Volcanoes

Dedication

*We dedicate this book to Gordon A. Macdonald (1911–78), a great volcanologist, teacher, and dear friend, who wrote an excellent textbook (*Volcanoes *– 1972) that served as the progenitor of this work, and also to the memory of all volcanologists who, motivated by concerns for their fellow human beings and by their desires to understand volcanoes better, came "too close to the flames," and paid the ultimate price.*

Rob Cook, Elias Ravian	Karkar, 1979
David Johnston	Mount St. Helens, 1980
Salvador Soto Piñeda	El Chichón, 1982
Alevtina Bylinkina, Andrei Ivanov, Yurii Skuridin, Igor Loginov	Kluchevskoi, 1951–1986
Alexander Umnov	Karymsky, 1986
Maurice & Katia Krafft, Harry Glicken	Unzen, 1991
Victor Perez, Alvaro Sanchez	Guagua Pichincha, 1993
Geoff Brown, Fernando Cuenca, Nestor Garcia, Igor Menyailov, Jose Zapata	Galeras, 1993
Asep Wildan, Mukti	Semeru, 2000

Volcanoes: Global Perspectives

Second Edition

JOHN P. LOCKWOOD
Affiliate Faculty, Department of Geology
University of Hawaii at Hilo, Hawaii, USA

RICHARD W. HAZLETT
Professor Emeritus, Pomona College
Claremont, California, USA

SERVANDO DE LA CRUZ-REYNA
Instituto de Geofísica, Universidad Nacional Autónoma de México
Ciudad de México, México

WILEY Blackwell

Edition History
John Wiley & Sons Ltd (1e, 2010) © John P. Lockwood and Richard W. Hazlett

Registered Offices
John Wiley & Sons, Inc., 111 River Street, Hoboken, NJ 07030, USA
John Wiley & Sons Ltd, The Atrium, Southern Gate, Chichester, West Sussex, PO19 8SQ, UK

Editorial Office
9600 Garsington Road, Oxford, OX4 2DQ, UK

For details of our global editorial offices, customer services, and more information about Wiley products visit us at www.wiley.com.

Wiley also publishes its books in a variety of electronic formats and by print-on-demand. Some content that appears in standard print versions of this book may not be available in other formats.

Library of Congress Cataloging-in-Publication Data

Names: Lockwood, John P., author. | Hazlett, Richard W., author. | Cruz-Reyna, Servando de la, author.
Title: Volcanoes : global perspectives / John P. Lockwood, Affiliate Faculty, Department of Geology, University of Hawaii at Hilo, Hilo, Hawaii, Richard W. Hazlett, Professor Emeritus, Pomona College, Claremont, California, Servando de la Cruz-Reyna, Instituto de Geofísica, Universidad Nacional Autónoma de México. Ciudad de México (CdMx), México.
Description: Second edition. | Hoboken, NJ : Wiley-Blackwell, 2022. | Includes bibliographical references and index.
Identifiers: LCCN 2021031545 (print) | LCCN 2021031546 (ebook) | ISBN 9781119478850 (paperback) | ISBN 9781119478843 (adobe pdf) | ISBN 9781119478836 (epub)
Subjects: LCSH: Volcanism–Textbooks. | Volcanology–Textbooks. | Volcanoes–Textbooks.
Classification: LCC QE522 .L63 2022 (print) | LCC QE522 (ebook) | DDC 551.21–dc23
LC record available at https://lccn.loc.gov/2021031545
LC ebook record available at https://lccn.loc.gov/2021031546

Cover Design: Wiley
Cover Image: © Bruce Omori

Set in 9.5/12.5pt SourceSans by Straive, Pondicherry, India

SKY10038992_112822

Contents

Volcanoes – Global Perspectives: Preface To The Second Edition

More than ten years have passed by since the First Edition of *Volcanoes* was published in 2010. Our world of volcanology has changed in exciting ways over this decade, and we have much important new material to discuss in this Second Edition. Volcanoes have continued to erupt (some 61 eruptions exceeding VEI 3 have taken place since 2010), and we are describing the largest of these and the ones that have had the most impact on society since the First Edition. This new edition contains more than 80 new photographs and figures to better illustrate volcanic features and processes more clearly, discusses or refers to more than 180 important volcanoes whose properties are summarized in an Appendix, and includes an updated Bibliography (with more than 1000 entries) that includes important papers describing recent eruptions and new findings.

Professor Servando de la Cruz-Reyna of the National University of Mexico has joined us as a co-author, and this edition benefits greatly from his long experience in applying statistical tools to analyze volcanic hazards and risk, as well as his invaluable Latin American perspectives on volcanic activity. In the first introductory chapter, Servando, Rick and I describe our personal experiences with specific eruptions that we have learned from, eruptions that have taught particularly important lessons for volcanology. A visit to St. Pierre, Martinique in 2019 stimulated a renewed appreciation for the impact of the 1902 Mount Pelée eruption on the development of volcanology, and led to the realization that we had failed to pay enough attention to the Peléean class of volcanic eruptions in our First Edition – that failing is now addressed. We have also greatly improved our discussion of pyroducts, the subsurface conduits of molten lava responsible for the areal distribution of pahoehoe lava flows, and of the genesis of the caves remaining when pyroducts drain after cessation of eruptive activity. Volcanologic research is improving the foundations of knowledge upon which all our science rests, and we briefly summarize the most important of these advances and new research tools developed over the past twelve years. The most productive of these new tools are remotely operated, and are constantly monitoring volcanoes and their impacts on the Earth's atmosphere from space and exploring new volcanic worlds beyond the bounds of Earth. We discuss the remotely operated vehicles (ROVs) that are now widely used to reveal the secrets of the most active volcanoes on Earth, those beneath the sea.

Although basic research is essential, the most important volcanological advances of concern to Society at large are those aspects that directly benefit Society – advances that enable humans to co-exist more harmoniously with the Earth's volcanoes. These are the broad fields of "Applied Volcanology," where volcanologists focus on better understandings of the hazards associated with volcanic activity, the means to evaluate and mitigate volcanic risk, and the best ways to share their knowledge with others – enabling people to live and work more safely with their volcano neighbors. Great strides have been made in these fields, and the past decade has seen major advances in our ability to better monitor active volcanoes and to provide advance warnings so that vulnerable populations can evacuate from dangerous areas. One measure of our success in these efforts is to compare the numbers of people killed by volcanic activity in the recorded period before and after 2010. From the advent of modern communication and better reporting of volcanic activity in 1850 (widespread use of telegraphy) to 2010, there were about 100,000 volcano-related fatalities, or about 6,250 fatalities per decade. In the 11 years since 2010, only about 1200 people lost their lives. This is of course related to good fortune (there have been no eruptions with VEI's in excess of 5 in this period!), but also reflect the fact that our capabilities to forecast eruptions and to evacuate the vulnerable have improved greatly. Of interest is the fact we identified 19 volcanologist colleagues who were killed by volcanic activity prior to the 2010 publication of our First Edition. Fortunately, we have no more martyrs to add to this list – which attests to the value of remote sensing tools volcanologists are using to monitor hazardous eruptions, but also perhaps to good luck!

"Applied Volcanology" is not limited to academic research. Earth Science teachers at all levels are applying their understanding of volcanology to better inform their students about volcanic activity and the nature of volcanic hazards. This will enable students who may never see an erupting volcano to better evaluate news stories about far-away eruptions, and to distinguish between overly sensational stories and factual reporting that puts facts in context. An entire new journal, the *Journal of Applied Volcanology* is now devoted to publishing important research focused on volcanic hazards and risk mitigation. Emergency managers, land use planners, and civic officials also have need to understand volcanic processes when their communities are threatened. We have endeavored again in this new edition to avoid overly technical discussions and unnecessary use of "jargon," with the important needs of civil authorities, teachers and students particularly in mind. We also hope that laymen and travelers will find our book useful for understanding the volcanoes they meet along the way, and to hear the exciting stories of fire and fury that all volcanoes have to share with passers-by who pay attention! Volcanologists mostly write papers and books focusing only on the scientific aspects of volcanoes they know, but some write Memoirs about their lives, and some write fiction based on stories volcanoes have told them. We include an appendix called "*Fun Reading*" to show what volcanologists write about in their "spare time"!

Our debt to colleagues who have shared their knowledge of volcanology with us remains deep. We cited over 100 of them in our Preface to the First Edition, and will not repeat their names here. We do however again wish to again acknowledge Pete Lipman, Jim Moore, and Chris Newhall, who reviewed sections of this book. Dave Clague and Ken Rubin kept us abreast of their exciting new discoveries of active submarine volcanism. Stephan Kempe provided us with important insights into the dynamics of fluid lava flow emplacement processes, and we really mean <u>IN</u>sights! Kempe's studies from <u>inside</u> pahoehoe lava flows around the world have revealed important details about the critical roles played by pyroduct conduits in subsurface lava transport. Shawn Willsey of Twin Falls Idaho shared valuable perspectives on Snake River volcanism. Anne-Marie LeJeune, then Director of the Volcanological and Seismological Observatory of Martinique was a wonderful host during a visit to St. Pierre and has provided new perspectives about the impact of the 1902 eruption on volcanology. Hugo Moreno-Roa (OVDAS) answered many questions about southern Chilean volcanism. Luis Gonzalez de Vallejo and Pedro Hernandez Perez provided important updates and photographs of the large, destructive 2021 eruption of Cumbre Vieja volcano on the island of La Palma. Scott Bachmeier (University ofWisconsin-Madison) and Walt Dudley (University of Hawai'i-Hilo) contributed importantinformation about 2022 Tonga eruption dynamics and its tsunami impacts. The 2018 eruption of Kilauea volcano changed many of our concepts about this volcano, and we appreciate the professional photography of Brad Lewis and Bruce Omori, who contributed many photographs. Matt Patrick of the USGS Hawaiian Volcano Observatory also contributed important photos of Kilauea features. Jake Smith again did magnificent work to compile the data necessary to update our list of volcanoes mentioned in text and the endpiece map, as well as kept track of the many new photographs used in this edition. He is also largely responsible for digital contributions to the e-book version of VOLCANOES. Andrew Harrison, Rosie Hayden, Rathi Aravind, Shyamala Venkateswaran and other Wiley editors in Oxford and India turned our rough manuscript submittals into the fine book you are holding today!

As always, despite our best efforts, we know that some factual errors or mechanical shortcomings may have crept past editors, and we will very much appreciate readers letting us know about any of these or about shortcomings you would like to see corrected in future Editions!

JOHN ("JACK") LOCKWOOD, VOLCANO, HAWAI'I
RICHARD ("RICK") HAZLETT, HILO, HAWAI'I
SERVANDO DE LA CRUZ-REYNA, MEXICO CITY
JUNE, 2022

JOHN P. ("JACK") LOCKWOOD

Jack Lockwood nearly died on the first active lava flow he ever saw (Hawaii – 1971), but the experience of working with molten rock was an epiphany that changed his life's focus to volcanology. He spent the next two decades working with the U.S. Geological Survey's Hawaiian Volcano Observatory and later as a USGS team leader responding to volcanic crises and disasters worldwide. His primary professional interest has been on what Thomas Jagger called *Humanitarian Volcanology* — the application of volcanology to the needs of Society. He left the USGS in 1995 to form an international consulting business focusing on volcanic hazards and risk assessments. He is now semi-retired, living with his long-term field partner and wife Martha on the lower slopes of Mauna Loa volcano.

RICHARD ("RICK") HAZLETT

Richard Hazlett has spent most of his career in geology teaching at Pomona and Occidental colleges and the University of Hawaiʻi at Hilo. His research in volcanology includes Neogene volcanism in California's Mojave Desert and eruptive activity at San Cristóbal (Nicaragua), Vesuvius (Italy), Krafla (Iceland), Kīlauea (Hawaiʻi) and Makushin (Alaska) volcanoes. While at Pomona he also devoted significant time to environmental education, including establishment of the Pomona College Organic Farm and senior editorship of the Oxford University Press Encyclopedia of Agriculture and the Environment. He presently is retired in Hilo, Hawaiʻi, where he maintains a research connection with the Hawaiian Volcano Observatory while continuing to learn folk guitar and develop familiarity with local reef fishes as an avid snorkeler and diver.

SERVANDO DE LA CRUZ-REYNA

Servando de la Cruz-Reyna has dedicated his life to research at the Institute of Geophysics of the National Autonomous University of Mexico (UNAM). For years he has advised the Mexican National Civil Protection System on different aspects of volcanic risk management and is one of the founder members of the Scientific Advisory Committee for Popocatépetl Volcano, and of other active Mexican volcanoes, receiving the National Award for Civil Protection in Prevention 2006. He has worked on modeling physical processes in active volcanoes, on the recognition of eruption precursors, on the statistical assessment of long-term volcanic hazards, and on the applications of those methods to volcanic risk management and social response, particularly on Popocatépetl, Colima, El Chichón, and Tacaná volcanoes. He has taught the graduate course on volcanic risks at UNAM for decades. Currently, he is editor-in-chief of Geofísica Internacional.

Preface To The First Edition

This book has a long history. It was originally conceived as a revision of Gordon Macdonald's classic book *Volcanoes* (Prentice-Hall, 1972), following his too-early passing in 1978. We had both worked with Macdonald, who friends called "Mac," and wanted to see his plans for a second edition of *Volcanoes* fulfilled. Originally John "Jack" Lockwood (JPL) planned a simple updating of Mac's text, and Richard "Rick" Hazlett (RWH) planned to contribute artwork to make a more attractive new edition. We quickly found that a simple updating of the original *Volcanoes* would not be sufficient, however, as much of Macdonald's writing reflected the uncertainties of his time, which meant a major revision would be needed. Over the years, under the guidance of several Prentice-Hall editors, the focus of our book changed; less and less of Mac's original writing remained, and a decision was eventually made by Prentice-Hall to abandon preparation of a second edition. Arrangements were then made for publication of this book by Wiley-Blackwell Publishing. Although Gordon Macdonald no longer is formally listed as a co-author of this book, his legacy of volcanic knowledge was heavily relied upon, and some of his original words remain in this text (with the permission of Prentice-Hall and Mac's family).

Rick joined the project as co-author in 1993. His long experience in teaching volcanology to students at universities in Hawai'i and California adds invaluable academic perspectives to this book.

When Gordon Macdonald wrote *Volcanoes* in 1970, the science of volcanology was poised at the threshold of a new era of discovery and understanding, but that threshold had not yet been crossed. In his influential 1972 book, Mac wrote that "Comparatively little progress has been made in understanding the fundamental processes of volcanic activity." How true those words were in 1970, but how untrue now! In the decades since the 1972 edition of *Volcanoes*, people have undoubtedly learned more about the causes and nature of volcanism than in all previous time: Inclusion of this new knowledge and placing it in a global framework has been the foremost challenge before us.

Revolutionary new tools and techniques have also been developed since Macdonald wrote the original *Volcanoes*. Our knowledge of volcanism at that time was almost entirely based on observations of subaerial volcanoes, since those were the only ones readily available for study. Manned deep submersible vehicles, originally used mostly for biological observations, have subsequently become available as "field tools," and have increasingly been deployed for direct observations of submarine volcanoes and volcanic terrain on the floors of the world's oceans. These observations, along with new side-scan sonar imaging techniques, Remotely Operated Vehicles (ROVs) and extensive research drilling of the oceanic crust, have at least quadrupled the numbers of known volcanoes around the world. Exploration of the Solar System over these years has now revealed that volcanoes are actually commonplace extraterrestrial features. Volcanic eruptions have taken place on the Moon, Mars and Venus, and active volcanoes (of a sort very different than those of Earth) have been observed on the moons of Saturn, Jupiter and Uranus.

The eruption of Mount St Helens in 1980 had a major impact on volcanology. Not only was this complex eruption one of the best documented in history, but it also served to change the perceptions of millions of North Americans, who learned that they too had active volcanoes in their backyard – just like the volcanologists had been saying all along! This eruption provided examples of numerous volcanic processes that had been poorly understood and never observed in detail before; illustrations from Mount St Helens are used liberally throughout this new edition. Four other major volcanic eruptions followed (or began) over the next 15 years, and were also well studied by volcanologists before, during, and after their principal activity – the long-lived East Rift zone eruption of Kīlauea that began in 1983, the Mauna Loa eruption of 1984, the Mt Pinatubo eruption of 1991, and the ongoing eruption of Soufrière Hills volcano, Montserrat – which began in 1995. Each of these five eruptions was very different from one another, and each provided important new information about "how volcanoes work" – information that we have relied on extensively.

While writing this book, we have carried on Macdonald's emphasis on descriptive rather than "interpretive" aspects of volcanology, although the processes that form volcanic features are also described where understood. In some sections we touch upon more theoretical aspects of contemporary volcanology, but only to provide an idea of some approaches that can be taken rather than to provide comprehensive treatments. Our bibliography points the way forward for those who are more deeply interested in theory. We have also unashamedly tried to emphasize "applied" aspects of volcanology where appropriate. The applied interfaces between volcanic activity, global ecology, and human society are summarized in Part V: "Humanistic Volcanology." That term was coined by Thomas Jaggar, founder of the Hawaiian Volcano Observatory, and was used by Gordon Macdonald in his writings. We have strived to continue this "humanistic" focus in our book, and are carrying on the chain of human contacts that lead from Jaggar to Macdonald, and now to us and to this book.

We are grateful to many colleagues who shared important insights and knowledge of subjects they know far more about than we do. Many of our colleagues have reviewed parts of the manuscript at various times and shared their ideas and constructive criticisms over the years, including Steve Anderson, Oliver Bachmann, Charley Bacon, Steve Bergman, Greg Beroza, Kathy Cashman, Ashley Davies, Pierre Delmelle, Dan Dzurisin, John Eichelberger, Bill Evans, Tim Flood, Patricia Fryer, Darren Gravley, Michael Hamburger, Ken Hon, Tony Irving, Caryl Johnson, Steve Kuehn, Ian Macmillan, Mike Manga, Doug McKeever, Calvin Miller, John Mahoney, Chris Newhall, Harry Pinkerton, Karl Roa, Mike Ryan, Hazel Rymer, Tim Scheffler, Steve Self, Phil Shane, Ian Smith, Jeff Sutton, Carl Thornbur, Bob Tilling, Frank Trusdell, and Colin Wilson. Having had so many well-qualified geologists comment on parts of this book has caused a minor problem: we've found that there is no universal agreement as to what should be included, and it is clear that no single book will "make everyone happy." We have learned from each of these reviewers, and have humbly tried to accommodate their oft-conflicting suggestions as best we could. Many other colleagues have contributed photographs for this book, or provided insights from their own expertise. These include Mike Abrahms, Shigeo Aramaki, Tom Casadevall, Bill Chadwick, Yurii Demyanchuk, Bill Evans, Dan Fornari, Brent Garry, Magnus Gudmundsson, Cathy Hickson, Rick Hobblit, Caryl Johnson, Stefan Kempe, Hugh Kieffer, Minoru Kasakabe, Takehiko Kobayashi, Yurii Kuzman, Paul-Edouard de Lajarte, John Latter, Brad Lewis, Andy Lonero, Jose Rodríguez Losada, Sue Loughlin, Yasuo Miyabuchi, Setsuya Nakada, Tina Neal, Vince Neall, Hiromu Okada, Paul Okubo, Tim Orr, Yurii Ozerov, Tom Pierson, Jeff Plescia, Mike Poland, Ken Rubin, Mike Ryan, Etushi Sawada, Lee Siebert, Tom Sisson, Don Thomas, Dorian Weisel, Chuck Wood, and Ryoichi Yamada. The late Tom Simkin of the Smithsonian Institution and five USGS colleagues (Pete Lipman, Jim Moore, Chris Newhall, Bob Tilling, and the late Bob Decker) deserve special acknowledgement for their wisdom shared with us over the years, and for the ideas we have purloined from their many seminal publications. We are indebted to support personnel at the University of Hawai'i, Pomona College, and the US Geological Survey, for encouragement and expert advice over the years, including Jim Griggs of the USGS and Dianne Henderson of the University of Hawai'i, who gave extensive help with preparation of photographs and line illustrations. Paul Kimberly of the Smithsonian Institution and Wil Stetner of the USGS provided the Dynamic Map files we used in the Volcanoes of the World map. (In the text numbers within square brackets following a volcanic site's name refer to that site's position on this map.) Ari Berland and Todd Greeley, both Pomona College undergraduates, and Jacob Smith of the University of Hawai'i at Hilo compiled extensive data bases, reviewed writing from a student standpoint, and prepared maps. Andrika Kuhle spent long hours compiling and organizing book figures. Julie Gabell's careful editing greatly improved parts of the manuscript. Our friends Maurice and Katia Krafft, who were tragically killed at Unzen Volcano in 1991, provided invaluable background information from their wealth of volcano knowledge, and loaned historical photographs, several of which are used in this book. Bob McConnin and Patrick Lynch of Prentice-Hall, and Ian Francis, Rosie Hayden, and Janey Fisher of Wiley-Blackwell provided critical editorial guidance, as did many other staff at Wiley-Blackwell. A sabbatical semester Lockwood spent at the University of Hawai'i at Manoa in 1988 gave important logistical support and stimulation, as did a research period at Pomona College in 2003. The US Geological Survey's Volcanic Hazards Program supported Lockwood for many years – enabling him to investigate volcanic eruptions and disasters in many lands, and to learn "under fire" from colleagues and foreign volcanologists. A 2002 sabbatical stay at the Alaska Geophysical Institute, and a 2006 sabbatical semester at the University of Auckland provided Hazlett with wonderful facilities and colleagues to aid in final writing.

I (JPL) wish to express gratitude to my wife Martha, who has been my able but unpaid field companion and assistant in the falling ash, mud, and sulphurous fumes of active volcanoes around the Pacific, and who has always kept on, even when paid assistants have faltered because of fatigue, boredom, or fear. She has also been a constant source of editorial and technical counsel as this edition has come to completion over the past several years, and has endured extensive "loss of companionship" over the final months as "The Book" took priority over normal marital responsibilities.

Part of the royalties from this edition will be used to establish a *G. A. Macdonald Student Volcanological Field Research Fund* at the University of Hawai'i, so that young men and women at the University will be better able to seek volcanological knowledge from the ultimate source – the volcanoes themselves.

JOHN P. ("JACK") LOCKWOOD

Jack Lockwood worked for the US Geological Survey for over 30 years, including 20 years in Hawai'i, based at the Hawaiian Volcano Observatory. In Hawai'i he monitored dozens of eruptions of Kīlauea volcano, and the last two of Mauna Loa. During non-eruptive times he deciphered the prehistoric eruptive history of Mauna Loa by geologic mapping, and became a leader of USGS international responses to volcanic crises and disasters worldwide. He has monitored eruptive activity of volcanoes as diverse as Gamalama, Nevado del Ruiz, Nyiragongo, and Pinatubo. Increasingly he has become focused on "humanitarian volcanology" – the application of volcanology to the needs of society. He left the USGS in 1995 to form a consulting business, Geohazards Consultants International, to continue international service. He is a commercial pilot, and with his wife Martha operates a ranch near the summit of Kīlauea.

RICHARD W. ("RICK") HAZLETT

Richard Hazlett is Coordinator of the Environmental Analysis Program and a member of the Geology Department at Pomona College in Claremont, California, where he teaches an upper-level course in physical volcanology. He has undertaken and supervised geologic mapping, geochemical studies, and stratigraphic analyses on many volcanoes worldwide, including a hazards assessment at San Cristobal volcano in Nicaragua, seismogenic landslide analysis at Vesuvius in Italy, study of blue-glassy pahoehoe and phreatomagmatic ejecta at Kīlauea, Hawaiʻi, and most recently, research on the late prehistoric history of Makushin, one of the most active volcanoes in the Aleutian Islands. His work has involved detailed examination of ancient volcanic terrains as well, focusing upon the Mojave Desert region in the US Southwest. Further interests include environmental science and *agroecology* – the development of sustainable agriculture by applying the principles of ecology to food production.

About the Companion Website

This book is accompanied by a companion website.

www.wiley.com/go/lockwood/volcanoes2

This website includes:

- Interactive volcanoes map
- List of prominent world volcanoes
- Figures from the book, in Powerpoint slides

PART I

INTRODUCTION

Volcanology is a specialized field of geology – the science of volcano study. *Volcanologists* are not only the scientists who study volcanoes (mostly geologists, geophysicists, geochemists, and geodesists), but include the devoted technicians who spend their lives monitoring volcanoes at observatories. To become a volcanologist, one must certainly study a great deal of geology and other physical sciences, but the title cannot be meaningfully earned only by reading books or bestowed by any university. Volcanoes themselves are the best teachers of volcanology, and the most respected volcanologists are those who have studied volcanoes in the field for many years. Volcanologists spend their careers seeking better understandings of volcanoes but must also be concerned about how their work will contribute to human social needs. Enabling people to coexist better with their volcano neighbors, sharing their knowledge with students and the public, devising means to utilize the tremendous stores of renewable energy stored within volcanoes, and perhaps saving lives along the way – these are all noble goals to strive for!

This part contains only one chapter, an important one that begins with introductory personal narratives written by authors to give a clearer understanding of what volcanic eruptions are like to experience first-hand. We then discuss some basic terminology and include a section on the history of this young science.

Volcanoes: Global Perspectives, Second Edition. John P. Lockwood, Richard W. Hazlett, and Servando De La Cruz-Reyna.

Companion website: www.wiley.com/go/lockwood/volcanoes2

CHAPTER 1

Eruptions, Jargon, and History

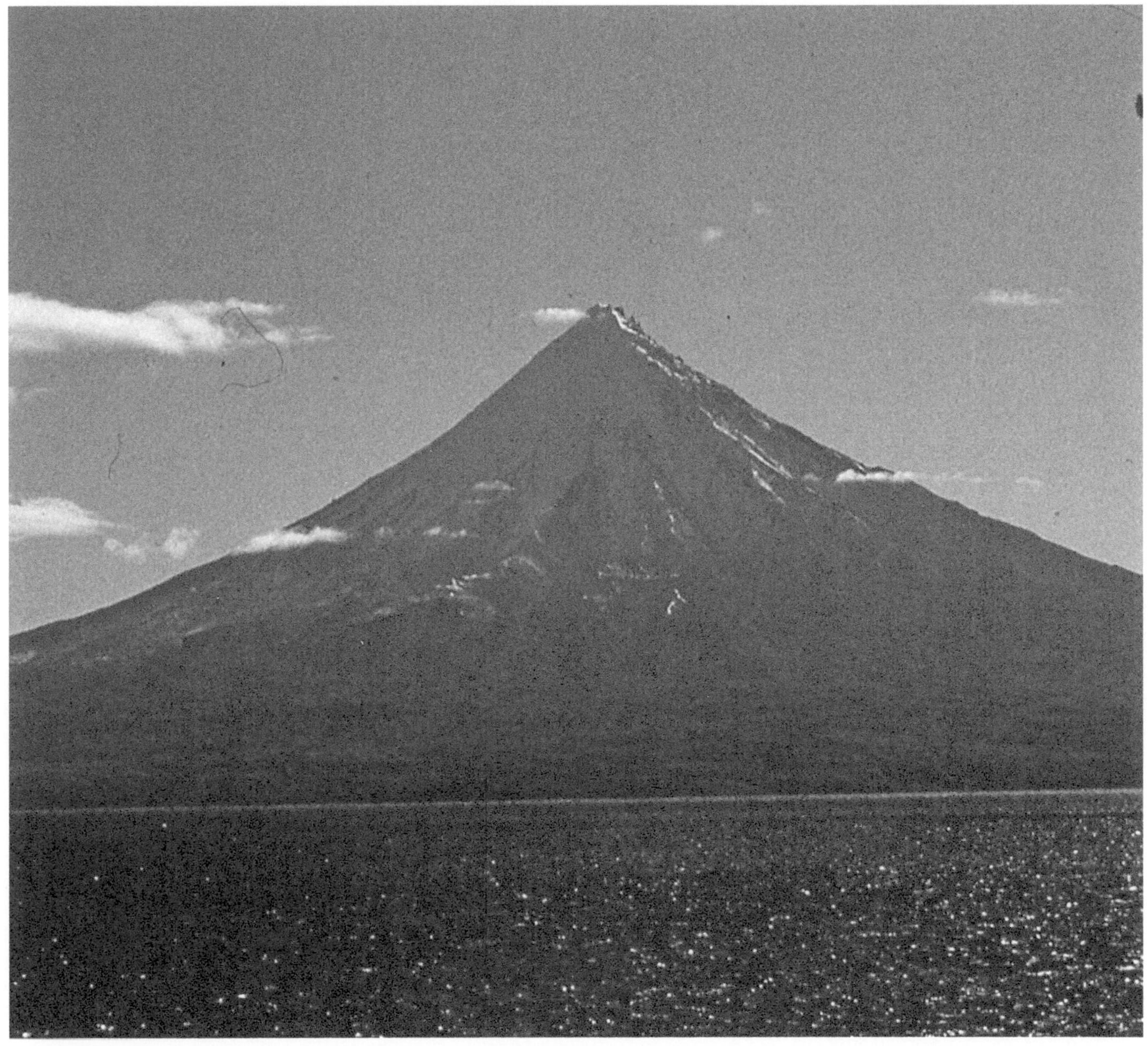

Source: Credit J. Lockwood.

Volcanoes: Global Perspectives, Second Edition. John P. Lockwood, Richard W. Hazlett, and Servando De La Cruz-Reyna.

Companion website: www.wiley.com/go/lockwood/volcanoes2

Volcanoes assail the senses. They are beautiful in repose and awesome in eruption; they hiss and roar; they smell of brimstone. Their heat warms; their fires consume; they are the homes of gods and goddesses.
(Decker and Decker 1991)

Volcanic eruptions are the most exciting, awe-inspiring phenomena of all of Earth's dynamic processes and have always aroused human curiosity and/or fear. Volcanoes, volcanic rocks, and volcanic eruptions come in many varieties, however, and to begin to understand them, one must absorb a great amount of terminology and information. We'll get to that material soon enough, but first let's explore what volcanoes are *really* like! The facts and figures in subsequent chapters could be boring if you lose sight of the fact that each volcano and every piece of volcanic rock that you will ever study was born of fire and fury, and that all volcanic rocks are ultimately derived from underground bodies of incandescent liquid called **magma**–molten rock. Every volcanic mountain or rock that you will ever see or touch once knew terrible smells and sounds that you must close your eyes to imagine!

French volcanologists loosely divide the world's volcanoes into two general types: ***Les volcans rouges*** (red volcanoes) and ***Les volcans gris*** (gray volcanoes). "Red volcanoes" are those volcanoes that are mostly found on mid-oceanic islands and are characterized by **effusive** activity (flowing red lava). The "gray volcanoes," generally found near continental margins or in island chains close to the edges of continents, are characterized by **explosive** eruptions that cover vast surrounding areas with gray ash. This is a rough classification for most volcanoes, although there are many that have had both effusive and explosive eruptions throughout their histories (or during individual eruptions). The volcanic hazards and risks posed by each of these types of eruptions differ greatly and will be described in detail in later chapters.

We hope that in this chapter you will gain some understanding of the look, smell, and *feel* of erupting volcanoes, and that this will put the material of the subsequent chapters in a more relevant light. To provide this, we three authors will describe our personal experiences during three major eruptions – one "red" and two "gray." This is our chance to describe the reality of dealing with major, destructive eruptions in personal terms, with the freedom to discuss our feelings as we deal with the human, political, and volcanic uncertainties that characterize all major eruptions.

"Rick" Hazlett will start off by describing his experiences during the 2018 eruption of Kīlauea, Hawai'i [17] – the most voluminous and destructive eruption in that volcano's history. Servando De la Cruz-Reyna will then discuss his human emotions and exciting experiences during the explosive eruption of El Chichón, Mexico [57] in 1982 – an eruption that caused the deaths of about 2000 people. "Jack" Lockwood will follow with his experiences during the 1991 Pinatubo, Philippines [135], eruption. Pinatubo was the second largest eruption of the entire twentieth century, and it had a major social impact in the Philippines, political repercussions, and global climate impact. In these and a few other places where we use the first person "I" with reference to personal experiences, we will identify ourselves by our initials, RWH (Hazlett), SCR (Servando De la Cruz-Reyna), or JPL (Lockwood).

A "Red Volcano" in Eruption – Kīlauea – 2018

In early April 2018, I (RWH) was working as a volunteer researcher at the United States Geological Survey (USGS) Hawaiian Volcano Observatory (HVO) on the island of Hawai'i, perched on the rim of the huge crater (caldera) atop Kīlauea volcano. This was a major place of worship for early Hawaiians, some of whom continue their ancient cultural practices here even today, honoring the Fire Goddess, Pele. In more recent times, Kīlauea has also become a major hub of world volcano science. The eruptions of Kīlauea are frequent, and most are gentle enough to study up close, though "gentle" is a relative term as you will see when you soon read the accounts of my co-authors! In the first five months of 2018, Kīlauea volcano was behaving in its typically dynamic but non-threatening way. Until early spring, there was no reason to believe that anything was about to change. . .

Unlike ordinary gray volcanoes with their centralized vents, Kīlauea has developed a far-flung internal magma transport system over a few hundred thousand years, allowing it to erupt fluid lava in multiple places, sometimes tens of kilometers from the summit. A complex magma reservoir system, extending several kilometers beneath the summit, frequently feeds lava to the floor of a broad central caldera, or collapse basin, in the form of persistent, sluggishly active lava lakes. Underground outlets can also divert the magma underlying the summit into one or two narrow rift zones – or zones of weak, fractured crust, stretching down the flanks of the volcano to the east and southwest. The East Rift Zone, by far the more active of the two, extends 130 km from Kīlauea's summit (elevation 1250 m) to the ocean floor 4–5 km below sea level (Figure 1.1)

For ten years prior to April 2018, Kīlauea had been erupting highly fluid lava simultaneously in two locations: A lava lake confined to a cylindrical pit about 200 m wide on the floor of Halema'uma'u Crater, which played to the delight of visitors viewing the scene from the Jaggar Museum overlook next to the observatory, 1.5 km away (Figure 1.2). Twenty kilometers to the east, lava also issued from a large spatter and lava shield, Pu'u'Ō'ō, a structure that had begun forming in 1983, and remained almost continuously active ever since. Pu'u'Ō'ō rose above remote forest in the uninhabited central part of the rift zone, but its vent frequently sent streams of lava to the south coast of the island, cascading down a 300–400 m tall slope. These flows destroyed dozens of homes and inundated roads on the East Rift Zone flanks where they spread out and gradually added hundreds of hectares of new land to Hawai'i Island. Despite the destruction, local residents and visitors alike took delight in observing the frequent displays of volcanic creation.

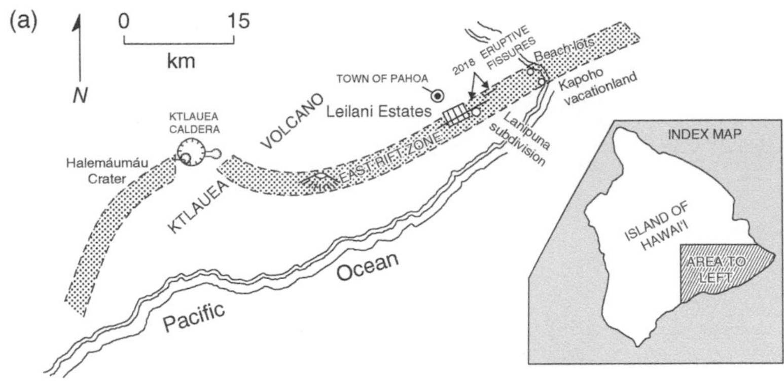

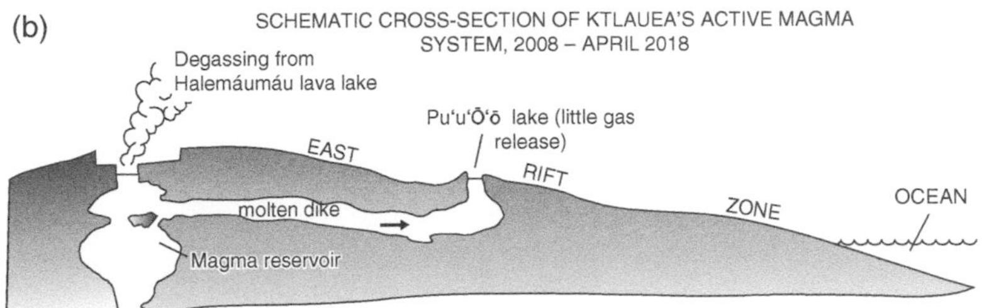

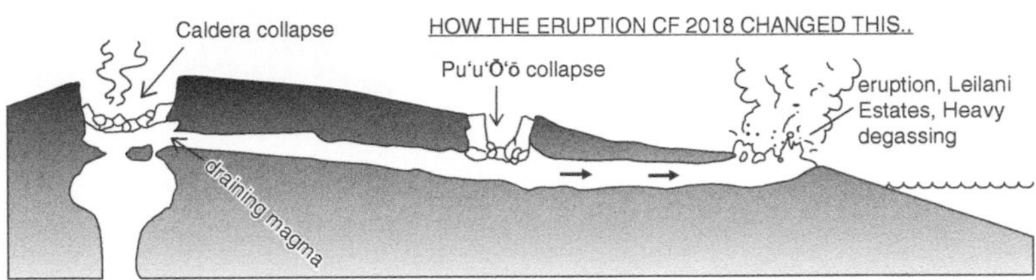

FIGURE 1.1 (a) Map of Kīlauea volcano [17], island of Hawai'i, showing major features related to the great eruption of May–August 2018; (b) cross-sections of the volcano before and during the eruption, drawn along the trace of the East Rift Zone

FIGURE 1.2 Halema'uma'u lava lake, as seen within the summit caldera of Kīlauea volcano [17] on 1 February, 2014, was continuously active between 2008 and 2018. It drained away rapidly at the start of the 2018 Lower East Rift Zone eruption. Source: USGS photo by Matt Patrick.

FIGURE 1.3 Puʻu ʻŌʻō vent area looking east, on 18 April, 2018. The vent consisted of a jagged, gently sloping lava mound surmounted by a cluster of craters with active lava circulating in the pit closest to the viewer. Source: USGS photo by Carolyn Parcheta.

The connection between summit lava lake and rift zone activity at Puʻu ʻŌʻō had been stable for many years: The Halemaʻumaʻu lava lake served as a chimney for gases escaping the underlying magma reservoir, behaving like a natural standpipe, with the fluctuating lake level reflecting magma pressure within the molten reservoir below. The Puʻu ʻŌʻō vent, 20 km to the east, behaved more like a drain for this system, tapping off magma from the summit reservoir that otherwise would accumulate there. Much less gas escaped from Puʻu ʻŌʻō as a result. From a distance on many days this flank vent appeared inactive, although molten lava was churning beneath the crust (Figure 1.3).

In mid-March, devices that measure the slope of the mountainside, called tiltmeters, indicated that pressure was pushing up the land all around Puʻu ʻŌʻō. Observatory scientists knew that a shallow pocket of magma must be accumulating there, and on 17 April, they warned the public that a fresh outbreak of lava was likely "somewhere" in the vicinity. As pressure rose beneath Puʻu ʻŌʻō, the lava lake in Halemaʻumaʻu, as though responding to a blockage, also began rising. On 21 April, it overflowed the rim of its vent pit, spilling out onto the surrounding crater floor. The observatory issued another eruption hazard bulletin three days later.

Then, on 30 April, the situation abruptly changed. The molten lava at Puʻu ʻŌʻō suddenly drained and the floor of the pit there began falling in. Swarms of earthquakes indicated that the pent-up magma was shifting underground along the East Rift Zone, like water rushing from a breaking dam, into the Puna lowlands ("Lower East Rift Zone" or LERZ) not far from the town of Pāhoa, an area where tens of thousands of people lived. The earthquakes were accompanied by a distinctive type of shaking called harmonic tremor which is produced by liquid (magma) intruding through cracks in fractured rock. HVO issued a public alert. It was time for residents to consider evacuating!

I knew that if a major eruption broke out in this heavily populated region of Hawaiʻi, I would be deeply involved in any emergency response. I was not afraid of what might happen, though I was certainly concerned – I had seen plenty of eruptive activity during a summit eruption of Kīlauea back in 1974 (see my account in the First Edition of this book) – but that had taken place in an uninhabited part of the Hawaiʻi Volcanoes National Park, where no lives nor homes were ever threatened. This was definitely different, and I realized that if lava broke out here, it would cause serious problems for many, perhaps thousands of people.

Leilani Estates (Figures 1.1a and 1.4) is a residential subdivision in the LERZ, close to the town of Pāhoa. Even though State and County officials and developers were well aware of volcanic hazards and the volcanic risk of authorizing development in this area – smack atop the East Rift Zone – this subdivision was approved and by the mid-1960s, housing began to spring up throughout Leilani Estates. Other nearby subdivisions also were approved and grew, including Lanipuna, Beach Lot Estates, and Kapoho Vacationland near the eastern cape, where the LERZ continues into the sea. Thousands of people now live in what was clearly seen by geologists and civil defense authorities as a hazardous area.

Mid-afternoon on 3 May, the first eruptive vent, later called "Fissure 1" opened in the midst of homes in Leilani Estates. Fortunately, most residents had safely fled by then, but no one could tell exactly where or when – even if – the eruption would break out next. Many evacuees worried about lost pets, friends, and their properties which might or might not be looted if not destroyed. No one knew when they could safely return home. Many were retirees, with scant social safety nets and life savings threatened. For

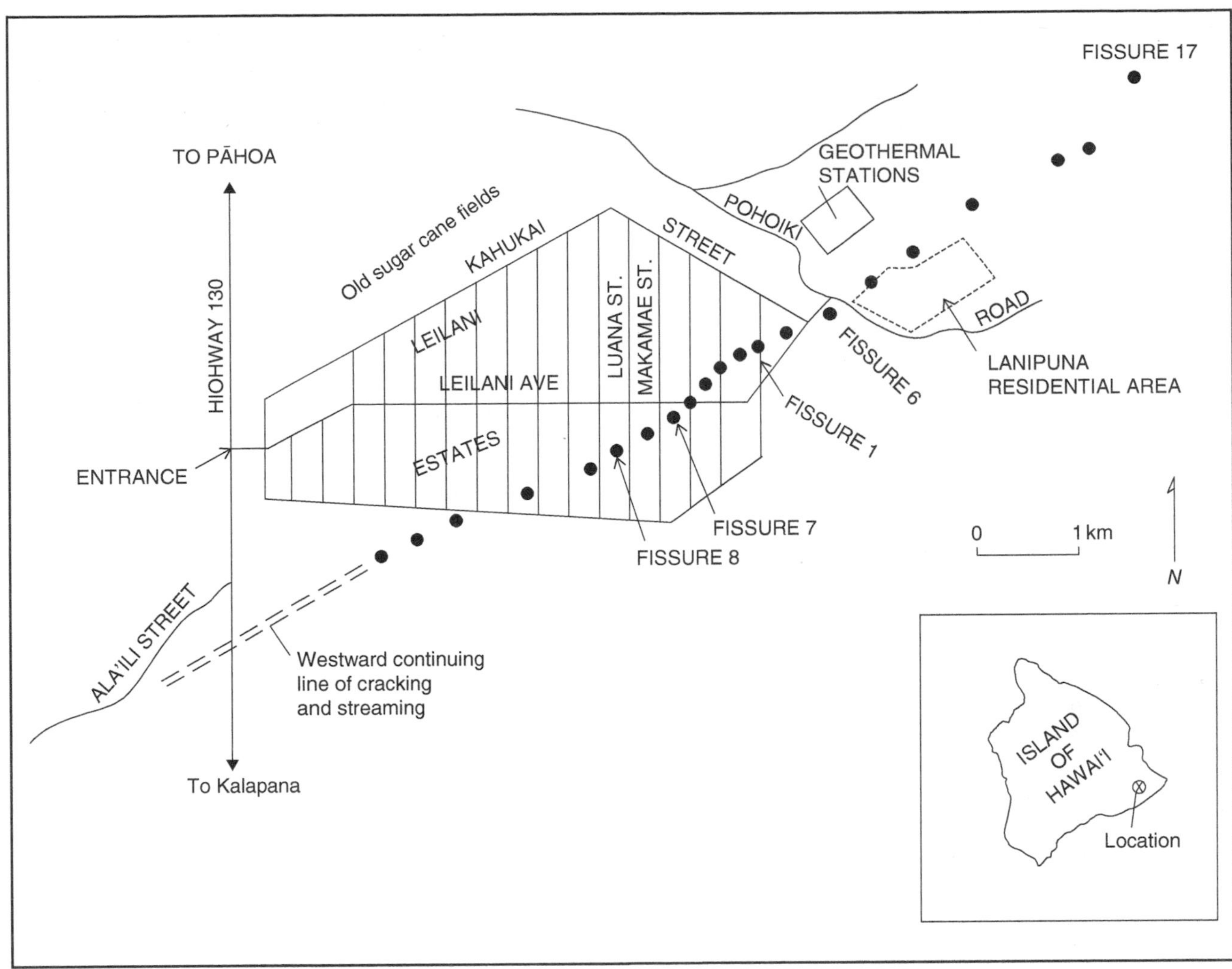

FIGURE 1.4 Map of Leilani Estates and 2018 eruptive fissures (black dots).

most disasters, the intense psychological stress of a hurricane or earthquake, typically comes and goes in short order – these are brief physical events. Imagine how a developing disaster with no end in sight would impact individual psyches, and no end was in sight for this eruption!

Adding to this stress, a magnitude (M) 6.9 earthquake jolted the island on 4 May. The south flank of Kīlauea shifted seaward as much as five meters when this happened, causing light to serious damage across several hundred square kilometers. Perhaps the earthquake occurred in response to the buildup of magma pressure in the East Rift Zone. Almost certainly, it opened fractures allowing magma to reach the eruption site more easily, fed from the bloated summit magma reservoir. By this time, the level of the lava in the Halema'uma'u lake was dropping rapidly.

"Fissure 1" was to be followed during the next three weeks by the opening of 23 additional eruptive fissures, each no more than a few hundred meters long, stretching in a 6-km-long set of closely spaced vents, both up- and downrift of the original outbreak. No order in the developmental sequence could be seen; fresh outbreaks of lava jumped up- or downrift relative to earlier-opened fissures, though the general trend was one of migration *down* the East Rift Zone, closer to the subdivisions at the coast (Figures 1.1a and 1.5). Hawai'i County Civil Defense, Police, and Fire Departments, all worked in close collaboration with the National Guard, HVO, and USGS geologists to assure that no lives would be lost and that residents could remove as much property as they safely could before homes were inundated by lava. They were remarkably successful doing so.

Like typical Kīlauea rift zone eruptions, this one began along numerous fissures, but eventually settled down to concentrated outpouring from a single vent ("Fissure 8") along the initial 24-fissure system in the middle of ill-fated Leilani Estates. A measure of my excitement during the first three weeks of this eruption can be reconstructed from journal notes I took during this early stage:

4–5 May: The Observatory asked me to partner with Carolyn Parcheta, a new staff geologist, to monitor eruptive activity in the field. We got going around 7 p.m. on Friday night, and did not knock off until 6 a.m. the next day. Our first objective was 'Fissure 6' in the thick woods at the east end of Leilani Estates, just south of a major geothermal plant. Entering the Civil Defense closed zone we met the team we were assigned to relieve at a street intersection in the dark forest. Silhouetted through the trees, repeated blasts of lava kept occurring, pitching bright yellow-orange globs of molten spatter up to a couple of hundred feet. Wild, gaseous bursts

FIGURE 1.5 Fissure 20 springs to life, 14–16 May, 2018, with lava gushing as much as 50–100 m above the mouth of the fissure that guides it to the surface. Hawaiian geologists sometimes call the line of lava fountains a "curtain of fire," a characteristic feature of red volcano eruptions. Source: USGS photo by Richard Hazlett.

reverberated every few seconds with a background noise of loud hissing. Our cameras could not do the stark scene justice. Our senses alternated between awe at the beauty of the eruptive activity, and shock about what was happening.

"The vent was no longer growing, and we proceeded to spend the next six hours simply watching activity die out here while making sure that the steaming, sulfurous cracks in the adjacent streets were not opening further. (I used my ruler a lot!) Everything seemed pretty quiet during the early morning hours of Saturday. We went patrolling for ground cracks in the abandoned Leilani subdivision, locating them where they broke street pavements and measuring their widths. By about 5 a.m., we were tired but remained alert because we knew that things could change quickly; Kīlauea's eruptions take breaks like this, sometimes lasting for a few days, then pick right back up again. Around 5:30 a.m., while measuring one crack, I heard a familiar, unwelcome noise coming out of the woods very close to us, behind some houses nearby (a sound resembling waves sloshing up on a beach or water coming out of a blowhole). I told Carolyn to turn off the car engine to hear this warning-sound better, and to see if we could ascertain the direction of its source. She agreed a new outbreak must have begun, though neither of us had a good idea where it was. Sulfur gas began filling the air all around us, so we donned our masks and drove further into the subdivision to find the new vent. We ran into a concerned local resident still in the process of evacuation, delayed in part because he feared looters. He said that he thought there were actually two new outbreaks nearby – 'somewhere'. With this tip, we moved along Kahukai Street – the north road, and sure enough up we soon came upon a dense cloud of wood smoke and fume rising from the trees ahead. No lava was erupting from this new crack yet, but we reported the location to the mobile command post in Pāhoa for a helicopter pilot to investigate. We then backtracked along the north road; there was no other way to exit. I kept looking down each of the south-trending streets that we intersected as we drove along, all the while feeling a pit in my stomach. And suddenly, looking down Makamae Street, *I saw it*; an explosive outburst of blood-red lava fountaining above the forest only about a half mile away. Carolyn turned onto the street and drove us toward this new vent. Taking care as we parked to turn the vehicle around for a quick getaway, we walked to within a few hundred feet of the roaring fissure (Figure 1.5) where she took a GPS measurement and photos. She was a bit braver than I about approaching the outpouring flow, but we both worried with good reason about the burning roadside power line. During our brief time there, we saw that the vent was becoming increasingly active – certainly more active than all previous openings we had seen in this eruption so far. HVO formally designated 'our' new openings Fissure 7 and Fissure 8."

Carolyn and I witnessed the ongoing activity of Fissure 8 directly across Luana Street in Leilani Estates late in the day on 5 May. We had to move our field vehicle back up the street from this site, not so much because of advancing lava but because the power poles around us were beginning to burn down, and live electrical wires were threatening. I saw a lot of methane flame emerging from openings in the edge of the active flow and heard a continuous series of muffled explosions in the surrounding woods;

grenade-like methane bursts. We avoided entering the surrounding forest – for multiple reasons. A photograph taken later in the eruption by Brad Lewis, shows a flow advancing on a Leilani street road as the adjacent vegetation was being consumed by fire and blue methane flames were emerging from road cracks (Figure 1.6).

The rest of May flew by in kaleidoscopic fashion. I witnessed new vents opening and homes burning, measured lava fountain heights (Figure 1.7), and met with stunned residents who were forced to evacuate and leave most belongings behind, just before their homes were destroyed. Multiple streams of lava poured southeast from the evolving set of fissures, covering scattered homes and some of Puna's most productive agricultural land (Figure 1.8).

As the various fissures, 1–24 opened in May, magma intruding the LERZ from uprift drove out pockets of older, still liquid magma stranded underground from previous eruptions and intrusive episodes. Aging magma cannot erupt on its own, having cooled, become more viscous, and lost the gas concentrations needed to help erupt. If driven out by fresh melt from below, however, older magmas tend to erupt explosively; sudden pressure reduction causes gases that remain trapped within them to gather into enormous bubbles that ultimately burst with great force through the viscous liquid. Fissure 17 opened in this way on 12 May, releasing a torrent of andesite lava, a rare composition in Hawai'i that can only form through the fractional crystallization of aging, stored magma. The andesite likely evolved from molten rock originally emplaced in 1955. In contrast, all the other fissures erupted only more fluid basaltic lava.

The explosive eruption of Fissure 17 led to one of the few injuries of the eruption. A large block, tossed ballistically hundreds of meters from a vent, tore through the roof of a house and injured a resident who had refused to

FIGURE 1.6 A Leilani Estates road is buried by an advancing pāhoehoe lava flow as methane gas being distilled from heated vegetation burns in road cracks. Source: Photo © Brad Lewis.

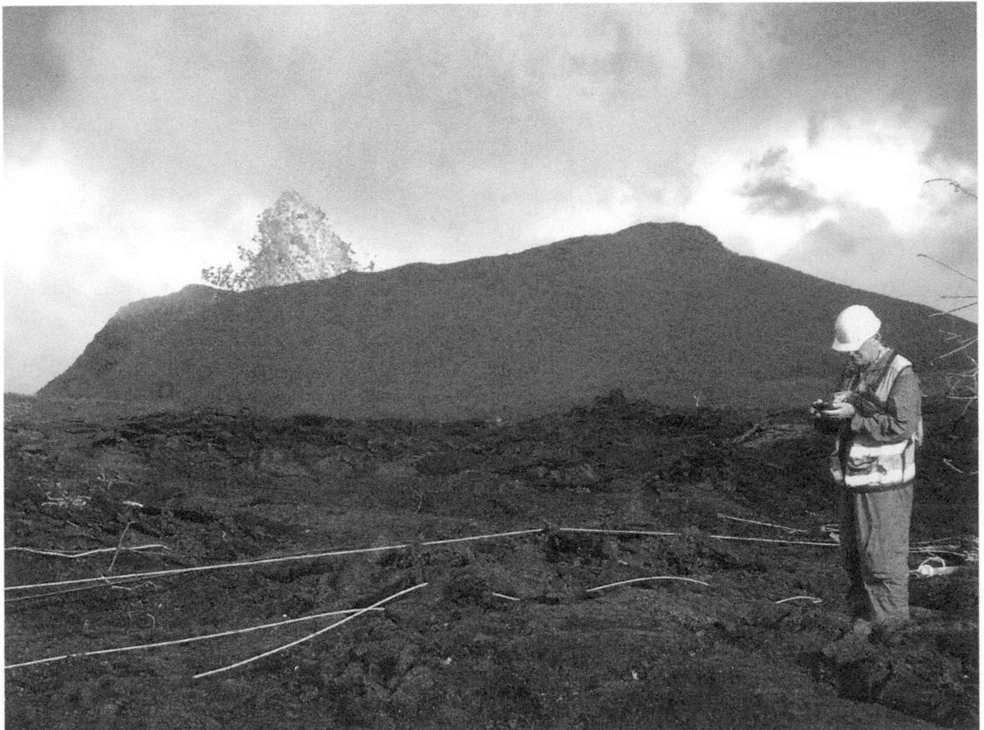

FIGURE 1.7 Trigonometry works! Measuring the height of a lava fountain within with a GPS and inclinometer. Fountains rise about 60 m above the base of a freshly forming cone (Fissure 8) in Leilani Estates. This cone has since been named Ahu'ailā'au by Hawaiian elders, meaning "Altar of Ailā'au" (a vindictive Hawaiian volcano god before Pele). Downed power lines and electric cables litter the landscape. Source: USGS photo by Samantha Gassett.

FIGURE 1.8 Lava flows course through the Malama Ki area of the LERZ on 21 May, 2018, on their way to building new land along the southeast coast of the island of Hawai'i. Source: USGS photo by R.W. Hazlett.

evacuate the area. Fortunately, the injury was not grave, though it well might have been. The concussive bursts from Fissure 17 blasts were strong enough to rattle and even shatter glass in neighboring homes, requiring geologists and other observers to wear ear protection within a kilometer or so distance of the vent.

On 25 May, volcanic activity in the LERZ entered its prolonged final stage, as Fissure 8 reactivated to become the dominant remaining vent of the eruption. The escaping magma had a composition more closely matching that of the Pu'u 'Ō'ō lavas, and the discharge rate (amount of lava released per unit time) accelerated. Hundreds of millions of cubic meters issued through a set of closely spaced lava fountains, around which a 55-m-high cone of spatter and pumice fragments grew (Figure 1.7). An outlet on the northeast side of the cone released a molten river that traveled as fast as 30 kph into a channel as much as 250 m wide that carried the lava all the way to the ocean 10 km to the east, reaching Beach Lots and Kapoho Vacationland subdivisions. Frequent small overflows from the channel built levees that both confined and deepened the lava river (Figure 1.9). Ironically, even though the eruption rate had increased, the threat to the remaining homes in Leilani subdivision had diminished owing to the way the flow path of the lava had stabilized.

6 June: "I took the 2–10 p.m. shift today with Brian Shiro, the HVO seismic system manager. (Brian had recently transferred to the Observatory from the Pacific Tsunami Warning Center in Honolulu.) As usual, I started out with a set checklist of places to investigate in the field, and then the unexpected happened to make life more interesting. The first 'unexpected' came with a drive to the southwestern end of the new rift [fissure] system, in a forested area of rural residences and farms along Ala'ili Road. Some observatory geologists are concerned that the eruption will soon spread from its current vent at Fissure 8 to the west, cutting the last open highway into south coastal Puna. I believed that there was little evidence to support this forecast, but Brian and I were surprised to discover quite a few new ground cracks crossing Ala'ili Road in a zone about a kilometer wide. We measured, oriented, and photographed the biggest ones. None had opened to more than about 1.5 cm wide, but their presence was certainly significant. We cannot tell when they opened. Possibly these were older and not related to a fresh, new dynamic development. Anyhow, this fracturing has now directed our eyes westward to be sure that we are not 'flanked' as this eruption progresses.

"The mid-afternoon overflight on this day brought deep, powerful shock: We witnessed the Kapoho Beach Lots/Vacationland subdivisions being consumed by lava Figure 1.10). A massive 'a'ā [rough, crumbly lava] flow poured rapidly through the subdivision, reaching beautiful Kapoho Bay on the night of 3 June. Now, it was rapidly filling this shallow bay (Figure 1.11). Residents had been evacuated just before their homes were consumed, but most of their belongings had been left behind in the rush. We watched helplessly as many beautiful homes including one belonging to a friend of mine, and another owned by the County Mayor – burst into flames and instantly disappeared underneath monstrous fiery masses of advancing 'a'ā, in places 15 m thick. Kapoho Bay and its wide fishpond were island treasures, but quickly became a peninsula of jagged hot rock extending over a kilometer into the ocean. We could only track with our GPS

FIGURE 1.9 A vast river of lava in this area up to 150 m wide escapes Fissure 8 and pours through Leilani Estates behind self-constructed levees. A small tongue of lava overflows a levee in the lower center view toward one home. (It never got that far.) Overflows like this helped construct levees as high as 20 m above the pre-existing landscape. Source: USGS photo by R.W. Hazlett.

FIGURE 1.10 Lava consumes homes in the Kapoho Beach Lots subdivision, 4 June, 2018. The helicopter pilot and I were both choked up to see this, each having friends who owned homes here. Source: USGS photo by R.W. Hazlett.

FIGURE 1.11 4 June, 2018. View westward across Kapoho Bay just after dawn. A rapidly moving ʻaʼa lava flow more than 10 m thick has already destroyed about 100 homes, and is now entering the bay. A total of nearly 500 homes would be destroyed by this flow, including all of these visible here. Source: USGS photo by Matt Patrick.

the northern perimeter of the flow because heat, gas, and laze [acidic vapor produced by lava entry into the sea] made flying conditions to the other side perilous. Through the billowing mist we could tell that the lava had made it all the way to the edge of the spectacular coral tidepools at Waiʻopae that everyone feared would be lost. A day and a half later, they are all gone (Figure 1.12). Lava is Nature's ultimate 'clean-slater'. A hurricane, earthquake, and even an atomic explosion will leave traces behind of what once was. But a lava flow buries and incinerates human-built structures slowly and utterly, though the social shocks of obliteration and personal losses echo in the aftermath. How strange from the air to see the patterns of streets, orchards, homes, and powerlines now disconnected in small patches across the district like random pieces of a jigsaw puzzle scattered across black asphalt (Figure 1.8). The lava is adding to the island in a beautifully powerful way and providing a future platform for new life, but that is not much comfort here and now."

As these exciting events were taking place 35–45 km east of Kīlauea's summit, the floor of Kīlauea's caldera collapsed in a stepwise fashion. The reservoir beneath Halemaʻumaʻu was pouring out its magma supply through far-away Fissure 8. My notes from 6 July describe this phenomenon:

"The southern part of Kīlauea caldera's floor continues to drop dramatically, in two ways – (1) by gradual subsidence day-by-day, broken by (2) sudden downward 'bolts' we are calling 'Type-A events', which can lower an area now covering several square kilometers as much as two meters *all at once*. This rapid, scraping drop generates big, shallow, long-duration earthquakes, generally in the magnitude 5-6 range. Each Type-A event is followed within minutes by a sudden jump back up in the area immediately overlying the magma chamber, and none of us really knows why. Geophysicists think that it must have something to do with alternate release and re-dissolving of gases in response to suddenly changing reservoir pressure. Or, perhaps it is elastic rebound in the crust. There is little empirical background to explain anything at this scale. No one ever expected a Hawaiian-type caldera to develop this way."

"Mysterious Type-A events are taking place like clockwork. Each event is about the same size, and each is spaced by about a day, 24–26 hours. Like a giant Old Faithful Geyser, you can plan your visit to the summit around them, which HVO does while it continues to evacuate equipment and gear from its creaking headquarters building on Uwekahuna Bluff (now spalling plaster from twisting I-beams). Movers in Hilo wait until a Type-A event takes place, then scurry on up, since for a period of 3–4 hours following each collapse, everything is eerily quiet and stable atop the volcano. After that, the forecastable earthquake swarm leading up to the next event, mostly in the magnitude 2–3 range, starts up once again, putting people on edge once more who live within a few kilometers of the mountaintop."

FIGURE 1.12 Lava enters the ocean across a broad front, nearly a kilometer wide in this view, where Kapoho Vacationland subdivision and its famous coralline tidepools (Wai'opae) existed a few days earlier. Over a 6-week period, this flow added over 3 km^2 of new land to the island of Hawai'i. Source: USGS photo by R.W. Hazlett.

FIGURE 1.13 Kīlauea [17] caldera floor collapse (foreground) during a Type-A, M-5 seismic event, observed on the afternoon of 6 July, 2018. The crumbling, dust-shrouded outer caldera and crater walls in the background are a couple of hundred meters high, 3 km from the camera. Source: USGS photo by R.W. Hazlett.

I took an afternoon shift viewing Kīlauea caldera from the overlook at the abandoned Volcano House Hotel, perched on the stable north rim. I wanted to experience a Type-A event close at hand and was certainly not disappointed. When it happened an hour and a half after arrival, it startled the hell out of me – the power of it was so overwhelming. It began with a noise like a sonic boom, followed by violent shaking that forced me to grab a rail next to the lounge overlook to avoid being thrown down. My colleague was "dancing" around with me. Inside the hotel, I could hear shifting furniture and a lot of clinking glasses, stored behind the bar. Almost at once, another noise rumbled in, resembling the rushing shush of an approaching wave. This was the sound of countless rockfalls cascading down the high caldera wall to the west (Figure 1.13). The big, widening maw of Halema'uma'u suddenly filled with an evolving plume of dust. This persisted for what seemed like the better part of a minute. Then all this activity stopped as suddenly as if someone had thrown a switch! For the next several hours, the summit was absolutely calm and peaceful – just another beautiful Big Island afternoon. A GPS station indicated that the south caldera floor had dropped 1.7 m during this one intense event."

Over 60 Type-A events with magnitudes >5 eventually took place, finally ending on 4 August with the new caldera floor over 600 m deeper than it was at the start of the eruption. The rims of a yawning pit, 6 km^2 in area, marked the perimeter of the drained, underlying reservoir. Throughout most of this collapse, from mid-May on, essentially no fresh magmatic ash erupted; this was primarily a *passive* collapse occurrence; just lots of noise and rock dust billowing into the air whenever the crust suddenly dropped.

Within 1–2 hours of each Type-A event, a surge or pulse of lava emerged from Fissure 8. The timing was not coincidental, geophysicists thought, though other surges and pulses, timed from a few minutes to 10–15 minutes apart, also could be seen in the

TABLE 1.1 Chronology of Events, 2018 Lower East Rift Zone Eruption of Kīlauea Volcano.

17 April – Hawaiian Volcano Observatory (HVO) issues warning of a possible new vent opening either at Pu'u'Ō'ō or along adjacent areas of the East Rift Zone
21 April – Building magmatic pressure causes the lava lake at Halema'uma'u Crater to overflow
24 April – HVO issues another volcanic hazard warning based on continuing swelling and earthquake activity around Pu'u'Ō'ō.
30 April – Pu'u'Ō'ō Crater floor begins collapsing as earthquakes propagate east into populated lower Puna area.
1 May – Lava lake begins draining at Halema'uma'u. HVO warns residents in lower Puna that an eruption is possible there in a matter of a few days.
3 May – At 5 p.m., Fissure 1 opens up in Leilani Estates subdivision.
4 May – A magnitude 6.9 earthquake, the biggest to strike Hawai'i in 43 years, takes place as the south flank of Kīlauea slips seaward as much as 5 m.
3–10 May – Short fissures continue opening in Leilani Estates with bursting eruptions of lava from old, stored magma trying to escape. Small flows fed.
10 May – Halema'uma'u lava lake now drained down 300 m, no longer visible from crater rim. HVO issues warning that new fissure outbreaks are likely downrift of Leilani Estates.
12 May – New fissure opens up downrift of initial set, as forecast.
15 May – HVO issues warnings of explosions imminent in draining Halema'uma'u conduit.
16 May – First Halema'uma'u explosions take place.
18 May – Series of Halema'uma'u explosions continues, while hotter, less viscous lava begins pouring from Lower East Rift Zone fissure vents. Larger than previous ones, the flows begin heading toward the sea.
23 May – Series of Halema'uma'u explosions continue, while first lava flows reach the sea southeast of Leilani Estates
27 May – Lava eruption focuses on reactivated Fissure 8 in Leilani Estates, which formed in early May. A large flow moves northeast, soon turning east toward Kapoho.
30–31 May – The main caldera floor begins subsiding around Halema'uma'u. The first of 62 collapse events ("Type-A") takes place with as many as 700 magnitude 4+ earthquakes per day, for the next two months. Summit gas emissions decline and explosions cease.
3–5 June – Lava reaches the ocean at Kapoho Bay, destroying adjacent residential neighborhoods and farmland. Activity continues unabated at Fissure 8. Type-A events continue at summit.
Mid-July – Lava eruption from Fissure 8 begins to pulsate as height of lava fountaining slowly declines.
2 August – The last of the Type-A events occurs at the summit.
4 August – Eruption has ceased at Fissure 8, except for sporadic bursts of spatter immediately around the crater vent, last visible the first week in September.

stream of emerging lava in the LERZ, possibly related to the way gases escaped the erupting conduit. In general, the greater the time interval between surges, the greater their intensity. We still do not understand their origin.

By the first week in August, Kīlauea had essentially exhausted itself. Lava continued to spurt weakly from the vent inside the Fissure 8 cone until for another month, but otherwise, the show was over. The eruption had not only provided a new, instructive view of what the red volcano Kīlauea could do, but it was also a remarkable opportunity to test new ways of studying volcanoes, from applied drone technologies to infrasound, RSAM analyses, communication-apps, flow-path software, and many other imaginative and useful approaches (Chapter 14). The great eruption of 2018 was certainly an historic event for Hawai'i, not only the most destructive to property, but one of the best managed in terms of public safety. Not a life was lost. During the 150 days of this activity, about 0.77 km^3 of lava poured out – mostly through Fissure 8 – covering 35 km^2 of land and adding 3.5 km^2 of new land to Hawai'i Island. The lava buried 50 km of roads and destroyed 716 homes. Total loss of property has been estimated at nearly $800 million dollars. Table 1.1 summarizes the 2018 events.

2021 UPDATE: Kīlauea's return to activity on 20 December, 2020

Kīlauea's dramatic summit collapse in 2018 (Figure 1.13) accompanied by rift zone eruptions was not unprecedented; major collapses or drainages of lava from the caldera almost as significant also occurred in (or shortly before) 1823, 1832, 1840, 1868, and 1924. In each case, the caldera eventually healed its "wounds" with outpourings of fresh lava that gradually filled collapse pits and built up new caldera floors, though long quiet periods sometimes passed before eruptions resumed. The collapse in 2018 was no exception. The new subsidence basin remained tranquil, degassing peacefully for 27 months, long enough for a small water lake to appear at the bottom supplied from the surrounding, higher-standing water table. But then on 20 December, 2020, a fresh fissure opened in the northern wall, disgorging molten lava that quickly evaporated the water lake and steadily created a pool of lava ultimately over 200 m deep (Figure 1.14). The eruption was very sluggish compared to all the excitement in 2018, and lasted

FIGURE 1.14 16 April, 2021 view of Halema'uma'u, the ever-changing collapse pit on the floor of Kīlauea Caldera. The down-dropped fault block in the middle ground, 80 m below the overlook, preserves part of the severed National Park Crater Rim Road – still intact though clearly inaccessible. The active lava lake lies 280 m farther down, and has filled Halema'uma'u to a depth of 215 m. The eruptive source vent on the northwest margin of the lake is marked by a plume of blue SO_2 gas. Source: USGS photo by R.W. Hazlett.

a little over five months before the volcano once again became peaceful ...But, not for long! On 29 September, 2021, the volcano sprang back to life in almost the same location, and continues to erupt up to the present (June, 2022). The lake of lava has deepened a further 65 m in the interim.

A Gray Volcano, El Chichón, Erupts in 1982

El Chichón Volcano [57] is located in the State of Chiapas (17.36°N, 93.23°W), a region of southeastern Mexico predominantly populated by the Zoque indigenous people. After six centuries of quiescence, El Chichón erupted explosively on 28 March, 1982, producing heavy ashfalls over southeastern Mexico, forcing the closure of numerous airports and roads. Moderate eruptions continued for a week, and then two major Plinian ("violently explosive"; Chapter 5) eruptions took place on 3 and 4 April. The explosive episode lasted about a week, and it is estimated that it caused nearly 2000 fatalities and the evacuation of about 20,000 people, as well as heavy losses of livestock, and major damage to coffee, cocoa, banana, and other crops. This was the worst volcanic disaster in the history of Mexico, and I (SCR) was there. Before the 1982 eruption, there was scarce information about previous volcanic activity in the region. Around 1928, felt earthquakes and the appearance of small fumaroles in a hill locally called "El Chichón" for a common, locally growing plant, motivated the visit of the German–Mexican geologist Federico Müllerried from the Institute of Geology of the National Autonomous University of Mexico (UNAM) who wrote three papers recognizing El Chichón as an active volcano (Müllerried 1932a, 1932b, 1933). However, no eruption followed the 1928 seismic swarm. Afterwards, El Chichón remained quiet and little-studied for about 40 years. Research revealed that the volcano had not erupted for six centuries, which gave people the general impression that it was dormant and therefore not very threatening. However, its proximity to an oil-producing region and potential for geothermal energy production prompted further geological study. Two geologists from the Mexican Electrical Power Company (Comisión Federal de Electricidad, CFE) felt earthquakes during their fieldwork on the volcano between December 1980 and February 1981 and described them in an unpublished internal CFE report (Canul and Rocha 1981) that was released only after the eruption ended.

But of course, at the time I was not aware of any of that. Until then, my background had been mostly in physics of the Earth's interior. I began switching to physical volcanology after observing the 1975–1976 and 1982 effusive activity of Volcán de Colima [51]. The first news about something happening in Chiapas arrived at our Institute of Geophysics (IGEF) at UNAM in Mexico City, 667 km to the northwest of El Chichón, on Monday 22 March, 1982. Local Zoque residents living in a remote jungle area with poor communications had been reporting earthquakes to authorities for weeks. They seemed to be increasing in magnitude. Around Tuesday 23 March, the Chiapas state government contacted UNAM with a request to investigate the phenomenon. At this stage, it was not quite clear at the IGEF if the felt seismicity was a tectonic swarm – not uncommon in the intensively fractured water-rich karstic crust of Chiapas (Figueroa 1973; Mota et al. 1984) – or volcano-related seismicity. The volcano sits in the middle of a massive tertiary siltstone and sandstone mountain range, underlain by Cretaceous dolomitic limestone and a sequence of Jurassic or Cretaceous evaporitic anhydrite and halite beds. It is isolated; its closest other volcano of geologically young age is 200 km away (Duffield et al. 1984). Apart from the Müllerried study, the only published information about El Chichón at the time was a brief geographical description in the International Association of Volcanology and Chemistry of Earth's Interior (IAVCEI) Catalog of Active Volcanoes of the World and Solfatara Fields (Mooser et al. 1958) and a single mineralogy-oriented paper (Damon and Montesinos 1978).

At the time, the IGEF did not have a department of volcanology and we belonged to the Department of Seismology. At UNAM, the staff involved with seismic studies was split into two separate institutes: (1) Geophysics, which focused on tectonic seismology and (2) Engineering, concerned with construction in earthquake-prone environments (IGEF and IINGEN, respectively). Mexico is a country with high tectonic seismicity and recording the aftershock activity that followed major earthquakes was a common task. That was done mostly using 15 kg MEQ-800 "portable" seismic stations each consisting of a battery-operated smoked paper helicorder, a vertical-motion seismometer, and two car batteries with a charger to power electronics for a reasonable time, plus various accessory equipment.

Survival Tips for Field Volcanologists

In emergency situations, always have an escape plan; know the lay of the land and available routes for making any exit needed. Look over your shoulders from time to time to be sure that you are not blindsided by something happening nearby; e.g. a hidden lava flow moving through the forest, a slope that is shedding small, frequent rockfalls, strange sounds. Near threatened structures watch out for "secondary" hazards, such as downed electric lines, potentially explosive fuel tanks or septic systems. Be personally prepared. . .have personal protection equipment handy, plenty of water, flashlight and batteries, gloves, strong boots, basic first aid kit, and survival snacks. Radio or cellphone communication capability is essential.

Two teams with participants of each institute (staff and graduate students) were organized with the plan to leave for Chiapas on 25 or 26 March with each team carrying five seismic stations for installation at various places around El Chichón. Telemetry was out of the question. Radio transmission requires expensive equipment, and it can be very difficult on short notice to obtain permission to use specific new radio-communication frequencies. Cellular telephones in Mexico were then a subject of science fiction. So, data from each portable field station had to be collected daily by people managing the station directly in the field.

In 1982, the IINGEN was also participating in a major CFE project: The building of the Chicoasén hydroelectric dam, a project that included the study of reservoir-induced seismicity. The dam is located about 49 km south of the volcano, and its seismic monitoring network included stations located from 27 to 62 km south of El Chichón. That seismic network was not intended for volcano monitoring nor for real-time monitoring of the dam. The helicorder paper records were collected and sent to UNAM every month for analysis. Only weeks after the eruption ended did we learn that shallow volcanic seismicity preceding the eruption by at least a month, was recorded there, unnoticed as it took place (Havskov et al. 1983). Detailed analysis of the seismograms took months to years to be interpreted in greater detail (e.g. Medina et al. 1990, 1992; Yokoyama et al. 1992; Jiménez et al. 1999; Legrand et al. 2015).

Since IINGEN already had some infrastructure in the dam area, south of the volcano, they would be covering that southern sector in the event of an eruption, while IGEF would be covering the northern sector. For speedy access, this meant that we would be flying either to Villahermosa 75 km north of the volcano or Tuxtla Gutiérrez, 72 km south of El Chichón, the state capitals of Tabasco and Chiapas, respectively, in an airplane chartered by the Chiapas government, and then by helicopter into the field. Unexpectedly, the day before the initial planned departure, a phone call from the Chiapas government informed us that the airplane and the helicopter were engaged and would not be available until the following week.

On the night of Sunday, 28 March, 1982, at 11:15 p.m., a sudden, powerful steam-rich explosion marked the onset of El Chichón eruption. The 26 March flight cancellations prevented us from recording some of the precursory seismicity of the volcano, and the eruption itself, but it also probably may have saved the lives of some of the mission participants, as several of the manned stations were to have been installed right on the flanks of the volcano.

The opening explosion produced an eruptive column about 18.5 km high (Matson 1984). Ashfall and ballistic lithics made many roads impassable in the region and forced the closure of the airports at Villahermosa and Tuxtla Gutiérrez. However, no pyroclastic flows or surges – deadly clouds of hot ash and other fragments spreading across the landscape – were produced that day. Nonetheless, around 100 lives were lost, mostly from roof collapses caused by ashfall and falling stony fragments, mainly in Nicapa, a small town 7 km to the north–northeast. There, a church roof collapsed on people taking shelter.

In response, the next day, the UNAM group separated into smaller teams of two or three that traveled to impacted or potentially threatened areas, using any available transportation. Petroleos Mexicanos (PEMEX), the national oil company, offered several vehicles to transport staff and equipment in a DC-3. But the plane could not fly farther than the city of Veracruz because drifting ash clouds forced the closing of the air space in the whole of southeastern Mexico.

My team consisted of three IGEF staff scientists and two graduate students. In the following days, many other scientists from the Institutes of Geophysics, Engineering, Geology, and other UNAM institutes (about 15 scientists, technicians, and graduate students from IGEF, and 10 more from the other UNAM institutes) as well as from several federal and local governments and other universities joined this operation (De la Cruz-Reyna and Martin Del Pozzo 2009).

In Veracruz, PEMEX provided us with pick-up trucks to carry our instrumentation into the field. As we approached El Chichón, the effects of ashfalls became increasingly evident (Figure 1.15). Villahermosa was in a noon darkness, and the green jungle around the area was mantled by the white–gray ash that extended over most of eastern Mexico (Figure 1.16). Even ships in the Gulf of Mexico reported ash falling on their decks.

We reached the vicinity of the new eruption on the afternoon of Monday, 29 March, and began to install and operate seismic stations around the north sector of the volcano that evening and early the next day. In Villahermosa, the state government, through its Ministry of Transport and Communications, offered valuable logistical help and personnel to assist setting-up the portable seismic network and drive vehicles. Systematic smoking and collection of seismogram records on the entire northern portable network was achieved by Wednesday, 31 March, just a few days after the eruption started (Havskov et al. 1983; Jiménez et al. 1999). Simultaneously, the IINGEN made a similar deployment of monitoring stations in the southern sector of the volcano, arriving from central Chiapas. A smaller number of portable stations were deployed in the southern sector, since the Chicoasén dam network was also recording the activity. However, the blocked roads and the lack of telephone or radio communications made it impossible to keep sustained contact between the widely dispersed teams (De la Cruz-Reyna and Martin Del Pozzo 2009).

Our first impression upon arrival in the region affected by heavy ashfall was the great confusion and disbelief that prevailed among the people. Many highways were closed by the poor visibility. Official information was scant, and the perception of the ongoing phenomenon among different officials and the general public varied widely, so that it was difficult to find coincident opinions. We set up a headquarters in the town of Teapa, about 36 km northeast of El Chichón, a town in which several roads surrounding the volcano converged. Collected smoked seismograms were delivered daily there and it was the place where information could be shared about other participants and the development of the crisis.

FIGURE 1.15 The road from Coatzacoalcos to Villahermosa on 29 March, 1982. Ashfall reduced visibility to about 100 m. Source: Photo by S. De la Cruz-Reyna.

FIGURE 1.16 Ash blanketed land near Villahermosa. Note the ash sticking on the cables, affecting power and telephone communications. Source: Photo by S. De la Cruz-Reyna.

It was evident from looking at the seismograms during the first days of seismic network operation that the eruption that began on Sunday, 28 March, was not over. The few stations installed between 29 and 30 March showed an almost complete calm in the first hours of Tuesday, 30 March, but large-amplitude tremors mixed with long-period (LP) earthquakes started up later that day, a phenomenon associated with shallow moving magma (Jiménez et al. 1999). The seismograms recorded continuing earthquakes through the next day, including a four-hour long period of explosive blasting, less intense than the opening salvo, but still worrisome (Yokoyama et al. 1992; Jiménez et al. 1999).

Neither a volcanic emergency plan nor a Civil Protection or Civil Defense organization existed at the time in Mexico. The national Civil Protection System was not created until several years later. At the time of the first eruption, no civil single decision-making

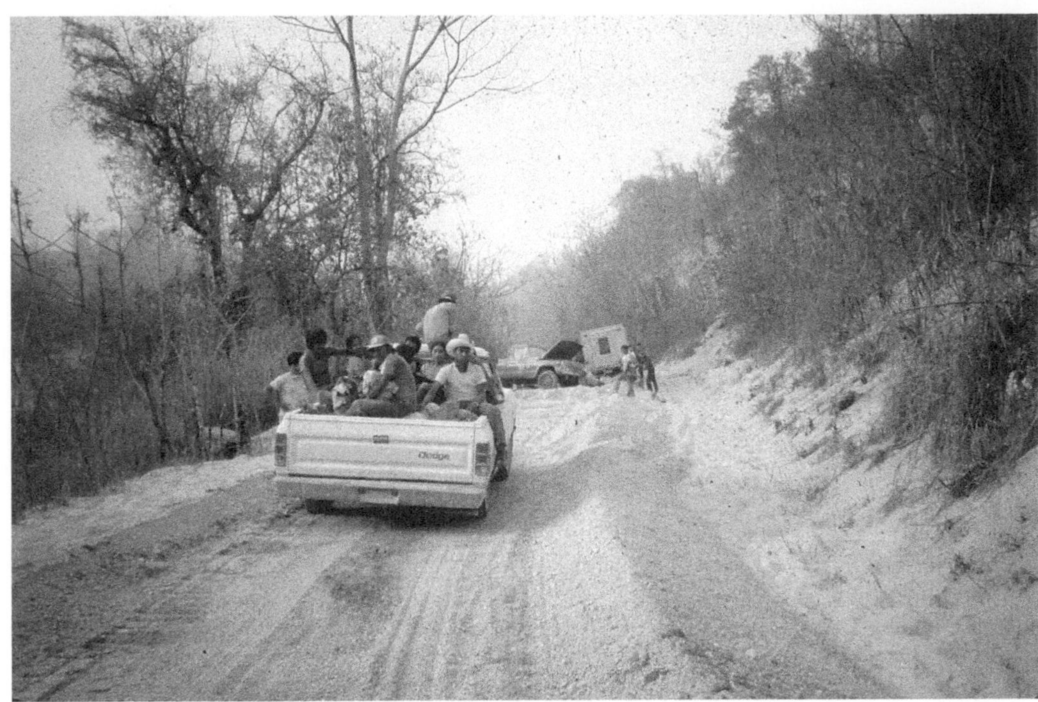

FIGURE 1.17 Traffic chaos in ash-covered roads. Cars damaged by thick ash deposits made two-way traffic almost impossible. Source: Photo by S. De la Cruz-Reyna.

institution existed that could manage the early stages of the crisis. As a consequence, the initial actions after the initial explosion on 28 March were uncoordinated and chaotic.

During the installation of the first seismic stations, and later, on the daily drives to each station to collect seismic records and service equipment, it was common to see ash-covered country roads jammed with vehicles with farmers trying to flee the area and vehicles driven in the opposite direction by people looking for relatives or trying to help an unplanned evacuation (Figure 1.17). Children lost in the confusion walked in despair along the roads. When possible, we stopped to drive them to shelter (Figure 1.18).

FIGURE 1.18 In the confusion, families separated, and Zoque people were very reluctant to evacuate unless families were complete. Source: Photo by S. De la Cruz-Reyna.

The only emergency plan existing at that time was the DN-3 of the Mexican Army, a selective military mobilization proposed in 1966 to help people in disaster situations. Unfortunately, that plan had two serious handicaps (which were later corrected). (1) The plan was mostly designed to deal with only commonplace disaster situations, notably floods, and relied heavily on aerial support by helicopters and fixed-wing aircraft. In addition, the general public and official perception of the hazards related to volcanic activity was prejudiced by the relatively mild eruptions of Popocatépetl [56] in 1919–1927, the birth of Paricutin [52] in 1943, and the 1975–1976 activity of Volcán de Colima [51]. Even the relatively powerful Colima volcano eruption of 1913 was not considered a "disaster" since no confirmed fatalities were reported at the time. (2) The plan also could only be launched by presidential order. Problem (1) precluded the full implementation of the plan, because the ash clouds made it impossible to transport by air the relief personnel and the rescue equipment. The logistics had to be changed to ground operations during the crisis. Problem (2) delayed starting the plan, as the president took a few days to issue the order. The full operation of the Plan actually began on Thursday 1 April (DN-III-E 1983).

Throughout the week of the eruption thousands were evacuated according to DN-3, first in a disorderly and then in a more organized way. The evacuees were transported mostly to the neighboring state of Tabasco. First 29 shelters were installed in schools, and later 8 more facilities were adapted in different cities. The plan evacuated a total of 22,351 persons (DN-III-E 1983). However, because of the confusion prevailing during the week of the eruptions, not all were removed in time, and some evacuees returned to their hometowns before the eruption ended. To the IGEF team on the northern sector of the volcano, it was clear that the main source of this confusion resulted from the lack of a central decision-making authority, expressing a single scientific consensus on the state of the volcano and on the possible evolution of its eruption. No state or federal civil authority assumed an overall lead in managing the crisis. The military central command responsible for implementing DN-3 lacked experience in volcanic phenomena, and the local 31st-Military Zone staff, familiar with the area and the Zoque people, were unhappy with how the situation was being handled. In addition, a separate scientific team from CFE arrived to observe El Chichón at about the same time as we did. They were mostly scientists with ample experience in geothermal exploration and exploitation, but they were as inexperienced dealing with active explosive volcanism as we were. Since they came from the state power company, the chief scientist of that team was appointed by the DN-3 commander as the sole scientific advisor for crisis management.

Communications between the now multiple scientific teams were off to a bad start because of this division-in-force. Our attempts to discuss scientific information were not welcome, and even worse, not all the authorities were willing to listen to us. Although the UNAM crews manning the seismic stations could at times exchange information in the Teapa headquarters, no organized meetings with the other groups (CFE or DN-3 command) were possible nor encouraged, meaning that each group obtained only a partial view of the eruption as it unfolded. Several attempts to talk with the DN-3 commander received the same answer: *We already have scientific advice and no more is needed.* In particular, on Thursday, 1 April, in another attempt to show our seismograms to the commander, the answer was: The CFE chief scientist says that the worst of the eruption has already passed, and no further evacuations are planned. In addition, the authorities reprimanded one of our graduate students for telling residents that the eruption might well continue.

We were not convinced that the "worst has already passed." Our seismic network indicated that the mountain could erupt again dangerously. Fortunately, we retained full support from local institutions, including The Ministry of Transport and Communications of the State of Tabasco, as well as the local PEMEX and CFE offices. Likewise, the local military command, in particular, the general commanding the 31st military zone carefully listened to our information, though his ability to act upon it was overridden by the DN-3 commander – for the duration of the emergency.

We kept operating the portable seismic stations trying to understand the volcano's behavior, particularly after the second, moderate eruption on the 30th March, after which the seismic activity showed some reduction, and the eruptive activity appeared to have stopped. The air cleared as the ash in the atmosphere settled down. However, episodic LP earthquake swarms and smaller amplitude tremors continued through Thursday and Friday – a warning sign of things come we feared.

Most of the villages near the volcano, such as Guayabal and Colonia el Volcán were nearly abandoned except for the few farmers who returned to watch over their crops and farm animals. Besides the ash, many of the roofs showed holes from ballistic impacts; rocks thrown out like cannon-shot by blasts that fell down onto the structures (Figure 1.19). On Friday, 2 April, despite our concerns, the DN-3 command communicated that the evacuees could return to their towns.

For a few hours on Friday, 2 April, we were able to get a view of El Chichón from Colonia el Volcán, 4 km to the southeast of the crater. Part of the crater dome was gone, expelled during the 28 March eruption. That afternoon only a few men from the village remained there, drinking alcohol to quench their fear. This situation was common in most of the villages surrounding the volcano. We learned that the evacuees were being escorted by authorities back home. In particular, a geologist of the CFE team, Salvador Soto Pineda, was commissioned by DN-3 command to conduct a group of soldiers and an unknown number of villagers back to their town: Francisco León, 5 km south of the volcano (Tilling 2009).

FIGURE 1.19 The damage of the initial major steam-blast explosion and the second moderate eruption was mostly caused by lithic impacts and ashfall. Source: Photo by S. De la Cruz-Reyna.

FIGURE 1.20 The initial explosion of the Plinian phase of Saturday, 3 April, 1982. Source: Photo by S. De la Cruz-Reyna taken from Ostuacán, 11.3 km of the volcano.

Two UNAM colleagues, a graduate student and I were the crew manning the northwest sector seismic stations. The last station that we needed to service each day was OSC, at a 11.3 km distance from El Chichón. We had placed it between the small village of Ostuacán and the foot of a small hill from which the volcano could be seen. Every evening we drove to that place, and on Saturday, 3 April, we were particularly interested to see the seismogram, because the other stations showed a significant reduction of the seismicity through the course of the day. In addition, the commander of the 31st military zone, the army general who had a good perception of the situation and was interested in our monitoring, had settled a provisional headquarters in the small Municipal Hall building of Ostuacán. We left the car in front of that building and walked to the seismometer site. It was almost dark when we finished replacing the smoked paper around the seismic recording drum, and I was adjusting the dials of the MEQ-800 helicorder when, at 19:35, the seismograph stylus suddenly began pounding strongly against the edges of the penmotor casing. I just kept staring at the stylus, wondering what was wrong with the equipment, when an impressive rumble and the trembling of the ground could be felt. It took us a bit to realize that a major blast was beginning. By quickly climbing the small hill, we could see and photograph the initial explosion (Figure 1.20). Then, minutes later, strong lightning illuminated the development of a massive eruptive column – a towering pillar of gas, ash, and other ejecta rising from the crater. An increasingly intense shower of light pumice began falling on us. The pumice fragments were small and warm.

About 20 minutes later, we could see a dim glow below the developing column. Looking attentively, we realized that it was a glowing cloud heading our way (Figure 1.21), slightly to our right. It took a moment to realize that it was a pyroclastic flow – a cloud of hot gases, ash, and other volcanic particles moving at high speed. We knew that we were standing at about 11 km from the volcano, but in the darkness and the confusion induced by the heavy pumice fall, it was very difficult to estimate the size and speed of that pyroclastic flow. We hurried back to the Municipal Hall and discussed the situation with the general. He instructed his communications officer to contact the main headquarters at Pichucalco by radio. The officer unsuccessfully tried for a long time to contact any other army group. It seems that the static electricity in the falling ash prevented any radio communication. We then decided to split our team, one person staying to record the eruption, and the others returning to Teapa to inform others about the situation. In the following minutes, the falling debris, pumice, and heavier lithic fragments increased in intensity. We could see dents appearing in the tin roof of the Hall, and then the lights went out. The one-floor school building in front of the Hall, which the army was using as supply storage, collapsed under the weight of accumulating ash, pumice, and stony fragments. Two soldiers emerged from the wreckage slightly wounded to report to the officer in charge.

At that moment, we did not believe that the pyroclastic flows could reach Ostuacán, but we were not completely sure. The few locals remaining in Ostuacán, apparently intoxicated, reacted strangely to this situation. They laughed and joked about it. After a while, the intensity of the fragments' fall decreased, and we decided to go back to check the seismograph. It was quite dark outside, and the flashlight beam only reached two or three meters. We opened the seismograph vault, and it was working fine. The seismograph had remained saturated with seismic noise, penned to the stops for about 20 minutes and then the signal gradually decreased its amplitude to produce a legible record. Since the road probably was covered by thick layers of ash, and the seismograph was recording variable signals, we took turns to sleep and to check the record.

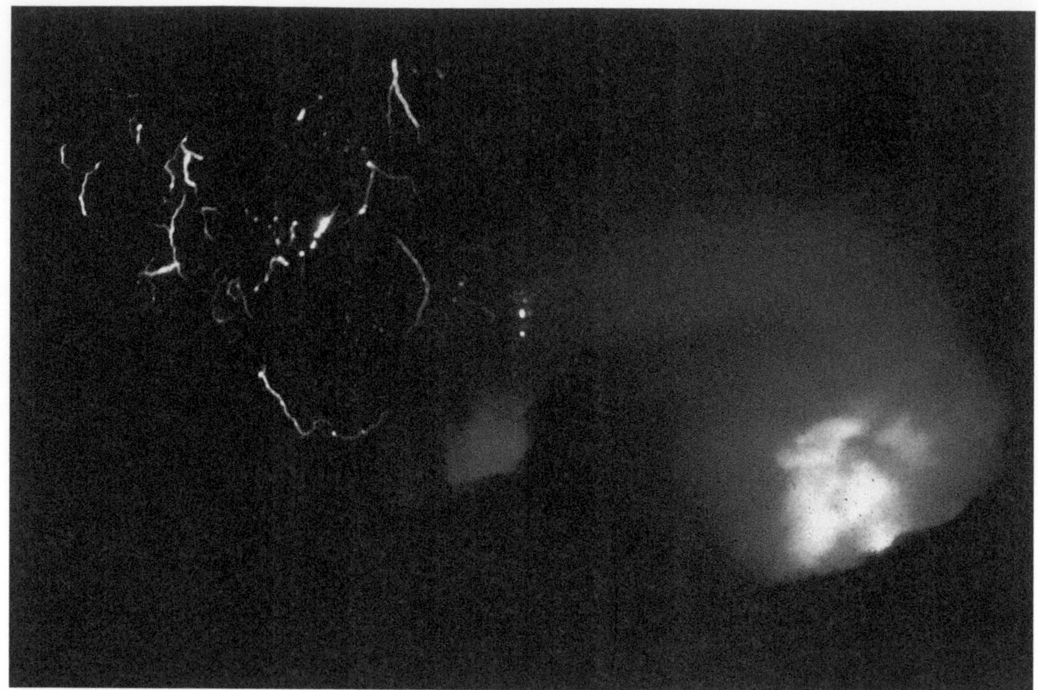

FIGURE 1.21 The glow in the lower right corner is one of the pyroclastic flows generated during the Plinian phase of the 3 April, 1982 eruption that was coming directly toward us. The Plinian plume is marked by continuous lightning. Source: Photo taken from Ostuacán by S. De la Cruz-Reyna.

A few hours of relative calm followed, but at 5:10 a.m. on Sunday, 4 April, another rumbling began, and the seismograph stylus started pounding again. This time the seismograph remained saturated for more than two hours. It was impossible to see anything, and in the darkness, the thought that many people might be dying that night was quite oppressive. That Sunday morning the sun did not appear. Total darkness caused by the ash cloud lasted until about 3:00 p.m. that day making it impossible to see what was happening beyond about 20 meters.

The following days and weeks consisted of frantic efforts to keep seismic stations operating and to appraise the extent and impact of the eruption. This time the seismograms showed a marked change in the nature of the seismicity. After the main eruption of Sunday, the continuous tremor signals and the LP events were replaced by volcano-tectonic earthquakes (VT) with clear P and S phases. Although, in the first hours, their very high rate made them to overlap in the helicorder records, in the following few days it became clear the rate of VT activity decayed in an almost exponential form, and the number of events reduced to background levels in about four weeks (Jiménez et al. 1999). This was interpreted as a result of the stress release caused by the eruption in a large volume of crust around the magmatic system, and thus acceptable evidence that the peak of eruptive activity had passed. At that time, we could explore the devastated area, but only partially, as most roads were destroyed by the eruption and by the increasing rain (Figure 1.22). At that time, we had much improved relations with the DN-3 staff and the 31st Zone general, and we could fly in army helicopters to reach devastated areas.

We later learned that the 3 and 4 April eruptions produced ash plumes 32 and 29 km high, respectively (Macías et al. 1997). Pyroclastic flows and surges, and shockwaves swept all flanks of the volcano devastating villages within 6–9 km around the volcano summit (Silva-Mora 1983; Sigurdsson et al. 1984, 1987; Carey and Sigurdsson 1986; Macías et al. 1997; Scholamacchia and Schouwenaars 2009). The eruptions killed a large but unknown number of people (later estimated between 1700 and 2300 from an Army census and the counting of reportedly missing persons among the evacuees), made more than 20,000 homeless, caused severe economic damage, mostly from cattle stock loss, and extensive damage to the coffee, cocoa, and banana plantations. The people of those villages not only lost their property, but they also witnessed first-hand how many of their fellow residents had been severely burned by the pyroclastic flows. The power of the pyroclastic flows and surges was evident in towns like Francisco León (Figure 1.23), where CFE geologist Salvador Soto and returning evacuees were killed at the time of the final eruptive phase, and in El Naranjo.

The March–April eruptions obliterated much of the pre-existing summit dome, forming a new crater about 1 km wide and 230 m deep. An estimate of the total tephra volume ejected by the eruptions was about 1.1 km^3 (Carey and Sigurdsson 1986). Although this amount of mass expelled (magnitude) is not among the largest in the world for volcanic eruptions, the extremely high emission rates (intensity) of the final explosive phases nevertheless had catastrophic consequences. Perhaps a better wording for this was expressed by Jaime Sabines, a Chiapanecan poet who wrote in his "Crónicas del Volcán," referring to the volcano and the 3 and 4 April eruptions "*. . .but on Saturday night, six days after his birth, he returned to do their thing, this time with more force, with abundance, with a terrible generosity.*"

FIGURE 1.22 The road from Pichucalco to Chapultenango was destroyed at several points. Source: Photo by S. De la Cruz-Reyna.

FIGURE 1.23 Remains of the concrete-reinforced church in Francisco León, a town located at the foot of the volcano, about 5 km SW from El Chichón crater. Some people remained in town and others returned after the lull of activity preceding the major Plinian phases – they all perished. Source: Photo by S. De la Cruz-Reyna.

The Magdalena River flows past the southwestern sector of the El Chichón volcanic edifice. Pyroclastic flow and surge deposits of the 1982 eruption formed a dam that blocked the water flow, forming a lake that accumulated 40 or 50 million cubic meters of hot muddy water in about 4–6 weeks (Medina-Martínez 1982; Macías et al. 2004). This situation was explained to the general of the 31st Zone. The DN-3 Plan was over, and the general took independent measures to evacuate that part of Ostuacán and other small towns downstream that might be flooded if this natural dam burst. Sure enough, the pyroclastic dam broke after midnight on 26 May, flooding with scalding mud the threatened riverbank corridor Figure 1.24). The breakout flow reached an electric power

FIGURE 1.24 The devastated area around the El Chichón summit as seen in 1990 – view to the northeast. In the background the lahar-filled Magdalena River drains off towards the Gulf of Mexico. The March-April eruptions obliterated most of the pre-existing summit dome, forming a new crater about 1 km wide and 230 m deep. Source: Photo by S. De la Cruz-Reyna.

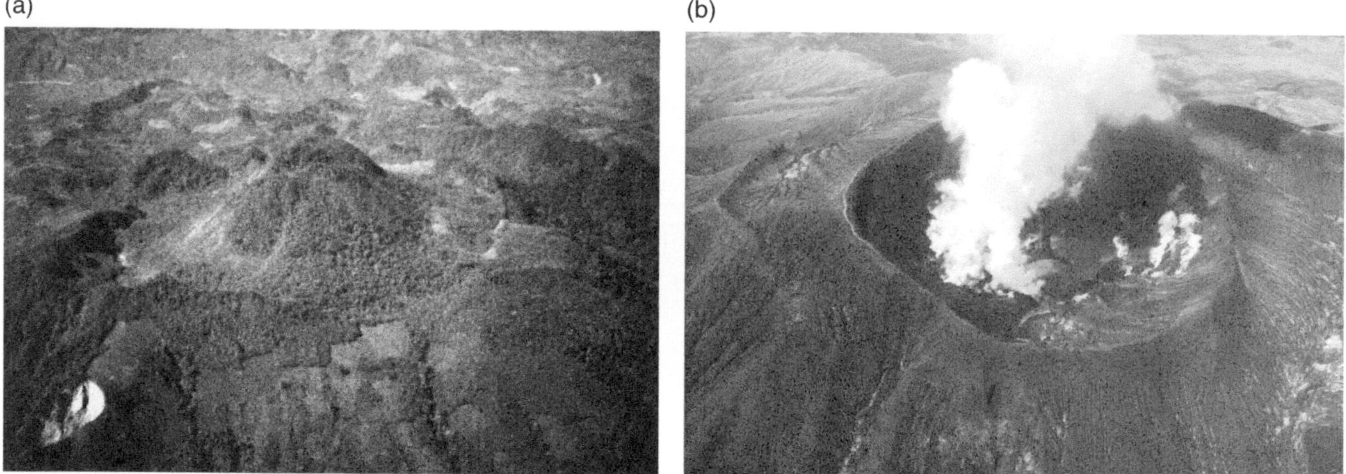

FIGURE 1.25 (a) Summit dome of El Chichón as it looked in early 1982. Source: Photo by Paul Damon. (b) The devastated summit area around El Chichón as seen in 1990. The crater has about the same diameter as the dome in (a). Source: Photo by R. I. Tilling.

facility under construction, the Peñitas hydroelectric dam, about 35 km downstream. Thanks to timely early warning, most people along the river remained safe. However, three employees at the dam were injured and one died. Since mid-September 1982, except for flurries of limited rockfalls from the steep walls of the 1982 crater in June 1992 and September 2017 (triggered by the Chiapas 8.2 earthquake), activity at El Chichón has been restricted to low-level, fluctuating hydrothermal activity in the new crater lake (Casadevall et al. 1984; Taran et al. 1998; Armienta et al. 2000, 2014). Figure 1.25a illustrates the summit dome shortly before the eruption, and in Figure 1.25b, a similar view of the post-eruption crater. After the eruptions, some land north of the devastated area was offered to survivors. However, many returned and rebuilt villages at their original sites.

The El Chichón disaster had several causes, among them the lack of awareness and preparedness among the public and authorities, in part resulting from the long mean recurrence time between major eruptions of the volcano (Espíndola et al. 2000). But a key reason was the lack of an organized procedure compelling the exchange of ideas among scientists, aimed at attaining a consensus with a set of consistent recommendations to accepting decision-making authorities. Perhaps one of the lessons learned from this disaster is that the burden of responsibility involved in forecasting the evolution of a complex phenomenon, such as volcanic eruption cannot be assigned to a single person. It is essential to consider the perspectives of all scientists observing the eruption, and focus the decision-making on convergent opinions. A National System of Civil Protection (SINAPROC) was committed by Mexican law in 1986 following a major earthquake to protect the people from the effects of natural catastrophes, and the DN-3 Plan has been upgraded by the Army to manage a variety of possible emergency scenarios. This in fact is an essential structural part of the SINAPROC. That law includes establishing a standing Scientific Committee for each potentially hazardous natural phenomenon, which must be consulted for crisis management and decision-making.

Another "Gray Volcano" in Eruption – Mount Pinatubo – 1991

Major eruptions commonly confront responding volcanologists with more than "simple" monitoring responsibilities. When human lives and critical social infrastructure are at risk during evolving volcanic crises, volcanologists may find themselves called upon to provide credible advice to government agencies responsible for complex life and death decisions. This was the role I (JPL) played during part of the Pinatubo crisis, as leader of a USGS team. Because the Pinatubo [135] eruption involved considerable loss of life, huge property losses, and had complex international political repercussions, my work there largely involved coordinating monitoring efforts, crisis management, and addressing socio-political aspects of the eruption. Those are aspects of the eruption that will be emphasized here rather than my limited field studies. For scientific descriptions of this major eruption, the magnificent study of Newhall and Punongbayan (1996) is an essential detailed resource. Wolfe and Hoblitt's paper in that volume is an excellent short summary of the eruption and its impacts.

Clark Air Force Base (Clark AFB), then the largest US Air Force (USAF) base outside continental United States, and the once proud "Pearl" of the USAF Pacific Command, was located about 10 km from the base of a heavily forested, deeply eroded mountain called Pinatubo. Although geologists had known that Mount Pinatubo was a dormant volcano, it was only one of dozens of dormant volcanoes along the Luzon Arc. Prior to 1991 no one at Clark nor in surrounding communities knew that this volcano would soon produce the second largest eruption of the twentieth century – an eruption that would have significant economic impact on the Philippines, would affect global climate, and would alter the future of US military bases in the Philippines. Pinatubo was the home of an indigenous tribe of traditional people called the Aeta, whose entire culture was intertwined with the mountain. The Aeta economy was based on hunting in Pinatubo's verdant forests, small-scale farming, and fishing in cool mountain streams. They believed in a benevolent god of the mountain called Apo Namalyari – a god who would become very angry with them in 1991.

Unrest at Mount Pinatubo, first marked by phreatic (steam blast) explosions in early April, increased in tempo for the next two months. Volcanologists from the Philippine Institute of Volcanology and Seismicity (PHIVOLCS) noted the threat, established early seismic monitoring equipment, and recommended an evacuation of Aeta people living near the summit. Long-standing cooperation between PHIVOLCS and USGS volcanologists facilitated the deployment of a small survey team to Pinatubo in late April to work with PHIVOLCS to bolster monitoring.

My direct involvement in the developing Pinatubo crisis began on 30 May, 1991 when I received an urgent early morning phone call from Tom Casadevall, the leading USGS expert on the impact of major explosive eruptions on aircraft safety. Tom had just received a FAX from Chris Newhall, leader of the USGS Pinatubo team, stating his belief that a major eruption of the volcano was imminent, that there was a significant potential threat to aviation, and that major US military facilities and personnel in the Philippines were in jeopardy. Local officers at Clark AFB were aware of the threat, but their superiors at the Pacific Air Force (PACAF) command in Honolulu and at the Pentagon were not convinced that a "mere volcano" could actually threaten the most powerful Air Force base in the western Pacific. Casadevall knew of my high-level Air Force connections from previous lava-diversion experiments in Hawai'i, and asked if I could arrange a meeting at PACAF Headquarters to convince staff that the threat was imminent and that evacuation of Clark should be considered. Casadevall visited the Pentagon to brief Defense Department civilian leaders about the crisis, and I met with U.S. Senator Daniel Inouye, Chairman of the Senate Armed Services Committee on 1 June to brief him on the situation. An emergency meeting was set up for 3 June. Chris Newhall flew in from the Philippines and we met at Hickam AFB on 2 June. That night we received the devastating news that a deadly pyroclastic eruption had just taken place at Unzen volcano [141] (Japan) killing over 40 persons, and that our friends Maurice and Katya Krafft and USGS colleague Harry Glicken were missing. Maurice had recently sent me a draft copy of a video explaining volcanic hazards and the risks associated with pyroclastic flows, and I had planned to show parts of that video at the next day's meeting with PACAF leaders. Maurice was at Unzen for filming, but was planning to travel to Pinatubo, where we had hopes of a possible rendezvous.

Chris began the 3 June PACAF meeting with a description of the Pinatubo situation and the pyroclastic flow risks to Clark AFB. I began my briefing by showing Krafft's video of actual pyroclastic flows they had filmed in Alaska and explained that Maurice and Katya had apparently been killed by such flows in Japan the previous day. This made for a very somber meeting – everyone present understood that Pinatubo was not playing games. After questions, a roomful of generals and colonels agreed the situation was dire and that action was needed – the USAF "Pearl of the Pacific" needed to be evacuated. The PACAF commander (General James Adams), concluded the meeting by delegating responsibilities for various operational aspects of a massive evacuation that lay ahead. Newhall and I were convinced that the needed evacuation would take place, and he continued on to Washington to coordinate efforts at the USGS headquarters. My family had a pre-planned East Coast trip, and I left immediately after the 3 June meeting for New York. I was at Princeton on 7 June, when a magmatic dome appeared on Pinatubo's surface, indicating magma had now reached the volcano's surface. There was no longer any question that the volcano was primed for a major eruption. We were in New Jersey as the eruption intensified, as the Clark AFB evacuation was executed, and when the climactic eruption occurred on 15 June. I returned to Hawaiʻi as soon as I could, and immediately prepared to travel to the Philippines to provide relief to the heroic USGS volcanologists who had just survived their "trial by fire."

More than 15,000 Air Force personnel and civilian retirees had been successfully evacuated from the base on 10 June, as major earthquakes intensified, and a spectacular explosive eruption sent ash over 19 km above the summit on 12 June. The evacuation zone around the summit was increased, and about 58,000 people fled the summit area before the climactic eruption. All non-essential personnel were evacuated from Clark, leaving only the USGS volcanologists with the base commanders and a small security team. The climactic eruption began in full fury just before 2 p.m. on 15 June. Volcanologists observed all seismic stations go silent as they were overwhelmed by the pyroclastic flows that were racing down Pinatubo's slopes toward them. Airfall ash was blanketing the base, pyroclastic flows were about to reach the base perimeter, and base commanders and the volcanologists had no choice but to flee for their lives – driving through near-zero visibility ashfall conditions to a college campus 23 km to the east, where other evacuees were waiting out the ongoing eruption. A typhoon was making landfall that day, and heavy rains fell across the area just as ashfall intensified. About 300 people were killed around the volcano that night as roofs collapsed under the weight of rain-soaked ash. The USGS team and air base commanders returned the next day to plow through ash and document the devastation. All of Clark's aircraft hangars and other buildings had collapsed, thousands of individual homes had been destroyed or seriously damaged, and all but one seismometer on Pinatubo had been destroyed. A riveting description of those chaotic days and the courage displayed by Air Force personnel and civilians is described by Anderegg (2000).

After my return from the East Coast, I packed emergency field gear and "respectable" business attire, then left for the Philippines on 24 June, making a stop at Hickam AFB for briefing about USAF command concerns and capabilities in the current crisis situation. I was then taken to the Philippine Consulate for a meeting with Consul Peter Chan who gave me an overview of the social and political complexities caused by the ongoing eruption, and issued me an "official" visa to expedite customs and immigration clearance. Then, back to the Honolulu Airport for flights to Manila via Seoul. I arrived in Manila late at night on 25 June, and was taken to the US Embassy for a briefing about diplomatic needs and protocols to be observed. The next day at the US Embassy, I met the Clark AFB Commander, General William Studer, who now commanded all military activity related to the eruption, including US Navy operations at Subic Bay, which had also been greatly impacted by the eruption and was the center of ongoing evacuation efforts. Studer briefed me on the extensive space-based intelligence assets available at Clark AFB, assets that proved to be critical for monitoring Pinatubo activity in the weeks ahead. I was then taken to the headquarters of the Philippine Volcano Observatory (PHIVOLCS) in Quezon City, where I met with the director and old friend Ray Punongbayan, who gave me a summary of losses to date (Table 1.2), explaining the logistical and political difficulties in coordinating monitoring and hazard responses in the three separate provinces that surround Pinatubo. After more high-level meetings the next morning, I finally left for the two-hour drive to the USGS-PHIVOLCS operations center that has been set up at Clark AFB.

TABLE 1.2 Mount Pinatubo eruption impact as of 26 June, 1991. Information from PHIVOLCS. The impacts continued to increase owing to the expanding areas of lahar devastation and related deaths in evacuation camps, and nearly doubled over the next year.

IMPACT OF THE JUNE 15 ERUPTION	
Families evacuated	26,646
Persons evacuated	133,862
Dead Missing Injured	289 39 1,500
Homes destroyed	25,078
Homes severely damaged	39,817

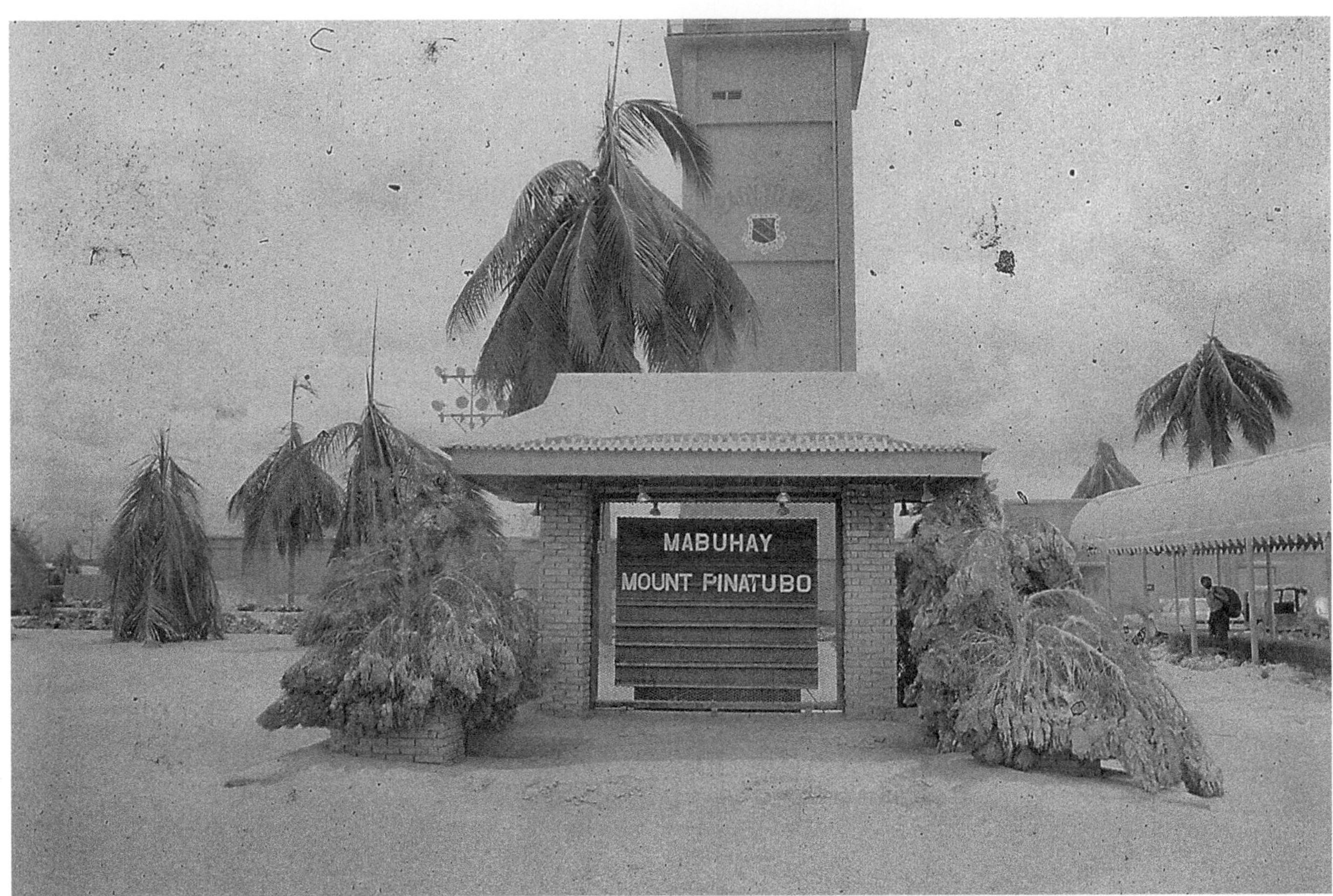

FIGURE 1.26 Headquarters of the Clark AFB in late June 1991. This is a color photograph, but there is no color to be seen anywhere – everything is covered with gray ash. The drooping, broken palm tree fronds are the universal signatures of explosive eruptions anywhere in the tropics where more than a few centimeters of wet ash fall. Source: USGS photo by J.P. Lockwood.

Light ash was falling from leaden skies as our embassy car entered the ruins of Clark air base on 27 June, 1991 – everything was buried by a half meter of tephra (Figure 1.26). I met with my weary, battle-scarred USGS and PHIVOLCS colleagues who had been under incredible pressure for the past fortnight dealing with uncertain eruption conditions and efforts to restore basic monitoring capabilities. They had named their ash-covered facilities "PVO" – the Pinatubo Volcano Observatory. A quick drive around revealed the almost total destruction of homes and large buildings at the base (Figure 1.27). The 1,500 or so Filipino employees who had worked at the base before the eruption were now unemployed, and the adjoining Angeles City, was now mostly buried by ash and rendered economically devastated.

A large caldera had apparently formed at the Pinatubo summit on 15 June, but continuous ash generation and pyroclastic flows pouring down upper flanks prevented any direct knowledge about caldera morphology or evolution. A helicopter flight the next morning allowed a view of an eruption plume boiling upward from the caldera (Figure 1.28). Near the base of the plume, small pyroclastic flows poured over low portions of the unseen caldera rim (Figure 1.29) and flowed short distances down upper slopes. Light ashfall continued to blanket the summit area, especially to the northwest, where initial pyroclastic flow impact was the greatest. These areas, in Tarlac and Zambales Provinces, were the areas where most Aeta evacuees had been settled in refugee camps and ongoing lahar (mudflow) damage was particularly severe. Clark AFB was located in Pampanga Province, and coordination between volcanologists and emergency agencies in these areas was especially difficult (Rodolfo 1995).

My dedicated USGS and PHIVOLCS colleagues at the Pinatubo Volcano Observatory (PVO) were outstanding as they worked in dangerous situations with limited helicopter support to restore seismic stations on the devastated slopes of Pinatubo. Restoration of these monitoring and telemetry stations was complex, but six stations were operating by 30 June. Station locations were uncertain in these pre-GPS days – leading at times to wildly divergent earthquake locations. Earthquakes with magnitudes exceeding M = 5 were continuously jolting the area and indicated expansion of the caldera. At one point it appeared a new ring fault might be developing east of the caldera – perilously close to Clark. Was the caldera about to enlarge with another major eruption? Harmonic tremor associated with magma was widely felt and caused furniture to dance across floors. Access to classified military satellite imagery eventually was arranged, and it was learned that the caldera was only slightly increasing in size, lessening our fears. The

FIGURE 1.27 Civilian housing at Clark AFB on 27 June. About 15,000 civilians had been removed from such homes during an emergency evacuation on 10 June as the Pinatubo [135] eruption intensified – just before the climactic 15 June eruption. Evacuees left almost all personal belongings and many pets behind as they boarded buses to USN refugee accommodations at Subic Bay – not realizing that that area was also about to be severely impacted by falling ash on 15 June. Source: USGS photo by J.P. Lockwood.

military satellite systems were invaluable, able to "see" through dense ash clouds, and enabled me to monitor changes in caldera morphology as the eruption continued. The ingenuity of Air Force technicians at Clark was amazing. They managed to find a "Cyclops" infra-red monitoring system that would normally be attached to an F-4 fighter jet and jury-rigged it to the base control tower with a remote display, so that volcanologists could evaluate eruptive thermal activity real-time and observe hot pyroclastic flows emerging from the rim of the developing caldera! On the early morning of 2 July, a particularly large eruption occurred, and

FIGURE 1.28 Basal area of sub-Plinian eruption plume boiling above Pinatubo caldera [135] This plume was rising about 10 km high, but we were flying too close to see the upper portions. Such plumes are actively changing every few seconds, with black jets shooting out of the base (temperatures too high for steam to condense), and cooler mixtures of gray steam and ash forming contorted, interfingering pulses of convecting tephra higher up. At night, these plumes are marked by continuous lightning, and red fire at their bases. Source: USGS Photo by J.P. Lockwood.

FIGURE 1.29 Small pyroclastic flow boiling over Pinatubo caldera [135] rim. These gas-rich clouds only traveled a few hundred meters downslope, and in some cases were "sucked back" into the base of the ascending plume as strong winds rushed upslope to feed ascending plumes. Source: USGS photo by J.P. Lockwood.

the Air Force Cyclops operator on duty called me to report that incandescent material was blanketing Pinatubo's upper slopes. Continuous large earthquakes were "clipping" seismic signals on all seismographs, harmonic tremor was very strong, and could be felt by us at Clark, 25 km from the active caldera. Continuous lightning illuminated the rising plume, and at 03:40, Manila Airport radars indicated rising eruption plume almost 17 km high with an umbrella cloud more than 10 km diameter! By 04:20, the tremor subsided, the plume disappeared, and everyone went back to sleep. Frequent large felt earthquakes became "normal" background for our work.

Eruptive activity continued, and radar measurements from Manila showed that sub-Plinian explosion plumes were rising sporadically to over 15 km above the volcano nearly every day. I took several reconnaissance flights with USAF and US Navy helicopters over the next few days to deploy field crews to reestablish seismic stations and survey the situation above Clark and on the populated areas below. During these recon flights, I was impressed by the huge volumes of unconsolidated airfall ash clinging to the volcano's slopes and by the even greater volumes of pyroclastic flow deposits that had choked all pre-existing drainages – to depths obviously exceeding 100 m. Like my colleagues, especially the late Dick Janda, I realized the terrible risks that were facing the hundreds of thousands of people living along Pinatubo drainages on the flat-lying agricultural lands below. Dick and I had both witnessed the terrible carnage downslope of Ruiz volcano in Colombia [71] a few years earlier (Chapter 14) when nearly 30,000 people were killed by lahars. I was in a very somber mood as I finished a flight and was contemplating the risks faced by people living on the flat-lying ground only a few meters above the level of existing rivers that drain Pinatubo. My 29 June field work includes this note:

> *"I realize now that after complaining about "what others didn't do" at Ruiz in 1985, I'm now in the same situation. What can be done to save lives in this area where there is no high ground to flee to?"*

Major lahars (mudflows) had been generated during the initial climactic eruption, transitioning from coarse debris flows upslope to hyperconcentrated stream deposits on the plains below. Even the minor rains that had occurred so far had begun to move unconsolidated ash downslope (Figure 1.30), but the worst of seasonal monsoon rains had not yet begun. Erosion of these deposits threatened vulnerable areas on the populated alluvial plains surrounding Pinatubo, where more than 250,000 people lived. More than 5 km^3 of unconsolidated pyroclastic flow deposits (up to 200 m thick in buried valleys) and about 4 km^3 of airfall ash mantling slopes was rapidly eroding, forming lahars and debris flows that were filling river drainages in and below Pinatubo. The airfall ash mantle was particularly vulnerable to erosion by heavy rainfall, and sediment deposition was blocking drainages and causing widespread flooding (Figure 1.31).

FIGURE 1.30 Incipient erosion of airfall tephra on the lower slopes of Pinatubo [135] on blast-denuded slopes about 10 km east of the former Pinatubo summit. Source: USGS photo by J.P. Lockwood.

FIGURE 1.31 Flooding along the Sacobia River below Mount Pinatubo [135] foothills. Light volcanic ash is falling, and ash has blanketed everything. Lahar deposits have filled river channels with sediment, forcing streams to flood adjoining villages and fields – tens of thousands of people have been displaced. Source: USGS photo by J.P. Lockwood.

Weather monitoring became critical for predicting lahar activity to warn people and order evacuations before their homes were impacted. The ingenuity of PHIVOLCS technicians was again amazing. Telemetered rain gauges were installed on Pinatubo's slopes, and rainfall intensity on Pinatubo's slopes gave invaluable warnings to trigger evacuations orders for threatened populations. The seasonal monsoon and cyclone-related storms that impacted the volcano caused incredibly heavy rainfall. At 19:30, on 9 July, a rain gauge on the east flank of Pinatubo reported that one inch of rain had fallen in 13 minutes! At the same time, Manila airport reported that an ash plume was ascending over 15 km high, adding fresh airfall ash to the picture. Lahar warning monitors were not yet in place, and the probability that massive lahars were cascading down all river drainages was extreme. That night and the next day were devoted to supporting PHIVOLCS and emergency management agencies in delivering house-to-house warnings to vulnerable populations. The hazards were two-fold: Lahars were eroding stream banks and undermining homes in upstream areas, and rapid sediment deposition from hyperconcentrated flows downstream filled existing channels and caused rapid flooding of streamside areas.

Aerial inspections of the summit area when weather conditions allowed showed that small pyroclastic flows were continuing to form by boilover from the bases of eruption plumes and were pouring down drainages (Figure 1.32). Thin volcanic ash deposits re-blanketed upper slopes after each of the frequent summit explosions. The thick pyroclastic flows that had buried pre-existing drainages were hot, probably over 500°C, and when their interiors were exposed by erosion or slope failure, violent steam eruptions formed – sending ash clouds to as much as 10 km height. These explosions provoked speculations about renewed eruptive activity on the middle slopes of Pinatubo and resulted in great civil concern. I spent a great deal of time with reporters assuring them that these phreatic explosions did not indicate renewed eruptive activity. Rumors and crazy risk mitigation ideas were inflated by the media, and my efforts were frequently directed at tamping down these stories. At one point it was suggested that active lahars might be bombed to divert them or to make diversion channels, and my advice was sought to help the Philippine Air Force plan such missions! Great diplomacy was required in such situations – because the ideas may have been proposed by important politicians!

Making sure that volcanological realities were understood by the many agencies that were involved in disaster relief was one of my most critical responsibilities. I represented our USGS team in what seemed like endless high-level meetings, with politicians, emergency service providers, and international lending agencies. Major international assistance came from several UN agencies,

FIGURE 1.32 Front of a small pyroclastic flow pouring down a ravine on the upper flanks of Pinatubo [135]. As our helicopter pulled away, the ash flow passed harmlessly below us. Note the denuded slopes – this area had been covered with dense forest before the 15 June climactic eruption. Source: USGS photo by J.P. Lockwood.

the World Bank, the Asian Development Bank, and many countries, including Australia, Canada, France, Germany, Israel, Japan, the Netherlands, United Kingdom, and the United States. Dozens of NGOs also were involved in relief efforts. A problem I soon encountered was the realization that these agencies are accustomed to dealing with disasters that happen suddenly – like earthquakes or floods and are then over. These agencies are accustomed to developing fixed budgets for funding restoration work and then moving on to deal with other disasters. The idea that volcanic eruptions can last for months or longer is not easy to budget for, and the concept that destructive activity can last long after eruptions end was completely beyond their traditional thinking. I needed to explain at every meeting that the impact of the Pinatubo eruption would last long following the eruption ended, and that property losses and fatalities would likely occur for several years as volcanic ash continued to erode from Pinatubo's slopes and cause devastation on the slopes below. The needs of the Philippines were long-term, and international assistance should reflect that fact.

One meeting I particularly remember was called on 7 July at a Philippine Air Force Base by General Fidel Ramos, then head of the Philippines National Disaster Coordination Council (Civil Defense). I sat next to Ramos and was impressed by his military background that allowed for no wasted time. He asked good questions and expected succinct, brief answers. He clearly understood my statement that the destructive impacts of lahars would be affecting this part of central Luzon for long after the eruption ended. He wanted ideas as to how the displaced populations would be able to coexist with Pinatubo if the lahar hazards persisted for several years? When would the area be safe again for economic rejuvenation? It turned out that General Ramos would be elected president of the Philippines in 1992 and would be later instrumental in the economic rejuvenation of the Pampangas area.

Dick Janda and I represented the USGS in a particularly critical meeting organized by the UNDRO (UN Disaster Relief Organization) and UNDP (UN Development Program) in Manila on 12 July. Almost 30 major disaster relief agencies were represented. These international agencies had immediately reacted to the Pinatubo disaster by sending staff resources and by committing large amounts of funding for relief and reconstruction, but the eruption had taken place almost a month ago, and we geologists were telling these agencies that the disaster was not yet over! Losses of lives and property could be expected to continue for several more years? The Philippine government representatives were told bluntly that "donor fatigue" had set in, that these agencies had other disasters to deal with (almost 2000 people had just been killed by flooding in China), and that further disaster relief would largely be the responsibility of the Philippine government and people. Fortunately, the ingenuity and resilience of the Filipino people was already a major factor in the recovery of the area. Even before I left the Philippines in late July, while the eruption continued and destructive lahars poured off Pinatubo, resourceful entrepreneurs were excavating lahar deposits to recover charcoal for sale (Figure 1.33), residents were excavating buildings that had been completely buried by ash and lahar deposits to recover materials to rebuild homes on higher ground, and industrial uses for volcanic ash were being explored. Mount Pinatubo sandy ash is now essential for concrete production throughout Luzon – modern Manila is being built with Pinatubo ash!

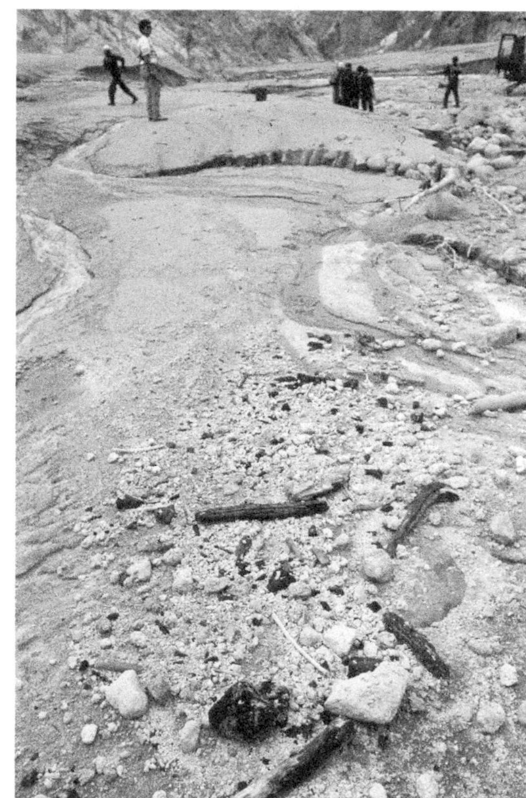

FIGURE 1.33 Eroding pyroclastic flow deposits on the upper Sacobia River, 10 km east of the Pinatubo summit [135]. Note the abundant carbonized wood fragments, which were already being mined downstream by displaced people for sale as charcoal. Source: USGS photo by J.P. Lockwood.

On 15 July, an eruption plume rose to 15 km above Pinatubo and intense tremor was felt by all of us at PVO. A morning flight around the summit showed that dangerous lakes were forming above ash-impounded river drainages, increasing the lahar hazard should ash dams break. On 16 July, Chris Newhall returned to assume oversight of PVO activities, and I packed up and departed for Manila that evening. I was committed to fieldwork in Kamchatka, and I needed to return to Hawai'i to prepare for that expedition. Leaving PVO and the dedicated USAF, USGS, and PHIVOLCS staff was bittersweet – the eruption was not over and I knew many crises and sleepless nights lay ahead. Light ash from the latest Pinatubo explosion began to fall that afternoon in Manila, the international airport closed, and my flight reservations were cancelled. 17 June was devoted to clearing ash from runways, but I was able to board a flight early the next morning. Fortunately, I had filled a large sample bag with Pinatubo ash from the PVO entrance to take along as "souvenirs" as I had left Clark two days ago. This proved useful at Narita Airport, when I was able to convince a Northwest Airlines supervisor that the ash would be extremely useful as fertilizer for his bonsai plants. I do not know if this was true – but was able to trade some ash for an upgrade to First Class for the flight to Honolulu! On arrival back in Hawai'i, I was able to see for the first time the spectacular atmospheric effects of the ongoing Pinatubo eruptive activity (Figure 1.34), but left the next day for fieldwork at Gorelli volcano [157]. This was too far north to admire Pinatubo atmospheric effects, but I was there to witness the August palace revolution that toppled the Soviet Union. I arrived with a USSR flag flying, but with my Russian volcanologist colleagues, saw the old Russian flag raised just before my departure!

FIGURE 1.34 "Pinatubo Sunset" as viewed from Hawai'i on 21 July, 1991. Such sunsets were observed around the world for weeks after the eruption. Source: Photo by J.P. Lockwood.

Epilogue – What Lessons Were Learned from the 1991 Eruption?

The lessons taught by the Mount Pinatubo [135] eruption were many. The most important were observations of the 1991–1993 global climatic eruption impact. Self and others (in Newhall and Punongbayan 1996) showed that the immense volumes of SO_2 injected into the stratosphere "produced the largest perturbation to the atmosphere since the eruption of Krakatau [128] in 1883." They showed that global cooling more than offset the warming effects of anthropogenic greenhouse gas emissions over this period. For me it was important to learn that military satellite monitoring systems could be invaluable for monitoring eruptive activity. This knowledge proved important for establishing subsequent cooperative civilian-military programs that led directly to the saving of thousands of lives during the eruption of Merapi volcano, Indonesia in 2010 (Chapter 14).

For volcanologists who deal with major disasters and for society in general, it is important to take a long-term view of disaster impacts. The social and economic scars of major eruptions *will* heal over time; the economic recovery of the Pinatubo area after 30 years is a good example. After being elected president of the Philippines in 1992, and after ownership of Clark AFB was transferred to the Philippine government, one of Ramos's early acts was to create an entity called the "Clark Development Corporation." The CDC declared the former American air base to be an "Economic Freeport" with slashed tax rates and investment incentives. Results of this decision have been amazing! Foreign investment poured into the area, and the former Clark area has become a major tourist destination. The thick 1991 volcanic ash deposits that buried the area in 1991 are no longer visible, and the area is now carpeted with five-star hotels, gambling casinos, duty-free mega stores, a golf course, and light industrial facilities. Workers who lost their jobs in 1991 have learned new skills to qualify for higher-paying work. Their children sought higher education and now have technological and managerial jobs. The Aeta, whose forest homes were destroyed and who suffered greatly in refugee camps are now returning to work in reforestation projects and jobs related to tourism. Recovery of areas to the west of Pinatubo, where lahar damage was greater, have been more limited, although ash has increased the fertility of agricultural lands. Skidmore and Toya (2002) and Cuaresma et al. (2008) refer to the "Schrumpterian Concept" of *Creative Destruction*, and make the case that natural disasters can serve to rejuvenate societies and spur economic growth. This has certainly proved true in the aftermath of the 1991 Mount Pinatubo eruption, and might be remembered by volcanologists for some consolation as they deal with the death and destruction associated with volcanic disasters. Human societies are indeed resilient, and they do heal over time. Legends about the Phoenix Bird arising from the ashes do come true, but this might take a long time. Pompeii did not begin to be restored for almost 2,000 years!

Conclusions from These Three Narratives

Some useful generalities:

1. All volcanic eruptions are different, the large ones evoke feelings of awe, excitement, and to a greater or lesser extent, concern and fear for those nearby.
2. Effusive red volcanoes such as Kīlauea [17] tend to have frequent, gentle eruptions that can be studied close up, whereas eruptions of the explosive gray volcanoes are less frequent but have far-reaching impacts.
3. Eruptions of gray volcanoes like El Chichón [57] and Pinatubo [135] are more dangerous than those of red ones, and much more caution is needed during close observation.
4. People's lives can best be protected in areas where eruptive activity is frequent (e.g. Java, Hawai'i). This is because in areas of relatively frequent eruptive activity, residents are better informed and are more likely to recognize early warning signs of impending activity, and volcanologists will be able to prepare more accurate predictions of future activity. Most important, local residents are more likely to accept the advice of public officials and comply with mitigation efforts such as evacuation.
5. The magma reservoirs beneath effusive volcanoes are typically complex and dynamic, and they can inject magma great distances at shallow depths underground, even tens of kilometers from the volcano's top. Flank eruptions are common and can be the norm. In contrast, the magma systems beneath typical explosive volcanoes are usually more localized beneath volcano summits.
6. Erupting lava and gases are the major concerns of effusive eruptions on red volcanoes, but simple field precautions can usually prevent disaster. In contrast, explosive eruptions of gray volcanoes typically involve serious hazards such as falling ash, pyroclastic ash flows (Chapter 7), and mudflows (Chapter 11). Eruptions may threaten extensive areas, and evacuations of large numbers of people are often necessary. This makes a major volcanic eruption as much a sociological and economic event as a geological one.

These eruption accounts illustrate some of the reasons that volcanoes are being studied and illustrate the sort of practical use volcanologists are trying to make of their knowledge. The modern science of volcanology combines well-documented field observations and instrumental surveillance of active volcanoes, as well as careful field mapping of older volcanoes to learn their histories. Laboratory studies that reveal how magma is formed, stored, and evolved beneath active volcanoes are also vitally important. The long-term goal of all volcanologists is to have a better understanding of "how volcanoes work." Only such understanding of volcanoes can lead to the knowledge and development of tools needed to fulfill our ultimate goal – the protection of human lives and property. Volcanologists seek to recognize hazardous areas and evaluate risk, warn of danger, and thus save lives. In addition to these benefits, volcano studies have revealed many other equally important connections between volcanic activity and the human experience. The final section of this book, entitled "Humanistic Volcanology," explains the relationship of volcanoes to our history, our climate, and to the metals and energy that we use and to the soils whose productivity provide our food. It is no exaggeration to state that without volcanic activity life might never have developed on Earth. Without the ore-concentrating roles of magmatic systems and volcanoes, technical civilization would be impossible. Volcanoes are infamous for their destructiveness, but they are also beautiful, have created much of the land we live on, and help sustain the world in ways that few of us fully grasp.

Generalizations being what they are, there are exceptions and elaborations needed for each of the statements made above, and we shall touch on them again throughout this book. For the time being, however, we have shared some of our personal experiences, and hopefully hinted at reasons why this science is an exciting humanistic as well as a scientific journey.

Some Basic Terminology

As humans develop specialized knowledge, whether it be about the baking of bread, ballroom dancing, brain surgery, or whatever, new words are coined to describe common features or processes. These words become "shorthand" to express thoughts in a few words rather than many. Volcanologists have their own shorthand jargon, though we strive to avoid complex terminology as this can be a barrier to effective communication with the general public. We here define some of the terms we will be using in this book – in narrative form rather than presenting them in a glossary. The Index will also direct you to further definitions when needed.

First, just what is a **volcano**? Most people first think of volcanoes as those beautiful, steep-sided, symmetrical, pointed cones like Mount Fuji] [146] in Japan, Mount Shasta [30] in California, or Mount St Helens [31] in Washington State (at least as it was before the catastrophic eruption of 1980!). But in fact, most volcanoes have shapes and forms that differ greatly from the postcard views and cartoon sketches we all learn to draw in school. Volcanoes come in many shapes and sizes, from small hills to the largest mountains on Earth, and to even larger mountains on other planets. Their shapes vary from majestic snow-covered cones (Figure 1.35) to inconspicuous hills, huge, lake-filled craters with volcanic rims, and vast lava-covered plains.

FIGURE 1.35 Osorno volcano, southern Chile [75] viewed from the west across Lake Llanquihue. This classic Andes stratovolcano has not erupted for over 150 years, but remains seismically restless. Source: Photo by Mary-Ann del Marmol.

In this book, we accept the common understanding of what a volcano is – a mountain or hill (an edifice) that develops when molten rock reaches (or closely approaches) Earth's surface and erupts. But while this serves as a practical definition, it has limitations. For example, the floors of the world's oceans are dominantly underlain by lava flows, but those volcanic rocks are mostly derived from elongated fissure vents within deep sea rift valleys in ways that commonly did not allow for the construction of conspicuous near-vent edifices. Vast areas of continents are also blanketed by incredibly voluminous outpourings of lava or hardened volcanic ash that buried pre-existing topography, but resulted in no conspicuous "volcanoes" that grew above the land. In an instance like this, the "edifices" built are not mountains or hills with craters on top, but rather are sprawling gently sloping volcanic *plains* that can be related to a common source vent or cluster of vents. It might seem a stretch to call features like this "volcanoes," but they are.

Some non-volcanic landforms such as the "mud volcanoes" of sedimentary basins and geothermal areas (Figure 1.36) and the "asphalt volcanoes" found on the seafloor above salt domes are also sometimes referred to as volcanoes, and can form very large submarine edifices (e.g. Fryer et al. 2000). Although they can also impact large subaerial areas [e.g. the devastating Indonesian mud volcano "Lusi" (Davies et al. 2008)], they are not true volcanoes in our sense, because they do not serve as vents for molten rock, and will not be considered here. Terrestrial volcanic eruptions mostly involve the escape of siliceous magma and gas, but other sorts of very different volcanic activity can also occur on the outer planets of the solar system, where the make-up of "rock" is quite unlike what we find on Earth (Chapter 12).

Igneous rocks result from the cooling of molten material originating inside the earth. They are divided into two related clans: **volcanic** (or **extrusive**) and **plutonic** (or **intrusive**) rocks. Volcanic rocks are products of erupted magma, whereas plutonic rocks are formed from magma that crystallizes underground. In general, volcanic rocks have sparse visible mineral crystals (**phenocrysts**) owing to rapid cooling of magma after eruption, whereas plutonic rocks are usually coarsely crystalline owing to slow cooling. Compositionally, most igneous rocks range from those like **basalt** typically containing low relative proportions of silica molecules (SiO_2), to those like **rhyolite** and **granite** that are silica rich. Rocks like **andesite** and **dacite** that have moderate amounts of silica are said to be **intermediate** in composition. As you will learn, silica content plays an important role in eruptive processes (Chapter 3).

Large bodies of magma underlie most volcanoes, and the terms **magma reservoirs** and **magma chambers** have been used interchangeably by many writers to describe them. Bachman and Bergantz (2008) would distinguish between the two terms on the basis of magma eruptibility. We prefer to define magma "chambers" as single discrete bodies of fluid melt and magma "reservoirs" as the overall magmatic system underlying a volcano or volcanic center – a system that may well consist of several separate magma

FIGURE 1.36 A small "mud volcano" in the Uzon [167] geothermal basin, Kamchatka. This one was formed by rising steam above a "mudpot." Other mud volcanoes, rising from sedimentary basins because of density contrasts, can spread out to cover vast areas. Source: USGS photo by J.P. Lockwood.

chambers and feeder conduits. Magma chambers may cool to form masses of coarsely crystalline intrusive rocks called **plutons**, and especially large plutons or closely spaced plutons may form extensive **batholiths** beneath volcanic belts. When exposed by later erosion, such batholiths may form majestic mountain ranges like the Sierra Nevada of California, their volcanic covers completely eroded away. Coarse-grained igneous rocks can also form at shallow depths directly beneath volcanoes. If directly related to overlying volcanoes, such rocks, whether coarse or fine-grained, are referred to as **subvolcanic** or **hypabyssal** rocks.

The passageways that supply magma from subterranean chambers to the volcanoes above are called volcanic **conduits**, and come in many shapes. When magma freezes in elongate fractures, the thin, sub-vertical tabular structures called **dikes** are formed (Chapter 4). Magma may also intrude laterally from magma chambers (or from dikes) to form the sub-horizontal structures called **sills**.

Eruptions may take place from single pipe like vents or from long fissures. The eruptions may be **explosive**, blowing out large amounts of fragmented (**pyroclastic**) debris (or "ejecta") consisting mostly of quenched magmatic lava fragments, commonly mixed with older rock material; or they may be **effusive**, erupting mainly fluid lava. Pyroclastic material is classified according to the sizes and shapes of fragments with the finest material, dust to fine sand-sized particles termed volcanic **ash**. Solidified ash beds are called **tuffs**, and **welded tuff** results where settling ash particles are so hot that they melt together. Coarser pyroclastic fragments include **lapilli**, **bombs**, and **blocks** (Chapter 7). Volcanic **breccias** are accumulations of large, blocky fragments embedded in finer, generally pyroclastic material. If the blocks are separated from one another so that they are not in mutual contact, then they are said to be "matrix supported." Otherwise, they are "clast supported."

In contrast to pyroclastic debris and breccias, lava flows are classified according to their surface features. Smooth flows, as mentioned above, are called **pāhoehoe**, and rubbly ones **ʻaʻā**. There are also **blocky lava flows** (Chapter 6).

Eruptions resulting from the direct action of magma or magmatic gas are **magmatic** eruptions. There are various kinds of magmatic eruptions, classified according to their relative explosiveness and volcanic products, including Hawaiian, Strombolian, Pelean, Vulcanian, and Plinian-type activity. Eruptions generated by the heating of water external to the magma (**hydrovolcanic eruptions**) may take place either in shallow water (Surtseyan-type eruptions), or on land, where magma interacts with shallow groundwater (Chapter 7). **Phreatic** eruptions, from the Greek word for "well," are dry-land steam-blast explosions that throw out only the solid fragments of surrounding older rocks. Similar eruptions in which the material ejected is partly or wholly magmatic are termed **phreatomagmatic** eruptions. Phreatomagmatic eruptions produce **maars** – wide, low-rimmed craters commonly occupied by lakes.

Eruptions are invariably driven by the expansion of dissolved gases (**volatiles**) within the magma as it nears the surface. Some eruptions show mixed explosive and effusive characteristics, in some instances taking place simultaneously on different parts of the same volcano. The pyroclastic material accumulating from explosive eruptions may be deposited by powerful, extremely dangerous ground-hugging ash clouds called **pyroclastic density currents** (PDCs). Two major categories of PDCs are **pyroclastic flows** and **pyroclastic surges**. Material blasted high above a vent can also simply rain down from above – **tephra**, **airfall**, or simply **fall** deposits.

Geologists recognize several distinctive types of volcanic edifices (Chapters 9–12). Those long-lived, large ones that grow from repeated eruptions through the same conduit system are referred to as **polygenetic**, while those that result from a single eruption are called **monogenetic**. Polygenetic volcanoes include **shield** volcanoes and **composite** (or "**strato**") volcanoes. Monogenetic volcanic edifices can be separate, individual volcanoes or distributed as structures at satellitic vents on the flanks of larger polygenetic volcanoes, and include **pyroclastic cones** (spatter, cinder, pumice, or ash), volcanic **domes**, and **lava shields**.

Shield volcanoes, such as those characteristic of Hawai'i and most other mid-oceanic islands, include the largest volcanoes on Earth, have very long lives, and are mostly made up of long, thin lava flows that form broad, gently sloping (generally 5–10°) shield-shaped mountains (Figure 1.15). Many eruptions occur on the flanks of these volcanoes tens of kilometers from their summits, commonly localized along radial fracture zones termed **rift zones**.

Composite volcanoes (Chapter 9) are the massive, steep-sided (30–35°), "pointy" mountains of classical shape like Mount Fuji [146] or Kronotsky [168] (Figure 1.13 These volcanoes (sometimes called "stratovolcanoes") are built up over long periods of time from hundreds or thousands of eruptions. They are composed of layers of pyroclastic material, primarily volcanic ash, interbedded with lava flows.

Of the monogenetic volcano types, **cinder cones** (sometimes also called **scoria** cones) are the most common subaerial volcanoes on Earth. They are made up of a bubbly type of lapilli called cinder, and have cup-shaped craters at or near their summits. Some are horseshoe-shaped and may be associated with lava flows. Cinder cones have short lives and may build up in a few days or a few decades at most. A volcanic **dome** is a jagged mound of lava that was very stiff and partly or even completely solidified as it escaped the vent. Domes may form gradually over a period of years, swelling from within and occasionally exploding or oozing out sluggish lava tongues.

In some areas, clusters of volcanoes and minor volcanic landforms exist around a central volcano called a **volcanic center**. The central volcano and surrounding edifices may share the same magma reservoir or may maintain separate conduits to their sources of melt, which in turn derive from a common source of heat. Especially large clusters of volcanoes, with or without volcanic centers, comprise **volcanic fields**.

Many volcanic centers are centered around **calderas**, large collapse craters generally many kilometers in diameter formed from the sudden withdrawal of magma and gases from a shallow underlying magma chamber (Chapter 10). The largest and structurally most complex calderas occur in continental settings. Much simpler and generally smaller ones occur on mid-oceanic islands.

There you have it – a starter potpourri of terms. Now let's learn about the history of volcanology.

History of Volcanology

The Ages of Superstition and Legends

[Japanese kanji: "Fire Mountain"]

Les volcans tremblent,
se gontflent,
s'ouvrent et explosent. . ..
Impuissant devant
les convulsions de la nature,

l'imagination populaire met en scène
des forces surhumaines,
venues d'un monde supernaturale

The volcanoes shake,
swell,
open and explode . . ..
Helpless before
convulsions of nature,

popular imagination envisions
superhuman forces,
coming from a supernatural world
[Maurice Krafft 1991]

Among all creatures subject to volcanic eruptions, human beings are unique in that they feel compelled to understand the reason why "fire" should come from Earth. The earliest explanations of these fearsome "fire mountains" were given in religious terms based on superstition, and usually invoked the actions of subsurface gods who were either displeased with the terrestrial world or were fighting among themselves. Myths and legends are very different ways in which traditional people describe and understand volcanic activity, but both can be associated with experiences living close to active volcanoes (Vitaliano 1973; Sigurdsson 1999). How volcano myths have developed (and continue to be created) is discussed by Cashman and Cronin (2008). Myths are basically fiction – the products of peoples' superstitions and fears, whereas legends almost always have a basis in actual *events*, embellished by storytellers as they pass down these original stories to succeeding generations, provide important clues about past eruptive activity. Volcano myths are widely circulated but are never true. A good example of a never-ending myth regards the efforts to throw virgins into erupting volcanoes to please the gods. This is a great myth for cartoonists and fantasy filmmakers to exploit, but is obviously false: Erupting volcanoes are not all that accessible, nor are virgins (arguably) all that plentiful.

The Valley of Mexico and its surroundings, located in the middle of the Trans-Mexican Volcanic Belt, has also been a center of social development for over two millennia, where several great city-states, including Teotihuacan, Tenochtitlan, Tula, and Cholula developed. The valley is bordered by two emblematic stratovolcanoes on its southeastern rim – the dormant Iztaccíhuatl [55] and the currently active Popocatépetl [56] (Figure 1.37a), volcanoes that generated many myths involving love, gods, and war. One beautiful, romantic myth that involves both volcanoes is told in many versions – here is one:

(a)

(b)

FIGURE 1.37 (a) Detail of the 1891 painting "*Valle de México visto desde el Tepeyac*" by Jose María Velasco (1840–1912), illustrating Popocatépetl [56] and Iztaccíhuatl [55] volcanoes, as seen from the NW. An alignment of monogenetic volcanoes in the foreground marked the limits of Mexico City at that time. Source: Reproduction authorized by the Instituto Nacional de Bellas Artes y Literatura, 2021; (b) A calendar illustration of the Popoca-Itza legend. Source: Jose Enrique de la Helguera (1910–1971) showing the Aztec warrior Popoca mourning his love – with the volcanoes Iztaccíhuatl and Popocatépetl in the background.

"A long time ago, when the Aztec Empire dominated most of central Mexico, frequent wars took place among the subject kingdoms. One of them, the Tlaxcaltecas, was at war with the Aztecs. Princess Iztaccíhuatl (Iztac white, cihuatl woman), the most beautiful daughter of the king, fell in love with one of the king´s warriors, Popoca. He asked the king for the hand of the princess, and the king conceded, under the condition that he wins the war. Popoca prepared his weapons and left for the war. He fought bravely and, after many battles, won the war, killing the enemy leader. Meanwhile, a local rival of Popoca told Iztaccíhuatl that the warrior had died in battle. Iztaccíhuatl, believing the lie, cried until she died of sadness. When Popoca returned victorious to claim Iztaccíhuatl's hand, he was overwhelmed by the news of the death of the princess. He then took her body in his arms and carried her to the top of a high mountain, holding a smoking torch. On the summit of the mountain, he knelt in front of the dead princess until the snow covered them. The Gods, moved by their tragedy, transformed them into great snowy mountains. Every time the heart of the warrior within the mountain recollects his love, he shakes the torch until it smokes, and that will repeat until the end of time."

Popoca's spirit now resides within Popcatepetl, and Izta's spirit is now present in the profile of her new home, Iztaccihuatl (Figure 1.37b).

Myths have little utility for volcanologists, but legends are different. The knowledge of traditional peoples living near volcanoes should not be underestimated. Conversations with longtime residents of any volcanic area will often reveal unrecorded legends that are still being passed down by storytellers today. Their stories are worth listening to, for no matter how embellished they may be by the "poetic license" of generations of storytellers, "grains of truth" from which the legends have sprung are found in most of these stories. Although traditional peoples did not have modern tools or knowledge, their stories of past eruptions are mostly based on actual observations, and attempts to reconcile these human accounts with modern volcanic knowledge can be a most fruitful source of information for volcanologists interested in understanding "prehistoric eruptions." As examples, I (JPL) was fascinated to find that native people in the Virunga volcanic belt (eastern Congo and Rwanda) knew very well which volcanoes had been active in pre-written history because they distinguished between "female" and "male" volcanoes. The lady volcanoes were those that had "emitted blood" (i.e. lava) in times past, whereas the male volcanoes were the ancient ones that had not erupted for a very long time. Their accounts matched scientific observations very well.

A Navajo woman in Grants, New Mexico, once told me a story to explain a long, narrow pāhoehoe flow (the McCartys flow) that extends more than 50 km between Gallup and Albuquerque. The woman apologized before sharing the legend saying, "Of course, you won't believe this – this is just old superstition," and then recited a detailed tale about a battle between an evil giant and a young brave: "When the giant (Ye'litsoh) was felled by a stone from the young man's sling, he fell to the earth and made the ground shake all over. The giant's blood poured forth as a red torrent that flowed like a river across the land. When the red blood dried it turned black – you can see this dried-blood river today." This eruption (the youngest in New Mexico) took place over 3,000 years ago – yet the events (earthquakes and eruption) that people witnessed are still vividly documented through story-telling – a testimony to the power of oral tradition.

Mount St Helens volcano [31] has erupted many times in the 4000 years prior to the well-described 1980 eruption (Chapter 8), and the legends of neighboring native American tribes (mostly involving love affairs between competing gods) record rich details about major explosive eruptions that pre-date 1980 (Cashman and Cronin 2008).

The indigenous Mapuche people of southern Chile and Argentina Patagonia live with dozens of active volcanoes in their midst. The volcanoes are only one feature of their surrounding physical universe (the Mapu), which includes all elements of the Earth, the Sky, and Life. Gods inhabit all of the Mapu, and one of the most important is Pillan, a volcano spirit who does not live in the volcanoes – he is the volcanoes. Present day Mapuche families are direct descendants of Pillan and his wife Wangulēn, and know the histories of the surrounding volcanoes, because the volcano histories are their histories. Mary-Ann del Marmol and I (JPL) spent three days in the company of Marta Melilet, a Mapuche scholar living near the small town of Melipeuco, Chile. I spent long nights with Marta, listening to stories about Pillan and his activities. We also spent a wonderful day circling beautifully snow-capped Llaima volcano [77] in Conguillo National Park (Figure 1.38) and learning more stories. To Marta, these stories were neither myths nor legends, these were realities because they happened to her when she was a part of Pillan. She told me about how Pillan rose above Llaima in lightning-filled clouds that would produce "culebras de fuego" (fiery serpents) that would rush down the volcano and then turn to stone. She told me that Pillan's wife Wangulēn could become very sad at times, and that her copious tears could flow off the volcanoes, forming rivers of mud and rocks that would inundate surrounding lowlands. After these floods Wangulēn's tears would freeze, forming the snow and ice of volcanic summits.

Seismic activity almost always precedes volcanic eruptions, and is commonly described in legendary accounts. Another example comes from Hawai'i, where "prehistoric" time conventionally refers to the period before European written accounts. Although there are many credible legends about pre-European eruptions of Kīlauea volcano [17], none had been known about prehistoric Mauna Loa [15] eruptions, until I (JPL) came across a previously overlooked story dealing with the origin of some littoral cones ("Na Pu'u O Pele") along the southwest coast of Hawai'i Island. Westervelt (1963) recounted a tale he had been told by an old Hawaiian man about the origin of these cones, a story of Pele's ire after she had been jilted by two chiefs who had spurned her romantic advances. In the story, Pele was so mad at the chiefs that she had "caused the earth to shake by stomping her feet on the

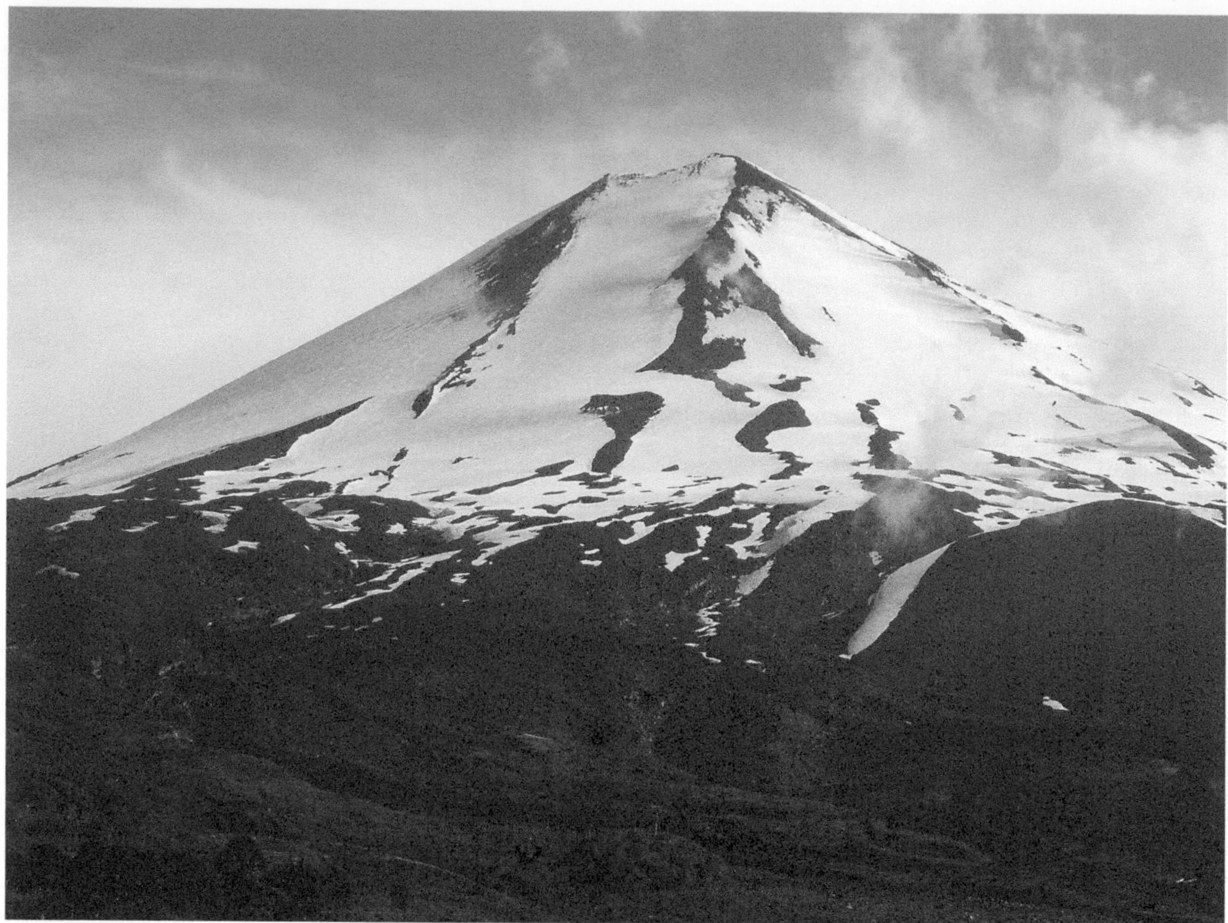

FIGURE 1.38 Llaima volcano [77] in southern Chile – one of the volcano god Pillan's manifestations. One of the most active volcanoes in Chile, it last erupted in 2009, sending ash across the border into Argentina. Source: Photo by J.P. Lockwood.

ground in anger" – before sending two lava flows to the sea, trapping the chiefs between them and turning them into the cones. Later mapping showed that two separate flows had indeed erupted at that time and that the "legend" was an accurate description of the sequence of events that formed the two cones. This eruption was radiocarbon dated at 300 BP, showing that Mauna Loa's humanly recorded history began long before Hawai'i was "discovered" by Captain Cook. Hawaiians understood basic volcano processes very well – long before Westerners came for formal study. They knew that their chain of island-volcanoes is younger to the southeast and interpreted this as evidence for the migration of Pele from west to east over time. They also knew that magma was stored beneath the summit of Kīlauea Volcano, and that this magma (i.e. Pele in their oral traditions) could travel along subterranean pathways to erupt on Kīlauea's lower flanks, often expressed as "battles" between Pele and her sometimes suiter, the pig god Kamapua'a. Interestingly, Pele is not the only volcano deity present in Hawaiian tradition. She was preceded by a more fearful male spirit, 'Aila'au, a destructive force whose name translates as "forest eater." The Bible is also a source of possible volcano legends. The Old and New Testaments, as well as the Quran, were all written in areas of the Middle East near where volcanic activity was occurring nearby, and each includes references to destruction by fire, or by "fire and brimstone." Youthful-appearing cinder cones in Syria's Golan region are less than 250 km northeast of Jerusalem, and similarly aged volcanoes associated with the Red Sea rift zone and Arabian Peninsula lie not far to the south. Geologic studies indicate that volcanoes very near Jerusalem erupted long before the time of biblical stories, but the writers of these words certainly knew about volcanic activity that had occurred nearby. In the story of Sodom and Gomorrah (Genesis 19: 24–26), it is written that:

> *"Then the LORD rained brimstone and fire on Sodom and Gomorrah, from the LORD out of the heavens," "And Abraham went early in the morning to the place where he had stood before the LORD.[28] Then he looked toward Sodom and Gomorrah, and toward all the land of the plain; and he saw, and behold, the smoke of the land which went up like the smoke of a furnace."*

In Exodus 13:21–22, which describes Moses' flight east from Egypt (toward active volcanic areas), it is written that:

> *"The Lord went before them in a pillar of cloud to guide their way by day, and in a pillar of fire to give them light at night."*

FIGURE 1.39 Mount Ararat [123], a dormant volcano in northeastern Turkey, viewed from the north in Armenia. This is the volcano where Noah's arc is reported to have landed after The Flood. Source: Sevak Aramyan/123 RF.

In Exodus 19:16–19

> *On the morning of the third day there were thunders and lightnings and a thick cloud on the mountain and a very loud trumpet blast, so that all the people in the camp trembled.* [17] *Then Moses brought the people out of the camp to meet God, and they took their stand at the foot of the mountain.* [18] *Now Mount Sinai was wrapped in smoke because the Lord had descended on it in fire. The smoke of it went up like the smoke of a kiln, and the whole mountain trembled greatly.* [19] *And as the sound of the trumpet grew louder and louder, Moses spoke, and God answered him in thunder.*

Other volcanoes figure in Genesis. As the Great Flood waters receded, Noah's arc landed on Mount Ararat [123], a dormant 5165-m-high volcano in northern Turkey (Figure 1.39). Genesis 8.5 states that "*The waters decreased steadily until the tenth month; on the first day of that month, the tops of the mountains became visible.*" These mountains were all volcanoes, but probably Noah did not appreciate the volcanological significance of his historic expedition! Some eruptions may be the source of legends that have lost their volcanic affinities, such as the possible volcanic destruction of the "Lost Continent of Atlantis" (Chapter 13).

The Quran also has a passage that possibly refers to volcanic eruptions, but this is likely copied from Old Testament accounts:

[Quran 99.1–2] *When the earth is shaken with earthquakes, the earth brings out its loads* (Figure 1.40). (Earthquakes precede volcanic eruptions.)

As classical Greek civilization spread across the Mediterranean 2500 years ago, a panoply of gods was devised to explain the natural world. **Hephaestus** (εφέατσος) was the son of the all-powerful god Zeus, but he had been thrown down from heaven by his father after an argument and was condemned to spend his days on (and within) Earth. He was badly injured on landing, and although lame and ugly (Figure 1.41), he was a gentle god and became responsible for all artisans, including weavers, sculptors, and blacksmiths. To forge metal, he sought fire on Earth and was henceforth associated with active volcanoes of the Mediterranean. When the Roman Empire vanquished Greek civilization 2300 years ago, they accepted many of the Greek gods as their own, but they gave them different assignments to suit their imperial needs. They renamed Hephaestus **Vulcan** and his duties were focused only on the forging of metal; he became responsible for manufacturing swords and armor. His principal forge was beneath the active island-volcano north of Sicily named "Vulcano [111]," which has lent its name as a descriptive term to all other volcanoes on Earth and to the science of volcanology. Roman philosophers made many important speculations about the origins of volcanoes (Macdonald 1972), but the decay and fall of the Roman Empire marked the end of an interest in natural explanations for natural phenomena. The Western world then plunged into a

١ إِذَا زُلْزِلَتِ الْأَرْضُ زِلْزَالَهَا

٢ وَأَخْرَجَتِ الْأَرْضُ أَثْقَالَهَا

FIGURE 1.40 Quran quote about possible eruption.

dark millennium where only religious dogma was allowed to flourish. Active volcanoes were considered gateways to hell and were thus to be avoided!

The critical importance of legends for documenting prehistoric eruptions was proven by the work of Blong (1982) who compiled the oral legends of people in the highlands of Papua New Guinea. The legends, told in many languages by native peoples over a vast area, told of a "taim tudak" (a "time of darkness") when homes collapsed and crops were ruined, causing great suffering. Later work (Blong et al. 2018) showed that this VEI 6 eruption took place offshore on Long Island by about 1660 CE, produced at least 10 km^3 of tephra, and was one of the Earth's largest eruptions in the past 600 years. (See Chapter 5 for a discussion of VEI; the Volcano Explosivity Index).

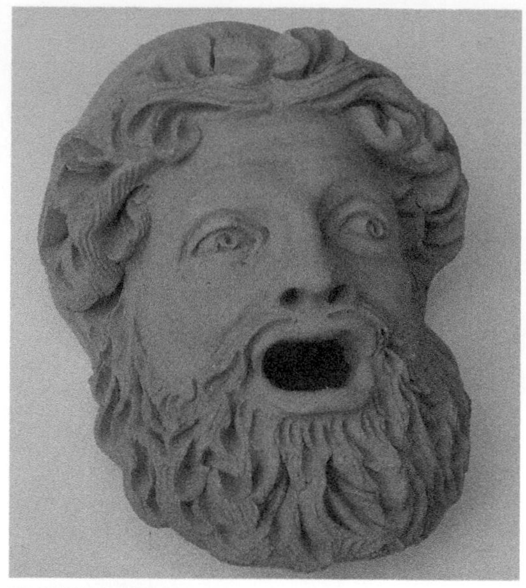

FIGURE 1.41 Hephaestus, Greek god of blacksmiths and other artisans, later named "Vulcan" by the Romans. Sculpture by R. Spada. Source: Photo J.P. Lockwood.

The Age of Enlightenment

Around 600 years ago, as the Renaissance dawned, people began to seek more logical natural phenomena to explain volcanic activity, but they encountered stiff opposition from conservative authorities who preferred religious interpretations. Nonbelievers in prevailing dogma could be condemned to death for heresy. Early alchemists (who spent most of their efforts attempting to transform materials into gold) attributed volcanoes to pent-up gases ignited by burning coal, sulfur, or oil, to electricity, or to frictional heating of air blowing through confined passages. Because of their inability to travel or to explore, they never made field observations. While their ideas might seem strange to us today, these experimenters were courageous to look beyond the supernatural in seeking explanations for natural phenomena, and they paved the way for the Age of Science that was about to unfold. By the mid-eighteenth century, academic discussion began to focus on an explanation that volcanoes are the products of erupted molten material formed deep inside Earth. At that time, French geologists, notably Jean-Etienne Guettard (1715–1786) and Barthelemy Faujas de Saint-Fond (1741–1819) recognized the volcanic origin of various cones and craters in the Clermont-Ferrand region of south-central France. They recognized the importance of erupting volcanoes in the Earth's history, and came to epitomize the *School of Volcanists* – those who believed that volcanic activity had been more common in the world than people imagined.

Abraham Gottlob Werner (1749–1817), a German professor of mineralogy from the School of Mines in Freiburg, developed a competing *geognosic theory* of geology, based on the biblical interpretation that was widely embraced by the religiously conservative "establishment" of the day. He posited that the earth had cooled solid long ago (forming granites and metamorphic rocks), but had later been covered by a primitive, global, ocean from which all stratified rocks and minerals making up the crust (including lava flows) precipitated. Fossils in some strata, though absent in lava flows, were good evidence of marine origin. Volcanoes were merely anomalies, associated with burning coal deposits. There were, Werner pointed out, extinct volcanoes in the Bohemian coal fields – evidence that burning coal had produced them. A powerful public speaker who held his post for 40 years, Professor Werner did not travel more than a few tens of kilometers from his hometown and so, like most of his predecessors, was hampered by lacking solid physical evidence for his hypotheses. But his logic, though misplaced, was internally consistent, and he was very persuasive. His theory fit in well with religious belief in a Great Flood, and Werner's perspectives became known as the **Neptunist School**. Meanwhile, resulting from these observations, an epic controversy arose between Werner's Neptunists and many of the leading scientists of the day. Even the great German poet Goethe (an accomplished naturalist in his own right, who had climbed Vesuvius [109] and observed erupting lava during a spectacular eruption in 1787) questioned the popular conclusions of Werner (see the epigraph for Chapter 3).

The leading opponent of Werner's theories was James Hutton (1726–1797), an influential Scottish geologist with great field experience, who had seen undeniable proof that many igneous rocks had been intruded into sedimentary rocks from below – an impossibility if the layered rocks were younger. Hutton and his fellow disbelievers were called **Plutonists** and were much disparaged by their opponents. Hutton was honest about the limits of his knowledge, and respected the limits of what is knowable, unlike some of his scientific predecessors. In 1788, he wrote: "Our knowledge is extremely limited with regard to the effects of heat in bodies, while acting under different conditions, and in various degrees." He knew that molten rock is active in Earth's crust but did not pretend to know why. It would take nearly two centuries and much additional scientific controversy before answers emerged, thanks to the development of new geophysical techniques for examining Earth's interior and to the discovery of plate tectonics in the 1960s (Chapter 2).

One of Werner's influential students was Leopold von Buch (1774–1853). Although originally a staunch Neptunist, he traveled extensively and made observations of volcanoes and lava flows in Italy and France that seemed to have no relation to the presence of coal beds required by Werner as a causative agent for volcanism. His further observations of lava flows in the Canary Islands in

1815 showed that they were too obviously related to volcanoes and not to sedimentary processes, and he finally broke with his professor and became an avid supporter of the Plutonist school. Unfortunately, he went too far with his conclusions: He developed a widely accepted "Craters of Elevation" theory which sought to explain the inclined lava flows around volcanoes as evidence that volcanic landforms were caused by internal magmatic intrusions uplifting and deforming originally flat-lying lava flows. That beds of lava and ash may have originally accumulated at their present dip angles, now known to be the case on young volcanoes, seemed unlikely to him.

The Emergence of Modern Volcanology

Volcanic action exhibits itself chiefly in the eruption or exhalation of heated matter in a solid, semi-liquid, or gaseous state, from openings in the superficial rocks that compose the crust of the globe. (Scrope 1862)

The modern science of volcanology began to develop in the early nineteenth century, after the fundamentals of geology had been established by geologists of the preceding century. Early observations made by curious non-scientists, including Johann Steininger, a teacher from Trier who explored explosion craters in the Rhenish Slate Mountains in western Germany, later proved critical to modern understanding. Steininger coined the term "maars" to describe the type of volcano he studied – a term still in use today, as noted earlier (Lutz and Lorenz 2013).

Credit for being the "first volcanologist" should perhaps go to Sir William Hamilton (1730–1803), as he is the first person to devote much of his life to studying active volcanoes and to providing perceptive eye-witness accounts of eruptive activity (Figure 1.42). Hamilton was a colorful, aristocratic politician, naturalist, and dilettante who served as British Ambassador to Italian kingdoms in

FIGURE 1.42 Sir William Hamilton in full regalia, with Vesuvius [109] visible through his window. A contemporary painting by David Allen. Other paintings show his head to be normal size! Source: https://commons.wikimedia.org/wiki/File:Sir_William_Hamilton_by_David_Allan.jpg

Naples from 1764 to 1800. Vesuvius [109] was characterized by intermittent effusive activity throughout most of this period, and Hamilton devoted great efforts to documenting this activity – published in a series of sometimes illustrated letters to the Royal Academy (Hamilton 1776–1779), of which he became a member. He climbed to the Vesuvius summit more than 20 times, and his descriptions of lava lake and flow activity are classic (see, e.g., his description of flowing ‘a’a – Chapter 6)..

A later aristocrat who devoted most of his life to the study of volcanic activity was George Poulett Scrope (1797–1876) (Figure 1.43). Scrope was a cosmopolitan, well-educated gentleman who made contributions to many fields in addition to volcanology; he is also known for his writings in economics and for his pioneering social work. He studied at both Oxford and Cambridge, under the pioneer geologist Adam Sedgwick. He was a friend of Charles Lyell, and like Lyell, he realized the critical importance of fieldwork. He was fascinated by active volcanism and spent his early years observing the active volcanoes of Italy, carefully observing eruptive activity at Stromboli [113] and Etna [112] and studying the eruptive products of Vesuvius and the Phlegrean fields. In later years, he went on to study older volcanoes near Rome and the extinct volcanic fields of France and Germany. Although born George Julius Thomson, he married the wealthy heiress Emma Phipps Scrope in 1821, changed his last name, and was elected to Parliament where he worked tirelessly to improve the welfare of England’s poor. While still a young man in 1826, he published the first-ever modern textbook in volcanology, entitled *VOLCANOES – The Character of their Phenomena, their Share in the Structure and Composition of the Surface of the Globe, and their Relation to its Internal Forces: with a Descriptive Catalogue of All Known Volcanoes and Volcanic Formations* (Scrope 1862). Scrope made many important observations, including a recognition of the role of water (steam) as the driving force of explosive eruptions. He also became a partisan in the raging debate over the craters of elevation theory, using his wealth of field studies to refute von Buch’s ideas. Scrope and his contemporary Charles Lyell together proved the essential role of careful field studies, and paved the way for the major advances in volcanology that were about to occur as a new generation of geologists began systematic exploration of active volcanoes around the world.

FIGURE 1.43 George Poulett Scrope. Source: Photo © Natural History Museum, London.

FIGURE 1.44 James D. Dana, “America’s first volcanologist.” Dana devoted his life to the study of volcanoes and mineralogy and was editor of the *American Journal of Science*. Source: Lithograph by Rudolph Hoffmann – after photo by Matthew Brady, Courtesy of Yale University Library.

James D. Dana (1813–1895) (Figure 1.44) studied under Benjamin Silliman at Yale, traveled to the Mediterranean after graduation, witnessed an eruption of Vesuvius, and had already authored his classic *System of Mineralogy* when he was selected (at age 25) to join the US Exploring Expedition, the “Wilkes Expedition,” for a four-year scientific exploration of the Pacific Ocean. This four-year voyage gave Dana a global perspective of volcanology and enabled him to recognize the evolution of young volcanic islands through old age and eventual submergence beneath atolls. During Dana’s month-long stay in the Hawaiian Islands in 1840–1841, he saw Kīlauea in eruption and met the “missionary volcanologist” Titus Coan (Chapter 6). This friendship was to prove seminal in fostering his life-long interest in volcanology. Returning to Yale University to teach, he married Benjamin Silliman’s daughter Henrietta, later became chairman of the Department of Geology and editor of the influential *American Journal of Science*, where he ensured that many classic volcano studies were published. After publication of his classic *Characteristics of Volcanoes* in 1890, Dana became generally regarded as America’s first volcanologist.

An important development was about to revolutionize volcanology – the development of a worldwide telegraph system in the last half of the nineteenth century. The telegraph made it possible for news of volcanic eruptions in far-away places to reach scientists in near real time and for expeditions to be mounted to investigate major eruptions. The cataclysmic 1883 eruption of Krakatau volcano [128] was reported around the world almost before the tsunami waves had receded, and major international expeditions were soon mounted to begin investigations of the tragedy. The report of the Dutch expedition to their colony (Verbeek 1885) was the most comprehensive documentation of a major volcanic eruption that had ever been prepared. The contrast between the reporting of the Krakatau eruption with the lack of documentation of previous major devastating eruptions (e.g. Asama [145] – 1783 or Tambora [134] – 1815) is revealing. Before the Age of Communications began less than 200 years ago, news of major eruptions never reached the scientific community.

Mount Pelée [87] rose with apparent innocence above the island of Martinique in the southeastern Caribbean in early 1902. A disastrous eruption of this volcano that would take place that spring would forevermore change the face of volcanology (Chapter 7).

FIGURE 1.45 Ruins of St Pierre, Martinique, after the devastating eruption of Mount Pelée [87] in May 1902. Mount Pelée in the ash-obscured background. Source: Library of Congress Prints and Photographs Division, Washington, D.C., USA.

Although the mountain had been "smoking" for a long time, and earthquakes had been shaking the island for months, residents of the important city of St Pierre did not realize that Mount Pelée was a dangerous volcano about to erupt. When the volcano did erupt violently on 8 May, 1902, almost 30,000 people lost their lives (Figure 1.45) The entire world was shocked by the tragic loss, and scientific commissions flocked to Martinique to investigate the tragedy. The Mount Pelée eruption was one of three large eruptions around the margins of the Caribbean in 1902 (devastating eruptions also occurred that year at La Soufrière-St Vincent[1] [86] and Santa Maria [58] volcanoes), and it was soon followed by another devastating of Vesuvius [109] volcano in 1906. These eruptions at the dawn of the twentieth century focused the world's attention on volcanic activity, had a major impact on the development of volcanology, and would change the careers of three people who were not volcanologists at the time. Three men in particular, Lacroix, Jaggar, and Perret, had never considered the study of volcanoes particularly important before 1902, but would go on to make major contributions to volcanology as they bridged the gap between two centuries.

François Alfred Lacroix, (1863–1948) (Figure 1.46) was a well-known French mineralogist who had never studied volcanoes before he was sent to Martinique after the 1902 disaster. He spent a year on the island, conducted critical interviews with eyewitnesses, and published the most detailed reconstruction of the events that preceded and accompanied the eruption. He was the first to recognize the nature of the deadly flowing clouds of hot gas and rock that he named "*nueés ardentes*," an internationally used term equivalent to "pyroclastic flows." [Local Portuguese had actually introduced the term as *nuvem ardente* way back in 1580 while witnessing a similar eruption in the Azores. Lacroix simply resurrected it in French (Wallenstein et al. 2018).] He founded the Observatoire Volcanologique de la Montagne Pelée and went on to study the major eruption of Vesuvius in 1906. Although mainly known for his contributions to mineralogy, he continued to advise the French Government about volcanic crises for the rest of his life. His meeting with Thomas Jaggar at the ruins of St Pierre was doubtless a factor in the evolution of Jaggar's career.

[1] There are two volcanoes named "La Soufrière" in the Lesser Antilles—this one, more specifically referred to as "La Soufrière-St Vincent," and another 300 km to the north on the island of Guadeloupe.

FIGURE 1.46 Alfred Lacroix (second from left) inspecting pyroclastic flow deposits above St Pierre, Martinique after the 1902 Mount Pelée [87] disaster. It was here that Lacroix coined the term "nueés ardentes" to describe the pyroclastic density currents that form such deposits. Source: Photo from the Krafft Archives.

Thomas A. Jaggar (1871–1953) (Figure 1.47) was a young Harvard geology professor in 1902 who had made important studies of many geologic processes – but had never seen a volcano. Following the tragic eruption of Mount Pelée in 1902, he was dispatched on an emergency mission to Martinique by the US Government. Although he only spent a few weeks there, the experience of walking among the ruins of St Pierre (Figure 1.45) and viewing the corpses of thousands of victims still lying in the ruins completely changed his life. As he wrote in his biography (Jaggar, 1956):

> *As I look back on the Martinique expedition, I know what a crucial point in my life it was and that it was the human contacts, not field adventures, which inspired me. Gradually I realized that the killing of thousands of persons by subterranean machinery totally unknown to geologists and then unexplainable was worthy of a life work.*

His career did indeed change and the rest of his life was devoted to the study of volcanoes. He made several volcano expeditions in the next few years: to Vesuvius (1906), to the Aleutian volcanoes (1907), to Hawai'i and Japan (1909), and to Costa Rica (1910) to gain an understanding of volcanic activity and hazards. His 1906 trip to Italy was especially important as he there met Frank Perret (see below) and Tempest Anderson (who had described the 1902 La Soufrière eruption), and again worked with Lacroix. He also learned about the important work of the historic Osservatorio Vesuviano, founded in 1841 (see Chapter 14). These expeditions convinced Jaggar that *expeditionary* volcanology (mostly after disasters) could never be successful for understanding the *processes* of volcanism. Without an understanding of volcanic processes, there was no possibility of ever *predicting* eruptions, and prediction had to be the primary goal of what he termed "humanistic geology." He made fervent public appeals for the establishment of permanent observatories to continuously monitor volcanoes and seismic regions.

After the 1906 San Francisco earthquake he succeeded in urging the Geological Society of America to pass a resolution calling on "Governments and private enterprise to establish volcano and earthquake observatories"; and, in a widely read article in *The Nation* (1909), he decried the preoccupation of geologists with studies only of the past ("the bones of Jurassic reptiles, and in finding out all about iron and coal") and urged young people to devote their lives to "humane rather than historical" science. After his visit to Kīlauea in 1909, he resolved to found a permanent volcano observatory there, and spent the next three years raising funds to accomplish his goal. He took leave of absence from MIT in 1912, left his family behind, and moved to Kīlauea to found the Hawaiian Volcano Observatory (HVO). He never returned to his academic appointment, left his first family, and spent the rest of his life assuring that the HVO record of volcano observation would be unbroken. He was a prolific writer on many subjects besides volcanology, an avid inventor of new devices for volcano monitoring, and an eloquent spokesman for his "humanistic" beliefs. He was *not* a very good observer, however, and his descriptions of Hawaiian volcanic eruptions are so entwined with his interpretations that they are

FIGURE 1.47 Thomas A. Jaggar, founder of the Hawaiian Volcano Observatory, in 1928 alongside a prototype of the amphibious landing craft he invented. Source: USGS photo from Hawaiian Volcano Observatory Archives.

of limited value today. As Gordon Macdonald told me (JPL) in frustration one day, "*Jaggar always thought like a promoter, not like a geologist – he never made a map to let us know what happened!*"

Frank A. Perret (1867–1940) also was touched (more indirectly) by the Mount Pelée disaster and subsequently became one of the most fascinating of the early twentieth-century volcanologists. Perret never finished college, nor had formal training in geology, yet his contributions to volcanology in the form of his lucid descriptions of explosive volcanic activity are timeless. Perret was self-educated as an electrical engineer, worked directly with Thomas Edison, and later founded his own company in New York to manufacture electric motors and batteries. When word of the 1902 Mount Pelée disaster arrived in New York, his only previous awareness of volcanism was associated with his viewing the "amazing sunsets" associated with the 1883 eruption of Krakatau as a young man. His health began to fail in 1902, however, and while on a recuperation visit to the Caribbean, he stopped by Martinique to visit the ruins of St Pierre. Like Jaggar and Lacroix before him, Perret's life was forever changed by the destruction wrought by Mount Pelée, and he devoted the rest of his life to the study of volcanoes.

To learn about active volcanism, he abandoned his business, traveled to Italy in 1904, and apprenticed himself to R.V. Matteucci, Director of the Vesuvius Volcano Observatory. He lived in Naples for 20 years, witnessed the devastating 1906 eruption of Vesuvius, and wrote a classic description of the eruption and its effects (Perret 1924). He later became affiliated with the Carnegie Institution in Washington, DC, where all of his volcano studies were published. Although plagued by ill-health, he traveled widely to active volcanoes around the world, and his frail, dapper figure, well-dressed with Van Dyke beard and straw hat (Figure 1.48) became a well-known sight during volcanic eruptions. He traveled to Hawai'i with R.A. Daly and Thomas Jaggar in 1909, made some of the first-ever quantitative measurements of molten lava temperatures, and was one of the first people to recognize the importance of

FIGURE 1.48 F.A. Perret with an improvised "seismometer," listening to subterranean noises at the Campi Flegrei, Italy [108]. Source: Photo from the US Library of Congress.

explosive eruptions in Kīlauea's past. When Mount Pelée returned to activity in 1929, Perret returned to Martinique the following year to observe the action and to advise local residents about threatening eruptions. He lived in a shack high on the flanks of the volcano to better observe the activity, and he was nearly killed by a nuee ardente in 1930. His nonchalant eye-witness accounts of this activity (see Chapter 7) make for some of the most exciting reading in the annals of volcanology.

Active volcanoes are found throughout the world, and it is no surprise that the major volcanologists of the twentieth century have come from countries facing the greatest eruptive perils.

Alfred Rittman (1893–1980) (Figure 1.49) was born in Basel, Switzerland, and there met the wealthy Swiss banker Immanuel Friedlander, a widely traveled amateur volcanologist who published the influential journal *Zeitschrift für Vulkanologie* (1914–1936), and privately founded the Institute of Volcanology based in Naples. Rittman was named the Director of Friedlander's institute, and there pioneered the use of petrographic, geochemical, and geophysical methods to better understand volcanic processes; his studies of the magmatic evolution of Vesuvius and Etna were major contributions to the foundations of volcanology. When Friedlander's institute closed on the eve of World War II, Rittman was offered a university position in Germany, but refused the appointment as he did not wish to associate with the Nazi Party. He taught for a while in Egypt, and later became Director of the Institute of Volcanology in Catania. His influential textbook *Volcanoes and their Activity* (1962) remains in use today. The English language reference cited here is a translation of the original *Vulkane und ihre Tätigkeit* (1936), which was also translated into French and Italian. His system for "Eruption diagrams" (see Chapter 5) is an excellent, but largely overlooked, graphical means to portray the nature of individual eruptions. He was one of the longest-serving presidents of IAVCEI, the International Association of Volcanology and Chemistry of the Earth's Interior – from 1954 to 1963.

There are more than 30 active volcanoes in Russia, all part of the Pacific "Rim of Fire" extending from Kamchatka down the Kuril Islands. The Soviet Union produced many great volcanologists in the twentieth century to study these volcanoes, but because most of their publications are in Russian, and because Cold War complexities made travel for Soviet scientists and international exchanges difficult, not much has been known about their work. Georgii S. Gorshkov (1921–1975) (Figure 1.50) was able to travel widely and became well known in the West. He began his volcanological studies with extensive field work along the length of the Kuril Islands, and then began to study the numerous active volcanoes of Kamchatka. Gorshkov was primarily a volcano seismologist, and pioneered the study of *teleseisms* (earthquake waves generated by distant earthquakes) to define the geometries of magma chambers underlying closer volcanoes. He documented the 1956 eruption of Bezymyanny volcano [170] (Gorshkov 1959), a volcano whose pre-eruptive behavior closely mimicked the pre-eruptive behavior of Mount St Helens as the volcano approached its 1980 eruption. Unfortunately, volcanologists in Washington did not appreciate the similarity in eruptive style and did not realize that the catastrophic eruption that was about to occur on 18 May had almost the same clearly recognizable eruption precursors that Gorshkov described at Bezymyanny.

FIGURE 1.49 Alfred Rittmann (1893–1980), pioneering Swiss volcano-petrologist who spent most of his life studying the magmatic evolution of Vesuvius and Etna. Source: Photo by G.A. Macdonald.

Japan is one of the most volcanically active areas on Earth and has more than 40 on-land volcanoes that have erupted in historical time – almost 70 if one counts minor islands and submarine volcanoes in its territory. From a public safety standpoint, however, earthquakes have been a more serious threat to human life and property, and Japanese earth science was traditionally focused on the field of seismology. When Thomas Jaggar founded HVO, he turned to Japan for expertise, and HVO's original seismometers were designed by Fusakichi Omori (1868–1923), the pioneering Japanese seismologist whom Jaggar had met in 1909. Omori gave Jaggar one of his seismographs and plans for a seismic vault, and HVO's Whitney Seismological Laboratory was constructed in accordance with these plans. Volcanology itself was not an important focus of Japanese science, however, until Takeshi Minakami (1910–1983) (Figure 1.51) was assigned to the Asayama Physics Research Laboratory on the slopes of Asama volcano [145] northwest of Tokyo in 1934. Minakami had trained as a seismologist at Omori's Imperial University, but at Asama, his interests in volcanology blossomed, and he can rightfully be called the "Father of Japanese Volcanology." Asama was extremely active while Minakami was there, and with great effort (he needed to carry heavy batteries upslope to his laboratory for several years as there was no electrical service), he pioneered the field of volcano seismology. He recognized two different classes of volcanic earthquakes, and analyzing them allowed him to successfully forecast several eruptions. His "Minakami Classification" of type "A" (deep, sharp) and type "B" (shallow, LP) volcanic earthquakes is now used worldwide to recognize the ascent of magma within volcanoes. Minakami's visit to Indonesia before World War II was critical to enabling Indonesian volcano observatories to function during the war years under military occupation. His many students and junior associates have gone on to make Japan one of the world's leading centers for volcano research. A historic visit of Minakami and his associates to HVO in 1963 for cooperative research studies set the stage for the close collaboration between Japanese and American volcanologists that continues today.

FIGURE 1.50 Georgii S. Gorshkov, pioneering Soviet volcanologist. Source: Photographer unknown.

FIGURE 1.51 Professor Takeshi Minakami, the founder of modern Japanese volcanology and leading volcano seismologist. Source: Photo courtesy of Shigeo Aramaki.

There are many recently deceased "pioneers" in volcanology whose contributions we should perhaps honor, but that list is long, and we must be getting on with the rest of this book. Suffice it to say that most of today's volcanologists owe their careers to the influence of great teachers who have passed away in recent years. These teachers include giants like Robert Decker, Peter Francis, Dick Stoiber, and George Walker – dear friends and incredibly productive volcanologists whose contributions to our science and to hundreds of their students are immense and long-lasting. It is also worth noting that while the history of volcanology has been male dominated and mostly Western-based, like much of modern science, participation of a more diverse humanity is presently underway; a sea change we welcome and encourage (See Epilogue). This is reflected in many of the references and observations we'll cite in the coming chapters.

Questions for Thought, Study, and Discussion

1. Outline the major differences between the eruptions of "red" and "gray" volcanoes. Why are these volcanoes designated with these colors?
2. Contrast the hazards of doing volcanological field work at "red" vs "gray" volcanoes.
3. Why is the definition of a volcano simply as a "mountain or large hill that erupts" inadequate?
4. Would you define "volcano" any differently than we have? Why, or why not?
5. What are the two largest kinds of volcanoes, and how would you recognize each? What are three smaller kinds of volcanoes, and how are they distinguished?
6. A steep-sided volcano consisting of many layers of hardened lava alternating with beds of cinder and ash suddenly rips open with a sluggish flow of molten rock pouring out of a crack extending down its flank. No ejecta are disgorged and volatile release is minor. (a) What kind of volcano is this? (b) What kind of vent opened up? (c) What kind of eruption is this?
7. Eruptions of highly fluid lava with very little associated pyroclastic material create what kinds of volcanic features?
8. Contrast the Neptunist and Plutonist schools of thought. What evidence finally allowed the Plutonists to overcome the Neptunists?
9. There are admirable goals and benefits to both "pure" volcanological research focused on the expansion of human knowledge, and to "applied" volcanology focused on the immediate social and economic needs of society. What is the proper balance between these two end uses of volcanology? What balance would *you* hope to strike if you pursue a career in volcanology?

FURTHER READING

Dvorak, J. (2017) *The Last Volcano: A Man, a Romance, and the Quest to Understand Nature's Most Magnificent Fury.* New York, London: Pegasus Books, 344 p.

Lubrich, O. & Nehrlich, T. (eds.) (2021) *Die Vulkane des William Hamilton, Naturreportagen von den Feuerbergen Ätna und Vesuv.* Darmstadt: Wissenschaftliche Buchgesellschaft, 272p. (Containing reproductions of the figures and text, translated into German by J. v. Koppenfels, of "Campi Phlegraei" by William Hamilton, 1776 and 1779).

McIntrye, D.B. and McKirdy, A. (2012) *Hutton: The Founder of Modern Geology,* 2nd edn. Berkeley, California: NMS Publishing, 80 p.

Parfitt, E.A. and Wilson, L. (2008) *Fundamentals of Physical Volcanology.* Oxford: Blackwell.

Scarth, A. (2009) *Vesuvius: A Biography.* Princeton, New Jersey: Princeton University Press, 342 p.

Sigurdsson, H. (1999) *Melting the Earth: The History of Ideas on Volcanic Eruptions.* Oxford: Oxford University Press.

Sigurdsson, H. et al. (eds.) (2015). *Encyclopedia of Volcanoes.* Amsterdam: Elsevier Press.

PART II

THE BIG PICTURE

This part consists of three chapters that describe the global environment in which volcanism operates on the surface of Earth, including the revolutionary impact of plate tectonics on earth sciences, and the nature of "magma" (molten rock below Earth's surface) – how magma is generated, and how it ascends to the surface. **Chapter 2** discusses the realization that Earth has a dynamic, ever-changing surface upon which volcanoes play vital roles, describes the various tectonic environments in which volcanoes occur, and how they differ depending on these environments. **Chapter 3** discusses how magmas form, how they cool to form volcanic rocks, and the all-important role of volcanic gases – important but often overlooked magmatic constituents. **Chapter 4** describes the physical properties of magmas, how they reach the surface of Earth, and the mechanisms that trigger volcanic eruptions – "why volcanoes erupt."

Volcanoes: Global Perspectives, Second Edition. John P. Lockwood, Richard W. Hazlett, and Servando De La Cruz-Reyna.

Companion website: www.wiley.com/go/lockwood/volcanoes2

CHAPTER 2

Global Perspectives – Plate Tectonics and Volcanism

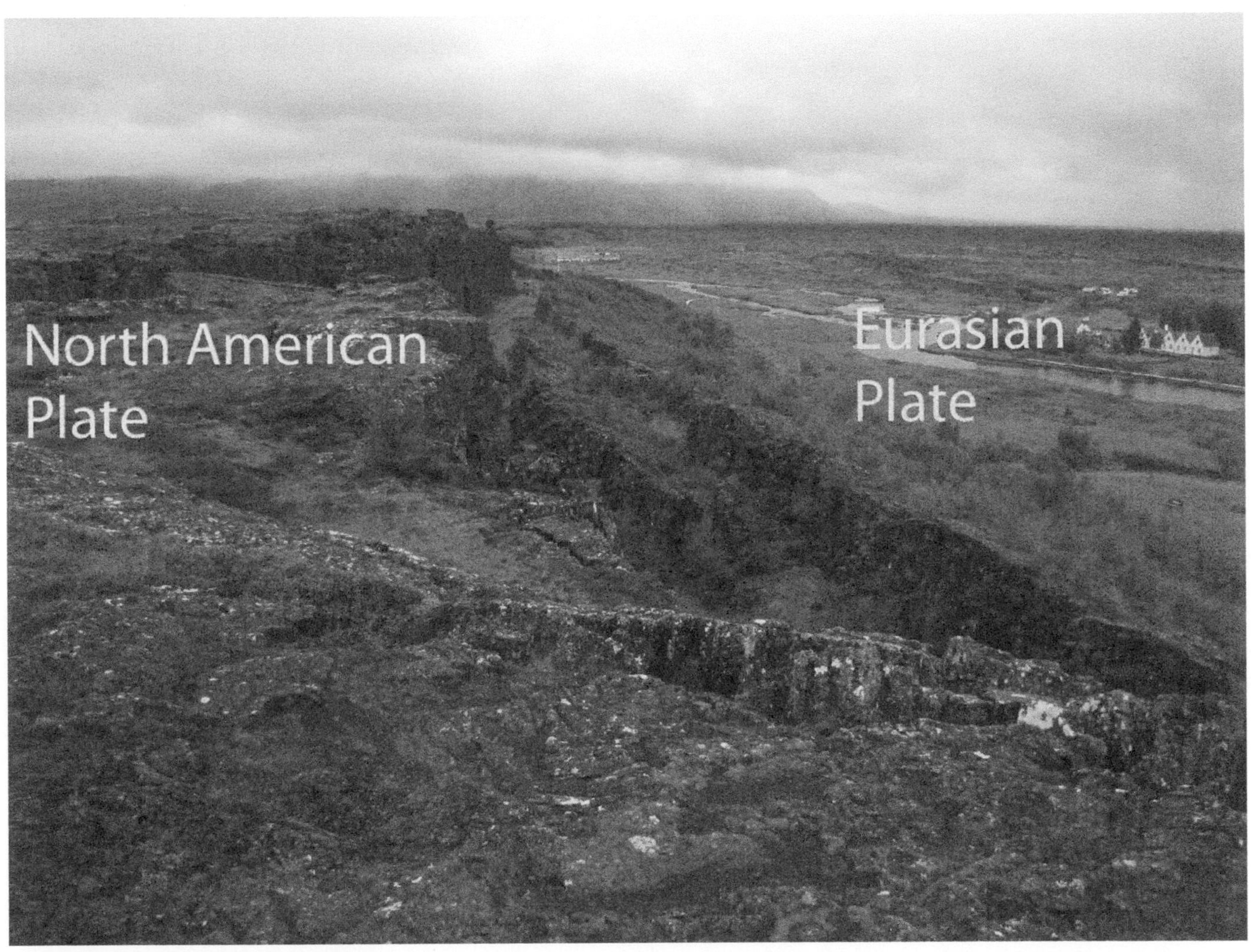

Source: Credit R. Hazlett

Volcanoes: Global Perspectives, Second Edition. John P. Lockwood, Richard W. Hazlett, and Servando De La Cruz-Reyna.

Companion website: www.wiley.com/go/lockwood/volcanoes2

A map of the Earth on which the position of volcanic vents is marked shows at a glance that their distribution can hardly be a matter of chance.

(Bonney 1899)

Birth of a Theory

In the decades following the triumph of the Plutonists over the Neptunists (Chapter 1), geologists established a broad consensus that there are three fundamental kinds of rocks making up Earth's crust and surface. James Hutton proved that **igneous** rocks form from the cooling of magma. (The term "igneous" ultimately stems from the Sanskrit root, *agni*, meaning "fire.") **Sedimentary** rocks are formed by the deposition and compaction of loose sediment resulting from weathering, erosion, and transport of rock fragments, or by the evaporation and chemical precipitation of saline waters. Either sedimentary or igneous rocks may further be transformed through compaction, heating, and mineralogical changes into **metamorphic** (changed) **rocks** by deep burial and deformation. By the middle of the nineteenth century, geologists also sensed that rock materials could in time be recycled through all three states – igneous, sedimentary, and metamorphic – in a process that eventually came to be called the **rock cycle**. The modus operandi of the rock cycle remained a mystery, however, as did the strange distribution of one great subclass of igneous rocks – the volcanic rocks.

For nearly two centuries after the birth of modern geology, no one could explain why volcanoes occur where we find them. In 1945, for example, Thomas Jaggar wrote the following about the chains of volcanoes fringing the Pacific Rim (Jaggar 1945, p. 1):

The Pacific girdle of fire is the most systematic series of partitions on the globe, dividing its surface by means of downgoing, upright, lava-filled fractures. Such a partition goes down perhaps 1800 miles to the core of the globe between Alaska and Patagonia, between Kamchatka and Java, divided into curved cracks in plan; no one knows "why"...

Then, with stunning rapidity and from sources wholly unexpected, the great "why" posed by Jaggar was answered. Like so many seminal leaps forward in science and technology, it took modern techno-industrial warfare to develop the new research tools and funding that allowed people to explore new fields and investigate areas that they might otherwise have ignored. In a mid-twentieth-century effort to improve ways of locating and tracking enemy submarines, researchers used sonar to map the bathymetry of the deep sea, and ship-towed magnetometers to plot ocean floor magnetic properties – important for navigational purposes. Several astonishing results emerged. Most spectacularly, investigators confirmed that the major ocean basins are crossed by a system of gigantic, undersea mountain belts called mid-ocean ridges (MOR). Parts of this enormous feature had been discovered a hundred years earlier by soundings in the central Atlantic, but the interconnectedness of it all had not been appreciated at the time. It wound around the whole planet, like the seam of a giant baseball, sending out branches here and there that could be related in a simple glance of the map to noteworthy features such as the Gulf of California and the East African Rift Valley.

FIGURE 2.1 Harry H. Hess, originator of the "seafloor-spreading" hypothesis. Source: Photo courtesy of Princeton University Library.

Scientific breakthroughs are often nurtured by coincidental factors, and such events combined to focus the life's work of one of the twentieth century's most influential geologists – Harry Hess (Figure 2.1). Hess was a graduate student at Princeton University in 1931, when he was asked to accompany the Dutch oceanographer F.A. Vening Meinesz on a submarine expedition to measure gravity fields in the Lesser Antilles. This was Hess's introduction to a lifelong interest in the origin of ocean basins, and the initiation of his wonderment about why oceanic trenches were marked by major gravity lows. His PhD dissertation on serpentinite belts of the Appalachians raised further questions about the role of ultramafic rocks in certain mountain ranges. Then, as a US Navy Captain during World War II, he discovered strange, flat-topped seamounts on the Pacific floor that he named **guyots** (pronounced "GEE-ohs"), whose summits all seemed to systematically deepen with distance from the MOR. The summits of many guyots were later found to be composed of coral reefs and associated lagoonal deposits, which develop in shallow water. Because of this, Hess reasoned, some unknown process active in the ocean floor must be submerging them deeper the farther one travels away from the ridge crest. Indeed, the *whole ocean floor*, not simply the guyots, deepened.

Likewise, new undersea mapping refined the understanding that some ocean basins are bordered by deep **marine trenches**. These were parallel to long, gracefully curving archipelagos characterized by active volcanoes, called **island arcs**, though in some places, the trenches are close to continental shorelines, as along the coast of South America.

Finally, researchers found that ocean floor lavas preserved a record of past reversals in Earth's magnetic field. **Magnetic reversals** happen when Earth's magnetic polarity (referring to the positions of the north and south magnetic poles) suddenly flip. The north magnetic pole suddenly becomes the south magnetic pole, and vice versa. Lava flows contain magnetic minerals such as magnetite, and when these flows cool, these internal magnets record ambient magnetic field directions and intensities. Lava flows thus record the history of Earth's magnetic field, and show the most recent field reversal about 700,000 years ago, with prior reversals occurring thousands of times in Earth's distant past. By the late 1950s, enough data had accumulated to begin showing a remarkable pattern of reverse-normal magnetic stripping parallel to and symmetrical about the MOR, recording a history of pole reversals going back nearly a hundred million years.

Sir Arthur Holmes, one of the greatest geology teachers and thinkers of the past century, provided a speculative explanation for all these later observations around the time that Vening Meinesz was doing his early gravity work in the late 1920s. In 1927, Holmes postulated that great cells of convecting hot rock well up from the deep earth to stretch and tug at the dense crust underlying Earth's continents. That deep crust then flows like taffy, pulling the continents along with it and in places breaking it into giant fragments. The denser, subcontinental crust forms the ocean floors, Holmes reasoned, and near continental margins it may convert through metamorphism to even denser eclogite, helping to drive the convection process below even faster and to explain the gravity data reported by Meinesz.

Holmes' speculation proved to be an important step toward our current model of global plate tectonics – a brilliant feat of induction. Unfortunately, without the evidence he needed to support his idea, he was too far ahead of his time to be taken seriously by the scientific community. The related observations of a well-known contemporary, Alfred Wegener, were being severely attacked by many influential professional geologists. Wegener had postulated in 1911–1912 that Earth's continents had drifted apart from one another over time. What else could explain the trans-oceanic coincidences in the distributions of many fossils, mineral deposits, and geographical features such as the "fit" of western Africa's coast with that of eastern South America? His **continental drift** theory suffered from the lack of a testable, even "reasonable" explanation, and was generally ridiculed. Wegener – a meteorologist by training – suggested that inertial forces caused by Earth's rotation were responsible for the separation of continental fragments and the opening of oceans such as the Atlantic. But geophysicists quickly demonstrated this could not possibly be happening, given the well-tested strength and mechanical properties of rocks. Continents did not "plow through the ocean bed," like ships moving through the sea, as Wegener suggested. No matter the stronger merits of Holmes' convection cells hypothesis, the rebuke of Wegener rung in his appropriately cautious ears.

Harry Hess enters the story again, however, by taking the newly discovered information of the 1940s and 1950s, brushing off the older model of Holmes, combining with his own oceanographic observations, and giving it a respectable update. Holmes was on the right track, Hess concluded. The paleomagnetic, bathymetric, radiometric, and gravity data were proof. In 1960, he informally printed and distributed to colleagues a new, well-integrated view of how the physical world works, arguing that the creation of new crust through magmatic intrusion and volcanic activity at the MOR provided a suitable explanation for Wegener's continental drift (Hess 1960). Moreover, stepping past Holmes, he proposed that the ocean floor sank back into the planet continuously wherever it entered the trenches, its elements to be recycled slowly in Earth's perpetually overturning interior (Hess 1960). The high-standing continents were simply passive passengers in the process, embedded *within* rather than plowing *through* the surrounding oceanic crust, which was continuously renewed at the MORs. The rate of production of new oceanic crust simply matched its rate of disappearance at the edges of certain ocean basins. He called his explanation, which was not formally published until 1962, "an essay in geopoetry." In a companion paper, Fisher and Hess (1963) also correctly proposed that volcanoes of arcuate island chains were the evidence of melting above the zones where the oceanic crust sank back into the deep Earth.

Hess's theory was testable, and in succeeding years verifications of his basic model have come from diverse fields, including paleomagnetics, seismology, paleontology, geochronology, and geodesy. His model, referred to as **seafloor spreading** by others, provided the fundamental conceptual framework for an integrated approach to the global analysis of tectonics and volcanism (Figure 2.2). Earth's crustal structure is very different from, and more complex than, that imagined by Hess, but his "geopoetry" remains the basis for a scientific revolution in the earth sciences, which at last has explained the pattern in volcano distribution puzzled at by T.G. Bonney (epigram), Jaggar, and many other pioneering geologists.

By 1968, the revolution in new understanding was complete, and the world's first holistic theory of how our planet works became popularly known as **plate tectonics**. "Tectonics" is derived from the Greek word for "builder," *tekton*; and the term "plate" was originally used by Vening Meinesz to describe areas of the ocean floor that are enclosed by prominent features such as the MOR, trenches, and the edges of continental shelves. The "type plate" identified by Meinesz was the Caribbean Basin. "Plate" now refers to any coherent area of Earth's shallow layering, whether continental or oceanic, that is capable of independent movement relative to other plates.

It is now generally accepted that Earth's outer surface is segmented into about a dozen major plates (and numerous minor ones) that are continuously being slowly moved about by mantle convection at speeds ranging from a few millimeters to several centimeters per year (Figure 2.3). These plates vary from less than 50 to perhaps 200 km in thickness and have their bases in a zone of low seismic wave transmission velocities, termed the **low velocity zone** (LVZ), in the upper mantle (Condie 1982). The plates consist of Earth's crust plus a layer of rigid upper mantle. Geophysicists call this combination of crust and upper mantle slice the **lithosphere** (from the Greek *lithos* "rocky" plus *sphaira* "sphere"). The plastic, partially molten layer beneath the lithosphere is the **asthenosphere** (from the Greek *asthenēs* "weak" plus *sphaira* "sphere") (Figure 2.4).

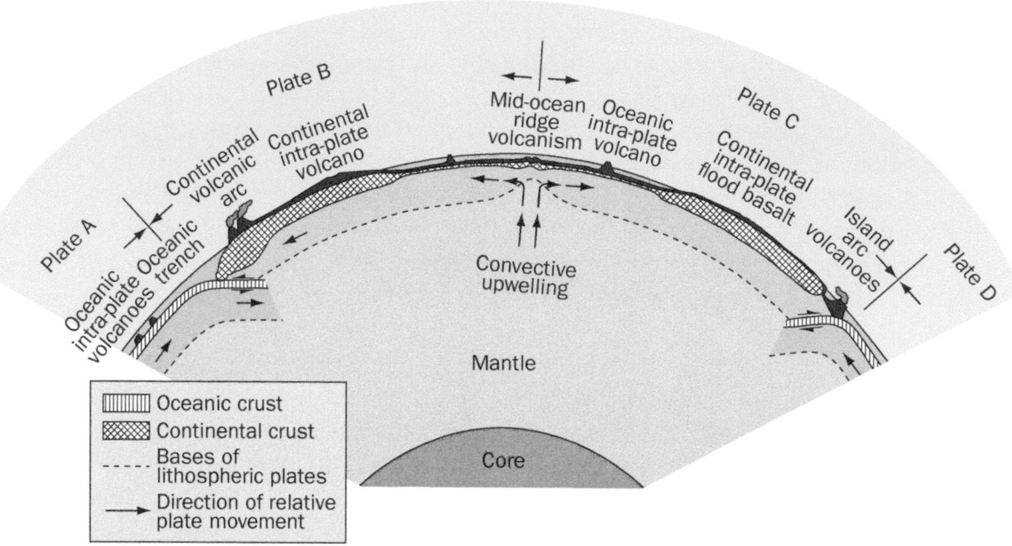

FIGURE 2.2 Schematic cross-section of Earth, showing locations of volcanoes relative to major plate tectonic features. Vertical scales are distorted.

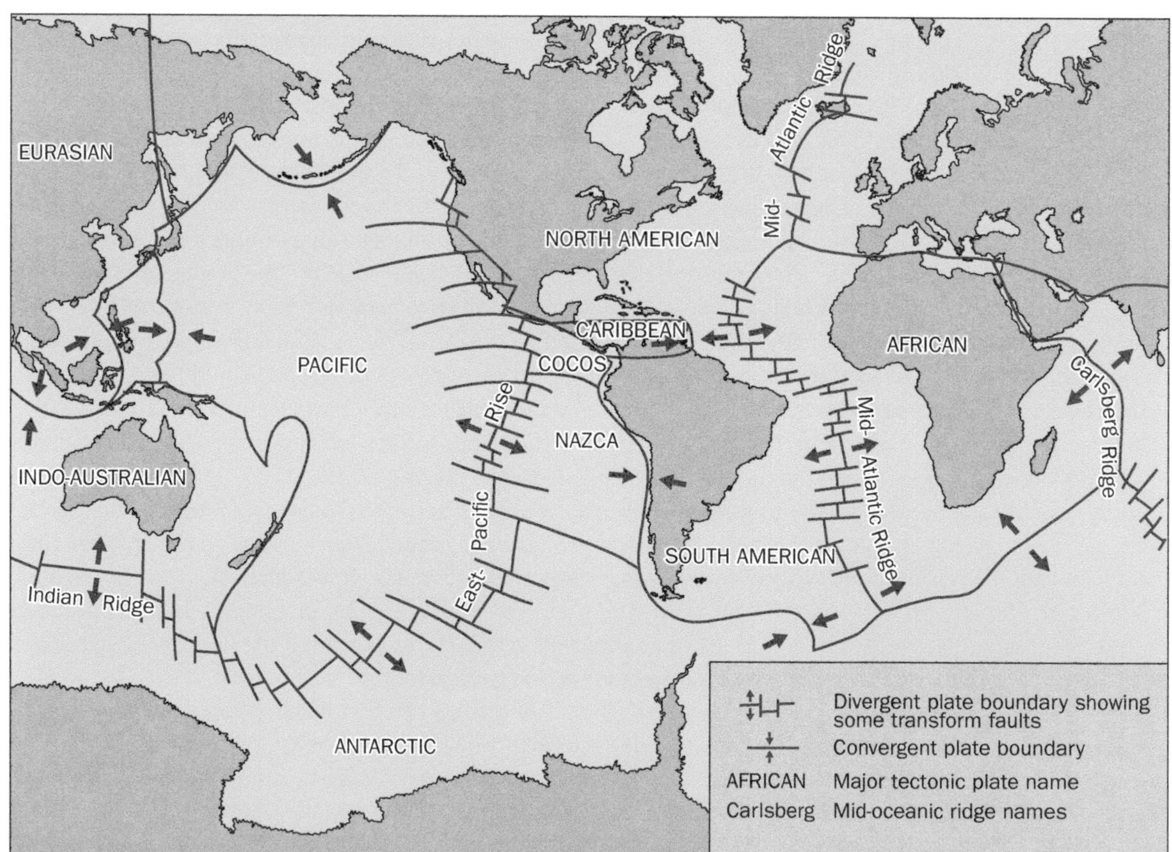

FIGURE 2.3 Earth's major tectonic plates and mid-ocean rises, showing divergent, convergent, and major transform fault boundaries. Effusive volcanoes tend to occur at divergent plate boundaries and explosive volcanoes tend to occur at convergent plate boundaries. Map is compiled from various sources, including Heirtzler (1968, p. 59) and McClelland et al. (1989). See the map at the end of this book for more details.

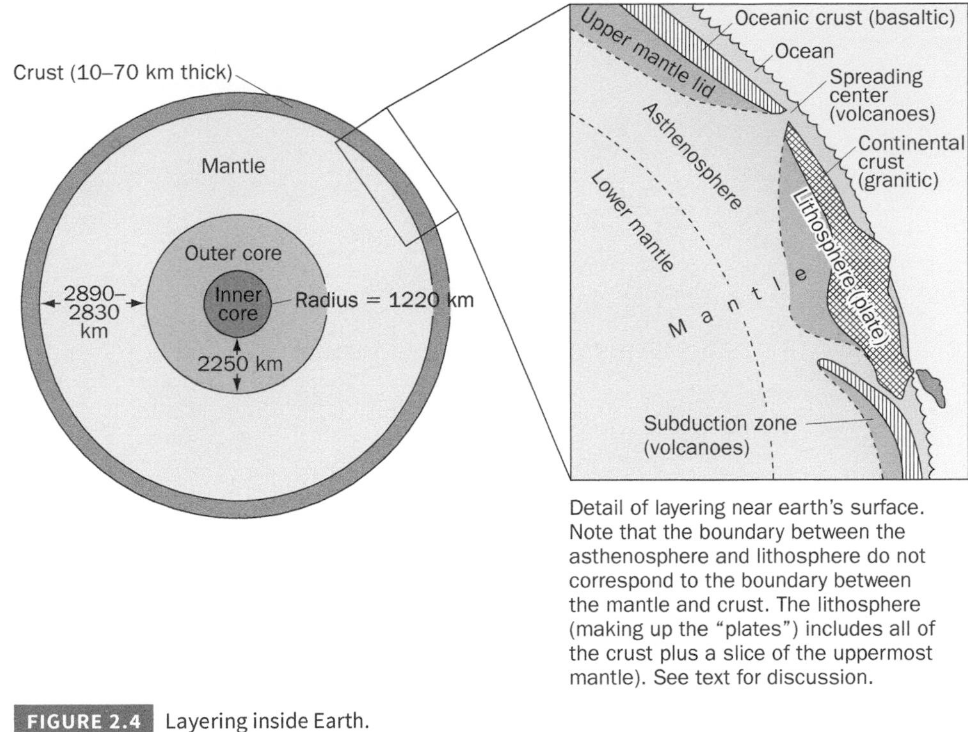

Detail of layering near earth's surface. Note that the boundary between the asthenosphere and lithosphere do not correspond to the boundary between the mantle and crust. The lithosphere (making up the "plates") includes all of the crust plus a slice of the uppermost mantle). See text for discussion.

FIGURE 2.4 Layering inside Earth.

Three fundamental kinds of plate interactions take place in response to underlying mantle convection: **divergent**, **convergent**, and **transform** plate boundaries. Because little, if any, volcanic activity takes place along transform plate boundaries (a form of strike-slip faults), we will not discuss them further.

Divergent plate boundaries exist where plates separate, as through seafloor spreading. "Spreading center" is a synonym for divergent plate boundary, but this term is a bit misleading since the formation of a new crust through volcanic activity occurs along a boundary *line* and not a *center*. Most divergent plate boundaries occur on the ocean floor, because the oceanic crust covers a larger percentage of Earth's surface than the continental crust, and because the lithosphere beneath the oceans is thinner and weaker than it is in continental areas. Most of the world's volcanic activity occurs along divergent plate boundaries, although it takes place largely unseen on the floors of the world's oceans.

Convergent plate boundaries exist where plates collide (Figure 2.5). Most of these boundaries lie near the edges of ocean basins because the contacts between continental masses and the sea floor tend to be structurally weak. Lithosphere weighted with dense, heavy oceanic rocks will generally plunge beneath continental lithosphere as continental and oceanic plates collide, since continents generally consist of lower density rocks.

In some regions, convergent plate boundaries trend across oceanic basins. The overriding oceanic plate is usually smaller than the plate sinking beneath it, and in most cases, it is rigidly attached to a nearby continental land mass. The body of water atop the overriding plate is termed a **marginal sea**. Numerous marginal seas, such as the Sea of Japan, for example, fringe the western Pacific Ocean.

Subduction (meaning "under-moving") is the term used to describe the processes whereby one plate dives under another. Deep marine trenches stretching from hundreds to thousands of kilometers typically mark the sites of subduction on Earth's surface. Individual trenches may be several tens of kilometers wide and as much as seven kilometers deep relative to the adjoining ocean floor. Some trenches lying close to large river mouths, such as the Columbia River in the northwestern United States, fill with sediment, and cease to be expressed as depressions in Earth's surface. They nevertheless remain "trenches" purely from the standpoint of tectonic behavior, since the underlying

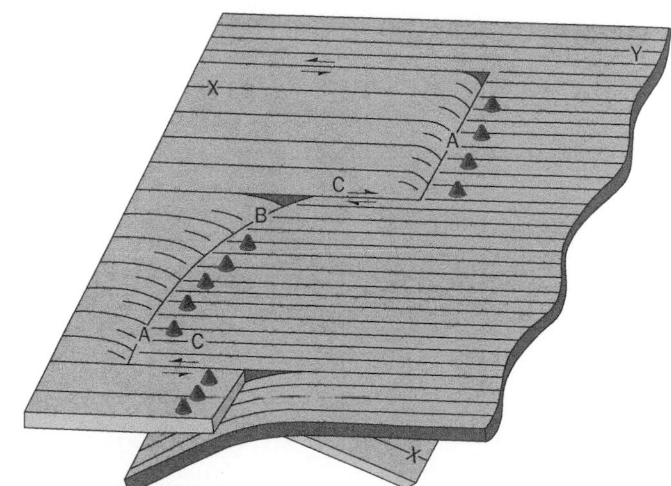

FIGURE 2.5 Various geometric configurations for boundaries between convergent plates ("X" and "Y"), showing most common volcano loci. A = subduction normal to plate convergence, B = Subduction oblique to plate convergence, C = Transform (strike-slip) fault boundary. Source: Modified from Gill (1981, p. 391).

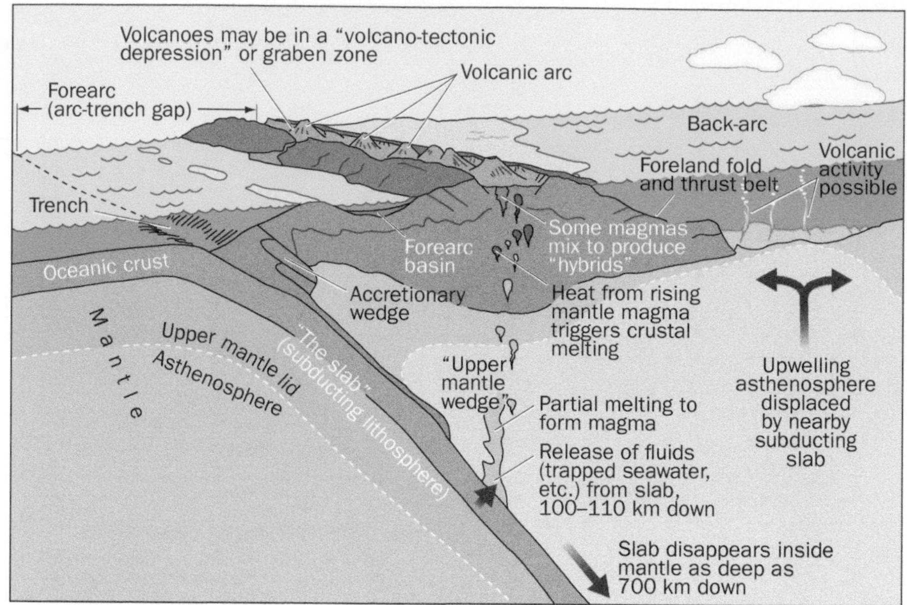

FIGURE 2.6 Oblique cutaway showing major geographical features and dynamic processes of a convergent plate boundary. There are, of course, many variations on this pattern. For example, back-arc regions are not submerged everywhere. Nor do coastal islands in many places mark the presence of an offshore accretionary wedge. In ocean areas, only the tops of the arc volcanoes may stick above the sea level, forming "island arcs."

bedrock continues to bend and sink along these loci, carrying considerable amounts of sediment mixed with seawater down into the mantle. Trenches appear arcuate on maps, simply owing to the fact that the intersections of curving surfaces on a sphere will form arcs.

The relationship of subduction to volcanic activity, as noted by Hess, will be explored further later. For the present, it is important to note that most oceanic trenches have parallel belts of volcanoes rising along the fringes of the overriding plates (Figure 2.6). The distance between the trench and the volcanic belt, termed the **arc–trench gap**, is a function of the steepness of the subduction zone, with smaller arc–trench gaps associated with more steeply subducting lithospheric slabs. This implies that processes responsible for creating subduction zone magmas must occur at nearly the same depth inside the earth. Plates that subduct at very shallow angles (less than about 15°) have no associated volcanoes, which is another important clue to the origin of magmas in this setting. Arc–trench gaps are a few tens to several hundreds of kilometers wide. If volcanoes grow upon a continental foundation, they form what are called **continental arcs**. If they form volcanic islands fringing a marginal sea, such as the Northern Mariana or Antilles archipelagos, they constitute **island arcs**. Volcanic arcs are typically no more than a few tens of kilometers wide. Volcanoes of continental and island arcs are similar, being mostly "gray volcanoes" characterized by steep slopes and explosive activity, but there are distinctive and important differences as well. In some regions, volcanic activity also occurs behind arcs, on the floors of marginal seas and within continental interiors. This **back-arc volcanism** is also plainly related to subduction, and tends to be widespread relative to the narrowly confined volcanism within arcs. In general, back-arc volcanism is less vigorous as well as more widely dispersed than arc volcanism. But, in fact, some of the largest eruptions known to have ever occurred have taken place in back-arc continental environments. The geometry and names of the major plates and their boundary zones, along with locations of Earth's most important volcanoes are shown on the map at the end of this book. From this map, it is indeed apparent that most volcanoes are located at or very near plate boundaries. We'll discuss the exceptions (intraplate or "hot spot" volcanoes) separately following the sections on plate boundary volcanoes.

Volcanoes along Divergent Plate Boundaries

Divergent plate boundaries are, for the most part, marked by an approximately 65,000-km-long discontinuous MOR system along which a new oceanic crust is being generated (Figure 2.3). The magma characteristically erupted here is the most "primitive" (least evolved) to reach the surface from the mantle – primarily as a highly fluid type of lava termed tholeiitic basalt (Chapter 3). Most other melts are contaminated to one degree or another by older rocks, or compositionally modified before eruption. Submarine volcanism is further discussed in Chapter 12.

The crest of the slower-spreading parts of the MOR, such as the Mid-Atlantic Ridge (25 mm/yr) is marked by an axial rift valley, the floor of which is distinguished by networks of subparallel fractures (rifts) – conduits for the escape of volatiles, hot mineralized water, and lava. The valley is a **graben**, a down-dropped block of crust bounded by faults resulting from tectonic extension. Because of the ongoing dilation, the basaltic accumulations are being continuously and slowly torn apart and spread out in this setting, so that few distinct volcanoes form, though there are a few spectacular exceptions. Those that do are primarily elongate, broad structures less than a few hundred meters high and complexly deformed (Fornari et al. 1987), and even these are mostly ephemeral features destined to be destroyed as seafloor spreading continues. Where spreading is much faster (e.g. the East Pacific Rise, at 80–120 mm/yr), axial valley development is subdued or non-existent, and there is even less of a chance for substantial volcanic edifice growth.

In many places, high-temperature underwater springs and vents (hydrothermal fields) persistently erupt along submarine divergent plate boundaries, sites of strange, deep-sea marine ecosystems (Karson et al. 2015). Their concentration varies according to the spreading rate: along slow-spreading divergent plate boundaries, one active hydrothermal field every 100 km is typical (Beaulieu et al. 2015). Where crustal heat and spreading rates are rapid, as many as one substantial vent area every 10–30 km may be observed (Palgan et al. 2017).

The MOR system emerges above sea level in just a couple of places – Iceland and the Afar region of Ethiopia. There, broadly elongate shield volcanoes cut by fissure systems parallel to the plate boundary are active, exemplified by Krafla [103] and Askja [102] in Iceland, and Erta Ale [121] in the Afar – which although 50 km across at the base rises only 600 m at its summit. Erta Ale is one of the most active basaltic volcanoes on Earth, in fact, having shown almost continuous lava lake activity since 1967.

Continental Rift Zone Volcanoes

Mechanically, these zones of active volcanism are related to the extension of continental plates by mantle upwelling, and they bear dynamic similarities to the divergent-plate volcanism of MORs. The associated volcanoes are of very different geochemical composition, however, apparently due to differing depths of melting, to different mantle source materials, or to magma contamination during ascent. Because continental rift zones are characterized by high heat flow, they are of considerable importance because of possible geothermal resources. Three examples of rift zone volcanism in continental settings will be described: one directly related to divergent-plate volcanism, and two from more stable continental environments.

The East African rift valleys are southwest-trending extensions of the Carlsberg Ridge-Red Sea spreading system and result from the pulling apart of the crust with the sinking of long, narrow blocks (grabens) between fault scarps. The grabens have been filled with thick sediments of largely volcaniclastic origin, and it is here that many of the oldest traces of early humans and their predecessors have been found, which is not an accident (Chapter 13). Adjoining the Red Sea (Afar Triangle in Eritrea), the rocks are tholeiite basalts similar to those of the MORs (Tazieff 1970), but further south, the lavas are moderately to very deficient in silica and rich in sodium or potassium, or both. Some of them are among the poorest in silicon and richest in alkalis of any volcanic rocks on planet Earth.

Another example is the Rhine Graben of southwestern Germany – a Y-shaped split in the crust associated with a gentle, broad up-arching above an area of continental extension, related to the intrusion of magma from below. Several volcanic fields are associated with it. The Vogelsberg field lies directly astride the eastern arm of the Y, and the Kaiserstuhl lies within its leg. To the east of the arms lie the volcanoes of the Hesse district, between the arms lie those of the Westerwald, and to the west lie those of the famous Eifel district, where over 200 monogenic volcanoes have developed, many associated with diatremes (Chapter 4) and maars (Chapter 10). Some of these volcanoes are of Holocene age, as are young volcanoes in central France (Chaîne des Puys) and northeastern Spain (Garrotxa volcanic field). Because of their long Quaternary eruptive histories, none of these volcanic areas should be considered extinct, and the odds are good that future eruptions will occur in these areas. As is true of most volcanoes associated with continental crustal extension, the volcanic rocks of these fields are alkalic and deficient in silica.

The Rio Grande Rift is a down-faulted depression and a series of deep sedimentary basins that extends over 1000 km from southern Colorado in the United States to northern Mexico. It is characterized throughout its length by normal faulting related to crustal extension, recently active volcanism, and high heat flow. Rifting began in this zone about 30 million years ago, and produced volcanoes varying in composition from basalt to rhyolite. Basaltic volcanism increased in intensity four to seven million years ago, and has continued into Quaternary time (Eaton 1979). Although the rift is believed to be underlain by a sill-like body of basaltic magma at depth (see Chapter 3), seismic activity and geodetic measurements suggest that no extensional rifting is occurring at the present time (Sanford et al. 1979).

Volcanoes along Convergent Plate Boundaries

Whereas divergent plate boundaries mark the places where a new lithospheric crust is created, convergent boundaries are the zones where the lithosphere is destroyed (or at least largely hidden from view) through the process of subduction. Over two-thirds of the world's known subaerial volcanoes occur in this environment, most of them in the circum-Pacific "Ring of Fire," both in continental, and in island-arc settings. On a gross, global scale, the Pacific Basin is shrinking as the Atlantic grows wider. This slow consumption of Earth's largest oceanic region at the expense of a younger, growing ocean on another side of the world has been going on since Mesozoic time and the breakup of the ancient supercontinent of Pangaea, creating the interlinked system of arcs called the Ring of Fire. The Ring, in fact, is not a complete loop – it more resembles a "Horseshoe of Fire." To the southwest, it begins on the North Island of New Zealand, extending northward through Melanesia into eastern Indonesia, the Philippines, Japan, and Kamchatka, eastward through the Aleutian Islands and southern Alaska, and southward along the western coast of the Americas to southern Chile (see the map at the end of this book) and the South Sandwich Islands, beyond which it terminates in a few lonely volcanoes close to Antarctica's Palmer Peninsula. To be sure, there are gaps in the line, such as that from Alaska to southern British Columbia, and northern California to Mexico, where transform faulting rather than plate convergence dominate. But even there, some volcanic activity takes place where side-stepping faults greatly reduce pressure on the underlying mantle.

The great Alpine–Tethys collision zone between the African and Eurasian plates is another major location for convergent plate volcanoes. In the Mediterranean, a dying sea where Africa is undergoing slow (in geological time frames) collision with Europe, the plate boundary consists of a complex zone of microplates along which are located the active volcanoes of Etna [106], Stromboli [107], Vesuvius [109], and Vulcano [111] – the namesake of volcanology. These volcanoes were known to ancient Mediterranean civilizations (Chapter 1), and it is from here that much of our volcano knowledge has been, and continues to be, gained. Older volcanoes are found further east in this collision zone, and include Elbrus [122], the highest mountain in Europe (5642 m), Ararat [123] (traditional landing place for Noah's Ark (Figure 1.39)), and the numerous and varied volcanoes of Turkey (Yürür and Chorowicz 1998).

The aforementioned discussion has been restricted to the active Holocene volcanoes of convergent plate boundaries, as these are the ones that can be seen and most easily studied. In contrast to most submarine volcanoes, however, which can retain their general shapes for over 100 million years in their protected locations beneath the sea (until their inevitable destruction at colliding plate boundaries), subaerial volcanoes are rapidly cut away by erosion. Their original characteristics must be inferred by studies of their deep roots or of their partially preserved products. The oldest subaerial volcanoes that retain their original constructional forms are of Pliocene age – probably none of them more than 10 million years old. Although this book focuses on the young, mostly active volcanoes we can presently see, no volcanologist should be so mesmerized by the present that the hundreds of thousands of volcanoes that have lived their moments of fire and glory in past times are forgotten.

A map showing the distribution of Mesozoic plutonic rocks in the circum-Pacific region (Figure 2.7) hints at the extensive volcanic activity of that age. Although erosion has removed most of the proof of direct connection between these batholiths and surface volcanoes, most if not all of these deep-seated plutons must have vented to the surface in many places and over the long, complex history of batholith emplacement, which can exceed 50 million years.

No two volcanic arcs are quite alike. To appreciate the variation in volcanism that may take place within and between them, it is instructive to discuss two well-studied examples, one from the northwestern United States and one from northeastern Russia.

"Cascadia" – The Cascade Arc

The Cascade volcanic arc stretches 1250 km from northern California into southern British Columbia (Figure 2.8). It results from the relatively slow (4 cm/yr) convergence at a highly oblique angle (N50°E) of two plates, the oceanic Juan de Fuca to the west and continental North American plate to the east. The southern terminus of the arc presently lies in the area of Lassen Volcanic National Park, California [40]. The northern boundary is less easy to define, in large part because of past glaciations. Isolated groups of young volcanoes continue from southern British Columbia all the way into the Yukon Territory. These northern volcanoes are not related to the low-angle (10–18°) subduction of young ocean floors responsible for volcanism in the main Cascade Range, however. If we use this subduction as a criterion for defining the Cascade arc, the northernmost possible extent of Cascadian volcanic activity appears to be the Garibaldi volcanic field [27], directly inland from central Vancouver Island (Souther 1990).

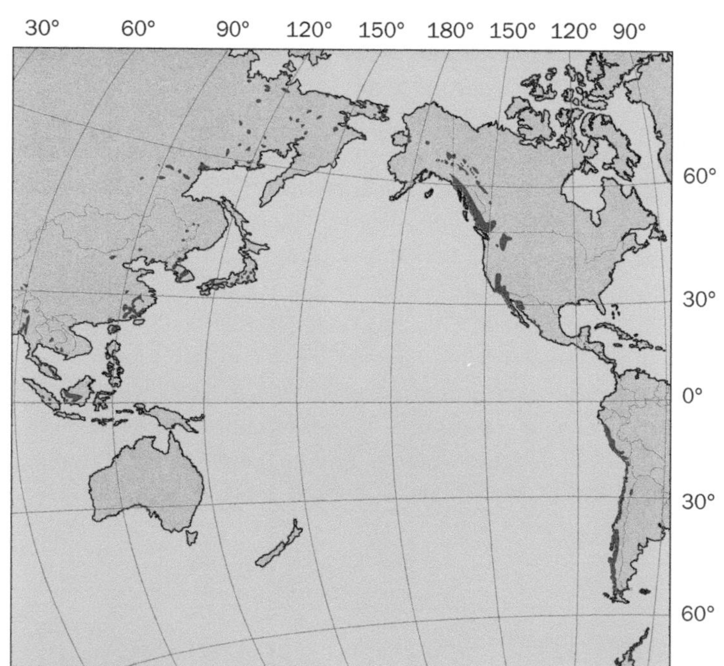

FIGURE 2.7 Distribution of Mesozoic plutons in the circum-Pacific region. Tens of thousands of now eroded-away volcanoes must have once been active above these plutons. Source: Bateman and Eaton (1967, p. 602).

The volcanoes of the Cascade arc include 30 large polygenetic composite volcanoes and volcanic centers, three young, topographically well-defined calderas, three

50-km-wide back-arc, mostly basaltic polygenetic shield volcanoes, and approximately 2300 small, monogenetic shield volcanoes and cinder cones of basaltic and andesitic composition, principally located in Oregon and northern California (Figure 2.9) (Luedke and Smith 1981; Guffanti and Weaver 1988; Hildreth 2007). The big polygenetic volcanoes range from knobby piles of explosive, intermediate to high-silica domes (e.g. Mount Garibaldi [27]) to near symmetrical edifices constructed mostly of low-silica lava flows (e.g. Mount McLoughlin [28]). Some of these polygenetic centers have erupted a wide compositional range of lavas and pyroclastic deposits (e.g. Mount St Helens [31]). Others are compositionally homogeneous (e.g. North Sister [35]). Although the small monogenetic volcanoes greatly outnumber the polygenetic ones, by far most of the volcanic material erupted in the arc has come from the latter (Hildreth 2007). Volcanism has taken place atop a foundation of older sedimentary and metamorphic rocks uplifted as much as 1500–2000 m above sea level. Crustal extension has faulted and down-dropped much of this terrain in the southern part of the arc.

At least 50 eruptions have occurred in the Cascade arc since the start of historical observations, around 1750. These have taken place at eight volcanoes, all but two of which are composite edifices. Interestingly, only three of these eruptions have occurred since 1900 (if we count the on-and-off Mount St Helens activity since 1980 as one eruption), underscoring how irregular volcanism can be within individual arcs. A large subduction-related earthquake centered beneath the northwest coast of Washington State late in the eighteenth century may have stimulated, or be related to the burst of ensuing nineteenth-century volcanism, but there is no way of proving this at present (Harris 2005).

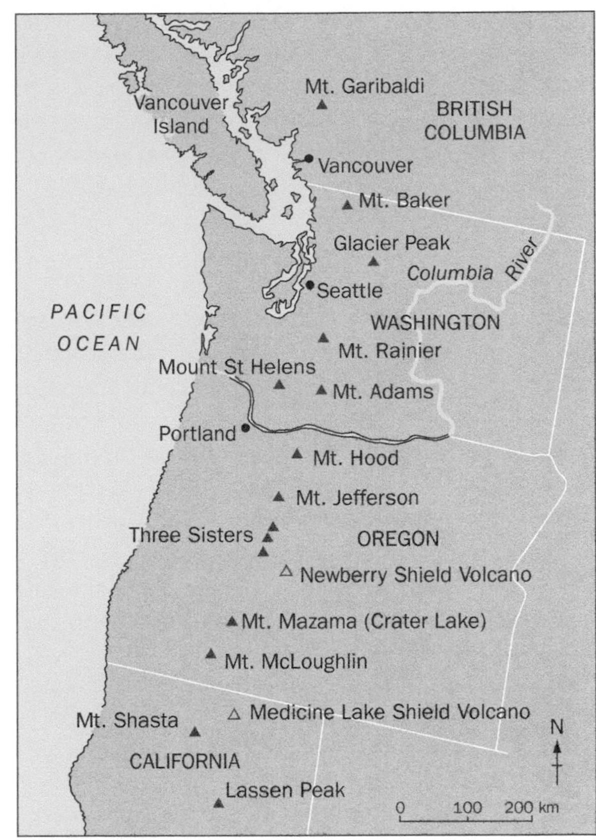

FIGURE 2.8 Large polygenetic volcanoes of the Cascade Range. Black triangles are composite volcanoes. Open triangles are shield volcanoes (Chapter 9).

FIGURE 2.9 Goosenest, near Mount Shasta, California [30]; one of thousands of small- and medium-sized monogenetic volcanoes in the southern Cascades. Goosenest is a shield volcano composed of andesitic blocky lava flows enclosing a central core of pyroclastic material, mostly cinder, bombs, and blocks, which protrude to form the summit cone. This type of volcano is uncommon or non-existent in other arcs. Source: Photo by R.W. Hazlett.

One can calculate the mean spacing between the large composite cones of the Cascade arc easily enough with a ruler and a map, which turns out to be about 85 km. But this is a misleading figure, for that spacing is by no means regular. Rather than forming a continuous trend, the alignment of prominent composite cones takes sharp jogs in places, possibly reflecting the presence of deep bedrock faults that have permitted melts to penetrate areas off the arc. In addition, there are prominent gaps in the distribution of volcanoes – the largest over 130 km long – notably north and south of the Mount Shasta [30] and between the volcanoes of the Mount Garibaldi group [27] and Mount Baker [34] in northern Washington. Presuming that the generation of magma is occurring *throughout* the underlying slab–mantle wedge, it is difficult to explain the presence of such gaps. Hildreth (2007) argues that they probably represent the presence of lithosphere that compositionally and structurally inhibits the ascent of melt all the way to the surface.

The predominance of monogenetic low-silica largely basaltic volcanism in the southern Cascade arc implies that subduction is generating a different kind of magma to the south than in the north. But this is not necessarily so. The difference probably is due to the fact that tectonic extension in the south permits the direct ascent and eruption of numerous small batches of mantle-derived basaltic melt that stall out in the thicker, more compressed lithosphere to the north (Hildreth 2007). Other arcs, such as the Indonesian and Central American, lack noteworthy monogenetic volcanic fields because they do not experience significant extensional tectonism. The volcanism that builds the big composite cones of the Cascades and these other arcs represents a more complex evolution of magma that will be discussed further in the next two chapters.

Volcanism in the Cascade arc appears to have been episodic during its 32-million-year history, but, by a rough estimate, eruption rates have declined six-fold during this time as plate convergence has slowed down (by a factor of 5) and the obliquity of convergence has increased (Verplank and Duncan 1987; Hildreth 2007). Let's contrast this situation with an arc that has arisen from more direct and rapid plate collision.

Kamchatka

The Kamchatka Peninsula (Eichelberger et al. 2007) extends from the eastern tip of Siberia southwestward toward Japan, enclosing the Sea of Okhotsk to the west. As in the case of the Cascades, the Pacific Plate is colliding with this region of continental crust, but at double the rate (8–9 cm/yr), and nearly head on, perpendicular to the trench, forcing a considerable amount of coastal uplift, notably in the region where the Aleutian–Komandorski island arc approaches the peninsula (Pedoja et al. 2013). The subducting plate is less hot than it is beneath the Cascades, being located much further from the MOR where it originated, and this may contribute to its very steep angle of subduction – about 50° – owing to somewhat greater density. Because most subduction-related melts tend to be generated above slabs that have sunken to depths of roughly 100–110 km, this also means that the arc–trench gap is substantially less in Kamchatka than it is in the Cascades (175–200 versus 320–450 km).

The average spacing between composite volcanoes or related calderas on the Kamchatka Peninsula in eastern Russia is a mere 29 km, though spacing here too is irregular, and includes some offsets and gaps. The arc extends about 750 km as a landward continuation of the much longer (1400 km) Kuril–Hokkaido island arc. Some 30 active volcanoes exist in Kamchatka alone, along with nearly 200 extinct or dormant ones. About 70 of these are large polygenetic volcanoes and calderas less than a million years old, and there are over a thousand monogenetic cones, domes, rings, and maars. Included in this group is Klyuchevskoi [171], one of the largest composite volcanoes in the world.

While only three geologically young calderas exist in the Cascade Range, the Kamchatka arc is noteworthy for its concentration of young calderas, most of which, unlike Crater Lake [32], Oregon, did not form from the collapse of a pre-existing composite volcano, but rather developed from the wholesale collapse of the crust above large, shallow magma bodies (Chapter 10). Eight large calderas occur, with one 150-km segment of the arc consisting of nothing but adjacent and overlapping calderas and their attendant edifices. This cluster includes Uzon Caldera [168] in Kronotsky National Park, an area active with hot springs and geysers and as scenically wonderful as Yellowstone National Park [47] in Wyoming; and the Klyuchevskoi volcanic center, which includes the active volcanoes Klyuchevskoi [171], Bezymianny [170], and Ploskii Tolbatchik [168] (Figure 2.10).

The Kamchatka arc consists of two parallel chains of volcanoes (Figure 2.11). The most vigorous and recent volcanism takes place in the eastern belt, which closely fringes the Pacific coast of the Peninsula. Roughly 200 km inland from this alignment is the 450-km-long Sredinny (Central) belt, which has a record of intensive effusive, shield-building activity within the past 2–3 million years, but is now apparently dead. The Sredinny Range could mark the existence of an older subduction zone that ceased to be active when an ancient island arc (the so-called Kronotski terrane) collided with and accreted to Kamchatka. Subduction beneath the peninsula continued, however, in the ocean basin to the east of the newly added landmass. That gave birth to the presently active volcanic belt (e.g. Avdeiko et al. 2006). The anomalously vigorous volcanic activity in the Klyuchevskoi–Shiveluch [171–172] group apparently marks the edge of the downgoing slab, where intensive upwelling of hot rock might be expected owing to the stirring of the mantle around the sinking margin of the plate. The slab subduction angle in this region is only around 35°, accounting for the westward offset of the Klyuchevskoi–Shiveluch volcanoes relative to the remainder of the active arc.

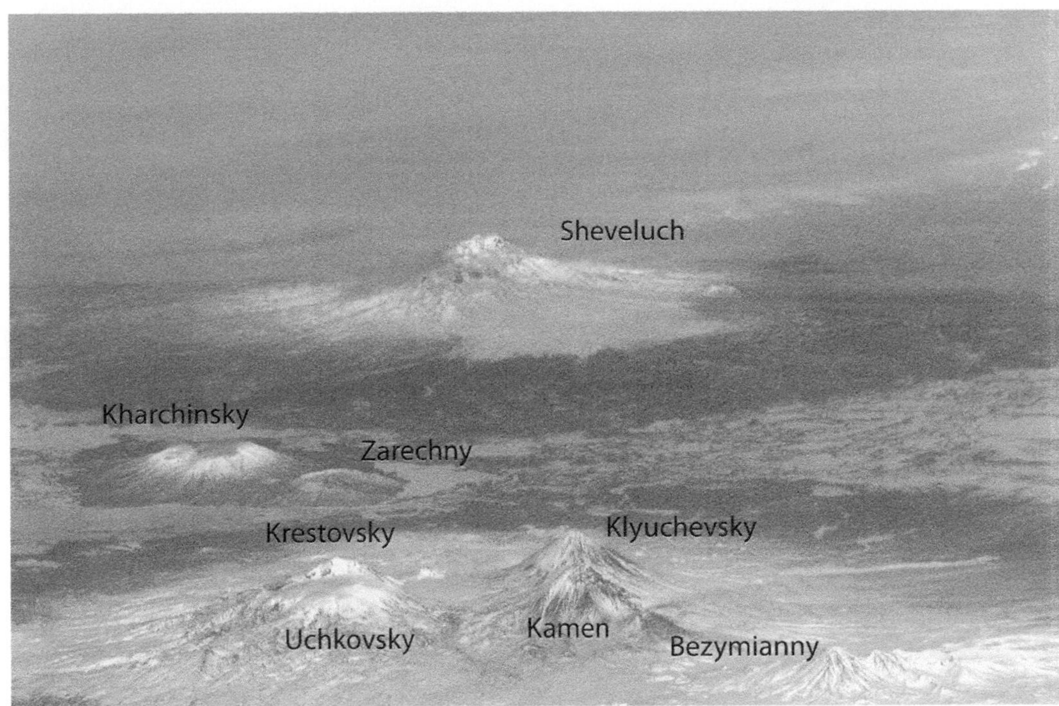

FIGURE 2.10 View of the Klyuchevskoi volcanic complex [171] from the International Space Station, including Shiveluch [172] Volcano further north (toward the upper margin of the image). Source: Photo courtesy of the NASA Image Science and Analysis Laboratory, Johnson Space Center, ISS-001-E-6505.

Some vigorously active volcanic arcs on thin or youthful continental lithosphere develop large grabens, tens of kilometers wide, hundreds of kilometers long, and hundreds of meters deep, striking along their axes. Arc volcanoes align along the lengths of these grabens, which as a result are often called volcano-tectonic depressions (Chapter 10). Examples include the Nicaraguan and Sumatran depressions. Arching of the overriding plate during convergence must account for formation of volcano-tectonic depressions, causing the broad crest of the uplifted arc to collapse like a gigantic keystone block. The Central Kamchatka Depression might appear to have formed in the same way – and indeed is bounded along its eastern side by a normal fault system, typical of volcano-tectonic depressions. But given that the current arc lies largely to the east of this feature, it is more likely that it is just an artifact of two parallel volcanic ranges growing adjacent to one another, augmented by mild back-arc extension.

Despite the differences in the concentration and explosiveness of volcanoes in Kamchatka and the Cascades, there is not necessarily a substantial difference in the total volume of volcanic produce in these two arcs within the past few million years. The difference lies more in the styles of volcanism than in the amount of fresh new volcanic material produced. Hildreth (2007) making first-order comparisons of volcanic output between the Cascades, Alaska Peninsula, Andean, and other continental volcanic arcs found rough parity and even greater Cascades volcanic output during Quaternary times than in other arcs with higher concentrations of large polygenetic cones and volcanic centers (Table 2.1). The

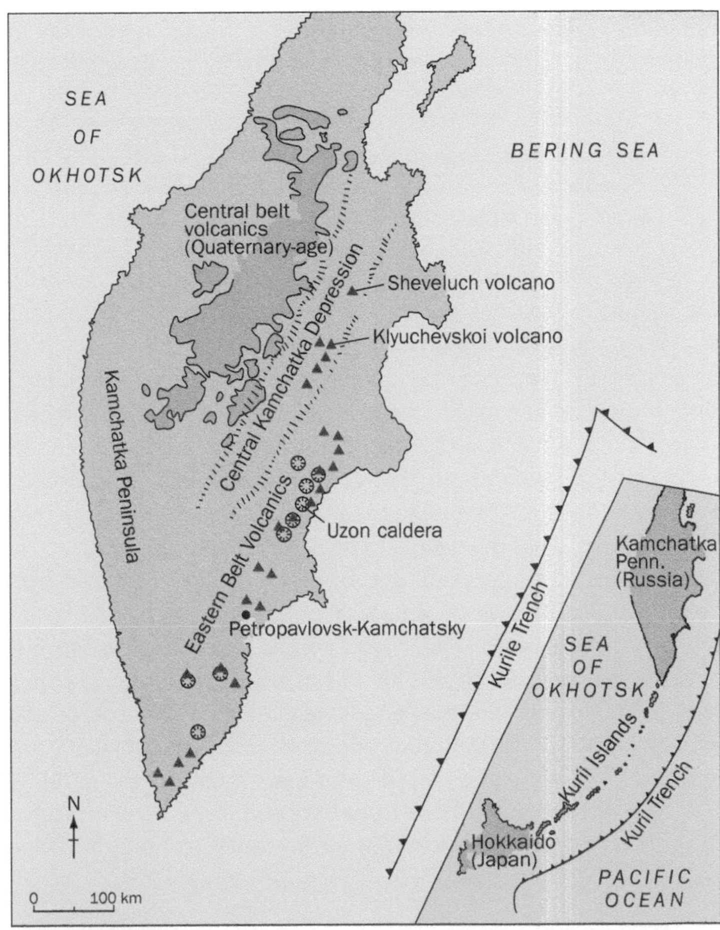

FIGURE 2.11 Major features of the Kamchatka volcanic arc. Dark triangles are large, geologically young polygenetic edifices. Hachured circles are young calderas. Source: Map modified from Fedotov and Masurenkov (1991).

TABLE 2.1 Comparisons of volcanic arc productivity during Quaternary times (the past 1.8 million years).

Arc (and research source)	Length of arc (km)	Number of major polygenetic volcanoes	Volume of material erupted (km³)
Cascades (Hildreth 2007)	1250	30	6400
Andean Southern Volcanic Zone (Hildreth 2007)	1400	80	5300
Alaska Peninsula (Hildreth 2007)	1150	55	2000–3000
Central America (Carr et al. 2003)	1100	39	3464
Northeast Japan (Arakami and Ui 1982)	1000	45	1480*

*Most of these volcanoes are less than 500,000 years old. If the rate of productivity in this time scale is expanded to include all of the Quaternary, then the productivity of this arc increases to almost 6000 km^3.
Source: Modified from Hildreth (2007).

comparison with island arcs is more tenuous, given the submergence of so much of their eruptive material, but it may be that because the lithosphere is thinner and less evolved in those settings that a larger proportion of intruding magma can erupt. For instance, recent studies of magmatism in the Aleutian arc suggest that it has been enormously productive, between 89 and 120 km^3 of igneous material – including plutonic rocks – per kilometer of arc length every million years (Jicha et al. 2006).

Episodicity and "Vigor" of Arc Magmatism – Real or Imagined?

From time to time, people believe that the amount of volcanic activity around the world is increasing, and some popular media popularize these notions. Those who believe in apocalyptic endings for Earth pay keen attention to such "trends." But are they real? Does volcanic activity occur in pulses worldwide? Is the world heating up to a fiery self-immolation? Newhall and Self (1982) note:

> *A careful study of the record of the past 100 years shows [an] apparent drop in global volcanism during periods in which [news]editors and scientists were preoccupied with other things, such as world wars, and apparent increases during periods of universally strong interest in volcanism, such as the years immediately following . . . [the eruptions of] Krakatau and Mont Pelée.*

Coats (1951) noted that major pulses in the number of reported eruptions in the remote Aleutian arc correspond to the times of exploratory expeditions. In other words, people documented volcanic activity whenever they were close at hand and paying attention.

Taking into consideration these very human factors, the *historical* pattern of global volcanic activity shows no evidence of increasing, decreasing, or even pulsating worldwide. However, bursts of volcanic activity do appear to occur in certain regions from time to time, which might be correlated to episodic plate motions – including sets of strong earthquakes – or releases of pent-up energy along volcanic arcs. Especially noteworthy are very explosive eruptions at three widely spaced volcanoes in the Caribbean region in 1902 and a spasm of volcanic outbursts in the Bismarck archipelago in 1972–1975 (Cooke et al. 1976). In another example, Prueher and Rea (2001) studied seafloor volcanic ash layers in the northwestern Pacific and found evidence for simultaneous, pulsating volcanism in the Aleutian and Kuril–Kamchatka arcs. Each episode of increased volcanic vigor lasted from about 10,000 to a few hundred thousand years, and was separated by somewhat more sluggish intervals lasting as long as a half-million years. Similar periods of intensive volcanism, or "flare-ups," each lasting a few million years, appear to have effected island arcs throughout the western and northern Pacific in late Jurassic, late Eocene, and early Oligocene times (Jicha et al. 2006; Beranek et al. 2017). Verdel et al. (2011) argue that a Paleogene flare-up in the Iranian volcanic arc happened when the underlying subducted slab steepened or "rolled back," allowing hot, magma-generating mantle to well up under reduced pressure conditions. This case is not unique; slab rollback, slab pull, slab buckling, rupture, and other subduction-related behaviors we do not have time to discuss have been invoked by earth scientists to investigate particular arcs in many parts of the world. Tectonic scenarios abound (e.g. Van der Meer et al. 2018).

Newhall and Self (1982) compared the volcanic activity in different arcs, and set up a standard of comparison between the various arcs that they call **volcanic vigor**. Volcanic vigor may be defined in several ways, whichever is most convenient for a particular researcher. Vigor may be:

1. the number of volcanoes active within the past 10,000 years in a volcanic arc, divided by the length of the arc; or
2. the total duration (say, during a century) in which documented eruptions have taken place in an arc, divided by the length of the arc; or

3. the total number of documented eruptions (say, during a century) above a certain level of explosiveness (VEI>3; see Chapter 5) divided by the length of the arc.

Newhall and Self suggested that there is a definite, albeit weak correlation between the dip of the subducting slab beneath an arc and volcanic vigor. "Sluggish" arcs tend to be associated with shallow-dipping slabs. In contrast, the steep plunge of the Pacific Plate along the east coast of Kamchatka correlates with one of the world's most "vigorous" volcanic arcs.

The thickness of the overriding plate also influences the vigor of a volcanic arc. Where upper plate thicknesses are less than 20 km (e.g. the Tonga–Kermedec and Izu–Mariana islands in the western Pacific), arc volcanism tends to be less vigorous – or at least less powerfully explosive – than along arcs with thick upper plates (25–35 km) and similar subduction angles. Newhall and Self further suggested that weak volcanism ensues where young, relatively warm oceanic lithosphere subducts beneath an arc. An example of this is the Cascade Range.

Intraplate Volcanoes

Not all volcanoes can be ascribed to the interactions of plate boundaries. Perhaps only three-quarters of them are so associated worldwide. The rest lie in locations often thousands of kilometers from the nearest plate edge, and must somehow have other explanations. Fortunately, the insights learned in studying plate tectonics, and a better grasp of Earth's interior properties, enables us to develop some reasonable hunches about what is going on.

By far, the largest group of "intraplate" (off-plate margin) volcanoes can be related to so-called **hot spots**. The term "hot spot" may be misleading, however, in that we do not really know if heat, or *heat alone* is responsible for the production of molten rock in each of these areas. Other phenomena, such as reduced pressure or increased volatile content, can also cause rocks at high temperature and pressure to melt (Chapter 3). Because there is no direct evidence that all hot spots are really all that "hot," we prefer to call them *melting sources,* but since "hot spot" is in general use, we shall accept it (warily) to describe these areas of intraplate volcanic activity. Such hot spots underlie areas of concentrated volcanic activity, generally accompanying uplift and seismic activity over broad areas, and are in some cases associated with the genesis of linear volcano chains thousands of kilometers long. Several dozen linear volcanic chains are present on Earth's surface, most of them on oceanic crust – especially trending across the Pacific (Figure 2.12).

The most famous of these volcanic island chains is associated with the "Hawaiian hot spot", a melting source presently centered beneath the southern half of the Island of Hawai'i near the center of the Pacific Plate. Extending northwest of the Hawaiian hot spot, a chain of over 100 volcanoes in progressively increasing states of erosion and submergence continues all the way to the Kuril-Kamchatka Trench, more than 5000 km away (Figure 2.12). The increasing age of volcanoes with distance from the hot spot led Canadian geologist J. Tuzo Wilson (1963) to propose that the volcanic chain developed as the Pacific Plate shifted over a fixed melting source rooted in the sublithospheric mantle beneath present day Hawai'i. Subsequent radiometric dating of sea floor lavas indicates that the rate of plate motion across the Hawaiian hot spot is about 8–9 m per century – double what it was 50 million years ago (Zheng et al. 2018) – and dating of volcanoes along the Hawaiian–Emperor chain largely validates Wilson's idea. Some noteworthy refinements have been made however: We now know that, relative to the deep earth, the Hawaiian hot spot has moved hundreds of kilometers southward over time (Tarduno et al. 2003). And the rates of lava production and volcanic growth have also increased significantly, perhaps by an order of magnitude (e.g. Robinson and Eakins 2006). The Hawaiian melting source is not as stable as Wilson first imagined; but it has remained remarkably persistent, nonetheless, given the dynamism seen in the rest of Earth's crust.

Jason Morgan (1971) suggested that a source for this melting might be analogous to the hot air mixed with smoke that rises from a chimney or smokestack. The stream of polluted hot air does not spread out at once as it exits the mouth of the chimney, but continues to ascend tens to hundreds of meters into a still sky, well-defined and narrow, owing to its great buoyancy relative to the surrounding, cooler air. Fluid mechanicists refer to this as *thermal plume* behavior. Similarly, according to Morgan, hot spots represent localities on Earth's surface where thermal plumes of unusually hot, plastically deforming rock well up from great depth and impinge upon the underside of the lithosphere. He referred to them as **mantle plumes** and suggested that they might originate from positions on the core–mantle boundary, 2900 km below Earth's surface.

In recent years, much controversy has revolved around proving that mantle plumes such as these really do exist. Some geophysicists argue that the deep mantle is under too much pressure to develop long-lasting plumes, and that hot spots may be explained by melting in a shallow, compositionally heterogeneous mantle. Melting could be caused by tearing at the base of the plates that shift ever westward in response to Earth's rotation, bringing to mind Wegener's early explanations for "continental drift" (e.g. Dogliani et al. 2005; Foulger et al. 2005; Scoppola et al. 2006). More recently, however, studies of long-period (far-traveling) seismic waves beneath several large hot spots have revealed strong evidence that they are indeed deep-rooted. Beneath both Hawai'i and Yellowstone [47], for instance, structurally complex thermal plumes appear to extend all the way into the lower mantle

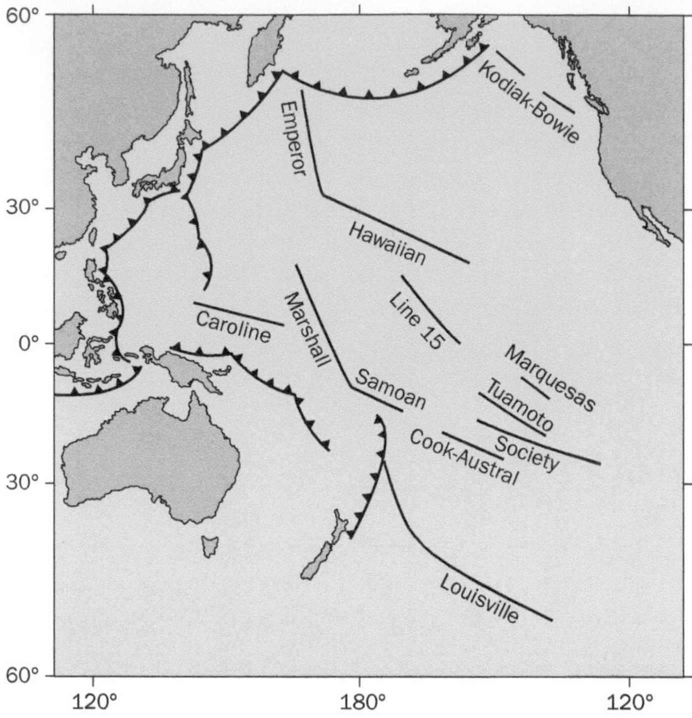

FIGURE 2.12 Linear hot spot volcanic island chains and seamounts in the Pacific Ocean. Most of the volcanoes along these chains are represented by atolls or submerged seamounts. Source: Compiled from Epp (1979) and Duncan and Clague (1985).

(e.g. Schmidt et al. 2012; Huckfeldt et al. 2013; Kerr 2013). Also relevant are certain gases released from hot spot volcanoes such as Kīlauea [17], including unusually large amounts of the heavy isotope helium-3 (Chapter 3) and osmium, elements thought to be concentrated in large quantities in the lower mantle and core. However, the significance of these gas compositions remains controversial (Lassiter 2006).

Hot spot tracks such as that of the Hawaiian–Emperor Seamounts (Figure 2.13) have some spectacular continental counterparts. For instance, the Snake River Plain is the track of the Yellowstone hot spot, with a chain of volcanic centers aging southwestward all the way to long-extinct calderas on the Oregon Nevada border (Camp 2013). Along the axis of the Great Dividing Range of eastern Australia, a conspicuous hot spot track links the 40,000-year-old Newer Volcanics [155] of Victoria with the 25-million-year-old Glasshouse Mountains volcanic center in Queensland (Cohen et al. 2004).

Other Long- and Short-lived “Hot Spot” Melting Sources

The Cameroon Line (Fitton 1980) is a 1500-km-long chain of Tertiary to Recent volcanoes that extends across oceanic and continental portions of the African plate from the Gulf of Guinea into Cameroon and Nigeria (Figure 2.14). Most of the volcanoes along this chain produce alkalic basalt lavas, many of which contain xenoliths of sub-crustal peridotite. Perpetually shrouded in clouds, Mount Cameroon [106] is the largest volcano in the chain, and rises to 4100 m above the West African coast. Presently, this is the most active volcano along the Cameroon Line, and has had nearly a dozen historic eruptions, most recently in CE 2000. The Cameroon Line is marked by over 30 maar lakes of various ages in western Cameroon, including Lake Nyos [107], infamous for its release of deadly CO_2 gas in 1986 (Chapter 14). In contrast to hot spot volcanic chains like the Hawaiian–Emperor Seamounts, there is no clear age progression along this volcanic chain, nor is there any indication of crustal extension as in the continental rift zone volcanic belts (Fitton 1980). Fitton considers the chain to mark a zone of weakness related to an inactive arm of a triple-junction spreading center related to the Mid-Atlantic Ridge. Other well-argued explanations have been offered too, including the presence of one or two mantle plumes, and “peculiar” lithospheric strain associated with a right-angle bend in the plate boundary to the west (e.g. Deruelle et al. 1991; Burke 2001).

In a few cases, hot spot mantle plumes may develop along the MOR and other divergent plate boundaries. Iceland is a spectacular example of one such hot spot. The reason that this island in the North Atlantic exists, in fact, is largely owing to anomalous heat-induced (thermotectonic) uplift coupled with high output of volcanic material. Likewise, the Afar hot spot in Eritrea and

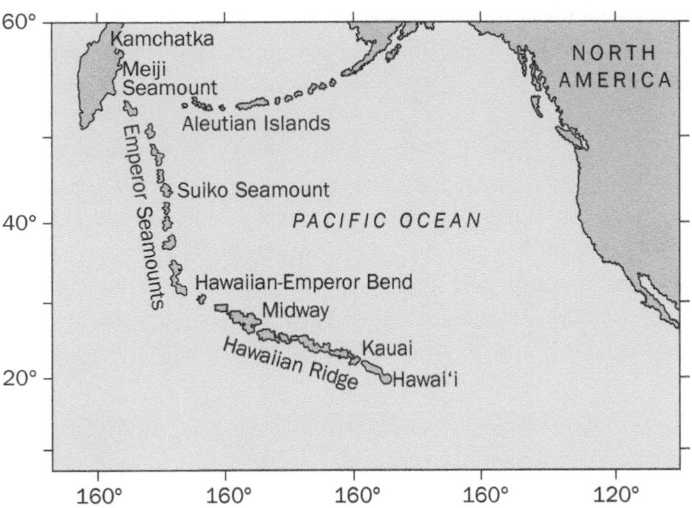

FIGURE 2.13 Location of the Hawaiian–Emperor volcanic island and seamount "hot spot" chain. Source: Modified from Clague and Dalrymple (1987).

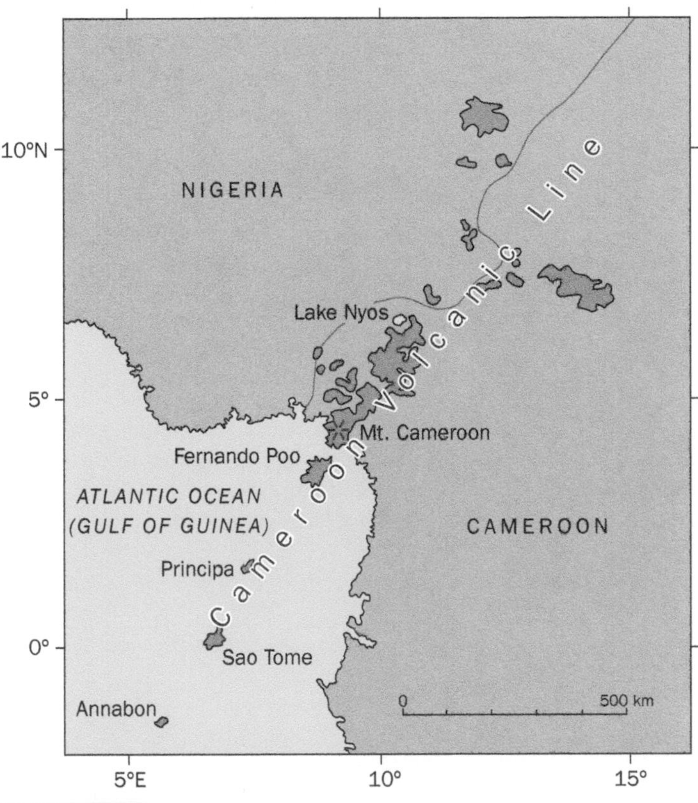

FIGURE 2.14 Distribution of Quaternary volcanic rocks along the "Cameroon Line," a 1500-km-long belt of intraplate volcanoes that cut across both the oceanic and continental crusts on the African plate. Source: Modified from Fitton (1980).

Ethiopia lies at the junction of three divergent plate boundaries, forming the Gulf of Aden, Red Sea, and East African Rift Valley. Volcanism at this *triple point* is much more vigorous than along adjacent plate boundary segments.

The Hawaiian hot spot has been active for a long time – at least 80 million years. The Yellowstone hot spot is not nearly as long-lived, dating back to about 16.5 million years. But this is nonetheless impressively persistent, given how dramatically geologic change has occurred in this same span of time elsewhere in the American West. Other hot spots, such as the ones underlying areas around Flagstaff (Arizona) and Auckland (New Zealand), are only a few tens or hundreds of thousands of years old. Will these thermal anomalies persist, or die out in short order? The geologic evidence suggests that many relatively short-lived "transient" hot spots have occurred elsewhere in the geologic past. Alternatively, these sites might simply represent areas of localized crustal extension that stimulate magma to form in the immediately underlying shallow mantle because of lowered confining pressure (Chapter 3; Koppers et al. 2003).

In some cases, volcanic activity can be ascribed to plate boundary interactions taking place many hundreds or thousands of kilometers away. Over the past 45 million years, India, once a large island like Greenland, has been in collision with Asia, owing to the subduction and consumption of the intervening prehistoric Tethyan sea floor. The elevated Himalayan Mountains, Tibetan Plateau, and Gangetic Plain are consequences of this collision. In central Mongolia's Hangay region, 2000 km from the leading edge of the Himalayan collision zone, a basaltic cinder cone field has grown over the past few hundred thousand years in response to collisional strain transferred far north into the Eurasian plate. Similarly, the Tibesti Volcanic Province, isolated deep within northwestern Africa, may be a far-flung expression of the tectonic collision of Europe with the African Plate (Deniel et al. 2015).

Questions for Thought, Study, and Discussion

1 If mantle convection slows down over great spans of future geologic time, how will this change what is seen at Earth's surface, both tectonically and volcanically?

2 Why do you think there are so few subduction zones fringing the Atlantic Ocean, while the Pacific Ocean is ringed with them? Why does the MOR divide the Atlantic Ocean floor symmetrically, while it lies well off to one side of the Pacific Ocean floor (mostly to the southeast)?

3 Why don't we see substantial volcanoes grow along the MOR systems? It is, after all, one of the most volcanically active zones on Earth.

4 Why are there substantial differences in the volcanic activity seen in the Cascades and Kamchatkan volcanic arcs?

5 What might explain why arc volcanoes are *spaced* above subduction zones? (Why does not a single, through-going volcanic ridge develop along the crest of an arc rather than the distinctively separated volcanoes that we see?)

6 How do hot spot volcanoes differ from those of volcanic arcs?

7 Why should the Cameroon Line not be regarded as a hot spot track in the same sense as the Hawaiian chain?

FURTHER READING

Condie, K.C. and Pease, V. (2008) *When Did Plate Tectonics Begin on Earth?* Boulder: Geological Society of America.

Foulger, G.R. and Jurdy, D.M. (2007) *Plates, Plumes and Planetary Processes.* Boulder: Geological Society of America.

Hill, R.L. (2004) *Volcanoes of the Cascades: Their Rise and Their Risks.* Guilford: Falcon.

Karson, J.A., Kelley, D.S., Fornari, D.J. et al. (2015) *Discovering the Deep: A Photographic Atlas of the Seafloor and Oceanic Crust.* Cambridge, UK: Cambridge University Press.

Oreskes, N. (2001) *Plate Tectonics: An Insider's History of the Modern Theory of the Earth: Seventeen Original Essays by the Scientists Who Made Earth History.* Boulder: Westview.

Sigurdsson, H. (1999) *Melting the Earth: The History of Ideas on Volcanic Eruptions.* Oxford: Oxford University Press.

CHAPTER 3

The Nature of Magma – Where Volcanoes Come From

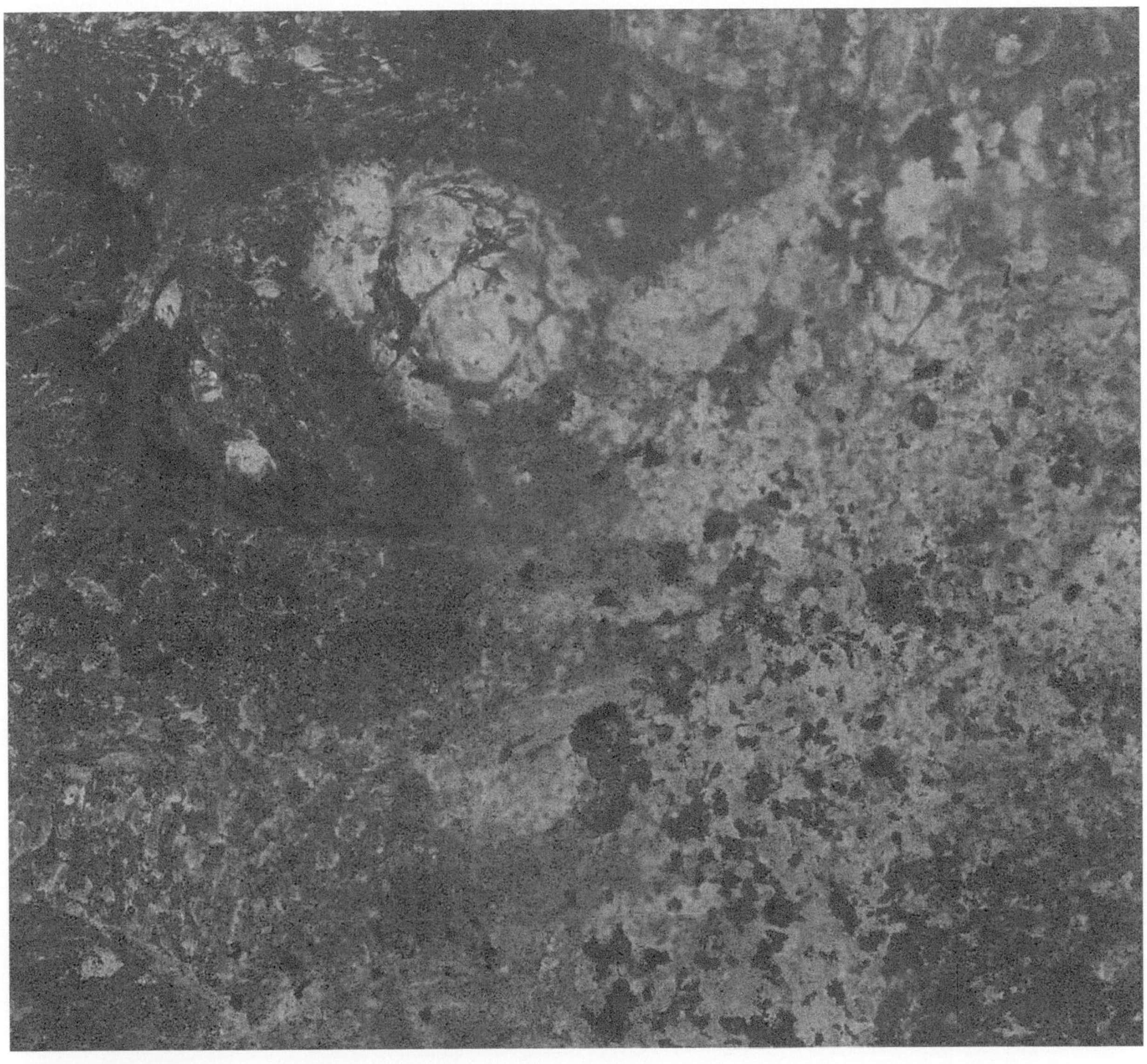

Source: Credit B. Omori

Volcanoes: Global Perspectives, Second Edition. John P. Lockwood, Richard W. Hazlett, and Servando De La Cruz-Reyna.

Companion website: www.wiley.com/go/lockwood/volcanoes2

Basalt, der schwarze Teufelsmohr, Aus tiefster Hölle bricht hervor, Zerspaltet Fels, Gestein und Erden, Omega muß zum Alpha werden. Und so wäre denn die liebe Welt Geognostisch auf den Kopf gestellt.

(Goethe 1827)

Basalt, dark devil's blackman, Breaks out of deepest hell, Splits rocks, stones and earth, And omega must become alpha And thus the dear old world is geologically turned upside-down.

(Goethe 1827)

Origins of Magma

Throughout human history people living near active volcanoes, including Goethe, have strived to explain the source of the terrifying fires volcanoes bring forth (Chapter 1). We readily observe that erupting lava is "hot" and thus the origin of that heat is critical to understanding the origin of magma – even though factors other than heat alone (pressure, composition, and volatile content) can cause hot rocks to actually melt. A lot of heat needs to be present for these other factors to be relevant, however, so the source of that heat is important to understand.

Four primary sources of heat have been active inside Earth since its formation: nebular, kinetic, gravitational, and radiogenic (Figure 3.1). Nebular and kinetic heating were significant when Earth first formed. Later, these faded away to be replaced by gravitational and radiogenic energy as the major contributors to Earth's heat.

Gradual cooling of the solar nebula allowed solid particles to condense, which in turn drew together gravitationally to form larger bodies, ultimately assembling planet Earth 4.6 billion years ago. Earth's full growth and the internal differentiation of its mass into a core, mantle, and crust may have taken place within only a 10–100-million-year period, though substantial additional material also slammed into the planet during a catastrophic cosmic disturbance 4.1–3.8 billion years ago. All impacting rocky objects not only brought heat to Earth acquired from their nebular formation, but also released tremendous amounts of heat energy (**kinetic heat**) on colliding with its surface. Impact temperatures were hot enough to melt impacted crust as it shattered, creating an unusual type of deposit rich in glassy fragments, termed **suevite** (Figure 3.2). Such temperatures exceed the hottest known magmas erupting today, in one known instance by over 1000°C (Timms et al. 2017). But they are short-lived compared with the high temperatures measured at individual active volcanoes. More importantly, the deep fracturing caused by large impacts, with craters hundreds of kilometers in diameter, is sufficient to trigger melting and magma generation beneath the pulverized crust, which

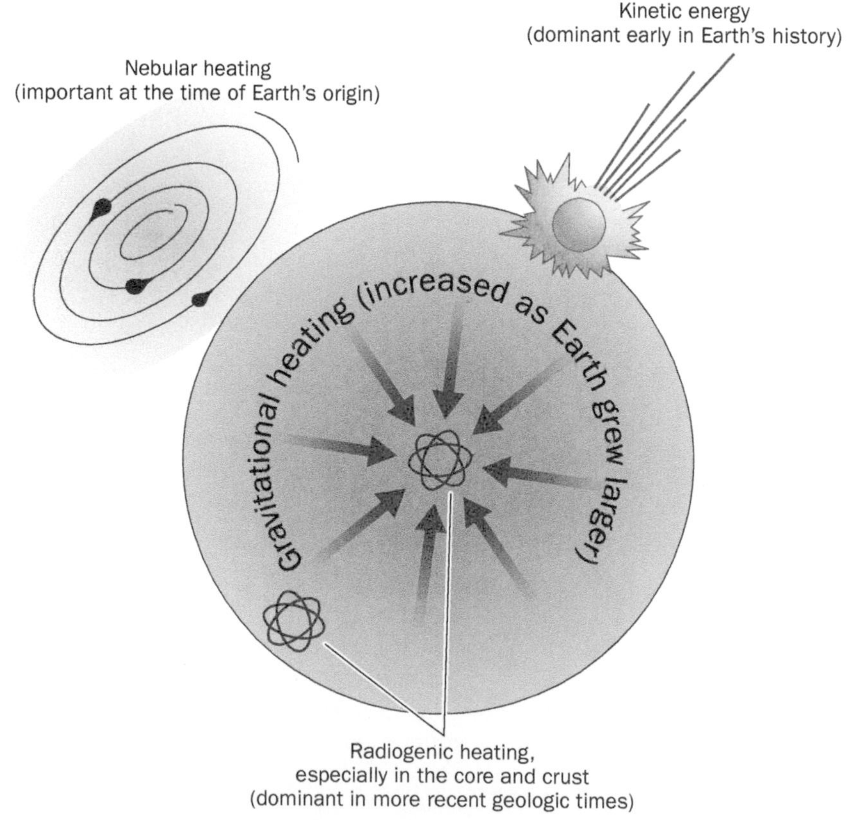

FIGURE 3.1 Sources of Earth's heat.

FIGURE 3.2 Suevite sample from the 14.5-million-year-old, 24-km-wide Nordlinger Ries meteorite crater in southern Germany. The dark irregular fragments are volcanic glass, the residue of continental crust that melted during impact. They are set in a bed of completely pulverized, fused rock dust with fragments of other, unmelted crustal rocks, mostly granitic, that blasted into the atmosphere then rained back down to blanket the surrounding landscape. Suevite is one of several rock types collectively called impactites. Source: Photo by R.W. Hazlett.

fills craters with upwelling lava flows of more ordinary terrestrial appearance. Those earliest impact craters, were not preserved, although a 3.5-Gy-impact crater and associated melt products have been discovered in Australia (Ohmoto et al. 2021). We'll explore a few ancient examples of this **impact volcanism** in Chapters 12 and 15.

The moon preserves the scars of the many impacts from space taking place during this early time. Scaling up to Earth's size, it is reasonable to believe that 20,000–25,000 craters exceeding a 20 km diameter once pocked Earth's surface, with approximately 40 craters exceeding 1,000 km diameter. A few even exceeded 5,000 km across; a starkly different world indeed (Ryder 2000). By 3.8 billion years ago, however, our planet had settled down to conditions conducive for sustaining a homeostatic atmosphere, oceans, and life, but giant meteor impacts continued at a less frequent pace, bringing heat and generating renewed melting (Lowe et al. 2014).

The oldest lavas covering the primordial earth's surface included one type, **komatiite**, not produced by volcanoes in modern times, suggesting the early existence of a hotter planetary interior. Geochemical studies indicate that komatiite flows erupted at temperatures in excess of 1600°C, some 400°C greater than the hottest lavas erupted today (Zimbelman and Gregg 2000). Komatiitic eruptions were widespread as the planet's thin lithosphere responded to major new additions of extraterrestrial debris and the intensive release of heat from its deep interior.

As impacts grew less frequent, Earth began to cool by radiation and release of entrapped gases. Inside the growing solid mass, gravitational compaction and redistribution of mass generated heat, but certainly not enough to keep the globe entirely molten, as might have been its original state. In fact, the geologic record reflects that the fraction of molten material inside Earth has significantly diminished over billions of years to only about 15% of the planet's present volume.

Research by the Comte de Buffon in France in the eighteenth century and by the famous British mathematician William Thomson (Lord Kelvin) in the nineteenth century attempted to constrain the age of Earth assuming that the planet could not be older than the time it would take for it to cool solid *from an entirely molten condition*. They presumed that Earth's interior cooled by simple heat conduction – the same phenomenon that makes the metal handle of a pan hot to touch on a stove top. Their calculations ranged from 75,000 (Buffon) to 40 million years (Kelvin). Because of the existence of active volcanoes, by this reasoning, Earth *must* be less than 40 million years old. Geologists, though, having documented many ponderously slow rates of natural change, had good reason to doubt this calculation. They argued that Earth must be much older, and that there must be an additional, unknown source of heat keeping the planet hot inside. That heat source – radioactive decay – was finally discovered by Henri Becquerel, the French physicist, in 1895.

Many elements radioactively decay to form less heavy particles. A byproduct of this decay, which occurs continuously inside Earth, is heat (Table 3.1). Since it takes billions of years for some elements in rocks to decay, this means that heat can build up and remain within a stony planet like ours essentially throughout its existence. At some point after four to four-and-a-half billion years ago, **radiogenic heating** superseded kinetic heat as the major source of Earth's thermal energy. That heat could not escape Earth's interior by simple conduction, however, as Buffon and Kelvin presumed. Something else must be facilitating its release because heat conduction is too slow a process, otherwise the planet would simply have remelted.

Scotsman Sir Arthur Holmes, whom we introduced in Chapter 2, provided a solution to this paradox with his 1927 hypothesis, proposing that convection rather than conduction, is the primary vehicle for releasing Earth's pent-up heat energy – as well as driving its tectonic plates. Holmes suggested that cells of hot rock ascend buoyantly from Earth's core–mantle boundary, spread out just beneath the surface, lose heat to the surface via radiation as well as conduction, and descend again, similar to water boiling in a pot. A complete cycle of overturning boiling water may take only a few seconds in the pot. For heated rock in Earth's mantle, the turnover may take as long as a couple of hundred million years. Several layers of "stacked" convecting cells may exist between surface and core as well, reflecting more complex layering in the mantle.

TABLE 3.1 **Relative contributions of heat escaping from present-day Earth from radiogenic decay, based upon geoneutrino studies. Note the discrepancy between the estimated sum total of radiogenic heat released and the total directly observed from measurements at Earth's surface. this discrepancy remains unexplained, but geoscientists are now confident that radiogenic decay is the primary heat source for driving plate tectonics and, ultimately, volcanism.**

Decay series	Primary location inside Earth of decay	Heat energy released (in terawatts, TW)
Thorium-232	Core	8.3
Uranium-238	Core	8.0
Potassium-40	Continental crust	3.0
Total estimated radiogenic heat release (from geoneutrino research)		19.0
Total directly observed heat release through Earth's surface		31.0

Source: Based on Araki et al. (2005)

It is difficult to imagine that solid rocks can move like a fluid, but indeed, over great periods of time at high temperatures, this is possible. Holmes was quick to point out that the mobility and plasticity of solid rock can be seen in the graceful folds of sedimentary strata and gneisses. Glaciers, too, are good examples of solid material moving in response to gravitational forces over long periods of time.

Holmes' insights helped to explain the discrepancies between geology and mathematics, although seismic investigations of mantle plumes indicate that the realities are not simply a matter of geometrically elegant "boiling pot"-shaped convection cells (Montelli et al. 2003). We still have much to learn about our planet's internal structure and internal dynamics.

The Physics and Chemistry of Melting

We now know that several factors are important in determining whether or not rock will melt to form magma: temperature, composition, pressure, and "water" (volatile) content are each vital, and their interactions governed by complex thermodynamic realities (Bohrson and Spera 2001). Various combinations of these factors, all related to early Earth history, convection, and plate tectonics, are responsible for magma production at a few distinctive levels and places inside Earth.

Most rocks are made up entirely of minerals, which are defined as unique arrangements of non-living matter having crystalline structures. A crystalline structure is an *orderly, symmetrical arrangement of atoms* inside a substance. Most of these atoms tend to bond with one another ionically, meaning that they come together because some atoms are deficient in electrons whereas others, to which they are drawn, have some electrons to spare. Atoms having an unequal number of protons and electrons are electrically charged, and we call them **ions**. Nature tends to neutralize electrical charge wherever it arises, so ionically bound minerals are the norm.

Ions that have the same charge (possess the same number of electrons out of balance with protons) repel one another, with higher charges leading to greater repulsive forces. Silicon, which is the most common **cation** – or ion lacking electrons – on Earth, also has an especially high charge. Silicon ions are highly repulsive of one another, but being very common, they also are found in the vast majority of rocks and minerals. Silicon bonds with oxygen, the most common **anion** (ion having excess electrons) on Earth, and so minerals and rocks tend to be made up largely of silicon ions fixed into position by intervening oxygens. Since oxygen atoms are much larger than silicons, four oxygen ions snuggle around each silicon ion in crystals, never more, never less. The result is a molecular arrangement shaped like a tetrahedron, the **silica tetrahedron** (Figure 3.3). Silica tetrahedra are not charge-balanced arrangements, despite the efficient use of space. Each oxygen ion, having two extra electrons, has a charge of −2, whereas each silicon, lacking four electrons, has a charge of +4. There is an excess negative charge of −4 in other words. To deal with this, nature employs a couple of strategies. One is to tuck additional cations in between the tetrahedra within crystal structures, so that the oxygens split their charges between the silicons and surrounding cations. This happens, for example, in the common volcanic mineral olivine. The other is to actually combine the tetrahedra so that each oxygen shares its charge with two adjoining silicons. Oxygen atoms that so partition their charges are termed **bridging oxygens**. In the common mineral quartz, all tetrahedra are interlinked and all oxygens are bridging, and the two-to-one proportion of silicons to oxygens – the only elements found in quartz – gives the mineral its chemical formula, SiO_2.

Silicon ions tend to be spaced more closely together in arrangements where tetrahedral oxygens are bridging than where they are non-bridging. This has important implications, because it means that minerals such as quartz tend to be less stable when heat

a) Four oxygen ions (charge −2), snuggle around one silicon ion (charge +4) in each silica molecule.

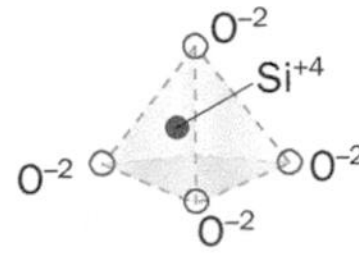

b) . . . This is shown using a "stick figure" here. The figure has a tetrahedral ("four-faced") shape.

c) The charge on the silica molecule is unbalanced:
$4 \times (-2) = -8$
$1 \times (+4) = 4$
$4 - 8 = -4$

d) To correct this, nature combines tetrahedra to build the crystal structures of silicate minerals, such as quartz.

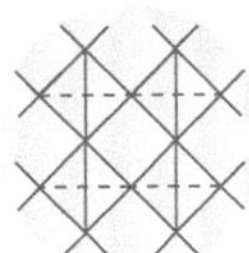

In every quartz crystal, each of the oxygens share its charge with two silicon atoms. That is, every tetrahedron is linked together through its apical oxygens to form a solid mass.

By sharing oxygens, the net negative charges are balanced out for each tetrahedron. The structure is stable:
$4 \times (-1) = -4$
$1 \times (+4) = 4$
$4 - 4 = 0$

e) In some minerals, such as olivine, not all oxygens are shared between tetrahedra. To address this, nature tucks positively charged ions (cations) in the spaces next to unshared ("non-bridging") oxygens. This neutralizes the net negative charges that otherwise exist and would destabilize the mineral.

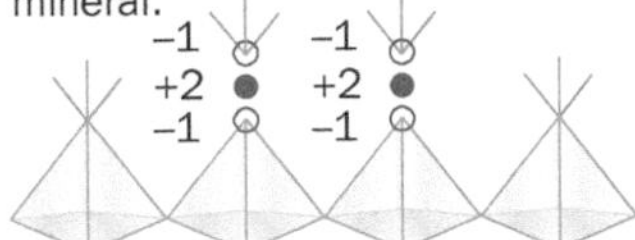

(The cations found in olivine are magnesium, which makes green olivine, or "peridot," and iron, which will make olivine black if present abundantly).

FIGURE 3.3 The silica tetrahedron.

energy is present than minerals such as forsterite (magnesian-rich olivine). If minerals are mixed together, their melting temperatures are suppressed even further – a phenomenon called **fluxing**. The heat energizes the ions, allowing them to break their bonds, and high-charge ions in close proximity to one another can do this with relative ease. The result is that the mineral melts. Since no two mineral species are quite alike, this means that a rock made up of different kinds of minerals *will not melt all at once, but rather mineral type by mineral type*. For instance, the melting point of pure quartz is 1650±75°C, while the melting point of forsterite, with its more widely spaced silicon ions, is around 1900°C. If fluxed in a rock also containing abundant H_2O-bearing minerals, however, quartz can start melting at temperatures as low as 700°C.

Norman L. Bowen, using laboratory furnaces at the Carnegie Institute in Washington DC, began studying this phenomenon by examining the feldspar mineral group around 1910 (Bowen 1913). He modeled and experimentally documented the melting and crystallization of granitic rocks and discerned the sequence, now known as Bowen's Reaction Series, in which the crystallization of cooling granitic magma is essentially just the reverse process of melting if considered as a closed system. He found that three kinds of mineral melting reactions take place. In the **discontinuous reaction series**, consisting of biotite (magnesium-rich mica), clino-amphibole, calcium pyroxene, and olivine, a mineral will partially melt at a specific critical temperature, reacting to form another structurally and chemically similar, but more stable mineral in the process. Hence, calcium pyroxene will break down to form olivine when temperature rises to a certain level in a melting rock, called a **peritectic point** by chemists. In the **continuous reaction series**, made up of the plagioclase mineral series, melt is *continuously released* and mineral compositions *continuously change* as temperatures rise. Sodium-rich plagioclase (albite) is stable at low temperatures in granite. At higher temperatures, calcium-rich plagioclase (anorthite) persists until melting is complete. At temperatures in between, plagioclase contains both calcium and sodium mixed together in temperature-sensitive proportions. If one can establish the pressure and H_2O content of a melt, the ratio of these two cations can be measured to determine the temperature at which a particular plagioclase existed at the time it acquired that composition (Kudo and Weill 1970). Minerals whose make-ups provide such information are called **geothermometers**. (There are also mineral **geobarometers** developed for pressure determinations.) Using geothermometers, it is possible to calculate the maximum temperature a rock reached as it was partially melted – or, conversely, to calculate the approximate eruptive magma temperature of volcanic rocks even millions of years old. The chemical variations and textures seen in plagioclases, in particular, provide important clues to the cooling histories of magmas and the causes of volcanic eruptions (e.g. Tepley et al. 2000; Shane 2015). One type of geothermometer, involving volcanic glass, can even be used to determine thermal histories of past eruptions (e.g. Helz and Thornber 1987).

Finally, Bowen noted that some minerals do not change their compositions or react out as they heat up and melt. When melting occurs, the liquid they release matches the chemistry of the original minerals. In the jargon of chemistry, they melt **congruently**. Potassium feldspar and quartz are examples of such minerals.

Bowen used his discovery about how molten rocks melt and crystallize to formulate a general theory that states how a range of igneous rock types can differentiate (form stepwise) from a single parent magma. **Magmatic differentiation** can most easily be explained by imagining that the crystals that develop early in a melt sink and accumulate in a layer of distinctive composition at the bottom of a magma chamber. The remaining melt crystallizes a newer suite of minerals as it cools, which in turn may settle out to build a bottom layer of another composition – and so on, the end product being a layer-cake of rock types that might have been mixed together to form a single blend had gravity not influenced the crystallization process. There are other ways that differentiation can take place, too, but this particular method, called **gravitational fractional crystallization**, seemed the most obvious to Bowen, and has stimulated thinking about magmas ever since.

So certain was Bowen about his theory that he wrote (Bowen 1928):

> *The only rival hypothesis [to differentiation is] the doctrine of the mixing of two fundamental magmas (basaltic and rhyolitic) but this has been found to fail so completely that the concept of differentiation has come to be regarded as a fact as well established as the observed rock associations themselves.*

However, current research shows that fractional crystallization is less important and magma mixing is more important for explaining the wide variety of igneous rocks than Bowen ever thought (Reid et al. 1993; Thornber 2003). It is not always wise to feel absolutely certain about your beliefs!

In a laboratory Bowen could remove and quench samples of melted rock whenever he chose. But under natural conditions, the molten fraction tends to separate before enough heat accumulates to melt rock *completely* at any given point inside the earth – at least outside of the core. The separating liquid migrates along crystal grain boundaries and through fractures that may open in the surrounding crust in response to stress, heating, and the buoyancy of the new magma. It also tends to be more silica (SiO_2) enriched than the unmelted residue it leaves behind. This process, intrinsic to the generation of all of Earth's erupting magmas, is termed **partial melting**.

Pressure tends to keep ions together in crystalline structures. Hence, with increase in pressure, there is an increase in the temperature at which a rock of any given composition and H_2O content will melt. Rocks having high silica content tend to melt at lower temperatures than rocks with low silica contents, given the same level of pressure.

Many rocks contain H_2O and other volatiles in pores, fractures, and even along grain boundaries between crystals. Each H_2O molecule has a slight charge imbalance from one side of the molecule to the other. Given this property, H_2O molecules are slightly attracted to the ionically charged surfaces of adjoining crystals. Given the heat, H_2O makes it easier for ions to diffuse and melting to occur than under dry conditions, hence H_2O-bearing rocks tend to melt at lower temperatures than dry rocks. The greater the lithostatic confining pressure on the magma, the more water can be absorbed (Table 3.2A)

Most melting takes place at three general localities inside the earth: the outer core, the asthenosphere, and in the upper mantle-lid (or "wedge") above subducting plates. Earth's outer core, 2900–5155 km deep, mostly consists of molten iron and nickel. It represents the largest known magma body in the solar system. Yet, being so deep, it feeds no volcanoes. Melting in the crust is largely due to heating from intruded mantle-derived magmas (Bergantz and Dawes 1994).

Earth's mantle, 70–2900 km deep, is mostly solid, but hot enough to convect, as mentioned earlier. It is made up of silicate rocks, in contrast to the core. At the core–mantle boundary, the rock has a higher melting temperature than iron and nickel. Yet, its solidness is only sustained under very high pressures. Where a typical specimen of hot mantle rock suddenly taken to the low pressure of sea level, the rock would melt at once into extremely fluid lava. However, at one level in the mantle, 100–640 km down, the pressure is *just low enough* and the temperature *just high enough* for partial melting to take place. This is the level of the **asthenosphere**, the soft bed upon which the rigidly overlying plates slide. Mantle convection cells reach their highest level in this layer. The partial melt fraction rises as basaltic magma, the main form of molten rock erupting from sea floor spreading centers, continental rift valleys, and other regions of tectonic extension (crustal stretching). At these loci, plate separation enables asthenospheric mantle to well up within a few kilometers of Earth's surface, inducing intensive partial melting. Most of the world's volcanic activity results from this **decompression melting**. Where mantle plumes ascend, a similar form of hypothetical melting occurs: the plume rock, bringing with it great heat from deeper inside the planet, enters shallow, lower-pressure regions, and begins to melt. The thermodynamic factors that govern the upward transport of that molten rock, with a summary of critical constants involved are given by Lesher and Spera (2015).

Somewhat different processes operate beneath volcanic arcs. Large volumes of trapped seawater and moist sediment ride downward into the earth where masses of heavy oceanic plate are subducted into the mantle. At a depth of around 100–110 km, the great heat of the surrounding mantle drives off the seawater – or various components of it within the sinking plate, making rocks in the overlying wedge of mantle less dense and more buoyant. The escaping fluids may induce partial melting, owing to the effect of H_2O on melting temperatures described earlier. Whether this happens or not, however, the upwelling, volatile-bearing mantle rock decompresses, which most certainly triggers partial melting. The resulting basaltic magma ascends into the overlying crust, assimilating parts of it while heating other parts enough to induce further partial melting, especially in silica-rich rocks. Granitic magma bodies result, accompanied by metamorphism in neighboring rocks that are sufficiently refractory to remain solid (Blatt et al. 2006).

Classification of Magma and Igneous Rocks

Geologists distinguish two subclasses of igneous rocks: **plutonic** and **volcanic** (Chapter 1). Plutonic rocks form by cooling and crystallization of molten rock deep beneath Earth's surface, whereas most volcanic rocks form by solidification of magma erupted onto the surface or emplaced at shallow depths beneath volcanoes. Some volcanic rocks form from fragmental material erupted during explosive eruptions (Chapter 7), but, as for the ones that cool directly from erupted magma, they may usually be distinguished from plutonic rocks in hand specimens by the lesser degree of crystallization, and the common occurrence of gas bubbles or **vesicles**. In contrast, plutonic rocks, which crystallize completely and almost completely degas during their formation, are said to be **holocrystalline**. In general, the larger the crystal grain sizes in plutonic rocks, the slower the cooling time required to develop them, although other factors are important as well.

When volcanic rocks contain obvious larger crystals separate from a finer-grained or glassy **matrix** where individual crystals are not observable without a hand lens, the rocks are said to be **porphyritic**, and the larger crystals are called **phenocrysts** (Figure 3.4). The rock matrix separating phenocrysts in volcanic rocks consists of a featureless or finely "sparkly" stony matter called **groundmass,** which is not so featureless when magnified in thin section! Some volcanic rocks lack phenocrysts altogether, consisting entirely of groundmass material. They are **aphyric**, meaning "crystal-free," and may represent the eruption of extremely hot magma, the gravitational setting of phenocrysts in the magma chamber before extrusion, or the filtering out of phenocrysts from portions of a flow because of shear-related processes during emplacement. Glass-rich volcanic rocks generally indicate rapid cooling, though some glassy rocks are non-volcanic and can form by fault movements deep in the crust (creating "pseudotachylite"); or are associated with meteoritic impacts and areas struck by lightning.

Apart from the "volcanic" versus "plutonic" distinction described earlier, geoscientists have also developed more refined approaches to classifying magmas and igneous rocks, according to specific need. Historically, the first efforts at classification were based upon direct observations – the practical description of the colors and textures of rocks seen with the unaided eye. For purely practical reasons, special attention was given to those rocks of interest in mining and construction. The word "basalt," for instance, can be traced back to the ancient Egyptian quarryman's "basanos." The growth of mineralogy as a scientific field, especially beginning in the eighteenth century enabled more precise rock classification, because observers recognized that certain kinds of minerals are characteristically restricted to certain kinds of rocks. But it was the development and proliferation of modern analytical instrumentation that really allowed a detailed understanding of the *elemental* make-up of materials and permitted their systematic classifications based on chemical compositions. Combined with experimental studies, such as those of Bowen, this has also enabled geoscientists to decipher with great precision the origins and ages of many igneous rocks.

FIGURE 3.4 Coarsely porphyritic andesite from a prehistoric flow in the Andes, on the upper north slope of Villarica volcano, Chile [76]. The plagioclase phenocrysts are likely of andesine composition (the Andean name gets around!). The coin is 2.5 cm in diameter. Source: Photo by J.P. Lockwood.

One generally used classification scheme for volcanic rocks is based on the ratios of the oxides of the alkalic elements

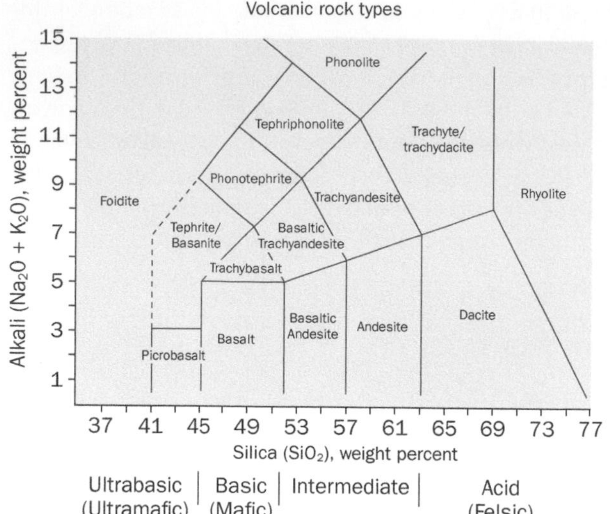

FIGURE 3.5 The International Union of Geological Sciences (IUGS) Total Alkali versus Silica (TAS) diagram for volcanic rocks. Note that chemical composition alone is the basis for this classification. Source: From Le Bas et al. (1986). Reproduced with permission of Oxford University Press.

sodium and potassium vs silica (Figure 3.5). The terms **ultramafic** and **mafic** (or sometimes **ultrabasic** and **basic**) refer to low-silica igneous rocks containing lots of iron and magnesium. **Felsic** rocks (sometimes referred to as **acid** rocks) are highly siliceous, and contain lower quantities of iron and magnesium but more potassium and aluminum than their mafic counterparts. Figure 3.5 does not distinguish between lavas and pyroclastic rocks of similar composition. That distinction is made by using the volcanic composition as a term-modifier; for example, **basaltic ash** and **basaltic lava**. So, ultimately, the textures, structures, *and* chemical compositions of particular volcanic deposits must be considered to render a complete rock identification. We shall explore some of these textural and structural aspects further in subsequent chapters.

For further information on the classification of igneous rocks, we recommend that you explore the International Union of Geological Sciences (IUGS) Subcommission on the Systematics of Igneous Rocks, available online and in most igneous petrology textbooks (e.g. Blatt et al. 2006). The presentation above simply illustrates the great variation in volcanic rock and magma compositions, and provides context for the names of rocks and minerals that we will find useful to discuss later in this text, beginning with the next section.

Principal Magma Types

Basaltic Magmas

Basaltic melt is the most common type of magma on Earth, as it appears to be elsewhere on the inner planets and moons of our solar system (Chapter 12). It is the product of decompression melting in the upper mantle, erupting at all spreading centers worldwide. It also erupts within and behind some volcanic arcs, and in areas of extending continental crust – places far from the closest plate boundary. The residue of the partial melting that creates basalt is **harzburgite**, a chemically depleted, olivine- and pyroxene-rich form of *peridotite* that occasionally appears as xenoliths (fragments) within basalt flows – especially in **alkalic basalts**. Harzburgite is inferred by seismic studies to form a high-density layer or zone underlying some active basaltic volcanoes (e.g. Okubo et al. 1997). The most widespread basalts on Earth are those erupted on the seafloor, on oceanic islands and in areas of shallow mantle upwelling, called **tholeiitic basalts**, after Tholey, in the Saarland (Germany). Tholeiites are notably richer in iron and lower in alumina than those along convergent plate margins.

Minerals typical of basalt include olivine, pyroxene, and calcic plagioclase, which are all stable at higher temperatures than most silicate minerals. This indicates that basalt magma must form at higher temperatures than other magma types. High-temperature melts have low viscosity – that is, greater fluidity. They also are less dense than surrounding source rock, are buoyant (Chapter 4) and can rise rapidly – in some instances traversing kilometers of crust in a matter of just a few days or weeks (Klein et al. 1987; Rutherford 2008).

As magma cools to crystallization (or *liquidus*) temperatures, viscosity increases because silica (SiO_2) tetrahedra begin linking together within the melt. Even before crystal nuclei develop, long silica-based polymer chains will form in the cooling magma. The more silica present within the melt, the more viscous it tends to be at any given temperature. Of all magma types, basaltic magma contains the least amount of silica, which contributes to its high fluidity during ascent.

Relative to other magma compositions, basaltic magma generally contains only small amounts of volatiles, ranging from about 0.5 to a little more than 1 percent of the total weight of the magma that reaches the surface. On the basis of vesicle volumes as related to experimentally determined magmatic solubility of water, Moore (1970) concluded that basalt magmas erupted on the deep ocean floors contain from 0.25 to 0.9 percent H_2O. These low values are typical. In contrast, some subduction-related erupting basalts may contain as much as 6 percent H_2O (Roggensack et al. 1997). Lack of volatiles, especially dissolved water, plainly plays a role in explaining the eruption behavior of most basaltic magmas (Wallace et al. 2015). Low volatile content explains why explosive basaltic volcanic eruptions are not frequent, nor typically large. While there are some exceptions, and basaltic ash and pumice do exist, basalt erupts primarily as highly fluid lava. An entire class of less-common alkalic basalts and related alkaline lava types (for example, hawaiites, phonolites, ankaramites, basanites), are characterized by higher amounts of

NaO and K_2O, form under different melting conditions than do tholeiitic basalts, and are frequently associated with the earliest and late stages of Hawaiian volcanism. They also form as intraplate mafic eruptions in many continental areas. Alkalic basalts are more fluid and rise to the surface faster than do normal basalts, form thinner, fast-moving flows. and commonly are hosts for xenoliths of ultramafic, mantle-derived rocks.

Andesitic and More Silica-Rich Magmas

Andesitic magma occurs primarily along convergent plate boundaries such as the Andes, from whence the name derives, where subduction has generated melting in the upper mantle wedge separating the down-going plate from the lithosphere above. The origin of andesite is still somewhat unclear. Geochemical evidence of both crustal and mantle contributions exists in typical andesite. Volcanic rocks that are chemically classed as andesites can vary considerably in physical appearance. Some are dark where finer-grained and glassy; others are lighter gray in color and contain numerous phenocrysts of plagioclase as in the porphyritic andesite from the Andes shown in Figure 3.4. A general consensus is that andesites must be derived from the upper mantle above dewatering, subducting slabs, initially perhaps as basaltic magmas that later become modified by assimilating material from the lower crust. Andesites can also form by the mixing of basalt with more silica-rich magma in magma reservoirs (Blatt et al. 2006). Andesites and basalts often erupt in close proximity to one another, further suggesting a genetic connection. In younger oceanic island arcs, basalt tends to be more prevalent (Devine 1995), while in older island arcs (e.g. Philippines and Indonesia) and continental areas, andesites usually predominate over basalts, suggesting that partial melting of silica and alumina-rich continental rocks are important to andesite production (Eichelberger 1978). This appears to be confirmed by the enrichment of elements such as lanthanum, barium, and potassium in many andesites – elements concentrated in continental crust and derived sediments (Rogers and Hawkesworth 2000).

Like andesites, **dacites** and their well-crystallized equivalents (*granodiorites*) primarily occur along convergent plate boundaries. Many early petrologists like Bowen thought that dacites, andesites, and granite itself formed primarily through gravitational fractional crystallization in basaltic magma. While it is clear that this process is important in some long-sustained mafic magma reservoirs (Gunnarsson et al. 1998; White and Urbanczyk 2001; Haase et al. 2006; Neal et al. 2019), the regional volumes of other more siliceous igneous rock types relative to mafic rocks seem to rule this out as the only mechanism to describe their origin (Blatt et al. 2006). Impossibly large bodies of basalt magma would be required to generate the great volumes of more siliceous granitic rocks found in continental mountain ranges. Some dacites and granodiorites may well derive from fractionated mafic melts, or from partial melting of lower crustal rocks, such as gabbro (e.g. Smith and Leeman 1987), but most result from crustal partial melting and the mixing of derivative silica-rich granitic magmas with silica-poor basaltic or andesitic magmas, as revealed by field observations (e.g. Wiebe et al. 2001; Harper et al. 2004) and compositional calculations.

H_2O and other volatiles lower magma liquidus temperatures. This is because these volatiles are generally ionized in solution at a high temperature and are apt to form temporary bonds with other elements, hence inhibiting the development of stable crystal nuclei. As molten rock approaches the surface, the escape of volatiles causes the melt to suddenly become more viscous and to start crystallizing new minerals. Evidence of this occurs in many lava flows and domes, which contain two distinctive populations of crystals – coarse phenocrysts formed by slow cooling deep underground, and numerous fine, generally needle, or lath-shaped crystals, termed **microphenocrysts** and **microlites**, which represent sudden crystallization from water loss during ascent to the surface. In many cases, suddenly stiffening magma will stop rising altogether and solidify at a shallow depth, such as beneath the growing dacite dome of Mount St Helens [31] in 2004 to 2006 (Cashman et al. 2008). Some of these very shallow magma bodies have the textural appearance of surface lava, even including minor vesicles, although if water is entrapped, rapid growth of large phenocrysts is possible. Geologists call them **hypabyssal** (meaning "very shallow").

In contrast to basalt that originates in the mantle, **siliceous magmas** form in the silica-rich crust at lower melting and liquidus temperature conditions (Clemens 1998; Chappell and White 2001). Because of lower temperatures and higher silica contents, these melts are more viscous than basaltic magma. As they rise, siliceous magma may cool to form extremely viscous "crystal mushes" that require later remobilization in order to erupt (Jackson et al. 2018). Moreover, since continental crust contains more bonded H_2O than the mantle, granitic melts have a considerably higher volatile content than basaltic melts (Table 3.2A). The viscosity of ascending magma may increase due to volatile losses as it rises closer to the surface. Consequently, proportionately much less granitic magma can erupt to form siliceous volcanic lava flows than mafic magma can erupt to form major basalt outpourings. And, if siliceous magma does erupt, it is apt to do so accompanied by significant explosive activity. Rhyolitic pyroclastic rocks are thus much more voluminous than rhyolitic lava flows, and the most powerful known volcanic eruptions are associated with high-silica magmas (Chapter 8).

Granites, dacites, andesites, and continental basalts (those that occur along continental convergent plate boundaries) are relatively rich in aluminum and calcium. Petrologists call them **calc-alkaline** or **high-alumina** rocks. **Alkaline rocks**, as mentioned earlier, contain higher amounts of alkali elements, notably potassium in continental areas, and sodium in oceanic settings.

Laboratory experiments indicate that alkaline compositions arise from small amounts of mantle partial melting at great depths – as much as 100–150 km – sometimes associated with subduction, as shown by Italian volcanoes, which are notably alkalic in composition. More typically, alkalic volcanic rocks are associated with areas of extending continental crust far from any plate boundaries, and with the initial and final stages of mantle plume (hot spot) volcanism, as mentioned earlier for alkalic basalts in the Hawaiian islands.

Carbonate-based Magmas

The rarest type of terrestrial magma is not based on silica chemistry, but upon carbonate. This magma, essentially molten baking soda, has erupted at about 50 highly localized sites, primarily in a few young or immature continental spreading centers, notably including the Rhine Graben of western Germany and the East African Rift Valley (Wooley and Church 2005). The principal product of such eruptions is a light-colored rock called **carbonatite**, occurring both as lava flows and in pyroclastic deposits. Carbonatites also are found as veins within some volcanic diatremes (Chapter 4), all located in areas of thick, otherwise stable continental crust.

The only historically active carbonatite volcano is in East Africa. The 2000-m-high basaltic composite cone of Oldoinyo Lengai [120] ("The Mountain of God"), in the Gregory Rift Valley of Tanzania, has intermittently erupted sodium-rich carbonatite ash and lava during its roughly 300,000–400,000 years of cone growth. In the latter half of the twentieth century, carbonatite eruptions filled its northern summit crater to the brim with mottled brown, white, gray, and black lava flows studded with spike-like spatter cones (Church and Jones 1995). The explosive summit eruption of 2007–2008 destroyed this fantastic landscape, however, leaving a gaping collapse pit in its place.

Although pure calcite ($CaCO_3$) has a high melting temperature (1339°C), alkali-rich carbonate minerals, such as those found in carbonatites, in some instances, mixed with minor amounts of nepheline and pyroxene, require much less heat to melt. Hence, carbonatite lavas are at the cold end of the magma temperature spectrum, typically around 550°C, only about half the temperature of erupting basalts. This means that molten carbonatite flows lack the incandescence of active silicate lava, resembling dark mud or porridge as they spill out (Figure 3.6). Carbonatite lava also is much less viscous, with a consistency resembling that of motor oil at the moment of eruption. Active flows may be only a few centimeters thick, and run in channels across which one can easily step. The flows can spread more rapidly than conventional lavas, but eruptive volumes tend to be small, so the lava rarely travels very far.

Chemists and carbonated beverage drinkers alike well know that the solubility of carbon dioxide increases with pressure on a liquid. Experimental studies indicate that for mantle-derived basalts originating at depths of greater than 80 km, as much as 40 percent of the bulk weight of the source magma may be dissolved CO_2 (e.g. Wyllie and Huang 1976), suggesting a connection between the upper mantle and carbonatites. Indeed, observations of gaseous emissions (CO_2, He, N_2, and Ar) from the northern crater of Oldoinyo Lengai show patterns essentially identical to emissions from divergent plate boundaries on the ocean floor, where mantle upwelling clearly facilitates volcanic activity (Teague et al. 2008; Fisher et al. 2009).

In many of its occurrences, carbonatite is closely associated with a few distinctive types of sodium-rich silicate rocks, including nepheline basalt, fenite, phonolite, and ijolite, which originate from small amounts of partial melting beneath the relatively thick lithosphere of continents. Neodymium and strontium isotope measurements and other mineralogical data suggest that, as the CO_2-rich silicate magma rises and the solubility of CO_2 drops, a carbonatite melt fraction is capable of separating to erupt independently (Dawson 1998; Ulrich and Sindern 1998). Evidence for this includes rounded nodules of silicate rock within some carbonatite bodies (Bell and Simonetti 1996). The phase separation likely takes place in the middle and shallow crust (Lee and Wyllie 1997; Baudouin et al. 2018), explaining how Oldoinyo Lengai and other similar, now extinct volcanoes have been able to produce these two very different kinds of volcanic material from single conduits.

Magmatic and Volcanic Gases

Although volatile compounds (dissolved gases in magma) are not normally considered as eruption products (all the attention is usually focused on lava and ash), they are an important magmatic constituent, and constitute the greatest *volume* of material erupted by volcanoes (Figure 3.7). The critical role of gases in explosive volcanic activity can easily be underestimated if only their mass is considered; their volume is much more important. As noted above, dissolved gases typically account for no more than about 1 percent of the weight of most basaltic lavas, which de-gas as they approach shallow depths, 3–4 percent of volcanic arc basalts and andesites, and 7 percent or more of highly silicic melts. The exsolution and expansion of gases as magma approaches the surface exert astounding forces. Bardintzeff and McBirney (2000) cite the example of one cubic meter of typical rhyolitic magma with dissolved gases stored deep underground, which can expand to 670 m^3 of fragmental material and gas upon reaching atmospheric pressure. It is no wonder that explosive volcanic eruptions take place (Chapters 7 and 8)!

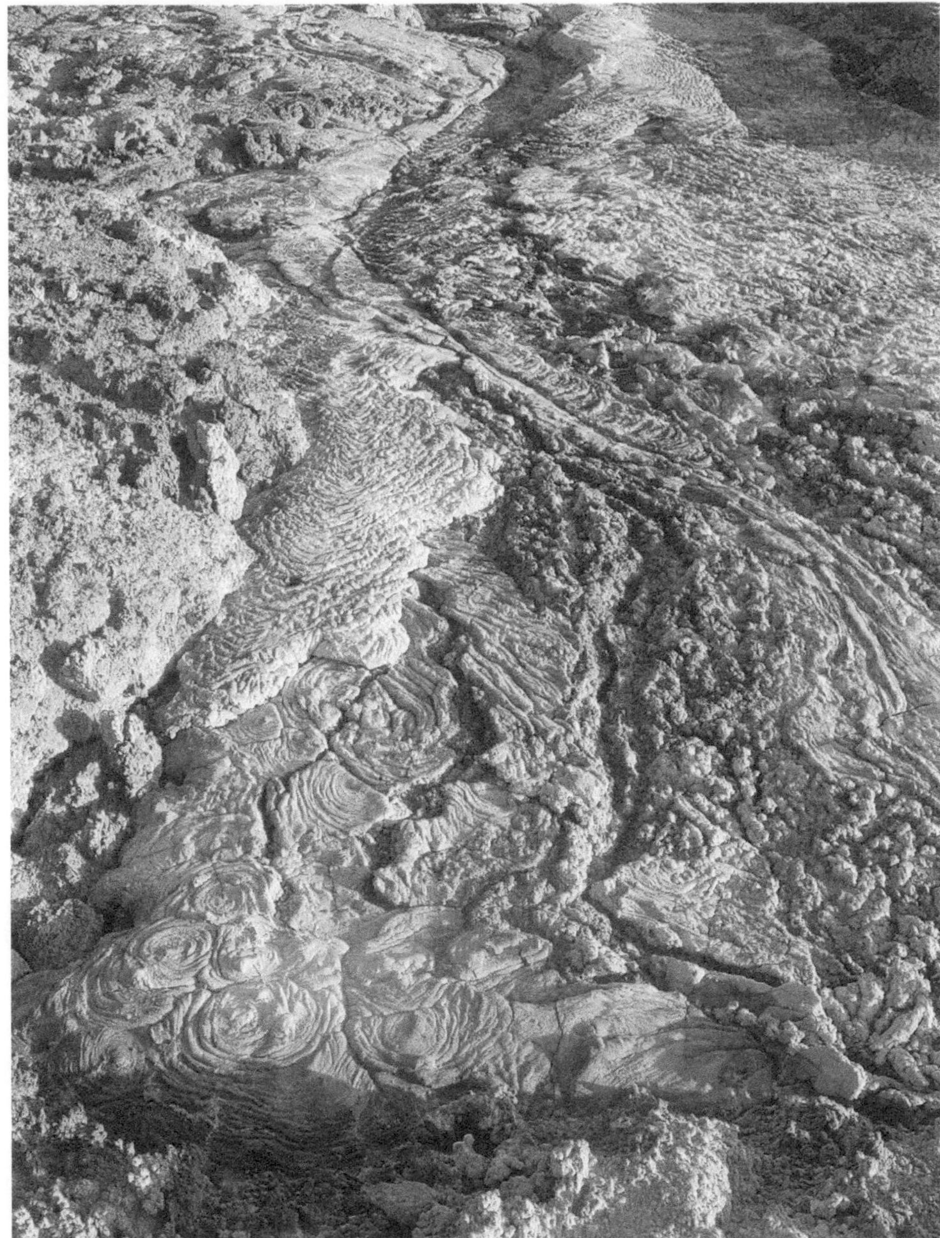

FIGURE 3.6 Fresh carbonatite lava flow, Oldoinyo Langai volcano, Tanzania [120]. Flow in lava channel about 2 m across, and indicates high emplacement fluidity by smooth surface textures. Source: Photo by Donald McFarlane.

Gases are the most important volcanic product in terms of influence on eruptive style and on world climate and atmosphere. Were volatiles not present within magmas, there would be far fewer volcanic eruptions on Earth, and those that did occur would simply involve the passive extrusion of lava, perhaps in response to the rare impacts of giant meteorites.

The amount of volatiles that can be dissolved in magma depends on the non-volatile magma components and on the magma's temperature and confining (lithostatic) pressure. Solubility increases with increasing pressure and decreases with increasing temperature for water. For sulfur, the pattern is different: Sulfur solubility increases as magma temperature increases – as much as 5–7 times each 100°C. This is a primarily why basaltic magmas are typically more SO_2-rich than the generally cooler rhyolitic melts. Differences in ferrous iron content between basaltic and rhyolitic magmas also influence the solubility of sulfur, with the correlation between the two ordinarily being positive (Haughton et al. 1974).

The standard unit of pressure in volcanology is the megapascal (MPa), where one MPa is equivalent to ten bars, or approximately ten times the pressure of the atmosphere at sea level. For example, Hamilton et al. (1964) showed that andesitic magma can contain 4.5 percent dissolved H_2O (by weight) at 100 MPa and 10.1 percent at 530 MPa. The solubility of water in basaltic magma in

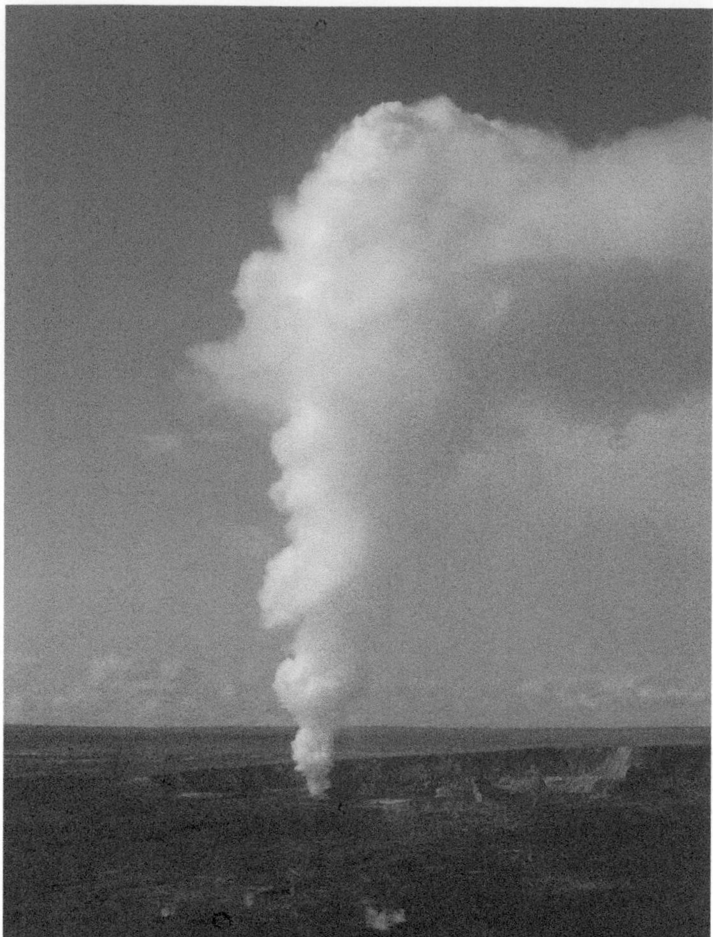

FIGURE 3.7 Volcanic degassing plume from Halema'uma'u crater, Kīlauea volcano, Hawai'i [17] in May 2008. This plume, consisting dominantly of water vapor, also contains around 10–15 percent SO_2. Halema'uma'u has often produced several thousand tonnes of SO_2 per day, which has impacted agriculture and residents downwind. Note that the rising plume rotates according to northern hemisphere Coriolis forces, and that it flattens and is deflected downwind when reaching its "neutral buoyancy level" (Chapter 4) about 1500 m above Kīlauea's summit. Source: Photo by J.P. Lockwood.

TABLE 3.2A **Solubility of water in anhydrous granitic melt at 800°C, averaged from four data sets reported in Behrens and Jantos (2001).**

Pressure (MPa)	Water content (weight %)
50	2.8
100	4.3
150	5.0
200	6.0
250	6.8
300	7.4
400	8.6
500	9.5

Source: Modified from Behrens and Jantos (2001).

this range is slightly less. **Henry's law** is the relationship describing how solubility of volatiles changes with pressure – which in the case of volcanoes, can be taken as mean depth in a magma column beneath a vent as well:

$$C_v = kP^{\beta} \tag{3.1}$$

where C_v is the concentration (mole fraction) of H_2O or gas dissolved in the melt, P is the pressure of the melt (MPa), k is a constant, which differs according to the kind of volatile being studied and the composition and temperature of the melt (Table 3.2b). "k" typically ranges from about $4.1{\cdot}10^{-6}\ Pa^{-0.5}$ at the lower end for siliceous magmas, to $6.8{\cdot}10\ Pa^{-0.7}$ for mafic magmas. β is another constant dependent on magma composition, with values from 0.5 in rhyolitic melts to 0.7 for basaltic liquids (Woods 1995). As volatiles

TABLE 3.2B **Change of water solubility in rhyolite melt at 100 MPa, experimentally determined by Yamashita (1999). Note that rising temperature (as through injection of a hotter melt from below) drives the water out of the magma as a vapor, which could press against the roof of a magma reservoir and increase potential for eruption.**

Temperature (°C)	Water content (weight %)
800	4.0
900	3.8
1000	3.6
1100	3.4
1200	3.3

Source: Modified from Yamashita (1999).

TABLE 3.3 **Sample compositions of volcanic gases in volume percent of gases released.**

Sample	1	2	3	4	5	6
H_2O	49.14	97.1	95.0	88.87	77.20	37.00
CO_2	23.41	1.44	4.53	6.64	11.30	48.90
CO	0.49	0.01	–	0.16	0.44	1.51
SO_2	25.94	0.54	0.02	1.15	8.34	11.84
H_2S	–	0.23	–	1.12	0.68	0.04
S_2	–	–	–	0.08	–	0.02
HCl	0.42	2.89	0.02	0.04	0.42	0.08
HF	–	0.26	–	–	–	–
H_2	0.49	0.70	–	1.54	1.39	0.49

1 Sample from basaltic eruption fume, Etna [121], 1970, convergent plate margin (Huntingdon 1973).
2 Sample from fumarole atop Momotombo volcano, Nicaragua [65], at 820°C, convergent plate margin (Symonds et al. 1994).
3 Average of four "low-temperature" fumarolic gas analyses, Shiveluch volcano, Kamchatka [172], convergent plate margin (Fedotov and Masurenkov 1991).
4 Gases emitted by an andesite flow at 915°C, Merapi volcano, Java [131], convergent plate margin (Symonds et al. 1994).
5 Gases from the lava lake of Erta Ale, Ethiopia [121], at 1130°C, divergent plate margin (Symonds et al. 1994).
6 Kīlauea [17] summit basalt flow, 1170°C, hot spot/mantle plume volcano (Symonds et al. 1994).

come out of solution, they expand – about 1600-fold in the case of an ordinary mole of water at atmospheric pressure. Henry's law quantitatively shows that the rapid exsolution of volatiles associated with reductions in magma pressure – the causes of volcanic explosions – must occur within the bodies of large volcanoes themselves rather than beneath them. Volcanic gases consist mostly of H_2O, CO_2, and SO_2 (Table 3.3).

Dissolved Water Content

Most of the "smoke" seen issuing from volcanic craters and fumaroles consists of water vapor. Effusive eruptions usually begin with the billowing of white steam from opening fissures before lava reaches the surface. Water vapor (steam) is invisible when its temperature exceeds its boiling point and only shimmering heat waves may be seen rising directly from high-temperature vents, but visible steam will reappear higher above, due to cooling and condensation. Most of the water emitted from fumaroles is **meteoric** – that is derived from heated groundwater. Studies of stable hydrogen isotopes in volcanic steam indicate, however, that re-heated groundwater alone cannot account for all of the water that issues from volcanoes – especially during major explosive eruptions. There is also a primary "magmatic" (or **juvenile**) component, at some volcanoes exceeding 50 percent of total steam emissions (Giggenbach 1987; Hedenquist and Aoki 1991). Oxygen isotope content is the primary means by which these two dissolved water sources can be distinguished (e.g. Friedman et al. 1974).

Carbon Dioxide

The various constituents of volcanic fumes have differing solubilities in magma which independently change as a function of pressure as well as temperature. That means that the volatile make-up of magma and the compositions of its gas bubbles change as the melt rises from its point of origin to the vent. Carbon dioxide is thought to be the dominant gas dissolved within mantle-derived melts rising through the lower to middle crust, and has been estimated to comprise as much as 0.7 weight percent of juvenile basaltic magma (Gerlach et al. 2002). The solubility of carbon dioxide drops drastically with reduction of pressure, however. A typical batch of molten basalt becomes over-saturated with respect to CO_2 even before it reaches the shallow crust, causing the exsolution of CO_2. Increases in CO_2 emissions thus commonly precede eruptive activity. Where magma rises rapidly to the surface, CO_2 may not be exolved gradually, and can contribute to explosive violence of eruptions, as noted by Allison et al. (2021) at Sunset Crater, Arizona.

In fact, the general depth from which a particular magma ascends can be inferred based upon the relative proportions of the gases it supplies through degassing vents at the surface. Magma rising from storage at shallow depths (e.g. – from <10 or 15 km) shows a more significant increase in SO_2 and/or water vapor degassing relative to CO_2 on the way up, than magma rising *directly* from much greater depths. This is because the shallower-sourced melt lost most of its CO_2 while positioning itself within the shallow crust in the first place. The "classic sequence" of gas emissions preceding a deep-sourced eruption is this: first CO_2 increase in volcanic emissions during the deep stage of ascent, and then a SO_2 + H_2O increase as the magma gets close to the surface. A shift to lower CO_2/SO_2 ratios is generally a cause of concern for volcano watchers.

These observations are certainly useful for determining the origins of magmas feeding active volcanoes, though variable factors such as shallow groundwater interaction with magmatic gases may need to be taken into consideration (e.g. Turner et al. 2013; De Gregorio et al. 2014; Lopez et al. 2017). Shallow groundwater can dissolve SO_2 released from nearby magma – a process called **sulfur scrubbing**. Resulting surface gas measurements could mislead observers as a result.

During the 2018 Kīlauea [17] eruption (Chapter 1), the gasses released from active erupting areas such as Fissure 8 had a CO_2/SO_2 ratio of ~0.3, while steaming cracks uprift from these vents (Fissures 9 and 10) had CO_2/SO_2 ratios between 1 and around 10. Gas geochemists in the field found that those areas with a ratio close to 1 were worrisome, because they suggested that the magma was quite close to the surface and could break out anytime, while areas with higher ratios were of even greater concern. Where steaming fissures crossed Highway 130 – an important designated evacuation route, virtually no sulfurous gas could be detected at all. But this could well have been due to sulfur scrubbing beneath this site. Other observations plainly showed that molten rock was not far from reaching the surface at that location (Allan Lerner, personal communication 2018).

Carbon dioxide emissions from volcanic arc and mantle-plume-related hot spot volcanoes tend to be notably higher than they are from volcanoes along mid-ocean ridges and in mature continental rift zones. In volcanic arcs, carbon dioxide enrichment could derive from several sources: Subduction and partial melting or the metamorphism of siliceous carbonate rocks and sediments can provide substantial CO_2 to arc magmas (Marty et al. 2001). Arc volcanic gases can also incorporate carbonate-rich fluids from crustal metamorphic and metasomatic reactions, many of them triggered by magmatic heating. A third source is simply the heating and breakdown of various carbonate minerals embedded in subducted oceanic plates and sediments, such as magnesite and siderite that directly precipitate from seawater or result from hydrothermal alteration of the ocean floor at spreading centers, and which remain stable to great depths and pressures (e.g. Ducea et al. 2005).

Where magma reservoirs and intrusions cool slowly in the shallow crust, carbon dioxide continues to leak out steadily even after substantial crystallization. In fact, a key indicator of shallow ascending magma is a slow but measurable diffusion of CO_2 and helium from seemingly stable ground, a phenomenon called **soil efflux**, which excludes other kinds of volcanic gas because of their higher solubilities in magma or groundwater. A particularly worrisome case of soil efflux began at Mammoth Mountain [43], a popular ski resort in east-central California in 1989, as molten rock ascended to within a couple of kilometers of the surface. Carbon dioxide above the intrusion infiltrated forest soils next to Horseshoe Lake, at the southern foot of the volcano. Ordinarily, the amount of CO_2 in the soil pores of healthy forests is no greater than about 1.5 volume percent; but at Horseshoe Lake soil pore concentrations quickly rose to 30–90 percent, poisoning tree roots and turning the forest into a wasteland. Plant morbidity in volcanic regions can be an early warning of magmatic activity below (Pfanz et al. 2004). Over a period of 15 years, some 40 hectares of forest died, and by the late 1990s, Mammoth Mountain was still releasing nearly 50–150 tonnes of CO_2 a day through passive soil emissions – comparable to the level of many frequently active volcanoes (McGee and Gerlach 1998). A doubling of CO_2 efflux emissions took place, accompanied by ascending seismic activity, from 2009 to 2011, a likely indicator of rising magma (Lewicki and Hilley 2014). Degassing peaked and then receded once more, but no one can say what may come next. The last eruption at Mammoth Mountain was a minor phreatic event about 600 years ago, yet gas emissions tell us that the volcano is still magmatically restless at depth and remains a potential threat to the 8,000 residents of the town of Mammoth Lakes, on its eastern slope.

Fumaroles that primarily emit carbon dioxide are called **mofettes**. Mofette gases typically have temperatures much below the boiling point of water and sometimes are nearly as cool as the surrounding air. CO_2 is a heavier-than-air gas, and may accumulate in closed basins and can drain down small valleys, causing an uncommon, but lethal volcanic hazard (Chapter 14).

Carbon monoxide is a minor constituent of some high-temperature fumaroles. The abundance of carbon monoxide relative to carbon dioxide is a function of the temperature and composition (oxygen fugacity) of the erupting magma and the issuing fume. CO_2 gradually decreases and CO concentration rises by some two orders of magnitude with a change in eruption temperature from 800 to 1200°C. The oxygen bonds in carbon dioxide break down, forming carbon monoxide with increasing heat (Heald et al. 1963).

Sulfur

After water and carbon dioxide, sulfur is the next most abundant volatile present in typical magmas (Table 3.3). Owing to multiple valences (sulfur may carry charges of 0, −2, +4, or +6) sulfur produces many kinds of volcanic gas, reflecting various degrees of oxidation and reaction with CO_2 and water. S_2, SO, SO_2, SO_3, H_2S, H_2SO_4, COS, and CS_2 are all present in volcanic fume (Heald et al. 1963), although the most common species by far are SO_2 (+4 sulfur valence) and H_2S (+2).

The apparent color of SO_2-rich fume clouds changes with the nature of incident light. Tiny SO_2 aerosol droplets scatter shorter wavelengths of sunlight, in the blue color range. As you look through a sulfurous fume cloud toward the sun, SO_2 droplets scatter these light waves away from you, and the fume takes on a brown hue. If you look away, with your back toward the sun, the aerosols scatter light toward you instead, giving the cloud a bluish cast. Those blue or brown colors always tell volcanologists that SO_2 is being emitted, before the gas reaches our sensitive noses! Of the manifold gases emitted by magmas, SO_2 is the most readily detectable. In addition to coloration, mentioned earlier, SO_2 is a strong absorber of ultraviolet sunlight, allowing it to be detected and measured using UV-spectrometers from safe distances (Chapter 14).

Interaction of SO_2 with groundwater during ascent through shallow crust will dissolve it, then re-emit the sulfur in the form of H_2S – the pungent molecule people associate with the odor of rotten eggs. The human nose can detect H_2S at concentrations as low as 1 part per billion (!), which is about 10,000 times less than the concentrations regarded as hazardous to health. No one enjoys the smell (except for gas geochemists – who like all gas smells!). Volcanic SO_2 emissions potentially have great environmental impact (Chapter 13). Pinatubo volcano [128] in the Philippines released an estimated 20 million tons of sulfur dioxide during its paroxysmal eruption on 15 June, 1991 (Chapter 1), most of which entered the stratosphere as a persistent aerosol haze. The amount of SO_2 released by continuously active volcanoes of the world is very large, and has been estimated at 4.6×10^5 t/yr from Kīlauea volcano [17] alone – although this is only about 4 percent of the typical yearly global volcanic output (Sutton et al. 2001).

Halogens

The halogens, chlorine, fluorine, and bromine are minor volcanic gas constituents. While H_2O, CO_2, and SO_2 are measurable as weight percents, halogens are measurable only in parts per million (ppm), with chlorine and fluorine rarely exceeding 5000 ppm in a silicate melt, and bromine typically less than 250 ppm (Carroll and Holloway 1994; Aiuppa et al. 2009).

Water vapor reacts with fluorine and chlorine to create HF and HCl, both of which are extremely acidic aerosols in volcanic fume. Where molten lava enters the sea, boiling of the salt water creates steam clouds rich in hydrochloric and sulfuric acid droplets. Hawaiians refer to this irritating, corrosive steam as **laze** ("lava" plus "haze"). With a pH of 1.5–2.5, raindrops condensing from laze steam will sting the eyes, and eat holes in the clothing of observers in a short time. If the laze contains HF, camera lenses and eyeglasses may also be damaged. It is definitely good practice to observe the interactions of molten lava and seawater from the *upwind* side!

Chlorine is more soluble in silicate magmas than CO_2 and sulfurous gases. Chlorine and fluorine only begin degassing from magma when it is just a few tens of meters from the surface. The general order of volatile solubility from least soluble to most soluble is $CO_2 < H_2O < SO_2 <$ HCl (& HF). While highly toxic (industrial chlorine gas was the first type of poison to be used against enemy soldiers in World War I), the halogens are typically quite scarce in magmas, and are in any case are very difficult to detect and measure.

Some long-lasting post-eruptive vents issuing only steam and sulfur with minor carbon dioxide are called **solfataras**, and can be easily spotted from a distance owing to the yellow coloration of the sulfur crystals lining the vents. Halogen-dominated fumaroles, similar to mofettes and solfataras do not develop since the concentrations of chlorine and fluorine tend to be so low in magmatic gases.

Some geochemists have attempted to use the ratios of sulfur (primarily SO_2) to chlorine in volcanic fume as an index of the readiness of particular volcanoes to erupt. But the difficulty of accurately measuring chlorine, in particular, makes this approach daunting. Another volcano monitoring approach involves measuring the total emissions of SO_2 alone, independent of chlorine (e.g. Edmonds et al. 2003). As noted earlier, the quantity of SO_2 released from fumaroles around a volcano will tend to increase as magma approaches the surface then slowly taper off following eruption to be replaced by H_2S, steam, and other gases. If no fresh batches of magma rise to revive fumarolic activity, the remaining magmatic gases gradually disappear as well leaving only steam vents behind which, in turn, slowly die as underground heat dissipates.

New volcanic gas species in very minor amounts continue to be discovered in volcanic emissions, recently including HNO_3 and NO_x (an active irritant in urban smog) both of which result from the thermal breakdown of atmospheric air near the surface of molten lava pools. Halogen oxides and halocarbons (including natural chlorofluorocarbons, or CFCs), can form through chemical reactions between volcanic fluids and enclosing rocks (Mather et al. 2004). One of the more interesting of these exotic gases is bromium monoxide (BrO), which has been detected in the eruptive clouds of several volcanoes, including Soufrière Hills [84] following the renewal of volcanic activity in 1995.

Like volcanic chlorine and commercial CFCs, BrO plays a role in depleting life-protecting stratospheric ozone (O_3) by causing the dissociation of ozone molecules into ordinary diatomic oxygens (O_2). In fact, though far less abundant than chlorine in volcanic emissions, BrO may annually contribute to as much as a third of all volcanogenic ozone depletion globally. So corrosive is BrO to ozone that local ozone holes can appear temporarily above volcanoes that have expelled a large amount of this gas (Bobrowski et al. 2003). Bromium and chlorine oxides injected into the upper atmosphere by a large Plinian eruption will attack ozone by the **Brx-Clx-Ox reaction**:

$$\mathrm{BrO} + \mathrm{ClO} \rightarrow \mathrm{Br} + \mathrm{ClO_2} \tag{3.2}$$

$$\mathrm{ClO_2} + \mathrm{M^*} \rightarrow \mathrm{Cl} + \mathrm{O_2} + \mathrm{M^*} \tag{3.3}$$

$$\mathrm{Cl} + \mathrm{O_3} \rightarrow \mathrm{ClO} + \mathrm{O_2} \tag{3.4}$$

$$\mathrm{Br} + \mathrm{O_3} \rightarrow \mathrm{BrO} + \mathrm{O_2} \tag{3.5}$$

(Note: The term M* refers to the energy and momentum provided by collision with a third particle.)

Perhaps 20–40 percent of all atmospheric ozone depletion is triggered by bromine compounds, from both natural and artificial sources. Natural biological processes and human manufacture, however, create by far the largest concentrations of bromine in the stratosphere. Volcanic sources of chlorine and bromine amount to no more than a few percent of the total (Russell et al. 1996).

Helium

Inert gases such as nitrogen, helium, and argon are among the least concentrated of volcanic emissions, although almost always present in trace amounts. Nevertheless, helium has been of special interest because it provides a useful tracer for the mantle contribution to volcanic exhalations. There are two isotopes of helium, ^{3}He and ^{4}He. Helium-4 is widely produced in Earth's crust by the radiogenic decay of thorium and uranium. Helium-3 however is only concentrated in the mantle as a residual element left from the planet's early formation. Only a minor amount has managed to escape from the deep interior, and that has done so by processes related to volcanism. At present, the ratio of ^{3}He to ^{4}He is about 1:100,000,000 in the air we breathe. Measuring fumarole or soil efflux samples with higher ratios therefore is a good sign that molten rock is welling up from the mantle. Around fumaroles and high-temperature springs in arc settings, vapors commonly record ^{3}He:^{4}He levels 5–10 times that of ordinary air, while at hot spot volcanoes such as those of Hawai'i and Iceland, the difference can be as great as a factor of 15. Mid-Ocean Ridge basaltic glasses commonly record an intermediate tenfold enrichment of ^{3}He relative to ^{4}He at spreading centers (Craig and Lupton 1981).

At Mammoth Mountain [43] prior to 1989, ^{3}He:^{4}He levels were 3.5–4 times atmospheric levels, as measured in fumarolic emissions. But in 1989, the onset of the seismic activity stimulated by the Horseshoe Lake intrusion accompanied a rise in this value to 5–6, where it has hovered ever since. Mammoth Mountain has erupted primarily rhyolitic material throughout its fitful history, but the high ^{3}He emissions indicate that it must have some sort of "chemical communication" with the mantle. Molten basalt working its way into the lower crust is not only causing partial melting of more silicic magma at shallower depths, but must be transferring mantle volatiles to it as well. A similar observation has been made at Yellowstone caldera, Wyoming, [47] where gas emissions from the Mud Volcano steam field yield ^{3}He:^{4}He ratios comparable with those of mid-oceanic hot spots. The presence of a thick continental crust is no impediment to the escape of helium-3 from the mantle (Sorey et al. 1998; Christiansen et al. 2002).

At Ambrym volcano [174], in the Vanuatu volcanic arc, active Ambrym volcano shows evidence of being fed from two distinct magma sources, based not only on basaltic lava compositions but also on contrasting He isotope ratios (Jean-Baptiste et al. 2016).

Mineral Sublimates and Fumarole Alteration Products

Persistent fumaroles, active sometimes for centuries, form atop the conduits of shallow, still partially molten magma bodies, and are marked by the presence of sublimate minerals, deposited by escaping gases. They are important features for our atmospheric gas composition because most of the gas released by volcanoes, perhaps by an order of magnitude, comes from slow, steady fumarolic emissions *between* eruptions, rather than during the eruptions themselves (Stevenson and Blake 1998). One of the best-known examples of a long-lasting fumarole field is the one on the floor of Solfatara Crater, in the suburbs of Naples, Italy. Known as the Forum Vulcani in Roman times, the solfataras here have remained continuously active for over 2000 years, and have been a constant draw for tourists and campers. Only one small volcanic eruption, toward the end of the twelfth century, has occurred at Solfatara throughout this long interlude.

Temporary fumaroles, lasting only a few weeks to months, may form at the mouths of cracks in the surfaces of cooling lava and ash flows. As gases escape from the fumaroles, they typically sublimate and condense brightly colored minerals. Stoiber and Rose (1970), in a survey of fumarole minerals forming around several Central American volcanoes, discovered that the chemistry of these sublimates varies according to the location of a fumarole relative to the eruptive vent as well as the time elapsed from the last eruption. Where halogen gases are present, we often find deposition of chlorides. The commonest are ammonium, aluminum, and ferric chloride, the first two forming bright white, and the last brilliant orange incrustations. Sulfates and hydrous sulfates are also common and include white anhydrite ($CaSO_4$), gypsum [$CaSO_4(2H_2O)$], alunite [K_2SO_4-$Al_2(SO_4)_3$], alum [$KAl(SO_4)_2$-$12H_2O$)], thenardite (used as a rat poison; Na_2SO_4), and epsom salt ($MgSO_4$-$7H_2O$). Hydrogen sulfide (H_2S) rising from depth is oxidized at the surface to sulfur dioxide or to pure yellow sulfur that often forms masses of beautiful delicate needle-like crystals (**flowers of sulfur**) around the mouths of solfataras. Minor amounts of metallic minerals may also be deposited, including metallic chlorides and sulfates, and oxides of iron, copper, lead, zinc, arsenic, antimony, and mercury. Many of these metals are transported as chloride complexes in the vapor state. Water combines with ferric chloride to give soluble hydrochloric acid and insoluble ferric oxide (Fe_2O_3) that is deposited on the walls of fumaroles as a lining of tiny, brilliant steel-gray to black plates of specular hematite. The fumarolic acids eat away the older rock lining at the mouth of the fumarole, and may leave behind residues of ochre-colored limonite (hydrated iron oxide) and, if solutions are highly acidic ($pH \leq 3$), milky opal. Other residues include lustrous black magnesioferrite ($MgFe_2O_4$) and yellow-brown goethite [FeO(OH) named in honor of the poet whose quote opens this chapter]. For those who marvel at minerals, fumaroles can produce a feast for the eyes!

Volcanic versus Anthropogenic Emissions

Many anti-environmentalist arguments point out that people "are not the only polluters; volcanoes also pollute the environment" – and even more so. But such arguments are spurious because the total pollutant loads, as in the case of bromine and chlorine mentioned earlier, are far less than supposed. For example, while volcanoes do release an estimated 130–230 million tonnes of CO_2 into the atmosphere annually, human consumption of fossil fuels, production of cement, natural gas flaring, deforestation, and related activities add as much as 40 billion additional tonnes, at an increasing rate from year-to-year (Gerlach 1991, 2011; Marland et al. 1998; Le Quéré et al. 2014). Similar studies comparing sulfur output of volcanoes to anthropogenic sources, primarily coal-fired power plants and heavy industries, indicate an order of magnitude difference with humans being the clear heavy-weights. We know that under natural conditions the atmosphere can easily reprocess the volcanogenic input, or life would have become impossible long ago. But coupled with the enormous human production of SO_2 and CO_2, the earth system can simply be overwhelmed. No wonder that earth scientists are so concerned about "the human volcano."

Questions for Thought, Study, and Discussion

1. *How* and *why* have sources of heating inside the earth changed since the time of the planet's formation?
2. What is the silica tetrahedron, and why might it be regarded as the basic building block of minerals?
3. Describe the roles of pressure and H_2O in melting.
4. What enables basaltic magma to erupt so easily, and why is it more difficult for granitic magma to do so?
5. What are the factors considered in the classification of igneous rocks and magmas?
6. What was N.L. Bowen's great contribution to our understanding of magma?
7. Why were the nineteenth-century scholars Buffon and Kelvin wrong in their calculations for the age of Earth?
8. What was Sir Arthur Holmes' great contribution to our understanding about the cooling of Earth?
9. How might measurements of SO_2 from a volcano be a useful indicator of its potential eruptive state?
10. What is the environmental significance of BrO? Do you think that we should be concerned about the impact of volcanic gases and the global environment? Why/why not?

FURTHER READING

Ebinger, C.J., Baker, J., Menzie, M.A. et al. (2002) Volcanic rifted margins. *Geological Society of America Special Paper 362*, 236 p.
Hibbard, M.J. and Hibbard, M. (2002) *Mineralogy: A Geologist's Point of View*. New York: McGraw Hill, 576 p.
Scholl, D.W., Kirby, S.H. and Platt, J.P. (1996) Subduction: Top to bottom. *American Geophysical Union Monograph Series*, 96: 384 p.
Sigurdsson, H. (1999) *Melting the Earth: The History of Ideas on Volcanic Eruptions*. Oxford: Oxford University Press, 272 p.
Young, D. (2003) *Mind over Magma: The Story of Igneous Petrology*. Princeton, New Jersey: Princeton University Press, 704 p.

CHAPTER 4

The Physical Properties of Magma and Why It Erupts

Source: R. Hazlett.

Volcanoes: Global Perspectives, Second Edition. John P. Lockwood, Richard W. Hazlett, and Servando De La Cruz-Reyna.

Companion website: www.wiley.com/go/lockwood/volcanoes2

Magma is not accessible to direct observation; its very existence is inferred, and its properties have to be deduced by circumstantial observations in nature and by laboratory experiments.

(T.F.W. Barth 1952)

We discussed how magma is formed and how different magma types are classified in the previous chapter, but here we need to add that of all the molten rock generated below and within Earth's crust only a small fraction – probably less than 10–15 percent – ever makes its way to the surface to form volcanic rocks. Most magma solidifies far below the surface to form *plutonic* rocks – never to see the light of day until cooled solid and exposed by erosion. This chapter explores the factors that enable magma to reach the surface and thus form the volcanoes, lava flows, and pyroclastic deposits that are the principal focus of this book. But first, some more background about magma temperatures and viscosity – critical properties affecting magma mobility. Basic thermodynamic properties are also essential to understanding magma evolution and transport mechanisms (Lesher and Spera, 2015) but are not discussed here.

Magma Temperatures

In the last chapter, we introduced the application of **geothermobarometry** – the study of mineral compositions to estimate original temperature–pressure conditions of igneous minerals at the time they cool and crystallize. Geothermobarometry can provide important insights about conditions that govern magma emplacement and related volcanic activity (Table 4.1). For example, Wark et al. (2007) employed the *Ti-in-quartz geothermometer* to document temperatures in the magma body that erupted to form the Bishop Tuff in central California, as recorded in the rims of quartz phenocrysts preserved within the tuff. The temperatures measured closely match those of earlier *Fe-Ti two oxide geothermometry* – in the range of 720–790°C (Hildreth 1981). Evidence was found that a small thermal pulse (ca. 60°) from injection of fresh mafic melt into the lower part of the magma chamber may have triggered the eruption of the Bishop Tuff – one of the most explosive North American eruptions of the past million years.

Optical pyrometers, developed for use in foundries over a century ago, were long used to estimate lava flow temperatures based upon radiated colors, which are also useful for visual estimates (Table 4.2). Volcanologists now use hand-held radiometers and thermocouples for more precise measurements. **Radiometers** measure temperatures based upon infrared radiation emissions from heat sources. **Thermocouples** (a voltmeter connected to high-melting-point wires – typically chromel–alumel) serve for direct measurements of lava temperatures, and involve insertion of bimetallic probes directly into or against the heat source – a difficult undertaking on active lava flows.

Lava begins cooling the moment it erupts, meaning that a single flow will exhibit a wide range of temperatures as it is emplaced. Using radiometry, Pinkerton et al. (2002) classified four "thermal components" of a typical active lava flow on Kīlauea volcano [17], including a flow core (greater than 1050°C), thin stretching crust (750–900°C), rigid solid crust (less than 750°C), and the flow margins (less than 175°C). Some greatly localized portions of the flow, such as active tumuli and openings in crusted over lava channels openings (Chapter 6) can be as much as 150°C hotter than the surrounding surface.

TABLE 4.1 Some important geothermobarometers.*

Type	Notes	Developers and important reference papers
Plagioclase geothermometer	Uses anorthite-albite ratios to determine temperature of plagioclase crystallization in magma	Kudo and Weill (1970); Stormer and Carmichael (1970)
Fe-Ti "two oxide" geothermometer	Estimates magma temperature based upon iron and titanium oxide mineral chemistry	Buddington and Lindsley (1964); Anderson and Lindsley (1988); Ghiorso and Sack (1991)
Al-in-hornblende geobarometer	Estimates depth of magma body emplacement based upon aluminum content in hornblende	Hammarstrom and Zen (1986); Anderson and Smith (1995)
Mg-volcanic glass geothermometer	Uses MgO content of volcanic glasses to estimate temperature at time of eruption and quenching	Helz and Thornber (1987)
Ti-in-quartz geothermometer	Magma temperature based on Ti content of quartz. Useful for granitic rocks formed at 10 kb pressure or less	Wark and Watson (2006)
Ti-in-zircon, Zr-in-rutile geothermometers	Magma temperatures based on zircon and rutile mineral chemistries in intermediate and silicic rocks. May be sensitive to pressure, even at less than 10 kb	Watson et al. (2006); Ferry and Watson (2007)

* For more information on geothermometers, barometers, and silicic plutonic rocks, see Anderson et al. (2008).
Source: Modified from Anderson et al. (2008).

TABLE 4.2 Correlation between the temperature of lava and its color.

Color	Temperature (°C)
White	> 1150
Golden yellow	1050–1100
Orange	900–1000
Bright cherry red	700–900
Dull red	550–625
Lowest visible red	475
Pizza oven temperature	260–315

Lava begins flowing at temperatures above about 700°C.
Source: Based on US Geological Survey, Cascade Volcano Observatory.

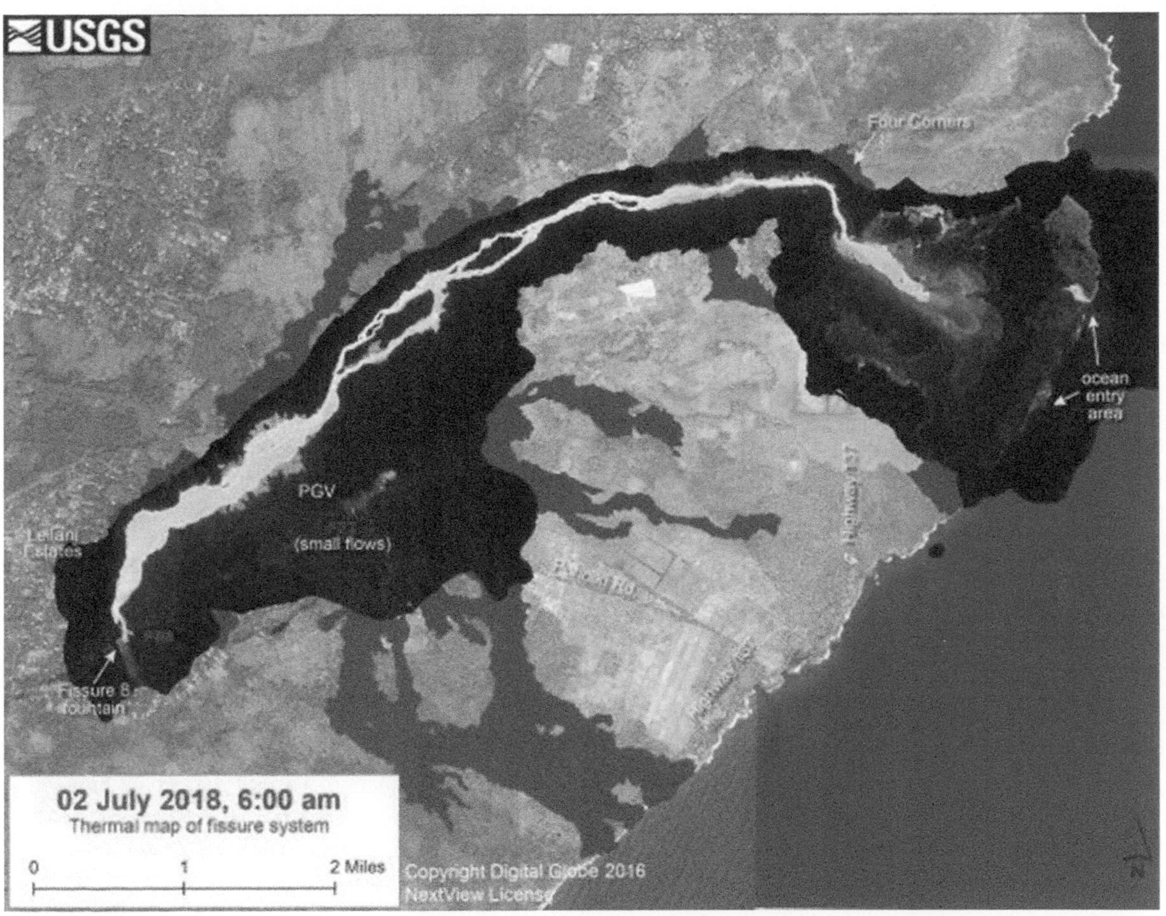

FIGURE 4.1 Composite image of thermal data superimposed on a satellite landscape. The brightest pixels show the hottest areas, with white being molten at the surface; black and gray solid, but at elevated temperatures (including hot water!). Source: USGS image by Matt Patrick.

Airborne heat-sensing **thermal cameras** have also been used to evaluate the movement and cooling of active lava flows, as during the 2018 eruption in Kīlauea's East Rift Zone. The images provide a precise record of the evolution and gradual cooling of the flows over the course of an eruption (Figure 4.1), information that has proved helpful for hazards assessment (Chapter 14).

Magma Rheology

Rheology is the study of how solids and liquids respond to **stress** – the application of *force per unit area*. In a broad sense, we can classify magmas (and lava flows) as being "elastic," "viscous," or "viscoelastic" in their rheological behaviors under differing conditions. An **elastic** substance is one like a rubber band; you can stretch it, but once you let it go (release the stress imparted by your fingers), it will return to its original size and general shape. It may seem counterintuitive, but solid rocks can also be regarded as being elastic bodies. If subjected to sudden blows, a rock will simply transmit shock energy through its body and return to the same state of rest as it had before being struck. This is how earthquake shock waves are transmitted through Earth; the "rubber band" stretching of rock takes place with the arrival of each seismic wave, but the energy then is transferred to the adjacent masses of rock as a compressive pulse or shearing movement along the line of travel, and the original rock mass returns to its rest position, unperturbed. There are limits, of course, to how hard and how fast you can strike a rock without it breaking. This limit defines the **strength** of a rock. **Faults**, the sources of earthquake shock waves, represent planes of rupture where rock strength was exceeded by powerful tectonic forces.

Viscous materials respond to stress by flowing rather than transmitting energy. The degree to which they do so defines their viscosity. Liquids of low viscosity flow readily; those of high viscosity flow only very slowly and with difficulty. As pointed out in the previous chapter, solids like glaciers and hot mantle rock can also flow, and might in fact be regarded as high viscosity fluids. A liquid of low viscosity, such as water, quickly spreads into a broad thin sheet when poured out onto a tabletop. A liquid of higher viscosity, such as cold molasses, poured out in a similar manner spreads slowly and never attains as thin and widespread a sheet as does the water. Still more viscous substances, like cold tar or shoemaker's wax, remain standing for a time as a steep-sided lump on the table, but over a period of a few days or weeks, gradually flatten and spread out. Even windowpanes and the hard slabs of stone benches in very old cemeteries deform and flow under the pull of gravity.

Two classes of viscous materials are of particular interest to us: the Newtonian and non-Newtonian fluids. **Newtonian fluids** are those like water in which the value of the viscosity does not change, irrespective of how fast the fluid moves across a slope:

$$\tau = \mu(du/dx) \tag{4.1}$$

where τ represents the "shear stress" – the stress imparted by a fluid in the direction of its flow going from the base of the current to its top (assuming that this surface is in full contact with the air); and du/dx is the changing speed of the flow also going from its base to top. Viscosity (μ) is a constant of proportionality in this equation. Newtonian fluids will maintain their fluid properties irrespective of how fast they are stirred or mixed (Hobiger et al. 2011).

A **non-Newtonian fluid** is one in which viscosity changes according to the rate at which shear stress is applied. An example is pudding in a cup. You can stir it, but a hole will open in the wake of the stirring implement that only gradually refills. Non-Newtonian behavior can be induced by mixing certain substances together. Take corn starch and stir it into a glass of water. At a certain critical point when the corn starch and water have developed a more or less uniform consistency, the solution undergoes **shear thickening** and it becomes non-Newtonian. "Shear thickening" means that as the mixture is stirred, its viscosity jumps and it stiffens. Put a finger slowly into the cup and turn it slowly, and it behaves like a liquid. But poke it rapidly, however, and it responds like a rubbery solid. Likewise, turn the cup over all at once and the mixture will stay stuck in the cup, only slowly slurping out as a single mass.

A **viscoelastic** substance is one in which, as the name implies, there is both viscous and elastic behavior. A viscoelastic substance will respond to stress by deforming, but after the stress is relieved, it will only partly recover its original shape and size. A significant type of viscoelastic behavior in volcanology is that of **Maxwell bodies**. You can make a Maxwell body by attaching a spring to a **damper**, which is essentially just an empty syringe in which you compress trapped air. The spring represents the elastic response of the system, while the damper represents the viscous response. Of course, Maxwell bodies do not all come with syringes and springs, but within their homogeneous masses, they combine behavioral aspects of both. If a Maxwell material is suddenly subjected to stress, it will respond instantly just like any elastic body, but then it will continue deforming beyond its **elastic limit**, at a constant, unrecoverable rate, meaning that when the stress is relieved, the Maxwell body will only *partly* recover its original shape and size. The elastic response (ε_{rev}) is defined by:

$$\varepsilon_{rev} = \sigma / E \tag{4.2}$$

and the viscous by

$$\varepsilon_{irrev} = t_1(\sigma / \mu) \tag{4.3}$$

with σ equal to the stress, t_1 the total time that the stress is applied to the body, and E a constant called the **elastic modulus**; more or less the "springiness" of the system.

The walls of some deep-seated magma conduits and of large, long-lived magma chambers consist of heated rock that softens and responds to pressures exerted by the magma viscoelastically. This is shown by the folded and sheared country rock associated with the margins of some deeply eroded granitic plutons, such as in the Sierra Nevada Range of California. The degree of deformation intensifies with proximity to the borders of these plutons. Such viscoelastic behavior may have an important bearing on the ability of the certain magma chambers to erupt, because it is more difficult for critical fracturing to initiate in their walls (Jellinek and DePaolo 2003).

Just as Newtonian bodies can change into non-Newtonian bodies, so too can elastic substances become viscoelastic and even viscous. The partial melting that generates magmas begins when solid elastic rock heats to the point where it becomes a viscoelastic mass until finally it melts and becomes fully viscous. The process works in reverse too. Consider the cooling skins of molten lava flows, or of fluid volcanic bombs tossed out of some volcanic vents. These bombs cool to the point of becoming viscoelastic as they sail through the air. They continue to cool and are deformed into aerodynamic shapes as they travel toward their points of impact, where sudden collision with the ground causes them to deform even further or to fracture.

Lava flowing out from eruptive vents is much more viscous than most liquids with which we are familiar. Flowing lava may appear to be very fluid, and descriptions may say that such lava "flowed like water." Actually, the hottest silica-based lava yet measured (at 1250°C) has a viscosity about 100,000 times that of room-temperature water. Rare carbonate-based lavas have lower viscosities than this, but still much greater than water. The false appearance of great fluidity results from the high density of lava, as much as three times that of water, which because of momentum on slopes causes it to assume the flow characteristics of a liquid of much lower density and viscosity.

Viscosity is the principal factor that determines the ultimate shape of lava flows, from thin pāhoehoe (Chapter 6) to extremely thick lava domes (Chapter 9). Field studies of lava flow morphologies is one of the best ways to evaluate the original relative viscosities of lava (Moore 1987). The viscosity of magma varies widely, depending on several factors, including its chemical composition (especially silica content), the amount and condition of included gas, the amount of solid load being carried, and its temperature. The presence of solid fragments (either mineral phenocrysts, microlites, or fragments of foreign origin) in magma also increases its bulk viscosity simply by increasing the frictional resistance to flow (Klein et al. 2018). Likewise, a rough bed beneath flowing lava flowing on a slope will impart friction that raises flow viscosity. The effects of gas are more complex. Vesicles may increase, or decrease viscosity dependent upon their shapes as well as the steadiness of flow. Spherical gas bubbles tend to make steadily moving flows more viscous, in a manner analogous to suds in a solution of soap in water; one can set a mass of soapsuds on a tabletop and have it remain there without flowing appreciably until the bubbles have burst. Vesicles elongated in the direction of flow, however, will reduce viscosity by facilitating shear. If the flow is unsteady, viscosity effects depend upon the ability of vesicles to change their shapes according to changing flow conditions. If vesicles cannot adjust their shapes fast enough to keep pace with changing flow conditions, flow viscosity tends to drop (Llewellin and Manga 2005).

The basic unit of measurement of viscosity in the cgs system is the **poise,** a factor of dynamic viscosity which is measured in units of grams per centimeter per second (g/cm/s). In the SI system, viscosity is measured in **pascal-seconds**, in which 1 pascal-second (Pa s) is equivalent to 10 poises. The viscosity of water is equivalent to 1×10^{-2} poises. By comparison, air has a viscosity of 17.4×10^{-6}; olive oil of 81×10^{-3}, tree pitch of 2.3×10^{8}, and glass of more than 10^{20} poises.

Temperature is the most important influence on viscosity. Given constant pressure, the higher the temperature of magma, the lower its viscosity. This is particularly evident in the behavior of lava flows. As they travel away from vents, they lose heat by radiation and conduction into the air above and the ground beneath, and the viscosity steadily increases. Macdonald (1972) reported that a flow on Mauna Loa [15] was found to be more than twice as viscous 19 kilometers down the mountainside as it was when it issued from the vent. On Etna [112], Walker (1967) found that the calculated viscosity of a small flow increased about 375-fold (from 0.4×10^{5} to 1.5×10^{7} poises) in a distance of only about 450 meters. He also found that the viscosity at which lava stops moving is between 10^{9} and 10^{11} poises, depending on slope angle. Direct viscometer measurements on flowing Kilauea lava yielded values of 380 Pa.sec (3.8×10^{3} poises; Chevrel et al. 2018).

Neuville et al. (1993) examined the viscosities of andesite and rhyolite at a range of temperatures. At 764°C, they measured a viscosity of 1×10^{10} poises in the laboratory for a sample of andesite, which dropped to 1×10^{3} poises at a temperature of almost 1400°C. For a sample of much more siliceous rhyolite, the range of viscosity values was from 1×10^{16} poises at 647°C to 1×10^{4} poises at 1643°C. The influence of dissolved water on the viscosity of magma is also important, as shown by the laboratory research of Shaw (1972) in which he demonstrated that a 4–5 weight percentage increase in magmatic H_2O can decrease the viscosity of a basaltic melt from 10^{9} to 10^{5} poises, depending on temperature. Table 4.3 gives viscosities of different melts as reported by Blatt et al. (2006).

TABLE 4.3 Some sample igneous melt viscosities.

Type	Viscosity (poises)
Basaltic melt at 1200°C	10^{2}–10^{3}
Andesitic melt at 1200°C	10^{4}–10^{5}
Rhyolitic melt at 1200°C	10^{6}–10^{7}
Erupting basalt	10^{3}–10^{4}
Erupting andesite	10^{5}–10^{7}
Erupting rhyodacite	10^{11}

Avard and Whittington (2012) studied the rheology of dacite lavas erupted in volcanic arcs, and concluded that heating due to internal shearing in flows (much like the heating you experience when you rub the palms of your hands together briskly) could reduce flow viscosity significantly, enabling ordinarily highly viscous flows to travel much farther on steep slopes than they could otherwise.

Every house has a plumbing system, and so does every volcano. . .

(Professor G.P.L. Walker addressing a classroom at the University of Hawai'i, October 1988)

Magma Ascent and Emplacement

It is now known that the level to which magma rises beneath (or within) volcanoes is largely controlled by local equilibrium between the density of the magma and the density of enclosing rocks. Ryan (1987a) noted that:

Magmatic fluids, generated at depth in Earth's mantle often come to rest at surprisingly shallow levels within the crust. It is, of course, remarkable that so long an odyssey should be interrupted just short of subaerial or submarine eruption.

Ryan then went on to introduce the concept of **neutral buoyancy** to describe the critical role of density in governing the rise of magma beneath volcanoes. The level of neutral buoyancy is the level where the density of magma equals the density of the surrounding country rocks, and is the depth at which magma may accumulate to form magma chambers. This concept is based on the realization that magma is a fluid (albeit normally a very viscous one), and will respond to the laws of gravity like any other fluid or plastic medium and will seek local mechanical equilibrium. If the density of magma is less than that of the surrounding rocks (positive buoyancy), it will attempt to rise; if its density is greater (negative buoyancy), it will tend to sink. The same principle was earlier observed by Francis (1982) who noted that "hydrostatic equilibrium" governed the injection of basaltic lava sills into layered Paleozoic sedimentary rocks of northern England. Walker (1989) applied the neutral buoyancy concept to a wide variety of volcanic settings, demonstrating its importance to understanding all volcanic processes that involve the storage and migration of molten rock.

Although neutral buoyancy is a well-established physical concept, ascending magma can stall and collect at levels not corresponding to its neutral buoyancy level. If the rising magma encounters an abrupt reduction in wall-rock density, as for instance encountering the crust–mantle boundary, it will slow down, accumulate, and even spread laterally for great distances along the contact, a process called **underplating**. More important for shallow magma bodies (<~10 km deep), may be the presence of groundwater that can advect heat away as it circulates, cooling and stiffening at least the upper portions of the molten mass. This is also important for producing geothermal power (Chapter 15). Magma may *seek* neutral buoyancy, but does not necessarily achieve it.

More is known about the generation, ascent, and emplacement of magma beneath oceanic volcanoes than about magmatic processes that take place below and within continental volcanoes, mainly because the processes involved are less complex, and the travel times from origin to eruption are usually much shorter (Table 4.4). In some parts of the world tectonic activity has also exposed the deep roots of oceanic volcanoes and their magma plumbing systems in places where oceanic crust has thrust up (**obducted**) onto the edges of continents at convergent plate boundaries (e.g. Dilek and Robinson 2003).

Many of the concepts about magma ascent and emplacement have been developed in Hawai'i, based on petrologic and geophysical studies. Some late-stage, low-viscosity alkali basalts ascend to the surface at high speed (several m/s), allowing these fluid magmas to carry mantle-derived xenoliths with them to the surface. Some of these peridotite fragments are cut by pyroxenite veins that are structurally similar to the veining in granitic migmatites (Wilshire and Kirby 1989). It may seem odd that brittle fractures and veins could form in the very hot, plastically deforming mantle, but perhaps rapid decompression may be involved,

TABLE 4.4 **Some estimated magma ascent rates beneath volcanic areas.**

Volcano	Type of volcano and tectonic setting	Rate of ascent (m/s)
Mount St Helens, Washington State [31]	Composite volcano, volcanic arc	0.007–0.15
Soufrière Hills, Monserrat [84]	Composite volcano, island arc	0.001–0.015
Mount Unzen, Japan [141]	Composite volcano, volcanic arc	0.002
Kīlauea, Hawai'i [17]	Hot spot volcano, Hawaiian mantle plume	0.03–1.7*
Kimberlite diatreme	Explosive volcanic pipes (see text below)	1.1–30

* See Klein et al. (1987).
Source: Modified from Rutherford (2008).

causing mantle rocks to snap apart like rapidly stretching putty. It may also be that these veins form by fracturing of larger mantle fragments after transport to cooler crustal levels. Magma under high pressures quickly exploits any pathways for escape and fills them. The rate at which it can rise from the mantle can be ascertained at Kīlauea, where volcanic tremor originating in the region of partial melting 40–60 km down may be tracked migrating upward into the volcanic edifice over a period of just a few months. These earthquakes most likely represent cracking of wall rock along the path of ascent by batches of melt (Aki and Koyanagi 1981; Chapter 14). In other cases, the magma entering the volcano apparently does so continuously and relatively passively, with little or no apparent seismic activity.

Although xenolith-bearing alkali basalts found at many continental monogenic volcanoes and on some oceanic volcanoes like Hualālai [14] (Hawai'i) must rise rapidly from their source areas and may erupt without significant pausing, this is more the exception than the rule (Clague 1987); most rising magma stalls and accumulates in magma reservoirs on the way up. These magma storage sites may prove temporary if magma density in upper portions of magma chambers decreases through compositional differentiation during storage and allows migration of residual melts to higher levels (see "Eruption Triggers" section). Two such magma storage levels are thought to exist under active Hawaiian volcanoes. One level lies at the mantle–crust boundary, the so-called **Mohorovicic (M) discontinuity**, where seismic waves passing through the mantle slow down upon entering the lower-density crust. The M-discontinuity lies about 25 km beneath the largest of the active Hawaiian volcanoes, Mauna Loa [15], and is somewhat shallower beneath Kīlauea [17].

At a shallower level, the rising magma stalls and accumulates at levels only a few kilometers beneath each volcanic summit. The magma reservoir underlying Kīlauea's summit lies 1.5–5 km down and is several kilometers wide (Figure 4.2). Seismological evidence suggests that it is not actually a spherical, liquid-filled void as depicted in some simplistic cartoons of volcanoes, but rather a spongy mass of plastic or viscoelastic hot rock permeated with channels of more fluid, eruptible magma (Ryan 1988; Dawson and Chouet 1999). Because the level of neutral buoyancy in large volcanoes like Kīlauea rises as the volcanoes grow and are compacted (become denser) at depth, it is likely that their shallow-level magma chambers gradually shift upward over time as well (Figure 4.3) (Decker et al. 1987; Ryan 1988).

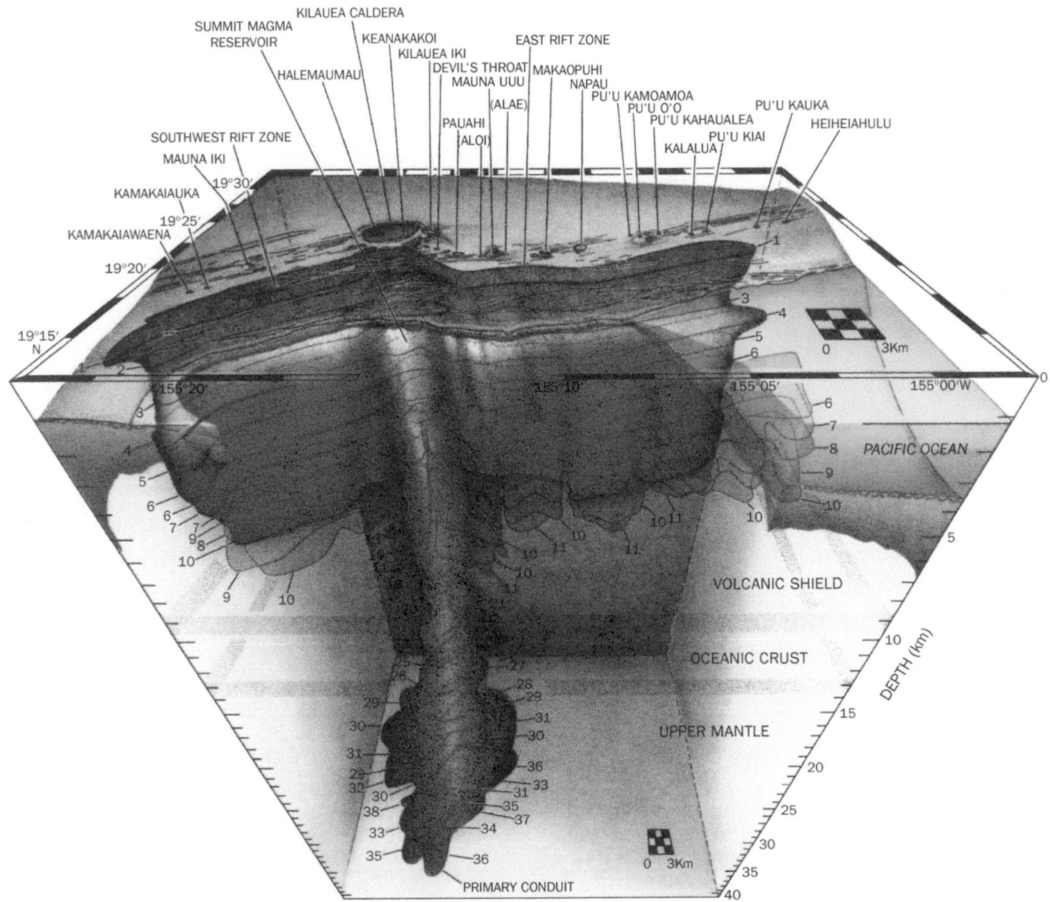

FIGURE 4.2 The structure of Kīlauea volcano's [17] magma system, as viewed from the south. This model is based on an analysis of over 25,000 earthquakes measured in this area between 1969 and 1985. Magma is seen to rise through complex channelways from its place of origin in the upper mantle, through the oceanic crust, and to its level of neutral buoyancy, where it accumulates in a shallow magma chamber beneath Kīlauea caldera and migrates along rift zones. Source: From Ryan (1988). Reproduced with permission of John Wiley & Sons.

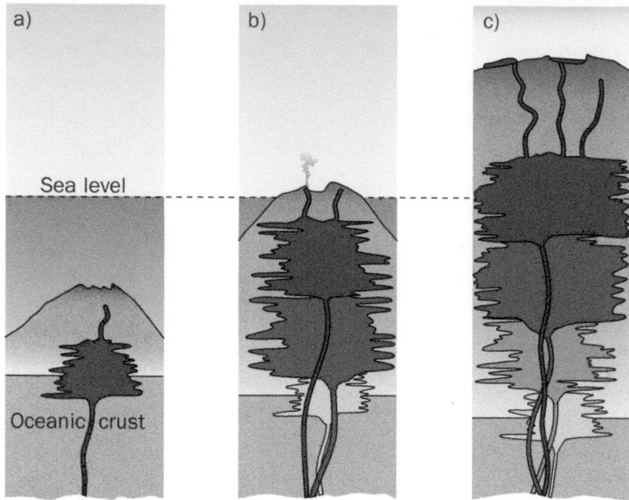

FIGURE 4.3 Magmatic evolution of an oceanic volcano. As surface eruptions build volcanic edifices higher, the density of the lower parts of a volcano increases, owing to compaction, mineralization, and accumulation of dense minerals. Active magma chambers (red) must thus ascend upward through growing volcanoes with time, always seeking their "Level of Neutral Buoyancy." (a) Submarine volcano; (b) Immature subaerial volcano; (c) Mature oceanic shield volcano, with summit lava lake and flank eruptive activity. Note that submarine slopes are steeper than subaerial ones, and that the base of the growing volcano subsides owing to isostatic adjustments. Source: Modified from Ryan (1987b).

Crustal extension and related faulting also greatly influence the pattern of magmatic intrusions within Kīlauea. While the north flank of the volcano is buttressed by the adjoining mass of Mauna Loa, the south flank is unsupported and slopes to the 5-km-deep ocean floor. As Kīlauea swells with magma, stresses accumulating within its southern flank have caused large-scale slumping and faulting across most of the mountainside – an area measuring roughly 60 × 80 km. An interplay of forces has developed that involves periods of (sometimes violent) seaward movement in the south flank, opening spaces within the volcano for more magma to migrate horizontally, essentially draining the source reservoir beneath the summit (Conway et al. 2018). The infilling of magma, and compaction of the underlying regions of the volcano, adds additional weight and pressure to the weak flank, which eventually responds by slipping further to accommodate yet another round of intrusions. It is impossible to discern cause and effect in this geological interplay; one form of behavior leads to the other in a feedback process. As a result, two narrow bands of extensional dilation, magmatic intrusion, and fissure eruptions, termed rift zones (Chapter 1), extend as far as 120 km from the summit of Kīlauea. Each marks the crest of the volcano's southern flank, and both are artifacts of gravity guiding the upwelling and lateral intrusion of melt as it approaches the surface. Similar rift zones are present on most of the world's oceanic shield volcanoes.

Although the summit magma chamber of Kīlauea actually lies within the body of the volcano, this is not the case for comparable magma chambers underlying most "gray" volcanoes. Perhaps this is simply because these volcanoes have smaller volumes than the Hawaiian shields. But – likely related to overall lower densities of continental crust – magma bodies in continental settings are also much deeper. The magma feeding the catastrophic eruption of Mount St Helens [31] on 18 May, 1980, for example, originated at depths of 7–14 km beneath the base of the mountain. All subsequent dome-building eruptions from 1981 to 1986 tapped magma stored at a shallower depth – less than 4 km, suggesting a close relationship with the channel feeding molten rock from a deeper-seated reservoir (Blundy and Cashman 2006). At Vesuvius [109], mineral chemistry and seismological data indicate that the magma reservoir also is "deep seated" – or was originally. Recent data suggest that it has shifted upward over the past 20,000 years some 9–11 km, with an ascent of 3–5 km between 79 and 472 CE – the dates of two very explosive eruptions. Scaillet et al. (2008), who provided these data, suggests that changing stress field conditions, including caldera collapse and gravitational spreading of the mass of Vesuvius, plus differences in volatile content of the magma could account for this remarkable change.

To think that all volcanoes must have single long-lived magma storage systems is probably erroneous. Turner et al. (2013) used geochemical data to show that three independent magma supplies fed eruptions at Bezymianny, Kamchatka [170], between 1956 and 2010. Similarly, two reservoirs feed magma to New Zealand's Tongariro volcano [179] (Arpa et al. 2017).

Koyaguchi and Kaneko (1999) provide evidence for ephemeral, rather than long-lived magma reservoirs beneath some volcanoes. These reservoirs may experience repeated partial solidification and remelting events, and can interact intensively with the surrounding crust as well as each other. Such appears to be the case at Ruapehu, New Zealand [178] – the most active volcano in Australasia (Gamble et al. 2003). Anna Myers (2007, p. 36) comments:

> *The picture is emerging, rather than having one big magma storage chamber, Ruapehu has a complex plumbing and reservoir system of relatively small-scale magma stores distributed throughout the crust beneath the volcano. Each magma batch evolves on its own timescale, assimilates surrounding crust, and then mixes with other batches. . . Ruapehu [has]. . . an "open" plumbing system, with the magma exchanging both heat and material with the surrounding crust.*

Similarly, Newberry [41] and Medicine Lake [39] volcanoes in the Cascade arc appear to be fed by multiple small magma storage sources spanning a wide range of compositions even more diverse than those seen at Ruapehu (Donnelly-Nolan 1988; MacLeod and Sherrod 1988; Figure 4.4).

At the other end of the spectrum from the numerous small magma sources of Ruapehu and the Cascade volcanoes are the large, discrete magma chambers underlying silicic calderas such as Yellowstone [47] and Long Valley [44] (from which the earlier mentioned Bishop Tuff erupted). These single magma bodies may have volumes greater than 5000 km^3, the size of the most

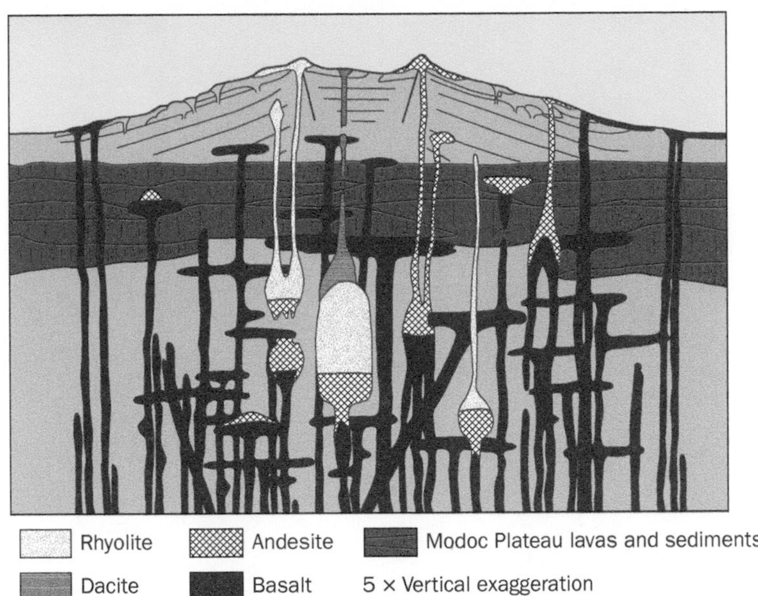

FIGURE 4.4 Cartoon sketch showing the inferred complexity of the magma reservoir system beneath Medicine Lake Volcano, California [39]. Seismic studies indicate that no single large magma chamber is present. The width of feeder dikes is greatly exaggerated. Source: From Donnelly-Nolan (1988). Reproduced with permission of John Wiley & Sons.

voluminous known volcanic eruptions (Chapter 10). The Yellowstone magma chamber underlies an area measuring 40 × 80 km – comparable with the city of Los Angeles, with a roof only about 8 km thick. Seismic data indicate that only about 10–30 percent of the chamber is presently liquid – the rest is a mushy bed of uneruptible phenocrysts and solid but very hot rock fully crystallized from the melt (Bachman and Bergantz 2008). Nonetheless, this vast pocket of magma is still quite active. Between 2004 and 2006, fresh mafic melt apparently flooded into the bottom of the chamber, causing the surface above to rise as fast as 7 cm/year – three times the historical rate of caldera uplift (Chang et al. 2007). Analogs of similar magma bodies – granitic plutons long crystallized and exposed by erosion – occur throughout the western United States, revealing varying degrees of interaction with mantle-derived melts and enclosing crustal rocks (Miller and Miller 2002; Žák et al. 2012). The largest silicic magma bodies are elongated lenses typically only a couple of kilometers thick, but from 10 to 20 km wide. Those associated with vigorous high-silica volcanism appear to lie mostly in the depth range of 4–10 km (Lindsay et al. 2001).

The development of **seismic tomography** ("slice picture") techniques, based on the observed variation of seismic wave velocities as teleseisms (waves from large distant earthquakes) pass through and below volcanoes, has allowed great refinement in our understanding of volcanic structures and magma reservoir geometries. Higher velocity zones indicate the presence of dense, crystalline rocks, whereas low velocity areas indicate the presence of high-temperature or partially molten rocks. Results show that some vigorously active volcanic centers such as Krafla (Iceland) [103] and Rabaul caldera (Papua New Guinea) [160] show low velocity anomalies, whereas others such as Taupo caldera (New Zealand) [180] and Hekla (Iceland) [92] show no clear anomaly structures at all, suggesting either extremely shallow or very deep magma reservoirs, or the lateral feeding of magma from distant sources (Finlaysen et al. 2003; Lee 2007).

One of the many questions in volcanology concerns where magma reservoirs and their overlying volcanoes develop. In the United States Pacific Northwest, many large composite volcanoes are positioned within grabens, which apparently formed as downdropped blocks of crust pulled open in gaps between the ends of parallel strike-slip faults. The extension developed within these blocks would make it easy for rising magma to form full scale reservoirs, simply by leaking into fractures incrementally opened by fault motion. It is not certain that this has happened in the Pacific Northwest, although in southern Greenland, plutons crop out throughout a deeply eroded region between the ends of large strike-slip faults whose motion appears to have permitted passive, simultaneous ascent of magma by pulling apart the crust (Harrison et al. 1990). In the Ivrea-Verbano zone in the Italo-Swiss Alps, mantle-derived melts accumulated at the base of rapidly extending metamorphic terrane in late Paleozoic times (Quick et al. 1992). Similar situations may have governed Miocene volcanism in the Los Angeles Basin, where tear faults related to the San Andreas plate boundary created tensional openings that resulted in extensive basaltic eruptions (McCulloh et al. 2002).

Further evidence to support faulting as a mechanism for accommodating upwelling magma comes from the Mopah Range of southeastern California (Hazlett 1990). There, detachment faulting occurred simultaneously with volcanism 16–23 million years ago. In this area of extreme crustal extension, volcanic vents and their feeder systems became aligned along the most intensively active normal faults in the upper plate of the regional detachment structure (Figure 4.5). The numerous alkalic basalt volcanoes of Europe (The Eifel volcanic field in western Germany, the Chaîne des Puys volcanoes of central France, the scoria cones of the Czech Republic and the La Garroxta volcanic field in northeastern Spain) were very active in Pleistocene time, producing monogenic cinder cones, lava flows and explosive maars whose locations appear to have been controlled by extensional tectonics. Although there have been no eruptions for the past 10,000 years or so in these countries, there are probably still pockets of magma beneath them, and no one should ignore the possibility of future volcanic activity in these areas.

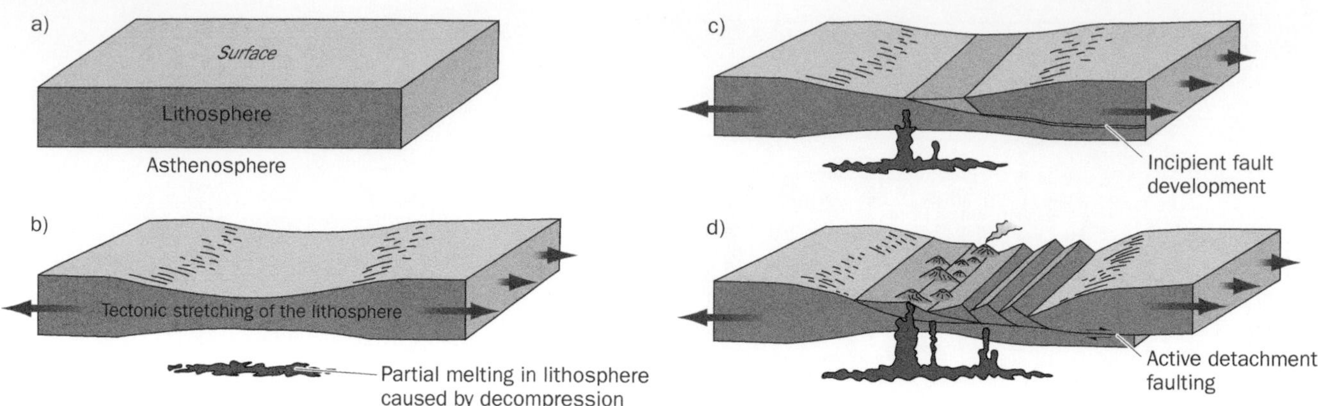

FIGURE 4.5 Schematic relationship between faulting and volcanic activity seen in the Mopah Range, California, and in other areas of rapidly extending continental crust.

"Frozen Magma" – Subvolcanic Intrusives

The relationships between plutonic rocks and their overlying volcanic rocks can be exceedingly complex. The great Sierra Nevada Batholith of California was apparently roofed in places by its own volcanic cover, yet this cover was itself intruded by later plutonic rocks and partly melted to form new igneous rocks, some of which solidified as plutons, and some of which probably reached the surface to form fresh, new volcanoes.

Steeply inclined or vertical pipe-like volcanic conduits and dikes (Chapter 1) commonly link magma reservoirs to overlying volcanoes (e.g. Takada 1988; Furman et al. 1992; Day 1993; Okubo and Martel 1998; Gudmundsson 2006). The dikes commonly radiate around conduits extending all the way up into volcanic edifices, or may form parallel to one another in long linear rift zones on volcano flanks. The large magma chamber beneath Platoro caldera (southern Colorado) remained active for over a million years after caldera collapse, and sent a dike swarm over 100 km to the south (Lipman and Zimmerer 2019). Individual dikes in this swarm only 1–2 m wide were as long as 20 km.

Where large magma chambers only a few kilometers deep inflate and lift their roofs, a pattern of inward-dipping concentric dikes, referred to as **cone sheets**, commonly develop (e.g. Ancochea et al. 2014). Magma drainage, faulting, and subsidence can generate another, almost vertically inclined set of concentric dikes enclosing the sinking roof block of the chamber, termed **ring dikes** (Johnson and Schmidt 2002; Browning and Gudmundson 2015). These special sorts of dikes are discussed further in Chapter 10. Volcanic eruptions may be fed through central conduits or wherever dikes intersect the surface (Figure 4.6).

Plugs are the solidified igneous rocks frozen within former volcanic conduits. They commonly resist erosion more effectively than the surrounding altered and fractured crust, causing the former conduits to rise as **volcanic necks**. Eventually, plugs become so viscous as magma cools and supply wanes that volcanic activity is no longer possible. Plugs and necks range in size from a few tens of meters to over a kilometer in diameter. Some necks rise as high as 500 m, and are usually steep-sided, often serving as pedestals for shrines, fortifications, and castles in Europe. Devils Tower (Figure 4.7) is a volcanic plug composed of a phonolite porphyry that intruded Mesozoic sedimentary rocks in southeastern Wyoming about 50 million years ago. The magma cooled very slowly, owing to the insulating nature of the enclosing rocks, forming columnar-jointed columns that average 2 m in diameter. Devil's Tower is considered sacred by neighboring Native Americans, who tell legends that the columnar striations were made by the claws

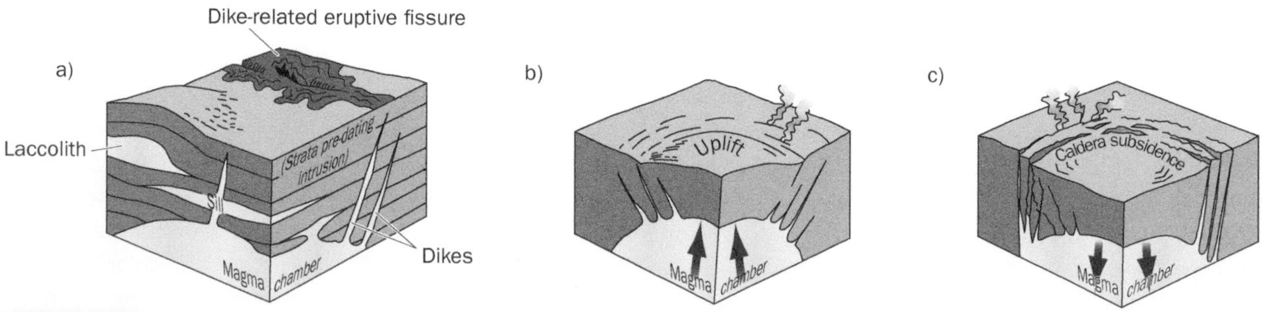

FIGURE 4.6 (a) Dikes, sills, and laccoliths are common intrusive structures associated with magma reservoirs; (b) Cone sheets can develop in the roofs of magma chambers accumulating in the shallow crust; (c) Ring dikes can develop in response to the withdrawal of magma from underlying chambers.

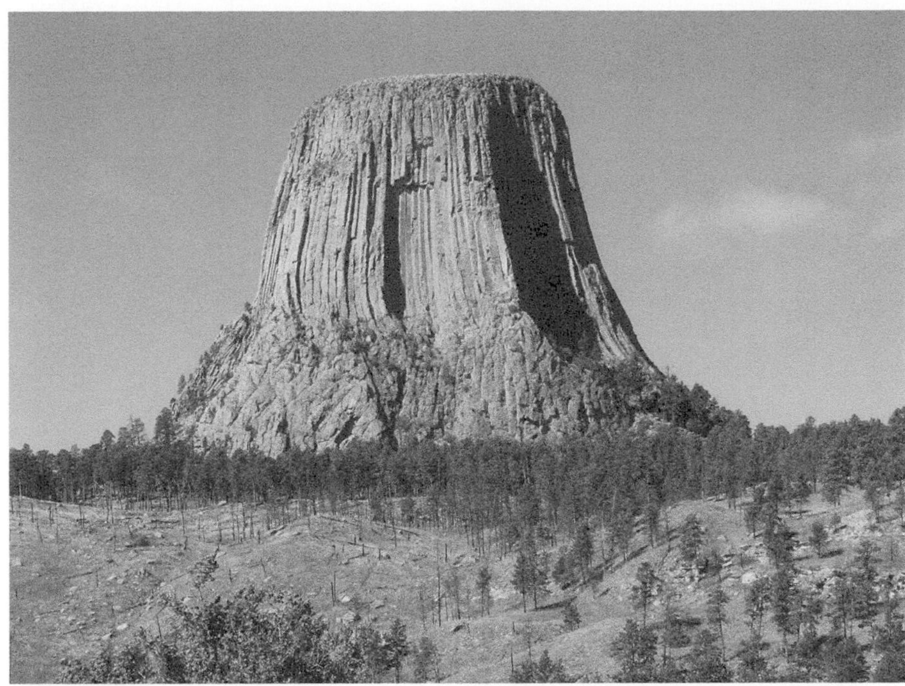

FIGURE 4.7 Devils Tower, southeast Wyoming. This 265-m-high volcanic neck comprises the remains of magma intruded into Mesozoic sedimentary rocks about 50 million years ago. The surrounding shales and sandstones eroded away long ago, but red sandstones of the Jurassic Sundance Formation enclose the base of the tower in the foreground. Source: Photo by J.P. Lockwood.

of a giant bear who was pursuing children up the tower. President Theodore Roosevelt made this the first U.S. National Monument in 1906 (River and Harris 1999).

Shiprock, New Mexico, is another spectacular volcanic neck that probably once did feed an explosive volcano and maar crater on a long-ago eroded-away surface (Figure 4.8). Dikes radiate in three principal directions from the base of this neck (Delaney and Pollard 1981). The dikes consist of hypabyssal *lamprophyre* (a dark, porphyritic volcanic rock), whereas the neck is mostly made up of tuff and breccia (Chapter 7). A volatile-rich magma quickly ascended from great depths to form this neck, fracturing bedrock around the conduit and filling those fractures with melt to produce massive dikes (Townsend et al. 2015). When the conduit-filling magma breached the surface, explosions ensued that apparently filled the pipe with a mixture of ash and rubble from both above and below. Fine veins of calcite crisscross the neck, evidence of late-stage carbon dioxide percolating through the breccia.

Explosive volcanic conduits such as Shiprock are called **diatremes**. They are distinguished from the conduits feeding shield volcanoes and probably most composite volcanoes, because of the predominance of explosive (generally alkaline) fill material. They are restricted to regions underlain at depth by Precambrian continental crust, and are of economic interest because some contain diamonds – an indication of their great source depths. Some, if not all, many represent the exposed conduits of surface maars (Chapters 1 and 7, Martin et al. 2007).

Hans Cloos (1941) published the pioneering work on diatremes, describing a group of 300 plugs and necks in Swabia, not far south of Stuttgart, Germany. These 15–18 million-year-old conduits, each a product of a single short-lived volcanic eruption, cut plateau-forming carbonaceous sedimentary rocks of Jurassic age, and are filled with a mixture of tuff, lapilli, finely ground sediment, and sedimentary blocks (Figure 4.9). The tuff consists of consolidated glassy ash. The blocks are of the same sedimentary rocks that make up the walls of the necks. These blocks are not arranged randomly, as they would be if they had fallen back into the conduit after being thrown into the air. Instead, many of them, particularly larger ones, are derived from the immediately adjacent wall rocks, which may be different from those a short distance above or below, and have simply moved outward and a little downward into the conduit. Flow structure is typically well developed in the tuff, with wedge-shaped fissures penetrating the conduit walls and the sedimentary blocks (White and Ross 2011). Many volcanic plugs show connections with underlying dikes. While some plugs are nearly circular in ground plan, others are oval, and in extreme cases they are gradational into fissure fillings; they become dikes at depth. Alignments of plugs are common, and strongly suggest their localization above deep-seated magmatic fissures. As magma in dikes approaches the surface, the unequal reduction of pressure along the crest of the dike encourages magma to focus its ascent at sites of least resistance, forming one or several narrow, conduits (e.g. Mastin and Pollard 1988). The conduits tend to become more pipe-like closer to the surface, because stresses in surrounding wall rocks approach uniformity at shallower depth.

FIGURE 4.8 View to west of Shiprock, a 500-m-high volcanic neck and associated dikes in northwestern New Mexico. The neck is a diatreme filled with welded tuff breccia that is more resistant than the surrounding soft sedimentary rocks, which have eroded away over the 25–30 million years since the diatreme was emplaced. Note the radially oriented, resistant dikes of minette lava that were emplaced during this eruption. Source: Photo by D.L. Baars.

Dike Features

Most dikes exposed by erosion extend as linear features for long distances, but some are broadly curved in plan, depending on the crustal stress field controlling their injection. In detail they commonly consist of numerous short fairly straight, parallel segments commonly arranged in *en echelon* patterns. The detailed courses of many are obviously governed by preexisting fractures. Real-time seismic and deformation monitoring of dike propagation show that dikes can open in hard crust very rapidly – migrating more than 20 km laterally in only two days during the 2018 Kīlauea eruption [17] (Lengliné et al. 2021).

Shallow dikes cutting pyroclastic debris tend to be regular and sharply bounded at deep levels within tephra cones, but may become very irregular or even feather out into the tephra at high levels, where confining stresses are low, and the material surrounding the intrusion is poorly consolidated (Figure 4.10). The loose cinder to either side of each intruding dike maintains its strength as a coherent solid simply through the interlocking of irregularly shaped fragments – a property called pseudocohesion (e.g. Sánchez et al. 2014). A spectacular example of a dike-fed fissure intersecting a small spatter cone was observed on the northeast rift zone of Mauna Loa on the Island of Hawai‘i in 1984. Where the dike intruded solid flows to either side of the cone, a distinct eruptive fissure opened at the surface. But where this same dike penetrated the loose spatter, a coherent fissure was unable to form, and a shallow graben developed instead.

Dikes are commonly bounded by selvages of glass, generally only a few mm or cm thick, resulting from quenching of magma against the cooler wall rocks. Some dikes are highly vesicular, but most are denser than the lava flows they may penetrate. Their density is partly due to escape of the gas bubbles before consolidation of slowly cooled intrusive magma, but more commonly dike density probably results from pressure of the overlying column of magma, in the same way that bubbles are prevented from forming in lava flows in deep water. Under lesser pressure at the tops of dikes, the escaping gas expands to form bubbles in the liquid, forming fragmented dikes and feeding lava fountains. Depending on magma viscosity, this process can begin at depths approaching a kilometer beneath a vent (Llewellin and Manga 2005).

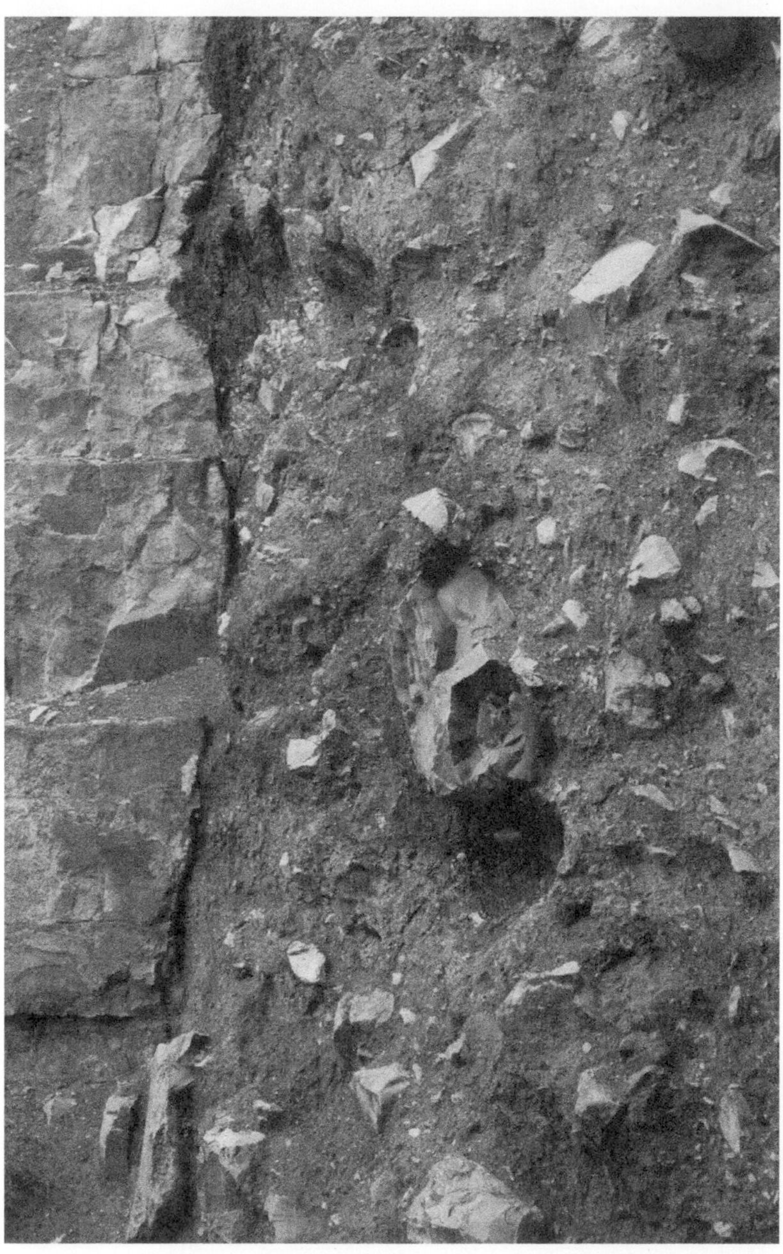

FIGURE 4.9 Detailed view of the tuffaceous breccia (right) comprising a diatreme intruded into solid limestone (left center) in Swabia, southwestern Germany. The height of the visible exposure here is about 3 m. Source: Photo by R.W. Hazlett.

Cooling of dikes by radiation and conduction of heat from their two approximately parallel boundary surfaces commonly results in well-developed columnar jointing approximately at right angles to the dike walls, and in the case of nearly vertical dikes, the exposed masses of columns may look surprisingly like great stacks of cordwood.

Although there can be no question that many dikes are the fillings of conduits that once fed surface eruptions, it is not often that one is exposed to reveal the connection with the lava flow it produced such as the outcrop shown in Figure 4.11. This is because lava usually drains down, or explodes out of a fissure during the final phase of eruptions, severing any connection with surface flows.

I (JPL) am probably one of the few people who have ever seen a dike forming at his feet – during a December 1974 summit eruption of the Kīlauea volcano, Hawai'i. Subsurface magma was radiating outward down the volcano's Southwest Rift Zone (SWRZ) (Figure 1.1), and surface fissures were spreading up and down the SWRZ. I was in line with the zone of propagating fissures that night, when a crack started to slowly open in the ash a few meters away, and a line of steam marked the new fissure. The steam would have been white in daylight, but was now colored a brilliant pink from the light of active fountains downrift. Within about 30 seconds, the red fumes turned darker, and a foolishly sampled whiff showed that the steam had turned to SO_2. The fissure rapidly opened to 10 cm or so, and the sulfur gases began to burn with blue flames. The roots of a small bush growing next to the crevice were being stretched taut across the widening crack. Suddenly, with a snap like a banjo string breaking, the root broke. I could hear a crackling, hissing, rumbling sound, and, out of curiosity, I straddled the rapidly opening crevice. I was amazed and fascinated to

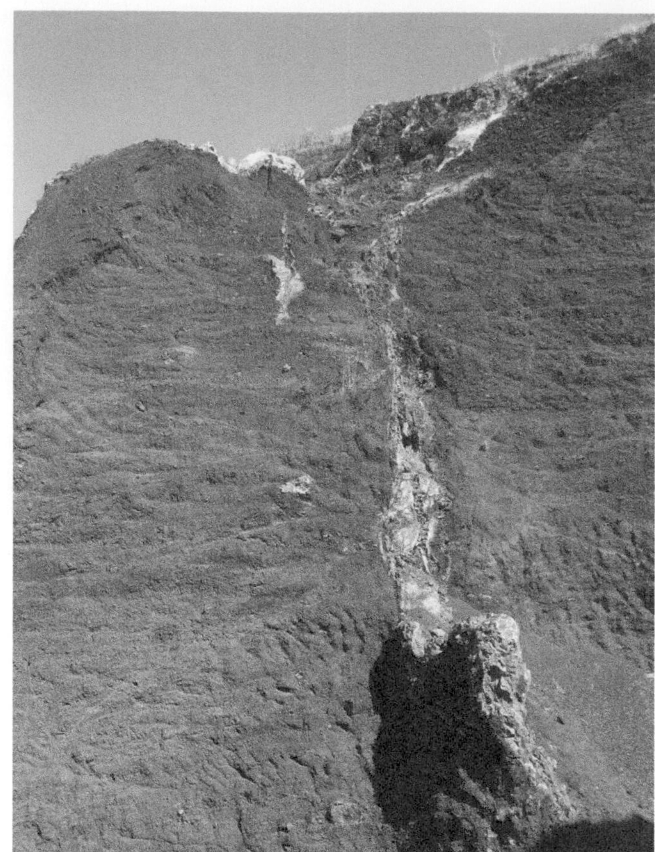

FIGURE 4.10 Basaltic feeder dike exposed in a cinder cone quarry near the village of Sauceda, 40 km south of Tequila volcano [50], Jalisco, Mexico. Note that in the latest stage of cone growth, molten lava and welded spatter coated the walls of the cone's summit crater. About 40 m of the dike is exposed in the quarry wall. Source: Photo by J.P. Lockwood.

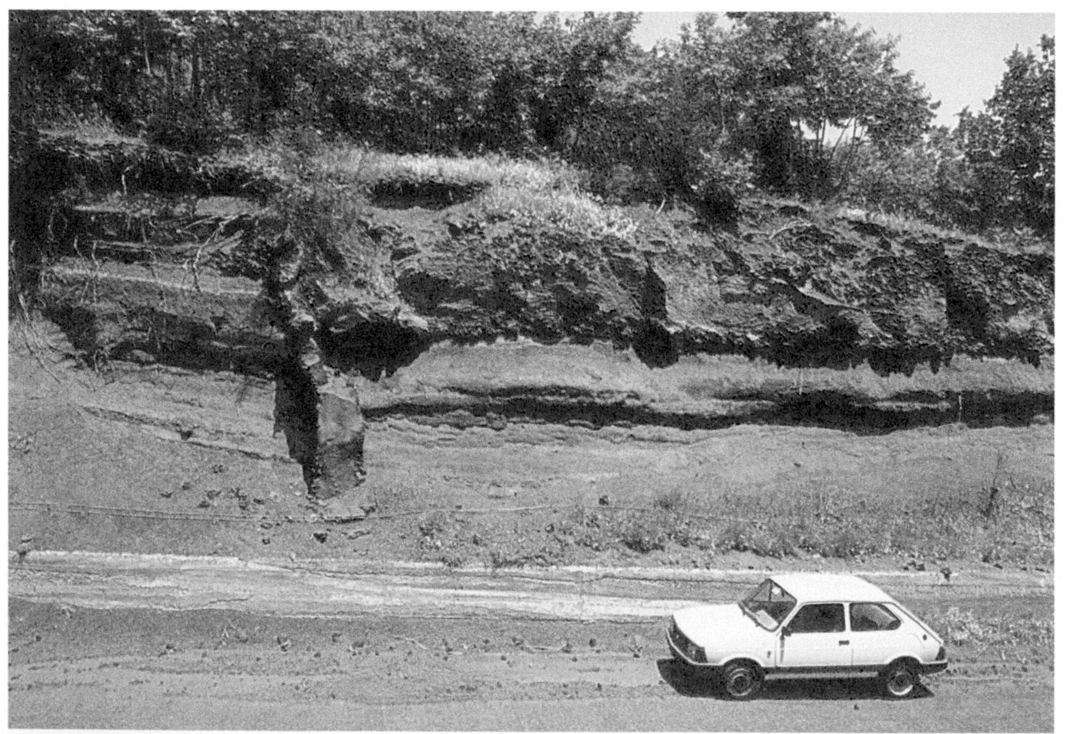

FIGURE 4.11 Cross-section of a Quaternary dike that supplied a fountain-fed `àā lava flow near Salto di Cani, north flank of Mount Etna, Sicily [112]. The dike intruded upper tertiary sediments along a fault that is downdropped to the east, and built a low spatter rampart on the upslope side of the erupting vent. Source: Photo by J.P. Lockwood.

see red lava slowly rising in the fissure about 10 m below me! I stood over the rapidly widening crack for only a few seconds – but what a thrill – watching the top of a lava dike ascending to the surface beneath my legs! The top of the dike began to spatter and eject blobs of lava, and I quickly moved away. Within a few minutes, lava fountains were shooting up from the dike to a height of more than 25 m!

Pseudodikes

Just to add a complexity, not all volcanic dikes form from magma intruding from below – some dikes in Hawai'i can be shown to have been fed from surface lavas that poured into open (or opening) fractures on the sides of volcanoes (Easton and Lockwood 1983). Vast quantities of lava have been observed to pour into open fractures during Hawaiian eruptions, and such "surface-fed" dikes may be much more common in volcanic terrains, at least those associated with fluid basaltic eruptions, than is generally recognized. We call these features **pseudodikes**. Elsewhere in the world, notably at the well-known Big Pumice Cut near Long Valley in eastern California, clastic sediment fills older fissures, creating sheet-like bodies resembling dikes from a distance, but packed full of material resembling ordinary stream gravel and sand.

Sills and Laccoliths

Less common than dikes, **sills** are also tabular intrusive bodies. They tend to be nearly horizontal in orientation, and instead of cutting across stratification as dikes typically do, they form parallel to, or intrude directly along the boundaries separating layered rocks of contrasting rigidity (Kavanagh et al. 2006). Such conditions prevail in layered roof rocks above shallow magma chambers or zones of intensive intrusion. A **laccolith** is a sill-like structure that is lens shaped with the overlying layering arched upward (Currier et al. 2017). It forms through repeated injections of melt along the same horizon (Menand 2007). Some researchers argue that sills and laccoliths can evolve into full-fledged magma chambers as they incorporate melt from sustained dike intrusions (e.g. Miller et al. 2005; Hawkins and Wiebe 2007; Gudmundsson 2012).

Like dikes, shallow sills may develop columnar joints, vertical in orientation in contrast to the horizontal columns of dikes. This can cause difficulty for geologists, for sills may be mistaken in cross-section for buried lava flows. Fortunately, a few useful criteria exist for telling them apart. For example, lava flows will not contain any inclusions of an overlying (younger) stratum. A sill, however, may dislodge fragments of an overlying stratum as it intrudes, incorporating them as xenoliths. Also, shallow sills may heat and thermally alter or mineralize wall-rock, both above and below. This, however, can only happen at the base of lava flows. In addition, many of the finer-scale structural features of lava flows (Chapter 6), such as flow joints, breccia at the tops and bottoms of `àā, caves and related pockets, zones of intensive vesicle formation, etc., are lacking in sills.

Triggers for Volcanic Eruptions – Why Volcanoes Erupt

It is becoming increasingly clear that the magma reservoirs underlying the world's active volcanoes are, for the most part, in a delicate state of gravitational equilibrium with enclosing rocks, and that many factors can contribute to their destabilization and to subsequent eruptive activity. Eruption "triggers" are mostly related to processes that increase the buoyancy and mobility of a magma body, and can be separated into three categories: processes taking place *below* magma chambers, processes *within* them, and processes taking place *above* them. The processes operating below and within a magma chamber unfold over long time periods, and typically set the stage for eruptive activity by transforming conditions, including melt temperature, volatile content, pressure, shape, and size. Such processes may control the eruption timing for volcanic systems that are already "primed." Major earthquakes may also trigger volcanic eruptions. Higgins (2009) suggests that a large eruption of the remote Tseax volcano in the Canadian Cascades might have been triggered by the M>8 megathrust earthquake of 1700.

Potential eruption triggering processes occurring *below* a magma reservoir are related to the rise of new material migrating from storage areas or magma generation sites in the mantle below, in most cases involving the injection of mafic melts into more silicic magma above. Summarizing the thinking of many researchers, Martin et al. (2007, pp. 89–90) comment:

> *Replenishment of a cool silicic reservoir with hot mafic magma may have a number or possible effects. The simple addition of a new volume of magma to a magma chamber has the potential to create an overpressure large enough to cause failure of the chamber walls and result in an eruption. Cooling of the mafic magma and heating of the Residual magma [more silicic melt above] may result in volatile exsolution which will also contribute to chamber overpressurization. Heating may also cause convective uprising and remobilzation of the resident magma, possibly resulting in eruption.*

As a case in point, Martin et al. (2007) suspect that intrusion of basaltic andesite melt into the bottom of the dacitic magma chamber of Santorini volcano, Greece, [114] triggered the onset of a three-year-long eruption cycle one month later (1925–1928). Intrusion of basaltic magma into more siliceous melt beneath Mount Pinatubo [135] set the stage for the 1991 eruption (Palister et also led to cyclical explosive eruptions at Makushin volcano [5] in Alaska during the early Holocene (Lerner et al. 2018).

Magma chamber overpressurization is an important process within magma chambers. It is the development of fluid and gas pressure within a magma chamber sufficient to fracture its walls allowing for a runaway feedback process culminating in eruption (Edmunds and Woods 2018). The feedback works like this: Upon initial fracturing of the chamber walls, loss of pressure causes additional volatiles to come out of solution (exsolve), which, because of the expansion of gases into the fractures, forces further fracture development and propagation, reduces pressure in the remaining melt, and causes volatile exsolution in a sustained, possibly even accelerating process. Fracture propagation is generally easiest in upward directions through the ceilings of magma chambers because it is here that the crust tends to be the weakest. An eruption occurs when – and if – these fractures reach the surface permitting trapped gases and entrained melt to escape.

The chemical and mineralogical differentiation of magma (Chapter 3) over time can serve as a significant "internally generated" trigger for volcanic eruption. As magmas cool, heavy minerals will form and sink toward the lower parts of magma chambers and lighter minerals together with exsolving volatiles will rise. In the uppermost part of the chamber, a lower-density, more buoyant, and silica-rich melt fraction will develop, potentially overpressurizing ceilings while seeking higher levels of neutral buoyancy. Other processes that may occur in magma chambers are related to the fragmentation of chamber walls and the sinking of dense blocks downward – which abet the processes of magma rise.

Potential triggers operating *above* magma chambers include a general weakening of roof crust because of hydrothermal and fumarolic activity, and short-term non-volcanic phenomena that can ever-so-slightly reduce confining pressure on the chambers or induce fracturing in their roofs. These "superficial" factors are diverse, and include the effects of earth and ocean tides, glacial melting, precipitation, and perhaps even variations in atmospheric pressure. Excessive precipitation has been shown to correlate with subsequent explosive activity of volcanic domes at Mount St Helens [30] (Mastin 1994), Piton de la Fournaise [125] (Violette et al. 2001), and at Soufrière Hills volcano [84] (Matthews et al. 2009).

Earth tides – elastic rise and fall of Earth's surface up to 40 cm caused by the gravitational attraction of the moon and sun – play a well-established role in triggering many short-term eruptions (Hamilton 1973; Mauk and Johnston 1973). Volcanic eruptions correlated with earth tides have been observed at volcanoes as diverse as Augustine [21], Fuego [60], Kīlauea [17], Mayon [137], and Stromboli [113]. Ocean tides have also been shown to affect the activity of submarine volcanic activity (Kasahara 2002). Earth tides are complex and have periodicities ranging from 12 hours to 19 years. The most important tides that have been shown to trigger volcanic eruptions are the **semidiurnal** tides that occur about every 12 hours, and the **fortnightly tides** that occur about every two weeks (when the moon is either full or in "new moon" status – times when the sun and moon are aligned with Earth). Some volcanoes show heightened eruptive activity at both tidal minima and tidal maxima (e.g. Fuego [61]; Martin and Rose 1981), some will respond to minima only, others are more sensitive during tidal maxima, while many others show no tidal responsiveness at all. No universal pattern exists, though correlations with fortnightly tides – when tidal stresses are greatest – are more common than semidiurnal ones.

FIGURE 4.12 Volcanoes have a higher probability of erupting when earth tidal stresses are at their maxima. This fanciful painting by Gong Futang shows a volcano goddess bursting from her volcano home when the moon is full! Source: J.P. Lockwood private collection.

Dzurisin (1980) demonstrated that Kīlauea preferentially erupts at times of fortnightly tidal maxima. As a result, Hawaiian Volcano Observatory (HVO) staff tend to be more alert when the level of summit inflation is high (Chapter 14) and the moon is full (Figure 4.12) In contrast, eruptions at Kīlauea's close neighbor, Mauna Loa [15] show no tidal correlation, perhaps because its magma reservoir is deeper. The three-week eruption of Gamalama volcano [140] (Mollucas island arc, Indonesia) in 1980 consisted of 34 individual Vulcanian eruptive explosions that devastated the central portions of Ternate Island and caused most of the island's population to evacuate for safety. A subsequent investigation (unpublished) showed that these well-timed eruptions corresponded well to semidiurnal tidal minima (Figure 4.13). The effect of tidal stresses on subvolcanic magma reservoirs probably depends in great part on the shapes and structural features of those bodies (vertically extensive reservoirs will react differently compared with horizontally extensive ones) and to preexisting regional stress fields. Given

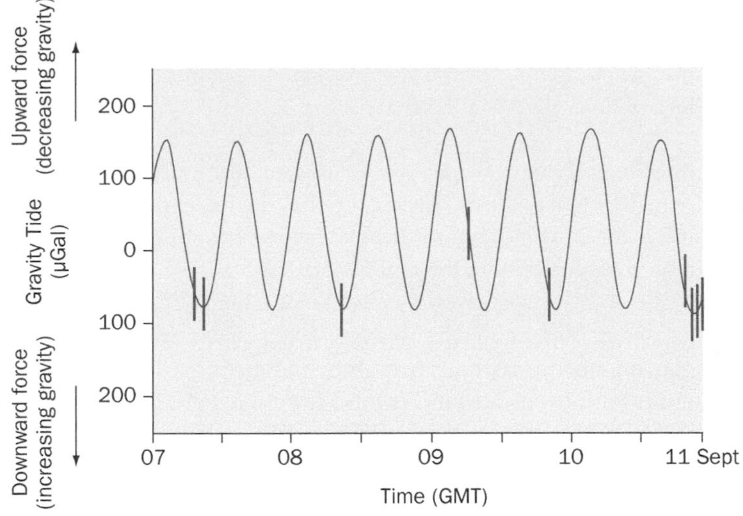

FIGURE 4.13 Semidiurnal earth tides for the period 7–11 September, 1980 and associated explosive eruptions at Gamalama volcano [140], Ternate Island, Indonesia. Eight of the nine eruptions during this time period occurred at or near tidal minima, which were the times of greatest compressive stress affecting the underlying magma chamber.

all the possibilities, and the fact that we cannot be quite certain how magma reservoirs are configured beneath most volcanoes, it would certainly be useful for volcanologists to bring along earth tidal plots or software to generate them when responding to eruptive episodes – just to see if patterns emerge that could be useful for forecasting future activity at those particular volcanoes (Chapter 14).

The reduction of overlying pressures on volcanic magma systems owing to lowered glacial loads have been shown to trigger eruptions in Iceland, the Andes and elsewhere (e.g. Lockwood and Lipman 1987; Mora and Tassara 2018). How small a pressure change can provide the critical "trigger" to initiate explosive decompression of a magma reservoir – how small a straw can "break the camel's back?" A hint comes from the timing of the climactic eruption of Mount Pinatubo [135] on 15 June, 1991 (Chapter 1). It was perhaps no coincidence that a major cyclone – Typhoon Yunya – passed within 75 km of Mount Pinatubo that morning, and that, at 11:00 a.m., the greatest pressure drop associated with the passage of the cyclone (6.3 mbar) was recorded at Clark Air Force Base (Oswalt et al. 1996). The nine-hour-long paroxysmal eruption of the volcano began less than three hours later, at 1:42 p.m. (Wolfe and Hobblitt 1996). Indications that slight changes in barometric pressure can affect the timing of eruptions has also been observed at Oldoinyo Lengai [120] (Fred Belton, personal communication 2007).

Of course, the passage of Typhoon Yunya did not "cause" the Mount Pinatubo eruption, but this may have provided the critical trigger that initiated an eruption that would have occurred anyway. Pallister et al. (1996) have used petrologic research to demonstrate that the primary factor that set the stage for this eruption was the apparent intrusion of basaltic magma into the base of a silicic magma chamber beneath the volcano, which began almost three months before the June eruption.

Repose Intervals

The **repose interval** of a volcano is the time that elapsed between eruptions. No two volcanoes are alike in this regard. Most have irregular repose intervals, but some have repetitious behavior on roughly regular schedules. Repose intervals may range from days to centuries. Jellenik and DePaolo (2003) postulate that the minimum time interval (T_{min}) between successive eruptions of a volcano is proportional to the volume of the magma chamber:

$$T_{min} = \Delta V_{min} / Q_{min} \qquad (4.4)$$

with ΔV_{min} the minimum increase in magma chamber volume required to pressurize it to the point of erupting, and Q_{mean} the chamber's average magma replenishment rate. This equation suggests that a volcano having a constant magma recharge rate will erupt at a regular interval, like Old Faithful Geyser; or if not regularly, then with the sizes of its volcanic eruptions increasing in direct proportion to the repose intervals between them. Smith (1979) pointed out that there is indeed a correlation between the energy released by eruptions and the repose times between the eruptions of many explosive volcanoes. The hiatus since the volcano's last eruption is a measure of the size of the next eruption to follow. Smith developed the following correlation law to show this trend:

$$\log \text{ repose interval}(\text{years}) = \log \text{ magma volume}(\text{km}^3) + 3 \qquad (4.5)$$

This relation is empirical, and was modified by Trial and Spera (1990) using a more recent data set for eruptions involving release of 1–100 km^3 of magma:

$$\log \text{repose interval}\left(\text{years}\right) = \log \text{magma volume}\left(\text{km}^3\right) + 2.5\left(\pm 0.5\right) \tag{4.6}$$

The repose intervals between extremely violent eruptions may be so long that persons living around a potentially active volcano may forget past eruptions, or lack a historical record of them. Newhall and Self (1982) note that the mean repose interval prior to 25 of the world's most violent, documented eruptions is 865 years. Looking at 21 of these most powerful blasts, they warn that 17 of the eruptions were the *first reported volcanic activity* at the volcanoes where they took place.

The repose interval equations are quantitative expressions of the critical role played by consistent magma chamber supply in triggering volcanic eruptions. But they only really pertain to volcanoes that are in what Wadge (1982) calls a **steady state of activity**; that is producing lava and pyroclastic material at an approximately continuous rate through time, be it through many short, small eruptions, a few large ones, or a mix of both. In fact, however, there are also many volcanoes that do not show steady-state behaviors (Lipman 1980), implying that their magma supply rates vary over time. Some, such as Kīlauea and arguably Vesuvius, show mixed steady-state and non-steady-state behavior in their histories. No one yet knows how typical steady-state behavior is for volcanoes. Chapter 14 explores this topic further from the standpoint of volcanic risk assessment.

Questions for Thought, Study, and Discussion

1. The intrusion of a high-temperature, volatile-poor mafic melt into the base of a low-temperature, volatile-rich silicic magma reservoir can trigger a highly explosive eruption, as in the case of the Bishop Tuff. Consider however what would happen if the reverse situation took place, with a more silica-rich melt introduced into the base of hotter mafic reservoir. (Evidence for this indeed exists in some historical cases.) Construct and explain a scenario.
2. Explain how the rheological behavior and viscosity of a lava change as it cools and solidifies.
3. Why do you suppose that slowly rising dikes begin to develop pipe-like conduits only when they get close to Earth's surface and not deeper down?
4. Dikes (and sills too for that matter) are planar bodies that tend to form perpendicular to directions of extension. How can this be demonstrated using a stick of chalk?
5. Why should a sill or laccolith not easily form where stronger, more rigid layers *underlie* a weaker, less rigid layer?
6. What are some non-volcanic analogs to magma overpressurization?
7. How do you explain the fact that many volcanoes erupt repeatedly through the same conduit rather than form new ones with each new eruption?
8. How would you interpret the historical record of a volcano that shows non-steady-state behavior for part of its existence, and steady-state behavior for the rest?
9. Why is the repose interval concept important for assessing volcanic hazards?
10. If you were going to try to forecast the readiness of a volcano to erupt, what factors and conditions related to the volcano would you like to know about and why?

FURTHER READING

Annen, C. and Zellmer, G.F. (2008) Dynamics of crustal magma transfer, storage, and differentiation. *Geological Society of London Special Publication 304*, 288 p.

Breitkreuz, C. and Petford, N. (2004) Physical geology of high-level magmatic systems. *Geological Society of London Special Publication 234*, 253 p.

Browne, B. and Saramek, L. (2018) Rates of magma ascent and storage (Chapter 9). In: *The Encyclopedia of Volcanoes* (ed. H. Sigurdsson, B. Houghton, and H. Rymer), 203–214. Cambridge, Massachusetts: Academic Press, Elsevier Publishing.

Duba, A.G., Durham, W.B., Handin, J.W. et al. (1990) The brittle-ductile transition in rocks: The heard volume. *American Geophysical Union Geophysical Monograph Series*, 56, 243 p.

Fagents, S., Gregg, T.K., and Lopes, M.C. (eds.) (2013) *Modeling Volcanic Processes: The Physics and Mathematics of Volcanism*, 431 p. Cambridge, UK: Cambridge University Press.

Faybishenko, B., Witherspoon, P.A., and Benson, S.M. (2000) Dynamics of fluids in fractured rocks. *American Geophysical Union Monograph Series*, 112, 400 p.

Masotta, M., Beier, C., and Mollo, S. (eds.) (2021) Crustal magmatic system evolution: Anatomy, architecture, and physico-chemical processes. *American Geophysical Union*, 264, 256 p.

Vetere, Francesco (ed.) (2020) Dynamic magma evolution. *American Geophysical Union*, 254, 208 p.

PART III

VOLCANIC ERUPTIONS AND THEIR PRODUCTS

This is the largest part of the book, comprising five chapters that are important for understanding volcanic eruptions and the many varieties of rocks and ash deposits that volcanoes produce. **Chapter 5** describes some of the classification systems that are used to describe and categorize different kinds of eruptions. **Chapter 6** discusses eruptive processes and volcanic products of the red volcanoes – those characterized mostly by lava flow activity. **Chapter 7** discusses the eruptive processes and volcanic products of gray volcanoes – those characterized by explosive activity that commonly blankets large areas of surrounding land with thick (and sometimes deadly) deposits of gray volcanic ash. **Chapter 8** discusses the eruptive mechanics of the largest explosive eruptions – including the very infrequent "Super Eruptions" that have the power to alter Earth's climate and human history.

Volcanoes: Global Perspectives, Second Edition. John P. Lockwood, Richard W. Hazlett, and Servando De La Cruz-Reyna.

Companion website: www.wiley.com/go/lockwood/volcanoes2

CHAPTER 5

Classifying Volcanic Eruptions

Volcanoes: Global Perspectives, Second Edition. John P. Lockwood, Richard W. Hazlett, and Servando De La Cruz-Reyna.

Companion website: www.wiley.com/go/lockwood/volcanoes2

To be useful, a classification must be as simple as possible; above all, it must be usable in the field.

(C.F. Park and R.A. MacDiarmid 1964)

One of the great achievements of the Age of Enlightenment has been the effort to categorize what we see in the natural world as a basis for better understanding. Many terms and concepts have been proposed that have not survived the test of time, and some older terms that *did* survive ought not to have because of the confusion and complexity they add to the present. This is especially true in a still-developing field such as volcanology. It is important to recognize that the most pressing need of the first volcanologists was to invent a language for communicating observations about eruptive behavior, both between themselves and to the authorities and public at large. Consistant classification is a question of public safety as well as scientific understanding (Chapter 14).

To complicate matters, the very term **eruption** is fraught with semantic ambiguity. For example, many volcanoes burst to life by generating a series of explosions, each separated by a few minutes of quiescence. But what if the time breaks are, say, several weeks, or several years in length? Are we still talking about the same eruption, or a series of separate ones? The Smithsonian Institution's Global Volcanism Program has suggested that if a *three-month time interval* has elapsed since the end of a particular eruption that eruption has certainly ended. While seemingly arbitrary, this seems like a reasonable time lapse given our record of how eruptions simmer down and come to an end at volcanoes worldwide.

Given this definition, eruptive activity that occurs after shorter periods of quiescence are termed **phases** (or **episodes**) of the same eruption. The term **phase** also refers to a *distinctive period of eruptive behavior* within a continuous eruption. Every volcanic eruption demonstrates varying behavior over its duration. Some eruptions may change from dominantly explosive phases to dominantly effusive activity as they progress, and all eruptions vary in phases of eruptive output – usually from initially high rates to zero as activity ends.

Long-lasting effusive eruptions of red volcanoes commonly have three phases, beginning with low **lava fountaining** along long linear fissures, later followed by higher **point-source fountaining** at restricted vent areas, then in turn by the sustained production of lava flows with or without low fountains. Explosive eruptions of gray volcanoes often begin with violent vent-clearing discharges of older rock, followed by sometimes cataclysmic eruptions of pumice and ash, commonly culminating with the rather sluggish growth of lava domes – again, three phases of activity. The transitions from phase to phase relate to critical thresholds in magma over-pressure and release of volatiles to which we will allude in the chapters to follow. In some cases, the phases of a single eruption may consist of cycles of changing, repeated behavior. An example is the East Rift Zone outbreak of Kilauea [17], which began in January 1983. During the first few years of this activity, each phase of the seemingly endless eruption began with sudden high fountaining and degassing of lava at the Pu'u'Ō'ō vent, which lasted just for a few days, then simmered down into quiet sloshing of molten rock in a crater vent that continued for several weeks.

On much longer time scales, eruptive patterns, often separated by years or even centuries, may be repeated, even with startling regularity. Such **eruption cycles**, for example, are one of the distinguishing features of Vesuvius [109], with extremely violent eruptions of similar character occurring 25,000, 22,000, 17,000, 15,000, 11,000, 8000, 3800, and 2000 years ago; a frequency of one violent eruption every 2000–5000 years. The most recent major outbreak is infamous for destroying Pompeii and Herculaneum in 79 CE. With this record, it seems inevitable that Vesuvius will erupt in the same way sometime in the next few millennia. More recently at Vesuvius, a cycle of less devastating eruptions – each consisting of a persistently mild, decades-long effusive phase culminating in a violently explosive phase – took place between 1631 and 1944. Individual eruptions in this sequence were separated by quiescent intervals lasting no more than 7 years. The regularity and rough predictability of this pattern inspired the establishment of the world's first volcano observatory (Chapter 14). The 1944 outburst apparently ended the three-century long period of rhythmic activity, and the mountain has remained ominously quiescent ever since.

How do we characterize volcanic eruptions? Many eruption characterization systems have been proposed over the past century, based on eruptive style, volume and composition of erupted products, and energetics. We here mention only some of those we consider most important.

Lacroix Classification System

The notorious volcano Vesuvius has the longest record of direct volcano observation, and it is here that some of the first scientific studies of eruptions were undertaken (Chapter 1). But there are hundreds of other potentially threatening volcanoes in the world, and early volcanologists soon recognized that the behavior of Vesuvius differed greatly from many, if not most, of them. They were faced with the challenge of developing a common language for describing volcanic eruptions that was *international* in scope. Professor Giuseppe Mercalli (1850–1914), also well-known for his work on earthquakes, started with his study of two famously active Italian volcanoes, Stromboli ("Type 1") and Vulcano ("Type 2"). He distinguished between effusive (molten lava)

TABLE 5.1 Lacroix system of classifying eruptive behavior (revised).

Class	Description
Hawaiian	Effusive eruptions of lava with little or no explosive activity apart from lava fountaining. Primarily basaltic. Originate at fissure vents. Associated with building of shield volcanoes and flood-basalt plains. VEI* = 0–2 (and higher for flood-basalt eruptions).
Strombolian	Moderately explosive eruptions producing cinder, bombs, and ash, which are initially incandescent as they leave the vent. Blasts often periodic, associated with bursting of very large gas bubbles in the vent. Typically basaltic and andesitic. Includes violent Strombolian behavior, such as the Vesuvius eruptions of 1906, which devastated Naples and killed over 100 people. VEI* = 1–3.
Vulcanian	Moderate to violent ejection of recently solidified lava. Associated with the clearing of conduits, often plugged with domes. In addition to angular blocks, large amounts of ash are produced. Eruption columns feature much lightning, and accretionary lapilli may accompany ash fall. Dense, ground-hugging pyroclastic clouds are possible with transition to Peléean behavior. If the vent-clearing eruption simply involves ejection of older rock fragments and lots of steam, it is said to be *Ultravulcanian*. Any composition where extrusion of viscous magma is involved. VEI* = 2–4. Named after typical behavior of the volcano Vulcano [105].
Peléean	Varied explosive activity including the formation and destruction of domes, spines, and the generation of destructive pyroclastic flows - termed *nuees ardentes* by Lacroix after the 1902 eruption of the namesake volcano, Mt. Pelée [87]. May include generation of sub-Plinian ash columns. Typically andesitic to dacitic compositions. VEI* = 3–5.
Mixed	We no longer recognize this too-broad class, which has not survived the test of time owing to lack of descriptive specificity.

Ranking is from least explosive to most explosive. *VEI – Volcanic Explosivity Index, described later in the text.

and explosive eruptions (ash and other pyroclastic products), and noted that "Mixed" eruptions showing characteristics of both can also take place (1907). Soon after, the French mineralogist Alfred Lacroix (1863–1948; Chapter 1), discussing matters with his Italian colleagues, proposed the more detailed classification of volcanic behavior that is still widely used today (Table 5.1; Lacroix 1908, pp. 74–81). He designated five kinds of eruptions, giving four of them the names of the volcanoes at which they often occur: **Hawaiian**, **Strombolian**, **Peléean**, **Vulcanian**, and "**Mixed**"- taking into consideration their relative effusiveness or explosiveness, and their characteristic types of ejecta, vents, and patterns of eruption. Lacroix correctly surmised that the principal cause of eruption variation was the viscosity of the magma involved. This was a start, and fortunately a good one, reflected by the fact that much of our contemporary approach to classifying types of volcanic activity derives directly from Lacroix's original work. But objections have sometimes been raised to these names on the basis that no volcano always exhibits the same sort of activity, although Lacroix himself emphasized that his system pertained to eruptive behavior, and not to individual volcanoes. In fact, two or more different types of eruptive behavior may develop simultaneously at different vents on the same volcano. For instance, vents at the summit of Stromboli have exhibited Strombolian, Hawaiian, and Vulcanian behaviors *simultaneously*. Most volcanoes do have characteristic eruption behaviors, but they may depart from that habit to varying degrees from time to time or as eruptive activity evolves.

Lacroix's classification system has by necessity also evolved, as other eruptive behaviors have been observed in parts of the world that Lacroix had never witnessed. These include "Plinian" and "Surtseyan" eruptions. To make matters more complex, the authors of many different textbooks and scientific articles have each proposed their own spin on the Lacroix classification scheme, because its qualitative nature allows some wiggle-room for interpretation and embellishment. In his book *Volcanoes*, Gordon Macdonald (1972) proposed 10 different eruption types! Classifications of eruptive behavior must remain qualitative and not rigid, as there is a spectrum of behaviors, with no rigid boundaries between them. Table 5.2 presents our suggestions for expanding the five types of eruptions that Lacroix described, although we realize intermediate behaviors exist, and terminology reflecting these transitional categories can always be used to refine descriptions.

Rittmann Diagrams

Following Lacroix, pioneering volcanologist Alfred Rittmann (1962) proposed a useful means of diagrammatically portraying the variations of behavior during individual eruptions through use of schematic figures he called "Eruption Diagrams" (Figure 5.1). He intended such diagrams to be used semiquantitatively, plotting readily observable eruptive parameters on vertical axes (such as eruption column height for explosive eruptions, or lava production rates for effusive eruptions) versus time on the horizontal axis. Such figures have proven especially useful for characterizing long-lived, multi-phase eruptions.

TABLE 5.2 **Additional eruption types not described by Lacroix. See Chapter 7 for more detailed descriptions.**

Class	Description
Phreatomagmatic	Violently explosive eruptions when rising magma encounters large amounts of water. Such activity can cause the formation of maars (Chapter 10), and characterize intrusive activity beneath volcanic crater lakes. These are some of the deadliest eruptions for volcanologists because activity may be unpredictable.
Surtseyan	A class of phreatomagmatic eruptions taking place in shallow water, especially along island coastlines. Often rhythmic, like Strombolian eruptions, with largely white eruption clouds of dense steam. In some eruptions, *cypressoid behavior* is exhibited, in which short, sharp blasts create jagged black "roostertail" ash-rich clouds shooting from the main steam-rich eruption clouds or directly from the water (Chapter 7).
Plinian	The most violent, catastrophic eruptions that can occur on Earth, with major global climate impact. Usually associated with the explosion of large andesitic to rhyolitic magma chambers and caldera formation. Eruption columns 20–55 km high penetrate the stratosphere, injecting large quantities of water and sulfur aerosols into the upper atmosphere. Airfall ash may blanket vast areas and voluminous pyroclastic flow deposits bury thousands of square kilometers surrounding new calderas (Chapter 8). VEI* = 4–8. Similar, but less intense explosive eruptions that may occur independently or follow major Plinian activity for months are termed *sub-Plinian*.

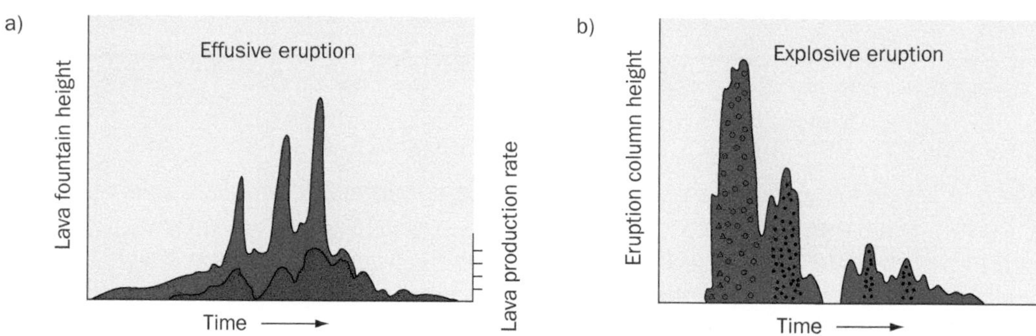

FIGURE 5.1 Schematic Rittmann "eruption diagrams" can be very useful for depicting variability of eruption styles during the course of eruptive activity, as shown in these hypothetical examples. (a) For an effusive eruption, the variations in the height of lava fountains (light red) and lava effusion rates (dark red) are plotted against time. (b) For an explosive eruption column height, variations are plotted against time. Episodes of vent clearing are indicated by triangles, pumice-rich pyroclastic flows are indicated by circles, ash eruption is shown by dots, and clear areas depict periods of fuming without ash emission.

Geze Classification Diagram

Geze (1964), provided a simplified version of the Lacroix scheme that remains broadly useful, classifying eruptions using a ternary diagram (Figure 5.2) in which the major magmatic products involved in eruptions, be they liquid, solid, or gas, are indicated. It is sometimes best to introduce students to volcanoes using the Geze perspective first, because it lends itself well to more in-depth development as the subject matter unfolds.

Walker Classification System

The Lacroix, Rittmann, and Geze schemes are most useful for characterizing eruptions in progress, but the field mapping and interpretation of *eruption deposits* require different criteria. Field geologists over a period of decades began to realize that kinds of volcanic activity not recognized by Lacroix must have left the deposits that they were describing.

To address these problems of nomenclature, and to add quantitative rigor to the system of classifying eruptions, Walker (1973) proposed a new classification scheme based upon factors that could be measured directly in the field. His work was in part derived from an earlier attempt by Tsuya and Morimoto (1963) to create an explosive magnitude index for volcanic eruptions, rather along the lines of the highly successful Richter magnitude scale already in place to study earthquakes. Walker's proposal was distinctly advantageous in that it could be applied to prehistoric or unobserved eruptions. Walker (1973) noted "Fewer than 10% of the explosive eruptions of the present century have been reasonably documented scientifically, and few volcanologists have the opportunity to observe more than 3 or 4 large explosive eruptions in their lifetime." He reasoned that the dispersal area (D) and degree of fragmentation (F) of air fall ejecta from any given eruption could be measured and correlated with at least some of Lacroix's traditional eruption types (Figure 5.3).

To measure air fall dispersal, Walker proposed isopach mapping. **Isopach maps** resemble ordinary elevation contour maps, but instead of representing equal-elevation positions, the isopachs represent positions of *equal deposit thickness.* For simplification, he proposed that only a single isopach be mapped for any given eruption, having 1/100 the thickness of the *maximum* air fall deposit thickness for that eruption – the "0.01 T_{max} isopach," as he put it. In other words, if an eruption has dumped a layer of ash that *at most* is 4 m thick, the T_{max} isopach will be designated on a map where ash deposits thin to 4 cm thickness (Figure 5.4). Outside this area, the deposit thickness would be less than 4 cm, thinning out to nothing. After delineating this isopach, the area enclosed by it (D) would then have to be measured. Extrapolation (surveying by "eye-balling") between eroded outcrops could be necessary, and for deposits older than a few tens of thousands of years, accurately determining fall dispersals could well be impossible. Walker was also careful not to count fall deposits associated directly with the construction of the cones and rims around volcanic vents.

To measure tephra clast sizes, Walker proposed that materials be sampled from a position along the 0.01 T_{max} isopach farthest from the vent (Figure 5.4), then sieved to determine the weight percentage of fragments having a diameter of less than 1 mm. This would constitute the **fragmentation index** (F) of the sample. Walker (1973) also proposed that F could be used to indicate the degree to which volatiles were involved in the eruption that produced the deposit. The greater the value of F, the more explosive and powerful the eruption, owing to the tremendous force of expansion as water transforms into steam. (Bear in mind that *power* is a concentration of energy – energy released as a function of time.) Lines 2a and 2b in Figure 5.3 represent different eruptive styles reflecting this variable.

By comparing the F and D parameters, Walker proposed several new eruption types to add to the Lacroix scheme (Table 5.2). There were certainly good reasons for doing so, based upon field observations. For example, Strombolian eruptions predominantly produce cinder, with F values less than 10 weight percent of total ejecta. Macdonald (1972) and Walker referred to these as "normal" Strombolian eruptions. But in some instances, the emissions of ash and other fine particles less than 1 mm in diameter are considerably greater, despite the fact that eruptive characteristics remain broadly similar. For example, at Paricutín [53], a 400-m-high cinder cone grew in Mexico in 1943–1952 (Chapter 14), and 70 percent of samples contain greater than 50 percent material finer than 1 mm *within* the 0.01 T_{max} isopach, where, if anything, one would expect to find generally coarser materials, because of proximity to the vent. Because of this, Macdonald (1972) referred to Paricutín as a "violent" Strombolian eruption, and Walker adopted this term. Observers report that there was indeed a larger amount of groundwater involved in the Paricutín activity than in shorter-lived "normal" Strombolian eruptions (e.g. Segerstrom 1950), seemingly vindicating Walker's assertion about the role of volatiles in fragmentation trends.

Surtseyan (phreatomagmatic) eruptions (Chapter 7) develop when magma invades shallow bodies of water, and may transition into Strombolian and Vulcanian eruptions as water is excluded from magma contact. The difference is not simply one of

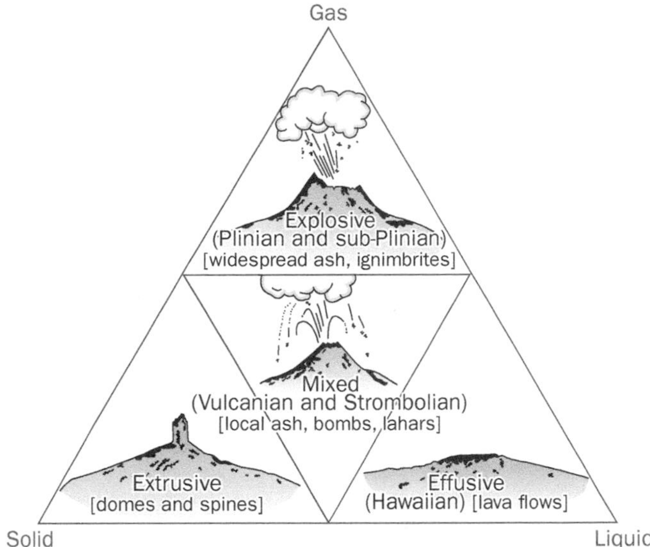

FIGURE 5.2 A simple classification of major eruption types, based on the proportion of gases, solids, and liquids erupted. Peléean eruptions would be intermediate between "Mixed" and "Plinian" types if plotted on this figure. Source: Modified from Geze (1964).

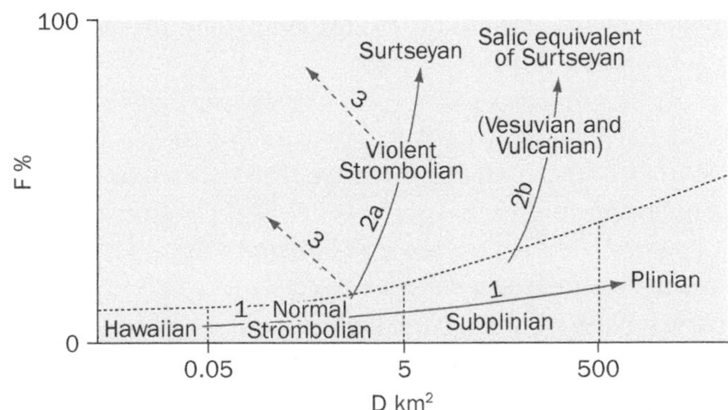

FIGURE 5.3 The Walker eruption classification diagram. The F axis is the percentage of ejecta in an air fall deposit with a grain size of less than 1 mm, as measured at the 0.01 T_{max} isopach (see text). The D axis is the area enclosed within the 0.01 T_{max} isopach for a given eruption air fall. Trend line 2a represents increasing explosive power in basaltic eruptions, and 2b the same for eruptions involving highly viscous, silica-rich magmas. Trend line 3 indicates weak eruptions that can only throw fine ejecta beyond their crater rims. Source: Walker (1973). Reproduced with permission of Springer Nature.

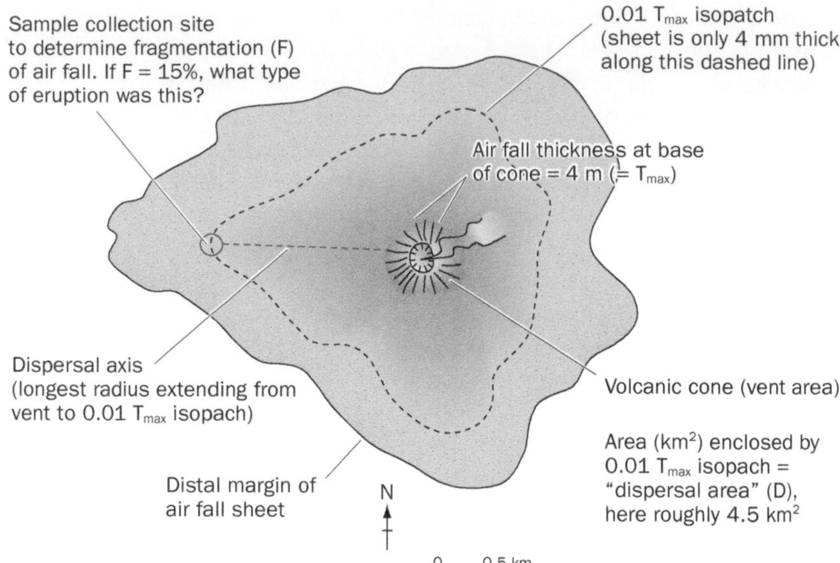

FIGURE 5.4 Schematic map of an imaginary volcano showing tephra information to be noted to apply the Walker (1973) classification system.

fragment size, but also of the shapes of fragments, especially ash particles. Walker (1973, p. 437) notes: "[Ash] particles have shapes related to their origins. In Strombolian/Hawaiian deposits they are often ragged and in part bound by smooth or rounded surfaces molded by surface tension. In Surtseyan, they are bounded by fracture surfaces and the inner walls of broken vesicles." They are steam-shattered, in other words.

Walker and Croasdale (1972) further proposed that an intermediate type of eruption be designated between normal Strombolian and Plinian activity. Walker cited as an example of this "new," intermediate eruption type, the Hekla [92] basaltic explosions of 1970 in Iceland, in which ash fell over 200 km from the vent fissure. As far as 18 km from the volcano, fall beds as much as 6 cm thick contained no more than 4 percent by weight fragments less than 1 mm in diameter. The deposit showed the distinctive scoriaceous (highly vesicular) appearance of a mammoth cinder cone eruption. This was a "super" Strombolian event in other words, or as Walker and Croasdale preferred to call it, given the unusually wide dispersal area of its ash fall, a **sub-Plinian eruption** (Figure 5.5, Table 5.2). Many, though certainly not all Vulcanian eruptions, have dispersal patterns in the range of sub-Plinian deposits. They differ in terms of the degree of fragmentation, with Vulcanian ejecta consisting mostly of broken, highly angular rock fragments, and sub-Plinian ejecta consisting mostly of broken bits of vesicular glass. Walker proposed that the involvement of steam, as in the case of violent Strombolian eruptions, also was important in explaining this difference in particle shape. With even greater amounts of water, he postulated a "salic [silicic] equivalent of Surtseyan" style of eruption – an extremely powerful phreatomagmatic type eruption for which he could not provide a specific example at the time. Self and Sparks (1978), however, provided just such a description a few years later, based upon careful study of several deposits including the 26,000-year-old Oruanui Tuff [181] in New Zealand and the 1870 eruption of Askja [102] in Iceland. Self and Sparks proposed replacing "salic equivalent of Surtseyan" with the term **Phreato-Plinian**. We shall consider Phreato-Plinian eruptions in much greater detail later (Chapter 8). They include the most fearsomely powerful volcanic explosions known - **Ultraplinian** eruptions.

The Walker classification scheme is limited by the fact that not all volcanic deposits, especially older ones, are amenable to the stratigraphic analysis he proposed. In addition, Walker recognized that events develop that can modify the ordinary patterns of pyroclastic fall deposition. Accretionary lapilli (Chapter 7) can be deposited from ash clouds, forming much larger clast sizes than associated ashfall. Rain can also flush ash out of an eruption cloud to produce a bed of fine-grained material anomalously close to a vent. Some blasts throw out large quantities of fine material formed in earlier pyroclastic eruptions, mixing deposits in a confusing way. Walker's scheme also ignores some "trademark" aspects of certain eruption styles that may be the only evidence one has to interpret past volcanic activity. Working in the remote Mopah Range of California's Mojave Desert, for instance, I (RWH) mapped several lenses of ash-poor, oxidized orange-red cinder up to 10 m thick and 100 m long sandwiched between early Miocene-aged basaltic lava flows. The cinder was a tell-tale sign of Strombolian activity, quite apart from pyroclast dispersal information, which after 18 million years of erosion was impossible to recover. Despite these shortcomings, Walker's 1973 classification scheme, as modified by Self and Sparks (1978) remains a good basis for approaching the classification of volcanic eruptions, especially younger, more explosive ones. D.M. Pyle (1989) recognized that fall depositional characteristics change exponentially with distance, and developed further improvements over the Walker methods for plotting field measurements. Pyle's methods are especially useful for the calculation of total fall deposit volumes.

Walker (1980) later defined five quantitatively measurable eruption parameters that are very useful in categorizing an eruption, and these terms are now used universally:

1. **Intensity**: The rate at which magma leaves a vent (i.e. its discharge rate), initially estimated in volume per second (m^3/s), and then later converted to mass per second (kg/s) based on density considerations. Eruptive intensity commonly varies with time during most eruptions, and eventually falls to zero as the eruption ends.
2. **Magnitude**: The total mass (or volume) of material erupted, measured in kilograms (kg), cubic meters (m^3), or cubic kilometers (km^3).
3. **Dispersive power**: The area over which the eruption products are spread by spending some time being transported high in the atmosphere before falling to Earth, measured in square kilometers (km^2).
4. **Violence**: For explosive eruptions, the distribution of products directly thrown away from a vent and scattered by *ballistic momentum*. (Think of "ballistics" as being like the trajectory of a baseball thrown through the air from one player to another.)
5. **Destructive potential**: The damage caused by an eruption to urban and agricultural lands, to vegetation, and other objects that can be used to measure impacts. In cold-hearted economic terms, this is measured in dollars or euros, numbers of injuries, fatalities, etc.

Volcanic Explosivity Index (VEI)

The **Volcanic Explosivity Index**, or **VEI** (Table 5.3), now universally accepted as a means to categorize the relative sizes of explosive eruptions, was developed by Newhall and Self (1982) at the suggestion of Robert Decker, then a Dartmouth College professor. The higher the VEI number, the more powerful the eruption. The VEI correlates the volume of volcanic ejecta and various other observed physical criteria, such as eruption column height and eruption duration. As we will see later, it is possible to infer some of these physical criteria based on a careful field study of past eruptive deposits. A somewhat parallel approach was proposed by Fedotov (1985), who devised a scale for explosive eruptions based logarithmically on magma discharge rate during eruptions.

Decker (1990) undertook a statistical study of the VEIs of eruptions to calculate the frequencies of eruptions having particular levels of explosivity. His research reinforced the observation that the smaller and less explosive the volcanic eruptions, the more frequently they occur. On a log-log plot of eruption frequency versus VEI, eruptions ranging from VEI 2 to 6 plot in a line with a slope of 0.5, meaning that each increase in eruption magnitude correlates with a fivefold decrease in frequency. For eruptions having VEIs of 6 to 7, the slope of the line increases to 1 (a tenfold decrease in frequency for every unit increase in VEI), and from 7 to 8, the slope increases to 10. From this analysis, he estimated that the average number of worldwide eruptions as a function of VEI magnitude is as follows (Figure 5.5):

VEI ≥ 2, 15 eruptions/year;
VEI ≥ 3, 3 eruptions/year;
VEI ≥ 4, 1 eruption every 2 years;
VEI ≥ 5, 1 every 10 years;
VEI ≥ 6, 1 every 50 years;
VEI ≥ 7, 1 every 450 years;
VEI ≥ 8, 1 every 300,000 years or more (Self 2006).

The upper limit of volcanic explosivity lies somewhere between 8 and 9 on the VEI scale, exemplified by the Toba super-eruption (Chapter 8). Toba [127].

General Applications

In practice, it is often best to describe different phases of an ongoing eruption using the revised Lacroix nomenclature (with its amendments from Walker, and Self and Sparks), while classifying the overall eruption after it ends in terms of its VEI or Walker pyroclast dispersal characteristics. Because the Walker and VEI systems are quantitative, they are more valuable for comparing different volcanic eruptions then the simple Lacroix classification, but the VEI system depends heavily on observations of ongoing eruption parameters. For prehistoric eruptions, whose only records are left by their deposits, VEI estimations are impractical and imprecise. Hence, we think it is critical to be able to recognize the *kinds of deposits* left by eruptions. As in the case of the Mojave Desert cinder lenses, it may be that the appearance of a few outcrops is the only evidence available for evaluating past volcanic activity. Field recognition and correct interpretation of style-dependent depositional characteristics (Cas and Wright 1987) are critical for analyses.

TABLE 5.3 The Volcanic Explosivity Index (VEI): Criteria for determination.

Criteria VEI →	0	1	2	3	4	5	6	7	8
Size description	Non-explosive	Small	Moderate	Moderate–Large	Large	Very large →			
Volume of ejecta (m^3)	< 10^4	10^4–10^6	10^6–10^7	10^7–10^8	10^8–10^9	10^9–10^{10}	10^{10}–10^{11}	10^{11}–10^{12}	> 10^{12}
Eruption column height (km)*	< 0.1	0.1–1	1–5	3–15	10–25	> 25 →			
Description of explosivity	← gentle, effusive →		← explosive →		cataclysmic, paroxysmal, colossal →				
Classification	← Hawaiian →	← Strombolian →	← Vulcanian →		← Plinian →	← Ultraplinian →			
Duration of continuous blasts (h)	< 1					> 12 →			
Injection of lower atmosphere (troposphere)	Negligible	Minor	Moderate	Substantial →					
Injection of upper atmosphere (stratosphere)	← None →			Possible	Definite	Significant →			

VEIs of the nine most powerful volcanic eruptions since 1400[†] *CE*			*Other well-known recent eruptions*		
Year	Volcano	VEI	1982	El Chichón, Mexico [56]	5
1991	Pinatubo, Philippines [128]	6	1980	Mount St Helens, USA [30]	5
1912	Novarupta, Alaska [18]	6			
1902	Santa Maria, Guatemala [57]	6			
1883	Krakatau, Indonesia [122]	6			
1815	Tambora, Indonesia [127]	7			
1660(?)	Long Island, New Guinea [150]	6			
1641	Parker, Philippines [132]	6			
1580(?)	Billy Mitchell, Solomon [154]	6			
1452	Kuwae, Vanuatu Islands [167]	6			

* For VEIs from 0 to 2, the column height is kilometers above the vent; for VEIs greater than 2, column height is kilometers above the sea level.

[†] Data from Briffa et al. (1998).

Source: Modified from Newhall and Self (1982).

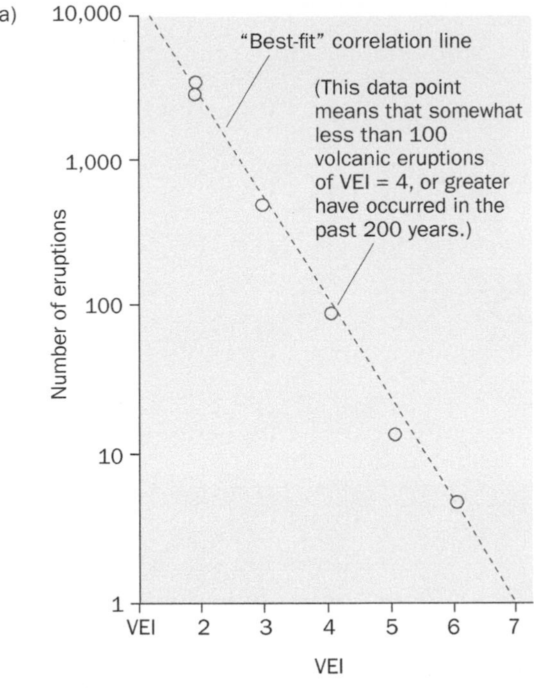

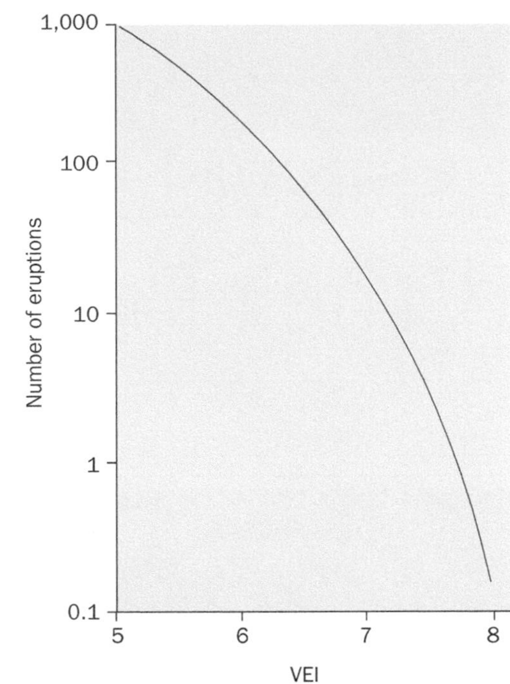

FIGURE 5.5 (a) Plot of frequency of eruptions of specific VEI and greater during the past 200 years, based upon statistical analysis of the eruption data base of Simkin et al. (1981) by Decker (1990); (b) Number of large eruptions over the past 10,000 years. Sources: Simkin et al. (1981) and Decker (1990).

Questions for Thought, Study, and Discussion

1 Discuss the merits and limitations in the application of the classification schemes of Lacroix, Rittmann, Geze, and Walker. When would it be most appropriate to use any one of these schemes in preference to all of the others? Could several schemes be usefully applied to describing and studying a *single* eruption? Explain how and why.

2 A series of explosions rock the summit of a composite volcano, throwing out numerous blocks of older lava amidst roiling clouds of ash and steam. An estimated $5 \times 10^{5.5}$ m^3 of ejecta are thrown out. Classify this eruption in as many ways and as precisely as you can.

3 Steam explosions throwing out occasional tails of dark, wet ash accompany the emergence of a new volcanic island out of the sea. Once the vent is sealed off, a series of rhythmic explosions throwing out clots of lava and dark clouds of ash begin to build up a young cone. Later, this explosive activity dies down as a fissure opens on one flank of the cone releasing a torrent of lava, with low lava fountaining occurring all along its length of the fissure. Where the fissure meets the sea, though, periodic steam blasts rip the shoreline, temporarily setting back the construction of new land. Classify this eruption. Is it multi-phase?

4 Can an eruption have both high intensity and low magnitude? Explain.

5 Can an eruption have both high intensity and show low violence? Explain.

6 According to Self (2006), the great Oruanui–Taupo caldera [181–180] eruption of 26,000 years ago had a VEI of over 8. If it occurred today, this eruption would have impacted most of the North Island of New Zealand, home to millions of people. How concerned should residents and authorities really be about the threat of eruptions like this?

FURTHER READING

Scarth, A. (2009) *Vesuvius: A Biography.* New Jersey: Princeton University Press, 352 p.

Thordarson, T., Self, S., Larsen, G. et al. (2009) *Studies in Volcanology: The Legacy of George Walker.* Perugia, Italy: IAVCEI Publications, International Union of Geodesy and Geophysics, 416 p.

Wood, C. (2009) "World Heritage Volcanoes: Global Review of Volcanic World Heritage Prospects: Present Situation, Future Prospects, and Management Requirements." *International Union for the Conservation of Nature World Heritage Studies*, 8, 62 p.

CHAPTER 6

Effusive Volcanic Eruptions and Their Products

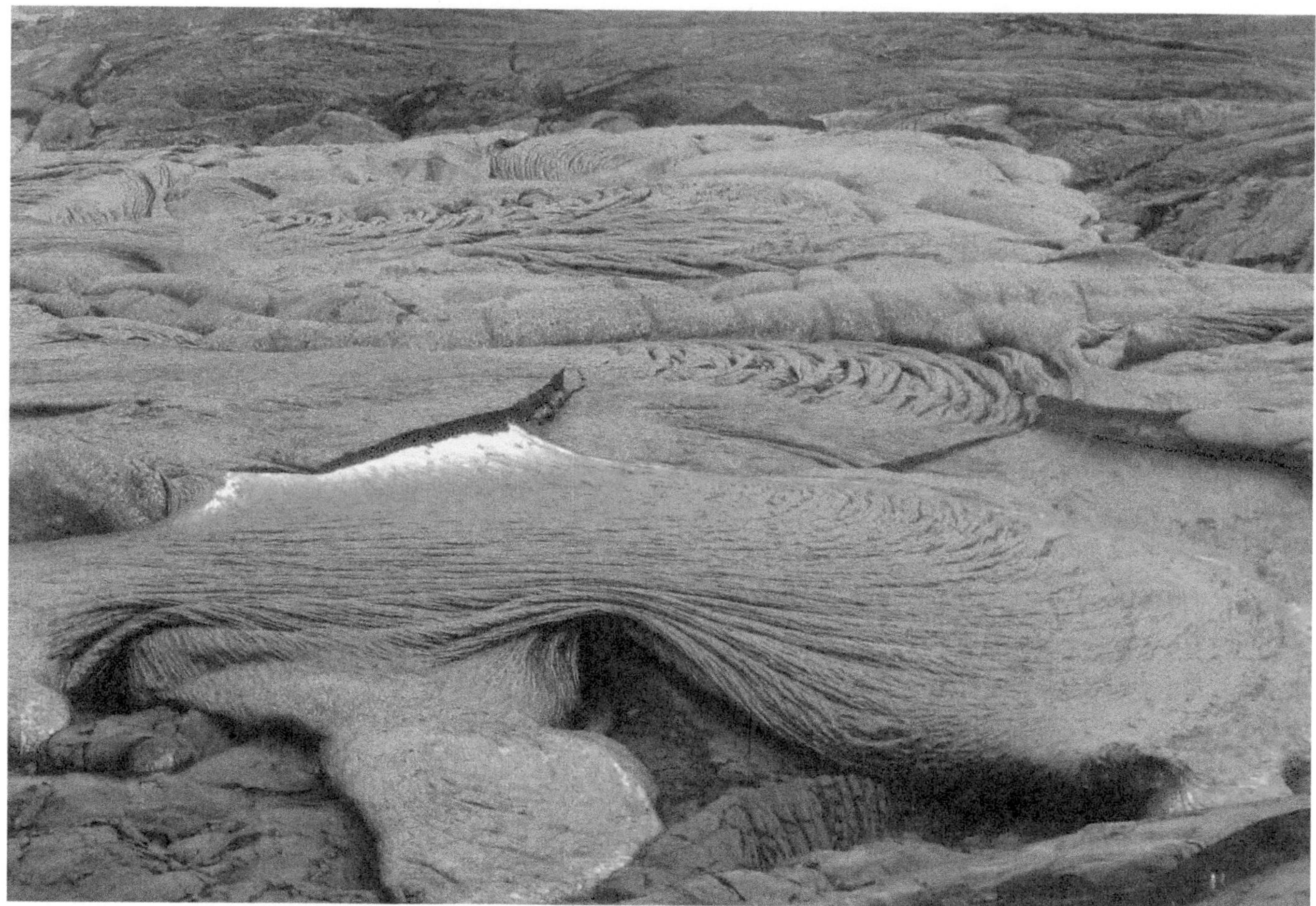

Source: Credit J. Lockwood.

Volcanoes: Global Perspectives, Second Edition. John P. Lockwood, Richard W. Hazlett, and Servando De La Cruz-Reyna.

Companion website: www.wiley.com/go/lockwood/volcanoes2

E Pele e
Ke akua o na pōhaku ʻenaʻena
ʻEliʻeli kau mai
[Oh Pele, Goddess of the burning stones,
let a profound awe possess me]

[Traditional Hawaiian oli (chant)]

The first thing that people picture or children draw when thinking about volcanoes is **lava**, the outpouring of fluid magma onto the earth's surface (Figure 6.1). Lava flows are common wherever molten rock reaches the surface without fragmenting explosively. Such flows come in many shapes and sizes, and feature many kinds of distinctive surface with differences mostly controlled by variations in magma viscosity and supply rates at the time of eruption. Lava flow types may be classified either according to their compositions (rock types, Figure 3.3) or by their distinctive appearances (Table 6.1).

Most of our emphasis in this chapter will be on basaltic eruptions and their lavas rather than on silicic lava flows, primarily because basaltic lavas are much more voluminous on Earth – and on the other rocky planets (Chapter 12). They are much more likely to be seen in action in places like Hawaiʻi or at frequently active volcanoes like Etna [112]. Siliceous eruptions also tend to be quite explosive, with their viscous, thick lava flows not usually traveling more than a few kilometers from vents. We will discuss siliceous eruptions and their products more extensively in the following two chapters.

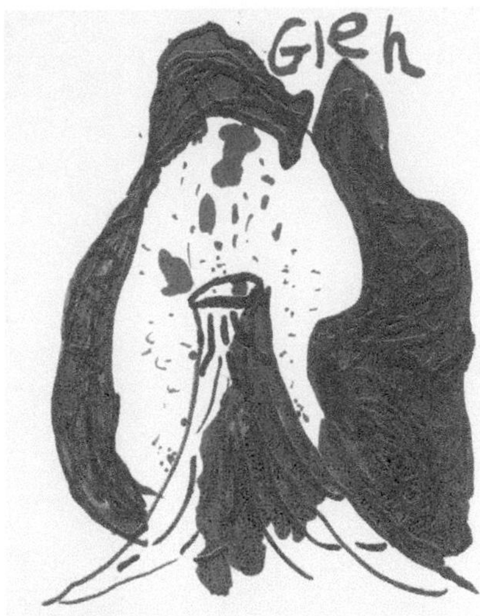

FIGURE 6.1 A "typical volcano," as envisioned by a 6-year-old child.

Hawaiian-type Eruptions

Alfred LaCroix proposed using the term **Hawaiian eruptions** to identify all forms of effusive volcanic activity that involve the eruption of fluid lavas, typically basaltic in composition, and usually involving both lava fountains and flows. A Hawaiian-style eruption begins where a dike of molten basalt breaches the surface. Swarms of sharp, shallow earthquakes precede the opening of a fissure in advance of the dike. Dilation of the ground above the ascending dike may form a linear graben from tens to over a hundred meters wide, perhaps stretching for kilometers. A fissure crack opens on the floor of the graben, parallel to its margins. Dense steam first issues, then blue SO_2-rich fume, sometimes pulsating shortly before the first blobs of red lava spurt out. Lava at first shoots into the air all along the length of the fissure, forming a curtain of incandescent, fountaining lava that may reach a few tens, or even hundreds of meters into the air. Such sheets of erupting lava present spectacular panoramas typically stretching from a few hundred meters to 10 or 20 km in length, the celebrated Hawaiian "Curtain of Fire" (Figure 1.5).

When one watches a curtain of fire in action, the heat and massive amounts of dancing liquid spatter being thrown upward give the impression that lots of lava is being produced, but this is not necessarily so. When fluid lava is fountaining during effusive eruptions, this is much less a display of erupting lava than it is a display of erupting gas. For typical lava fountains, more than 90 percent of the new material being erupted (by volume) is actually gas, chiefly water and SO_2. Each blob of fountaining lava is a highly inflated mass of spongy, bubble-filled molten rock much like popcorn or Styrofoam in texture. Even more significant to evaluating lava output is the fact that a lot of the molten spatter is simply being recycled – much of it has already been ejected, fallen back into the pools of lava at the bases of fountains and shot up again several times. Some of the gushing lava falls back around the sides of erupting fissures, quickly building steep-sided ridges of spatter, termed **spatter ramparts**, typically 5–10 m high (Figure 6.2). The spatter clots are highly fluid and usually spread out after impact, collecting to form loosely welded aggregates called **agglutinate** (Figure 6.3; Rader and Geist 2015). If the fissure trends across a slope, spatter ramparts will develop only on the upslope side of the vent since fountain-fed lava flows from the vent will incorporate and transport spatter as it flows away on the downhill side.

TABLE 6.1 **Relationship of the surface types of lava flows to their compositions and source vents.**

Roughness classification	Common composition	Source vent structures
Pāhoehoe (smooth, hummocky surface)	Basalt	Shield volcanoes, cinder cones, composite cones
ʻAʻā (covered with loose, spiny rubble)	Basalt, andesite	Shield volcanoes, cinder cones, composite cones
Blocky (lithic)	Basaltic andesite, dacite	Cinder cones, block lava shields
Blocky (glassy)	Dacite, rhyolite	Domes

FIGURE 6.2 Spatter rampart in formation. Kīlauea [17] east rift zone eruption – 2003. Source: Photo © Brad Lewis.

FIGURE 6.3 Welded basalt spatter (agglutinate) deposited on the inner wall of an eruptive vent during the 1969 eruption of Kīlauea volcano, Hawai'i [17]. Source: Photo by J.P. Lockwood.

Newly formed ramparts are inherently weak edifices, and may slump and collapse where mobilized agglutinate oozes from their sides or bases. Tongues of oozing agglutinate may meld into pāhoehoe, or more typically 'a'ā flows ranging up to 5–15 m thick and hundreds of meters long. Such secondary flows commonly raft patches of agglutinate on their surfaces, some of them meters in diameter, which did not coalesce into homogeneous liquid during flow. The resulting rock has been termed **clastogenic** ("fragment-formed") **lava** (Cas and Wright 1987; Sumner 1998).

Hawai'i has no monopoly on "Hawaiian-type" eruptions. The largest Hawaiian-type eruption in recorded history was the voluminous Icelandic Eldgjá [95] fissure eruption in ~935 C.E., which produced 19.6 km^3 of fresh basalt, comparable in volume to some of the smaller "flood basalt" flows building up Earth's prehistoric continental lava plains and plateaus. This was followed by the smaller but still impressive Laki [96] fissure eruption (12.3 km^3), which broke out in the same region in 1783–1784. Of the two major Hawaiian-style eruptions since publication of the First Edition of this book, one took place in Hawai'i (Kīlauea-2018), and the other, larger one, again in central Iceland (Bárðarbunga, 2011–2012). Bárðarbunga [98] erupted beneath the Vatnajökull glacier and in 180 days produced about 1.5 km^3 of basalt, which covered an area of 85 km^2. Krýsuvík-Trölladyngja volcano [100] on Iceland's Reykjanes Peninsula began erupting on 19 March, 2021 (for the first time in nearly 800 years!), and provided spectacular "Hawaiian-type" eruptive activity for foreign visitors and residents of the nearby capital city Reykjavik, with lava fountains over 400 m high. It is perhaps simply a historical accident that LaCroix did not designate Hawaiian-type eruptions "Icelandic" instead, though such eruptions are far more frequent and approachable in Hawai'i.

FIGURE 6.4 Spatter cone forming on Kīlauea [17] volcano's East Rift Zone, April 1983. Lava fountains about 40 m high. The cone is kept open by a flowing river of lava that carries away spatter falling on its surface. Source: USGS photo by J.P. Lockwood.

Many Hawaiian eruptions cease within a few hours or days, but some may continue for much longer, even persisting through various phases of activity for decades. After the first few hours, lava fountaining will become concentrated at certain points along the initial fissure where flow rates and fountain heights rapidly increase, while lava production ceases everywhere else. The concentration of lava production at one or just a few points establishes a feedback mechanism that increases eruption rates further. Greater discharge at a single point erodes and streamlines underlying dike walls, enlarging vents, and facilitating flow. As fountaining increases in vigor and height, larger amounts of **reticulite** (porous, pumice-like scoria consisting of intricate glass filaments) and **Pele's hair** (long threads of volcanic glass) will form in addition to frothy spatter. Instead of an elongate spatter rampart, a single much taller volcanic edifice (**spatter** or **lava and pumice cone**) begins to grow (Figure 6.4), capable of reaching more than a hundred meters in height over a few months of activity.

The construction of a prominent edifice and concentration of fountaining at one locality along a fissure may be regarded as the second phase of a Hawaiian eruption. Many short-lived Hawaiian eruptions do not last long enough for this to happen, but those that do may transition to a third phase, with continuing enlargement of the vent diameter owing to erosion of the vent walls and collapse into the underlying dike. As degassing becomes more efficient with widening vents, fountain activity will decrease. The surrounding cone can be partially consumed when this happens, and lava may pool up over the vent to form an **active lava lake**, overflows of which begin to construct a gently sloping lava shield on the remains of the earlier, steeper-sided agglutinate cone. High lava fountaining is no longer possible because of associated longer time for degassing, which favors production of pāhoehoe rather than ʻa'a lava flows (Figure 6.5).

Lava Lakes

Lava lakes can be either *active*, forming directly above eruptive vents and usually confined within coeval spatter cones, or pre-existing craters (Chapter 10; Coppola et al. 2016) or *passive*, as lava from vents located elsewhere is ponded in pre-existing craters or depressions encountered along the flow path.

Active lava lakes constantly churn as new lava wells up from below and cooler surface lava become denser than surrounding molten rock and founders. The thin semi-plastic silvery crusts on these lakes are in continuous slow movement across the lake surface, owing to convection of underlying lava. Duffield (1972) observed that the movements of lava crust plates across the Mauna Ulu lava lake (Chapter 1), mimicked the movement of tectonic plates across Earth's surface. Indeed, there are striking parallels to be seen, including examples of "sea-floor spreading" as new crust forms, "transform faulting," and "subduction," as older crust founders beneath younger (Figure 6.6).

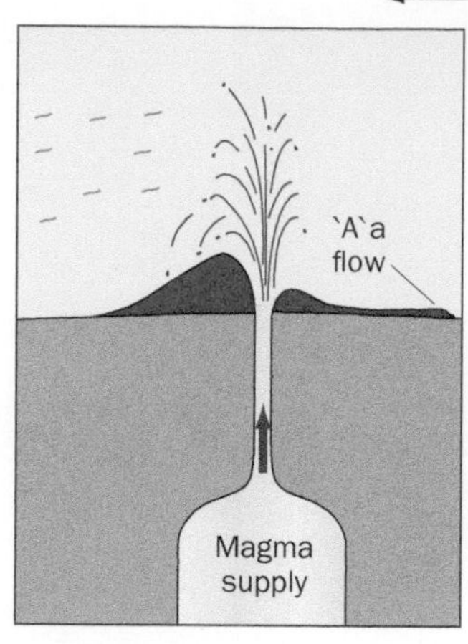

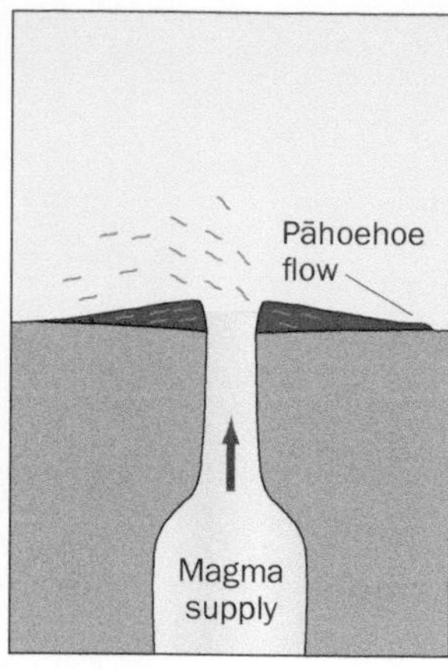

FIGURE 6.5 Effect of vent diameter on eruption style for Hawaiian-style fluid lavas. For constant magma supply rates narrow ascent will result in higher ascent speeds, which allow less time for bubble migration and degassing. Narrow vents (a), typical of early eruption phases, favor high fountaining, formation of spatter cones, and production of "shelly" pāhoehoe (Swanson 1973) and ʽaʽā flows (Table 6.1). Wide vents (b), that develop by vent erosion with time, favor formation of lava shields, lava lakes, and production of "tube-fed" pāhoehoe flows. Source: Modified from Swanson (1973).

FIGURE 6.6 Mauna Ulu lava lake surface circulation in 1970, East Rift Zone of Kilauea volcano [17]. The crustal movement mimics plate tectonics and "seafloor-spreading" centers on a much smaller scale, including a transform fault. The plastic crust being carried across the lake by underlying molten lava is less than a centimeter thick where formed, but thickens laterally as it moves from the spreading center. The active zone of upwelling melt is as much as a meter wide. Source: USGS photo by Wendell Duffield.

The supply of new lava to these lakes is constantly changing, and the balance between upwelling new lava and sinking of old lava crusts back into the lake often accompanies the process of **drainback**, where lake levels may drop as lava pours back into the source vents. The balance may be static, at which times lava lake levels remain constant for long periods; or excess drainback may cause sudden, rapid lowering of the lake's level. Increasing lava supply may cause the lava lake to overflow crater rims as sheet-like surface flows, or may feed long-sustained pāhoehoe flows through subterranean conduits in the submerged walls of the vent lake ("pyroducts") that may transport lava many tens of kilometers downslope.

Between 2008 and 2018, the active lava lake in Kilauea's Halemaʻumaʻu Crater (Chapter 1; Figure 1.2) exhibited episodic **gas-piston behavior** (Patrick et al. 2016). The circulating lava lake spattered as it released gases, normally through tears in its crust

FIGURE 6.7 View of Kīlauea Iki lava lake from the east. The source vent for the 1959–1960 eruption that fed this lava lake is at the far western margin of the lake, and is marked by a tall spatter cone and devastated forest zone downwind (southwest) of the high lava fountains from the actual vent. This passive lava lake is 1.5 km long, over 100 m deep, and provided a natural laboratory for studying how basalt lavas cool and differentiate over a long period of time. Source: Photo by J.P. Lockwood.

localized along the south side of the crater wall. But suddenly, all spattering would stop, and the lake level would well up by as much as 20 m. Not for long, however ; an explosive spattering and degassing would eventually resume, and the lake would quickly drop to its ordinary level once more. Geophysical observations suggest that these oscillations are somehow related to gas exsolution and volume changes that originate in the shallowest part of the vent (Patrick et al. 2016).

In some areas, lava may fill neighboring craters or low areas to create **passive lava lakes**, which though not positioned directly over vents, may behave in similar fashions to active-lakes as long as they continue to be fed fresh lava. But while small lava fountains often play across the surface of passive lava lakes, the gas pressure is not sufficient to cause substantial fountaining.

In November and December 1959, a spectacular eruption east of Kīlauea volcano's [17] caldera filled the prehistoric Kīlauea Iki pit crater to a depth of 135 m with olivine-rich tholeiite basalt (Figure 6.7) and for the next decades provided an excellent natural laboratory to study the cooling behavior of a very thick molten lava body. For a few days after cessation of the eruption on 20 December, 1959, the new crust was thin and unstable, but by late January geologists were walking on the hot crust to measure changes in its level. By April 1960, the crust was strong enough to support heavy equipment, and a program of scientific drilling began to measure the rate of crustal growth and the physical and geochemical properties of the crust and diminishing melt. Drills penetrated the crust 26 times during the next three decades (R. Helz 2012, personal communication). Kīlauea Iki is now the best studied natural example of basalt solidification processes in the world, and has yielded a treasure trove of over 200 scientific papers in petrology, geophysics, and geochemistry. These studies have enabled petrologists to study the ways in which basaltic magma changes in composition and mineralogy over time as temperatures drop and heavy minerals (principally olivine) grow and sink gravitationally within it (Helz 1980; Helz and Thornber 1987; Jellinek and Ross 2001).

The crust of the lava lake at Kīlauea Iki initially thickened very rapidly (over a meter a month), during which time radiation was the principal means of cooling. As the crust insulated the underlying basalt melt from radiative cooling, however, conductive cooling became predominant, with the rate of crustal thickening proportional to the square root of time:

$$\text{Thickness} = \text{n}\sqrt{t} \qquad (6.1)$$

where n is a constant dependent on environmental factors, principally rainfall, and t is time, in months.

This relationship held for the first three years at Kīlauea Iki, but in the years following the eruption, lake crust thickened at a somewhat faster rate. The boundary between solid and melt at the base of the crust became increasingly complex, both because of the formation of plastic, largely crystalline "mushes" (which were neither solids nor mobile fluids) and because of compositional inhomogeneities within the zone of crystallization. No "melt" (in the sense of material too fluid to drill through) existed in Kīlauea Iki after 1981, although a large, partially molten, incandescent plastic core still existed within the center of the lake (Figure 6.8). Crystallization of this plastic core proceeded from both above and below, and researcher Rosalind Helz (2000, personal communication) calculated that the Kīlauea Iki lava lake completely solidified sometime around 1994 or 1995, about 35 years after the eruption.

Lava lakes can be formed in ways other than through vent erosion during Hawaiian eruptions and the passive entrapment of lava within pre-existing craters. In some cases, largely degassed magma chambers stope all the way to Earth's surface with little eruptive fanfare. The top of the active magma chamber feeding Nyiragongo [118] has appeared many times in the deep central crater of that volcano (Figure 6.9). One or more lava lakes marked the exposed top of Kīlauea's magma chamber for most of the time between 1823 and 1924, and again from 2008 to 2018 (Figure 6.10).

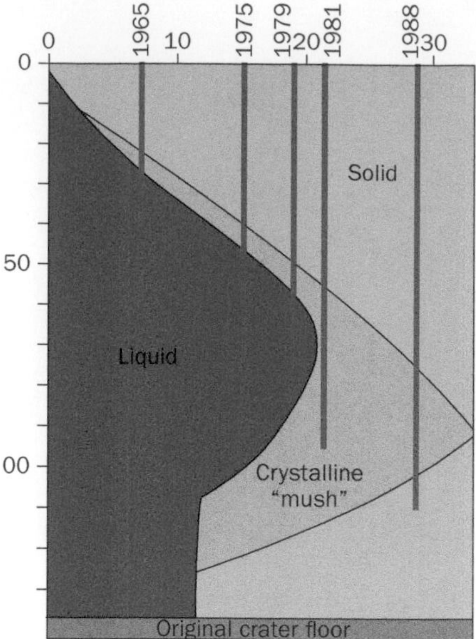

FIGURE 6.8 The cooling of Kīlauea lava lake – 1959 to 1988. Source: Data from various USGS studies.

Strombolian-type Eruptions

Strombolian eruptions, like Hawaiian, are common forms of basaltic volcanic activity. The mechanical difference is that gas exsolution in Hawaiian eruptions takes the form of numerous, small, escaping vapor bubbles that cause the magma to fountain continuously or passively flow out, whereas in Strombolian eruptions, very large bubbles develop deep in the vent, leading to rhythmic moderately explosive discharges as they escape (James et al. 2008; Pering et al. 2017). The explanation for the contrasting styles of vesiculation generally relates to variable magma ascent rates; slow rising magma has a chance to develop larger bubbles, fast rising magma does not (Parfitt 2004; Gonnerman and Manga 2007). Why some volcanic areas favor more rapid ascent of basaltic melts relates to tectonic setting, composition, magma supply rate, and gas content. Slightly more viscous mafic melts contribute to reduced ascent rates and favor Strombolian eruptions.

Seismic and Doppler radar studies have been made of the bubble formation which causes the bursting behavior so characteristic of Strombolian eruptions. Evidence suggests that individual bursts may originate as deep as several hundred meters within the magma conduit. Long-period seismic frequencies record the initiation of explosions up to a second or two prior to anything becoming visible at the surface (Rowe et al. 2000). At Etna's SE crater [112], bubble formation at conduit depths of 500 m has been inferred, with eruptive blasts following approximately 2–3 minutes later (Dubosclard et al. 2004).

FIGURE 6.9 View of Nyiragongo [118] summit crater, 1994. This lava lake is nearly 1 km in diameter. The active fountains are about 50 m across, although the undulating crust showed the molten lake extended beneath the entire crater. Source: USGS Photo by J.P. Lockwood.

FIGURE 6.10 Halema'uma'u lava lake, Kīlauea volcano, Hawai'i [17] ca. 1894. This lake was almost continuously active for nearly a century. Note the circular form and bounding lava levees – which are formed by short overflows from a rising lake surface. Source: Photo by J.A. Gonsalves.

Strombolian activity generally, though not always, takes place at open vents where moderately fluid lava stands at a high level in the vent throat, often just a few tens of meters below the rim (Capponi et al. 2016). Some of the clots thrown up by blasts strike the walls of the vent near the rim while they are still molten and form spatter agglutinate; others cool during their flights through air, and strike the surrounding ground in an essentially solid condition. Commonly they are still glowing red when they leave the vent, but have become black by the time they impact. If the magma level rises, more of the ejecta will exit the vent in a partly fluid condition, piling up around the vent as spatter. The eruption then seemingly grades into one of Hawaiian type, but rhythmic jetting of incandescent material at the vent continues every few seconds or minutes, much in contrast to ordinary Hawaiian lava fountaining.

The type-locality for Strombolian eruptions, Stromboli volcano [113], is located off the southwest coast of Italy. Its eruptions have been nearly constant, and consistent in character, since before the first Greek colonists settled the region over 2500 years ago. The most typical activity consists of explosive ejections of incandescent cinder and spheroidal or fusiform bombs thrown to heights of a few tens to several hundred meters above source vents. This activity may or may not be accompanied by discharge of a lava flow. If lava does emerge, it is generally somewhat more viscous than that of Hawaiian eruptions and forms somewhat shorter and thicker flows. Most ejecta are nearly solid when they strike the ground but larger materials are semi-molten and flatten appreciably, at times forming "cow-dung bombs" as much as a meter in diameter (Chapter 7).

Stromboli's ordinarily mild activity is punctuated at intervals of a few months to a few years by episodes of more violent eruption (VEI greater than 2) lasting no more than a few days each, in which showers of incandescent cinder and bombs are thrown to heights of as much as a thousand meters and great black ash clouds rise above the volcano. Bombs may fall at a distance of several kilometers from the vent. The general nature of the ejecta is the same as in the milder eruptions, except that glassy ash becomes much more abundant and occasional blocks of old rock are found in the ejected debris. The violent activity usually lasts for only a few hours or days and is commonly followed by a short interval of quiet during which only fumarolic activity occurs. Major eruptive activity in 2002–2003 involved damaging tsunami generation and renewed outbreaks on Stromboli's flanks (Calvari et al. 2008).

Cinder cones also develop from Strombolian eruptions that generally persist for only a few weeks or years – an outstanding example being Paricutín volcano [53] in central Mexico, which grew between 1943 and 1952 (Chapter 9). Cinder cones often cluster together in volcanic fields. Paricutín, for example, is merely the youngest in a group of several dozen similar cones, while over 600 of them are scattered across the San Francisco volcanic field in central Arizona.

Flood Basalt Eruptions

Many volcanologists classify flood basalt eruptions as a distinct type of effusive volcanic activity, although in fact they share the same general features as ordinary, lower-volume Hawaiian eruptions. The difference is largely one of scale; a lot more comes out of the ground when flood basalts erupt. But they also never build large, discrete polygenetic volcanoes, which may seem ironic considering the vast amount of magma that feeds them. Instead, they create wide lava plains and tablelands.

It is fortunate that no major flood basalt eruption has occurred in human history, because discharge rates are so great, and flood basalt flows move so fast, that it might be difficult to evacuate a threatened region in time. Whereas the volumes of typical large Hawaiian or Icelandic basalt flows range from less than one to a few km³, individual flood basalt flows range in volume up to 2000–3000 km³. A 55-km-long lava flow erupted in Hawai'i in 1855–1856 is impressive in comparison with other historically observed flows, but prehistoric flood basalt flows as long as 750 km are known. The most detailed studies of flood basalt volcanism have focused on the Columbia River Basalt (CRB) province in the northwestern United States (Tolan et al. 1989). There, between 17.5 and 15.5 million years ago, lava poured out across an area exceeding 150,000 km² – almost equivalent in size to the whole of Washington State (Figure 6.11). The sources of most of these flows are preserved in the Chief Joseph Dike Swarm of northeastern Oregon. Dikes in the Monument Dike Swarm several hundred kilometers farther west also fed numerous flows. Remnants of large spatter ramparts and pumice cones have been found where many of these dikes served as eruption feeders. Wright et al. (1989) estimate that the initial eruptive fissures of individual flows typically stretched 70–200 km! Given their huge volumes, the late-stage inflation of such flows (more later) probably took much longer than in the case of their ordinary Hawaiian counterparts – perhaps as long as several years as opposed to days or months (Swanson et al. 1975; Thordarson and Self 1998; Vye-Brown et al. 2013). Large as the CRB lava fields are, they are dwarfed by the much more voluminous late-Cretaceous Deccan Volcanic Province basalts of southern India. The products of these very largest flood basalt eruptions are called **large igneous provinces** (LIPs), and are discussed further in Chapter 9.

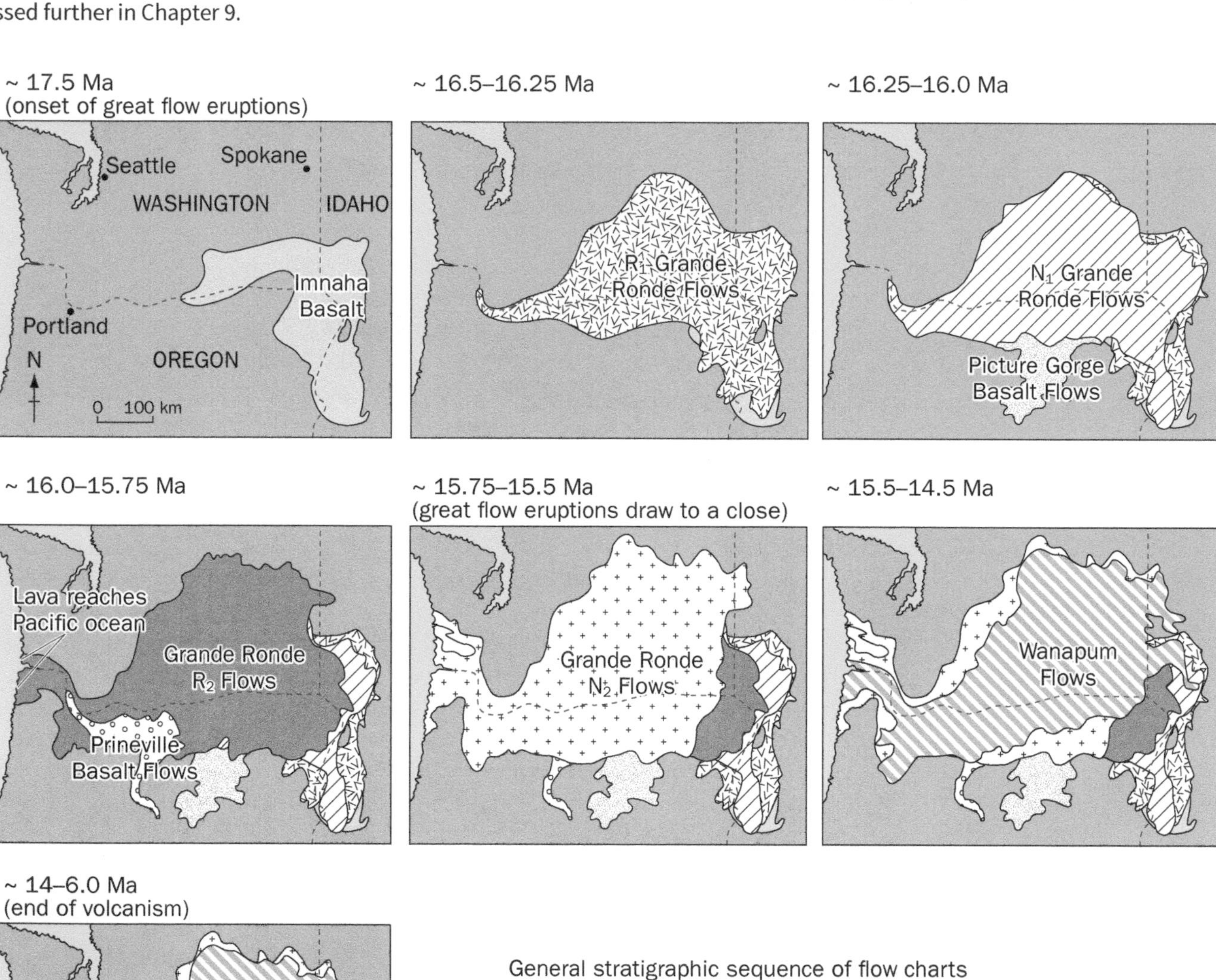

FIGURE 6.11 Rise and decline of Columbia River basalt volcanism. Ma = millions of years ago. Source: Modified from Tolan et al. (1989).

Pāhoehoe and 'A'ā

Lava flows vary significantly in morphology and many early attempts to classify them took place when geology was still a young science, including one by Italian Mario Gemmellaro in 1858 who designated nine (!) different categories based on field appearances (see Harris et al. 2017 for an excellent discussion). People living in basaltic volcanic terrains noted long ago that some lava flows were easy to cross on foot and some were not, and terms were soon devised to describe them on this entirely practical basis. Hawaiians (to whom the distinction was especially critical, as they usually went barefoot) called the smooth-surfaced flow type **pāhoehoe** and the rough-surfaced type **'a'ā** (Table 6.1). Icelanders called the two types **helluhraun** and **apalhraun**, respectively (Thorarinsson and Sigvaldason 1962). The rough-surfaced type was called **marubi** in Japan and **malpais** (bad country) in Mexico. When C.E. Dutton introduced the Hawaiian terms into the scientific literature in 1884, they met with strong opposition from British geologists. Bonney (1899, p. 79) scoffed that the terms are "the barbarous [expressions] of an insignificant and uncivilized race in a small archipelago in the North Pacific," and preferred the (more civilized?) terms **slaggy** and **clinkery**. T.A. Jaggar went to Greek civilization for his now-forgotten terms **dermolith** and **aphrolith** (1917). In the end, the Hawaiian words won out, and **pāhoehoe** (properly pronounced PAH-hoy-hoy) and **'a'ā** (ah-AH) are universally employed (Figures 6.12 and 6.13). These terms are presently used not only to describe the surface appearances of cold flows, but also actively moving flows.

The contrast between active 'a'ā and pāhoehoe flows is stark, and we'll dwell on their physical appearances in greater detail below. For now, though, it is important to grasp that both types of lava for an individual eruption are compositionally the same. Their differences relate to viscosity; *the relative stiffnesses* of the molten lava as it flows. Pāhoehoe is the more fluid form, while 'a'ā develops under more viscous conditions. Temperature is a key factor in controlling viscosity; the hotter the lava, the lesser the viscosity. But other factors are also at play, related to conditions of eruption and topography. Let us explore these further:

The Pāhoehoe–'A'ā Transition

Active pāhoehoe flows like a thick porridge beneath a thin, pliable crust. 'A'ā, being stiffer, rips its surface apart through its own forward motion, producing its tell-tale clinkery crust. While there are some "transitional" flows featuring aspects of both 'a'ā and pāhoehoe, most flows are either of one type or another. An individual flow, followed downslope from a vent, may show transitions between the two that appear over short distances. Some critical dynamic thresholds are clearly important to explain such sharp changes in physical appearance during emplacement.

Early during a typical Hawaiian-style eruption, high vigorous lava fountaining will create fountain-fed flows, as described earlier, in which the falling spatter loses temperature before coalescing into flowing lava at the base of the fountains. That heat loss is critical, and the result is that fountain-fed flows often move as 'a'ā. Later, as eruptive vigor and fountaining wanes, spatter falls from lesser elevations, or lava may emerge directly from a vent. Little air cooling occurs before the flowage begins, and hotter, less viscous pāhoehoe results. One result of this change in lava production is that where one sees pāhoehoe and 'a'ā produced by same eruption in mutual contact, the pāhoehoe is usually younger and overlies older 'a'ā.

Shear stress is the concentration of force in a material causing it to stretch out like a deck of cards being spread across a table. When stress rates increase as fluid pāhoehoe pours faster down a steep slope, the highly fluid molten interior will accelerate while the crust lags, generating considerable shear throughout the flow and decreasing its ability to move as a coherent stream. Instead of chilling into ordinary pāhoehoe crust, the solidifying portion may abruptly tear into detached fragments, forming 'a'ā (Figure 6.14).

FIGURE 6.12 Molten lava oozing out from a 30 cm-wide section of uplifted crust on an active pāhoehoe flow during a 2016 Kilauea volcano [17] eruption. Source: Photo by J.P. Lockwood.

FIGURE 6.13 Pāhoehoe and ʽaʽā flows formed during different phases of the 1972 Kīlauea eruption [17]. Where both lava types are produced during the same eruption, it is most common for the pāhoehoe to be younger. Source: Photo by J.P. Lockwood.

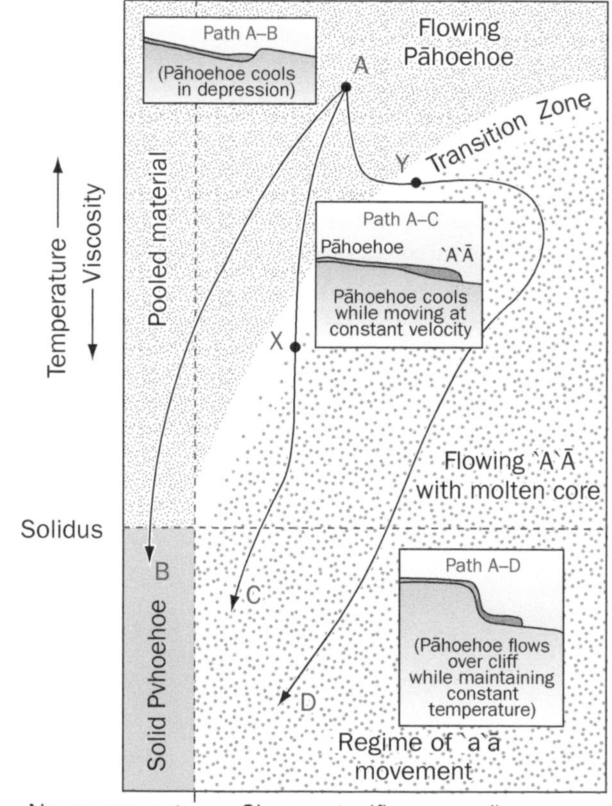

FIGURE 6.14 The rheological basis for the pāhoehoe–ʽaʽā transition. For path A–B, flowing pāhoehoe comes to rest while in a fluid state and continues to cool until solidification. For path A–C, flowing pāhoehoe continues to move as its temperature drops below a critical point (determined largely by composition); at point X, it can no longer flow as a fluid and by rupture cooler parts of the flow will convert to the ʽaʽā form. For path A–D, the pāhoehoe maintains a constant temperature, but converts to ʽaʽā at point Y where the flow's velocity suddenly increases (as by flow over a cliff). Theoretical aspects of the pāhoehoe–ʽaʽā transition are clearly discussed by Hulme (1974); field aspects by Peterson and Tilling (1980). Sources: Hulme (1974) and Peterson and Tilling (1980).

On gentler slopes at uniform grade and constant and slower flow speeds, the commonly observed change of pāhoehoe to ʽaʽā results largely from an increase in viscosity not only due to cooling, but also to loss of gas, and increasing degree of microlitic crystallization. (*Microlites* are tiny mineral grains that develop in glassy lava.) Many flows show transitional textures across part or even all of their surfaces, including fragmented plates of spiny pāhoehoe crust interspersed with patches of true ʽaʽā clinker.

FIGURE 6.15 Source of pāhoehoe flow that "leaked" out of ʽaʽā flow after the 2002 Nyiragongo [118] eruption. This pāhoehoe flow appeared one day after the ʽaʽā flow had come to rest. Source: Photo by J.P. Lockwood.

Once flowing lava has transitioned to fragmental ʽaʽā, it can never revert back to pāhoehoe, although the liquid cores of ʽaʽā flows occasionally break out from flow interiors with buildup of pressure or change in slope to spread and cool as pāhoehoe (Jurado-Chichay and Rowland 1995; Kauahikaua et al. 2003). Good examples of this were seen during the 2002 eruption of Nyiragongo volcano [118], Democratic Republic of Congo, where many homes were destroyed in the city of Goma up to two days after the ʽaʽā flows had come to rest. The fluid cores of the immobile ʽaʽā flows, gradually building up pressure from continued internal flowage of lava from upslope, emerged unexpectedly to feed destructive lobes of pāhoehoe (Figure 6.15).

Pyroducts, Conduits, and Caves

A bowl of porridge turned onto a slanting table will create a viscous, homogeneous flow that surrounds or overtops obstacles in its way. But molten pāhoehoe flows are not homogeneous, and their emplacement as they burn through forests, fill craters, and surround buildings is much more complicated than a spilled-porridge model implies! During long-lived eruptions, most of the "action" in an active pāhoehoe flow takes place in conduits beneath solidified crust, and flowing lava is mostly unseen, analogous to blood circulating through veins in your body. Such subterranean conduits allow lava to flow long distances with little heat loss, and are one of the factors responsible for the gentle slopes of typical "red volcanoes." If lava drains downslope from these conduits in the waning stages of eruptive activity, however, elongate caves may be formed. Reverend Titus Coan observed active subterranean lava flows in action in 1843 on the upper slopes of Mauna Loa volcano [15], and suggested the term **pyroduct** to describe the subterranean "rivers of fire" he witnessed.

Coan was a pioneering Congregational missionary in Hawaiʻi, and a well-educated naturalist who provided excellent descriptions of all the Hawaiian eruptions that took place between 1832 and 1880. His interest in the scientific aspects of volcanic activity was piqued by his meeting with a young J.D. Dana (Chapter 1) when the Wilkes Expedition visited Hawaiʻi in 1840–1841. Coan described subsequent eruptive activity of both Kīlauea and Mauna Loa volcanoes for the next four decades in letters to Dana, who published them in the *American Journal of Science*. These letters give us some of the earliest detailed observations of many volcanic phenomena that had never been directly witnessed before. His reports on the 1843 eruption of Mauna Loa volcano (Coan 1844) were the first direct observations of any eruption on this giant Hawaiian shield volcano, and are particularly important because they accurately describe the sustained transport of molten pāhoehoe beneath solid crust. After being awakened in his Hilo home on 10 January by a ruddy glow from far-away Mauna Loa, he decided to explore this major eruption firsthand, and made an arduous three-day climb through the rainforest to the open flat saddle between Mauna Loa and Mauna Kea [16]. Accompanied by a fellow cleric, but soon abandoned by his porters, he ascended the north flank of Mauna Loa's Northeast Rift Zone the next day, traversing the length of an active pāhoehoe flow and describing in great detail what he saw:

The lava on which we were treading gave indubitable evidence of powerful igneous action below, as it was hot and full of seams, from which smoke and gas were escaping. But we soon had ocular demonstration of what was the state beneath us; for in passing along we came to an opening in the superincumbent stratum, of twenty yards long and ten wide, through which we looked, and at the depth of 50 feet, we saw a vast tunnel or subterranean canal, lined with smooth vitrified matter, and forming the channel of a river of fire, which swept down the steep side of the mountain with amazing velocity. The sight of this covered aqueduct or, if I may be allowed to coin a word, this pyroduct – filled with mineral fusion, and flowing under our feet at the rate of twenty miles an hour, was truly startling. One glance at the fearful spectacle was worth a journey of a thousand miles. We gazed upon the scene with a kind of ecstasy, knowing that we had been traveling for hours over this river of fire, and crossing and recrossing it at numerous points. As we passed up the mountain, we found several similar openings into this canal, through which we cast large stones; these, instead of sinking into the viscid mass, were borne instantly out of our sight upon its burning bosom. (Coan 1844)

In describing the 1880–1881 Mauna Loa lava flow in his autobiography, Coan (1882, pp. 332–334) again defined "pyroducts," and specifically mentioned their role in insulating lava from heat loss:

As the lava first rushes down the steeper inclinations it flows uncovered; but its surface soon hardens, forming a firm, thick crust like ice on a river, and under this crust the torrent runs highly fluid, and retains nearly all of its heat. In this pyroduct, if I may so call it, the lava stream may pour down the mountainside for a year or more, flowing unseen, except where openings in the roof of its covered way reveal it.

A concise definition of "pyroduct," a marriage of Greek and Latin roots, is any internal channel, or conduit within a lava flow that feeds molten lava to an advancing flow front, or across the developing flow field downslope. This insulation of flowing lava in subsurface pyroducts is why pāhoehoe flows can travel such great distances and build such broad, gently sloping shield volcanoes. Measuring temperatures of flowing lava between holes in the crust has shown that heat loss during pyroduct transport is very slight, as little as 1.2°C/km of travel through conduits hidden within an active flow (Witter and Harris 2007). Coan's observations were not widely believed at the time, mainly because no formally trained scientist had ever had the opportunity to study an active lava flow firsthand. In fact, J.D. Dana, the influential Yale professor and editor of the *American Journal of Science*, challenged Coan's conclusions and wrote that what Coan had seen were actually deep volcanic fissures in the flanks of Mauna Loa – propagating downslope to feed the terminus of the 1843 flow (Dana 1852). But, the missionary was correct, and the professor was wrong. Nothing can sink an academic theory faster than good field observations!

Terminology

The term "pyroduct" was not widely adopted, despite Rev. Coan's sensible proposal. Instead, other words appeared as scientists began to describe similar features on other volcanoes. Perhaps the most used alternative today is "lava tube," although other expressions have also been introduced (Table 6.2). We reject usage of such terms including "tubes", "pipes", or "tunnels" to characterize what Coan described so well, and feel that "pyroduct" retains the practical value he intended because it is genetically descriptive. Other terms in volcanology also employ the Latin prefix "pyro-" including "pyroclastic" and "pyrocumulus," so in addition to historical precedent, Coan's "pyroduct" is hardly a semantic outlier in this branch of science.

Geometrically, "tube" is a flawed word because pyroducts are rarely tubular (cylindrical) in cross-section. Some are shaped like vertical slots with keyhole-shaped cross-sections, and others are nearly horizontal flat-roofed passageways, with their shapes continuously changing at multiple levels. Primary pyroducts near vents commonly evolve into slots or sheet-like lenses at distal parts of flows. "Pipes" describe man-made structures that carry fluids under pressure. Pyroducts are only rarely pressurized, and normally are only partially full of fluid. "Tunnels" are man-made structures that commonly facilitate foot or vehicle traffic, not molten lava!

When pyroducts are actively transporting molten lava, we refer to these structures as **lava *conduits***, but if molten rock drains away after cessation of eruptive activity to form long, sinuous caverns, we refer to these drained pyroducts simply as ***lava caves*** – not as *lava tubes* This follows the developing usage of volcanospeleologists – those specialists who explore, map, and describe the caves found in lava flows (e.g. Kempe 2012; Kempe and Kempe 2016). We are well aware that some small caves in lava flows have formed by processes that have nothing to do with the partial drainage of pyroducts, including erosion along faults or by coastal wave activity, but nevertheless refer to them all by the non-genetic, non-geometrical term "lava caves."

TABLE 6.2 Highlights in the development of terminology for subsurface lava conduits and caves.

Publication year	Reference	Description
1774–1775	Olafsen	"The running lava flowed through this channel like a river."
1779	Troil	"The upper crust sometimes cools and solidifies, even though molten matter keeps running underneath; in this way large caves form, the walls, floors, and ceiling of which are composed of lava and where a lot of dripstones of lava occur."
1799	Rosenmüller and Tillesius	They refer to caves (hölen) that "served for channels of the earthfire molten rivers."
1803	Kant	Kant states that lava caves (die Hölen) "dried from the outside while still being fluid inside" and that when "fluid finally dries up and retracts, then caves are created."
1844	Coan	Coan observed subterranean lava rivers during the 1843 eruption of Mauna Loa [15], correctly determined their genesis, and coined the term **pyroduct** to describe them.
1852	Dana	Dana used the term **tunnel** as he discredits Coan's observations.
1857	Coan	Coan again correctly described the genesis of lava conduits in the 1855–1856 Mauna Loa [15] flow, but referred to them as **subterranean pipes**.
1862	Scrope	Scrope (1862) refers to "caverns formed beneath the surface of a lava stream," with "pseudo-stalactitic projections made by the subsistence of the liquid."
1872	Scrope	Scrope referred to these features as **hollow gutters** and **caverns**.
1882	Coan	In his biography, Coan called these features both **subterranean ducts** and **pyroducts** in his recollections of the many Hawaiian eruptions he witnessed from 1840 to 1881. He wrote that "under this crust the torrent runs highly fluid, and retains nearly all its heat."
1884	Dutton	Dutton described then cold pyroducts of the 1880–1881 Mauna Loa [15] lava flow, and called them **lava tunnels**.
1896	Powell	Powell refers to these caves as **volcanic pipes**.
1909	Brigham	Brigham described these structures as **subterranean pipes**, and dismissed Dana's incorrect criticism of Coan's views with the words: "Those who have never seen a lava flow cannot well understand its action. I believe Mr. Coan's briefest account conveys a better idea [of these features] than the most elaborate theorizing of those who have never seen one."
1919	Jaggar	Jaggar used the terms **tunnel** and **tube** interchangeably to describe pyroducts of the 1919 Mauna Loa [15] eruption. This may be the first use of the term **tube** to describe these features.
1941	Anderson	Anderson popularized the term **lava tube** in his classic study of the Modoc Plateau lava caves.
1950	Perret	Perret favored the term **lava tunnel** in his important memoir.
2008	Numerous	"Lava tube" has unfortunately become the most commonly used term both for lava-filled conduits, and for drained pyroducts (caves) although they are still also referred to as "tunnels" or "pipes" in some publications.

Pyroduct Formation

On steeper slopes (near source vents) pāhoehoe flows are typically fed by well-defined channels, and if lava production rates are relatively constant and the eruption is long-lived, these channels will commonly crust over to form subsurface conduits; Coan's "pyroducts" (Figure 6.16). The formation of conduit roofs evolving from open channels can follow two processes: (a) narrowing of the channel rims by freezing of lava levees along channel walls, and (b) the accretion of plates of crust that are skimmed off channel surfaces where flow obstructions are encountered. Once a conduit roof segment is established, that roof forms a blockage for crustal fragments moving downstream and the roofed-over area will rapidly propagate upstream as more crustal fragments plate onto the upstream roof edge. Conduit roofs are also commonly thickened by new lava that may pour over them from overflowing channels or fresh extrusions of lava upslope.

Where pāhoehoe flows reach more gentle terrain, open air channel development mostly ceases and instead most lava is supplied by gravitational inputs beneath inflating lava sheets (Hon et al. 1994; Kempe et al. 2021). Such flowage tends to be concentrated along the most efficient pathways, which evolve into persistently active subsurface conduits as eruption continues. Sudden

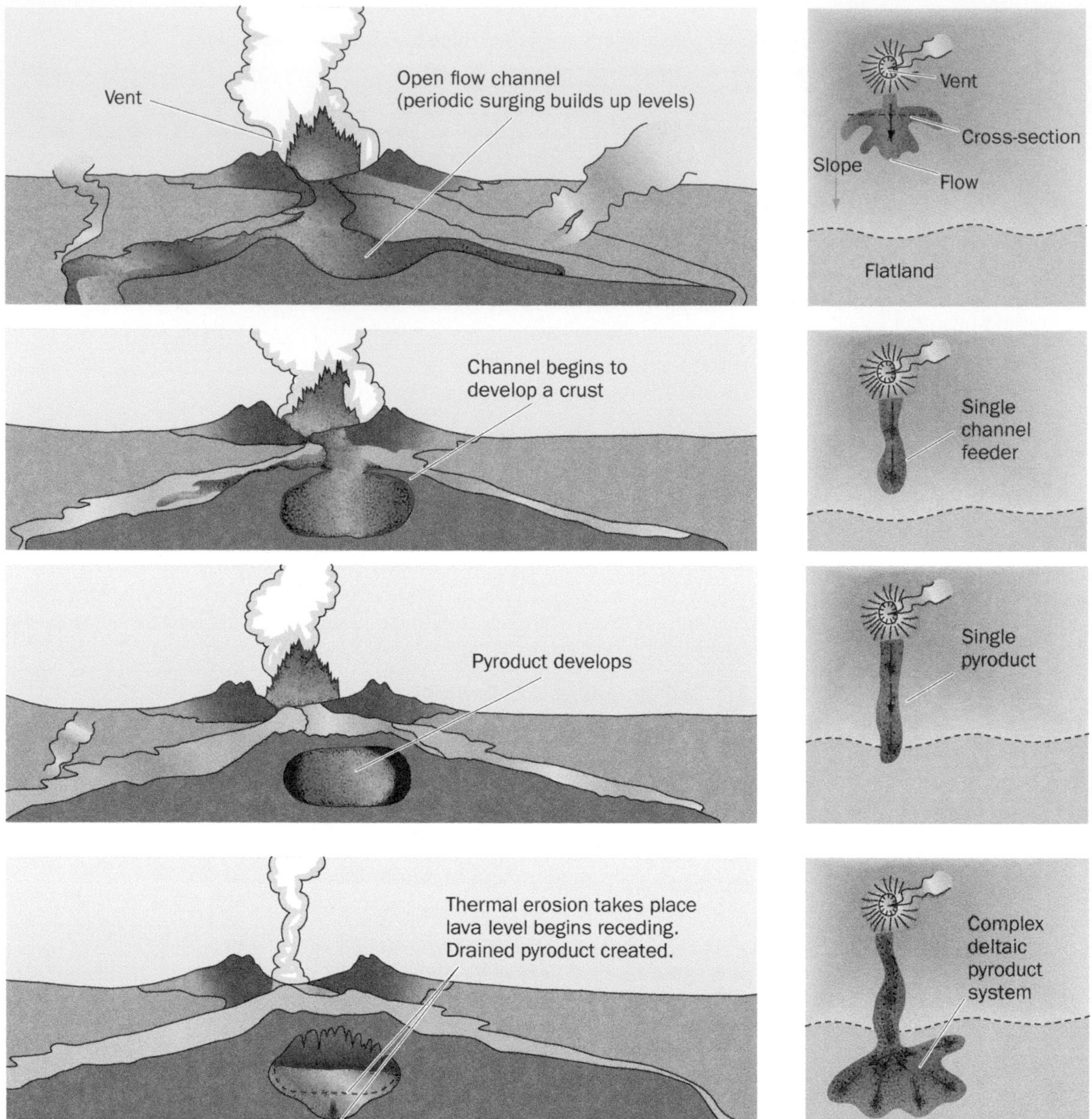

FIGURE 6.16 Mechanism for pyroduct formation, as an open lava channel crusts over during flow (view upstream toward source vent) and lava supply drains. Complex conduit systems may also form beneath inflating lava crusts on more gentle slopes, unrelated to surface channels. On plan views (side bars) dashed arrows indicate subcrustal flow.

extrusions of fresh molten lava may occur practically anywhere through the chilled crust overlying pressurized conduits in the lower reaches of an active flow, and in places the repeated raising and lowering of pressurized roofs can lead to the formation of circular accumulations of broken rock known as **shatter rings** (Figure 6.17; Kauahikaua et al. 2003). Where openings in the crust are small and develop nozzle-like geometry, escaping high-pressure gases from a flow conduit can blow spatter through them to pile up in steep-sided agglutinate pinnacles, or **hornitos** (Spanish for "little ovens"). The presence of the conduit can sometimes be traced by following a chain of widely spaced hornitos across a landscape (Figure 6.18). Sluggish movement may continue at a flow front even after an eruption has ended as the main flow channels or centralized conduit upslope continue draining into the narrower, more intricate "flow inflation" conduits at lower elevation.

Most likely all large pāhoehoe flows develop subsurface conduits to transport lava beneath their crusts as they advance. When active, their presence can only be inferred in most cases, unless **skylights** (large holes) in overlying crust reveal their presence below, as Titus Coan discovered (Figures 6.19 and 6.20). Skylights in a conduit roof provide useful opportunities for quantitative flow measurements and lava sampling. By throwing markers such as branches into an opening, which won't burn due to lack of atmospheric oxygen, the speed of flow can be measured by clocking the time required to travel to the next skylight downstream, and show that lava travels much faster when it is internally confined than it does when spreading directly across the surface. Skylights and cave entrances commonly remain to show the presence of partially drained conduits long after a flow has cooled. Where pyroducts did not drain and remain choked with lava, their presence can be observed only in erosional cross-sections (Figure 6.21).

FIGURE 6.17 A shatter ring formed in 1991 by the repeated flexing of a roof above a major active lava tube 5 km below the Kupaianaha lava lake, East Rift Zone, Kīlauea volcano, Hawai'i [17]. The ring is 58 × 46 m in plan view, and its rim ranges from 3 to 5 m high. Source: USGS photo by Tim Orr, looking to the southeast.

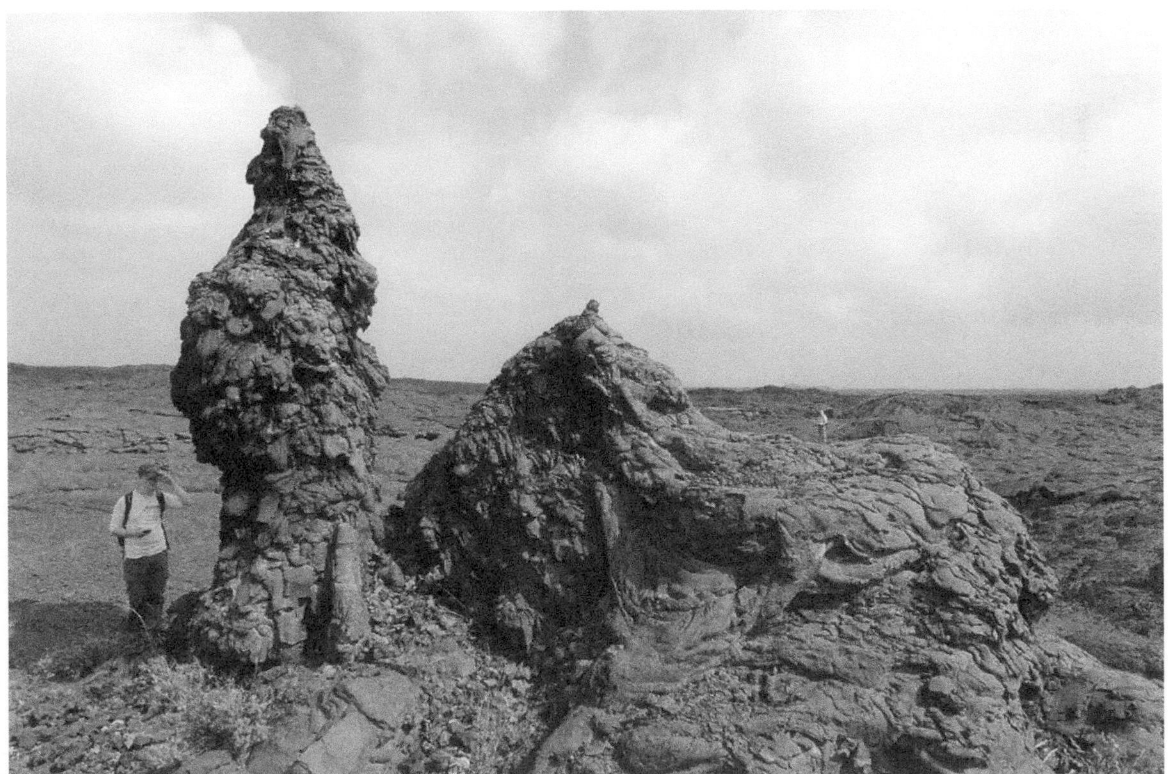

FIGURE 6.18 "Pele's Driblet Cone," lowermost and largest of a set of hornitos that formed on the west flank of Kīlauea volcano [17], above an active pyroduct in an early nineteenth-century pāhoehoe flow. Visitors named this feature over a century ago, but subsequent eruptions and relative inaccessibility have directed popular attention elsewhere. Source: USGS photo by Carolyn Parcheta.

FIGURE 6.19 The "Waha Mano" (Shark Mouth) window into a pyroduct of the 1859 lava flow, Mauna Loa, Hawai'i [15]. The red color was caused by high-temperature oxidation as atmosphere-contaminated gases poured out of this window, melting rock to form the stalactites visible on the pyroduct ceiling. Source: Photo by J.P. Lockwood.

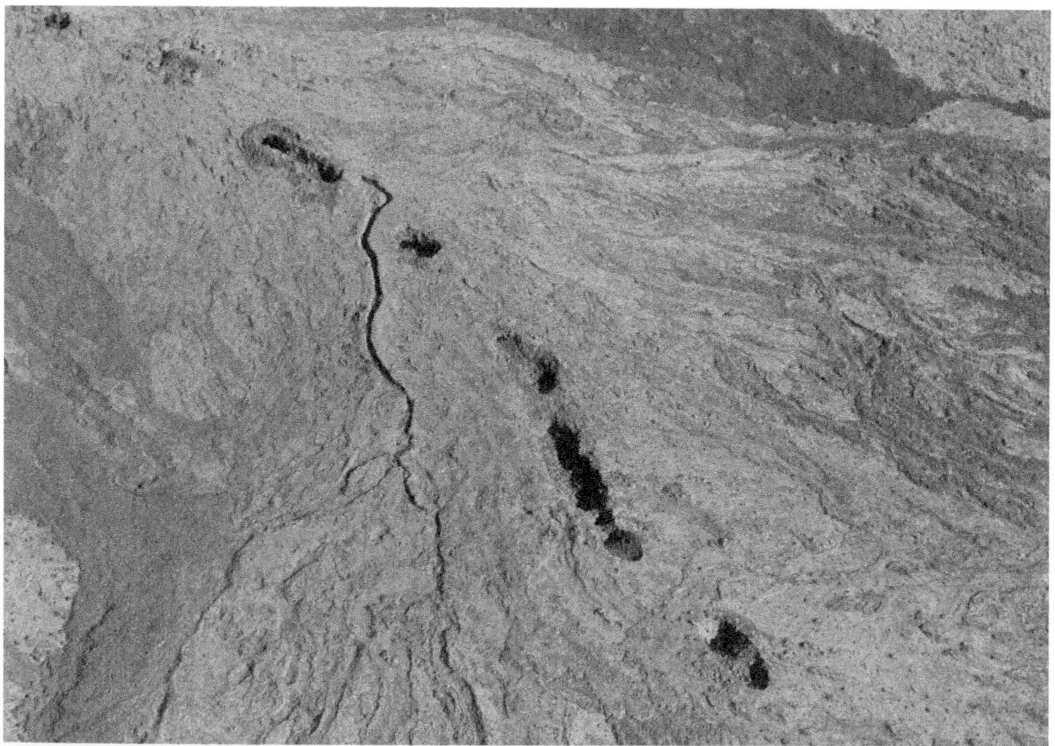

FIGURE 6.20 Series of "windows" or "skylights" above a prehistoric pyroduct on the southeast slope of Mauna Loa [15]. The skylights are 4–5 m in diameter. Source: Photo by J.P. Lockwood.

FIGURE 6.21 Cross-section of a filled pyroduct at Makapu'u Point, on the eastern end of Oahu Island, Hawai'i. Note the concentric banding, that records the narrowing of the pyroduct as the lava supply waned and lava began to freeze inwards. Source: Photo by J.P. Lockwood.

Exploration and detailed mapping of lava caves formed from pyroduct drainage (e.g. Kempe 2002) has revealed the complex dynamic history of these long-lived conduits as they evolve. They commonly deepen by thermal and mechanical erosion of their floors with time as lava continues to pour through them, and may develop complex intertwining passageways at multiple levels.

How exactly does this erosion take place? Some comparison with downcutting by water streams provides a clue. In that case, the rushing water uses sediment to abrade and scrape away the underlying rock. Similar abrasion by blocks embedded in the molten lava, perhaps knocked loose from walls and ceilings upstream as lava level declined, and simple plucking of rock fragments from the bottom of the current could take place – a form of *mechanical erosion*. But heat is a key factor too. Beneath a moving lava stream, heat cannot easily escape from the underlying rock. It becomes soft and plastic; easy to scrape or pry away mechanically. This form of **thermal erosion** is also thought to be important for creating the deep, slot-like canyons or rilles followed by certain flows on other planets (Schenk and Williams 2004). Kauahikaua et al. (1998) inferred downcutting rates of around 10 cm/day lasting for around 2 months in a conduit on the south flank of Kīlauea. In addition to heat, factors important for thermal erosion include character and strength of the substrate as well as temperature, volatile content in the molten lava, flow rates, turbulence, and flow duration (Williams et al. 1999).

Evidence for declining lava level inside deepening conduits is well preserved on the walls of lava caves. Commonly this includes bands in the walls, the tops of each showing lava stream levels at various stages of downcutting. From time to time, subsidence may pause, perhaps due to pyroduct obstructions downcurrent. The lava may freeze against the wall for a while, developing a smooth-sided accretionary shelf. The molten lava may even crust over entirely. As the level of the flow continues to decline, this will leave a second, deeper cave passage below, and the freshly formed crust becomes a **secondary ceiling**. Some large lava caves even have three levels of passages, stacked one atop the other with openings in their ceilings and floors permitting access from one to the other. Stacked passageways can be hazardous for exploration, since roofs may be thin and vulnerable to collapse.

Secondary ceilings may also form where large skylights break open the roofs of the master conduit. Sudden quenching of the molten current flowing through the pyroduct allows thin crusts to form across the flowing lava.

As erosion or drainage take place, the walls of the emerging cave can partly re-melt, developing glassy crusts that reflect light in cavers' headlamps (e.g. Kempe 2013). Long strips of re-melted lava may streak the vertical walls to either side of a cave passage. These may incline slightly or show deflection in the downslope passage direction, reflecting the movement of the liquid lava and hot gases down the conduit.

Just as water streams develop falls where they undercut resistant ledges by removal of weaker material beneath, **lava falls** (Figure 6.22) can also form in lava caves. Similar to waterfalls, lava falls cut back in the up-current direction after they begin forming, and may feature enormous plunge pools where the hot, fast-moving lava rushes to the bottom of each fall. Concentration of heat and escaping gases above some plunge pools may induce ceiling collapse directly above, domes that can propagate all the way to the surface forming new, hot skylights. Turbulence during flowage and de-gassing activity may produce delicate spatter that can adhere to pyroduct walls (Figure 6.23).

FIGURE 6.22 "Lava fall" within the upper Kazumura Cave, Kilauea volcano, Hawai'i [17]. This 10-m-high lava fall demonstrates the downcutting power of flowing lava when turbulent flow erodes a conduit floor, cutting through several older flows. Source: Courtesy of Peter and Annie Bosted.

The ceilings of lava caves are typically marked by lava stalactites (Figures 6.24 and 6.25; Kempe, 2013). These may resemble icicles, ending in twisted protrusions resembling bunches of grapes. They are formed by the freezing of molten lava dripping from the ceiling of the conduit. Sometimes, this is liquid left by the rapid lowering of the lava level, but most commonly, like the parallel streaks on cave walls, they are formed by a re-melting of the solid roof rock by the heat of burning gases, or by drips from pockets of molten lava above. If liquid continues to drip from the stalactites onto the floor of the cave after lava has stopped flowing, it will solidify to build up lava stalagmites (Figure 6.26).

Lava caves may extend as single, well-defined passages, or braid just as some streams do. Some have small primary side passages, generally toward their ceilings, that typically form as braiding shallow conduits during flow inflation. The master passage happens to be the one route that most lava eventually occupied to begin downcutting as the eruption continued.

Atkinson et al. (1975) describe the longest pyroduct-fed lava flow on Earth, in pāhoehoe flows from the 19-million-year-old Undara volcano [156], Western Queensland, Australia. The Undara lava cave system has an inferred length of over 160 km, but only parts of it are accessible for spelunkers today. On the island of Hawai'i, the 65-km-long Kazumura Cave is among the longest lava caves on Earth. Its main passage of 41 km in length can be traversed along most of its 1100 m of elevation change, though numerous lava falls, as tall as 20 m, require rappelling in places. Culturally, Kazumura and many other lava caves were important to early peoples as sheltered places to hide, cook, and sleep, for storage, burial, water supply, and religious worship. More recently, some lava caves in Hawai'i were designated as Civil Defense shelters during the Cold War.

Ultimately, pyroducts end downslope by feeding lava to advancing flow fronts through complex distributaries, analogous to the fanning of streams in river deltas, or through **breakouts** of molten lava that repave fresh surfaces farther upslope. In any case, molten lava may spend hours or days underground within a conduit before re-emerging, if indeed it erupts back onto the surface at all (Figure 6.27).

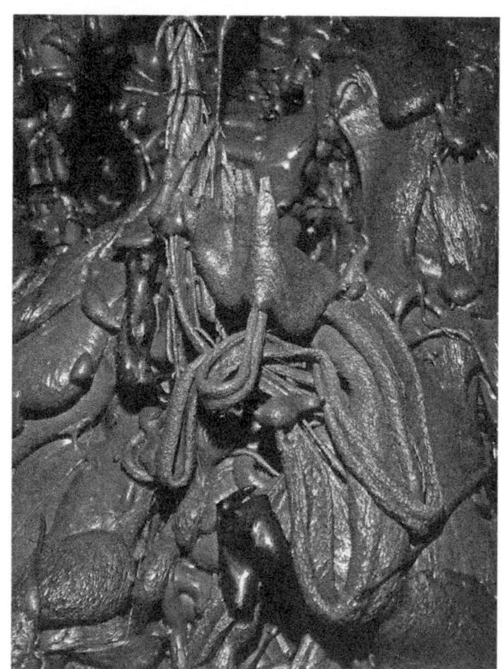

FIGURE 6.23 Spatter on the wall of a lava cave on the north slope of Hualālai volcano [14] associated with lava agitation at a "lavafall". Note the extreme fluidity of the spatter as it dripped down the wall, and also the variation in color, which demonstrates the changes in oxidation conditions as atmospheric oxygen varied within the active conduit. The area of this photo is about 14 × 20 cm. Source: Courtesy of Peter Bosted.

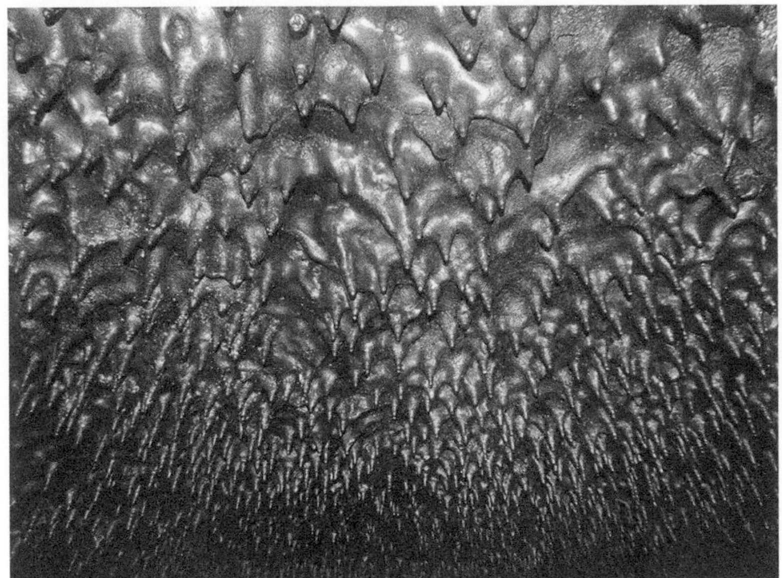

FIGURE 6.24 Stalactites formed by accretion of molten lava to the ceiling of a pyroduct on the flank of the eighteenth century Ai La'au shield, Kilauea volcano, Hawai'i [17], as lava levels fluctuated during flowage. Individual stalactites are 5–10 cm long. Source: Courtesy of Stephan Kempe.

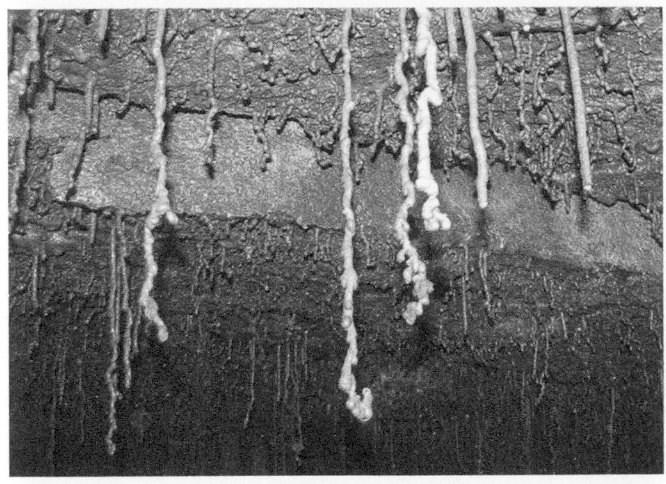

FIGURE 6.25 Less-common cylindrical stalactites (to 25 cm length), formed by secretion of molten lava from fluid pockets within the lava cave ceiling. The contorted ones are called lava helictites. Source: Photo by J.P. Lockwood.

FIGURE 6.26 Lava stalagmites, formed by the dripping of fluid lava from stalactites above. These are uncommon, as they are usually carried away by flowing lava on developing lava cave floors. Source: Photo by J.P. Lockwood.

Pāhoehoe Surface Structures

Many pāhoehoe flows have relatively flat upper surfaces – extending horizontally for kilometers – "flat enough to ride a bicycle on" (although it would be a bumpy, slow ride!). More typically, the upper surfaces of pāhoehoe flows are marked by a characteristic "mini-topography"; with relief on a scale of a few meters, including discrete hillocks called tumuli, uplifted blocks called lava rise terraces, broad depressions, and steep-walled pits unrelated to caves, termed inflation pits (Figures 6.28 and 6.35). Lava channels are commonly present in the upper reaches of a flow, and as described in the previous section, may be roofed over in places to form pyroducts.

FIGURE 6.27 The end of a long subterranean trip – from Earth's mantle, to magma chambers beneath Kilauea's [17] summit, down the East Rift Zone, and through shallow conduits to the coast! The erupting end of a lave conduit in the background feeds a lava channel at the ocean coast in 2009. The channel has begun to roof over by accretion of frozen lava at its edges, and became covered shortly after this photo was taken, extending the source pyroduct directly to the ocean entry. Source: Photo © G. Brad Lewis.

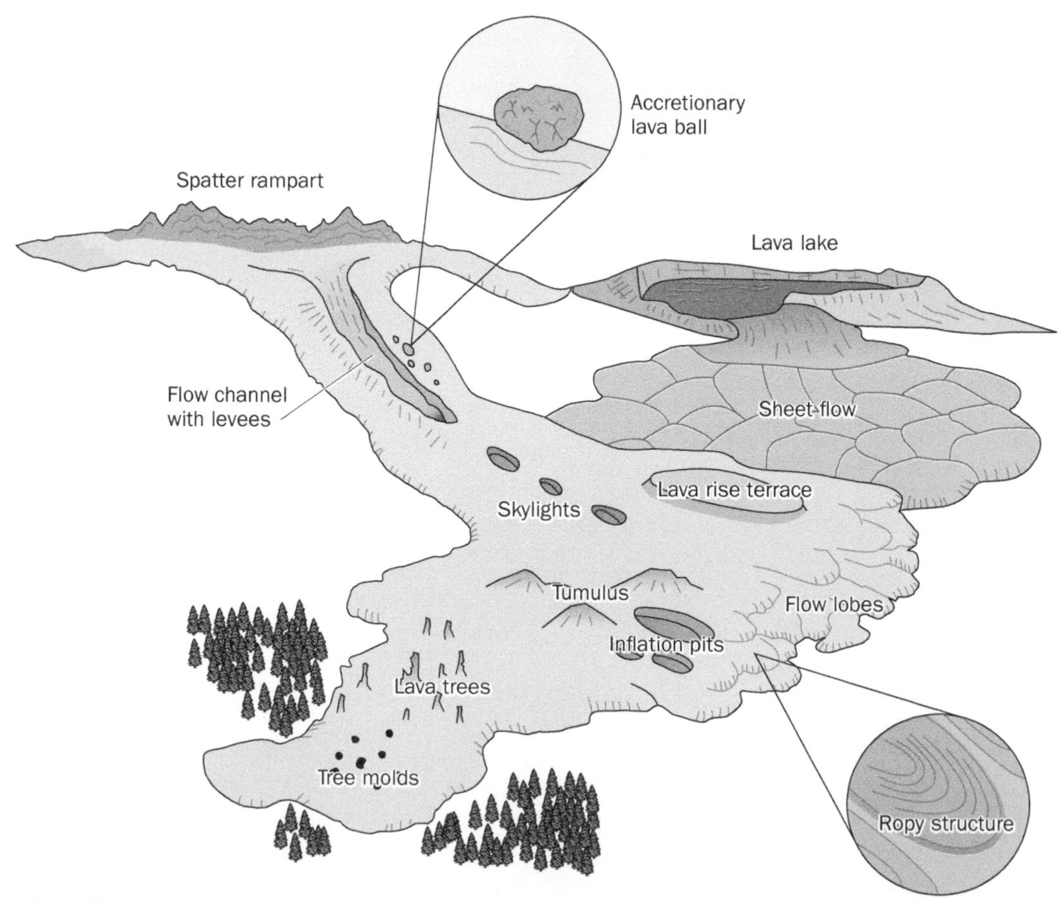

FIGURE 6.28 Distinctive surface features of pāhoehoe flows.

Such irregular features are characteristic of most pāhoehoe flows, especially those resulting from basaltic eruptions characterized by the relatively steady production of lava over long time periods. In some high-discharge effusive eruptions, however, extremely rapid extrusion onto nearly level ground causes the lava to spread out as laterally extensive, interfingering **sheet lobes** (Figure 6.32). Cracks separating low convex plates meters in diameter may be the only features marking the surfaces of these "hummocky parking-lot" flows.

The initial emplacement of pāhoehoe flows may only be the beginning of their development since molten lava from source vents often continues to inject the interior of flows sometimes for weeks or months after they develop rigid upper crusts and stop spreading across the landscape. As a result, like bread rising in an oven, the lava flow **inflates** (Hon et al. 1994; Walker 2009). This, in fact, is when much of the meso-topography described above may develop. Gradual inflation, combined with scattered breakouts from pyroducts can thicken a flow after initial emplacement by as much as several times. Features originally at the surface – including fine-scale pāhoehoe ropes (see below) and objects engulfed by the original flow, such as charred tree trunks, old cars, and bathtubs – will remain on that original surface as the flow's inflates dramatically beneath them. Other objects, fixed to the original ground beneath the fresh lava, may be encased more deeply, or even covered by the flow as it swells up around them (e.g. Figure 6.30).

Survival Tips for Field Volcanologists

When crossing irregularpahoehoe terrane , make the effort from time to time to climb to the top of a tumulus rather than walk around it. The air will be much cooler, refreshing breezes will be stronger up there, and the high ground makes a good vantage point to plan the easiest traverse path ahead.

Multiple breakouts and lobe extrusions from a flow front advance the flow incrementally. The smaller lobes, termed **pāhoehoe toes** (Figures 6.29 and 6.31), may measure only a few centimeters across. Larger ones may span tens of meters in width. This process repeats countless times, so that a cross-section of a typical pāhoehoe flow will appear as a great stack of flattened sandbag like masses, many of them hollow from later drainage of the conduits that fed them (Figure 6.32; Gregg and Keszthelyi 2004).

While some surfaces of stationary pāhoehoe flows inflate, other portions will **deflate**, especially near a source vent as lava drains away and rapidly loses its dissolved gases. Deflation in the proximal (higher elevation) part of flow is to be expected unless the rate of lava supply (vent discharge) is maintained or even increased. Most flows show evidence of simultaneous flow deflation near their vents with corresponding flow inflation at their lowland termini. Even at the scale of a single outcrop, only a few meters across one can see examples of inflation in one area and deflation in another, reflecting local redistribution of subsurface pāhoehoe in the final stages of solidification.

Everywhere one looks across the surface of a fresh pāhoehoe flow, rinds of glass provide a medium gray to black surface that may be highly reflective and take on a silvery, almost polished metallic appearance in sunlight. The rinds may only be a few millimeters thick, and are rarely thicker than a centimeter except where flows have been quenched by water. Surface glass grades inward into the microcrystalline interiors of flows through zones of mixed glass and microlites. Exposed to the elements and subject to devitrification, the glass breaks down readily, and a pāhoehoe flow develops a dull flow surface that transitions to tan and then red-brown colors over hundreds or thousands of years.

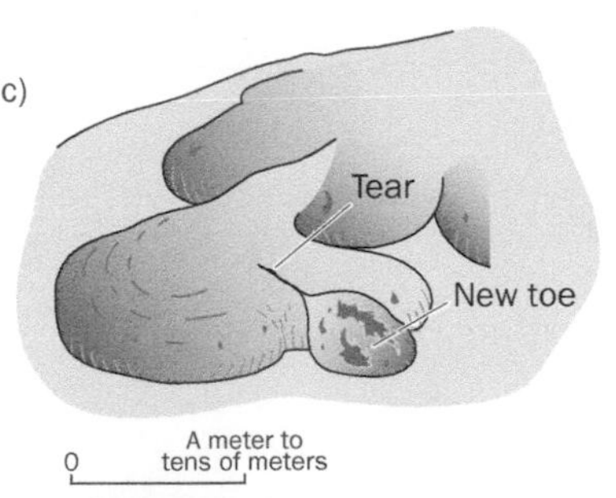

FIGURE 6.29 The fronts of pāhoehoe flows commonly move by spreading in lobes that freeze, then split as molten lava pressure rises inside, to feed new lobes.

Tumuli

Nineteenth-century geologist Reginald Daly introduced the term **tumulus** (plural **tumuli**) to refer to the steep-sided mounds found on surfaces of many pāhoehoe flows. The word had earlier been used to describe native prehistoric burial mounds in Europe and the eastern United States, and before that was applied by early Romans to identify the rounded, convex shields of their legionnaires.

Tumuli may be gentle domes or rugged obstacles with steep to overturned sides difficult to climb over and easier to walk around. They range in size to 15 m or more in diameter and commonly 1.5–3 m high (Figure 6.33), but in places are much larger; most are oval in ground plan, but some are nearly circular. Many are rent by deep open fissures at their tops – openings that formed as their once-flat surfaces were broken apart by intruding lava below. The tumulus crust commonly splits apart during uplift, and molten lava may squeeze up through the fractures to feed small flows that dribble down the sides, or they may explosively fail (Figure 6.34). The molten lava beneath large tumuli may drain away laterally, leaving grottos large enough to admit people (Halliday 1998). Not all tumuli form by uplifting of the original surface,

FIGURE 6.30 Infiltration of a wall of unmortared stones by an inflating pāhoehoe flow enclosing and partially inundating the compound of the thirteenth-century Waha'ula Heiau, a prehistoric Hawaiian sacrificial site on the southern coast of Kīlauea volcano [17]. Note the fingers of pāhoehoe that have oozed between the stones, showing how fluid this lava was when it was molten. Source: Photo by R.W. Hazlett.

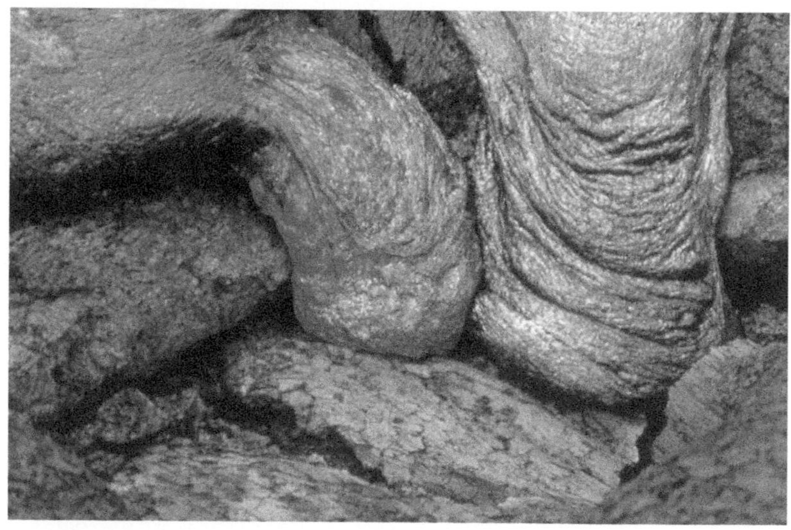

FIGURE 6.31 Flowing pāhoehoe toes on the flanks of Mauna Ulu, Kīlauea volcano, Hawai'i [17] in April 1974. The incandescent, fast-moving toe on the left is about 20 cm in width. The toe on the right is about 3 minutes old, and has developed a flexible, black glassy crust about 1 mm thick over a molten interior. This toe is still moving sluggishly, bounded by "strike-slip faults" in the crust along margins. Source: USGS photo by D.W. Peterson.

however. Some may form as lava is withdrawn from surrounding areas and the overlying crust subsides. In fact, it may be impossible to judge which surfaces went up in absolute terms and which went down – even if one is standing nearby as the surface slowly deforms!

Anderson et al. (2012) note (p. 1):

> *"Tumulus formation requires the proper combination of [lava] cooling and effusion rate. If cooling is too extensive and effusion rate is too low, the crust will provide too much resistance to bending. If cooling is too limited and effusion rates too high, crusts will not develop or have insufficient strength to resist fracture and subsequent breakouts . . . [T]umuli are rarely found over well-established [lava conduits] that typically have rigid walls/overlying crusts that exceed 2 m in thickness and provide too much resistance to bending. Silicic flows lack tumuli because the viscosity gradients within the flow are insufficient to concentrate stress in a localized area."*

FIGURE 6.32 Multiple pāhoehoe flow lobes formed during a single eruption of Kīlauea volcano [17] in 1972. Three separate lobes are shown here, each separated by a zone of red oxidation. Note the typical hollow voids in the overlying flow – caused by the coalescence and expansion of exsolved gases during flow. Source: Photo by J.P. Lockwood.

FIGURE 6.33 3 m. high tumulus on a 2003 pāhoehoe flow, Kīlauea volcano, Hawai'i [17]. This structure was formed by upwelling lava that broke open on the once horizontal, now uplifted surface and oozed out at the top. Note well-developed "ropy pāhoehoe" on the uplifted surface. Source: Photo by J.P. Lockwood.

Inflation Pits

Inflation pits are common on some pāhoehoe surfaces, and have been erroneously referred to as "collapse pits." These features may be less than a meter across or tens of meters in diameter (Figure 6.35). Most have near-vertical or overhanging rims, and they may be several meters deep. They certainly *look* as though they formed by collapse processes, but most form by differential uplift of the surrounding crust. I (JPL) have stood on inflating pāhoehoe flows as these pits slowly develop over a period of hours, and while visually persuaded they were subsiding, instrumental measurement showed in fact that the (hot) crust on which I was standing was going up – while the floors of the pits remained stationary.

FIGURE 6.34 "Exploded tumulus" on the southern flank of Kilauea volcano's [17] East Rift Zone. The displaced block (left center) is about 2.5 long, 1 m high, and weighs an estimated one tonne. It was carried about 5 m from the side of the tumulus by the force of the molten lava bursting forth. Source: Photo by J.P. Lockwood.

FIGURE 6.35 Inflation pit on the 1843 pāhoehoe flow, Mauna Loa volcano, Hawai'i [15]. Although these features indeed look like they formed by collapse, they actually form as the surrounding pāhoehoe flow is inflated by the subsurface injection of molten lava, leaving the original, uninflated crust behind. Source: Photo by J.P. Lockwood.

Surface Textures

Geologists have termed the surfaces of pāhoehoe as "ropy," "corded," "elephant-hide," "shark-skin," "entrail," "festooned," "filamented," "shelly," and "slabby" (transitional pāhoehoe). Some surfaces combine aspects of different types, for example, "shark-skin ropy pāhoehoe." Each term usefully conveys an image of physical appearance. It is beyond the scope of this book to discuss them all in detail. But consider the three following, distinctive examples.

> **Survival tips for Field Volcanologists**
>
> When crossing active pāhoehoe lava flows with well-solidified crusts, the surface may seem relatively cool, and there might be no indication of flowing lava below. Tumuli can slowly form and build up internal pressure from internal molten material. If such tumuli begin to make cracking sounds, better stand aside, as they can explode with no other warning!

Ropy pāhoehoe A characteristic feature of pāhoehoe is the occurrence in some areas of crust that is wrinkled and twisted into forms resembling the folds in heavy cloth, or parts of coils of rope (Figure 6.33). These forms, called **ropy structure**, result from the dragging and twisting of the thin, hot, plastic crust of the flow by movement of the liquid lava underneath. Friction results in the edges of narrow lava streams moving less rapidly than the middle. Where they form on the crust of such a stream, the lava ropes are bent into curves that are convex in the direction of the stream movement. Such a directional indicator does not provide a good sense of the overall motion of a lava flow, however, since many displays of curved ropy structure result from small eddies and other chaotic local movements. Statistically significant measurement of many rope structures must be made to gain an idea of the overall flow direction of an old lava flow in terrain where the original lay of the land is no longer evident. Ropy structure is glassy and somewhat fragile. Even so, pāhoehoe ropes may still be found preserved in outcrops hundreds of millions of years old (Figure 6.36).

Entrail pāhoehoe While ropy pāhoehoe tends to form on gentle to moderate slopes, cascading pāhoehoe on steep slopes may form **entrail pāhoehoe** instead. The name alludes to the startling appearance of this lava type, like a great heap of intestines, which consists of masses of overlapping pāhoehoe toes, each of which grew quite elongate owing to the pull of gravity on the slope (Figure 6.37).

Shelly pāhoehoe Near overflowing vents, rapid outpouring of gas-rich, low-viscosity pāhoehoe leaves a surface of fragile, glass-rich plates, in many places twisted into large folds. This crust overlies numerous void spaces, some as deep as several decimeters. To walk across such lavas without sturdy boots, long sleeves, tough trousers, and gloves, is to risk deep cuts, as one inevitably will break through surface plates. It feels and sounds like walking on eggshells, hence the name: **shelly pāhoehoe**. This lava is so fragile that it does not survive for more than a few thousand years before breaking down.

Shelly pāhoehoe describes well a type of lava Swanson (1973) termed "fountain-fed pāhoehoe," since it is typical of pāhoehoe close to lava fountain sources where considerable gas is entrained in lava. It is extremely dangerous to attempt crossing while

FIGURE 6.36 Proterozoic ropy pāhoehoe about 1.9 billion years old preserved on Flaherty Island near Sanikiluaq, Nunavut, Canada. Source: Photo by Richard Bell.

FIGURE 6.37 Entrail pāhoehoe, formed where fluid pāhoehoe lava tongues flowed down a cliff face in 1972, Kīlauea volcano, Hawai'i [17]. Source: Photo by J.P. Lockwood.

still active. Woe on volcanologists who break through these shells – they will never forget the smell of burning leg hair as high-temperature air rushes up pants legs! As pāhoehoe flows further away from vents, sufficient gas will be released that flow surfaces generally become stronger and more stable. Where molten pāhoehoe travels for any distance underground via pyroducts, it generally is extensively de-gassed and forms the smooth-surface flows Swanson termed "tube-fed pāhoehoe."

Survival Tips for Field Volcanologists

Channel levees are normally elevated above the surrounding terrain and the flowing lava below, and make excellent vantage points for viewing, but care must be taken to approach them from the upwind side, and to be alert for signs of instability. The sides of active lava channels, whether in pāhoehoe or 'a'ā flows, are potentially hazardous as they are often unstable and subject to sudden collapse. Also, because lava in these channels is generally agitated and not covered by lava crusts, they are extremely hot environments with copious sulfur fumes close to erupting vents. An additional risk are strong whirlwinds of heated air rising next to lava channels, which can fill the air with pieces of sharp volcanic glass. Some volcanologists call these "lav-nados." I (JPL) have witnessed "lavaspouts" form over new lava lakes that sucked up thin crust fragments and molten lava several meters above lake surfaces.

Environmental Conditions of an Active Pāhoehoe Flow

To witness actively flowing pāhoehoe may be a little disappointing at first – especially from a distance and during daylight hours. Most of the active parts of the flow are covered by apparently stationary, silver-gray crust, with molten lava only visible in a few places, mostly near flow margins where lobes of blood-red to orange fluid ooze from cracks in plates of upraised crust here and there. There is almost no conspicuous noise, except for the occasional creaking or rasping sound as one section of crust moves relative to another. Barely audible may be a continual "clicking" noise, caused by tiny chips of glass spalling off cooling, contracting surfaces.

The surfaces of flowing pāhoehoe cool very quickly, and a black crust 2–3 cm thick forms within a half hour or less. Such crusts can be walked on carefully with heavy boots, although they are very hot, can sag underfoot, and may cause boot soles to catch fire if one stands in one place too long. Beneath those thin black crusts on active flows molten lava continues to flow, and fluid pressure often breaks those crusts and allows lobes of fresh melt to ooze out unexpectedly. The splendors of an active pāhoehoe flow are greatest after twilight. Whereas "red rock" is hard to see during the day, "fire" will be visible everywhere at night, and myriad cracks in the lava show incandescent yellow orange lava below. Molten rock reflects off vapors at night, creating a dull red color in the sky. Numerous active outbreaks of molten lava not visible from afar in sunlight, may be seen in the dark.

Active pāhoehoe flows are continually releasing hot gases, especially near source vents. The flows lose their volatiles from exposed melt and from cracks that penetrate the hot crust opened during movement and cooling. An especially active portion of a spreading flow surface can easily be spotted from a distance by a combination of heat-shimmer and emerging brown to bluish-gray fume. Geologists quickly learn to approach active pāhoehoe flows from the upwind direction, since the heat and choking fumes blowing downwind from such flows are often unbearable. I (JPL) came close to death when standing near the margin of a large expanse of ponded pāhoehoe a few hundred meters below the Mauna Ulu vent of Kīlauea in 1974. The wind suddenly shifted direction and caught me in a plume of hot, sulfur-rich gas. I tried to run, but soon found that my throat had closed shut – an automatic physiological reaction to limit lung damage from breathing SO_2.

Not only the ground below, but the sky above an active pāhoehoe flow may show dramatic change as an eruption progresses. People who fight forest fires are well aware of the phenomenon of cloud formation caused when hot, moist air rising high above the fire condenses. These are *flammogenitus* ("fire-begotten") *clouds*. Similarly, vapor clouds characteristically develop above open lava streams and erupting vents from the condensation of rising hot air. They are called **pyrocumulus clouds** to distinguish their origin (Figure 6.38). Where these convectively expanding clouds reach sufficient altitude and size, they transition to **pyrocumulonimbus** clouds, capable of generating heavy rainfall below. Such rainfall may become somewhat acidified if it drops through clouds of sulfur dioxide and other volcanic gases, a process called **atmospheric scavenging** (Cuoco et al. 2013). In 1987, I (RWH) led a field party across a wide pāhoehoe flow several days old to visit the Kupaianaha vent on Kīlauea. The excursion began under a sunny, tropical sky. But by noon it was raining, and after examining the vent, the team became immersed in the midst of a lava-heated steam cloud too dense for even reading a field compass because of condensation inside the compass glass. By "dead-reckoning" several kilometers through this natural sauna, the group safely found its way back, warm, and thoroughly soaked. But this was certainly good luck. In places where heavy rain or wave splash falls on openly flowing lava, steam may be scalding, even deadly.

FIGURE 6.38 Small "Pyrocumulus" cloud formed over a major ocean lava entry along Kilauea volcano's [17] south coast in January 2013. Such clouds commonly form over lava flows when convective heat causes condensation of water vapor over heat sources, and can cause intense rainfall when they become unstable. Source: Photo © Bruce Omori.

The World's Most Fluid Pāhoehoe

Very fluid pāhoehoe can move several tens of kilometers per hour down steep slopes, and may leave residual flows only a few centimeters thick. During the 1977 eruption of Nyiragongo volcano [118] in central Africa, exceptionally low viscosity, silica-poor nephelinite lava left residual flows less than 2 cm thick on the steep upper slopes of the volcano (Tazieff 1977). During the 2002 eruption, the margins of residual lava flows surrounding small islands of older land (kīpuka) were less than 1 mm thick in places near source vents (Figure 6.39), although the "high lava marks" on adjacent burned trees showed the flow had initially been about 2 m thick as it rushed downslope. The thin flow margins consisted of glass quenched on the cold soil. The fact that voluminous lava had traveled down the mountainside without leaving more material behind shows the lava had no yield strength, and was in fact an exemplary Newtonian fluid (Chapter 4). During the same eruption, fast-moving, 2–3-m-thick fluid flows rushed past banana plants, burning off leaves, but not toppling the fragile plants. In comparison, the more viscous, "highly fluid" basaltic pāhoehoe flows in Hawai'i and at Piton de la Fournaise volcano [125] in the Indian Ocean are typically a few decimeters thick on their leading edges, and thicken within a few hours through flow inflation. Extremely fluid nephelinite lavas drained from Nyiragongo's summit crater again in May 2021, destroying homes in the outskirts of Goma.

Pāhoehoe, Forests, and Fossils

Where pāhoehoe sweeps through forests, it can freeze around living trees, sometimes forming casts around them. These will remain standing as the surrounding, more liquid lava drains away (deflates), forming ghostly hollow pillars of basalt called **lava trees** (Figure 6.40; Lockwood and Williams 1978). During the 1962 eruption at Aloi Crater on Kīlauea, lava trees as tall as 8 m were formed – the highest yet reported. Lava trees are fragile, geologically ephemeral features, and are never preserved on older flows (Figure 6.41).

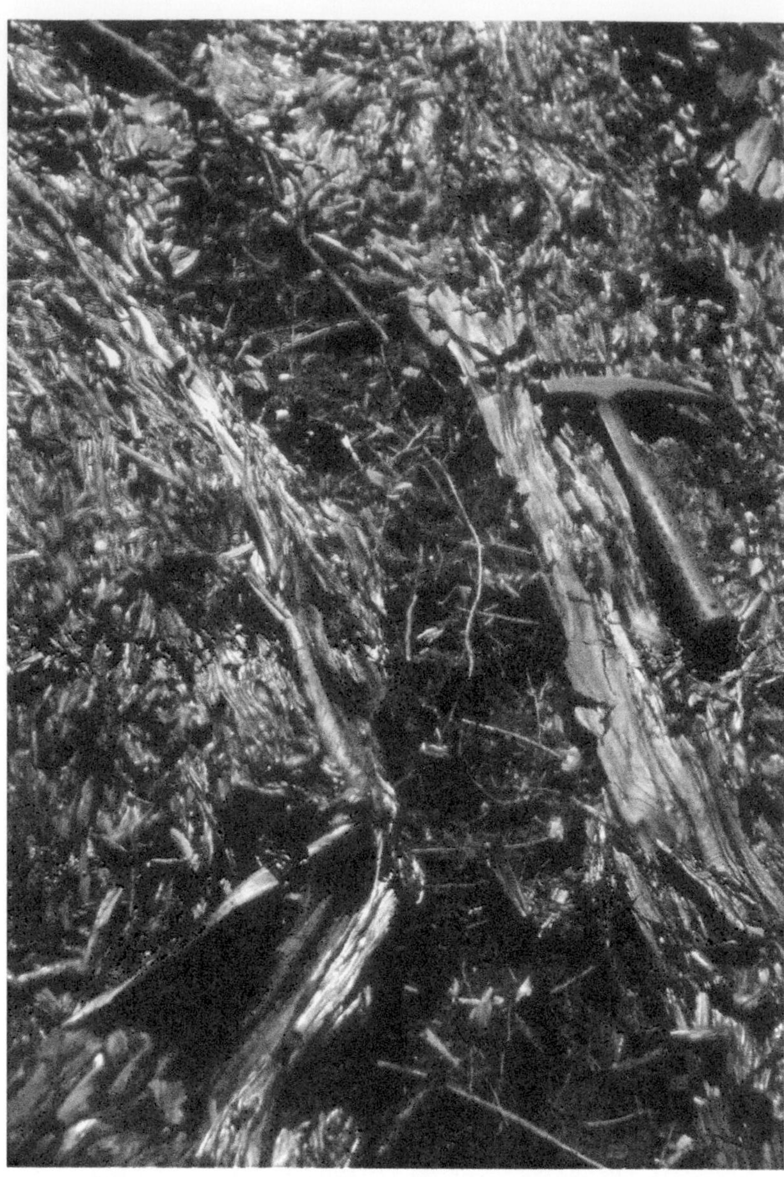

FIGURE 6.39 Small Kīpuka surrounded by 1 mm thick pāhoehoe flow, Nyiragongo volcano [118] eruption, 2002. Source: Photo by J.P. Lockwood.

Elsewhere, where deflation does not occur, the trunks of trees that burn away completely in the hot lava may remain preserved as cylindrical holes in the flow surface, called **tree molds**. Unlike lava trees, tree molds are much more durable, and are commonly preserved in once-fluid lavas thousands of years old. Some tree molds project vertically through the upper surfaces of flows (Figure 6.42); they are more commonly found sub-horizontally, representing trees that were knocked down by the flow, encased in frozen selvages, and carried along as their contained wood carbonized and burned away, or was destroyed by later biological activity. Tree molds as much as a meter wide and 4 m deep occur in some Mauna Loa [15] lava flows, and where obscured by vegetation can pose a hazard to travelers.

Lava trees and tree molds commonly preserve features that enable one to determine the direction of lava movement. Flanges or collars around the outside of a lava tree may slope and taper in the direction of original flow movement. And within a tree mold, the passage of lava around the mold when it was still a standing trunk may leave a tell-tale seam where the parting lava converged on the opposite, downcurrent side (Figures 6.40 and 6.42; Lockwood and Williams 1978).

Not only whole tree trunks, but delicate fossil impressions may be preserved in pāhoehoe too. In the walls of tree molds, one can occasionally find a detailed lattice of centimeter-scale ridgelines, the finer molds of cracks that formed in heat-shrunken wood, even to the scale of paper-thin partitions of lava. On the bases of flows, delicate impressions of plant stems and fern fronds can also be preserved (Figure 6.43). The lava forms thin selvages of glass, in part from quenching by water boiled out of the organic matter. The selvages prevent molten lava from filling in the holes left as the enclosed organic matter burns away.

Animal fossils are rare in lava, but a possible mold of an entrapped prehistoric mini-rhinoceroses was found within Columbia River basalt flows in Washington State (Chappell et al. 1961), and elephants killed by extremely fast-moving lavas on the slopes of a)

b)

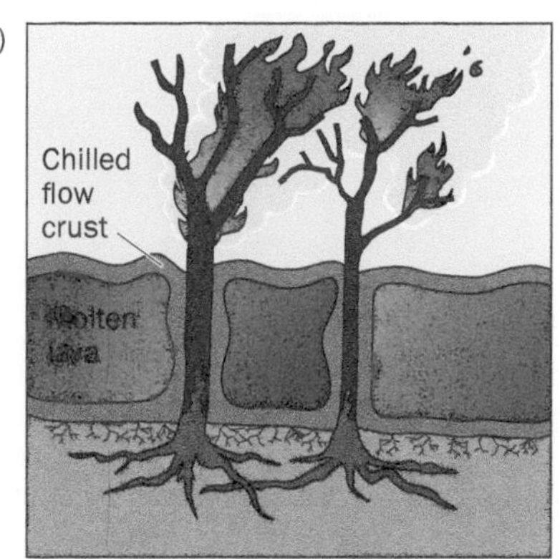

c)

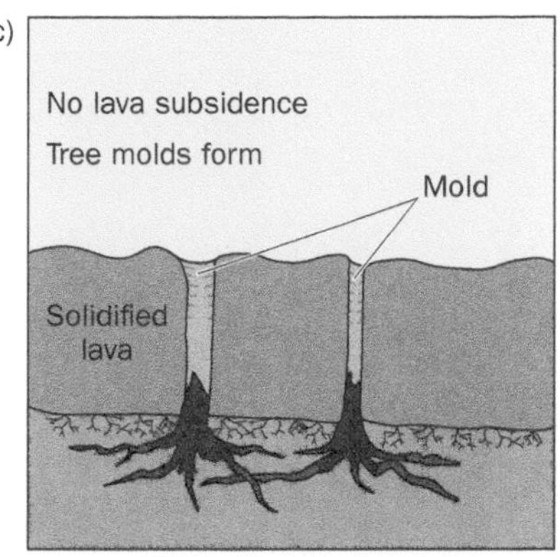

d)

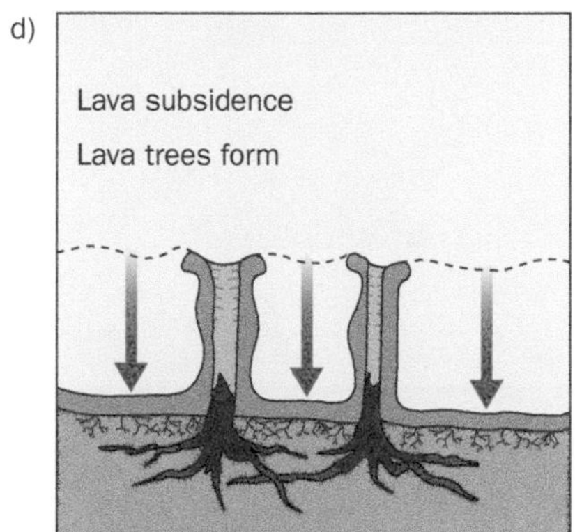

e)

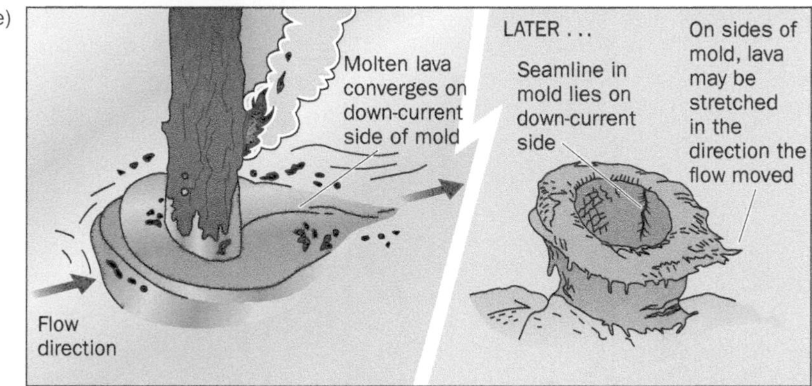

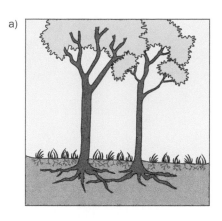

FIGURE 6.40 A pristine forest (a) is entered by lava (b). The lava subsequently chills around tree trunks, leaving deep molds (c); or, if after chilling, the lava recedes (d), it creates standing hollow pillars called lava trees. (e) Lava trees and tree molds both can preserve seam lines that serve as indicators of original flow direction.

FIGURE 6.41 Lava trees formed during the July 1974 eruption of Kīlauea volcano, Hawai'i [17]. These lava trees are about 3 m high, and indicate the thickness of the lava flow that surrounded these trees, before the molten flow interior drained away downslope and the crust subsided to the present level. Source: Photo by J.P. Lockwood.

Nyiragongo volcano [118], Zaire, in 1977 left perfect molds, down to the details of their trunks (Figure 6.44). I (JPL) tried to locate these "elephant molds" in 1989, but found out that farmers had discovered they could be excavated to reach underlying soil at shallow depths, and all had been planted with banana trees. The numerous clumps of bananas scattered about the field of pāhoehoe indicated where a herd of elephants had perished in 1977.

FIGURE 6.42 Lava tree mold preserved within a 1974 flow from Kilauea volcano, Hawai'i [17]. Note tree mold seam (arrow), which shows that flow was moving from left Lava tree mold preserved within a 1974 flow from Kilauea volcano, Hawai'i. Note tree mold seam (arrow), which shows that flow was moving from left to right at the time the tree (now burned away) was engulfed by flowing lava. Source: USGS photo by J.P. Lockwood.

FIGURE 6.43 Imprint of palm frond and other plant remains at the base of a 4000-year-old pāhoehoe flow, Hilina Pali, Kīlauea volcano, Hawai'i [17]. Source: Photo by J.P. Lockwood.

'A'ā Surface Structures

'A'ā flows are characterized by exceedingly rough, rubbly surfaces, in contrast to the relatively smooth surfaces of pāhoehoe flows. These surfaces consist of loose, angular, jagged fragments, many of which may be covered with tiny sharp spines. Vicious-looking angular spires of 'a'ā may project above flow surfaces (Figure 6.45). 'A'ā is commonly extremely difficult to cross without losing balance; leather gloves are well-advised, as falling can lead to some serious scrapes and cuts. It is difficult to convey with words an adequate idea of the roughness of a typical 'a'ā flow. It must be seen and walked on to be appreciated fully. Gordon Macdonald (1972) described the perils of 'a'ā well, when he wrote:

> *While mapping the active Hawaiian volcanoes, we had to cross and recross many kilometers of 'a'ā. The loose fragments rolled under our feet, and we fell frequently. Repeated lacerations of our hands soon taught us to wear gloves, and heavy work boots were often cut to ribbons in a week of hiking.*

An 'a'ā flow advances as the draining molten **flow core** in its interior overrides clinkery rubble shed from an oversteepening flow front. The rubble, which is termed **autobreccia**, originates as the lava surface chills and tears itself to shreds as it cools. Where overridden, the rubble accumulates as a **basal breccia** pad, matching in general appearance the surface breccia layer above, with

FIGURE 6.44 Bone-filled elephant cast in 1977 pāhoehoe near the base of Nyiragongo volcano [118], Democratic Republic of Congo. This pygmy elephant was about 2 m long. Note the mold of the elephant's trunk at the top of the photo (arrow). Source: Photo by Katia Krafft.

FIGURE 6.45 Spiny ʽaʽā surface of 1999 Mount Cameroon [106] lava flow, Cameroon. Source: Photo by J.-B. Katabarwa.

the flow core sandwiched in between. Though not a perfect analog, the moving tracks of a caterpillar tractor provide a useful image of the overall process (Figures 6.46 and 6.47). Because of their higher relative viscosities, active ʽaʽā flows may seem unimpressive as they advance very slowly across flat terrain. There may be little obvious movement of the flow, other than episodic cascades of lava fragments down the frontal face of a large, thick flow. Their molten interiors are mostly masked during the daytime by their overlying carapace of dark fragmentary material. At night, however, their appearance changes dramatically, as illustrated by another quote from Gordon Macdonald (1972, pp. 86–88) describing the movement of a large flow from Mauna Loa's Southwest Rift Zone in 1950:

> *The rate of advance was only about 30–40 feet an hour. The flow front, 50 feet high and half a mile broad, was a steep bank of reddish-brown to black clinkery rock buried by a heap of clinkery fragments that was accumulating at its base. For short periods it was quite motionless and appeared dead, except for small amounts of sulfurous fume and the peculiar odor of hot iron, resembling that of a foundry, that characterizes active basaltic lava flows. An incessant grating and cracking noise resulted from the shrinking and shifting of blocks on the flow surface and an occasional boulder tumbling down the flow front. At night, myriad red "eyes" glared out through holes in the dark cooler cover. During such quiet times, the amount of heat radiating from the flow was so small that it was possible to go right up to the flow front, and even climb part way up it. The upper molten mass of the flow interior moved more rapidly than the lower part, which was cooling and growing rigid against the ground. As a result, the flow front grew gradually steeper until eventually it became unstable at some point and a chunk of the dark clinkery rock began to separate from the mass of the flow behind it. Sometimes blocks leaned slowly forward as the crack behind it grew wider, until*

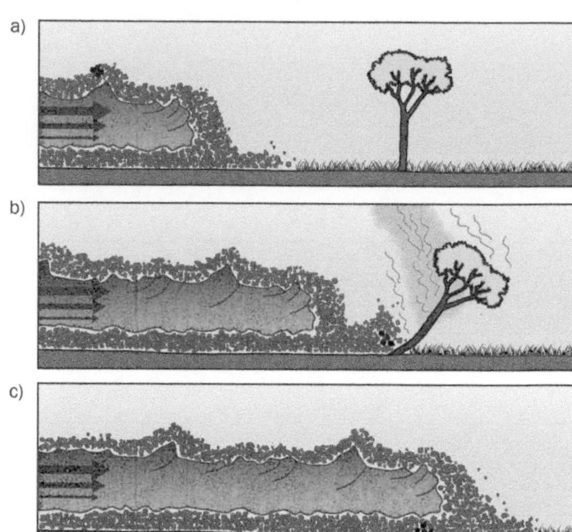

FIGURE 6.46 An ʽaʽā flow buries its own surface rubble as it advances. Note the changing position of some "reference blocks" (black) as they tumble over the flow front and are buried as the flow advances – propelled by the fluid lava within.

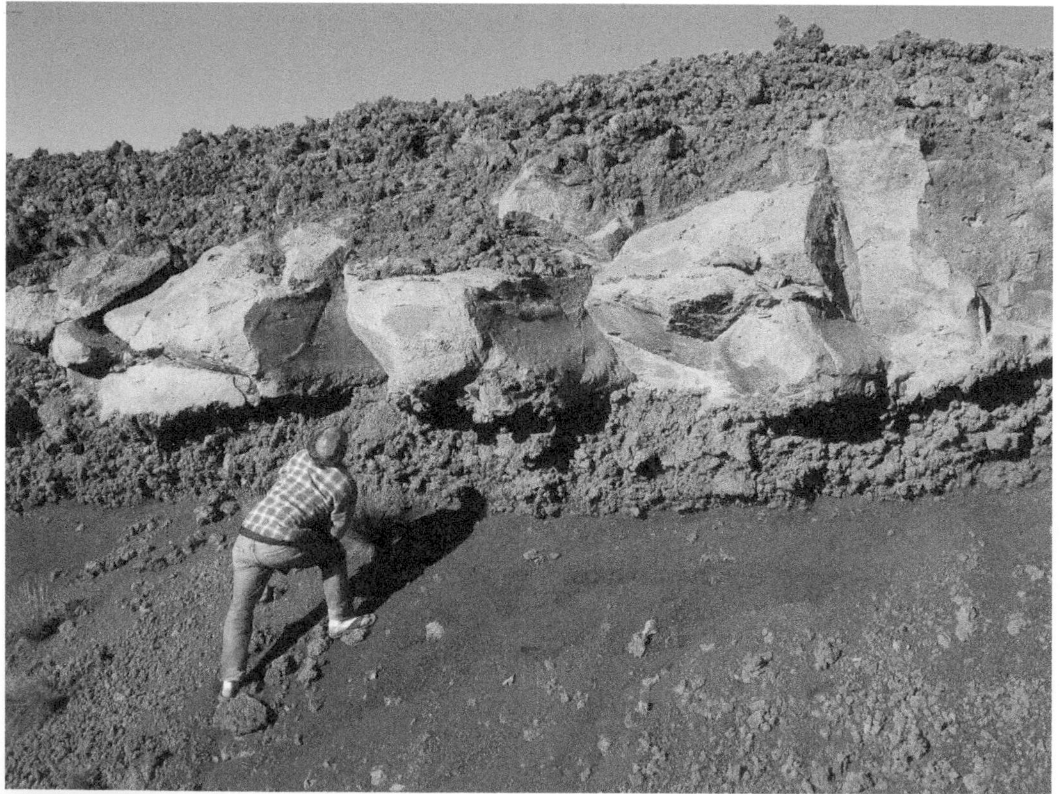

FIGURE 6.47 Prehistoric ʽaʽā flow cross-section, Mauna Kea volcano, Hawaiʻi [16]. Note the fragmental ʽaʽā above and below the central core. Because atmospheric oxygen is present beneath cooling ʽaʽā flows, underlying soils are commonly oxidized to a brick-red color.
Source: Photo by J.P. Lockwood.

finally it tore free and tumbled down the slope. At other times they started to slowly slide forward and downward along the forward edge of the flow – buoyed by unseen molten lava beneath. From the brightly glowing edge of the separation plane little streams of red-hot sand trickled down, formed apparently by the crushing and granulation of the incandescent lava. Eventually, the separation of the block became complete, and it also tumbled down the steep flow front. The surface left on the lava mass by the blocks separating from it glowed a bright orange-red with a temperature estimated to be about 900°C. The blocks also, where they broke open as they tumbled down, were brightly incandescent on the inside, but when the growing surface was exposed to the air, it cooled quickly and became darker, and within less than a minute most of them had become completely black. Yet, when a corner of such a block was knocked off with a hammer, the inside often was still cherry red a few millimeters below the surface. This illustrates well the low heat conductivity of the lava. The process of collapse was repeated over and over again all along the flow front, fragments rolling down to add to the bank of loose material at its foot. At the same time, the middle and upper parts of the flow crept almost imperceptibly forward. The main mass of the flow was a very viscous paste-like material, so viscous that it was impossible to push anything into it, but still sufficiently fluid to flow. As it oozed forward, it carried along on its back a cover of clinkery blocks, and at the same time it buried the fragments that had been accumulating at the foot of the flow front. A pall of reddish-brown dust hovered over the advancing flow.

William Hamilton, British ambassador to the Kingdom of Naples (Chapter 1) witnessed several 'a'a flows moving down the flanks of Mount Vesuvius [109] (Figure 6.48). His description of a fast-moving Vesuvius ʽaʽā flow in March–April 1766 (quoted in Phillip 1869, pp. 70–71) is timeless:

The lava had the appearance of a river of red-hot and liquid metal, such as we see in the glasshouses, on which were large floating cinders, half-lighted, and rolling one over another with great precipitation down the side of the mountain, forming a most beautiful and uncommon cascade. [. . .] It flowed like a torrent, with violent explosions and earth-shakings; the heat was such as to forbid a nearer approach than 10 feet. It ran with amazing velocity in the first mile, with a rapidity equal to that of the River Severn at the passage near Bristol. [. . .] The lava began to collect cinders, stones, &c, and a scum was formed on its surface, so that the whole appearance was like that of the river Thames, after a hard frost and great fall of snow, when beginning to thaw, carrying down vast amounts of snow and ice. The lower end of the current was covered with red-hot stones, a kind of wall 10- or 12-feet high, which rolled on irregularly and slowly about 30 feet in an hour.

FIGURE 6.48 A Mount Vesuvius [109]1761 a'a flow, as engraved by Pietro Fabris for Sir William Hamilton (Hamilton 1776, reprinted 2013). Source: Welcome Collection, Attribution 4.0 International.

As in pāhoehoe flows, central feeder channels and even rare pyroducts may form in 'a'ā. But 'a'ā flows lack the complex distributary pyroducts characteristic of pāhoehoe, and the central channels seen on the surfaces of some, if not all 'a'ā flows are only the exposed crests of broader molten flow cores. Lipman and Banks (1987) provided evidence of this relationship while studying an 'a'ā flow during the 1984 Mauna Loa eruption. They observed that as surges of lava traveled down the channel, the margins of the flow to either side swelled up as much as a meter or more. When the surge passed and the level of the molten lava in the channel dropped, the flow margins settled back down as well.

Accretionary Lava Balls

Accretionary lava balls (**bombes en enroulement**) are distinctive features found on the surfaces of many basaltic 'a'ā flows. They are typically egg-shaped, smooth-sided boulders 1–3 m in diameter that consist of low-density internal cores enclosed by glassy rims. Most originate within spatter cones formed by lava fountains during the early phases of long-lived eruptions. As described at the beginning of this chapter, spatter cones are incredibly unstable and fragile during growth, with large masses frequently dislodging and tumbling into the roiling lava they enclose. The agglutinated fragments are substantially cooler than the molten vent lava in which they suddenly become immersed, and they immediately acquire selvages of quenched glass. Because the bulk density of these porous, lava-coated spatter masses is much less than surrounding melt, they immediately bound back up to the surface, where they are rafted or tumbled end-over-end downstream in flow channels. The lava balls may float with as much as half their volumes above a conveying current. Some are destined to drift into large diameter pyroducts, and may be expelled from skylights downstream. Others are swept by surges up out of channels onto the surrounding flow surfaces; and, indeed, one of the most common places to find them is on the banks of pāhoehoe lava channels, where they may be recoated and made even smoother by later surges of lava (Figure 6.49). When a pāhoehoe flow transforms into 'a'ā, the lava balls will continue to ride along buoyantly above flowing 'a'ā, sometimes carried as far as 20 km from their points of origin (Figure 6.50). Accretionary lava balls can also form as fragments of channel walls fall into lava rivers and are coated by glassy rims.

Toothpaste Lava

Along the margins of some slow-moving 'a'ā, the clinker flow carapace may break open, allowing viscous core lava to extrude as aptly named **toothpaste lava** (Figure 6.51). Toothpaste lava may be considered an uncommon intermediate form between pāhoehoe and 'a'ā (Sheth et al. 2011). Precise conditions required to for it to form are unknown, though slightly greater flow viscosity related to a few weight percent higher silica content than is found in ordinary basalt is certainly one factor. The surfaces of each toothpaste extrusion are characteristically spiny and grooved, reflecting the pastiness of the lava as it oozed from the 'a'ā flow interior.

'A'ā Speed and Heat

Despite being a more viscous form of lava, 'a'ā flows often advance much more rapidly than pāhoehoe. This largely relates to a correlation between the lava discharge rate and the speed of advancing flow fronts; *the greater the eruption discharge, the larger the flows that emerge and the faster they move*. As mentioned above, episodes of high lava fountaining (high discharge) tend to create 'a'ā, while lower, less vigorous fountains (low discharge) generate pāhoehoe. A good example is the 1859 Mauna Loa eruption. During its high fountain phase, 116–235 m^3 of lava erupted each second and 'a'ā flowed away from the vent at an average speed of 267 m/hr.

FIGURE 6.49 Accretionary lava balls on prehistoric Mauna Loa [15] lava flows, Hawai'i. (a) Accretionary lava ball on the edge of a lava channel in a 250-year-old pāhoehoe flow. Note the welded spatter interior of this lava ball after the thin lava "shell" was broken. (b) Accretionary lava ball on a 200-year-old 'a'ā flow, about 25 km from spatter cone source (note hammer for scale). Source: Photos by J.P. Lockwood.

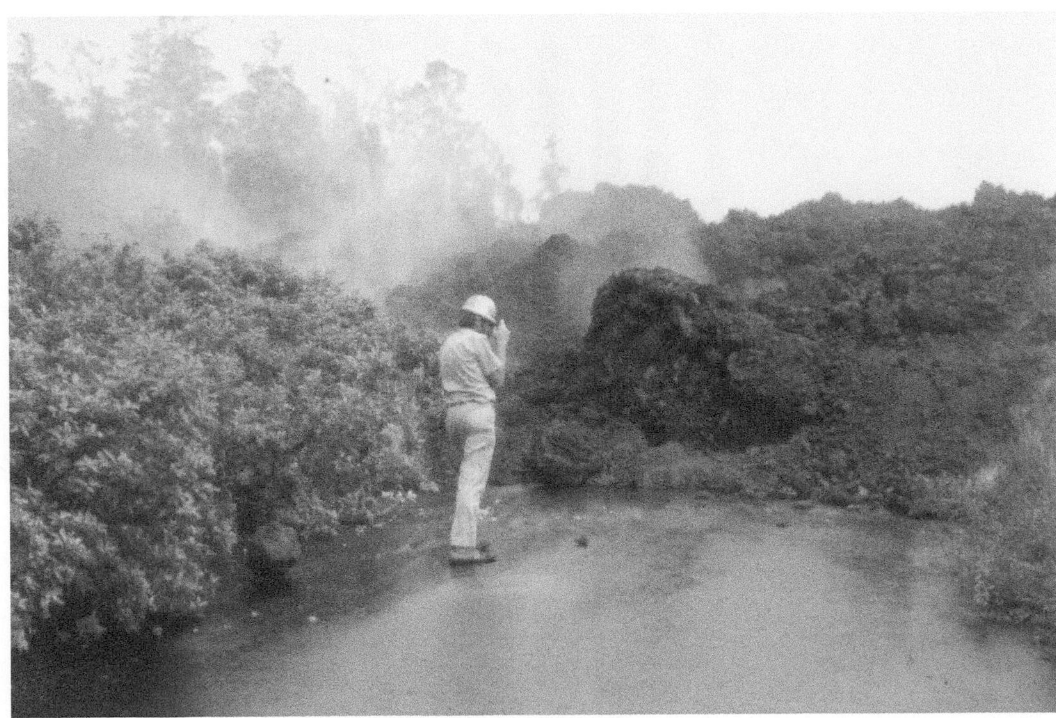

FIGURE 6.50 Rapidly advancing 'a'ā flow in Royal Gardens, Hawai'i – 1983. The flow, about 4 m thick, is advancing at about 200 m/h down the road surface. An accretionary lava ball is being pushed along in front of the flow. Source: Photo by Ben Talai.

FIGURE 6.51 Toothpaste lava issued from an 'a'ā flow of otherwise ordinary appearance, Kīlauea volcano [17], with field notebook for scale. The flow has a basaltic andesite chemical composition—unusually silica-rich and more viscous than most Kīlauea lavas. Source: Photo by R.W. Hazlett.

During the later pāhoehoe-generating phase, only 5–9 m^3/sec of lava erupted, flowing away at a lazy 7 m/hr (Rowland and Walker 1990; Riker et al. 2009). It is not unusual for an advancing pāhoehoe flow front to creep along at only a meter every minute, though speeds as fast as 30 km/hr may be observed in confined channels and conduits within the interior of the flow well upslope.

Other variables, of course, can also be important for determining how fast a lava flow moves, especially temperature and topography. Pāhoehoe entering a forest, for instance, slows down significantly as it interacts with moist ground and tangled vegetation. But even on a steep mountainside 'a'ā frequently wins the race. A pāhoehoe flow front tends to stop-and-go as individual buds, toes, and entrail lobes form, cool, and inflate sequentially; while the front of a fast moving 'a'ā flow is almost continuously on the move, with an ongoing shower of tumbling fragments that are noisily buried as the flow advances. 'A'ā flows have been observed moving as fast as 10 km/h down steep slopes, outpacing ordinary pāhoehoe flows (Figure 6.52). Even by daylight, incandescence is conspicuous at the restless snouts of these flows.

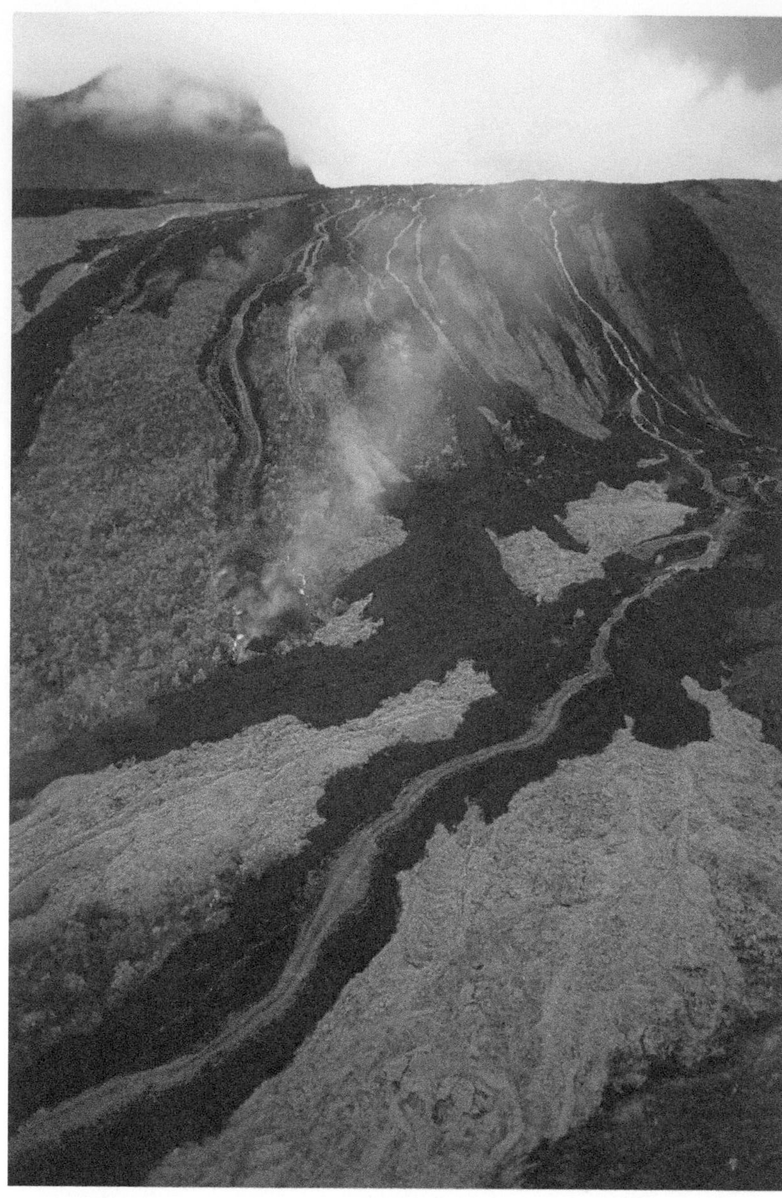

FIGURE 6.52 ʽAʽā flows descending down the east flank of Piton de la Fournaise volcano [124], Reunion Island in 2005. Note the well-defined central feeding channels – typical for ʽaʽā flowing on steep slopes. Source: Photo © Paul Edouard Bernard De Lajartre.

The radiant heat output of fast-moving ʽaʽā flows is much higher than it is for pāhoehoe, because flowing pāhoehoe will be largely covered with plates of non-incandescent crust, whereas such heat-shielding crusts may not form on the ʽaʽā. I (JPL) learned this dramatically on a pre-dawn morning flight with Donald Peterson (then scientist-in-charge at HVO) over Mauna Loa during the brief but spectacular 5 July, 1975 eruption: We were flying over a very active ʽaʽā flow that was cascading down Mauna Loa's north flank. The entire flow was incandescent and radiating heat energy. We were over a kilometer above and to the upwind side of the fast-moving (500 m/h) ʽaʽā, and I sensed no danger. The night air was cool (2°C on my Outside Air Temperature gauge), when suddenly I noticed "water" dripping off our wings. The "water" turned out to be paint, and we left that area very quickly! On landing, we found that dark paint on the plane had blistered, and we had burned several small holes in the fabric-covered wings. At the time we were unaware of Stefan's law, which states that radiant heat transfer is proportional to temperature to the fourth power.

$$Q = nT^4 \tag{6.2}$$

where Q is the rate of radiant heat release, T is the temperature of the surface releasing the heat, and n is a constant.

Stefan's law means that even a small increase in the temperature of a radiative surface will lead to a very great increase in the amount of heat released. A highly incandescent lava flow can heat objects above them far more than one that is even slightly less incandescent. Dark paint is a better "black-body" than light paint – so absorbed more radiant energy and blistered.

Methane Explosions

Both pāhoehoe and 'a'ā will "cook" overridden vegetation, leaf litter, roots, and other organic matter as they move across a landscape – especially in forests. The resulting distilled gases, especially methane, may travel underground many meters from the edge of moving flows, burning as they mix with oxygen to produce bright blue flames (Chapter 1, Figure 1.6). But, in certain cases, the gases can accumulate in subsurface pockets beneath areas that are seemingly safe, several meters from flow edges, where they may build up to explosive concentrations.

Survival Tips for Field Volcanologists

The distal margins of young 'a'ā flows are commonly hazardous places to climb, because the clinkery rubble here is generally loose and unconnected to interior solid portions of the flow. The middle and near-vent sides of these flows are generally much safer places to climb to flow tops, because the flows are thinner, and the fragments are commonly welded to one another or to the interior portions of the flows.

This phenomenon is most common and violent alongside advancing 'a'ā flows, perhaps because of their greater thicknesses and heat retention than pāhoehoe. Methane explosions never occur where a lava-invaded forest grows atop older, weathered 'a'ā flows, since such substrates have few cavities large enough to develop explosive concentrations of gases. A flow moving across older forested pāhoehoe flows, however, is a dangerous place to be. Numerous void spaces occur, largely related to partial drainage of pyroducts preserved within the overgrown flow. Methane can build up in these cavities, forming explosive cocktails when mixed with the proper amounts of air. Sometimes the detonations are small and merely cause muffled rumbling noises underfoot, but at times, violent explosions can blast craters up to 5 m in diameter, scattering shattered tree trunks through the air like javelins, and throwing blocks and dust over 100 m high. These explosions can commonly be heard several kilometers away, and may sound like cannon fire in the distance. We call these violent (and very hazardous) events *methane bursts*, although other gases besides methane are likely involved.

I (JPL) have seen trees 50 cm in diameter and large blocks hurled tens of meters into the air by methane bursts near advancing flows. I was watching an advancing pāhoehoe flow one night on Kīlauea, when a large, ancient tumulus behind me exploded violently. Unfortunately, my colleague (Chris Gregg) was standing on this tumulus, taking pictures. I heard him scream, and, as I turned around, I saw in the eerie red light that Chris had been thrown several feet in the air, and was now coming down, along with a massive amount of pāhoehoe blocks from the shattered tumulus. Chris and the rocks landed in an intertwined mass behind me, and he was moaning horribly. I feared the worst, but after he extricated himself from rocks and a crumpled camera tripod, he quickly regained his cheery composure. After a few minutes of rest and thanks to Pele, we continued with our observation duties, with renewed appreciation for this unpredictable hazard.

Survival Tips for Field Volcanologists

Sometimes it is safer to walk directly on the surface of the advancing flow (assuming the crust is sufficiently thick and not too hot!) than to walk in the forest at the edge of an active flow. Methane blasts can occur many hours after the advancing flow has come to rest.

Vesicles

Vesicles are "frozen gas bubbles" preserved in solid lava (Figure 6.53). They mostly form because of reduced pressure on the melt as magma ascends, but also in part due to the crystallization of minerals, which increases the percentage of volatiles in the remaining liquid to above saturation levels. Most gas bubbles begin forming in the magma prior to eruption, but many others form during eruptions, especially when molten lava ingests steam and gases from overridden water bodies or vegetation. In thicker, lower viscosity flows, the bubbles rise buoyantly, resulting in the concentration of vesicles in upper parts of flows. This may be useful for distinguishing flow top directions in deformed older terrains (Sahagian 1985). Thin flows cool too quickly to allow much movement of the bubbles, and are commonly quite uniformly vesicular throughout.

Vesicles vary widely in shape, size, and distribution within flows. Spherical to spheroidal vesicles are most typical of pāhoehoe, whereas subangular to angular vesicles are more typical of 'a'ā flows, where high viscosity during the final stages of flow motion twists and distorts them (Figure 6.53). The sizes of vesicles range from sub-millimeter to large voids tens of centimeters across. Rarely, nearly spherical vesicles up to 30 cm across are found with walls that are botryoidal (smoothly lumpy in a manner resembling a bunch of grapes), sometimes with small, sharp seams of lava projecting into cavities.

Ingestion of steam from moist earth or water below pāhoehoe flows commonly results in linear or roughly cylindrical groups of vesicles extending up into the lava from flow bases. Small tubular voids, known as **pipe vesicles**, can also form, usually less than a centimeter in diameter and projecting upward several centimeters to a couple of meters into the flow. The upper ends of

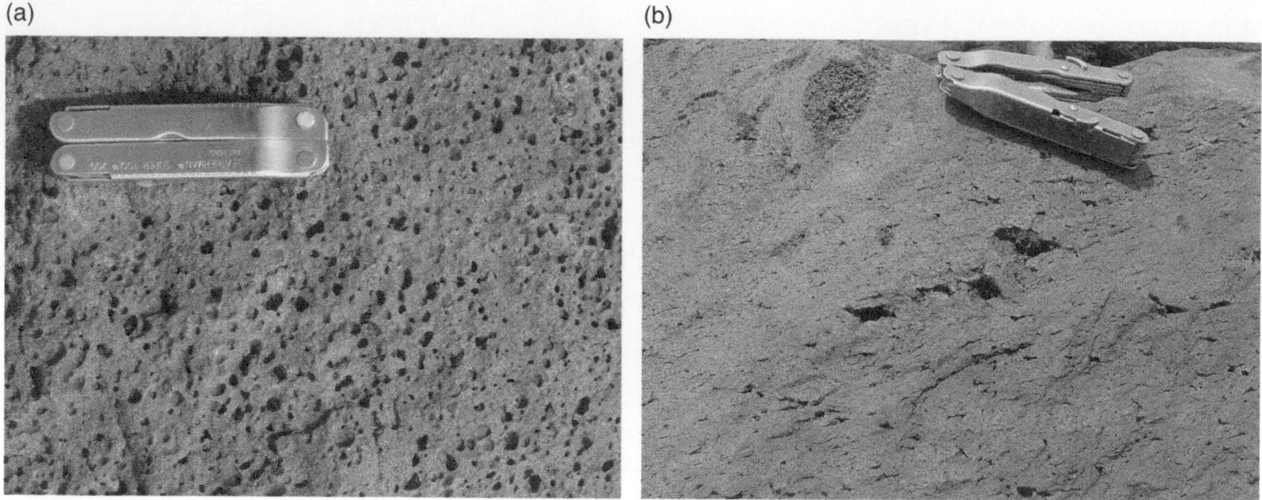

FIGURE 6.53 Distinctive shapes of vesicles in ʻaʻā and pāhoehoe. (a) Typical subrounded to spherical vesicles in pāhoehoe lava. (b) Typical angular and subangular vesicles in ʻaʻā lava. Note cognate xenolith near knife – such "ghost xenoliths" represent fragments of associated, earlier-cooled lavas that have been incorporated in younger molten material during flowage. Source: Photos by J.P. Lockwood.

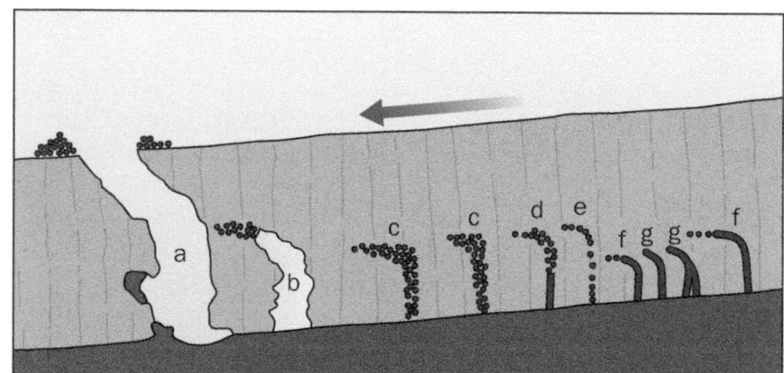

FIGURE 6.54 Cross-section showing the various kinds of pipe vesicles that may be found within a lava flow: (a) A large spiracle, or "mega-vesicle" blasted completely through the flow to its surface, with some underlying sedimentary material scooped up and included; (b) a spiracle passing into a vesicle cylinder (swamrs of dots); (c) vesicle cylinders; (d) a pipe vesicle (solid black mass) terminating upward in a vesicle cylinder; (e) a vesicle train (single line of spaced vesicles); (f) pipe vesicles passing into vesicle trains; and (g) pipe vesicles. (c) to (g) are proportionally exaggerated in comparison to the spiracles. Source: Modified from Macdonald (1972).

pipe vesicles commonly are bent downslope in the direction of flow movement (Figure 6.54). Where the lower skin of the flow is glassy and impermeable, the steam may burst upward into the overlying fluid lava explosively, creating an irregular cylindrical opening called a **spiracle**. These range in size from a few millimeters to 10 cm in diameter. Generally, they terminate within the body of the flow, but in the Pedregal lava flow on the outskirts of Mexico city, some of them pass entirely through the flow and are more than 30 m in vertical length. Mud from the underlying ground surface is in places carried up into the spiracle by the rising steam.

Extreme examples of steam erupting through active lava flows may also be seen in places such as Iceland, where the 63-km-long Younger Laxá flow smothered a boggy wetland valley near the town of Husavik. Steam blasting from the wet earth below drilled numerous vent openings in the flow as it came to rest, throwing up large amounts of spatter from the still molten flow interior – and, where blasting was more intense, finer pyroclastic material too. Over 6500 vent structures, termed **rootless cones**, developed there (Figure 6.55; Boreham et al. 2018). The cones are called "rootless" because they lack any direct, underlying magmatic conduits. The hornitos introduced earlier (Figure 6.18) are examples of rootless cones associated with pyroduct roofs, though the terms "hornito" and "rootless cone" are often used interchangeably.

Under less violent conditions, gases and aerosols trapped in vesicles during cooling eventually diffuse through microcracks in the solid rock. Before dissipating, however, some sublimate minerals may be deposited on vesicle walls, commonly light-colored needle-like zeolites. Zeolite precipitation often commences even before an eruption ends, and may continue for days or weeks following. On a somewhat longer timeline, hydrothermal solutions percolating through the rock readily accumulate within vesicles, where they slowly precipitate their cargo of dissolved silica-rich matter derived from the breakdown of minerals along their paths of ascent. Many vesicles fill up completely with hydrothermally precipitated mineral matter forming masses known as **amygdules** (from the resemblance in shape of many of them to almonds).

FIGURE 6.55 Rootless spatter cones or hornitos formed where lava flowed over wet bogs in the Laxá flow field, Aðaldalur Valley, Iceland. Source: Photo by R.W. Hazlett.

Joints

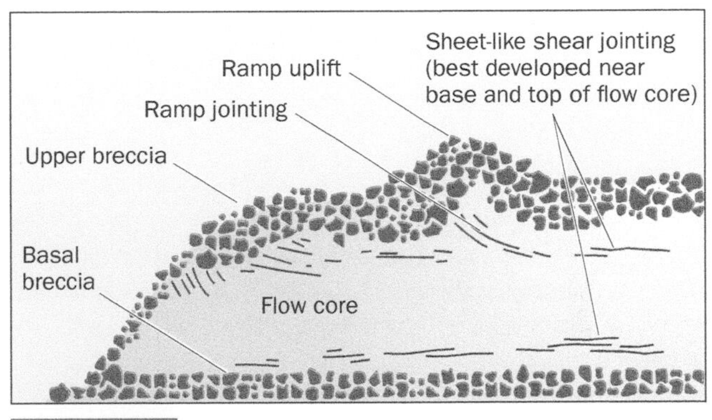

FIGURE 6.56 Shear-related jointing in cooling 'a'ā flows.

Almost all lava flows are cut by cracks known as **joints**. Joints have several different origins. In older flows most represent tectonic deformation of the earth's crust, mass wasting, or weathering long after their eruption. But those in younger flows originate almost entirely during flow emplacement. Joints parallel to the tops and bottoms of 'a'ā flows due to shearing during movement may be readily visible in outcrops (Figure 6.56). They generally develop as lava continues to deform for a short time after the viscosity becomes too high to sustain liquid or plastic flow.

Columnar joints are the most spectacular examples of emplacement-related flow jointing (Figures 6.57, 6.58, and 6.59). As lava flows cool, they shrink, setting up tremendous internal stresses. Laboratory studies by Lamur et al. (2018) showed that these joints form suddenly about 100° below flow solidification temperatures, from 840° to 890° for cooling basalts. The most prominent joints form at approximately right angles to cooling surfaces, which are usually the tops or bottoms of flows but may be on sides as well, as for example where a flow cools against the wall of a former river canyon or gorge. The joints tend to develop in three directions at roughly 60° to each other and form multi-sided columns. The columns commonly are hexagonal in cross-section, but range from four- to eight-sided (e.g. Phillips et al. 2013).

Cooling surfaces independently generate their own sets of columnar joints, which work their way stepwise into slowly solidifying flow interiors of all but the thinnest of lava flows. Over a period of days, weeks, or months (depending upon the thickness of flows), joint sets propagating from opposite cooling surfaces may meet to form single, through-going columns. Extension begins as stress builds up at the tip end of a partly developed joint, followed by sudden fracturing of the rock a few centimeters farther inside. The rock is momentarily relaxed after each spasm of propagation, but tension readily begins to accumulate with ongoing cooling until the breaking point once again is reached. The cyclical nature of this process results in paneled joint surfaces (Ryan and Sammis 1978). Where exposed by later erosion, the face of each panel typically features a feathery structure, termed a **plumose pattern**, the fine lines of which spread apart in the direction the panel opened over a period of microseconds and terminate in sharp, wedge-shaped edges termed **hackles** – the line of inception of fresh cracking (Pollard and Aydin 1988).

The episodic growth of columnar joints in cooling flows generates shock waves that may be heard as muffled "cracking" sounds at flow surfaces. That the downward growth of one column ultimately perfectly matches the upward growth of another growing

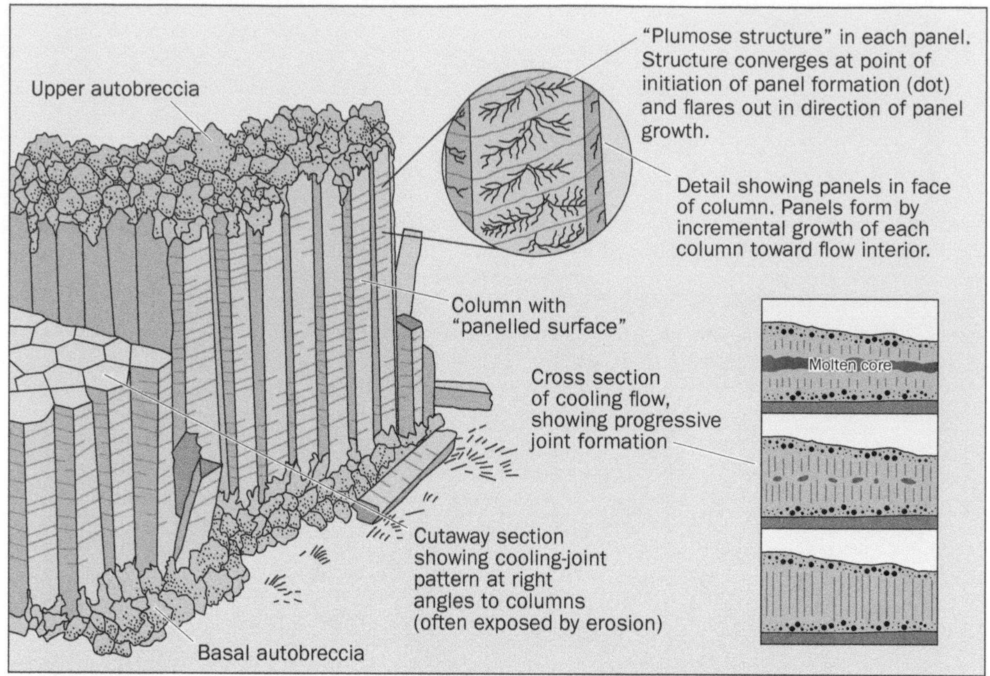
"Plumose structure" in each panel. Structure converges at point of initiation of panel formation (dot) and flares out in direction of panel growth.
Upper autobreccia
Detail showing panels in face of column. Panels form by incremental growth of each column toward flow interior.
Column with "panelled surface"
Molten core
Cross section of cooling flow, showing progressive joint formation
Cutaway section showing cooling-joint pattern at right angles to columns (often exposed by erosion)
Basal autobreccia

FIGURE 6.57 Columnar jointing in center of a very thick 'a'ā flow. Similar jointing forms as thick pāhoehoe flows and lava lakes cool.

a)
b)

FIGURE 6.58 Devil's Postpile National Monument, California. A late Quaternary lava flow was impounded along the Middle Fork of the San Joaquin River and cooled very slowly, forming well-preserved columnar jointing along its base. Subsequent glacial erosion removed overlying parts of this flow, as well as part of the deep interior, providing excellent exposures of the columns. (a) 15-m-high columns. (b) Glacially polished cross-section of typical columns, exposed at the present top of flow. Most columns are six-sided, but four-, five-, and seven-sided columns are also present. Source: Photos by J.P. Lockwood.

from below testifies remarkably to the fact that both sets of columns are responding to a common field of tensional stress. During the 1970s, seismometers operated on the surface of Kīlauea Iki lava lake recorded the development of cooling joints and other cooling-related microearthquakes.

Flood basalt lava flows a few tens of meters or more thick commonly develop multiple tiers of cooling columns (Figure 6.60). Above a thin basal zone with poorly developed inconspicuous jointing comes a zone with well-developed vertical columns from less than a meter to as much as 20 m long known as the **lower colonnade**. Lower colonnades commonly encompass the basal 10–30 percent of an especially thick flow. Next upward comes a zone in which the columns are much narrower and commonly have a variety of orientations, in some places forming fan-like rosettes or chevron-like structures. This zone is known, using another architectural term, as the **entablature** (Tomkeieff 1940). Entablatures in many flows make up as much as 60 percent of the flow mass. At the top of the flow comes a thinner zone, the **upper colonnade**, in which the columns are again wide but much less perfectly developed. In some flows, entablature is lacking, and in others there may be multiple sets of entablature and colonnade jointing present. Where lava enters water bodies, accelerated rapid quenching generally occurs and even more complex jointing arrangements can develop, including formation of numerous intersecting, curviplanar cracks that enclose rock masses that Forbes et al. (2012) calls "pseudopillows."

FIGURE 6.59 Columnar jointing at Giant's Causeway, Northern Ireland. Legend has it that these Paleocene basaltic columns formed a pathway that enabled giants from Ireland and Scotland to do battle across the Irish Sea. Source: Photo by J.P. Lockwood.

Other Types of Lava

Block Lavas

'A'ā fragments come in many shapes, ranging from rounded to angular to platy. Some are spiny and rough, and some are bounded by smooth fractures and are called **block lavas** (Figure 6.61; Harris et al. 2017). Macdonald (1972) and other authors consider block lavas to be an entirely separate category of flow distinct from pāhoehoe and 'a'ā. Spiny fragments can be found in some block lava flows, however, and some smooth blocks can be found in many spiny 'a'ā flows. Many flows consisting of conventional spiny 'a'ā throughout most of their length become blocky near their distal ends. Because of the gradations between all 'a'ā types, we will here consider block lava flows to be an 'a'ā textural variant, reflecting somewhat higher viscosity during fragmentation than typical spiny 'a'ā. Block lavas are characteristic of basaltic andesites and silicic lava flows, and are commonly associated with flows emanating from domes. They generally contain a larger amount of autobreccia relative to their solid cores than do spiny 'a'ā flows and blocks tend to be notably less vesicular.

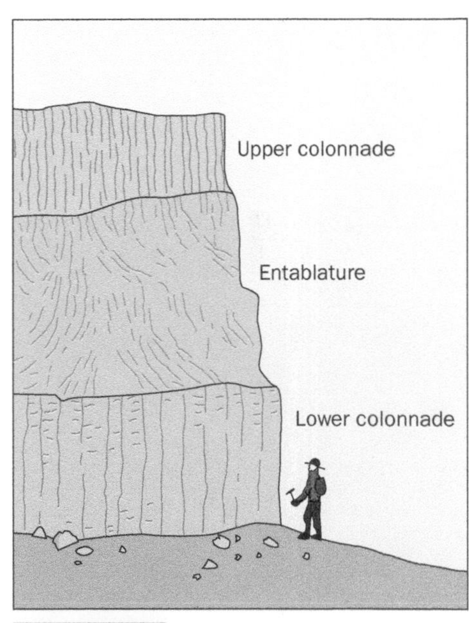

FIGURE 6.60 Idealized flood basalt cooling joint pattern.

Siliceous Lava Flows

Because of their higher viscosities, siliceous lavas typical of flows erupted from many continental volcanoes are thicker, shorter, and move much more slowly than the basalt flows erupted from oceanic volcanoes. Whereas a typical pāhoehoe flow may be hundreds or thousands of times longer than it is thick, a rhyolite or dacite flow may be only ten or a hundred times so. Many volcanologists in English-speaking countries term such fattened, high-silica flows **coulees**. But this word, which simply means "flow" in French, is applied to *all* compositions of lava in French-speaking parts of the world, so its use internationally has potential for confusion. The high viscosity of siliceous flows gives them steep flow fronts commonly over a hundred meters high – an order of magnitude greater than typical basaltic 'a'ā flows (Figure 6.62). Dacite block lava flows in Crater Lake National Park, Oregon [32], approach 400 m in

FIGURE 6.61 Basaltic andesite block lava flow on the flank of Pleistocene Brown Mountain volcano [29], near Fish Lake, southern Oregon. The thickness of the flow ranges up to approximately 75 m. Source: Photo by Stanley Mertzman.

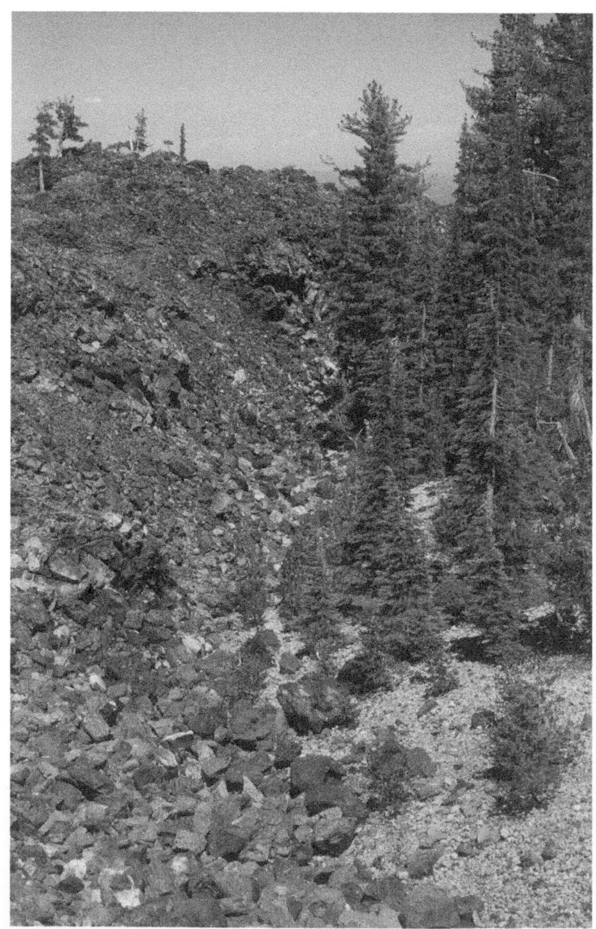

FIGURE 6.62 Edge of the thousand-year-old Glass Mountain rhyolite flow, Medicine Lake volcano [39], northern California. The steepness of the flow margin is indicative of the great viscosity of the lava as it erupted. Source: Photo by R.W. Hazlett.

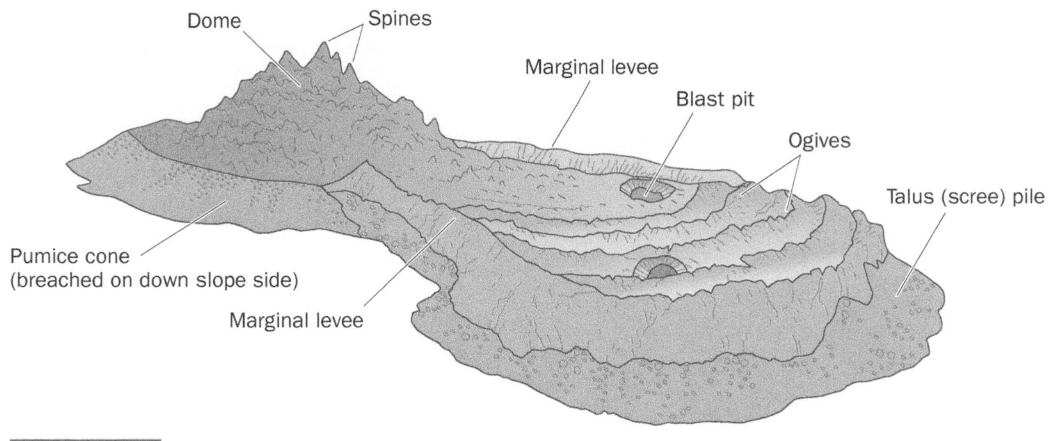

FIGURE 6.63 Typical features of a high-silica (rhyolite or dacite) lava flow.

thickness! High-silica flows rarely extend more than a few kilometers from their vents, though there are some spectacular exceptions. The Elephant Back rhyolite flow in Yellowstone National Park [47], for instance, is 16 km long (Christiansen 2001), and some rhyolite flows in neighboring Idaho comprise as much as 200 km^3 of lava (Bonnichsen and Kaufmann 1987).

The surface of a typical high-silica flow commonly wrinkles into ridges and troughs, termed **ogives** (similar to the ogives on glaciers), with amplitudes of 10–20 m or more (Figure 6.63). The ogives result from the freezing of the flow front while melt continues to drain from the vent. Compression forces the surface of the flow near its terminus to buckle and warp as the additional flow mass accumulates behind the stagnated, dam-like flow terminus. In this sense, then, ogive formation is analogous to the late-stage inflation of pāhoehoe flows. Near the vent, movement of the center of the flow often continues after consolidation of the flow margins, forming **levees** as the level of the flow upslope subsides with spreading of the lava downslope. Spines may puncture the flow surface in many places, especially near the vent, which may be marked by a towering lava dome. Blast pits, too, commonly pock mark the flow top, as the flow explosively releases pent-up gases.

Siliceous Glass

Siliceous lava flows tend to be quite glassy, and many rhyolite flows consist almost entirely of glass. This is because the high viscosity of siliceous melts suppresses the diffusion of ions and growth of microlites as degassing occurs prior to eruption. The most common type of high-silica volcanic glass is **obsidian**, a typically black, commonly bubble-free glass that readily fractures to develop conchoidally ribbed surfaces. **Vitrophyre** is a phenocryst-rich variety of obsidian. Knapped obsidian was a favorite source of arrowheads and spear points in early cultures worldwide. Many late prehistoric trade systems, notably in western North America and the Middle East, grew around the use of this material. The glass is brittle enough to flake and shape using horn, bone, or shell fragments. Although most obsidian is black, it occurs in many colored as well as transparent forms. Iron oxidation during late-stage flow and cooling may lead to red-brown colored glass (**mahogany obsidian**) that occurs in stand-alone masses or as streaks and patches embedded in ordinary black obsidian. The compositions of obsidian range from high-silica dacite to rhyolite. The darkness of obsidian arises primarily from the iron, and to a lesser extent the magnesium content of the magma.

Many obsidian bodies contain **spherulites**, which are spheres of radiating fibrous or needle-like, microscopically fine crystals (Figure 6.64) from only a few millimeters to over a centimeter in diameter. They result from the nucleation and incipient growth of crystals in hot, solid obsidian developing at close to the minimal cooling rate for glass formation (Lofgren 1971). Melts quenched more rapidly will not become spherulitic. The

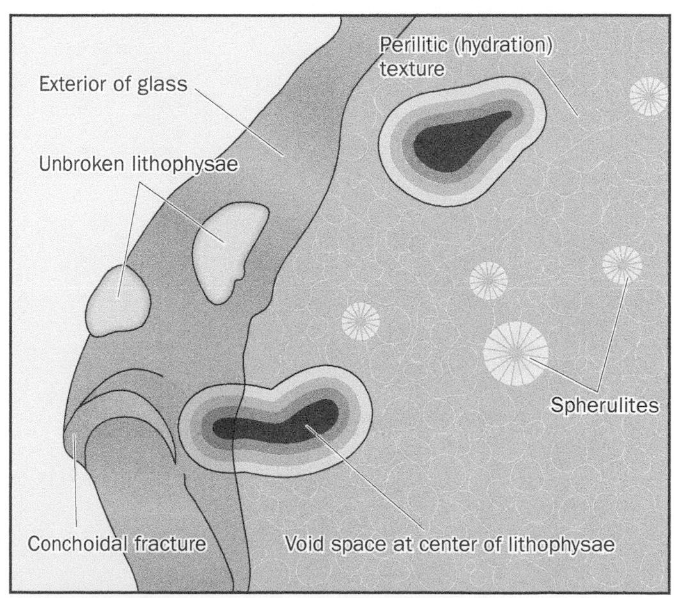

FIGURE 6.64 A sketch of spherulites, lithophysae, and perlitic texture in obsidian. Features are shown at natural scale, though perlitic fractures and spherulites may be microscopically fine.

minerals making up spherulites are commonly forms of high-temperature, low-pressure silica such as tridymite and cristobalite, plus more ordinary quartz and alkali feldspar (especially sanidine) – the last minerals to crystallize in Bowen's Reaction Series. Feldspar may be the only mineral present in some spherulites.

In some samples of obsidian, large, vesicle-like pockets up to several centimeters across are retained and lined with microscopic crystals through gas sublimation to form light-gray **lithophysae** (Figure 6.62). Some of these pockets may remain mineral-free. The pockets likely form as tensional stresses pull apart the stiffening magma, taffy-like, while it quenches solid Smaller, ordinary vesicles are typically not present in obsidian, because the quenching process inhibits exsolution.

Dating of young Prehistoric Lava Flows

Understanding a volcano's past history is critical for an evaluation of future volcanic hazards and risk. Prior to the 1940s, however, no accurate method existed for dating prehistoric eruptions less than a few hundred thousand years. Then, Willard Libby at the University of Chicago developed a technique for age determinations that involved the natural radioactive decay of ^{14}C, an isotope of carbon. Libby received the Nobel Prize for his revolutionary research.

Radiogenic isotope ^{14}C continuously forms as solar radiation interacts with nitrogen gas high in earth's atmosphere. Incident solar neutrons break nitrogen atoms, the most common substance in air, into ^{14}C and hydrogen atoms. Although the ^{14}C atoms are constantly decaying, their replenishment results in a semi-constant atmospheric ratio of ^{14}C to other, more common carbon isotopes (^{13}C and ^{12}C). When an organism is alive, the carbon in its tissues reflects this ratio, but upon death, radioactive decay lowers that ratio at a constant rate. Libby identified the decay rate for ^{14}C (half-life: 5730 +/− 40 years), which allowed the calculation of time elapsed since death based upon the proportions of carbon isotopes. Radiocarbon dating of carbonaceous material preserved beneath lava flows or within pyroclastic flows is the principal means of dating eruptions less than 40,000 years old.

When organic material is buried by molten lava, the heat rapidly distills off volatile components and all vegetable matter is usually burned quickly to ash – no carbonaceous material remains. In reducing environments, however, where access to atmospheric oxygen is restricted, elemental carbon (charcoal) may be formed and preserved. Charcoal can also form slowly at lower temperatures, and cannot be "burned" to ash at temperatures lower than about 500°C. Because of complete burning, above-ground vegetation is rarely preserved beneath lava flows, but carbonized plant rootlets may be present in soils beneath these flows. The

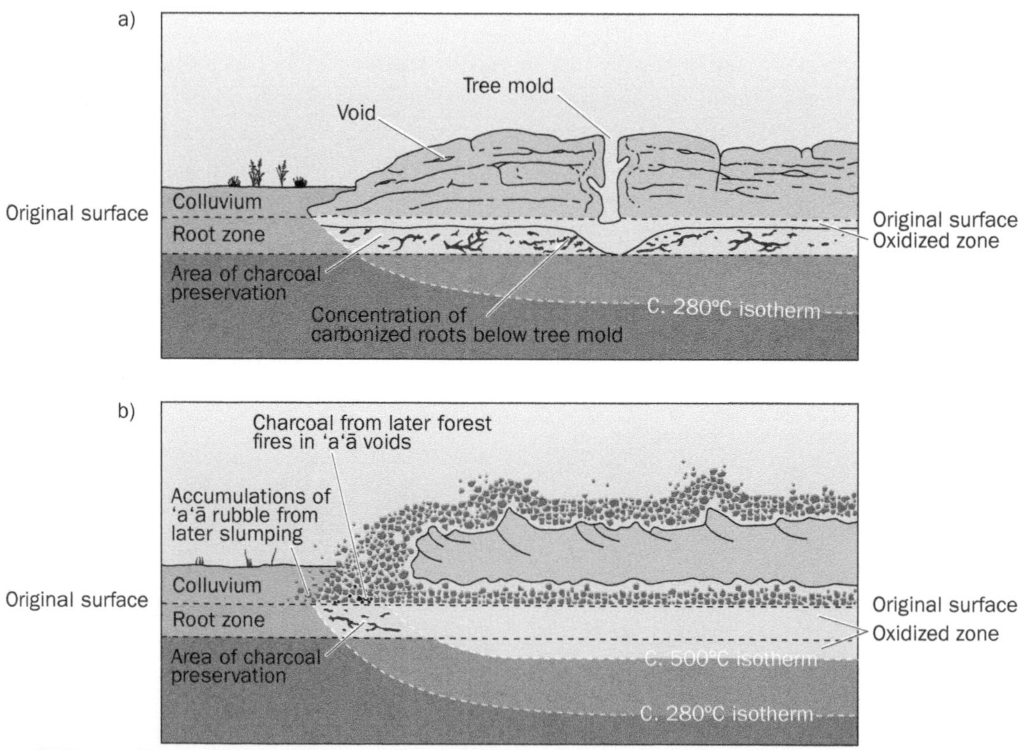

FIGURE 6.65 Sketch showing preservation environment for charcoal beneath (a) pāhoehoe and (b) 'a'ā flows. Note that charcoal may be preserved anywhere beneath pāhoehoe flows where oxygen circulation is limited, but only beneath the margins of 'a'ā flows – in areas where temperatures did not reach the combustion temperatures for charcoal. Source: Modified from Lockwood and Lipman (1980).

preservation of charcoal beneath Hawaiian lava flows was investigated by Lockwood and Lipman (1980) and means to predict where it can be found described (Figure 6.63). Carbonized rootlets can be found beneath any parts of pāhoehoe flows, as the impermeable pāhoehoe cover commonly prevents the atmospheric oxygen circulation required for complete combustion of buried organic matter. In contrast, the high permeability at the bases of 'a'ā flows usually permits the circulation of atmospheric oxygen, and results in the complete combustion of charcoal to ash residue. Carbonaceous material can best be collected beneath the margins of 'a'ā flows (Figure 6.65), where heat from the cooling lava was high enough to form charcoal, but not high enough to burn it completely away. Charcoal may also be preserved in pyroclastic density current deposits (Chapter 7), where trees are swept up and carbonized in the hot ash. Where organic mater was never present or not preserved beneath young lava flows, "Cosmogenic Exposue Dating" may be a useful tool. Trusdell et al (2018) used concentrations of the cosmogenic radionuclides ^{3}He and ^{36}Cl on the surfaces of volcanic ejecta and lava flows to date explosive events at the barren summit of Mauna Loa volcano.

Questions for Thought, Study, and Discussion

1. What aspects of eruptive behavior distinguish Hawaiian from Strombolian eruptions, and why do these two eruptive styles differ, even when erupting compositionally identical material?
2. What generally determines where a pāhoehoe flow will deflate or inflate?
3. What is the role of pyroducts in the spreading of pāhoehoe and the building of volcanic landforms?
4. Why should the ground around the margins of active lava flows be avoided in forests?
5. How are tree molds and hollow lava trees useful flow direction indicators?
6. What textures and structures would allow you to distinguish in cross-section a typical pāhoehoe flow, an 'a'ā flow, a high-silica rhyolite flow, and a flood basalt flow?
7. What primary physico-chemical factor accounts for the difference in the various kinds of siliceous volcanic glass?
8. Why is the presence of datable charcoal so much more widespread beneath pāhoehoe flows than it is beneath 'a'ā and blocky flows?

FURTHER READING

Calvari, S., Inguaggiato, S., Puglisi, G., Ripepe, M., and Rosi, M. (2008) *The Stromboli Volcano: An Integrated Study of the 2002–2003 Eruption*. American Geophysical Union Geophysical Monograph, vol. 182, 397 pp.

Hon, K., Gansecki, C., and Johnson, J. (2000) *Lava Flows and Lava Tubes: What They Are, How They Form*. DVD, Volcano Video Productions, Volcano Hawai'i, 75 minutes.

Lewis, G.B. and de Lajartre, P.E.B. (2007) *Red Volcanoes: Face to Face with Mountains of Fire*. New York: Thames and Hudson, 144 p.

Manga, M. and Ventura, G. (2005) *Kinematics and Dynamics of Lava Flows*. Geological Society of America Special Paper 396, 218 p.

Shackley, M.S. (2005) *Obsidian: Geology and Archaeology in the North American Southwest*. Tucson: University of Arizona Press, 264 p.

Witze, A. and Kanipe, J. (2014) *Island on Fire: The Extraordinary Story of a Forgotten Volcano [Laki, Iceland] that Changed the World*. Pegasus Books, Berkeley, California, 224 p.

CHAPTER 7

An Overview of Explosive Eruptions and Their Products

Source: Photo from the Popocatépetl volcano monitoring system kindly provided by the Centro Nacional de Prevencón de Desastres (CENAPRED).

Volcanoes: Global Perspectives, Second Edition. John P. Lockwood, Richard W. Hazlett, and Servando De La Cruz-Reyna.

Companion website: www.wiley.com/go/lockwood/volcanoes2

He looketh on the earth, and it trembleth: he toucheth the hills, and they smoke.

(Psalm 104:32)

Volcanic explosions differ greatly from chemical or nuclear ones – which involve complex changes in molecular or atomic structure. Volcanic explosions are much simpler – they are all due to the expansion of volatiles, as liquids convert to gaseous state. The most important volatile liquid to consider is H_2O – which expands some 1600 times in volume as water converts to steam (at atmospheric pressure), and is the driving force for most types of volcanic explosions. Carbon dioxide is another important driver for some volcanic explosions – but only increases in volume about half as much as water when converting from fluid to gas. This chapter will focus primarily on the *products* of explosive eruptions; the focus of the next chapter will be on the *processes* that form these products.

Whereas the red volcanoes described in the preceding chapter dominantly produce lava flows during typical eruptions, the gray volcanoes are characterized by the production of immense volumes of pyroclastic material that can blanket thousands of square kilometers surrounding and downwind from these volcanoes. The dominant eruptive products (by volume) during explosive eruptions are the gases that drive explosions (Chapter 3), but only the ejected ash and rocks remain behind to record such activity. Accordingly, we will begin this chapter with a discussion of **ejecta** – the various kinds of materials that can be blasted out of volcanoes by gases – especially during early phases of long-lived eruptions, then will focus on the various sorts of explosive eruptions that produce these materials. Potentially deadly pyroclastic flows and surges (pyroclastic density currents – PDCs) can be produced by different types of explosive eruptions, and will be briefly reviewed at the end of the chapter.

Ejecta Classification

Volcanic explosions produce fragmentary debris; everything from large, house-size boulders to finely crushed rock material, and to magma explosively sprayed into a mist of molten droplets and fragmentary glass and crystals. Ejecta is a catch-all term for anything blasted out of the craters of an exploding volcano. The word is used in both a singular and a plural sense, probably improperly. (The singular is **ejectum**, but it rarely appears in volcanological literature.) Synonyms include **pyroclasts** and **pyroclastic material**. The term "pyroclast" derives from Greek and Latin roots meaning "fire-broken" rock.

Volcanic ejecta may be classified in different ways – by their sources, modes of emplacement, general compositions, sizes, shapes, or by the degree of post-depositional consolidation. The different classification systems are useful for different purposes. Considering them first from the standpoint of source, H.J. Johnston-Lavis in 1885 proposed the following classification, still in use today: **Juvenile (essential) ejecta** are derived directly from molten magma, although the magma may have solidified first before fragmenting – for example, by forming a crust across a pool of periodically bursting lava in the crater of a volcano. Since magma-derived bombs are thrown out in a molten condition and cooled rapidly in the air or on the surface of the ground, they are partly or entirely glassy. **Accessory (cognate) ejecta** are older volcanic rock fragments that were formed within the same volcano. Some accessory ejecta are coarsely crystalline, formed from parts of magma bodies that crystallize at depth, in part as minor intrusions within the volcano, and perhaps in part on the outer edges of magma chambers. Accessory ejecta may be torn from the walls of the vent by gases during explosive discharge, or may be floated upward in the rising magma and explosively blown free at the surface. Fragments of non-volcanic rocks or of non-related volcanic rocks are called **accidental ejecta**. Such fragments typically show some evidence of thermal metamorphism owing to the great heat imparted to surrounding rocks through the walls of the magmatic conduit. Limestone fragments, for example, may be recrystallized to marble, a common metamorphic rock, and silicate rocks converted to a dense, dark material called **hornfels**. Hydrothermal solutions also commonly alter older wall rocks, and hydrothermally altered lithic fragments are very common around some vents.

Because the various kinds of ejecta may be difficult to classify without detailed microscopic and chemical studies (especially when it comes to comparing accidental with accessory fragments), volcanologists often simply refer to glassy ejecta as **vitric**, and all non-glassy ejecta as **lithic** – regardless of genesis.

The most widely used and useful classification of ejecta is that based on fragment size, which is easily measured in the field for larger particles, or by laboratory sieving for finer material. The sizes of clasts in volcanic deposits of all types are measured directly in millimeters (mm), but are commonly plotted according to their "**phi (ϕ) numbers**" – a shorthand for grain size ranges used by sedimentologists. In addition to grain size, volcanic ash can be further classified through a microscopic study of ash composition (Table 7.1). Larger ejecta fragments can also be subdivided on the basis of their shapes, which reflects their physical condition at the time they were ejected (Figure 7.1).

Modes of Ejecta Emplacement

Ejecta may be emplaced *ballistically* (where explosive, horizontal forces are involved, and fragments impact the ground at angles sharply from the vertical) or by vertical *fall* from overhead clouds (trajectories only influenced by gravity and prevailing winds). Finer-grained, lower-density ejecta tends to be emplaced as fallout, while larger, higher-density fragments are commonly emplaced

TABLE 7.1 Particle size scale for volcanic tephra.

Volcanic Clast Name	Blocks (angular) Bombs (rounded) >64 mm			Lapilli 2–64 mm					Coarse ash 0.063–2 mm				Fine Ash 0.009–2 mm			Dust <0.009 mm	
ϕ No.	−8	−7	−6	−5	−4	−3	−2	−1	0	1	2	3	4	5	6	7	8
Maximum diameter (mm)	256	128	64	32	16	8	4	2	1	0.5	0.25	0.125	0.062	0.031	0.016	0.008	0.004

Source: Based on Schmidt (1981).

TABLE 7.2 Relationship of common kinds of ejecta (Figure 7.1) to modes of emplacement.

Type of ejecta	Primary modes of emplacement
Ash	Fallout and PDC
Lapilli	(Depends on type of lapilli)
a) Accretionary lapilli	a) Fallout
b) Cinder (scoria)	b) Ballistic
c) Silicic pumice	c) Fallout and PDC
d) Reticulite, Pelée's hair, Pelée's tears	d) Fallout
Blocks	Ballistic and PDC
Bombs	Ballistic
Spatter agglutinate	Ballistic

ballistically. A ballistic particle is one that is thrown in an arc-like trajectory, like a ball tossed through the air, from a point of release to a point of impact (Figure 7.2). Ballistic fragments are especially abundant (and dangerous) products of smaller explosive eruptions. They have been responsible for numerous eruption-related casualties among visitors in the vicinity of active craters (e.g. Oikawa et al. 2016).

Ash and larger pyroclastic fragments that fall vertically to the ground from eruption clouds (without ballistic trajectories) are called **airfall**, **fallout,** or simply **fall ejecta**. The term **tephra** (Greek for ash) is sometimes used to refer to resultant deposits. The time spent adrift can range from minutes to days. More powerful volcanic explosions generate vast amounts of fall pyroclasts, but will throw out ballistic ejecta in the proximal areas of vents too. (Table 7.2).

Finally, substantial amounts of all types of ejecta may be deposited through the activity of PDCs, with most pyroclastic fragments spending little if any time passing through the open air at all.

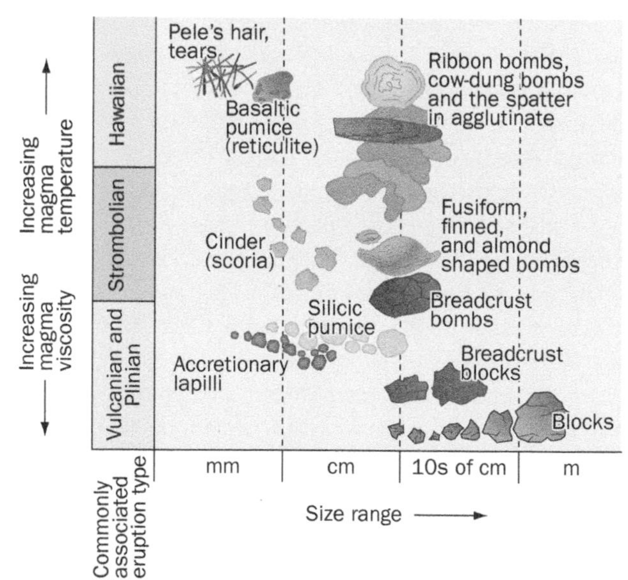

FIGURE 7.1 Shapes and sizes of the different kinds of volcanic ejecta, as a function of eruptive conditions.

A Closer Look at Ballistic Ejecta

Recall from Chapter 6 that Hawaiian eruptions can be mildly explosive and generate prodigious amounts of glassy spatter that accumulates around vents to form agglutinated spatter cones and ramparts. In an episode of high fountaining, spatter may be thrown a couple of hundred meters from a vent, rarely further, with most of the ejecta falling much closer to vents.

Spatter can also erupt during some Strombolian eruptions, but ejecta mostly consists of cinder and larger **volcanic bombs**, which are viscous masses of lava ejected in fluid or plastic states. Bombs are commonly rounded, subrounded, or streamlined

FIGURE 7.2 Night view of lava fountains showing ballistic trajectories of spatter bomb fragments, as a "littoral cone" formed at the south coast of Kīlauea volcano, Hawai'i [17] – January 1996. Fountains are about 20 m high – 10 s exposure. Source: USGS photo by Carl Thornber.

FIGURE 7.3 Spindle bombs from prehistoric Strombolian eruptions on Mauna Kea volcano, Hawai'i [16], showing aerodynamically modified shapes as liquid lava traveled through the air before solidification and impact. The scale is 30 cm long. Source: Photo by J.P. Lockwood.

owing to travel through the air. Particular varieties, including **spindle** or **fusiform** bombs, are identified according to their particular shapes (Figure 7.3). Uncommon, rounded forms, termed **cannonball bombs**, may be found scattered around some basaltic vents, and might owe their surprising symmetry to the transport of magma blobs within open conduits (Di Piazza et al. 2017). Bombs range from less than a centimeter to over a meter in length. The fluidity of volcanic bombs varies greatly and is responsible for their shapes (Figure 7.4). Some land while still quite molten and will flatten out or shatter on impact. Others solidify before they reach the ground, and can burrow deeply into soft volcanic ash.

The fluid cores of bombs commonly contain high concentrations of gases that continue to exsolve after eruption, stretching and cracking their chilled crusts to create a characteristic **breadcrust texture** (Figure 7.4). Some breadcrust bombs are deeply fractured, and some may completely disintegrate as they continue to expand after impact leaving a pile of rubble instead of a coherent bomb. Some will even accumulate enough gas pressure to explode in midflight – and have been known to burst even hours after impact! Following the 1980 Vulcanian eruption of Gamalama volcano [140] in Indonesia, I (JPL) found craters caused by bomb impact blasts as far as 2 km from the source vents. The rainforest had been shredded around these craters, leaving angular lava fragments driven into the trunks of surrounding trees.

Volcanic blocks are another form of ballistic ejecta, thrown out as solid, angular fragments of older rock. Blocks may be plucked from the walls of the conduit or rim around a vent during an eruption, or may be fragments of non-volcanic rock

a)

b)

c)

FIGURE 7.4 Ballistically emplaced lava bombs, showing effect of fluidity. (a) "Cow-pie" bomb formed as fluid spatter impacted the ground during a 1969 eruption of Kīlauea volcano, Hawai'i [17]. (b) Classic breadcrust bomb from the rim of Bolshoye Tol'batchik volcano, Kamchatka [163]. This bomb, which erupted in 1975, shows the crustal surface fracturing that occurred after impact as the bomb's fluid interior expanded. (c) A more viscous 1.5 m-diameter breadcrust bomb that shattered from internal fracturing after impact on the rim of Gamalama volcano, Ternate Island, Indonesia. Source: USGS photos by J.P. Lockwood.

from deeper sources. Some blocks are impressively large, as much as meters in diameter, and may be thrown hundreds of meters from violently exploding craters (Figure 7.5). **Bomb sags** form where ejected blocks land and depress soft, underlying muddy ash layers (Figure 7.6). Such structures (which despite their names are usually caused by falling blocks, not bombs) are especially common in muddy tephra deposits within a few hundred meters of vents, and may be expressed to depths of decimeters beneath points of impact. Ballistically emplaced volcanic bombs and blocks are not only produced by magmatic eruptions. Phreatic activity and hydrovolcanic eruptions in areas of hot springs and geysers can also eject lethal blocks and bombs (e.g. Figure 14.2).

Concentrations of blocky rubble embedded in ash or other matrix material are termed **breccias**. Not all volcanic breccia clasts are necessarily ballistically ejected blocks, however, but may simply be rubble swept out of a vent by a PDC, or even result from collapse of a vent or caldera wall during simultaneous deposition of pyroclastic material (see, e.g. Lipman 1976; Chapter 10).

A Closer Look at Fall Ejecta

Pyroclastic fallout from an eruption cloud is a lot like falling snow – it will drape (**mantle**) a landscape to uniform depths over broad areas as it settles out, irrespective of the topography (Figure 7.7). Such uniform blanketing contrasts with the variable thickness of PDC deposits, which accumulate on valley floors and in depressions, and are thinner on high-standing terrain. A bed of fallout ash

FIGURE 7.5 Large block ejected from Galunggung volcano [129] crater during a 1982 sub-Plinian eruptive phase. The block was thrown clear of the rising ash column and landed almost 1 km from the crater rim – behind me – I (JPL) never saw it coming! Source: USGS photo by J.P. Lockwood.

FIGURE 7.6 Bomb sag caused by impact of large block in wet tephra during the formation of the prehistoric Laguna maar, Gamalama volcano [140], Ternate Island, Indonesia. Source: USGS photo by J.P. Lockwood.

Survival Tips for Field Volcanologists

Approaching an active gray volcano (or any volcano with the potential for explosive behavior) requires special field gear. A strong hardhat is essential. Human skulls are quite fragile, and even small blocks falling from above can be lethal. Hardhats will also keep fine tephra out of your hair! Particle masks and eye protection are important to carry in one's field pack – when ash begins to fall, you will need them!

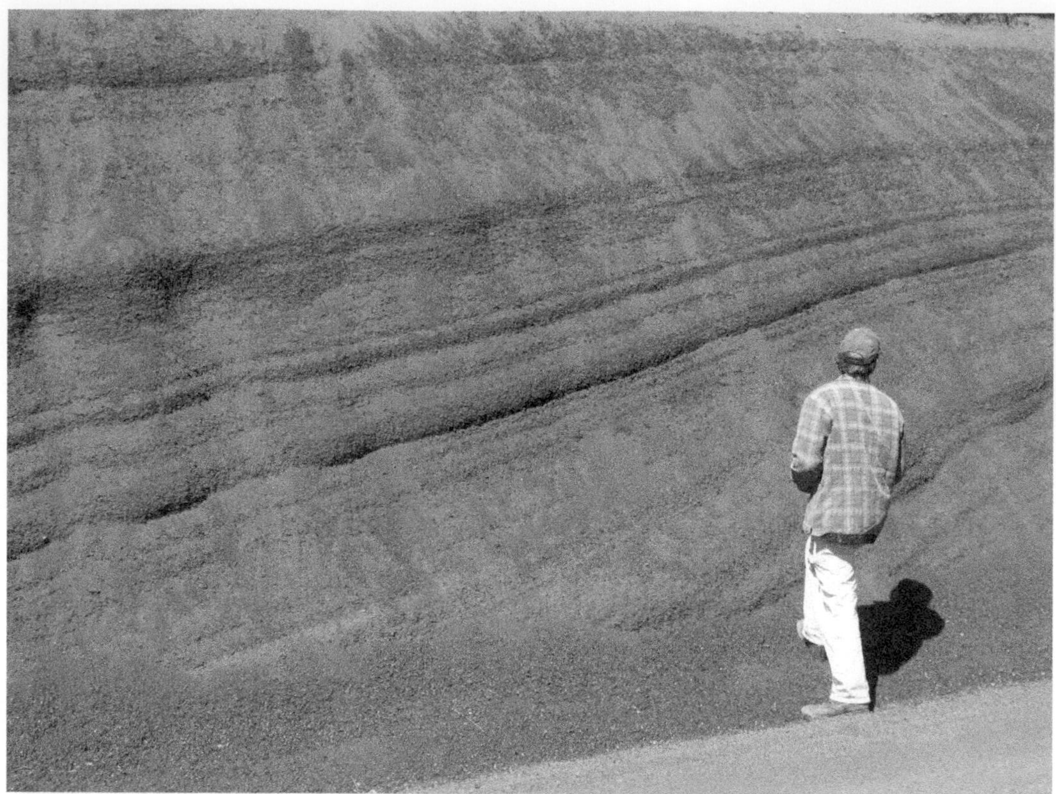

FIGURE 7.7 Fallout scoria and ash deposits downwind from a prehistoric eruptive vent on Mauna Kea volcano, Hawai'i [16]. Note the uniform thickness of individual layers, which mantle pre-existing topography – hallmark of airfall tephra deposits. Source: Photo by J.P. Lockwood.

may be as thick on a ridgeline as it is in an adjacent valley, though post-depositional slumping may thin layers on steep slopes, and wind can blow away the loose material from ridge crests and hilltops for months or years following an eruption. During tephra fall, strong winds may also affect the thickness of deposits around obstructions. Fallout beds gradually thin toward the distal areas of deposition, and isopach maps showing thicknesses of specific layers can be a valuable tool for identifying their sources (Chapter 5).

Ash

Volcanic ash is in no sense a product of burning, as are other "ashes." It is simply pulverized or finely fragmented magmatic material, with individual grains less than 2 mm in diameter. The forces that cause this explosive fragmentation are poorly understood (Zimanowski et al. 2003). Like other forms of ejecta, ash particles may be essential, accessory, or accidental, and either solid or liquid when they are thrown out, though because of their small size, the liquid ones cool and solidify very quickly. Sometimes, when the magma contains numerous crystals, the solid bits may be blown free of the melt to chill as **crystal ash**, each particle consisting of a single crystal or fragments of crystals with traces of adhering magmatic glass. Ash consisting of more than one type of material is customarily given a compound name: **vitric-crystal ash**, vitric-crystal-lithic ash, etc.

By far the most common variety of ash is vitric ash formed by the explosive disruption of liquid lava as gas expands in an open vent conduit. The gas exsolves to form a froth as the magma wells up from deep in the conduit, and as the bubbles continue to grow, the froth is literally torn apart as the molten rock approaches the surface. Some of the larger clots of froth remain as lumps of pumice, but most of it disintegrates to the point that all that remains are the septa separating individual bubbles, chilled to glass, and carried upward by the out-rushing column of gas. The shapes of the individual ash fragments clearly reveal their origin in the concave surfaces that once were the walls of bubbles. The curved and forked fragments of vitric ash are often referred to as **shards**. Some shards may be lightly coated with glass, suggesting that they were immersed again in molten material before being ejected for a final time.

Ash and **tuff** (solidified ash) beds often contain some fragments of larger size, quite apart from crystals. Those in which moderately to very abundant lapilli-sized ejecta (Table 7.1) of lithic material or pumice are scattered through the finer matrix are called **lapilli-ash** or **lapilli tuff**. Older, hardened beds containing blocks are known as **tuff-breccia**, and those containing bombs may be called **tuff-agglomerate**.

Near vents, each individual explosion during an eruption results in a shower of fragments that fall as a layer over the adjacent countryside. Each layer is the result of a separate explosion. Such beds are commonly well sorted according to the size of the fragments, and show a size gradation both vertically and laterally. Particles are usually larger near the base of the layer, because the larger fragments ejected by any one explosion have faster settling speeds than the small ones and so strike the ground first. For the same reason, the largest fragments tend to fall closer to the vent, while smaller ones are swept farther by the wind. Occasionally, ash beds are found in which the grain size increases upward. This can signify two diametrically opposed things: The gradation may result from increasing intensity of explosive activity, with larger fragments being thrown higher and drifting farther from the vent so that at any one place the fragments falling to the ground are larger than those preceding them; or it could result from decreasing strength of explosion, the magma being less completely blown apart so that the fragments formed are somewhat larger in later stages than they were earlier. Some ashfall beds also show fine internal laminations made up of slightly different particle sizes that can be related to abrupt fluctuations in the heights of eruption clouds (Walker 1980). Typically, these laminations are traceable laterally a variable distance away from the vent, representing the differing strengths of the eruption that produced them.

Fallout ash deposited by phreatomagmatic eruptions tends to be more conspicuously laminated than dry ashfalls, with groups of thin, subparallel layers separated by pyroclastic surge bedding (described later). Wet ash fall beds may also be slightly vesicular, owing to the steaming of moisture trapped by the hot ash, or from underlying heat emitted by previously accumulating material. The presence of a modest amount of moisture can actually strengthen a bed through the collective effect of the surface tension of water films around particles. Such wet cohesion improves the ability of some layers to resist wind and gas-blast erosion, and to cling more effectively to slopes. But if too much moisture is present, wet ash on slopes, particularly on slopes >10°, will remobilize and slide. Internal laminations fold and rumple as they slump downslope, much like a carpet being slid into a wall. This phenomenon is called **soft sediment deformation**. Soft sediment deformation rarely occurs in dry ash fall layers, which will slump and mix under the influence of gravity on steep slopes, but lack the internal cohesiveness to retain any semblance of original stratification.

Areal extent of ashfall from extremely powerful eruptions can be worldwide – literally! Extremely fine ash ("dust") from the 1883 eruption of Krakatau volcano [128] may have circled the earth three times. Long range ash distribution can also have important economic impacts. For example, the valuable bauxite ore deposits on the non-volcanic Island of Jamaica may derive from the fallout and weathering of ash, generated over millions of years by Antillean arc volcanoes lying approximately 1500 km upwind (Corner 1974).

The lateral movement of ash in the atmosphere depends on high-level wind speeds. Some of the 1.1 km^3 of ash from the 1980 Mount St Helens [31] eruption in Washington State drifted 1000 km to the east in a little less than 10 hours, indicating a high-altitude wind speed of nearly 100 km/h. Far larger than the Mount St Helen's blast, the 1912 Novarupta [19] eruption in Alaska released some 25 km^3 of ash and pumice, which darkened skies over an area of 260,000 km^2. The ashfall reached a thickness of about 30 cm at the village of Kodiak, 160 km from the site of the eruption. The even stronger explosion of Tambora [134] in the eastern part of Indonesia in 1815 ejected over 100 km^3 of ash, enough to bury the whole Island of Manhattan a kilometer deep!

Ash layers interbedded with ancient rocks of other types often are of great importance geologically because they provide a widespread time datum of unrivaled precision. A single ash bed traceable over tens of thousands of square kilometers represents an event that took place within a period of only a few days – a mere instant in the eons of the earth's history. It furnishes a means of correlating the associated rocks in time – rocks that may have formed under very different conditions and consequently be of very different character and contain very different types of fossils in different parts of the ash-covered region. As an example, a distinctive ash bed of Ordovician age widespread over eastern North America and Northern Europe is a critical marker horizon for dating a gigantic eruption that occurred some 450 million years ago – before the formation of the Atlantic Ocean (Huff et al. 1992).

The correlation of rock units by dating of interbedded ash layers is called **tephrachronology**, and was pioneered by Thorarinsson (1967), who was able to correlate late prehistoric and early historic volcanic and other historical events over large parts of Iceland. In the United States, fallout ash from the great eruption of Mount Mazama [32] (Chapter 10) has been traced as far east as central Montana and northward into British Columbia and Alberta, establishing the synchronism of associated rocks in the Cascade Range, Columbia Plains, Rocky Mountains, and western Great Plains (Wilcox 1965; Sarna-Wojcicki et al. 1983; Foit et al. 2004). Ashfall erupted at the time of the formation of the Bishop Tuff in the northern end of the Owens Valley, California about 740,000 years ago has been identified as far east as central Nebraska (Izett et al. 1970). Ash from the eruption of Glacier Peak, Washington [42], about 12,000 years ago (Mastin and Waitt 2000), has also been traced eastward as far as Montana; and the Pearlette ash, erupted during the late Kansan Ice Age, is found from western Texas to southeastern South Dakota, and possibly westward into Nevada. Tephrachronology has assisted in the dating of prehistoric fault movements (e.g. Anderson and Hawkins 1984) and in archaeology, paleontology, and climate change studies (Rapp and Hill 2006; Dorn 2009).

Ash layers, commonly altered to clay minerals, may appear very similar in hand specimen, and microscopic or geochemical methods may be required for correlation, although other features such as thickness, color, and stratigraphic position may be helpful (Westgate and Gorton 1981). The compositions and refractive index of any preserved glass fragments and relative abundances of the phenocrysts may also be important. For instance, the Mazama ash can be distinguished from the otherwise very similar Glacier Peak ash by the presence of crystals of augite, which are lacking in that from Glacier Peak (Powers and Wilcox 1964). An ash layer erupted from Mount Rainier [36] about 2000 years ago can be told from the Mazama ash by the higher refractive index of its glass (Wilcox 1965).

A challenge for geologists studying ash distribution is the possible change in ash composition with varying distances from the point of eruption. Most volcanic ash, consists of a mixture of phenocrysts and glass fragments. The phenocrysts generally contain considerably less silica, and more iron, magnesium, and calcium, than does the glass, are more dense, and fall faster from the eruption cloud. As a result, ash that falls at greater distances from vents typically contain less phenocrysts and may be richer in silica. This change of bulk ash composition by aerial sorting has been called **aeolian differentiation**.

Lapilli

Lapilli are pyroclasts in the range 2–64 mm (Table 7.1). The singular form of lapilli is **lapillus**, derived from a Latin word meaning "little stone." Like ash, lapilli may be essential, accessory, or accidental, ejected either in a liquid or a solid condition. Very fluid lapilli may weld together to form spatter, or agglutinate. There is no consistency in the terminology applied to accumulations of lapilli. Fisher (1961) and Schmid (1981) have suggested that consolidated masses of lapilli be called "lapillistone." To be more specific, perhaps it is best to call masses of rounded lapilli, **lapilli agglomerate**, and those of angular lapilli, **lapilli breccia**.

The rough clots of highly vesicular basaltic lava produced by Strombolian eruptions, called **cinder**, are typically lapilli-sized. The term **scoria**, the meaning of which has come to overlap with cinder, tends to be more loosely applied to any coarse, highly vesicular pyroclast (including some pumice deposits), and appears to be favored in English-speaking countries outside of North America. Cinder is ballistic in its deposition, leaving vents in a liquid or plastic state while solidifying during flight before impact. Volcanic cones consisting largely of cinder are described further in Chapter 9.

Explosive eruptions produce vast amounts of finely divided ash particles that form the principal solid components of eruption clouds. All of this ash will eventually fall to Earth as discrete particles, but in many cases, ash particles will coalesce to form layered ash balls called **accretionary lapilli**, or ash **pisolites** in older literature. These distinctive lapilli are commonly thought to form around water droplets, or by the adherence of moist ash particles in ash clouds to form muddy raindrops, which develop in much the same way that hailstones form around ice or particulate nuclei in thunderstorms (e.g. Gilbert and Lane 1994; Heiken 2006). "Hailstone" accretionary lapilli commonly are well-indurated and marked by nearly concentric fine layering. When observed to fall from ash clouds, they may appear damp, and often accumulate as aggregates of semi-spherical lapilli enclosed in fine ash matrices (Figure 7.8). They typically are 3–10 mm in size, but can reach more than 2 cm in diameter. Some muddy accretionary

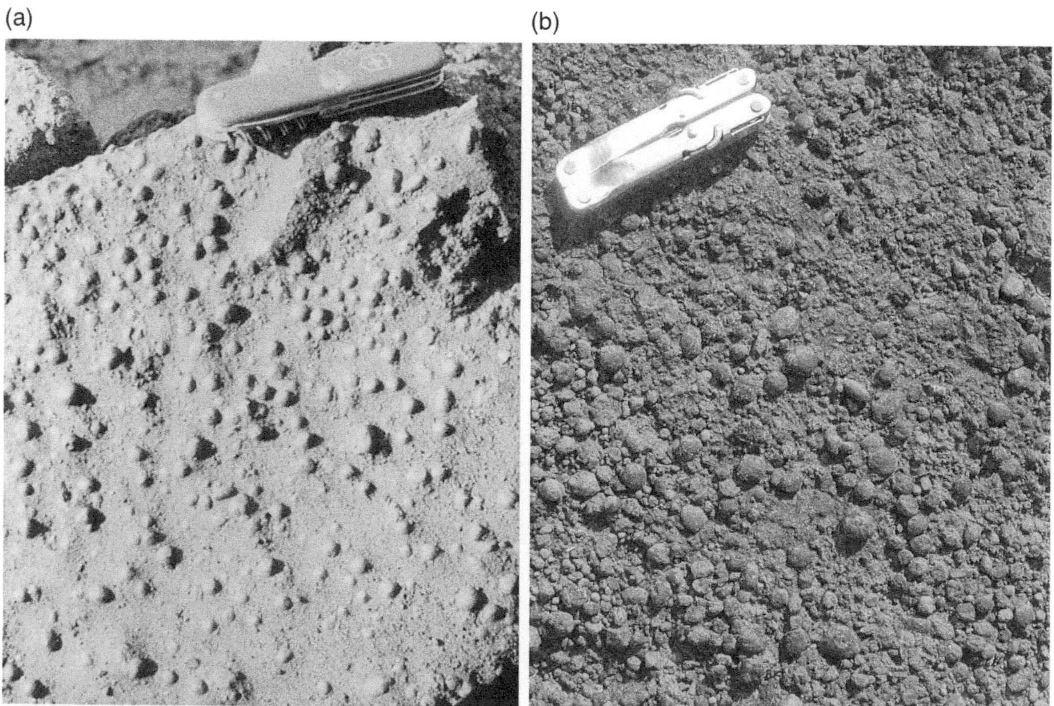

FIGURE 7.8 Accretionary lapilli deposits. (a) Most common type, where accretionary lapilli were deposited simultaneously with ashfall (Laguna maar explosion debris, Gamalama volcano [140], Ternate Island, Indonesia). (b) Concentrated deposits, where accretionary lapilli fell out of eruption cloud separately from ashfall (1790 Keanakakoi Ash, Kīlauea volcano, Hawai'i [17]). Knives 8 cm long. Source: USGS photos by J.P. Lockwood.

lapilli continued to grow after landing on loose ash during the 1924 eruption of Kīlauea volcano [17], when they were observed to roll along on loose ash after impact and enlarge like rolling snowballs (Stearns 1925).

Accretionary lapilli can also form by electrostatic attraction, forming loose spherical aggregates of fine ash that have no muddy matrix, and apparently maintain their cohesion electrically (Gilbert et al. 1991; Schumacher and Schmincke 1995; Telling and Dufek 2012). Such "electrical" accretionary lapilli were observed to fall during an eruption of Galunggung volcano [129] in 1982, and disintegrated into unconsolidated ash piles when they impacted hard surfaces – like volcanologists' hardhats or outstretched hands. They were preserved only where they dropped onto extremely loose, "fluffy," uncompacted fine tephra that was falling near-simultaneously. The tephra entombed the lapilli *within it* by backfilling of penetration holes with loose ash (Figure 7.9).

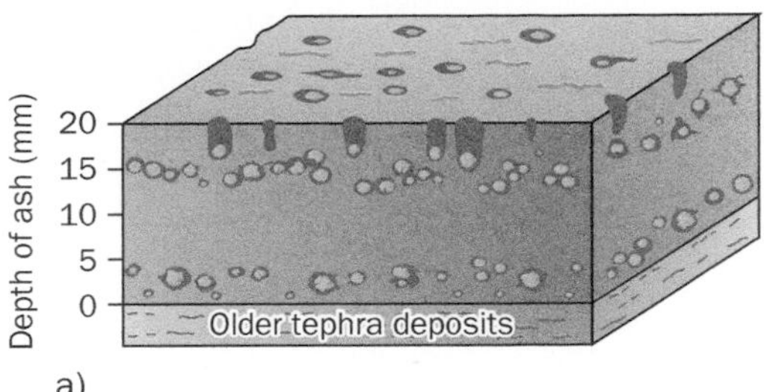

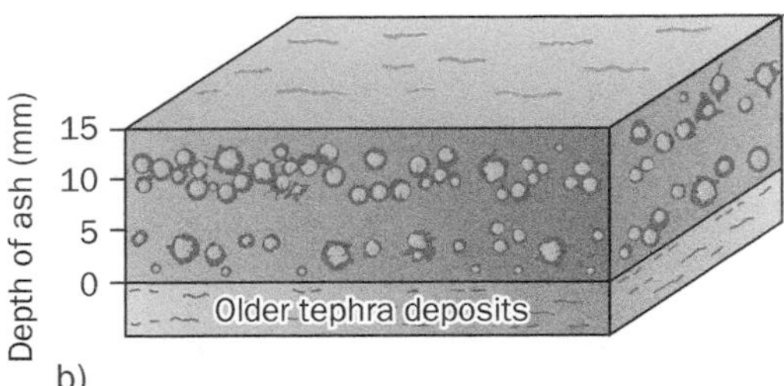

FIGURE 7.9 Formation of accretionary lapilli during the 13 August, 1982 eruption of Galunggung volcano, Indonesia [129]. (a) Accretionary lapilli have fallen in two periods, one contemporaneously with fine ash and one following the last ashfall. These last lapilli have drilled down up to 5 mm deep in the loose ash. (b) After rainfall compaction, the younger lapilli are now overlain by the older ash. Source: Sketch by M. Summers from field notes by J.P. Lockwood.

Pumice – Highly Vesicular Lapilli

The fourth-century BCE Greek philosopher Theophrastos of Eremos called pumice "worm-eaten wood." He conjectured that it was the product of sea-foam interacting with fire, a reasonable hypothesis at the time (Heiken and Wohletz 1985). Pumice occurs widely as interlayers and patches with obsidian in lava domes and flows (Chapter 6), but here we discuss it only as ejecta. Individual pumice pyroclasts range from less than a centimeter to as much as several meters in size – exceeding the size range of lapilli as strictly defined (Table 7.1), though more as an exception than the rule. Specimens occur in airfall deposits as well as clasts in pyroclastic flows. Some low-density basaltic scoria, **reticulite**, is even less dense than ordinary siliceous pumice, but the bubbles in reticulite are mostly unsealed and interconnected.

The existence of pumice is the strongest evidence we have that magma can froth much like the foam on a fresh stein of beer, as melt exsolves gases in the upper portion of a magma conduit. Pumice fragments may be regarded as bits of quenched, glassy magmatic foam. Most have high silica content, reflecting the fact that high-silica magmas are more viscous than basaltic magmas, do not release their entrapped volatiles as efficiently, and tend to produce very explosive and gaseous eruptions.

Because pumice has low-density and sealed vesicles, pumice clasts can float on water – the only stone known to do so. During large pumice-producing submarine or littoral eruptions, such as the 1883 Krakatau eruption [128], fields of thick floating pumice typically accumulate over large areas (Chapter 12). Some floating pumice fields are so thick that they can support the weights of people, wildlife, and downed trees. It is even likely that floating pumice has played a role in distributing flora and fauna to new lands, especially in explosively volcanic archipelagos like Indonesia (Symonds 1888; Simkin and Fiske 1983; Bryan et al. 2004). Pumice clasts gradually become water-logged, however, and over a period of months most fragments will sink, although individual rounded clasts commonly wash ashore thousands of kilometers from their sources.

The solid portions of pumice clasts consist of volcanic glass blown into thin cellular walls and strands, typically silver-gray in color. This color might be due as much to the thinness of the glass and the internal reflection of light within its numerous vesicles as to its composition (Klug and Cashman 1994). In fact, were the glass in an ordinary clot of pumice re-melted to form a bubble-free solid, most would become ordinary black obsidian, as shown gradationally in mixed pumice-glass domes and the fiamme found in many ignimbrites. Nonetheless, pumices of variegated color also exist. Watanabe (1986) describes prehistoric orange pumice beds at Aso caldera, Japan [143]. Imai et al. (1993) examined highly oxidized, sulfur-rich yellow pumice erupted in 1991 from Mount Pinatubo [128] in the Philippines. Paulick and Franz (1997) analyzed multi-colored pumices from the 5000-year-old Meidob volcanic field [115] in Sudan, where brown, buff, black, and ordinary gray pumices originated during the same eruptive sequence. They discovered that the color differences relate to differing concentrations of tiny magnetite and hematite (iron oxide) crystals dispersed in the volcanic glass, which, in turn, may be related to the mixing of two separate magmas with distinctive ferric iron abundances.

Some pumice clasts show compositional banding – remarkable evidence for magma mingling (Smith 1979). Three to five percent of the June 1912 Novarupta [19] pumice at Katmai, Alaska, for example, is made up of streaks of dark andesite within tan-colored rhyolite, with some mixed andesite and dacite samples from the upper part of the associated PDC deposits (Fierstein and Wilson 2005). Such pumice clasts indicate that during the process of eruption, melts of different viscosities occupied the same vent. The tabular or streaky structure of components in many mingled composition pumices suggests that the uprush of gas from the top of a magma column stretched and sheared inhomogeneous melt into fragments of varied composition. Continued expansion of gases within the molten glass expands the pumice and with aerodynamic drag during descent, may explain why so many pumice fragments are rounded.

Measured pumice porosities range from 64 to 85 volume percent, with considerable variation seen even in the products of single eruptions (Houghton and Wilson 1989; Gardner et al. 1996). This range relates in some critical but as yet poorly understood way to magma viscosity – or more particularly, the viscosity of the molten films that separate gas bubbles at the moment the pumice forms (Klug and Cashman 1994). Molten pumice with bubble wall viscosities of less than about 10^9 Pa s can continue expanding after ejection from the vent (Thomas and Sparks 1992), increasing the porosity of the final samples and causing the chilled crusts of pumice clasts to inflate and crack, forming breadcrust structure. At even lower viscosities (less than 10^5 Pa s), the expanding gases may break their enclosing films, merge, and escape altogether, causing the fragile structure of the pumice to collapse as it exits the vent. The low end of the pumice porosity range – 64 percent – therefore represents the lowest possible pore volume of specimens that ordinarily do not collapse during eruption. "64 percent" may also be regarded as the approximate *minimal volume percentage* of gas bubbles needed for pumice-forming explosions to commence at the top of magma columns (Gardner et al. 1996).

Pumice vesicularity also circumstantially relates to eruption intensity, as suggested by the work of Allen (2001) who studied the Kos [116] ignimbrite, a Plinian eruption deposit in the Greek Aegean. During the early phreatomagmatic periods of low to moderate discharge from the vent, pumice clasts were small, the size of fine lapilli, and moderately vesicular. All were rhyolitic in composition. As discharge (mass flux) increased and the eruption gained a Plinian character, rhyolitic pumices up to the size of blocks erupted with notably higher vesicularity. At the climax of the eruption, as the Kos caldera collapsed, banded pumices (rhyolite plus andesite) appeared, showing deep evacuation and mingling of melts within the magma chamber.

The initial sizes of pumice fragments can also be correlated with gas bubble volume and the relative permeability of the magma to escaping gases. Klug and Cashman (1996) concluded that pumices as small as 1 cm across develop if the magma is mostly gas bubble impermeable and the volume of bubbles in the melt is less than 70 percent. Given an increase in bubble volume to 90 percent in a highly permeable magma, pumice clots exceeding 10 cm in size can easily blast out of the vent. Gas permeability is generally boosted by rapid eruption rate.

Both ash and pumice are commonly erupted *together* during large eruptions, and usually have the same composition. The proportions of each may relate to variable ability of gas bubbles to coalesce within the magma in the moments leading up to an eruption, indicating that vesiculation of shallow magma often does not occur homogeneously. The melt in an erupting conduit may be visualized as having some areas in **volumetric equilibrium**, in which the volume of gas bubble formation is matched by the total volume of gas escaping from the magma (Figure 7.10). In this situation, typical for basaltic magmas, molten rock is erupted together as spatter and flowing lava. **Volumetric disequilibrium** prevails for erupting siliceous magmas, leading to explosive discharge of pumice and ash. If the formation and expansion of gas bubbles exceeds the rate of gas release, then bubble walls burst into small, curved ash shards. On the other hand, if the rate of gas release exceeds that of bubble expansion, then the melt becomes foamy and is blown into pumice fragments. The capacity of the magma to release its pent-up gases depends upon many factors, foremost the temperature, viscosity, and total volatile content (Klug and Cashman 1996; Blower 2001; Mueller et al. 2008).

Reticulite (Figures 7.11 and 7.12), uniquely basaltic in composition, is the most vesicular of all pumices, with porosities as high as 98 percent. Reticulite is distinguished by the close-packing of polyhedral bubbles with open walls (Mangan and Cashman 1996). In intermediate- to high-viscosity siliceous melts, surface tension in the walls of vesicles maintains spherical shapes. But in low-viscosity magma, bubbles expand to impinge against one another so that polyhedral walls are ruptured, giving reticulite a permeability that makes it *sink* in water almost at once. Individual bubble sizes are much smaller than those that may be found in siliceous pumices, typically only a fraction of a millimeter in diameter. In contrast, vesicles in rhyolitic pumice may be as much as several centimeters in size.

Reticulite has the typical color of basaltic glass (**sideromelane**), representing the internal play of light on material that is significantly more iron and magnesium-rich than that of siliceous pumices. Most reticulite forms during Hawaiian-style lava fountaining, from the frothing of clots of spatter. The addition of a mere 0.05 weight percent H_2O to melt may be all that is required for a Hawaiian-type eruption to shift from one of quiescent lava effusion to lava fountaining and reticulite production, with an increase in bubble nucleation rate of three orders of magnitude (Mangan and Cashman 1993). Observations of reticulite falls near

Equilibrium (spattering)

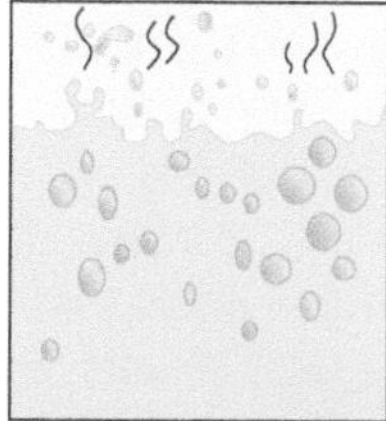

Gas volume production in melt = gas volume release from melt

Disequilibrium (Ash eruption)

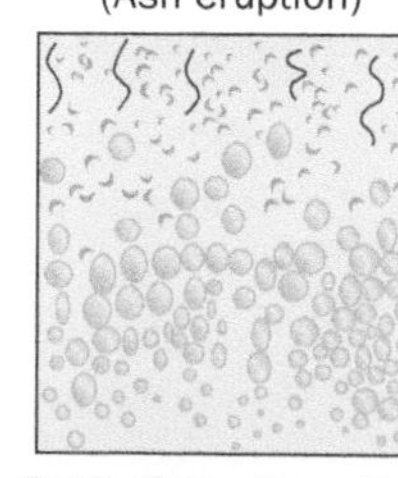

Rate of gas formation in melt exceeds rate of gas release

Disequilibrium (Pumice eruption)

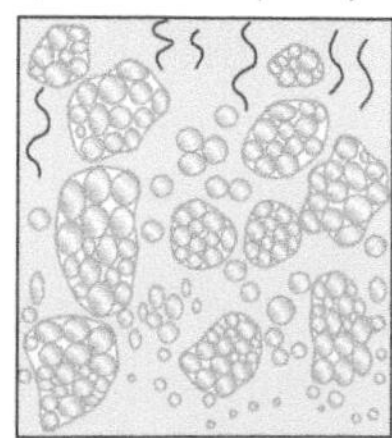

Rate of gas release from melt exceeds rate of gas formation in melt

FIGURE 7.10 The volumetric equilibrium concept of escaping magmatic gases.

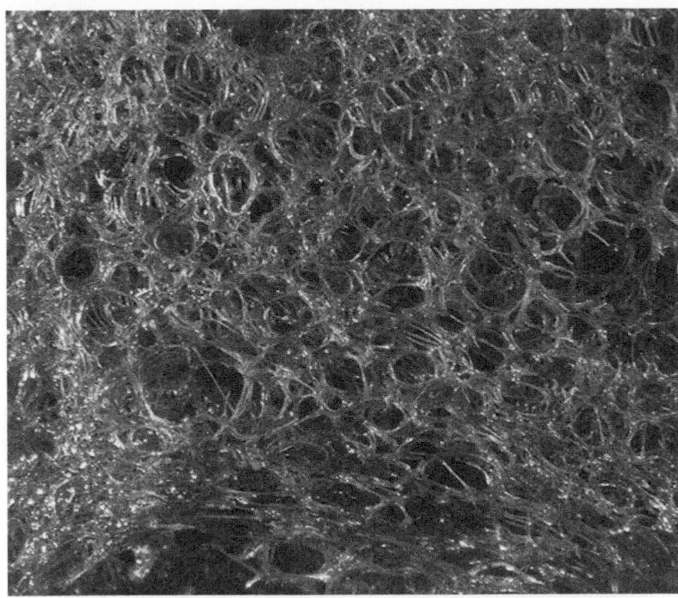

FIGURE 7.11 Internal structure of basalt reticulite from a 1969 eruption of Kīlauea volcano, Hawai'i [17], showing delicate network of glass filaments. Note that there are no closed bubbles, in contrast to pumice. J.D. Dana called such material "thread-lace scoria." View about 4 mm across. Source: Photomicrograph by Ben Gaddis.

FIGURE 7.12 Fresh reticulite fall deposit formed by 2018 lava fountaining, 50–75 m high, in Kīlauea [17] lower east rift zone. Note range in tephra sizes, in some instances related to welding of pieces to one another during ejection and aerial transport; and in the case of smaller samples related to further fragmentation upon impact. Directional indicator is 15 cm long. Source: Photo by Richard Hazlett.

lava fountains reveal that a range of fragment sizes develops, and that many larger fragments shatter and break into finer pieces as they strike the ground. Other forms of pumice are not quite as delicate as reticulite, but shattering must still play a role in modifying fragment sizes, especially during the transport of pumice in pyroclastic flows.

Sorting and Layering in Pyroclastic Deposits

The study of layering and layering order in natural rock deposits is called **stratigraphy**. Thicker layers (greater than 1 cm) are termed **beds**, and thinner ones (less than 1 cm) are called **laminations** (or laminae). It is possible for a single bed to contain internal laminations, or to thin out laterally with distance into a sequence of laminations. There is also a rough correlation between grain size and layer type in ejecta deposits, with finer particle sizes commonly making up the thinner layers.

The way in which different particle sizes in pyroclastic deposits are distributed is termed **sorting**. Perfectly sorted layers consist of particles of the same size, composition, and density. Moderately and poorly sorted layers show progressively wider ranges of fragment sizes mixed together (e.g. Figure 7.12). The sorting of ejecta deposits directly reflects the way in which it was deposited. Fall layers, for example, tend to show excellent sorting, while those of PDCs tend to be poorly sorted. Different sorting patterns reflect very different emplacement mechanisms for pyroclastic falls, flows, and surges. The excellent sorting of fall deposits reflects gravitational transport in air with very little turbulence. Particles of different weight and density fall according to their terminal velocities, with the heaviest particles dropping closest to the vent and falling faster from eruption clouds than lower-density ejecta.

Many well-sorted layers show gradations of grain sizes (Figures 7.13 and 7.14). If grain sizes are smaller upward, then the bed is said to show **normal grading**. If grain sizes coarsen (are larger) upward, then the bed displays **reverse grading**. A deposit that

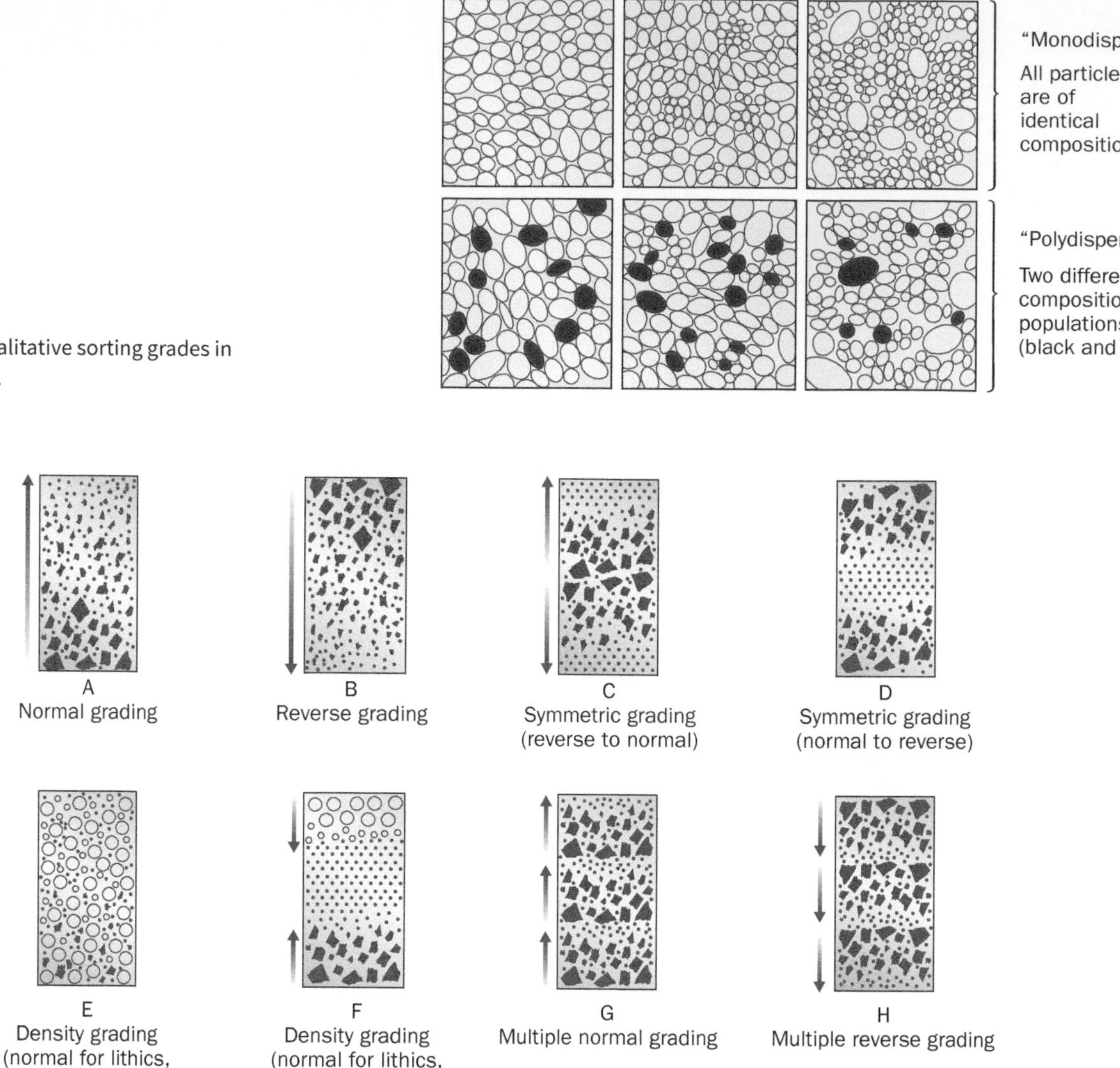

FIGURE 7.13 Qualitative sorting grades in pyroclastic deposits.

FIGURE 7.14 Types of grading commonly observed in pyroclastic deposits. Black particles are relatively high-density lithic fragments. Open circles are relatively light-weight pumice clumps. Dots represent ash particles. Source: From Fisher and Schmincke (1984). Reproduced with permission of Springer Nature.

neither shows grading, nor any internal layering, is called **massive**. The presence of grading in a layer may indicate that the conditions of deposition or the nature of source materials were changing as the layer accumulated. For example, as described for ash beds earlier, normal grading often indicates a decrease in the intensity of the eruption, while reverse grading illustrates just the opposite. However, other factors, too, can explain these same grading features, including shapes and densities of clasts. Dense clasts will fall to earth more quickly than low-density ones regardless of size and will end up at the bottom of beds. This is the process Professor George Walker called *the lady's handbag effect*.

Explosive Eruption Styles and Their Products

Following the above discussions of what is erupted during explosive eruptions, we turn to brief descriptions of the several principal types of explosive activity, roughly in order from least to most violent eruptions, along with descriptions of the rock deposits associated with each type. In doing so, we'll use three terms to describe the way in which gases and/or ejecta emerge from active vents: "volcanic clouds," "plumes," and "columns": A **volcanic cloud** is any visible emission of gas and particles from a volcano (including aerosols and vapors as well as solids). **Plumes** are volcanic clouds that are easily scattered and spread by ordinary winds. An eruption **column**, in contrast, is a volcanic cloud with great thermal and kinetic energy that ascends convectively from explosive vents and is not easily disrupted by winds until it reaches high altitudes. Eruption columns characterize the most powerful volcanic explosions, and if sustained for long enough, generally feed wide-ranging plumes from their crests.

Volcanic explosions vary in intensity from the weak spattering that commonly accompanies the eruption of very fluid basaltic lava (Chapter 6) to cataclysmic blasts that throw debris many kilometers into the atmosphere. Our usual conception of an explosion is a sudden violent outburst of very short duration – essentially a single brief impulse, like that of a booming cannon. Some volcanic explosions are like that, but most are of a longer duration, including continuous ejections of gas, ash, and rock fragments that persist episodically for several weeks or months. As Chapter 5 indicated, the classification of these eruptions depends upon multiple aspects that we often cannot directly observe and measure in the field. International efforts to improve upon our current approach to categorizing volcanic explosivity, in particular, continue (e.g. Bonadonna et al. 2016), but for our purposes, we'll stick with the "classical" framework that we have already introduced (Tables 5.1 and 5.2). A complexity is that the eruption types discussed are not necessarily unique to any one eruption. Eruptions commonly change behaviors as they evolve, and intermediate behaviors exist. We are describing Peléean eruptions as a separate category, although others may consider these eruptions as intermediate between Vulcanian and Plinian end-members.

Vulcanian Eruptions

Vulcanian eruptions are commonly driven by shallow explosions involving the expansion of water and carbon dioxide. They range from **Ultravulcanian** blasts, which only eject solid, older rock material, not directly related to the magma, or heat-source triggering the eruption, to blasts of ash and freshly solidified lava blocks commonly derived from recently emplaced domes or plugs and mixed with some incandescent ejecta from the molten rock underneath. This latter debris is the trademark product of "ordinary" Vulcanian eruptions. The ejection of angular lava blocks that travel far from the ascending ash clouds above these eruptions is a (very dangerous) characteristic (Figures 7.15 and 14.2).

Vulcanian deposits are characterized by large amounts of poorly sorted blocks and lapilli-sized angular lithic fragments mixed together with ash. Sorting improves with distance from a vent, and even farther out, the ash fall deposits of Vulcanian eruptions may be indistinguishable in appearance from those of non-Vulcanian blasts unless examined microscopically. When viewed under a microscope, Vulcanian ash grains commonly appear as highly angular bits of pulverized, pitted rock matter mixed with lesser amounts of vitric ash shards (Figure 7.16). Very large Vulcanian blasts may also include lumps of pumice, which are more common in Plinian eruption deposits. In fact, many, if not most, Plinian explosions commence with a Vulcanian phase, typically associated with "vent-clearing" activity that ejects older country rock and may precede later magmatic eruptions.

Likewise, many, if not most ordinary Vulcanian eruptions begin with an Ultravulcanian phase. Ultravulcanian outbursts commonly last no more than a few hours, starting with opening of a new, typically short fissure across the summit or flank of a volcano, or less commonly in an older hydrothermal area away from any young volcanic cone. A blast pit forms at a point along the fissure, often near its midpoint. Dark showers of accidental debris accumulate around the new vent to form a low block or block-and-ash cone. These deposits lack any vitric ash, spatter, or bombs whatsoever.

FIGURE 7.15 Vulcanian explosion from the summit crater of Volcán Popocatépetl. Mexico [56] – 17 June, 2019. The lightning-laced explosion cloud reached about 3 km above the summit during this brief eruption. Ballistic blocks from the destroyed summit dome can be seen impacting the ash-covered slope, and incandescent ejecta are seen just above the crater rim. Source: Photo courtesy of CENAPRED; Centro Nacional de Prevencion de Desastres.

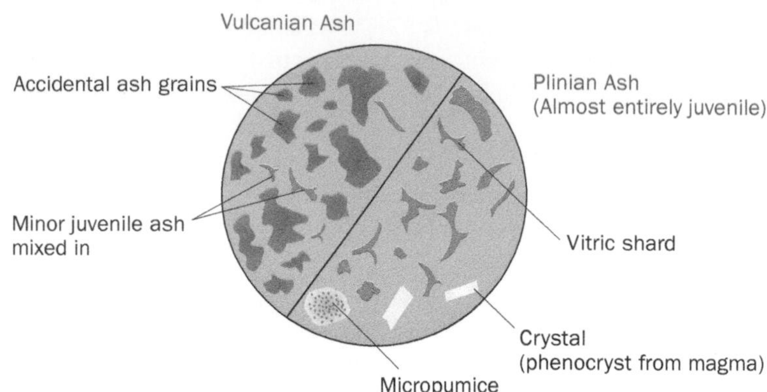

FIGURE 7.16 Contrasting compositions and textures of volcanic ash from typical Vulcanian (left) and Plinian (right) eruptions.

A good example of an especially destructive Ultravulcanian eruption is provided by Bandai volcano [150], on northern Honshu, Japan. Bandai is an 1800-m-high volcanic complex of overlapping andesitic composite cones. The dominant central one, 50,000-year-old Ko-Bandai, last erupted large amounts of juvenile material 25,000 years ago. Twenty-five thousand years seems like a long time, but Ko-Bandai is still a living volcano. It produced three major phreatic blasts between 3000 BCE and 806 CE. Then, shortly after dawn on 15 July, 1888, came the fourth in this series. Like the previous disturbances, the 1888 eruption derived its power from magmatically heated steam building up pressure at shallow depth, within a hydrothermal field that had been quietly developing on the upper northern flank of the volcano for many years. The only warning of impending disaster for local villagers was an ominous rumbling coming from the mountain, which preceded by only a half hour a moderately strong explosion centered in the north slope hydrothermal field. Continuous and strong earth tremors commenced with this first explosion, probably generated by the sudden release of subterranean gas pressure and the consequent high-pressure flow of a large amount of escaping vapor. While the ground was still trembling 15 min later, another moderately strong explosion shook the landscape, followed over the next few hours by 15–20 more, all focused on the same general locality. The blasts tossed blocks and finer stony debris to a height of 1300 m above the mountain's disintegrating northern flank, and a black, dust-laden pall of steam drifted up an additional 5 km. Condensing steam mixed with ash produced a shower of scalding hot mud, which cascaded all the way to the foot of the mountain. As the blasts continued, a large debris avalanche tore away the north flank of Ko-Bandai, quickly burying several villages. Farther downslope, it changed into a large lahar, wreaking more havoc and killing almost 500 people as it entered adjacent stream valleys. Later in the day, when the dust cleared, a new, steep-walled, dust-filled amphitheater could be seen, extending 2300 m into the volcano. The old hydrothermal field had vanished, quite literally, into thin air. Approximately a cubic kilometer of debris had been blown out, consisting entirely of old rock, some of which had partly decomposed to white clay before the eruption owing to intensive corrosion by strong hydrothermal acids.

Vulcano [111], the island-volcano providing the name for Vulcanian eruptions, is located in the Lipari islands north of Sicily, and its 1888 eruption – coincidentally the same year as Ko-Bandai's outburst – typifies activity that begins with an Ultravulcanian phase and becomes Vulcanian as time progresses. The first records of activity at Vulcano date from about the fifth century BCE soon after Greek fishermen settled the adjacent island coastlines. Between then and 1800 CE at least 10 explosive eruptions occurred. During the nineteenth century, activity began with a small eruption in 1831, followed by nearly 40 years of complete repose. Vulcano reawakened in 1873, with weak explosions that were repeated with increasing vigor in the years 1876–1879 and 1886. In each instance, the explosions threw out an abundance of fine ash mixed with angular fragments of rock. Some blocks were hot enough to glow dully at night but showed no signs of having been molten when ejected. Then, on 3 August, 1888, phreatic explosions became especially violent, tossing out rocks weighing up to several tons. Some heated debris flew 1.5 km and destroyed wooden boats anchored in the harbor at Porto di Levante (Figure 7.17). The fragments were of pre-existing, magmatically heated rocks; none were of fresh lava. After a 2-week-long hiatus, explosions began again on 18 August and continued at brief intervals over the next 19 months. Up to this point, the activity would still be called Ultravulcanian.

The resumption of the eruptive episode on 18 August brought a change in the sorts of material ejected. Fragments of freshly cooled, glassy lava were also now apparent, mixed in with the numerous blocks of older rock. The magma had risen significantly closer to the surface. Among the juvenile ejecta were breadcrust bombs having pumice cores enclosed in thin, cracked crusts of dense obsidian glass. Some bombs were still sufficiently plastic to flatten out very slightly without breaking when they plopped to the ground. The Italian volcanologist/seismologist Guissepe Mercalli, who suggested the term "Vulcanian" to Lacroix, described the general appearance of the eruption clouds rising during the 1888 explosions at Vulcano. His observations (1907, pp. 132–133) typify Vulcanian eruption clouds seen worldwide:

> *In the less violent explosions, large ejecta were lacking and the jet consisted of a dense gray mass of lapilli, sand, and ash that rose slowly, taking the form of a great cauliflower or giant mushroom . . . The strongest [explosions] commenced with a pine-tree [shaped] cloud that was absolutely black in daylight, culminating in arrow-like projections that rose very rapidly, within a few seconds reaching heights of many hundreds of meters, while from the flanks and summit of the cloud separated black streaks of stones and fine detritus. Large black rocks shot higher than the cloud, and within the*

FIGURE 7.17 Flank of Fossa di Vulcano, the active cone of Vulcano [111], Lipari Islands, that looms over the nearby harbor of Levante. It is easy to see, given this perspective, why this volcano is regarded as a threat! Source: Photo by David Buesch.

cloud darted lightning flashes, followed by short sharp claps of thunder, quite different from the rumbling that accompanied the beginning of the explosion. Then the cloud expanded in dense globes and volutes, finally building up to a height of 3 or 4 km, and becoming gray and then whitish as it gradually freed itself of heavier solid materials.

Mercalli's vivid descriptions are excellent, but he assumed incorrectly that when the eruption clouds turned color from black to white, it meant that the clouds were losing "heavier solid material." In fact, when black Vulcanian clouds turn to gray and white, it usually indicates that superheated (invisible) water has cooled enough to condense into visible white steam.

The bright lightning flashes within the ash clouds or between the clouds and the ground are characteristic features of Vulcanian eruptions. Volcanic lightning in an ash cloud results from differences in electrical potential due to the initial explosive fragmentation, and later to collisions between turbulently suspended ejecta particles. Lightning will strike when the electrostatic potential reaches a certain critical magnitude, starting as a nearly instantaneous, visible current of electrons that rushes all at once toward an area that is highly positively charged.

Seismic detection of volcanic lightning during the 1992 eruptive episode at Mount Spurr [24], Alaska, showed a strong correlation between the strength of lightning bolts and the amounts of eruption tremor and gas released (McNutt and Davis 2000). Lightning commenced 21–26 min after the start of each eruption, indicating that the explosive levels of static charge difference developed well within the rising eruption clouds rather than at the vent. Each lightning bolt was launched in the upper parts of the eruption clouds toward the positively charged matter that had fallen, or was in the process of falling to the surface (McNutt and Davis 2000). During eruptions of Redoubt volcano [23], not far from Mount Spurr, Hoblitt (1994) was able to correlate the numbers of lightning bolts directly with the magnitude (VEI) of explosions. Behnke and McNutt (2014) further applied very high-frequency lightning mapping instruments to confirm initially undetected Vulcanian ash plumes released by Redoubt – a definite plus for warnings of aviation ash hazards.

Some Vulcanian eruptions produce small-to-moderate-sized PDCs. If, in addition to pyroclastic density currents, volcanic activity is sustained and includes dome growth (Figure 9), the eruption style grades into Pelèean.

Peléean Eruptions

As mentioned in Chapter 1, the 1902 eruptions of Montagne Pelée (Mount Pelée [87]) on the island of Martinique, had major repercussions on the geologists who worked there at the time, and altered the future of volcanology. Alfred Lacroix, who witnessed much of the varied eruptive activity declared **Peléean** eruptions as one of his four major eruption styles (Chapter 5). As the archetype of this class, the 1902 and later eruptions of Mount Pelée are worth our describing in some detail.

Although earthquakes on Martinique and fumarolic activity had been noted at Mount Pelée's summit for several years, explosive activity had not been noted until late April 1902, when phreatic blasts and rising steam clouds were observed taking place at the summit. On 2 May, a red glow appeared atop the mountain, indicating magma had reached the surface. A light ashfall began over a wide area (Chretien and Brousse 1988). The volcano must have been undergoing major deformation at this time, but there was no equipment to monitor it. On 5 May, major lahars were generated from the volcano's upper slopes, reaching the sea and claiming the first lives. The climactic eruption occurred on the morning of 8 May, when observers at sea saw an ominous "black cloud" emerging from the south side of the summit crater (Scarth 2002). It is probable that voluminous pyroclastic flows were boiling over the crater rim at this time, although Fisher et al. (1980) have proposed that column collapse processes were involved. Sub-Plinian clouds were generated about the same time, and rose to an estimated 10 km above the surface (Figure 7.18). Pyroclastic flows rushed down the Rivière Blanche toward the sea, funneled by a low ridge north of the city of St Pierre. Though the denser parts of the PDCs hugged the bottom of the river valley, the lighter ash and gas-rich fractions overtopped the low ridge and tore directly through the city (Figures 7.19 and 7.20). Although little solid material was contained in this cloud, the force of this dilute, high-temperature blast demolished buildings and killed an estimated 28,000 people within seconds. Only exceptionally thick stone walls withstood the forces of the pyroclastic flow (Figure 7.21). These effects were graphically summarized by Macdonald (1972). Dome-building ensued shortly after this catastrophe and continued that spring and summer. The situation remained dangerous, and 3,000 more people died from pyroclastic flows generated by dome collapses and explosions. In October, a gigantic spine developed in the crater as viscous magma extruded (Figure 7.22), an unstable natural monument to the disaster.

Several generalities emerge from the Mount Pelée [87] eruptions that can be used to describe subsequent Peléean eruptions – such as those of Hibok-Hibok [138] (1948–1952; Macdonald 1972), Unzen volcano [141] (1990–1995; Yamamoto et al. 1993), Soufriere Hills volcano [84] (1995–2013; Druitt and Kokelar 2002), Merapi [131] (2010; Surono et al. 2012) and the ongoing eruptions of Sinabung volcano [126] (Nakada et al. 2019). Those generalizations show that Peléean eruptions typically last for several years, they involve viscous magmas, invariably involve dome growth and collapse, and are always accompanied by devastating pyroclastic flows. In contrast to the much larger Plinian eruptions (Chapter 8), they are much less powerful, usually not exceeding VEI 5. Magmas are typically andesitic, while those typical of Plinian eruptions are more silicic.

The recent reactivity of the Soufriere Hills volcano [84] on the Caribbean island of Monserrat has provided an opportunity for the best documentation of any Peléean eruption to date. The volcano had not erupted for over 400 years before the summer of 1995, when seismic unrest and fumarolic activity heralded its reawakening. The British Geological Survey founded what became the Montserrat Volcano Observatory in 1995 after initial summit phreatic explosions began, similar to those of Mount Pelée early in 1902. Monitoring activity has continued ever since, presently overseen by the University of the West Indies in Jamaica. As a major new dome began growing at the volcano's summit, concerns about pyroclastic flow threats increased and threatened populations were evacuated. Nonetheless, in 1997, a major PDC destroyed the capital city of Plymouth and killed 19 persons (Druitt and Kokelaar 2002). This severely impacted tourism and agriculture and the island's economy collapsed as authorities enlarged

FIGURE 7.18 Sub-Plinian eruption plume rising above the summit of Mount Pelée [87] in June 1902. Such a plume was observed following the initiation of pyroclastic flows on 8 May. Source: Courtesy of US Library of Congress. Photographer unknown.

FIGURE 7.19 Ruins of St Pierre, Martinique following the pyroclastic flow of 8 May, 1902. Note that very little ash was deposited by this PDC. Source: Courtesy of US Library of Congress. Photographer unknown.

A terrible volcanic Explosion—Mont Pelée in Eruption, May, 1902, Martinique.
Copyright 1902 by Underwood & Underwood.

FIGURE 7.20 Photograph taken from sea on 8 May, 1902, showing the burning city of St Pierre in the foreground and a "co-ignimbrite cloud" (Chapter 8) rising from the main pyroclastic flows closer to Mount Pelée [87]. Source: Courtesy of US Library of Congress. Photographer unknown.

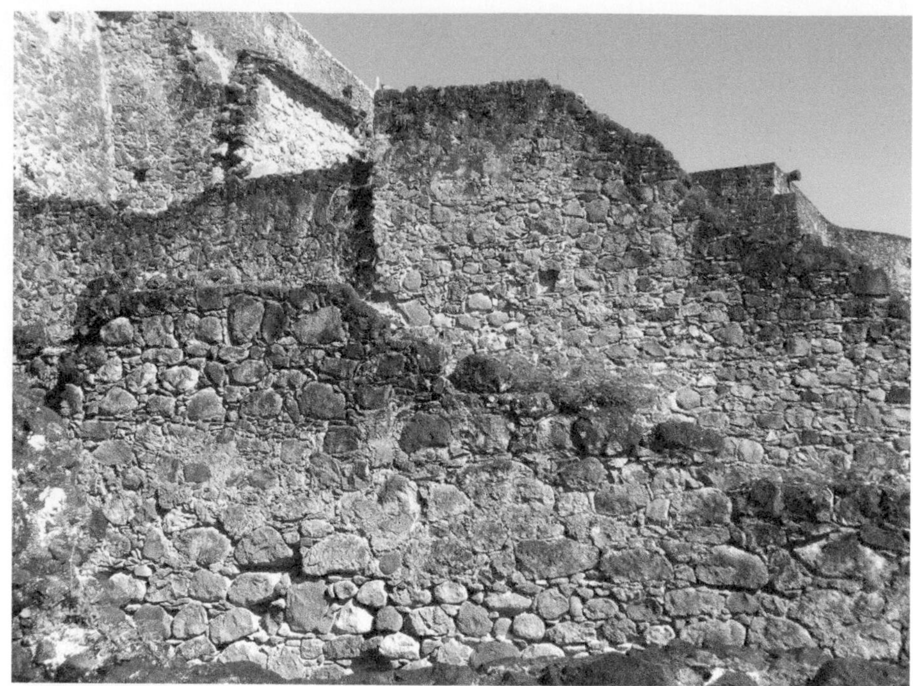

FIGURE 7.21 The only structures to survive the force of the 8 May pyroclastic flow that devastated St Pierre were the heavy stone walls of warehouses near the harbor that stored wooden casks of rum. All warehouse personnel were killed, and the rum casks exploded with the heat, adding to the conflagration. Source: Photo by J.P. Lockwood.

evacuation zones. More than half of Montserrat's population left for other islands, but it could have been much worse. A fresh surge of dome growth took place in 2008–2009, accompanied by numerous PDCs and sub-Plinian plumes rising as high as 12 km above the volcano (Wadge et al. 2014). Occasional dome failures have continued since then, with Vulcanian blasts generating small pyroclastic flows. The total eruptive volume over the more than ten years of eruption is impossible to know exactly, because of complex dome growth history and the fact that so much pyroclastic material has accumulated offshore, but likely it exceeds 0.6 km^3.

The other "La Soufriere" volcano in the Caribbean is 400 km to the south, on the island of St Vincent, and has been marked by repeated Peléean eruptions. The 6 May, 1902, eruption, which took place only a few hours before the catastrophic eruption of Mount Pelée on Martinique (Chapter 1), killed 1700 people. About 17,000 people were evacuated from the island during a 1979 eruption, and Peléean activity that began in December 2020 at La Soufriere's summit lasted only four months. These eruptions have all involved construction and collapse of summit crater domes, sub-Plinian ash plumes, widespread ashfall, and generation of destructive pyroclastic flows and lahars – all typical Peléean behavior.

FIGURE 7.22 A classic lava spine rising about 300 m above Mount Pelée's [87] summit dome in late 2002. Source: Courtesy of US Library of Congress. Photographer unknown.

Plinian Eruptions

The prototypical Plinian eruption is that of Vesuvius [103] in 79 CE, the eruption that buried the Roman city of Pompeii. This eruption type is named after Pliny, a Roman naturalist who witnessed, but did not survive the eruption. Plinian eruptions occupy the upper end of the VEI scale, and are the most energetic and potentially devastating of all volcanic activity, especially because of the voluminous PDCs that are generated. **Phreato-Plinian eruptions**, which are closely related, are essentially just Plinian eruptions that break out in shallow bodies of water. They tend to be even more powerful than ordinary Plinian eruptions because of the added explosive energy provided by larger volumes of incorporated steam.

Individual Plinian eruptions produce prodigious amounts of juvenile ash and pumice, with subsidiary lithic fragments. A typical eruption culminates in a towering eruption column of hot gas and steam mixed with millions of tons of ejecta blasted above the vent at speeds of up to hundreds of meters per second. Sheer inertial momentum sustains this column initially, but expanding steam and entrained, heated air sucked into the base of the column give it the convective boost it needs to climb kilometers higher into the atmosphere (Bursik et al. 1992). Eruptive columns with higher water and steam contents (greater than 5 weight percent H_2O) are more likely to start convecting than those that are drier, or have much higher proportions of CO_2 and other magmatic gases than of water (Sparks and Wilson 1976). Pliny's nephew (Pliny the Younger) witnessed a rising convection column above the summit of Vesuvius from 35 km away on 24 August, 79 CE, and thought the column resembled a rapidly growing pine tree of a local species (*Pinus pinea*) with a tall narrow trunk supporting a densely spreading crown or "umbrella" of roiling ash clouds. The umbrella developed where the column flattened out as it finally lost positive buoyancy in the stratosphere. The term **Plinian column** is used to describe these boiling clouds of ash and steam that rise more than about 15–20 km above volcano tops, and **sub-Plinian** for clouds that rise to lesser heights (Figures 7.23 and 7.24; Cioni et al. 2015). **Ultra plinian** describes those rare columns that rise above 40 km or so into the stratosphere (Figure 7.24). Eruptions of Krakatau [128] (1883) and Tambora [134] (1815) may be the only examples of this latter category within the past few hundred years. Most of what is known about Plinian eruptions come from studies of their distinctive ejecta deposits, and this will be the focus of the remainder of this section.

Plinian eruption deposits found throughout the world typically consist of four primary sub-units, here listed from bottom-to-top (Figure 7.25):

1. well-sorted basal pumice fall beds up to several meters thick, commonly containing angular fragments of older "vent-clearing" material; overlain by
2. cross-laminated pyroclastic surge beds; capped by
3. pyroclastic flow deposits that can be as much as several tens of meters thick. These latter deposits are typically a poorly sorted mixture of ash, pumice, and minor lithic fragments, with concentrations of angular lithic rubble at their bases and larger pumice fragments toward the top; and
4. a capping layer of well-sorted, unstratified ash up to several tens of centimeters thick (termed "co-ignimbrite ash").

Jurado-Chichay and Walker (2001) related this "standard" Plinian depositional sequence to eruption dynamics, beginning with the basal pumice fall layer that represents the initial, heavier-particle fallout from the plume as it begins to spread from the top of the eruption column across the surrounding landscape (Figure 7.26). A typical pumice fall bed is distinguished by an inversely graded lower part, reflecting a waxing in column strength; a uniformly sorted middle section showing steady state column activity; a horizon of coarsest pumice clasts showing greatest intensity of eruption; and, finally, a normally graded upper part that represents post-climax weakening of the column. The column then collapses (Chapter 8), forming PDCs that bury the pumice fall bed. Finally, the eruption ends, the fine ash lingering in the air above the freshly veneered volcanic terrain settles out to develop the co-ignimbrite layer.

Would that the world were so simple, but in many places – even where exposure is complete – the stratigraphic layering described above is not fully expressed in outcrops. One basic explanation is simply geographical – *unique factors control distribution of airfall and PDC deposits.* Prevailing winds at the time of eruptions guide plumes and their resulting airfall patterns, while directional eruption foci and even fine-scale topography greatly affect pyroclastic current deposition and distribution (e.g. Watt et al. 2015). Fallout deposits may accumulate on one side of a volcano while pyroclastic flows and surges charge down the other, leading to local stratigraphic sections that are unrepresentative of the full eruption sequence (Carey and Sparks 1986). Such, for example, is recorded in the 79 CE Vesuvius deposits, in which air fall took place primarily to the south-southeast, while pyroclastic currents poured down both the southern and western flanks.

Another geography-related depositional factor has to do with proximity to eruption sites. In general, the complexity and thickness of Plinian eruption deposits increases as one approaches source vents (Figure 7.25). For a typical deposit, the most distal material simply consists of a thin bed of fine, well-sorted tephra, which may be distinguished from Vulcanian ash only through microscopic examination. Within a few tens of kilometers of vents ash particles noticeably coarsen, and the basal layer of air fall pumice first appears. Pumice lapilli may be no larger than bird seed when first spotted, but they too coarsen ventward, with mean diameters exceeding the size of grapefruit within a few kilometers of sources. PDC deposits appear between the basal and air fall layer closer to the vent, documenting the reach of pyroclastic currents during the eruption. Within a few kilometers – the so-called **proximal area** – the stratigraphic layering more completely reflects the complex history of vent clearing, eruptive column development and collapse. Sorting deteriorates significantly, and blocks of ballistically ejected lithic material may stand out in the deposit. Local concentrations of poorly sorted lithic fragments might actually exceed those of all other pyroclastic materials in total volume and weight. The proportion of fine ash typically decreases, in no small part due to reincorporation of finer ejecta into the active convective column during eruption (Valentine 1998). Interlayering of the two types of PDCs (pyroclastic flows and pyroclastic surges) may also be present in alternating or random sequence, indicating rapid and localized transitions in particle concentrations

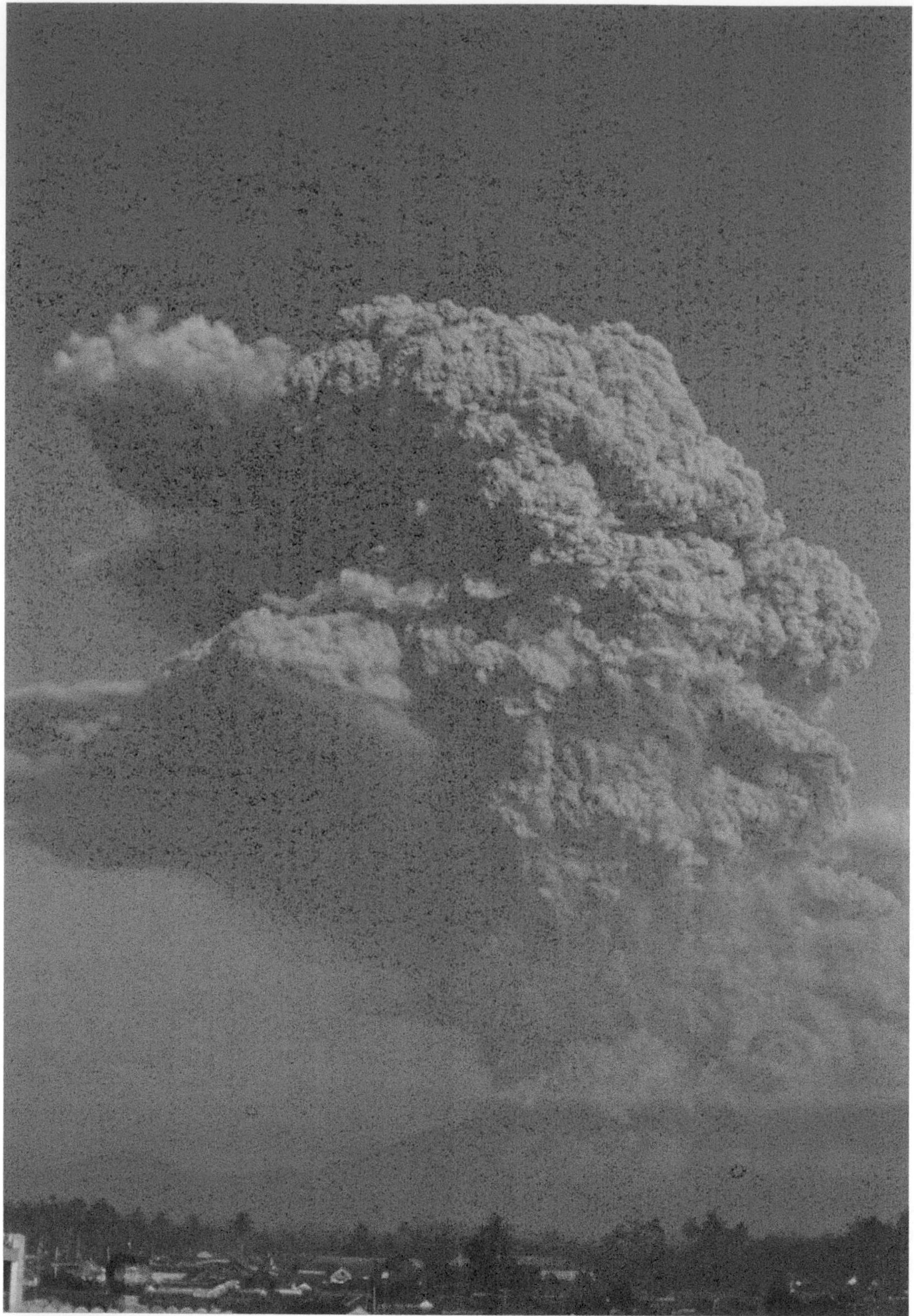

FIGURE 7.23 Sub-Plinian eruption of Galunggung volcano, Indonesia [129] – August 1982, photographed from the city of Tasikmalaya, 17 km to the southwest. This convective, ash-laden column has risen to about 12 km above the erupting crater. Source: USGS photo by J.P. Lockwood.

as air was entrained into a collapsing eruption column. Rowley et al. (1981), described the unconsolidated, 20 m thick near-vent layer left by the 1980 Mount St Helens [31] eruption as a **proximal bedded pyroclastic flow** (PBPF) deposit. There is considerable lateral variation in proximal deposits, variations that provide important clues about the complex dynamics of eruption at Plinian eruptive vents.

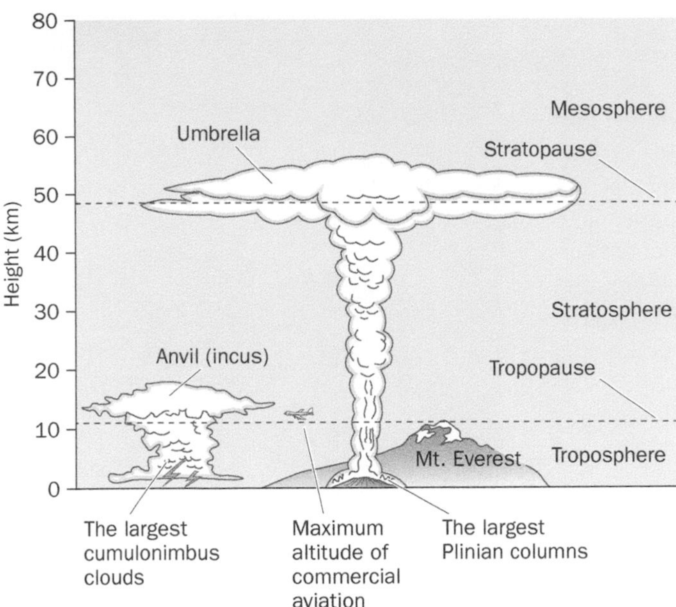

FIGURE 7.24 Comparative heights of Plinian columns and atmospheric clouds.

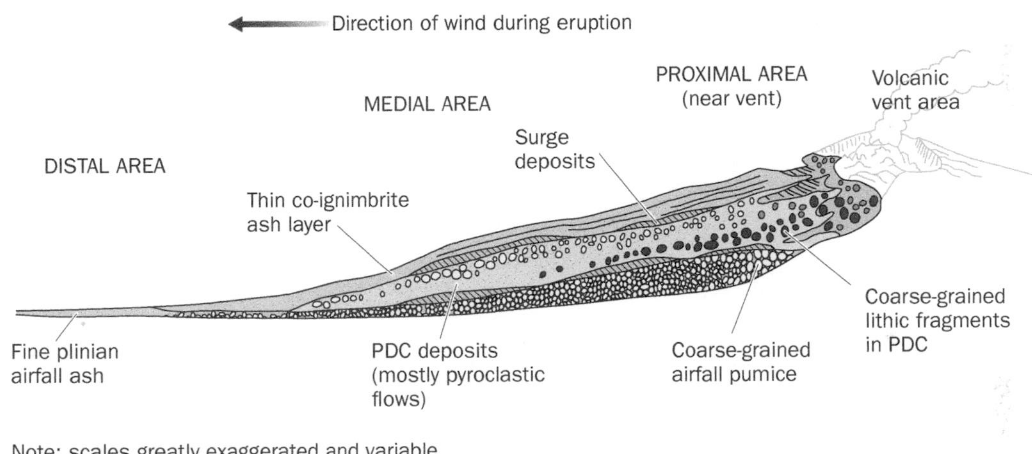

FIGURE 7.25 Proximal to distal anatomy of a typical Plinian eruption deposit.

Finally, not all Plinian eruptions unfold in a "standard" pattern. For example, in some eruptions gas thrust waxing in the eruption column is so rapid that mean air fall pumice sizes change abruptly, rather than gradationally upward. The 1947 eruption column from Hekla volcano [92] in Iceland, for example, ascended nearly 30 km in 20 min (Thorarinsson 1954). Fluctuating discharge rates might also oscillate erratically. Carey and Sigurdsson (1987) estimate that during the 79 CE Vesuvius [109] eruption, pumice discharge ranged from 7.7×10^7 to 1.5×10^8 kg/s and column height from 26 to 30.5 km. Waning can be abrupt, recorded by the absence of graded bedding and the presence of sharp contacts between deposited pyroclast sizes. Rapid shifts in eruptive column behavior can develop as vents increase in diameter during eruption and as exit speeds of ejecta and gas change. Pyroclastic discharge rate is also critical: A large mass of tephra suddenly injected into the column may force it to collapse even as eruptive discharge rate increases (e.g. Carazzo et al. 2012).

The topography of the surrounding landscape can also greatly influence PDC dynamics. Valentine et al. (1992) noted that, if a large amount of tephra hits the ground within the rim of an enclosing caldera, considerable amounts of pyroclastic material will pour back toward the vent where it is swept up and recycled in further discharge. The added mass may be sufficient to hasten column collapse, and will certainly make eruption dynamics more complicated and erratic. On the other hand, if tephra is shed exclusively outside a caldera rim little material can be recycled, and a convection column may continue to play without significant change as long as other factors, including vent diameter, mass discharge, and exit speed are steady (Woods 1988).

The classic Plinian eruption of 79 CE was mostly over in less than a day. The initial pumice fall put down a layer as much as 1.4 m thick across the small city of Pompeii over a period of about 18 h, enabling most of the population (about 90 percent of some 20,000 residents) to escape by foot and cart, though how many eventually reached safety is unknown. Many evacuees carried mattresses, pillows, or other protection on their heads to escape injury from pelting, though from archaeological remains, it is clear that some were felled horribly in their tracks. Some 12 hours after the development of the eruptive column, PDCs were already underway, initially directed toward the port of Herculaneum a few kilometers west of Pompeii (De Carolis and Patricelli 2003). When the first pyroclastic current finally reached Pompeii itself, the city's defensive wall managed to restrain it, but the accumulating ash and pumice provided a ramp on the upslope side of the wall for subsequent currents to overtop with ease. Magnetic studies of the PDC deposits indicate that the grid of city streets locally controlled the movement and subsequent depositional fabric of the basal parts of the incoming torrents of ash, pumice, and gas (Figure 7.27) (Gurioli et al. 2005).

The few people remaining in Pompeii at the time that had not already been buried or escaped certainly perished at this stage.

Pliny the Younger described the final phase of the 79 CE eruption in his Second Letter to the historian Tacitus, who witnessed it as he fled his uncle's estate near Vesuvius with his mother:

> *A dense dark mist seemed to be following us, spreading itself over the country like a cloud. "Let us get off the highway," I said, "while we can still see, for fear that, should we fall in the road, we will be crushed by the crowds that are following us." We had scarcely sat down when night came upon us, not such as we have when the sky is cloudy, or when there is no moon, but that of a room that is shut up, and all the lights put out. You might hear the shrieks of women, the screams of children, and the shouts of men; some calling for their children, others for their parents, others for their husbands, and seeking to recognize each other by the voices that replied . . . It now grew rather lighter, which we imagined to be the forerunner of an approaching burst of flames rather than the return of day; however, the fire fell at a distance from us; then we were immersed again in thick darkness, and a heavy shower of ashes rained upon us, which we were obliged every now and then to stand up to shake off, otherwise we would have been crushed and buried in a heap . . . At last this dreadful darkness was dissipated by degrees, like a cloud or smoke; the real day returned, and even the sun shone out, though with a lurid light, like when an eclipse is coming on. Every object that presented itself to our eyes (which were extremely weakened) seemed changed, being covered with deep ashes as if with snow.*

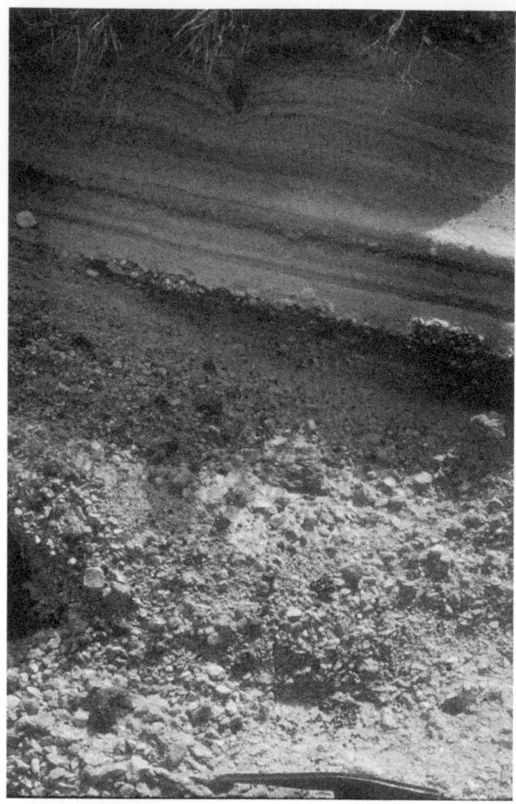

FIGURE 7.26 Plinian deposits from the 22,000-year-old eruption of Monte Guardia [110], Lipari island, Italy. Pyroclastic surge and fall deposits overlie pumice-rich, fine-depleted air fall. Note bomb sag in the uppermost layers of this meter-thick section. Source: Photo by R.W. Hazlett.

While most Plinian eruptions are associated with silicic magmas and their volcanoes, they are not unknown at some basaltic volcanoes. Nairn and Cole (1981) and Sable et al. (2006) describe distinctive deposits from a basaltic Plinian eruption of Tarawera Volcano, New Zealand [184] in 1886. The 1790 Kīlauea [17] eruption that largely destroyed the war party of the Hawaiian chief Keoua may have also been Plinian in character (Swanson et al. 2014). And Masaya Volcano [66], in west-central Nicaragua, has produced a set of powerful Plinian-type basaltic eruptions during the past 6500 years. Like Kīlauea, Masaya appears to alternate between spasms of explosive activity and intervals of more approachable lava lake eruption – for reasons yet to be understood (Perez et al. 2009).

Directed Blasts

One of the most powerful explosive phenomena associated with volcanic activity are **directed blasts**, incredibly destructive currents of hot gases and entrained material that can move at near-supersonic speeds across large areas on the flanks of structurally weak volcanoes, when magma chambers explode almost instantaneously, rather than releasing their energy gradually through constrictive summit vents. These highly destructive blasts cause widespread mayhem, but leave relatively sparse deposits, so are only recognized around young volcanoes where eye-witness accounts or devastated landscapes are left to record their effects. The 18 May, 1980, Mount St Helens directed blast swept across an area of about 600 km^2, but its varied deposits were much less than 1 m thick in most places (Hoblitt et al. 1981). Directed blast mechanisms will be discussed in detail in the next chapter.

a)

b)

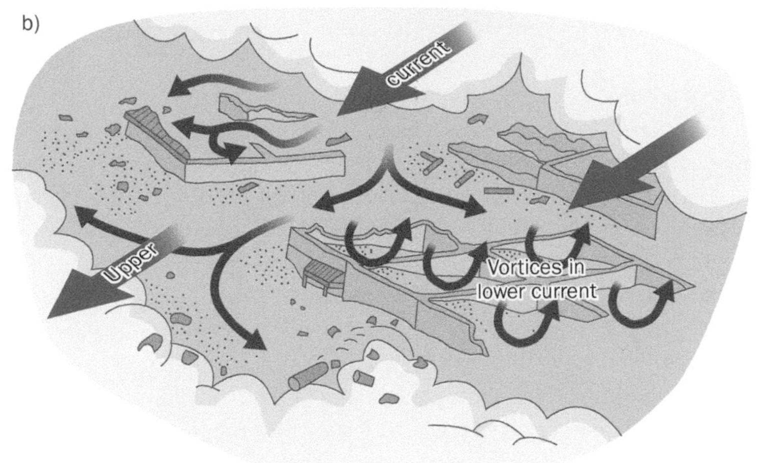

FIGURE 7.27 The 79 CE eruption of Vesuvius [109] impacted Pompeii in several ways: (a) An early air fall pumice phase began collapsing roofs within the first 6 h of eruption. According to forensic work by Luongo et al. (2003), 38 percent of all fatalities in the town were caused by roof and building collapse. About half of the victims (mostly women and children) were indoors, the other half in the streets. (b) Subsequent pyroclastic currents raced over the town. The upper parts of the currents were not deflected by buildings, but work by Gurioli et al. (2005) indicates that movement in the lower parts of the currents was significantly modified by city structures. Sources: (a) Modified from Luongo et al. (2003); (b) Modified from Gurioli et al. (2005).

Hydrovolcanic Eruptions

Hydroexplosions result from the sudden generation of steam where water comes in contact with hot rock or melt, as when molten lava rises through water-saturated rocks or is extruded into a lake or the ocean. But they need not accompany volcanic eruptions: Hydroexplosions unrelated to magma ascent are common in hydrothermal areas, such as the geyser basins of Yellowstone National Park (USA) [47], where super-heated meteoric groundwater reservoirs may slowly pressurize to critical thresholds as mineralization and silica precipitation seal surrounding crusts. Large hydroexplosions have formed more than 20 large (>100 m diameter) explosion craters in Yellowstone over the past 16 ka, with debris rims up to 30 m high (Muffler et al. 1971). The Mary Bay hydrothermal explosion crater complex on the shore of Yellowstone Lake measures 2.8 × 2.4 km in diameter, and may be one of the largest of such features in the world. Morgan, Shanks, and Pierce 2009 write that explosions large enough to create craters >100 m in diameter "*might be expected to occur every 200 years*" within the Park. Such an explosion in visited areas could cause significant loss of life, and would be likely to occur without warning. Geyser area steam blasts have also been frequent in New Zealand geothermal areas, including one 6000 years ago in Rotokawa that spread debris across an area 4 km wide. Focal depths of such explosions may be as deep as 450 m (Brown and Lawless 2001).

Phreatic and **phreatomagmatic eruptions** are both end-member types of volcanic hydroexplosions, the former involving steam only, and the latter, steam + magma. Phreato-Plinian eruptions are especially large and powerful phreatomagmatic eruptions whose dispersal areas (0.01 T_{max} of Walker 1973) exceed 50 km^2. Studies of ash deposits of the tragic 1790 phreatomagmatic eruption of Kilauea volcano (Swanson et al. 2015; Swanson and Houghton 2018; Figure 7.8b) indicate that it was Phreato-Plinian in nature, an unusual event for such an ordinarily "mild-mannered" volcano.

Explosive eruptions triggered by the heating and rapid expansion ("flashing") of external, non-magmatic water into steam lead to the development of characteristic volcanic or sub-volcanic features, including diatremes, maars, ash and tuff cones, (discussed in Chapters 4 and 9). Such features may additionally form from the exsolution and expansion of carbon dioxide from deep mantle sources. Where lava flows into lakes or the ocean at the shore, littoral cones can grow (Chapter 9). Where lava or hot ash blanket marshy terrain, numerous rootless cones and "pseudo craters" may also develop as secondary constructs from scattered steam blasts. Blocks ejected during phreatic eruptions can cause great local damage.

Ultravulcanian phreatic blasts merely discharge older rock material, while phreatomagmatic eruptions also eject vitric debris. For instance, some of the maars in the Rhine region of Germany are surrounded by low block-and-ash rims constructed from fragments of the surrounding sedimentary rocks. The explosions that formed them may have been wholly phreatic; but the presence of magmatic ejecta in some otherwise similar cones nearby suggests that magmatic gases played a part in the explosions that formed most of the cones. The low, wide rims surrounding Ubehebe Craters [45], ash cones at the northern end of Death Valley, California, consist largely of accidental ejecta, but also contain a few volcanic bombs, showing that although the explosions that formed them may have been very largely phreatic, magma and magmatic gases were also involved.

The explosiveness of all types of eruptions increases where rising magma comes in contact with groundwater at shallow depths. The effect is most conspicuous in the case of Hawaiian or Strombolian-type eruptions that otherwise would be only mildly violent. The nature of the Laki [96], Iceland flood-basalt eruption, described briefly in Chapter 6, strongly suggests that its explosiveness was at least partly due to the boiling of shallow groundwater. Another example is the 1960 flank eruption of Kīlauea, which began with the opening of fissures and the sinking of a shallow graben in an area where groundwater lay only about 25 m below the surface. During the first few days of the eruption, violent, black ash-laden steam was ejected, with much more explosive activity than that of typical Hawaiian eruptions.

Submarine eruptions in shallow water or near the margins of oceanic islands usually involve explosive hydrovolcanic activity (Figure 7.28), and commonly precede the formation of new volcanic islands. Explosive submarine eruptions involving hydrovolcanic activity will also be discussed in Chapter 12, but since many of these eruptions transition the ocean-land boundary, we shall provide one good example here, that of Capelinhos at the western end of Fayal Island in the Azores in 1957–1958 (Cole et al. 2001). The Capelhinos [89] eruption began in September 1957 when small earthquakes were felt and brownish-yellow, discolored water appeared a kilometer offshore. Within two weeks, incandescent fragments were being ejected from the sea explosively, and a new island quickly formed. By late October, the new island stood 80 m above sea level, but eruptive activity was episodic and island growth was irregular as explosive phreatomagmatic activity, characterized by jagged, **rooster-tail,** or **cypressoid** ejections of black, ash-laden steam jets, continued from several independent vents. Convecting "cauliflower" clouds of white steam rose kilometers over the entire eruptive zone as vents migrated eastward, toward Fayal. As vents became better sealed from seawater, activity became less explosive and generated Strombolian fountaining and minor lava flows. By 1958, the emerging tuff cones had

FIGURE 7.28 Hydrovolcanic "Surtseyan" eruption cloud rising above the submarine Kavachi volcano [164], western Soloman Islands (11 October, 2016). Note the ballistic fragments emerging from the explosion cloud and impacting on the water. Source: Photo by Alex DeCiccio – Wikipedia Creative Commons. https://commons.wikimedia.org/wiki/File:Kavachi_Eruption.jpg

joined the island to form a new peninsula. Tephra blanketed most of Fayal, destroyed hundreds of homes, ruined agricultural fields, and led to a major population exodus. Over 2 km^2 of new land accreted to Fayal, but the pyroclastic rocks are poorly consolidated and weak, and over half this land eroded away within the next half century.

Less than a decade after the Capelhinos eruption, a very similar outburst took place in the shallow sea off the southwest coast of Iceland, leading to the growth of the new island of Surtsey [90] between 1963 and 1967. This well-studied event gave rise to the name for the characteristic style of phreatomagmatic activity associated with eruptions through shallow water bodies – **Surtseyan eruptions** (Thordarson 2000).

Phenomena associated with Surtseyan eruptions are not restricted to oceanic volcanoes, and good examples are also found wherever volcanic activity occurs beneath fresh-water lakes (Godchaux et al. 1992) or water-filled volcanic craters. Renewed eruptive activity beneath these crater lakes can violently eject their waters in a Surtseyan manner, as New Zealand geologist Peter Otway and a colleague discovered on snow-covered Ruapehu volcano [178] in 1971. Ruapehu is a dangerous volcano, with a summit crater lake known for a history of hydrovolcanic explosions accompanied by destructive lahars (Chapter 11). Otway's team had been dropped off by helicopter near the summit to monitor activity, and were routinely surveying when part way through their work they noticed puffs of steam rising from the middle of the grayish-green, sulfurous lake. The steaming patch suddenly exploded, and jets of black ash-laden steam and large angular blocks shot upward. Peter, who was on the snow-covered rim of the crater, pulled out his camera to take a snapshot, when suddenly the explosions increased in violence; Ruapehu's crater lake had begun a series of hydrovolcanic explosions, and the New Zealanders were directly in the path of ejecta. Each had only a few seconds to grab ice axes and dig in, lying flat to the ground and holding on for dear life. A torrent of muddy water cascaded down, mixed with angular blocks that made sharp thuds as they impacted nearby. The water fortunately did not scald, and neither man was swept away during their 3 minutes of terror. John Latter and an assistant were surveying on the other side of the crater outside the impact apron, and were able to photograph the explosions from 300 m away (Figure 7.29). Analysis of his movie film showed that the cypressoid ejecta jets blasted out at speeds up to 150 m/s toward Otway. Latter was convinced that his field partners had been killed. Down below, observers knew that an eruption had started as the seismographs were going wild. A helicopter pilot was called, and maneuvered above clouds to see a column of convecting ash and steam obscuring the summit. He feared he had just "lost four geologists," but approaching closer, he found the mud-covered survey team and courageously rescued them from further danger. As Surtseyan eruptions go, this one was relatively small – had pyroclastic surges been generated, the field party would have been wiped out. The summit glacier was left mantled with a black, thin bed of muddy ash, and the ejected water poured down the mountainside. The resulting mudflow split a restaurant in half, though no one was injured.

Nuclear engineers concerned with reactor safety have long been interested in the potential for explosions as coolant water mixes with molten material, finding that the explosiveness depends critically upon the volumetric ratio of melt (the fuel) to water: the **fuel–coolant interaction ratio** (FCI) (Sandia Laboratories 1975). Extending this research to magma–water interactions, experimental work further suggests that hydrovolcanic eruptions appear to be most violent when one part magma mixes thoroughly with three parts water (FCI = 1:3; Wohletz and McQueen 1984). Magma rarely interacts with "pure" water, and factors such as vent geometry, fuel–coolant interaction and suspended aqueous sediment load can play critical roles too (White 1996). In any case, Strombolian eruptions appear to grade into Violent Strombolian and finally Surtseyan eruptions with increasing FCI.

The range in magma–water fuel–coolant ratios is reflected in the vesicularities of individual pyroclastic fragments. Explosions triggered purely by anhydrous magmatic degassing tend to produce pyroclasts such as pumice, vesicularities of which are clustered narrowly toward higher percentages (70–80 percent of sample volumes). The steam-rich magmatic blasts of the 1800-year-old Rotongaio Ash, New Zealand, in contrast, produced ejecta with vesicularities across a broad spectrum, from 20 to 60 percent. This seems to be typical of eruptions in which expanding steam, rather than magmatic degassing, initiates an explosion (Houghton 1993). The lower level of vesicularity in hydrovolcanic juvenile ejecta is a function of the power involved in gas expansion. Because the coefficients of expansion of magmatic gases such as CO_2, SO_2, and H_2S are much lower than those of water flashing to steam, it is possible for numerous non-hydrous bubbles to form in the magma before it explodes. Expanding steam, on the other hand, shreds much of the melt before such initial fine-scale vesiculation can take place (Figure 7.30).

The shapes and mixtures of ash and fine lapilli fragments ejected by Surtseyan blasts can provide important clues about the phreatomagmatic eruption process. For example, fragments showing mixed angular and subrounded edges with mud coatings are likely to have been recycled by falling back into a vent for later re-ejection a second or even a third time. Subrounded grains indicate ejection from melt that has not yet developed a crust, whereas angular grains result from solidification followed by explosive fragmentation of a crust (Houghton 1993). Because collapse of conduit and crater walls is much more common in wet eruptions than dry ones, perhaps due to the enormous lateral pressure pulses associated with shallow steam blasts, there tends to be an increase in the accidental lithic contents of hydrovolcanic deposits with increasing magma–water interaction (Romagnoli et al. 1993). In fact, variation in the lithic concentration of layers often provides better evidence of slight changes in magma–water interaction than simple fragmentation (Figure 7.31).

FIGURE 7.29 Hydrovolcanic explosion in the summit crater of Ruapehu volcano, New Zealand [178] – 8 May, 1971. (a) Initial blast from lake surface about 3 s after initiation – "rooster-tail" ejecta trails radial to explosion source rise about 300 m above crater lake surface. (b) Photograph taken 14 s after (a). This is a new, larger blast that has ripped through the first explosion column, and later sent a steam column 8 km high. This is the blast that nearly killed observers on the south rim of the crater. Source: Photos by J.H. Latter.

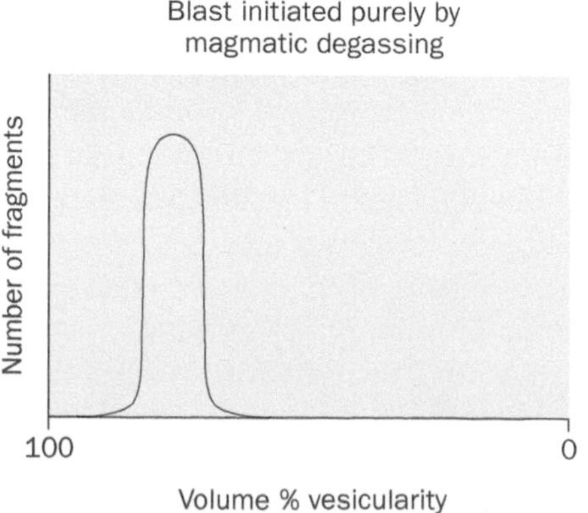

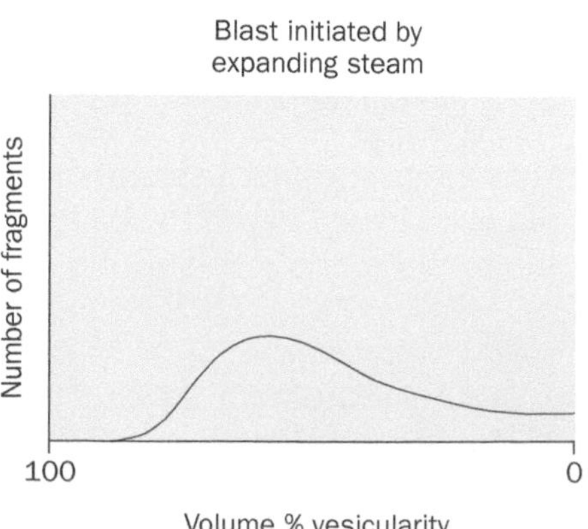

FIGURE 7.30 Contrasting vesicularity and fragmentation of volcanic material according to explosion source. Source: From Houghton (1993).

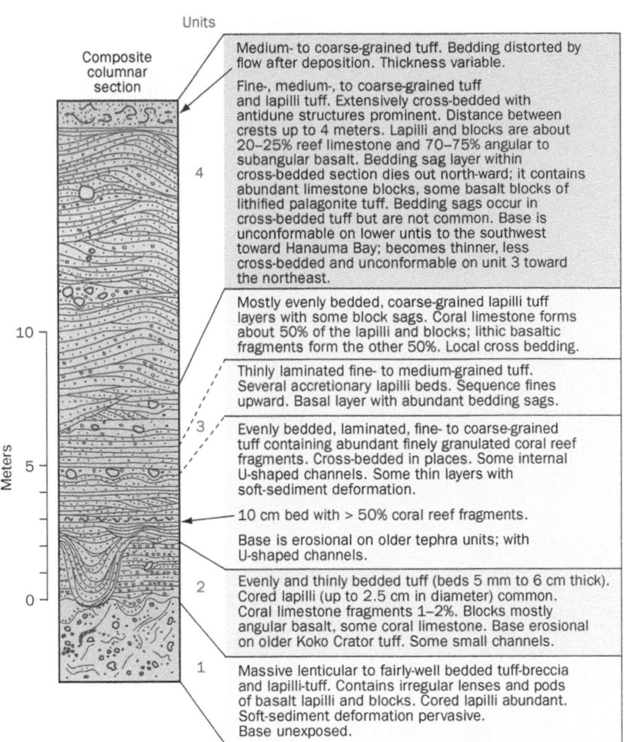

FIGURE 7.31 Stratigraphy of the Koko Head tuff cone deposits (Oahu Island, Hawai'i) – products of explosive hydrovolcanism and surge processes accompanying the formation of a littoral tuff cone. Source: From Fisher (1977).

Pyroclastic Density Currents (PDCs)

Pyroclastic density currents (PDCs) are potentially deadly, ground-hugging torrents of volcanic ejecta mixed with gases that can pour down volcano flanks on many different scales, from minor ones only a few hundred meters long and a few tens of meters wide, to devastating outpourings that can blanket thousands of square kilometers in deposits hundreds of meters thick – they are potentially the most devastating phenomena generated by large explosive eruptions. Because of their violence, geologists cannot directly observe the internal processes involved, but can only infer them from external field observation of active currents, through modeling, and by careful study of their deposits.

As a simple illustration of the phenomena involved, pouring a handful of flour onto the ground might impress (and perhaps distress!) you by how far and how easily the powdery white mass travels. The fine, light-weight particles trap air as they approach the surface, mixing with it to produce a highly mobile, air-plus-solid mixture that rushes across the floor very much like a moderately viscous fluid. This experiment is a quick and easy demonstration of how PDCs work (Colin Wilson, personal communication 2006). Instead of flour, of course, a typical PDC consists of various kinds of ejecta, including, blocks, ash, and in many cases, pumiceous lapilli. Many PDCs are extremely hot; hot enough to glow incandescently (hence the French name nuées ardentes coined by Alfred Lacroix (Chapter 5)). The heat helps buoy and mobilize the whole mass through the expansion of trapped gases. The mixture of gas, air, and solid matter varies in these ground-hugging eruption clouds, from rarefied masses having the consistency of dusty sandstorms, to denser media that spread out like fast-spreading sand or flour avalanches choked with abundant lithic blocks.

PDC deposits are commonly classified as either of **pyroclastic flow** or **pyroclastic surge** origin, although the processes that form them are complexly interrelated, and gradational interfaces between these end-members are common and difficult to classify rigorously (Branney and Kokelaar 2002). The internal complexities of pyroclastic surge deposits (cross-bedding and internal erosion features; e.g. Figure 7.32) suggest that deposition of these PDCs involve turbulent flow, whereas pyroclastic flow deposits are more homogeneous and likely reflect more laminar, sheet-like flowage.

In more powerful volcanic eruptions, PDCs transport ejecta of all sizes. Topography exerts a strong influence on the course of these surface-bound currents, which generally follow major valleys and dissipate against opposing mountain slopes, although extremely powerful currents can surmount high ridgelines. The deposits left by PDCs are distinctively different from airfall and ballistic accumulation, and different kinds of PDCs leave different kinds of deposits, as illustrated in Figure 7.32. It is worth spending a few minutes studying this figure before we move into exploring pyroclastic flows and surges in greater detail. But perhaps throughout it is wise to bear in mind the words of Carey (1991, p. 39) who concluded "It is . . . unrealistic to hope that all flowage deposits can be conveniently classed into either category and that a single set of transport and depositional processes would apply to each case." Nature throws us many instructive ambiguities!

It is also critical to note that PDCs are propelled by huge volumes of extremely hot gases and that these gases can separate from denser portions of the clouds and can cause great destruction in areas where little ash is deposited. The 28,000 victims of the 1902 Mount Pelée [87] eruption were killed by hot gas – not ash (Figure 2.19) – as were victims of the 1991 eruption of the Unzen volcano [141] and most of those felled by PDC clouds at Merapi [131] in 2010 (Chapter 8, p. 15).

Air-fall deposits

Ash fall

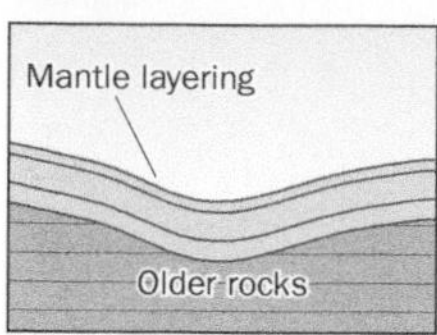

Pumice or cinder (scoria) fall

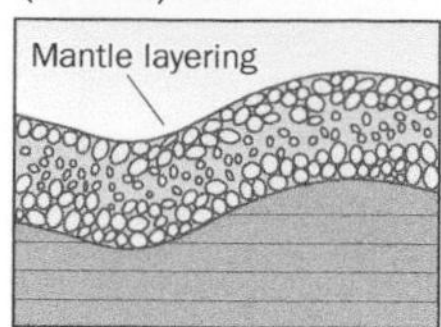

Phreatic or phreatomagmatic fall

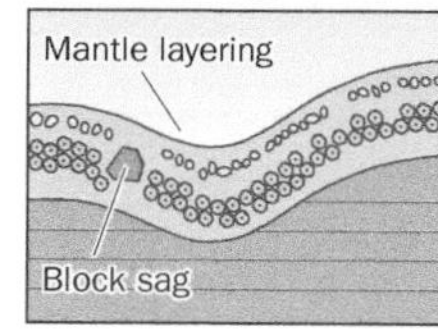

Welded pumice fall

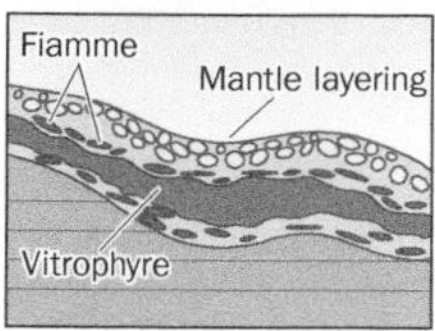

Pyroclastic current deposits

Surges

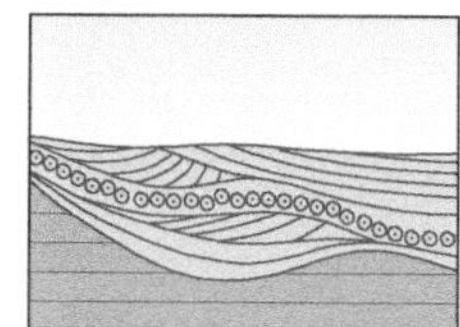

Block-and-ash flows

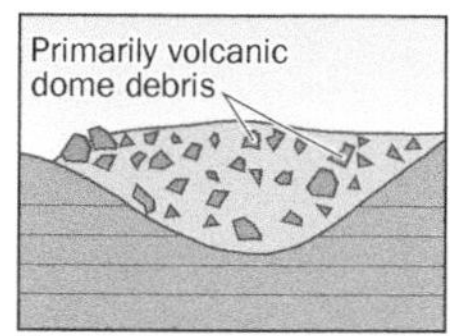

Ash-flow ("unwelded ignimbrite")

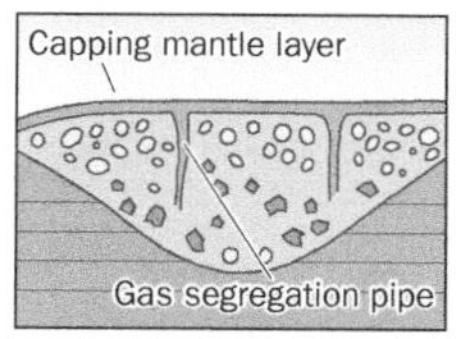

Ignimbrite

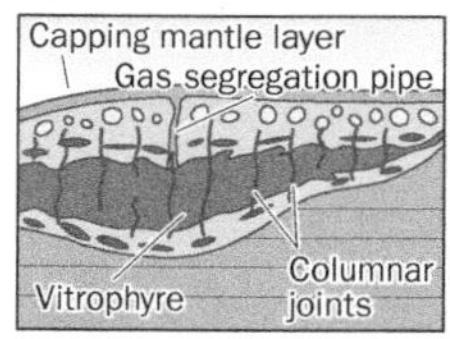

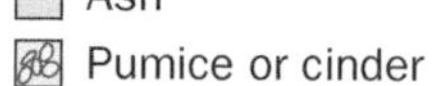

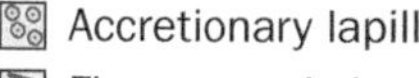

FIGURE 7.32 Some diagnostic features of pyroclastic deposits. Source: Modified from McPhie et al. (1993).

Pyroclastic Flow Deposits

Two general categories of pyroclastic flows exist; **block-and-ash flows**, which are especially characteristic of Vulcanian eruptions, and the much larger and energetic **ash flows**, which more commonly occur as a result of Plinian fountaining. Block-and-ash flow deposits tend to be highly localized around volcanoes, while ash flows lay down much thicker, more widespread sheets of ejecta, most commonly around highly explosive silicic volcanic centers. In fact, volumes of individual flow deposits range over six orders of magnitude (Smith 1979) – from the low-range block-and-ash flows (mostly less than 1 km^3), to a single massive ash flows sheet exceeding 1000 km^3 in volume.

Block-and-ash flows

Block-and-ash flows are a category of pyroclastic flows that are derived from the explosive disintegration of growing domes (Chapter 9) or from the margins of steep-sided active silicic lava flows. Their deposits differ from most other PDCs in that they typically include little or no pumice and contain abundant angular blocks derived from source domes, embedded in glassy volcanic ash (Figure 7.33). Unlike many ash flows, they also show little or no welding. Lithic clast sizes grow finer and less abundant toward the distal ends of deposits, similar to ordinary rock avalanches. Flow margins are very steep, but the tops are typically gently sloping and broad with hummocky surfaces of poorly sorted to extremely poorly sorted material. When a typical block-and-ash flow is emplaced, its sides commonly come to rest before its medial mass does, creating **lateral levees** as much as several meters high as the flow continues draining downslope. On gentler ground, the front of the flow also tends to slow before the central mass, resulting in a bow-like terminal mound rippled from compression ridges by the time everything comes to rest. In map view, block-and-ash deposits have a smooth, lobate, or spatulate shape, with the "spoon handle" extending upslope. Block-and-ash flow deposits formed from the explosive collapse of actively growing domes tend to have a larger proportion of fine ash relative to blocks than in deposits left by gravitational collapse of older domes (Freundt and Bursik 1998). These deposits rarely exceed 2 km^3 in volume or spread more than a few kilometers from their sources. We will tell you more about their origin in the next chapter.

Ash flows and ignimbrites

Ash flows are extremely powerful PDCs consisting largely of ash and pumice mixed with variable amounts of lithic material. Their deposits are distinguished from block-and-ash flows primarily because they tend to be more pumiceous (Brown and Andrews 2015). The term **ignimbrite**, derived from Latin ("fire-rain"), was introduced by Marshall (1935) who originally applied it to describe only those ash flow tuffs in New Zealand that showed a high degree of **welding**; the fusing together of very hot ash particles and other ejecta shortly after emplacement to create a hard, brick-like, or glassy mass. Since then, the word has acquired broader connotations and now generally refers to all ash flow deposits irrespective of their degree of welding. Roughly speaking, "ash flows" (termed "nuées ardentes" by Lacroix; see Chapter 1) refers to an eruptive phenomenon, and "ignimbrites" (or "ash flow tuffs") to the deposits, though a search of the literature will show ambiguity and evolution in these terms. We take them as approximate synonyms.

Ignimbrites from individual eruptions can form great sheets across hundreds or even thousands of square kilometers. The largest ash flow sheets form gently sloping uplands around source calderas tens of kilometers wide. Some are so thick – up to hundreds of meters – that they cover all the older highs and lows in the topography across which they erupted, and are said to be **landscape burying** (Figure 7.34). Marshall (1935) and Macdonald (1972) referred to these enormous accumulations as **rhyolite plateaus**. Good

FIGURE 7.33 Block-and-ash flow along the Gendol River, Merapi volcano, Indonesia [131]. Note the matrix-supported angular blocks from the summit dome and the concentration of large blocks at the base of the flow. Similar pyroclastic flows and lahars have killed thousands of people on Merapi's lower flanks in recent times. Source: USGS photo by J.P. Lockwood.

FIGURE 7.34 Flat landscape typical of areas buried by ash flows – the Valley of Ten Thousand Smokes, formed by the June 1912 Novarupta [19] eruption on the Alaska Peninsula, in what is now Katmai National Park. This ignimbrite bed covers approximately 100 km^2, and is as much as 210 m thick. Source: USGS photo by Robert McGimsey.

FIGURE 7.35 Laacher See [105] pyroclastic deposits exposed in the Schmitz Quarry, Mendig, Germany 3 km south of the Laacher See maar. The base of the 15-m-thick section cropping out above the dark quarry debris is a lithic-rich base surge deposit formed during initial vent-clearing phases of the eruption. This deposit contains fragments of older basalt substrate and other pre-eruptive material. A thin beige Plinian pumice airfall layer caps this basal deposit, with finely laminated, gray surge bedding forming the thickest layer at the top. Source: Photo by J.P. Lockwood.

examples are the Yellowstone Plateau [47] in Wyoming, and the tableland around Lake Taupo [180] in New Zealand. Ross and Smith (1961) wrote a classic resource for the field identification and petrographic interpretation of ignimbrites – a paper still useful today.

The main unit of many, if not most, ignimbrites is characterized by a massive, poorly sorted mixture of pumice lapilli, lithic fragments and juvenile ash, though owing to density contrast, pumice fragments may be concentrated toward the tops of individual sheets with lithic fragments toward the bottom. Variations in component clasts and textures can be quite informative about the emplacement history of an ignimbrite. Some deposits show faint traces of bedding, either from a sudden change in the supply of material to a gradually accumulating bed from laminar (sheet-like) flow during deposition, or from brief intervals in deposition during which fall ejecta accumulates atop flow surfaces (Wilson and Hildreth 2003). Changes in sparse lithic fragment composition may be all there is to show of a break in eruption between two pulses of ash flow deposition. From a distance of even a few meters, this compositional difference may not be obvious. One has to back away further – or observe the outcrop at a certain angle of lighting or under dry conditions – to observe the subtle layering that may be present. What in fact are two stacked ash flow units may appear to be a single massive sheet in some outcrops (Wilson and Hildreth 2003).

A good example of the sorting and grain size variations possible in ignimbrites may be seen in the deposits of the Laacher See [105] eruption Figures 7.35, 7.37). Laacher See was a very large, VEI = 6–7 eruption about 13,000 years ago that formed a caldera 3 km in diameter in the Eifel District of Western Germany. We do not often think of Central Europe as being volcanically active – but Quaternary volcanism has occurred here. The Laacher See eruption produced both Plinian and phreatomagmatic deposits and spread tephra over much of Europe, from Spain to Germany. Both pyroclastic flows and surges formed near the vent (Figure 7.35). The Plinian flows were generally warmer (300–500°C) than the "water-cooled" phreatomagmatic ones (less than 200°C) but are unwelded. Many of these flows were confined to steep-sided valleys, which they filled to overflowing within a few kilometers of the vent. Freundt and Schmincke (1986) identified three zones in these valley-confined flows – "proximal," "medial," and "distal" – each displaying a unique set of depositional features. They also identified eight characteristic marginal deposit types, each showing the variable influence of changing flow parameters and topographic confinement on the sorting of loads at the edges of the PDCs (Figure 7.36).

In many ignimbrites, the abundance of lithic clasts makes portions of them resemble ordinary block-and-ash flows. Extreme examples have been recorded in which lithic fragments make up *nearly half* of an ignimbrite, reflecting widening of the vent by blast erosion, collapse of conduit wall rock during eruption, or most catastrophic of all, caldera collapse. Three cubic kilometers of the

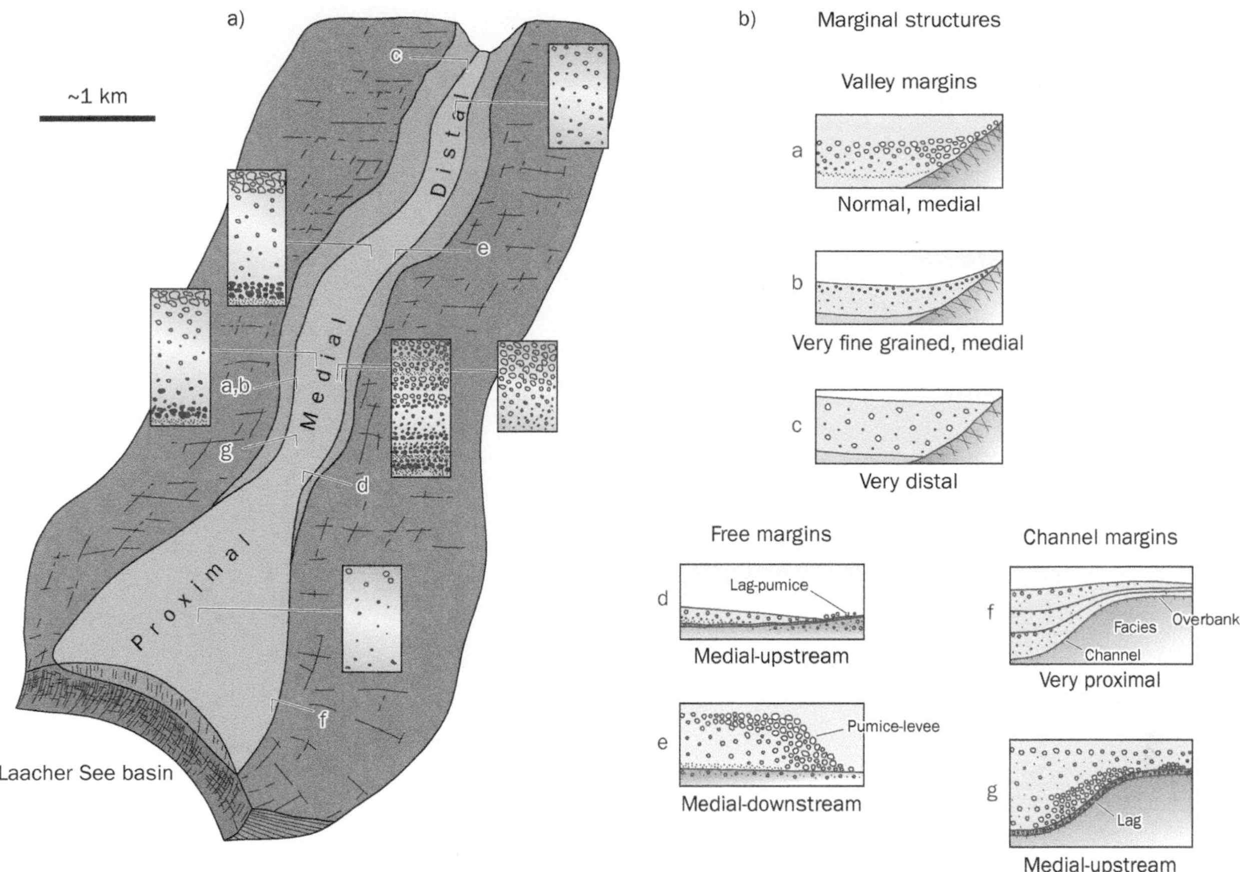

FIGURE 7.36 (a) Proximal to distal variation in valley-fill pyroclastic flows at Laacher See, Germany [105]; (b) marginal deposits, with blue letters keyed to schematic localities shown in (a). Pumices are represented by open circles. Source: From Freundt and Schmincke (1986). Reproduced with permission of Springer Nature.

60 km^3, 160,000-year-old Kos [116] ignimbrite in the Greek Aegean, for instance, consists of lithic fragments (Allen 2001). Such lithic clasts make up 6 km^3 of the Mazama Tuff [32] in Oregon (Bacon 1983), and 7 km^3 of the Bishop Tuff (Hildreth and Mahood 1986). In each case, there is a dominant concentration of lithic breccia in single layers near the base of the ignimbrite, with most rubble in fines-depleted facies concentrated near the vent (Figure 7.37). Proximal breccia layers such as these, 1–20 m thick, are termed **co-ignimbrite lag breccias**. Both lithic and poorly vesicular juvenile blocks may be present in the breccias, which may feature narrow, laterally localized accumulations of larger fragments called **breccia trains**. Individual lithic blocks often exceed 50 cm in diameter, and may approach several meters in size. Although some larger breccia-rich deposits stretch distally as far as 20 km from their source, most are restricted to within just a few kilometers (Wright and Walker 1977).

In addition to the sorting characteristics described above, ignimbrites typically show evidence of the escape of gases trapped within them at the time of emplacement. **Gas segregation pipes** are narrow, vertical "trails" left by the concentration and ascent of hot air and other volatiles through loose, settling ejecta. They are commonly marked by distinctively coarser sorting and discoloration, and in places are cemented by siliceous sinter. Escaping gases and steam help cool an ignimbrite, though to what degree is not well known. Interaction of released high-temperature volatiles with the atmosphere contributes to oxidation reactions, which commonly redden the surface and uppermost parts of ash flow sheets.

Ignimbrite welding is shown in thin sections by the fusing together of the molten edges of ejecta fragments. As vesicles close up, the gases within are expelled and may stream up into overlying gas segregation pipes or disperse more generally. Flattened, compressed pumices fuse under mounting pressure and temperature into irregular lenses of jet-black obsidian, a feature named **fiamme** ("flames") by Italian volcanologists. In cross-section, fiamme appear as conspicuous black lenses, often with fuzzy ends, in a gray or brown matrix. Fiamme are usually elongated in directions parallel to deposit surfaces. Geologists call this distinctive streaked structure **eutaxitic** (Figures 7.38 and 7.39).

McBirney (1965) described an unusual occurrence in several Central American ash flow tuffs in which fiamme are restricted to areas where the ash flows came to rest on river-laid sand and gravel that may have been saturated with water at the moment of burial. McBirney suggests that water vapor filtering upward through the hot ash and pumice beds was absorbed by glassy fragments, locally lowering their melting temperatures sufficiently to allow the glass to remelt and aggregate into fiamme.

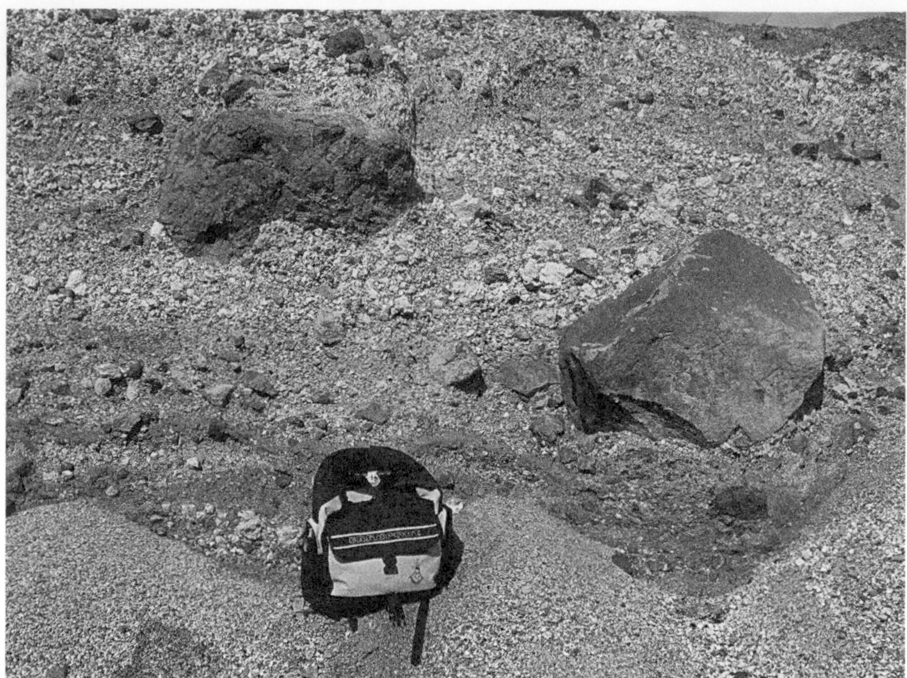

FIGURE 7.37 Detail of basal vent -clearing breccia at the base of the Schmitz Quarry section shown in Figure 7.35, above., illustrating the violence of the initial Lacher See phreatomagmatic activity. Large angular basalt blocks, smaller fragments of slate and indurated soil, blasted from pre-eruptive vent walls are in a matrix of juvenile pumice. Backpack is 40 cm high. Source: Photo by J.P. Lockwood.

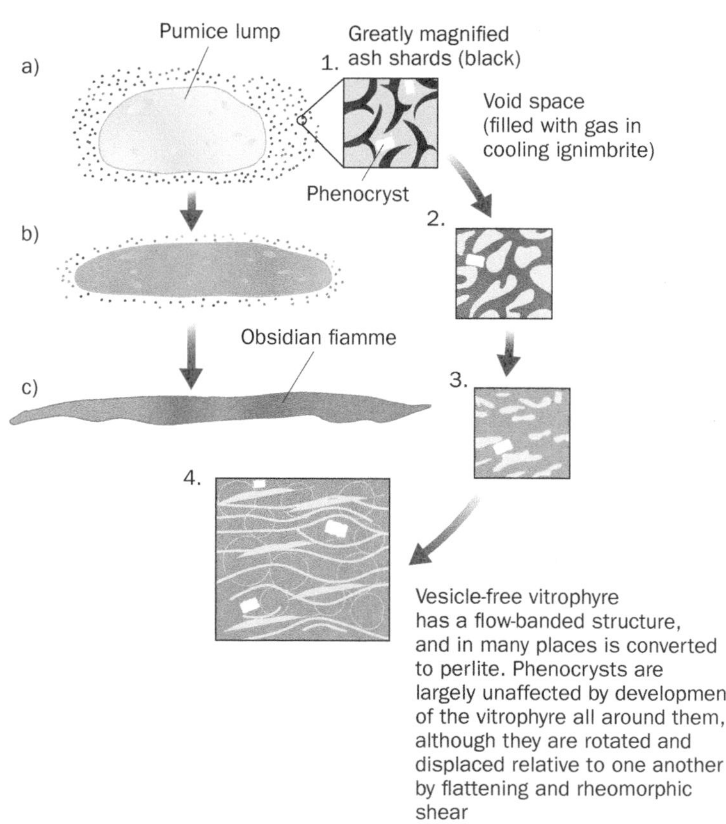

FIGURE 7.38 Progressive welding and flattening of pumice (a–c) and ash (1–4) fragments in ignimbrite.

The zone of most intense welding is usually somewhat below the middle of pyroclastic flow deposits, where heat was retained for the longest time following eruption. There, a layer of **vitrophyre** (phenocryst-bearing volcanic glass) up to several meters thick commonly develops in the days and months following an eruption, and may develop columnar jointing on cooling. The boundaries of the original ash particles welded together are often still clearly discernible under the microscope, as intricately curving

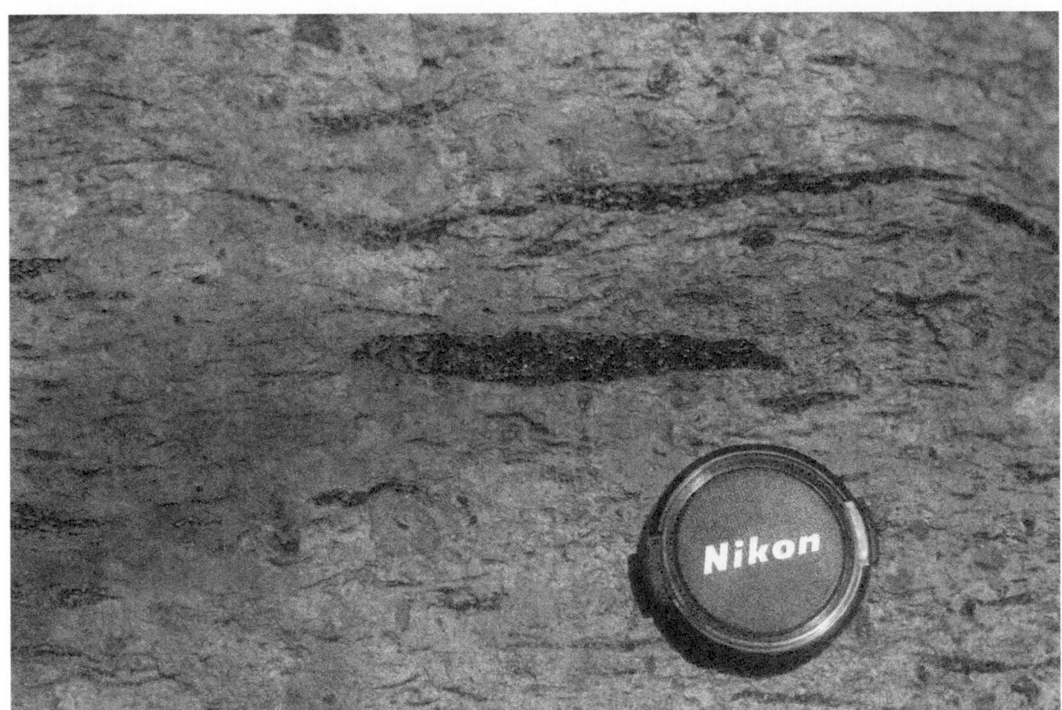

FIGURE 7.39 Fiamme in 40,000-year-old dacitic-welded ignimbrite on flank of Kuttara volcano [154], Hokkaido, Japan. These elongate glassy clasts (fiamme) were once sub-angular to subrounded pumice fragments that were compressed and stretched during internal flow and compaction of the ash flow matrix. Source: Photo by J.P. Lockwood.

and convoluted lines marked by stains or granules of iron oxide. A vitrophyre layer may be very extensive and retain rather uniform thickness over broad areas; but in other instances, vitrophyres are markedly lenticular and may pinch out, for example, from 10 m to 30 cm or even to zero, within horizontal distances of less than a hundred meters.

The degree of welding in ignimbrites varies with several factors, the principal of which are the thickness of the flow, its temperature when it came to rest, and the amount of gas it contained. Laboratory experiments show that the minimum temperature of welding of "dry" glass shards is probably about 750°C, but slight welding may take place in the presence of gases at temperatures as low as 535°C (Smith et al. 1958; Sheridan and Wang 2005). A landscape-burying ash flow tuff may be unwelded where it is thin over hills and densely welded where it is thick over valleys. That thickness is not the only factor is shown by the fact that thin flows may be thoroughly welded, whereas some thick flows show little or no welding at all. The Walcott Tuff in Idaho is only 7 m thick, but is densely welded, whereas the Battleship Rock ignimbrite in New Mexico shows no dense welding in a cooling unit 80 m thick (Smith 1960). In some deposits, the degree of welding decreases with distance from sources, reflecting a cooling of the ash during emplacement. But in the Owens Gorge area of the Bishop Tuff, welding actually increases 20–40 km away from the vent, mostly because the tuff accumulated to greater thickness in these areas. The tuff erupted as multiple ash flows in quick succession. While the earliest ash flow encountered rugged terrain, which slowed it down, it smoothed out the landscape to make it easier for later flows to travel farther (Wilson and Hildreth 2003). Calculations show that in 100-m-thick sheets with emplacement temperatures of 850°C, dense welding may be completed within a week; in caldera-fill deposits, 1 km thick with initial temperatures of 650°C, welding may continue for about a year (Bierwirth 1982).

When ash flows erupt in rapid succession, the stack of flows may at first have about the same temperature throughout the entire mass and cool as a single body, constituting what Smith (1960) coined a **cooling unit**. Freundt and Schmincke (1986) coined a parallel term, **eruptive unit**, to designate a sequence of closely related pyroclastic flows not hot enough to weld. In the San Juan Mountains of Colorado, the Bachelor Mountain [38] Rhyolite is a series of ignimbrites that issued from vents associated with a caldera near the town of Creede. At a distance from the vents, the formation consists of three separate members, but close to the caldera, it consists of a single cooling unit almost 1 km thick (Steven and Lipman 1976). The accumulation of ignimbrite layers near the vents was so nearly continuous that one did not have time to cool appreciably before the next one piled atop it, but only occasionally did exceptionally voluminous flows reach the distal areas (Ratte and Steven 1967). Other groups of ignimbrites that accumulated within some of the calderas of the San Juan region formed single cooling units as much as 1.5 km thick that are almost entirely densely welded (Lipman 2006).

Smith (1960) and later researchers (e.g. Christiansen 1979; Reihle et al. 1995) have distinguished several kinds of cooling units. A solitary ash flow with a normal cooling profile, as shown in Figure 7.40, constitutes a **simple cooling unit**. Heat loss is somewhat more rapid through the top of the ash flow than through the base, given the relative thermal conductivities of convective air versus

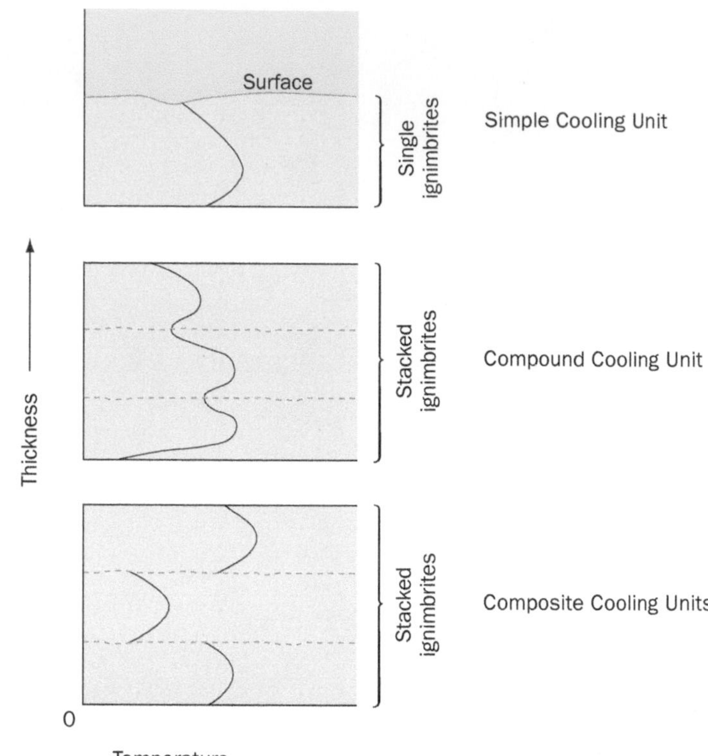

FIGURE 7.40 Cooling curves show the temperatures inside an ignimbrite just as it has settled and begun to cool. The curves can be reconstructed to varying degrees of precision based upon degrees of welding observed in the field and paleomagnetic studies. The three end-member cooling unit types are shown above.

ground at ambient temperature. A **compound cooling unit** is shown when ash flows stack together rapidly, as in the case of the San Juan, Colorado ignimbrites mentioned previously, with **welding maxima** and **minima** marking the cores and contacts of the separate ash sheets. Wright (1981) provides an excellent study of a compound cooling unit, the Rio Caliente ignimbrite in Mexico. **Composite cooling units** are separate but related ignimbrites that accumulate more slowly, each showing separate cooling profiles. The contacts between each ignimbrite in a composite stack may be highlighted by the development of soil horizons, or even deposition of sedimentary layers.

This identification of cooling units can be challenging, and it is wise to be cautious. Wilson and Hildreth (2003) mapped Bishop Tuff welding in the Owens Gorge area, where, as mentioned earlier, multiple ash flows accumulated in quick order. Welding does not correspond neatly to the contacts between the separate ash flow sheets in this deposit, but cuts across them in places. In some exposures two welding bands appear in the same vertical section, while in others, welding merges into a single broad zone. There is no clear correspondence between welding and the total thickness of the depositional package, and the densest welding, somewhat inexplicably, appears to correspond to steeper slopes. Wilson and Hildreth ascribe this pattern to irregularly increasing overall thickness of the ash flow stack during the eruption, after cooling had already commenced in the lowermost deposit. They also note that emplacement temperatures changed during the eruption, not only due to flux in primary magma temperatures, but also to variable cooling during eruption and transport of ejecta. For example, ash flows generated by convective column collapse are likely to be cooler than ash flows from the same host magma generated by boil-over (Chapter 8). Paleothermometry studies showed that the lowermost ignimbrite of the Bishop Tuff cooled from magma at a temperature of 723–737°C, while the magma of the uppermost sheet was emplaced at 749–790°C (Wilson and Hildreth 1997). Intensive volatile streaming moistened the ejecta near the surface of the tuff as the uppermost ash flow settled into place, further reduced melting temperature, softening the glassy ash particles and causing welding to take place at very shallow levels, largely unrelated to thickness of overburden.

The colors of ignimbrites vary according to composition and degree of welding. Poorly welded silicic deposits are light gray or even white in color. Ignimbrites emplaced at high temperatures will develop characteristic pink tops – products of oxidation in contact with air during gradual cooling (Figure 7.41). With progressive degrees of welding, they first acquire a pink or pale brown coloration, then orange-brown, red, and with complete fusion to glass, a black or glassy black tone. In general, hotter ash flows will tend to form darker, more strongly colored tuffs. Highly welded ignimbrites also develop columnar jointing as they cool. The joint planes tend to be less planar than those of columnar basalt, and they are commonly more widely spaced. For instance, cooling columns in the Los Chocoyos Ash, Guatemala [60], and the Bandelier Tuff, New Mexico are spaced as much as a meter apart. The more strongly welded the deposit – the better developed the jointing.

Welding is not a phenomenon unique to ash flows. Fallout deposits may rarely be welded from compaction of very hot material near vents. The dacitic 3600-year-old Thera and Therasia tuffs of Santorini, Greece [114] are outstanding examples (Sparks and Wright 1979). In other places, lava flows or other very hot deposits may weld underlying fall material. For example, in south-central Washington, where the Pomona Basalt (an extensive lava flow of the Columbia River flood basalts) overlies a mantle

FIGURE 7.41 Pyroclastic flow from the "Super-eruption" of Lake Atitlan caldera [60], Guatemala, about 84,000 years ago. The pink top of his unit is characteristic of high-temperature pyroclastic flow deposits, and is formed by oxidation during cooling. Source: Photo by J.P. Lockwood.

bed of vitric tuff, the tuff is welded at the contact with the basalt in many places, to thicknesses of as much as a meter. The texture and minerals found in the tuff are identical to those of welded ignimbrite (Schmincke 1967). Similar welded air fall tuffs are recorded beneath ignimbrites in Arizona (Enlows 1955), in Yellowstone National Park (Boyd 1961), on Tenerife in the Canary Islands (Soriano et al. 2002) and adjacent to a flow of rhyolite lava in Nevada (Christiansen and Lipman 1966).

Rheomorphic Ignimbrites

As a freshly, very hot ash flow deposit settles, it is held in place by the microscopic interlocking of countless tiny, highly angular pyroclasts – a sort of tenuous strength found in all deposits of loose material called **pseudocohesion**. But as welding ensues this interlocking weakens, and the whole ash and pumice bed begins sliding even on very gentle slopes, lubricated by internal films of molten fluid. The most densely welded vitrophyre-forming portions flow like sluggish lava, shifting and shearing over days or even weeks following an eruption in a manner akin to the gravitational creep of soils on slopes or to the flow of glacial ice. Rittmann (1958, 1962) coined the term **rheomorphism** ("deformation-shaping") to describe this late-stage movement, which can develop a very distinctive set of structures (Branney and Kokelaar 1994), but which also may create a structure closely resembling an ordinary high-silica lava flow (Henry and Wolff 1992). Rheomorphic adjustments may extend the original length of an ignimbrite by several hundred meters, while reducing its initial bed thickness as much as several meters.

Rheomorphic welded zones and vitrophyres are characteristically laminated, with planes of lamination highlighted by concentrations of flow-oriented crystal fragments, contrasting vesicularities, and even different colors resulting from hydrothermal alteration and weathering that was guided by lamination structure. The laminations record viscous flow in the form of very tight, irregular folding, including **sheath folds**, which are curving folded structures stretched in the direction of flow (Branney and Kokelaar 2002). In some deposits intricate sets of laminations also merge into shear zones resembling those found in obsidian domes and ʻaā flows that have slowly cooled inward from their margins while still moving. In some instances, the presence of large, stretched vesicles may play a role in the development of shearing (Pioli and Rosi 2005). Sets of laminations and related shears can form fabrics resembling the **S-C tectonites** of structural geology, in which the angle of intersection of laminations and shears reveals the sense of movement responsible for developing the structure (Figure 7.42; Soriano et al. 2002).

a)

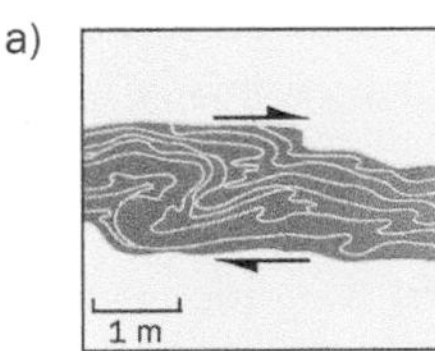

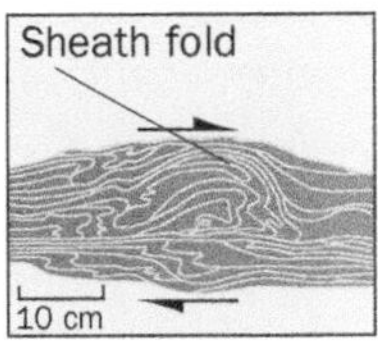

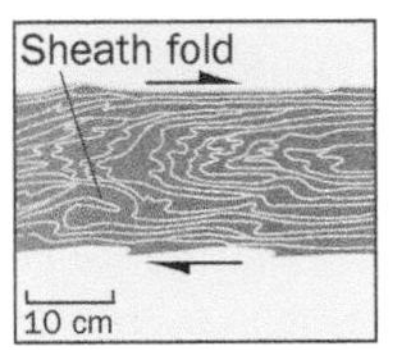

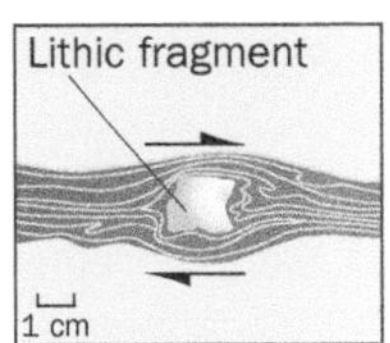

b)

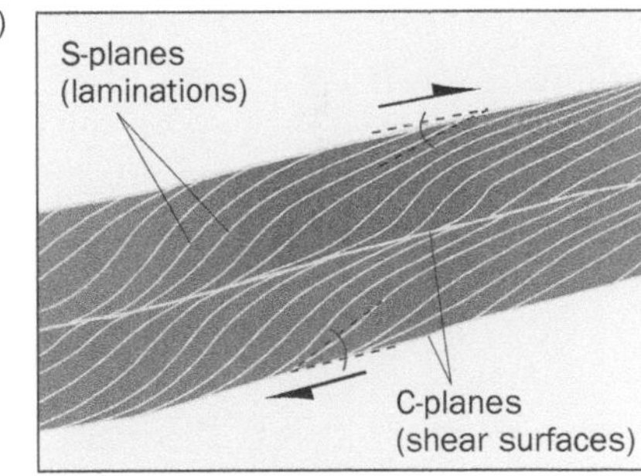

"S-C tectonite" fabric formed by flow laminations and shear zones in rheormorphic deformation. Dark headed arrows show sense of shear responsible for rotating and shear laminations (irregularities are smoothed out in this representation in order to illustrate concept)

FIGURE 7.42 (a) Contorted laminations within rheomorphic glassy ignimbrite in Nevada. Arrows indicate sense of shear responsible for creating contortions. Source: Based on Christiansen and Lipman (1966); (b) "S-C" fabric formed by flow laminations and shear zones in rheomorphic deformation. Dark headed arrows show sense of shear responsible for rotating and shearing laminations. Irregularities are smoothed out in this representation to illustrate concept.

Continual flow of the rheomorphic core of a deposit will disrupt the overlying lesser-welded portions of the flow that are riding piggy-back on its top. Pumice fragments may be stretched and pulled apart by late-stage deformation, with some grains undergoing rotation to open up void spaces within the deposit (Schmincke and Swanson 1967). The shearing, flowing mass beneath may intrude the loose materials atop to create laminated, dike-like masses that can extend all the way to the surface of the ignimbrite. These **autointrusions** may appear as ridges or curving parapets of glassy rock as much as several meters high. Autobreccias may also develop in the border between the densely welded center of the ignimbrite and the looser superficial material, or even at the contact with the unwelded base of the deposit.

The rare, welded air fall layers that show rheomorphic behavior tend to be especially potassium-rich, including peralkaline rhyolites and phonolites. Such compositions have lower glass melting temperatures and viscosities than more common calc-alkaline silicic materials. Soriano et al. (2002) describe welded rheomorphic pumice fall beds as much as 50 m thick around the rim of Las Cañadas caldera [104] in the Canary Islands. They introduced the term **welding sequence** to describe the well-preserved transition, generally over intervals less than 8 m, between unwelded air fall and "lava-like" material.

Late-Stage Mineralization of Ignimbrites

In addition to the post-depositional changes discussed earlier, other subtle changes may occur in ash flows long after the sky clears following eruptions, taking place unseen beneath the surface of fresh volcanic tephra. A pyroclastic flow, especially an ash flow, will discharge large amounts of gas and steam as it cools and compacts. The build-up of vapor pressure in pockets within the cooling ejecta, notably after heavy rainfalls, can trigger small phreatic blasts creating craters as wide and deep as tens of meters, weeks, or even months after the deposit settles. Thick ash flows can remain hot and be re-mobilized to generate secondary PDCs as much as four years after primary emplacement, as was learned after the Mount Pinatubo [135] eruption (Torres et al. 1996). This long cooling history and associated slow degassing of pyroclastic flows involves the transport and precipitation of a large amount of ions dissolved in hot vapors. Cooling and condensation lead to the growth of new minerals as both sublimates and precipitates, a process called **vapor-phase crystallization**, which acts to cement the flow. This is an important first step toward hardening it into tuff. The new crystals most commonly form in the open spaces between shards or fines-depleted coarser fragments, or in pores of incompletely collapsed pumices. In the densely welded portions deeper down, where gases are completely trapped as the ash flow welds all around them, pockets form that become lined with crystals upon cooling, including the lithophysae described in Chapter 6.

Cristobalite and tridymite, two low-pressure, high-temperature forms of SiO_2 are among the most abundant minerals formed by vapor-phase crystallization. The silica often hardens ("**sinters**") the ash around gas segregation pipes making them resistant to erosion. The surfaces of some ignimbrites are pock-marked with fumarolic mounds up to 3–4 m high and 10–20 m across. Notable examples include the Sherwin Grade area of the Bishop Tuff and the hummocky tableland of the Mamaku Tuff west of Rotorua, New Zealand [183]. The mounds are the egress points of gas pipes that have resisted wind and water erosion since the first years following eruption, when the airfall layers atop each ignimbrite were especially susceptible to removal. In more spectacular cases, silica-hardened gas pipes stand out as sharp, narrow pinnacles, tens of meters high in deeply eroded tuff beds. Examples include the Mazama Tuff in Crater Lake National Park, Oregon [32], and the Bandalier Tuff within Jemez Caldera in New Mexico [49].

Other important vapor-phase minerals include alkali feldspar, celadonite (which can turn an ash flow dark green as it hardens into tuff), and zeolites such as clinoptilolite. Zeolites are especially important minerals because they can easily absorb dissolved ions by bonding with H_2O. This made zeolite-bearing ignimbrites an attractive medium for the once-proposed storing of high-grade nuclear waste in the Yucca Mountain area of southern Nevada. The assumption was that any radioactive leakage from storage containers would be impeded by zeolites in the tuff. Laboratory experiments verify that zeolites readily absorb toxic radionuclides, but they can also disintegrate into colloids easily transportable by groundwater, tempering optimism about these "wonder-minerals" as potential sealants for soluble waste (Wilshire et al. 2008).

Devitrification, or the natural crystallization of glass, is a second type of mineralization associated with cooling ash flows. While vapor-phase crystallization takes place during the slow cooling of an ash flow, devitrification persists long after the flow has cooled to ambient temperatures. Indeed, glass alteration may continue for tens of millions of years following an eruption, and original glass becomes increasingly uncommon in older pyroclastic rocks.

Pyroclastic Surge Deposits

Pyroclastic surge deposits reflect extremely turbulent flow during emplacement, and are characterized by evidence of mechanical erosion at the bases of depositional units. Internal erosion channels, unconformities, and cross-bedding are also characteristic features. Surge deposits associated with hydrovolcanic activity are characterized by abundant shards of quenched glass. Surge deposits associated with Vulcanian activity commonly contain abundant lithic blocks and breadcrust bombs.

Crowe and Fisher (1973) and Wohletz and Sheridan (1979) studied the makeup of pyroclastic surge beds around Ubehebe Craters [45]. They found that deposits near the main vent consist of thin layers of coarse ash showing very gentle cross-stratification. They proposed that this structure developed wherever fast-moving, turbulent gases blew ash grains into low, regularly spaced piles, or dunes, much as Moore (1967) had observed at Lake Taal [136] in the Philippines following the 1965 phreatomagmatic eruption there. They designated this type of surge deposit a **waveform facies**, with the pyroclastic dunes called **sand waves** (Allen 1982; Cas and Wright 1987). The geological term **facies** refers to an "environment or conditions of deposition." The depositional "environment" of the waveform facies is one of high energy particle blasting. Farther from the vent, Crowe and Fisher also mapped lower energy "massive" and "planar" surge facies (Figure 7.43).

Like normal wind-blown dunes, a typical pyroclastic sand wave has a gradual **stoss side** facing the direction from which the wind, or current traveled, and steeper **lee side** facing the direction the current moved. The sharp, ridge-like dune **crests** are oriented at right angles to current direction. Thin beds and laminations, representing pulsations in deposition form parallel to the lee slope and readily help define each sand wave in cross-sectional exposures (Figure 7.44). Contrasting particle sizes highlight layer

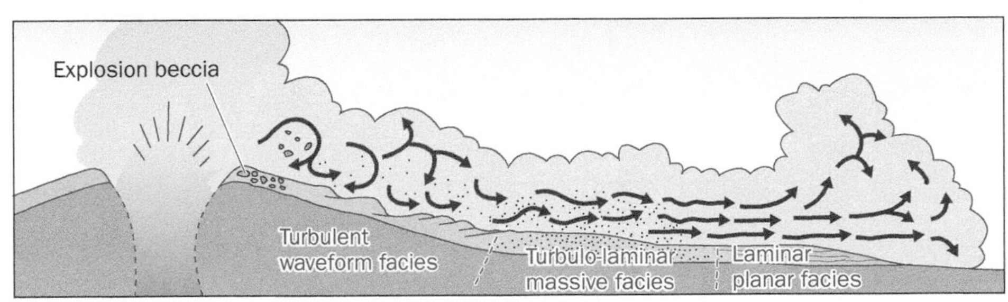

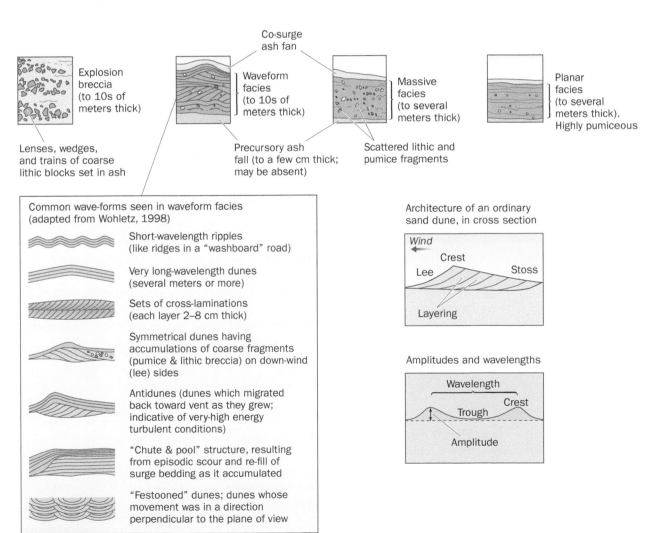

FIGURE 7.43 Dunes and the development of surge facies in a blast surge deposit.

FIGURE 7.44 Pyroclastic surge bedding in a late prehistoric PDC deposit in the Pollena Trocchia area, western flank of Vesuvius volcano, Italy [109]. Note the cross-bedding and erosional truncating of earlier beds, which testifies to the violence of surge emplacement processes. Source: Photo by Richard Hazlett.

boundaries, which are commonly truncated by **erosional surfaces** defining the stoss side of the sand wave. The erosion results from the plucking or rolling of fragmental debris, often set into motion by impacting fine particles swept along in the surge cloud. The material eroded from one side of the growing dune quickly re-accumulates in the calmer pocket of air on the other, lee side, adding a fresh layer and causing the crest of the wave to advance in the down-current direction. **Unconformities**, which are simply buried erosional surfaces, may be found truncating layering in many parts of surge deposits. Scouring and deposition frequently shifts from place to place as dunes try to adjust to changing conditions. Abrasive, particle-laden winds can erode earlier-formed beds and pre-existing land surfaces to depths of as much as several meters, only to refill erosional depressions with fresh ejecta a moment later, especially vigorous on steep slopes where speeds and turbulence may be greater.

To describe the shape of sand waves or dunes, we customarily refer to two measurements; **amplitude**, or wave height taken from the base of a dune to its crest, and **wavelength**, as measured from one wave trough to the next. Amplitudes and wavelengths range widely within surge deposits. They vary not only in response to changing conditions during surge deposition, but in response to the surfaces on which they form. For instance, at Ubehebe Craters, it can clearly be seen that the dune wavelength/wave height ratio in any given area relates directly to the angle of the depositional slope; the ratio increases on steeper slopes (Crowe and Fisher 1973).

More about PDC Sorting

In general, **most** pyroclastic surge and flow deposits can be distinguished from each other by their sorting and particle size ranges. Murai (1961), Fisher (1964), Passega (1964), Sheridan (1971), Walker (1971), and Buller and McManus (1973) were the first researchers to attempt to characterize the sorting of pyroclastic deposits in terms of origin. They concluded that: *pyroclastic falls are better sorted than pyroclastic surges, which in turn are better sorted than pyroclastic flows.* Lirer and Vinci (1991) examined 414 different pyroclastic deposits and concluded that although there is some overlap in sorting, effective discrimination on the basis of sorting alone can be done based on mean phi values. For example, deposits with -3φ sizes are most likely to be fall-related; 1φ to be pyroclastic flow-related, and 7φ pyroclastic surge-related (Figure 7.45).

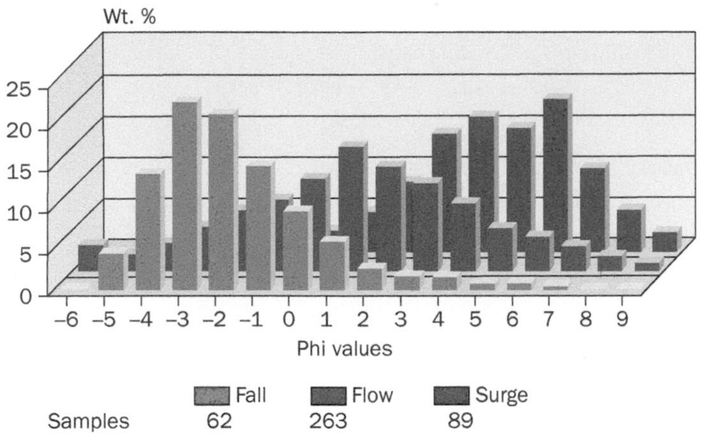

FIGURE 7.45 Histogram presented by Lirer and Vinci (1991) showing mean grain size phi -values for pyroclastic falls, flows, and surges, from a population of 414 eruption deposits. Coarser grain sizes to left. Source: Lirer and Vinci (1991). Reproduced with permission of John Wiley & Sons.

TABLE 7.3 Characteristics of principal pyroclastic deposits.

Pyroclastic deposit type	Origin	Deposition mechanism	Deposit characteristics
Fall ejecta	Gravity settling from overhead ash clouds	Accumulation by gravity settling through air or water, potentially affected by prevailing winds or water currents.	Typically, poorly consolidated layers of uniform thickness that drape pre-existing topography. Size sorting is moderate to good.
Pyroclastic surge	Ash column collapse, phreatic or phreatomagmatic eruption	Deposition from low concentration, relatively low-temperature mixtures of ash, broken rock fragments and gases. Surges are characterized by turbulent flow.	Usually poorly sorted and poorly consolidated. Dune and internal erosion structures are typical and result in irregular thicknesses over short distances. Commonly contain ballistically emplaced blocks – bomb sags are common in proximal areas. Accretionary lapilli common.
Pyroclastic flow	Boiling of magma chambers, ash column, or dome collapse. Most commonly associated with caldera-forming eruptions	Deposition from high-concentration, high-temperature clouds of mixed ash, gas, and juvenile magma fragments. Characterized by laminar flow.	Poorly stratified, with thicker deposits filling valleys, and thinner or absent deposits on highlands. Commonly shows poor grading, with denser fragments near bases. Evidence of high-temperature origin include welding, deformation of glassy pumice fragments, pink-oxidized tops, and gas pipes.
Directed blast	Violent explosions of suddenly depressurized magma chambers, typically triggered by massive landslides or sector-collapse	Extremely complex, and may involve deposition by pyroclastic surges, pyroclastic flows, and airfall. May be preceded by initial erosion by high-velocity hot gases activity.	Recognized by extreme violence of deposition mechanisms. Massive accumulations of uncharred logs may attest to devastated forest in young deposits. Deposits widespread and may accumulate in separate basins. Incorporation of eroded material from underlying substrates common in basal deposits.

Summary

Pyroclastic deposits and their differing origins may be grouped into five principal classes: (1) ballistic ejecta; (2) fall ejecta; (3) pyroclastic surges; (4) pyroclastic flows; and (5) directed blast deposits (Table 7.3). Recognition of these deposits is critical to understanding the volcanic history of many areas dominated by gray volcanoes.

Questions for Thought, Study, and Discussion

1 Describe the criteria we use to classify volcanic ejecta. Why are these criteria practical?
2 What are the main types of ballistic ejecta?
3 What criteria are used to distinguish the various kinds of volcanic ash?

4 How can we distinguish in ancient deposits the layers left by "wet" versus "dry" ash falls?
5 How might tephrachronology be useful in determining the age of latest faulting in a geological cross-section exposed by erosion or road cut?
6 What distinguishes Vulcanian from Plinian eruption deposits? Describe the characteristic features of each.
7 How do volcanic plumes differ dynamically from eruption columns?
8 How do the distal deposits formed by Plinian eruptions differ fundamentally from those in proximal locations to a vent?
9 Why does the full sequence of deposits produced in a typical Plinian eruption not always appear in even well-exposed cross-sectional outcrops?
10 What are directed blast eruptions, and why are their deposits so hard to discern in the geologic record?
11 In what fundamental ways do pyroclastic surge deposits differ from pyroclastic flow deposits?
12 What are "cooling units," and what knowledge do we gain by distinguishing them?
13 How do the dynamics and deposits of hydrovolcanic eruptions differ from those of Strombolian eruptions?

FURTHER READING

Branney, M.J. and Kokelaar, B.P. (2002). *Pyroclastic Density Currents and the Sedimentation of Ignimbrites.* London: Geological Society of London.

Fisher, R.V. and Schmincke, H.U. (1984). *Pyroclastic Rocks.* New York: Springer-Verlag.

Heiken, G. and Wohletz, K. (1992). *Volcanic Ash.* Oakland: University of California Press.

Kokelaar, B.P. (1983). The mechanism of Surtseyan volcanism. *Journal of the Geological Society of London* 140: 939–944.

Scarth, A. (2002) *La Catastrophe – The Eruption of Mount Pelée, the Worst Volcanic Disaster of the 20th Century.* New York: Oxford University Press, 246 pp.

CHAPTER 8

A Closer Look at Large-Scale Explosive Eruptions

Source: Credit J. Lockwood.

Volcanoes: Global Perspectives, Second Edition. John P. Lockwood, Richard W. Hazlett, and Servando De La Cruz-Reyna.

Companion website: www.wiley.com/go/lockwood/volcanoes2

The stupendous and terrific character of these catastrophes, the rarity of their display, and the dreadful extent of injury often resulting from them to the lives and property of the inhabitants of the surrounding country, make them the subject of general remark during and long after the period of their development.

(George Poulett Scrope 1862)

Given the extreme eruptive power of great explosive eruptions, their pertinence to global climate and atmospheric chemistry, and their implications for volcanic hazards analyses, we consider it important to explore the dynamics of these exceptional volcanic phenomena and other large explosive eruption types further in this chapter. We begin with a discussion of how to measure the sizes of these eruptions by studying their deposits, then describe Plinian eruption and pyroclastic density current (PDC) dynamics. We will also here describe the special class of explosive eruptions widely termed "Super-eruptions" by scientists and the media. These extremely violent eruptions, with VEI greater than 8 (Chapter 5), have never occurred within human-recorded history, and none are likely to occur within the lifetimes of those who will read this book, as they are exceedingly rare events in Earth's history. They involve the same physical processes described here for more typical Plinian eruptions, and we will refer to them again in Chapter 10 in our discussion of giant calderas, as these are the "super-volcanoes" that produce "super-eruptions."

Plinian Eruptions

Measuring the Sizes of Past Plinian Eruptions

Recall from Chapter 5 the important terms **magnitude** (the total mass erupted throughout the duration of the activity) and **peak intensity** (the maximum rate of magma discharge, measured in kg/s). Carey and Sigurdsson (1987) attempted to constrain the factors responsible for the peak intensity and magnitude of Plinian eruptions. They investigated 45 mostly prehistoric Plinian eruption deposits and discovered that their bulk masses range from 2.0×10^{11} to 6.8×10^{14} kg – three orders of magnitude difference. Estimated eruption intensity values also vary over three orders of magnitude, from 1.6×10^{6} to 1.1×10^{9} kg/s. They concluded that a positive correlation exists between explosive eruption intensities and their magnitudes. These values in turn relate directly to the volumes of the magma bodies being tapped.

Eruption magnitude determination is a relatively straightforward mapping exercise. The area and range of thickness of a deposit can be ascertained for younger eruptions by constructing an isopach map of fall deposits (Chapter 5; Sottili et al. 2004). Area (m^2) and thickness (m) are simply multiplied by the mean density of the fragments composing the deposit (kg/m^3) to arrive at the total mass (kg). There are, of course, enormous challenges in practice, and volume calculations can be quite challenging to determine (Pyle 1989). Many deposits are highly irregular in distribution. Erosion removes lots of material soon after eruptions, especially thin distal fall materials. Furthermore, the underlying bedrock topography, necessary for precisely determining thicknesses, may not be known. But under most circumstances for Plinian eruptions less than a few tens of thousands of years old, reasonably accurate estimates of mass can be made.

The estimation of eruption intensities involves an entirely different field approach, however, called **isopleth mapping** (Urbanski et al. 2003). Isopleths show contours of the *maximum diameters* of particular kinds of ejecta, primarily lithic fragments that do not break into smaller pieces upon impact, which pumices have the habit of doing. Sieving is required to sort out the largest fragments, which can be done in conjunction with determining overall particle size distributions. The resulting fragment-size dispersal information can then be used in models developed by Sparks (1986) and Carey and Sparks (1986) to identify eruption columns of particular intensity.

Recall from Chapter 7 that Plinian eruptions are distinguished by development of towering columns of heated gases, ingested air, and tephra – ash and pumice in enormous quantities, that rise 15 kilometers or more above a vent. Rarely, as at Sakurajima Volcano, Japan [142] in 1914, two or more eruptive columns from the same volcano or volcanic area may be active simultaneously (Todde et al. 2017).

The fundamental difference between an explosion cloud that just "fountains" and one that develops into a "column" is that the eruption column gains mechanical energy as it rises, fundamentally by ingesting air not far above its base, which grows hot and convects buoyantly upward, like heated air escaping from a furnace chimney. The explosion cloud must be sufficiently hot, dilute (relative to fountaining clouds), and not too turbulent – thereby enabling surrounding air to be sucked in effectively. Once a steady vertical flow of hot air, gases, and volcanic debris gets going, it can ascend as much as several tens of kilometers – an order of magnitude greater than fountaining sustains (Chapter 7).

The Rise of Plinian Columns

A Plinian column can appear immensely complex as it grows above an erupting volcano (Figures 8.1 and 8.2; Kaminski and Jaupart 2001). Bulbous "cauliflower" projections churn rapidly upwards, and shoot laterally out of the column. Many are initially black with ash as they emerge from cloud interiors – because they are above 100°C in temperature, and visible steam cannot form – until they cool somewhat and turn to white or gray. The column may be ringed by bizarre skirt clouds that form and disappear quickly. Electrical charge imbalances within and between a column and the surrounding ground create near-constant lightning – which can form a near-continuous din. The roaring of the eruption and constant cracking of electricity can make conversations difficult; and, if close enough, the "whoompwhoomp" sounds of falling blocks ejected from the base of a column can be heard.

When Sparks et al. (1973) first proposed their model for the formation of Plinian eruption deposits, much less was known about how Plinian columns are generated and dissipated than we know now. Subsequent well-observed Plinian and sub-Plinian eruptions (e.g. Mount St Helens [31], 1980, El Chichón [57], 1982, Mount Pinatubo [135], 1991, Mount Spurr [24], 1992, Soufrière Hills [84], 1996) have taught us much, but volcanologists have also created their own model eruptive columns and plumes, both numerically with computers and experimentally in tanks of water (which are reasonably safe research environments). Modelers have greatly benefited from the concepts of **fluid dynamicists** – the people who study how gases and liquids flow. Not only Plinian columns but the weaker and somewhat more variable Vulcanian clouds have been intensively studied (see, e.g. Clarke et al. 2002).

Plinian and sub-Plinian eruption columns consist of three vertically stacked regions (Figure 8.3), each dynamically distinctive, with variable ascent speeds and density relative to the adjoining atmosphere. The **gas thrust region**, immediately above the vent, is denser than air but does not collapse despite lifting a great load of pyroclastic material, owing to the powerful inertial discharge of its mass (Figure 8.4). Entrainment of air enables the gas thrust region to undergo transformation at higher elevation into the **convective region,** which remains lighter-than-air up to the altitude of neutral buoyancy (H_B).

At that level, atmospheric density and density within the rising column match, owing to lowered temperature, ingestion of ever larger volumes of air, and dispersion of ejecta as the column expands in response to reduced high-altitude air pressure. The column still retains momentum, however, and will continue to rise a bit further to its maximum height (H_T), all the while spreading to form the **umbrella region** (Figure 8.5). As a representation of this momentum, Sparks (1986) observed that the ratio $(H_T - H_B)/H_T$ for most Plinian eruption columns lies in the range 0.25–0.3.

The transition zone from a gas thrust into a convective column is a critical one, acting as a filter for fragments that fall out of the eruption column while the gas thrust loses momentum, but before it starts accelerating again convectively with the ingestion of a critical amount of rapidly heated air (Suzuki and Koyaguchi 2012). Woods (1988) and Wilson (1993) refer to the gas thrust lifting

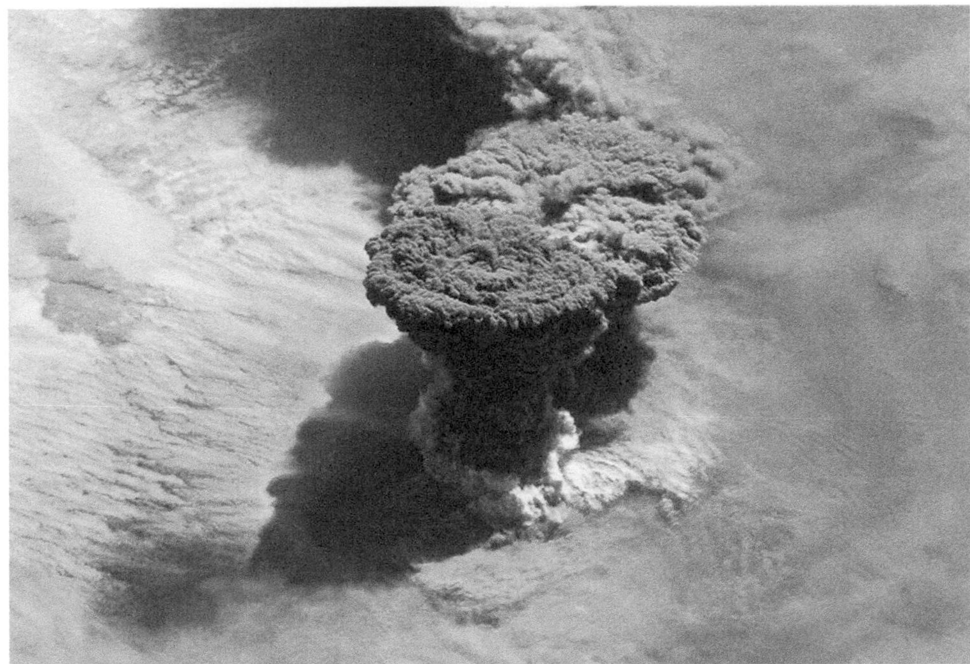

FIGURE 8.1 Eruption of Raikoke volcano [161], central Kuriles as photographed with a telephoto lens from the International Space Station on 22 June, 2019. A sub-Plinian eruption column has risen about 13 km high before spreading laterally at the level of atmospheric neutral buoyancy. Sulfur-dioxide-rich ash clouds from this eruption spread harmlessly eastward across the northern Pacific although numerous aircraft flight paths were diverted. Source: NASA image iss059e119254.

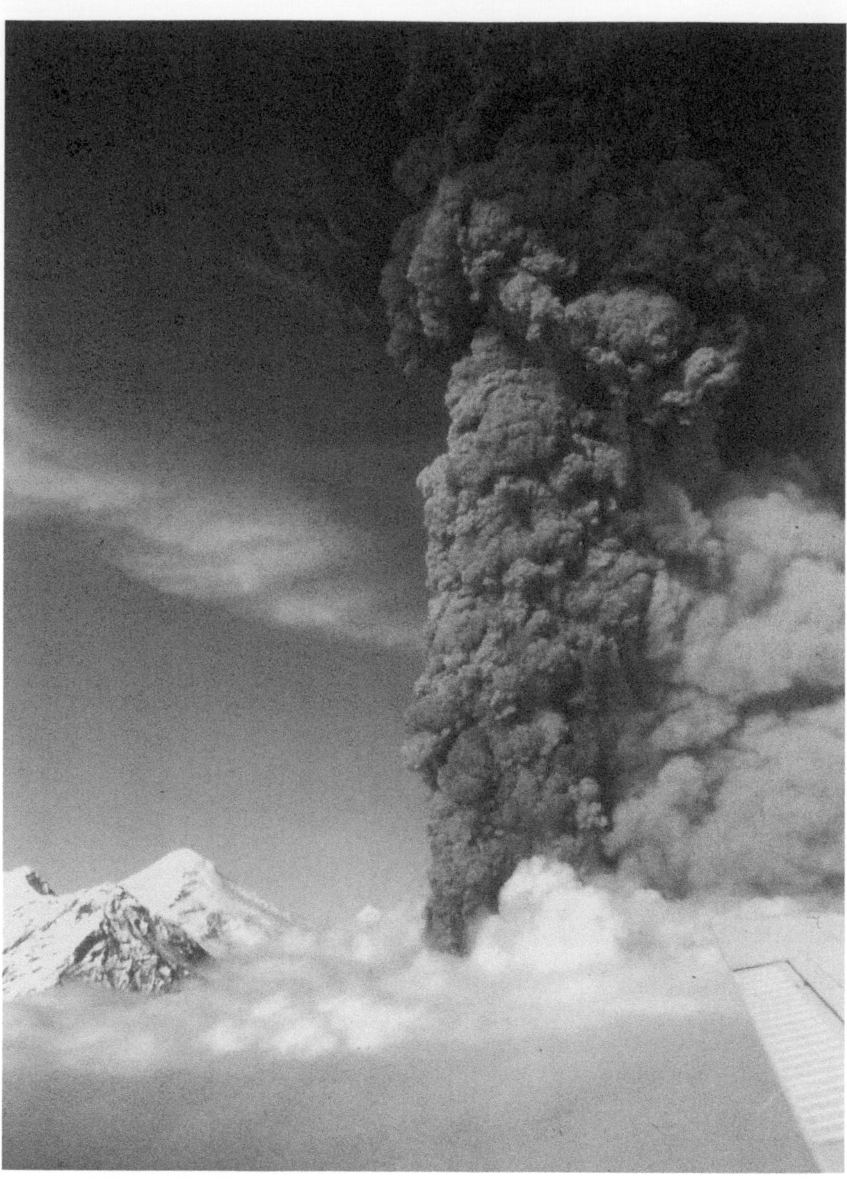

FIGURE 8.2 Sub-Plinian eruption column and overlying umbrella cloud from Crater Peak vent, Mt. Spurr volcano, Alaska, 18 August, 1992. Source: USGS photo by Robert McGimsey.

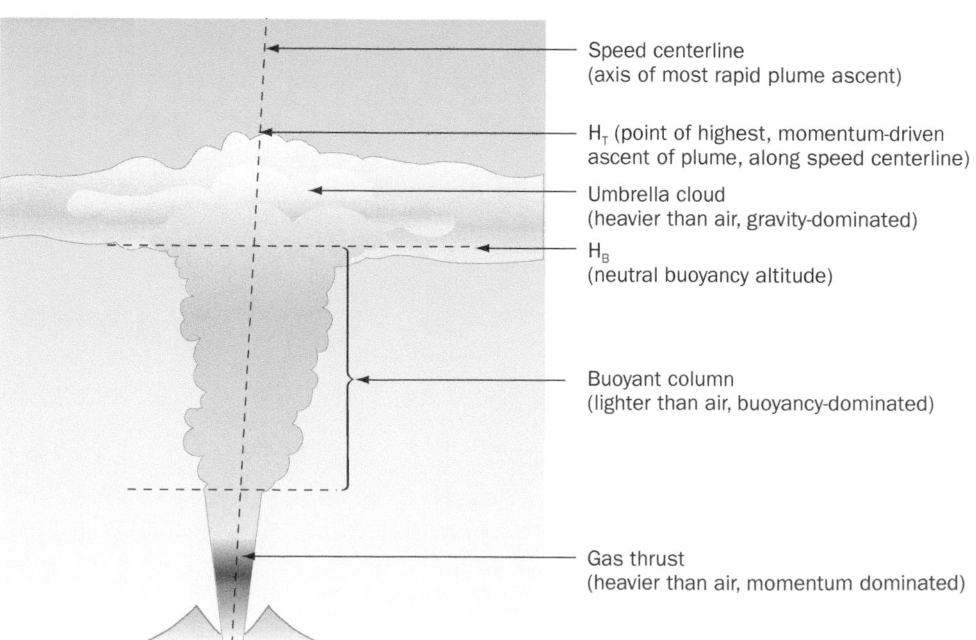

FIGURE 8.3 Plinian column structure.

FIGURE 8.4 Plinian eruption column rising from the decapitated summit of Mount St Helens volcano [31] on 18 May, 1980, about three hours after the onset of Plinian activity – view to east. The convecting column has risen at this time to about 20 km above the volcano. "Gas thrust" upward motion dominates the right side of the column, but minor downward movement of billowing ash is occurring on the left side; a partial column collapse. Source: USGS photo by Austin Post. Reproduced with permission of Elsevier.

of ejecta as **simple buoyancy**, and higher, air-related convective lifting as **superbuoyancy**. The transition between these states of eruptive buoyancy is related directly to two aspects of an eruption; the speed at which material leaves the vent (m/s), and the mass discharge of material from the vent (kg/s). For example, eruption columns that erupt rapidly but contain low concentrations of particles fail to become superbuoyant, because they cannot incorporate and heat air from the surrounding atmosphere very effectively. Likewise, if an eruption column leaves the vent slowly and contains a high concentration of particles, it is bound to collapse before ascending very far. The precise definition of a **(pyroclastic) fountain** is an eruption cloud that fails to achieve superbuoyancy or one that collapses from an earlier superbuoyant state (e.g. Dellino et al. 2014). The best combination of circumstances for maximum column buoyancy is a moderate rate of ascent, with a moderate level of mass discharge (certainly less than $10 \cdot 10^{9.5}$ kg/s; Figure 8.6). The water vapor content of columns also plays a role in their ascent, with wetter clouds able to capture air slightly more effectively than dry ones mixed solely with magmatic gases. Even vesiculation within a conduit may prove

FIGURE 8.5 Top of sub-Plinian eruption column from Galunggung volcano, Indonesia [129] – 16 August, 1982. The rising cloud has reached a point of neutral buoyancy at about 12 km altitude, and is beginning to spread out to form an umbrella cap. Source: USGS photo by J.P. Lockwood.

critical: If developing pumice fragments trap excessive amounts of exsolving magmatic gas because of their unusually high impermeable vesicle content, then a column will lack the total gas concentration in its initial thrust to do anything more than fountain (e.g. Michaud-Dubuy et al. 2018).

Temperature differences, density of erupted material, and most of all, rate and intensity of magma discharge are key determinants for the heights of eruption columns and the levels at which they flatten out into umbrella clouds (Sparks 1986; Ogden et al. 2008). This is illustrated in Figure 8.7, which also equates volume to mass of magma being discharged. Note the slight inflection points in the sets of lines drawn for both tropical (20°S–20°N) and temperate latitude cases. These inflections are caused because the atmosphere does not change gradationally all the way out to the fringe of Space, but is itself stratified into layers having slightly different physical and chemical properties. The atmosphere is thicker, less dense, and more moist at tropical latitudes than in temperate, making it slightly easier for columns to rise there.

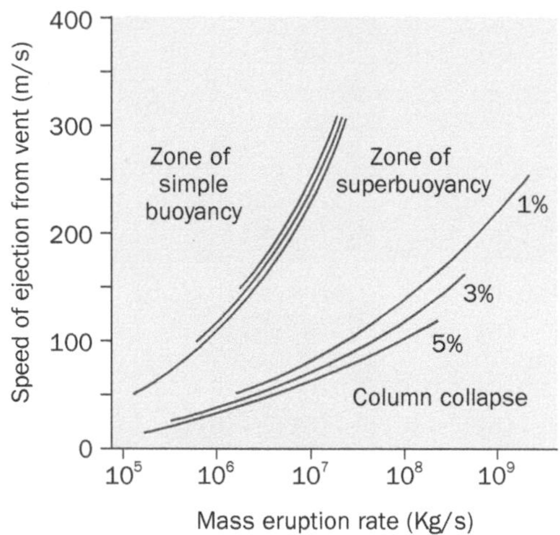

FIGURE 8.6 Relationship of eruption discharge, speed of discharge, and water vapor content of eruption column (in 1, 3, and 5 weight percents) to buoyancy properties. Convective columns will become established only in the zone of superbuoyancy. Source: Bursik and Woods (1991).

Of further interest is how *fast* convective columns ascend. Ascent rate depends not only on a combination of buoyancy factors, but also on initial momentum of the gas thrust region. The written description of Pliny the Younger of the opening phase of the 79 CE Vesuvius [109] eruption (p. 193) suggests that the kilometers-high column appearing above the Bay of Naples developed in just a few minutes. He was not exaggerating. Eyewitness accounts of the initial Mount St Helens [31] 1980 Plinian column by airline pilots indicate that the cloud was 18 km high in about 5 min (Rosenbaum and Wait 1981). Modeling by Sparks (1986), reinforces the point, made earlier, that entrainment of air, buoyancy, magmatic heat, and pressure–temperature changes in the atmosphere all play essential roles in determining the ascent speeds of columns. The results of their theoretical studies can be diagrammed to show how ascent speed changes in a column as it rises. The nest of curving lines shown in Figure 8.8 each relates to a different numerically predicted column. Column (a), for example, erupts from the vent at an ascent rate of 300 m/s; a very powerful eruption even by Plinian standards. By the time it rises to 5 km, the top of the column has slowed to about 140 m/s, and at 10 km, near the tropopause at high latitudes, it has slowed to about 100 m/s. Plots of two observed eruption column ascents, one for tropical Soufrière volcano [84] in the Caribbean (Sparks and Wilson 1982), and the other for temperate latitude Mount St Helens (Sparks 1986) are also shown. The matches with turbulence modeling are fair, though note the acceleration of the Mount St Helens column above 10 km does not match theoretical models well – perhaps a combined influence of strong upper level winds and unaccounted-for stratospheric properties (Zimbelman and Gregg 2000; Ishimine 2006).

Most particles that exceed 20 mm in diameter, excepting highly porous pumices, drop out of the sides of columns as they convect upward (Sparks et al. 1992). For the most powerful ultra-Plinian columns, as much as 60 percent of the bulk weight of

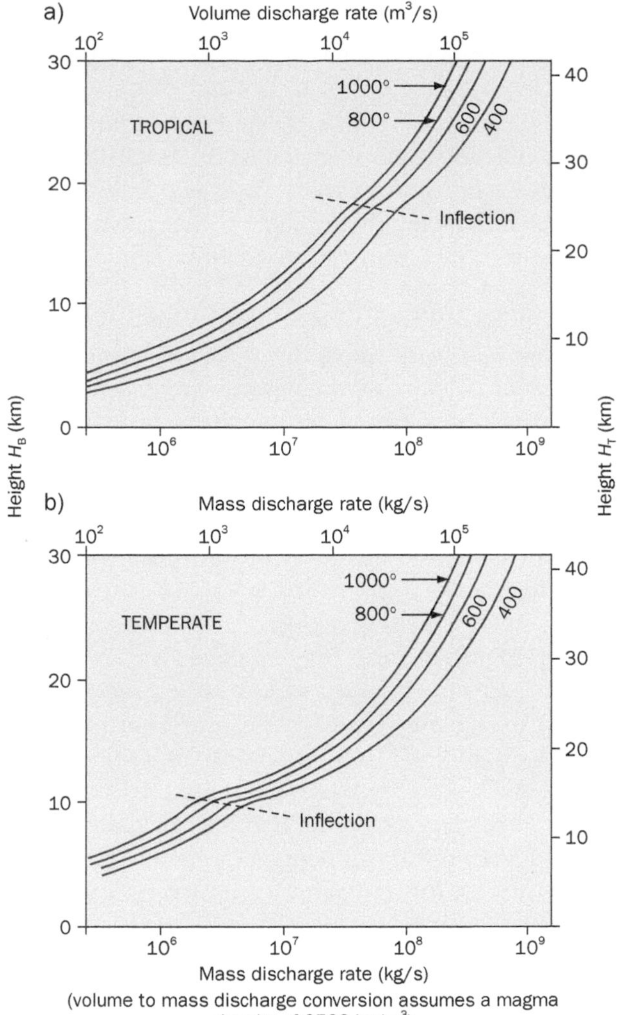

FIGURE 8.7 Changes in neutral buoyancy levels (HB) with changing volumetric or mass discharge rates, for Plinian columns of varying source magma temperatures. Note the differences for eruptions under tropical (a) and temperate (b) conditions. Source: Sparks (1986). Reproduced with permission of Springer Nature.

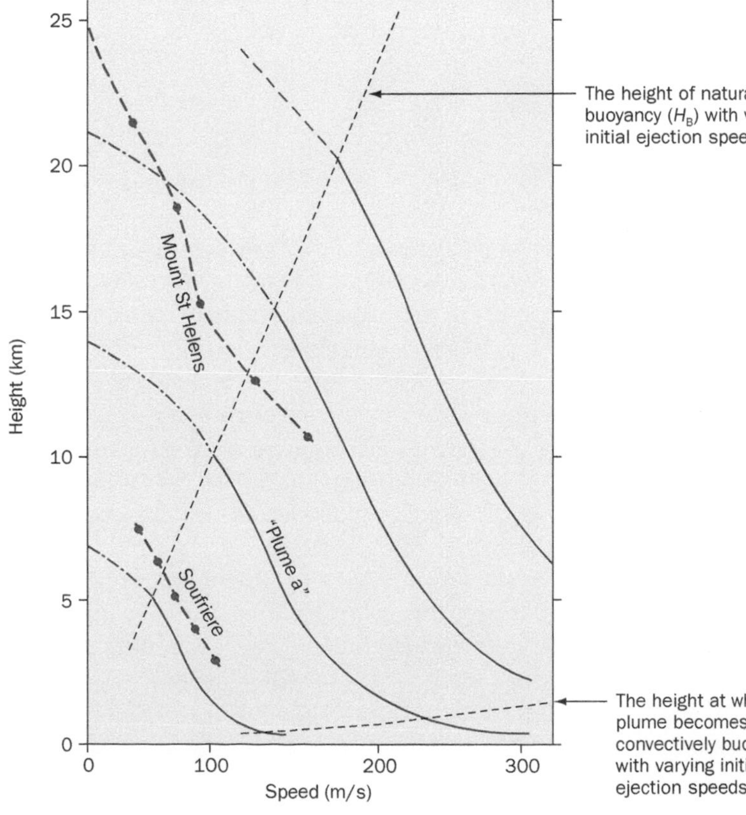

FIGURE 8.8 Theoretical variations of eruption column speed with height in the column, as measured at the centerline axis for columns of differing initial ejection speed (given at height = 0). Data for two historical eruptions are also plotted. Source: Soufrière, from Sparks and Wilson (1982), and Mount St Helens from Sparks et al. (1986).

solids can be very fine ash (less than 63 μm) capable of spreading hundreds of kilometers from the column axis in an expanding umbrella cloud (Walker 1981).

Observations and laboratory experiments indicate that the radial speed of umbrella expansions decrease as a function of radii, so that as speed declines, progressively smaller particles fall out to contribute to the growing tephra dispersal field below. In other words, the dispersal of fall ejecta can be regarded as a "map" of a slowing eruption umbrella. Carey and Sparks (1986) calculated estimates of original column heights for prehistoric eruptions. The highest eruption column is that estimated for the 186 CE Taupo [180] phreato-Plinian eruption in New Zealand – 51 km (Walker 1980), which is described further in the following text.

Wind, of course, will modify the simple radial distribution of tephra deposits according to grain size around a vent. If a strong atmospheric wind is blowing, the upwind spread of the umbrella cloud will cease where the spreading speed of the umbrella matches that of the opposing wind. This position is called the stagnation point. At a right angle to the wind direction, the umbrella will spread as though moving through still air, while downwind the cloud will, of course, propagate faster and farther (Wilson 1993). The distance that the center of the eruption column is displaced by wind from its position directly above the vent at H_B is termed the column's **axial displacement**, and this is directly related to the stagnation point, depending upon the altitude and the wind speed (Figure 8.9). This means that airfall deposits can serve as paleo-wind direction indicators. Carey and Sparks (1986) used the degree of ellipticity of wind-distorted airfall isopachs to estimate the high-altitude windspeeds occurring during a number of well-studied explosive eruptions. In some cases, this can indicate the yearly season in which eruptions have occurred.

In their 1989 study, Carey and Sigurdsson (1989) found that most of the 45 Plinian deposits they cataloged were preceded by pumice falls, but only when this early pumice exited the vent at calculated intensities in excess of 2.0×10^8 kg/s were major PDCs and caldera collapse also likely to occur. Bursik and Woods (1996) have more recently concluded that the most extensive pyroclastic flow activity develops during eruptions with intensities greater than 10^9 kg/s, with column activity persisting for approximately 20 min to over 30 h. PDCs are certainly possible at lower-intensity levels, but these are likely to be small, with attendant caldera formation improbable. Carey and Sigurdsson (1989) have also compared the calculated intensities to the magnitudes in their data set, coming up with minimal eruption durations from 1.1 hours for a simple (observed) Plinian pumice fall, to 208 hours for the gigantic caldera-forming Los Chocoyos [60] ash and pumice eruption in Guatemala (Rose et al. 1987).

Intensities and magnitudes, it turns out, correlate closely enough to be significant (Figure 8.10). Larger magma chambers develop higher overpressures, and so produce higher-intensity eruptions (Valentine et al. 1992). Ogden et al. (2008) further suggest that steady levels of high overpressure will stimulate oscillatory column rise and collapse, accounting for the periodic generation of PDCs recorded in many Plinian eruption deposits. Another important variable influencing eruption rate, quite independent of magma overpressure, is **vent diameter**. Wilson et al. (1980) modeled the relationship between vent radius (r) and mass eruption rate (M) as follows:

$$M = \left[\pi r^4 g \rho_m \left(\rho_{cr} - \rho_m\right)\right] / 8\mu \quad (8.1)$$

where ρ represents densities for both magma (m) and crust (cr), and μ is magma viscosity. Mass eruption rate is thus extremely sensitive to vent diameter, varying as the fourth power of the vent radius.

Field observations of graded air fall deposits suggest that many Plinian eruptions increase significantly in intensity shortly after initial venting, as indicated by reverse grading in pumice fall beds. Carey and Sigurdsson (1987), for example, inferred a greater than one-order of magnitude increase in the intensity of the early to middle stages of the 79 CE Vesuvius eruption, based upon differences in pumice sizes in the reversely graded Pompeii pumice fall. Perhaps the most reasonable explanation for this commonly inferred increase is that vents widen during eruptions. Erosive widening of a vent by particle abrasion and hydraulic weakening from fluctuating gas pressures certainly occurs in all explosive volcanic outbursts, as shown by the abundance of accessory and accidental fragments in their deposits. In some instances, a substantial increase in accidental ejecta corresponding to the sudden widening of a vent can even induce at least short-term column collapse (e.g. Shea and Houghton 2012). This is especially true for long-sustained, high-intensity eruptions. Higher-intensity eruptions also have greater magnitudes, which must be supplied by larger volume magma chambers.

Not all Plinian fall deposits show reverse grading – many are ungraded or show normal grading. Systematic enlarging of vents, plainly, is not the only factor involved in affecting deposit characteristics. A further complication is the recirculation of pyroclastic material. Ejecta falling within a few kilometers of the base of a large convective column may, depending upon its coarseness, largely be reincorporated by the column via powerful atmospheric updrafts acting at the column edge (Figure 8.11). There is also the complicated turbulence resulting from the fall back of pyroclasts into the top of the convective region from the umbrella cloud itself. Despite these caveats, the basic results of modeling provide a useful picture of the overall dynamics of Plinian eruption columns.

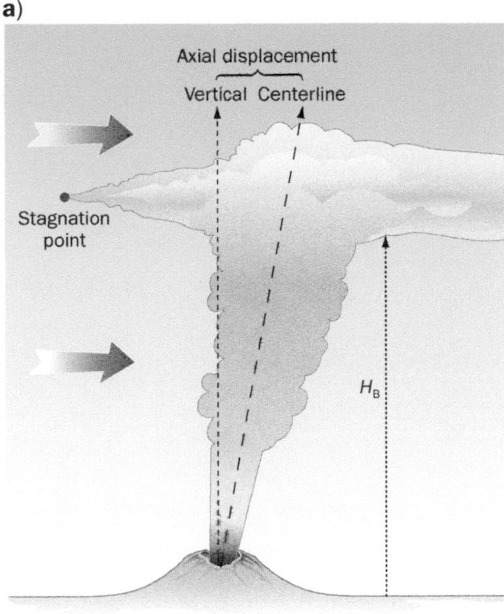

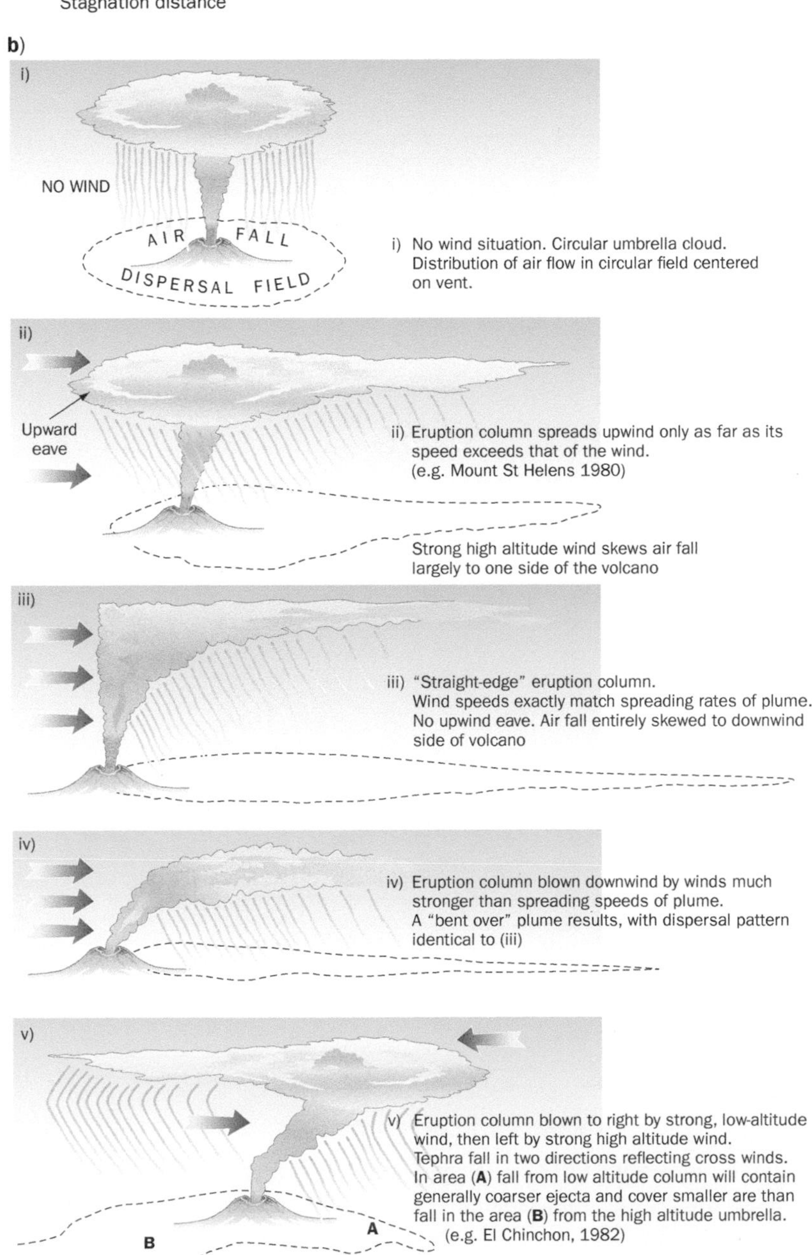

b)
i)
NO WIND
AIR FALL
DISPERSAL FIELD
i) No wind situation. Circular umbrella cloud. Distribution of air flow in circular field centered on vent.
ii)
Upward eave
ii) Eruption column spreads upwind only as far as its speed exceeds that of the wind. (e.g. Mount St Helens 1980)
Strong high altitude wind skews air fall largely to one side of the volcano
iii)
iii) "Straight-edge" eruption column. Wind speeds exactly match spreading rates of plume. No upwind eave. Air fall entirely skewed to downwind side of volcano
iv)
iv) Eruption column blown downwind by winds much stronger than spreading speeds of plume. A "bent over" plume results, with dispersal pattern identical to (iii)
v)
v) Eruption column blown to right by strong, low-altitude wind, then left by strong high altitude wind. Tephra fall in two directions reflecting cross winds. In area (A) fall from low altitude column will contain generally coarser ejecta and cover smaller are than fall in the area (B) from the high altitude umbrella. (e.g. El Chinchon, 1982)
A
B

FIGURE 8.9 Effects of wind on Plinian columns and air fall dispersals: (a) Geometric parameters. (b) Specific cases.

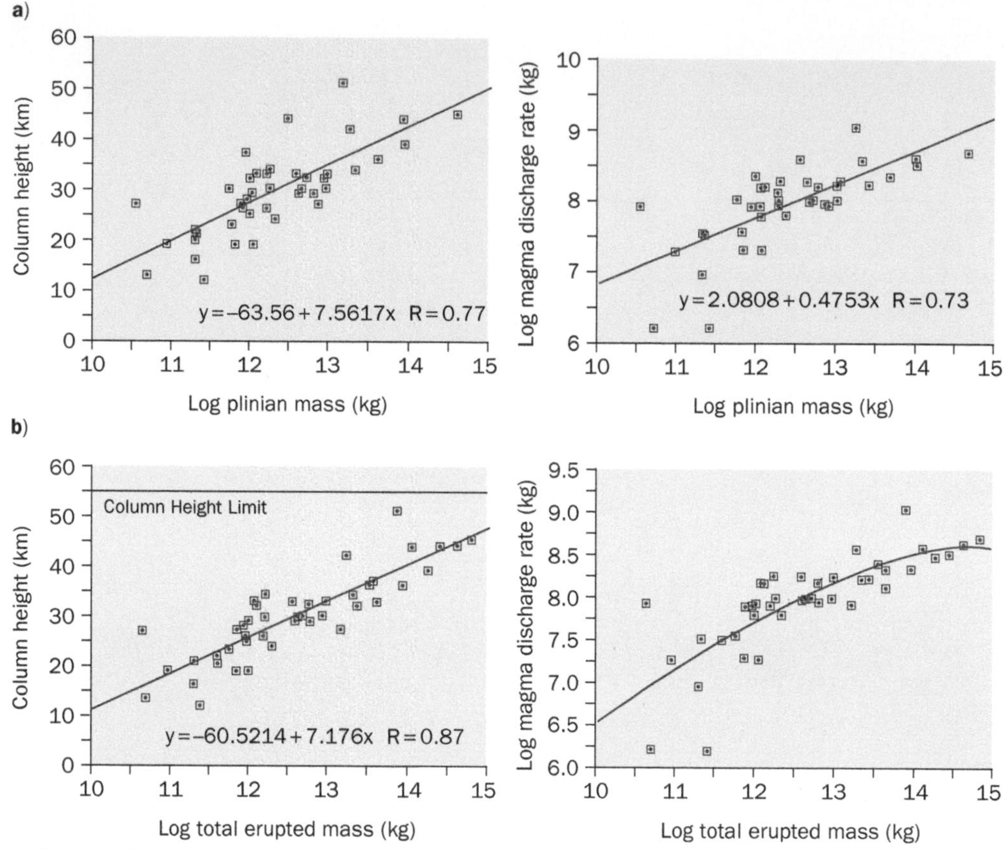

FIGURE 8.10 Correlations of various eruption parameters calculated for 45 Plinian deposits. Source: Carey and Sigurdsson (1989). Reproduced with permission of Springer Nature.

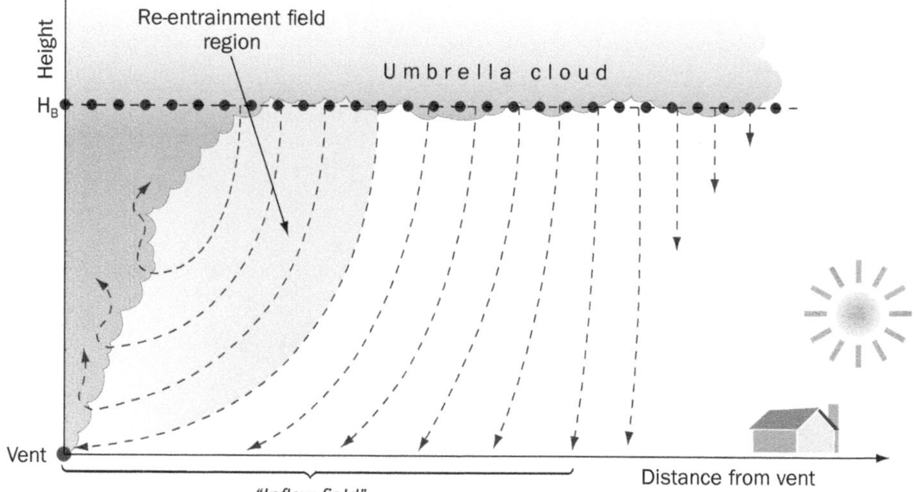

FIGURE 8.11 Re-entrainment of falling tephra into Plinian columns. Air rushing back toward the vent to replace that which is sucked up into the column deflects fall pyroclast trajectories (dashed lines) so that some particles may become re-entrained and recycled – repeatedly. Source: Modified from Bursik et al. (1992).

The Fall of Plinian Columns

As already mentioned, Plinian eruption column may partially collapse when the ratio of solid particles to gases and trapped air greatly increases density at any fixed-intensity level, potentially generating PDCs on volcano flanks. Historically recorded accounts of Plinian column collapse indicate that it may take just a few minutes for some columns to fall apart, although more typically, it can take several hours and involve numerous cycles of growth and collapse (Scott et al. 1996; Clarke et al. 2002). Some columns may collapse only on one side, or collapse in a sloppy, piecemeal fashion.

Wilson et al. (1980) calculated the physically important factors controlling column collapse (Figure 8.12). Note that the speed at which material exits the vent (which should not be confused with eruption intensity and mass discharge rate) is also critical (Neri and Dobran 1994). If discharge is supersonic – possible in some extreme Plinian eruptions – the standing shock waves generated can suppress ingestion of air, leading to column collapse above (Suzuki and Koyaguchi 2012). Water vapor content also plays a key role in governing column collapse, and consideration of this factor becomes especially important in comparing phreato-Plinian with drier Plinian eruptions (Wilson and Walker 1987). At a given level of intensity, steam-rich columns ingest far more air than dry ones do, and so might be expected to lack the critical particle–gas density balance needed to climb very high; they cool down too fast. Plumes whose conditions lie close to the dividing line between pure convection and collapse may experience short-lived pulses of collapse and recovery, as observed at Mount St Helens during its 1980 eruptions (Rowley et al. 1981). Sparks et al. (1978, 1997) concluded from density considerations that most column collapses begin at levels not exceeding about 10 km altitude, given a reasonable range of vent radii from 50 to 600 m.

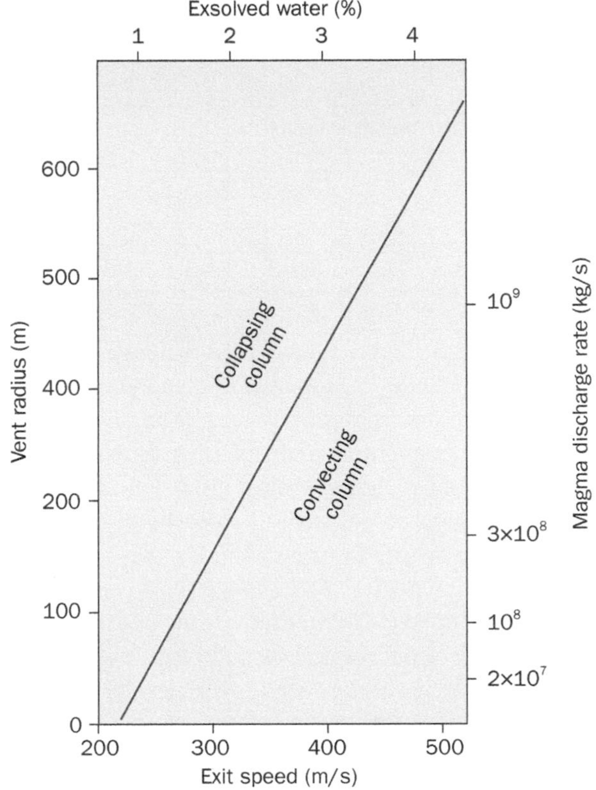

FIGURE 8.12 Factors governing the collapse of convective columns.

Carey et al. (1988) set up a series of experiments to investigate how eruption columns collapse. Their basic idea was simple: They filled a large tank with slightly salty water, acting as a proxy for Earth's atmosphere. Into this they injected lower-density fresh water through a narrow jet at the tank's bottom containing various concentrations of fine particles – analog Plinian columns, which they filmed. They varied the parameters of injection, including particle concentration and size and rate of injection to observe differences. Because the bulk (overall) density of a column of rising hot gases increases with particle concentration, they were able to simulate eruption columns ranging from highly buoyant (low particle concentration) to neutrally buoyant (high particle concentration). They compared their work with theoretical equations developed by fluid dynamicists (see, e.g. Turner 1986) to determine the extent to which their experiments reflected theory. Four basic modes of column collapse were observed, each dependent upon different levels of particle concentration and to a lesser extent particle size. The most dilute columns shed a continuous veil of enclosing fine particles after reaching their maximum heights, each creating a deposit of uniform width around the vent. With increasing particle concentration, the cascading veils became more prominent and small eddies of particle-rich column material began appearing along the column edges, slowly propagating downward and merging with the falling ejecta. Carey et al. (1988) noted eruptive analogs of their experimental observations from the April 1979 eruptions of Soufrière volcano [86] on the Caribbean island of St Vincent:

> *These [dilute] flows originated high in the eruption column about 10 min after the eruption began, when the column had reached 18 km in height . . . They descended as a curtain down the slopes of the volcano and out to sea. People who were overrun by the flow reported no sensations of high temperature, and there was no destruction associated with their passage. A thin layer of ash was the only trace of this type of activity. Such dilute flows provide a new mechanism for the deposition of fine ash in the proximal regions of explosive volcanoes.*

In their water tank experiments, when Carey et al. (1988) suddenly shut off eruption discharge through the nozzle at the bottom of their tank, the supported column of particle-rich fluid collapsed all at once, creating a capping, fine-grained fall bed. The heavy ash falls described by Pliny the Younger at the end of the 79 CE Vesuvius eruption doubtlessly accompanied analogous column collapse.

Pyroclastic Density Currents (PDCs)

As noted in previous chapters, PDCs are the deadly, ground-hugging torrents of hot ejecta mixed with gases that can pour down volcano flanks and potentially spread across surrounding landscapes well beyond. They can be produced by a range of eruption types (Table 8.1).

TABLE 8.1 Relationship of PDCs to styles of eruption.

Eruption type	Commonly observed pyroclastic currents
Hawaiian, Strombolian, Ultravulcanian	None. Ballistic ejection of fragments only
Vulcanian	Block-and-ash flows, base surges
Plinian	Ash flows, ground and ash-cloud surges
Hydrovolcanic	Base and blast surges

All PDCs are highly dilute, heterogeneous mixtures of gas and fragmental material. Most only flow responding to gravity as they sweep downslope across underlying topography, but some may be driven by momentum and explosive impulse, at least initially. Their high content of expanding gases makes all of them highly fluid ("autofluidized"; Chédville and Roche 2018). Complete gradations exist between PDC styles, and Branney and Kokelaar (2002) have shown that PDCs can change their characteristics during flow. As discussed in Chapter 7, there are two primary PDC types – the relatively higher particle concentration pyroclastic flows (typical of Plinian eruptions), and lower-density pyroclastic surges (common products of hydrovolcanic activity). What is "high" and what is "low" in terms of solids concentrations remains largely a relative question, however, since no one has been able to measure these values directly. Work with sediment cascades in shallow liquids and other modeling suggest that pyroclastic flows may contain "tens of percent solid particles" mixed with gases, while the percentage of suspended solid material must be as low as 1–3 percent for surges to form (Sparks and Wilson 1976; Bursik and Woods 1996). To initiate a PDC, most erupting gases must not separate from pyroclastic material, and conditions must be favorable for the incorporation of air. This requires rapid admixing of fragments with gases, which can take place either within vigorously vesiculating magma at the tops of conduits, or in and around the lower levels of a collapsing convection column. If fall ejecta around the column accumulate fast enough to thicken a deposit at a rate of at least 1–5 cm/s, then it may begin trapping air and magmatic gas in large volumes, resulting in pyroclastic currents (Figure 8.13). Too much water vapor in a Plinian or phreato-Plinian eruption column can hinder PDC development, however, owing to the buoyancy of steam. The critical level of "too much" steam in this regard depends in large part upon the speed with which material is ejected from a vent, but work by Koyaguchi and Woods (1996) suggests that in many cases less than about 15 weight percent water vapor is required for a column to remain fully buoyant.

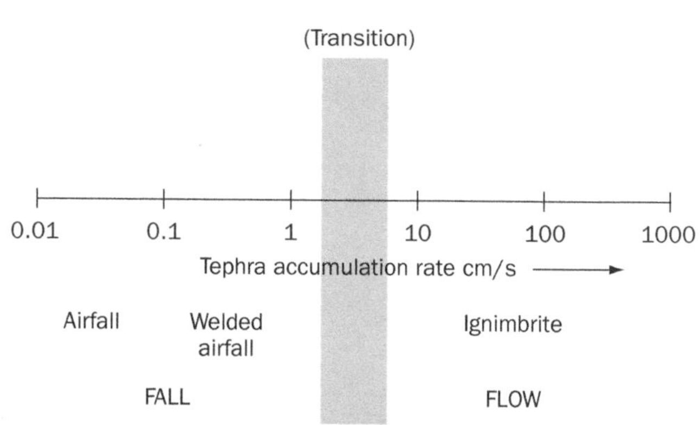

FIGURE 8.13 Theoretical relationship of ashfall (tephra) accumulation rates to generation of pyroclastic flows. Flows will form when ashfall, as around the margin of a collapsing column, is especially heavy. Source: After Wilson (1993).

The Origins of PDCs

Pyroclastic flows and related pyroclastic surges (see later in this chapter) can originate in several ways – all of which involve the rapid release of pressure on high-level magma bodies and subsequent volatile exsolution, the rapid admixture of expanding gases with fragmental rock material, the entrainment of surrounding air, and gravitational flowage downslope. Multiple types of volcanic activity can generate PDCs. The most frequent pyroclastic flows and surges are generated by the repetitive collapse of growing volcanic domes (Figures 8.14 and 8.15a). Large dome collapse events may depressurize underlying magma chambers and trigger explosive blasts with major PDC activity, as has been well demonstrated during the long-lived eruption of Soufrière Hills volcano, Montserrat [84] (Figure 8.16; Chapters 7 and 9). The sudden decompression of cryptodomes as shallow magma chambers caused by sector collapse (Figure 8.15b; Chapter 11) are uncommon, but lead to far more violent explosive activity. The most dangerous sources of high-volume PDCs (primarily pyroclastic flows) are Peléean eruptions or major Plinian column collapses (Figure 8.15c; Sparks and Wilson 1976) and/or vent "boilover" phenomena – pyroclastic fountaining without necessarily generating convective eruption columns (Figure 8.15d; p. 242). Smaller PDCs are also associated with initial phases of directed-blast phenomena (Figure 8.15e), and can also be generated by the secondary mobilization of primary PDCs or the collapse of thick silicic flow fronts (Figure 8.15f; Stoiber and Rose 1970).

Because pyroclastic flows are characterized by lamellar, non-turbulent flow at their bases, they generally do not substantially erode the underlying surface, in contrast to highly turbulent pyroclastic surges. Pyroclastic flows are beautiful to watch (from safe distances!), so long as they are not flowing toward populated areas. They are extremely "well-lubricated" by their high gas content,

(a)

(b)

FIGURE 8.14 Merapi volcano, Indonesia [131]. (a) Profile view from south, showing fuming dome at summit. (b) Close-up view of growing dome, October 1982. The dome was growing at a rate of almost 100,000 m^3/month at this time, and sending small pyroclastic flows and rockfall avalanches down the Gendol River several times per day. The magmatic SO_2 emissions from actively growing domes is usually intense, and the observers in this photo were gasping for better air! Source: USGS photos by J.P. Lockwood.

and move almost silently as they speed downslope. Small pyroclastic flows that I (JPL) have witnessed at close range at Galunggung [129], Java make almost no noise as they travel by – only faint whispering sounds – much like the rustling of leaves. Their beauty is tempered, however, by the knowledge that they are commonly extremely hot inside, and even momentary exposure to their hot gases will result in near-instant death. The people surrounded by dilute pyroclastic flow gases during the 1991 eruption of Unzen volcano [141] (including our friends Maurice and Katia Krafft and Harry Glicken) were burned beyond recognition with only a

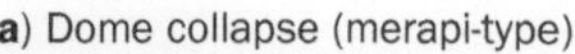
a) Dome collapse (merapi-type)

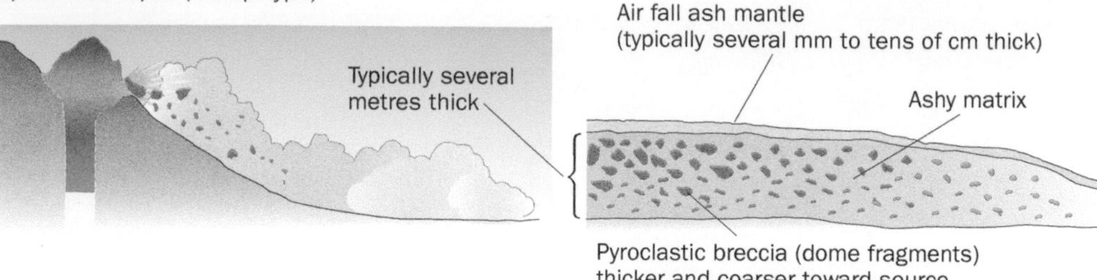
Typically several
metres thick
Air fall ash mantle
(typically several mm to tens of cm thick)
Ashy matrix
Pyroclastic breccia (dome fragments)
thicker and coarser toward source.
(Note: this sample shows crude reverse grading)

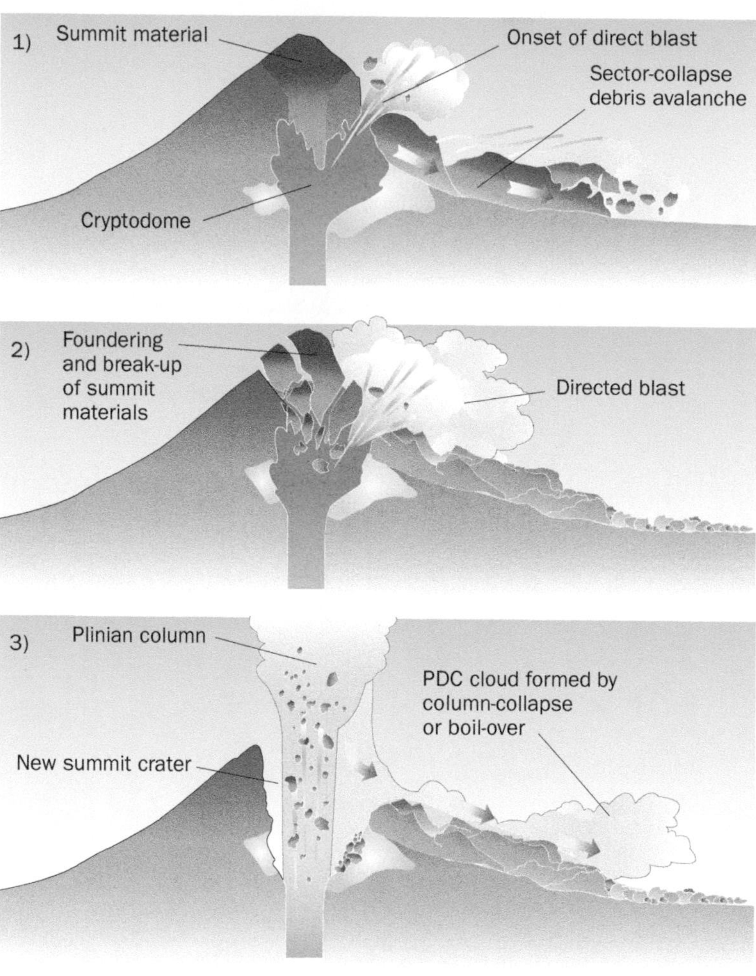
b) Sector-collapse, followed by magma decompression, directed-blasts and
Plinian activity, as at Mount St. Helens in 1980
1)
Summit material
Onset of direct blast
Sector-collapse
debris avalanche
Cryptodome
2)
Foundering
and break-up
of summit
materials
Directed blast
3)
Plinian column
PDC cloud formed by
column-collapse
or boil-over
New summit crater

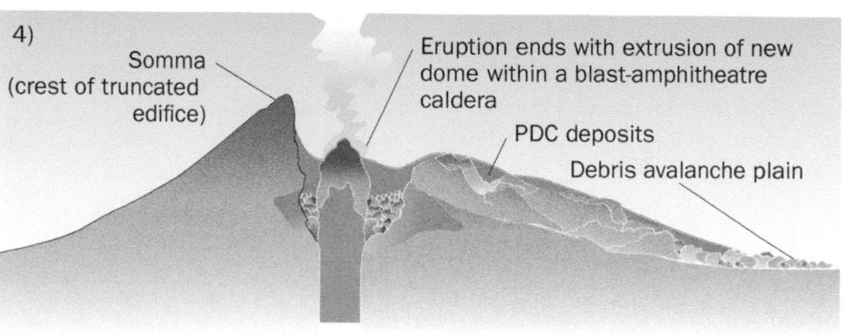
4)
Somma
(crest of truncated
edifice)
Eruption ends with extrusion of new
dome within a blast-amphitheatre
caldera
PDC deposits
Debris avalanche plain

FIGURE 8.15 Various PDC formation mechanisms showing associated deposits.

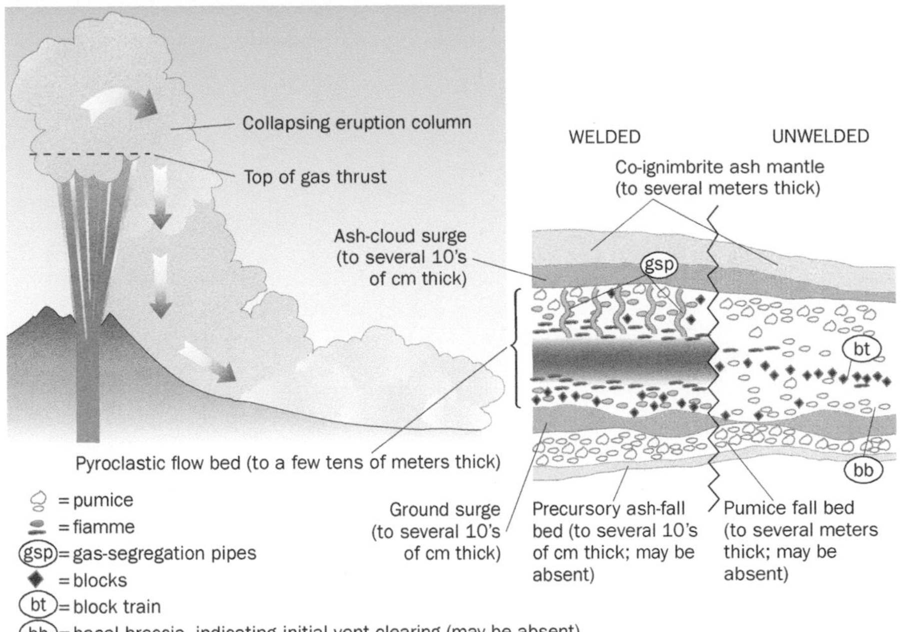

e) Cryptodome explosion

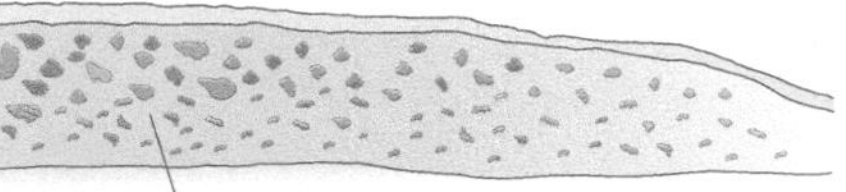

f) Silicic flow front collapse

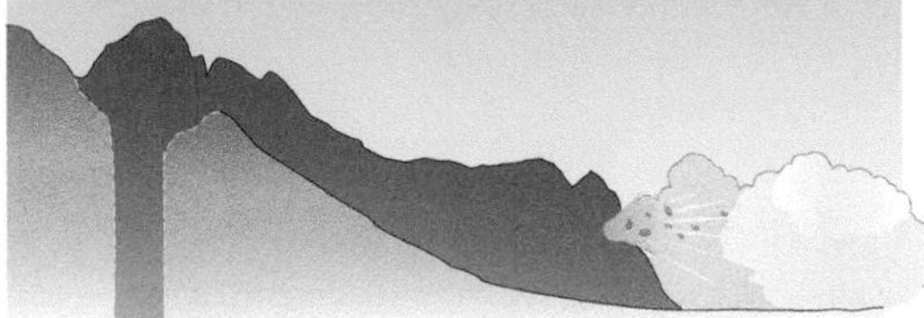

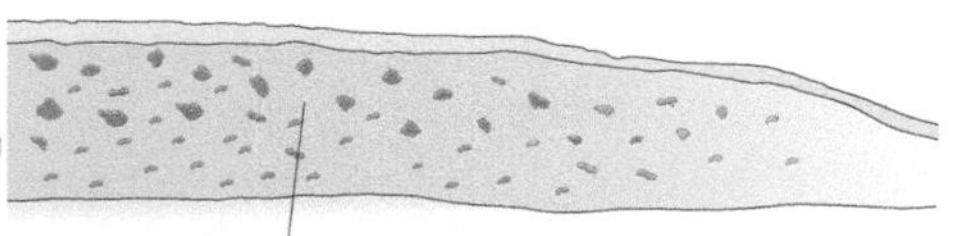

Similar to dome-collapse flow bed, above, but may show higher concentration of ash, fewer blocks

FIGURE 8.15 Continued

FIGURE 8.16 Soufrière Hills volcano, Montserrat [84], viewed from the northeast, showing the growing summit dome on 9 February, 1999. Voluminous pyroclastic deposits extend 3 km down the Tar River valley and several kilometers offshore on submarine slopes. Pyroclastic flows reached the sea on both east and west coasts of Montserrat and caused great destruction since 1996. Source: Photo by Simon Young, British Geological Survey, © NERC.

few seconds exposure; those at the very margins of the flows only survived for a few days, as their throats and lungs were horribly seared while they gasped for air during passage of the currents. Victims at the margins of the 2010 Merapi [131] pyroclastic flows died instantly while seated on chairs in their homes, even though no ash – only hot gases were involved.

Far more fortunate was Frank Perret (Chapter 1) during the 1929–1932 eruptions of Mount Pelée [87] on the island of Martinique. To view the growth and collapses of Pelée's active summit dome, Perret had a single-room wooden shack installed as a field station on a ridge 2 km below the volcano's summit. The station overlooked the valley of the River Blanche, through which most pyroclastic flows he witnessed traveled harmlessly below him to the sea. On the evening of 15 April, 1931, Perret was in his shelter, when:

> *I had one of those harrowing experiences which the volcanologist has to expect in close-range observation of an actual eruption. . . A huge spine had grown up near the west side of the dome summit. It was unstably poised, and its fall seemed imminent. Just at the close of the swiftly passing tropical twilight, in the dead calm between day and night winds, an unusual sound brought me to the cabin door. The whole mass had fallen, leaving a great scar on the dome from which poured forth an ash cloud of inky blackness, expanding upward as it rushed down the talus slope . . . At the same instant, an explosion on its eastern summit shot out a second avalanche with rising cloud white as snow; two mighty parallel columns, ominous, terrifying, moving straight toward the station. I shall not attempt to describe all the sensations I felt, nor the thoughts that swept through my mind at that moment. My first thought was of instant flight. With a distance of 2.5 kilometers between me and the crater, it might be four minutes before the clouds reached me, but a moment's reflection indicated that this could not be enough to escape the wide path of the cloud. I decided to risk the protection of the frail shack. Doors, windows, cracks, and holes were hastily closed and the onset awaited. Escapes from many former perils helped to allay my fears, but there was still the thought that this might be my last. I recall a sense of utter isolation; awe in the face of overwhelming forces of nature so indifferent to my feeble self. The track of the avalanche lay to the side of the station or these lines could never have been written. The chief dangers were heat and gas from the cloud. There was still a minute left. I peered out from the rear of the station. A sublime spectacle! Two pillars of cloud a thousand feet in height, apparently gaining in speed every instant and headed straight for my shelter. As I darted within, the blast was upon me – not a terrific shock as the reader might think, but swirling gusts of ash-laden wind, bringing*

a pall of darkness that might indeed be felt. The dusty air [that] entered every crevice of the shack was hot but not scorching. I felt the gases burning and parching my throat and then came a feeling of weakness (carbon monoxide?). It all lasted for half an hour, but it was nearly an hour before the feeling of suffocation was relieved by a kindly wind.

Painful as the experience had been – and still was – I had been granted an opportunity, from a situation only 200 or 300 meters from the direct path of the avalanche, to observe that marvelous manifestation of volcanism. During the whole time of its passage. . . there was a condition of absolute silence, broken only by the occasional slide of older lava blocks as they were thrust aside in its resistless sweep. (Perret 1935, pp. 61–62)

Survival Tips for Field Volcanologists

Whenever approaching an active volcano, always choose routes that "keep to the high ground." Unexpected hazards like PDCs, lahars, and dense CO_2 accumulations will flow to low ground. Not only are ridge tops safer than valley bottoms – there are more likely to be cool breezes up there!

As Perret's account indicates, pyroclastic flows have very low densities at their dilute margins owing to their low content of solid material, and will in some instances flow around fragile structures without damaging them. I (JPL) remember finding a soda bottle that remained upright on a charred table outside a home at the edge of a Galunggung pyroclastic flow in 1982. Although not toppled by the hot ejecta cloud, the glass had begun to melt and deform. Closer to the interiors of pyroclastic flows, the interactions between the turbulent, particle-laden hot gas cloud and buildings may be quite complex and destructive (Chapter 7; Doronzo 2013).

Small pyroclastic flows may pour down a mountainside at a relatively "slow" 20–30 m/s, while very large flows may reach speeds of 50 or even 100 m/s across wide areas (Freundt and Bursik 1998). The climactic Taupo, New Zealand [180] pyroclastic flow possibly attained supersonic speeds of as much as 500 m/s close to its vent (Legros and Kelfoun 2002)! While mostly guided by gravity and topographically confined, fast-moving ash flows have enough momentum to overtop ridges hundreds of meters high that stand in the way. For example, the 50 km^3, 3400-year-old Aniakchak ash flow [10], Alaska, surmounted slopes as high as 700 m, 20 km from its source vent (Miller and Smith 1977).

Boilover Pyroclastic Flows

Boilover pyroclastic flows originate in the tops of magma conduits in a zone of magma boiling that produces relatively dense mixtures of gases and particles, mixtures that lack the buoyancy to ascend far above crater rims. Wolfe (1878), as quoted in Ross and Smith (1961), described the 1877 Cotopaxi [68] pyroclastic flows thus: "*It is indeed one of the singular features of this eruption that the lava [pyroclastic flows] poured out of the crater not in one or several streams, but symmetrically in all directions, over its lowest edge as well as its highest points*" and "*that it flowed out over the highest crater edges like the foam from a boiling-over rice-pot [überwallenden Reistopfes].*" Taylor (1958) described the formation of boilover pyroclastic flows in his observations of the 1951 Mount Lamington [159] eruption:

On several occasions the discrete explosive events of the spear-head type gave way to a mass effect. Successive explosions fountained rapidly and extensively from many parts of the crater floor and filled the crater bowl with a massive, convoluted cloud of fragmental lava and gas. The cloud usually showed little tendency to rise. The heavy, yet buoyant mass seemed to behave as a layered hydrostatic column raised (sic) in the bowl of the crater. The heavier fractions poured out through the low gaps in the crater wall; the lighter fractions poured over the crater rim.

Macdonald (1972) refers to boilover pyroclastic flows as *overflowing glowing clouds*; Fisher and Schminke (1984) speak of them as *low-pressure boiling-over* flows. It is indisputable that many pyroclastic flows form in this manner, and that both column collapse and boilover mechanisms may operate during the same eruption. For example, Rowley et al. (1981) showed that "most pf's [pyroclastic flows] originated [as] bulbous masses of ash, lapilli, and blocks [that] rose only a short distance above the inner crater and spread laterally" during the 1980 eruptions of Mount St Helens [31], though some also resulted from column collapse. Pyroclastic flows did not commence at Mount St Helens until nearly 4 hours after a Plinian column had risen to more than 20 km altitude at the beginning of the 18 May eruption. In contrast, during subsequent eruptions, flows began to spill out of the crater *before* explosive columns were well-developed. In another instance, the PDCs of the 15 June, 1991 climactic eruption of Mount Pinatubo [135] originated from gravitational collapse of the basal portions of a Plinian column coincident with heavy tephra fall (Scott et al. 1996), whereas small PDC's later clearly formed by boilover (Chapter 1, Figure 1.29). In the eruptions of Soufrière Hills volcano, Montserrat [84], many pyroclastic flows have developed from crater overflows of low eruptive columns, and not from column collapse (Druitt and Kokelaar 2002).

In contrast to boilover flows from wide conduits where magma can release gas at shallow depths; deeper boiling in narrower conduits favors the development of the high-speed ejecta jets that result in the rise of pyroclastic material hundreds of meters above vents and the development of column-collapse-style pyroclastic flows – pyroclastic fountaining in other words. Lower volatile content in the boiling magma froth might well be another factor in these situations; the greater density of such mixtures would inhibit development of ascending Plinian clouds and predispose direct overtopping of crater rims by boiling, frothy, gas-rich magma that feeds PDCs.

Depositional Mechanics of Pyroclastic Flows

Pyroclastic flows fed by Plinian eruptions, whether formed by column collapse or boilover processes, exhibit similar characteristics once they leave source areas. How such flows come to rest and form their deposits is not well known. Some researchers have interpreted massive pyroclastic flow beds as resulting from the *en masse* settling of fragmental loads (Carey 1991). This indeed happens near the distal ends of many PDCs, and perhaps throughout most of the reach of smaller, denser currents such as block-and-ash flows in their final moments. Rheologically, such sudden deposition is characteristic of a type of non-Newtonian fluid (Chapter 4) called a **Bingham body** in which below a certain yield stress, the moving mass behaves almost like a solid. Individual particles do not change their positions much relative to one another within the transporting current prior to sudden deposition. Internal shearing can also occur. Particles near the bottom of the moving mass tend to stop first, frozen into place by frictional drag, while particles higher up in the deposit keep moving, setting up strong shear stresses within the shifting mass.

The notion that larger pyroclastic flow bodies such as ash flows settle by *en masse* deposition has been challenged by Branney and Kokelaar (1997, 2002), who argue such massive deposits develop primarily by **aggradation**, a process of gradual accumulation and build-up of fragments into a thick layer as dilute mixtures of gas and fragments move erratically across the depositional surface. Good cases can be made for aggradation during deposition of the Mazama Tuff around Crater Lake [32], Oregon, and Alaska's Valley of Ten Thousand Smokes ignimbrite, especially by the common texture called **imbrication** (Mimura 1984; Fierstein and Hildreth 1992). Imbrication, a typical feature of pebbles in river bed deposits, is the stacking of platy or elongate fragments swept along by a fast-moving current under turbulolaminar or laminar conditions (Figure 8.17). The clasts pile atop one another so as to slope back toward the source of the current. A process similar to aggradation and imbrication takes place in the traction deposition

FIGURE 8.17 Imbrication of pumice lapilli in the Tumalo Tuff, a Pleistocene ignimbrite west of Bend, Oregon. Pyroclastic flow moved left to right during deposition. Scale in centimeters. Source: USGS photo by J.P. Lockwood.

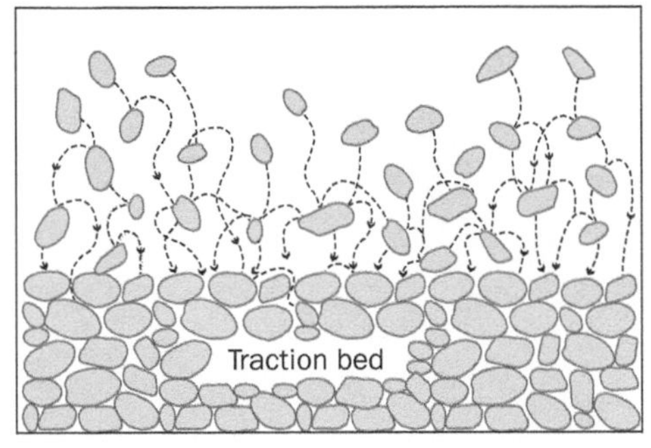

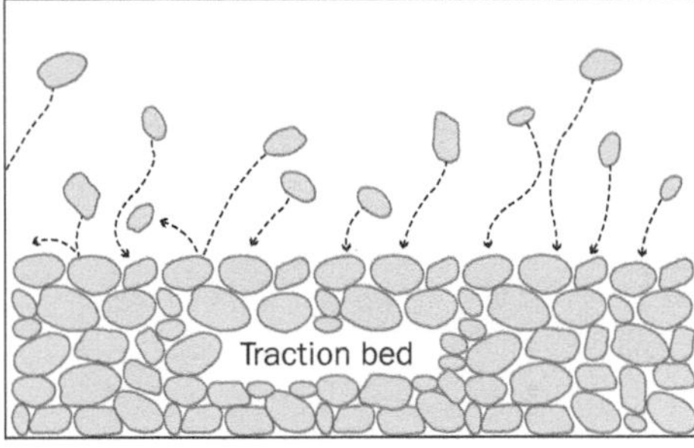

FIGURE 8.18 PDC fabrics are a function of the way in which the constituent particles settle from their gaseous transport medium.

of pyroclastic surges, suggesting that there is, in fact, a complete spectrum in transport behavior between pyroclastic flows and pyroclastic surges (Druitt and Sparks 1982; Druitt 1996).

If aggradation, imbrication, and basal traction bed development occur in both pyroclastic flows and surges, we are left once again dealing with the question of what accounts for the striking difference in the appearance of flow and surge deposits. Brown and Branney (2004) suggest that the critical defining factor has to do with the suspended particle concentration just above the surface of the traction bed right at the moment of deposition. If the particles are so closely spaced that they *interact and hinder one another* in their final settling, a massive texture characteristic of pyroclastic flows will develop. If fragments are *too widely spaced to collide and lock one another into position* in the moments before settling, surge-style bedding forms instead, similar to that of windblown dune fields in which individual sand grains spend awhile skipping and jumping ("saltating") great distances across a gradually aggrading surface before finally settling into place (Douillet et al. 2013). Brown and Branney envision flow texture-determining particle interactions as taking place in a narrow gradational zone separating the emplaced basal bed from the still moving current cloud above (Figure 8.18).

Ash Elutriation and Fines Depletion

Where an ash flow streams across the crest of a hill or ridge, a pocket of air may be trapped on the lee side into which coarser pumice fragments fall, and the lighter-weight material continue to be swept on (Figure 8.19). The resulting pumice-rich wedge may extend for tens of meters before pinching out in the direction of ash flow movement. Walker (1983) referred to such deposits as being **fines-depleted**, because they lack the high concentrations of ash seen throughout most of the rest of the ignimbrite.

Another common agent of fines depletion is the escape of gases and trapped air through the top of a settling ash flow. The finest particles of ash are filtered out (**elutriated**) by the gas, which tends to escape via gas segregation pipes (Chapter 7). Well-sorted, tell-tale concentrations of coarse pumice and lithic fragments that lagged behind during elutriation allow one to distinguish gas pipes quite readily in eroded cross-sectional cuts (Figure 8.20). They may extend vertically for many meters. Most are far better sorted than the surrounding ash flow deposits, illustrating the effectiveness of a high-density gas stream in filtering grain sizes. Many of them occur in clusters reflecting high concentrations of escaping gases from deeper in the deposit, in some cases derived from buried streams, marshes, and ponds. Some have more unusual origins. In the immediate aftermath of the 18 May, 1980 eruption of Mount St Helens, gas segregation pipes developed in pyroclastic flows from the embedded carbonized remains of stems and branches, pieces of the trees blown down and swept up by the pyroclastic current. The heated wood released gases into the overlying bed as it was being distilled to charcoal, gases that were concentrated enough to carry bits of charcoal upward for tens of centimeters (Druitt 1992). Most segregation pipes extend all of the way to flow surfaces, where the gas escapes through fumaroles following an initial burst of fine ash – a miniature eruption in and of itself. Some of these bursts are strong enough to scallop out shallow craters in the soft beds of late-stage fall deposits. Later hydrothermal alteration of included fragments by the concentrated flow of vapors gives some of them a distinctive ochre or red color. As large pyroclastic flows move downslope, tremendous amounts

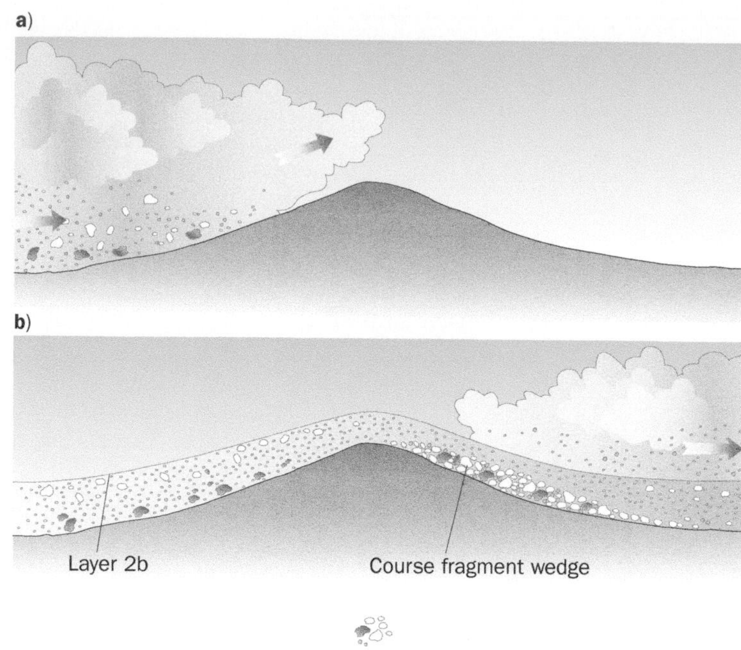

FIGURE 8.19 The launch ramp effect for development of fines-depleted coarse particle concentrations on the leeward side of topographic highs crossed by ash flows. Source: Based on Walker (1983).

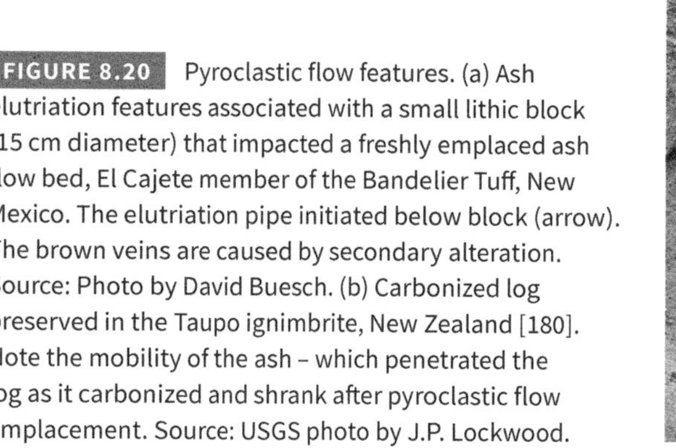

FIGURE 8.20 Pyroclastic flow features. (a) Ash elutriation features associated with a small lithic block (15 cm diameter) that impacted a freshly emplaced ash flow bed, El Cajete member of the Bandelier Tuff, New Mexico. The elutriation pipe initiated below block (arrow). The brown veins are caused by secondary alteration. Source: Photo by David Buesch. (b) Carbonized log preserved in the Taupo ignimbrite, New Zealand [180]. Note the mobility of the ash – which penetrated the log as it carbonized and shrank after pyroclastic flow emplacement. Source: USGS photo by J.P. Lockwood.

of material can periodically elutriate and jet from their advancing fronts, which is alternately (or simultaneously) entraining and releasing great amounts of surrounding air owing to their high turbulence and speed. These **flow head elutriation jets** consist not only of fine ash, but also lumps of pumice and even occasionally denser lithic fragments. A spectacular example of fines depletion related to this process, somewhat similar to the Mount St Helens example cited above, occurred during the eruption of the Taupo ignimbrite. Where this enormous ash flow overrode moist woodland, plants and trees were swept up and mixed in with the ash and pumice. Instantaneous combustion released large quantities of vapor that gushed upward through the ash. The elutriating ash-vapor mixture jetted from the advancing flow front and surface of the flow deposit, leaving a zone of concentrated charcoal logs and rounded pumice clasts behind (Walker et al. 1980).

Ash flows may not continuously deposit material as they travel across surfaces. Such non-deposition is termed **bypassing** (Brown and Branney 2004). Bypassing, accompanied by significant erosion of the ground where turbulent surge processes are involved may be common over distances of several kilometers on proximal slopes, where PDCs can be expected to move most rapidly. Scouring and erosion during bypassing probably provides a larger quantity of accidental and accessory ejecta to ash flows than is commonly appreciated. Later pyroclastic deposits, however, may bury bypassed areas, making their identification and interpretation difficult (Brown and Branney 2013).

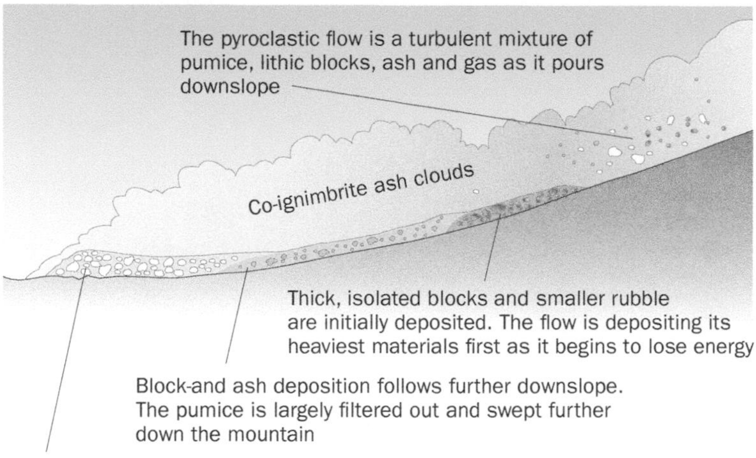
The pyroclastic flow is a turbulent mixture of pumice, lithic blocks, ash and gas as it pours downslope
Co-ignimbrite ash clouds
Thick, isolated blocks and smaller rubble are initially deposited. The flow is depositing its heaviest materials first as it begins to lose energy
Block-and ash deposition follows further downslope. The pumice is largely filtered out and swept further down the mountain
Lobate pumice beds accumulate at the terminus of the flow

FIGURE 8.21 Depositional facies of the Soncor pyroclastic flow, Láscar Volcano. Chile [82]. Source: Based on Calder et al. (2000).

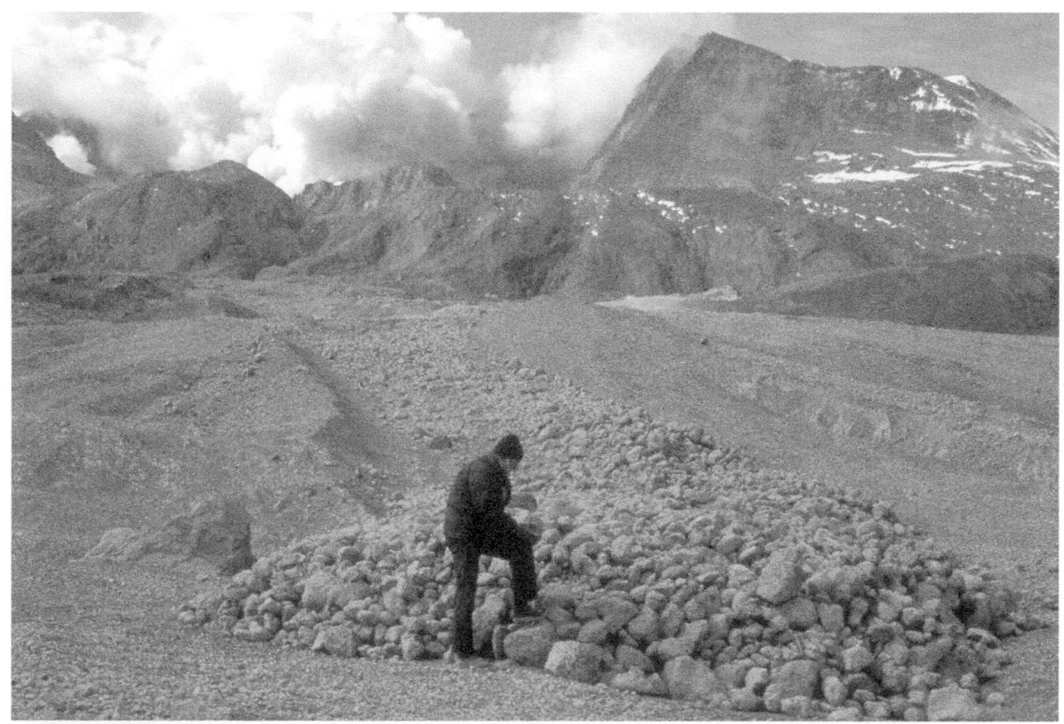

FIGURE 8.22 A remarkable concentration of pumice blocks marks the terminus of a small, late-stage pyroclastic flow erupted from the crater of Mount St Helens (MSH) [31] on 18 May, 1980. A grain dispersive sorting mechanism may have been at play to account for this concentration. View toward the MSH crater source, 5 km away. Source: USGS photo by Don Swanson.

Work by Calder et al. (2000) on the relatively small Soncor ignimbrite from Láscar volcano [82] in northern Chile reveals that as the initially turbulent ash flow traveled, it laid out a bed of coarse breccia closest to its vent with finer lithic breccia farther downslope, and finally, lobate pumice deposits at the terminus (Figure 8.21). The formation of basal breccias that thin distally, and pumice-rich upper layers and distal margins, is common in many ash flows (Chapter 7; Figure 8.22). Experimental work by Choux and Druitt (2002) provides some insight into the nature of particle separation during dilute, turbulent transport. They used a water flume in which particles of two distinct densities, one light (pumiceous) and the other heavy (lithic) were immersed. They varied the volumetric particle concentration in experimental runs from between 0.06 percent and 23 percent. At concentrations of less than a few percent, light and dense particles responded in the same way to the currents – they had **hydrodynamic** (or **aerodynamic**) **equivalence**, meaning that they had the same settling speeds. Above about 5 percent particle concentration, the behaviors of light and dense particles became separate. Heavier particles developed settlement patterns resembling the co-ignimbrite lag breccias mentioned in Chapter 7, while lighter particles accumulated atop the deposit and swept off to pile up at the distal end in a manner reminiscent of the Soncor pumices.

FIGURE 8.23 Pyroclastic flow pouring onto the sea off south coast of Montserrat Island, following a collapse of the Soufrière Hills volcano [84] summit cone in September 1996. Note co-ignimbrite plume ascending over flow base, and discolored water from previous flows. Source: Photo by Simon Young, British Geological Survey.

Co-ignimbrite (Co-PDC) Ash Plumes

As a pyroclastic flow travels, the gas-particle mixture at the top of the flow and at its terminus becomes lighter than the surrounding atmosphere owing to the fallout of larger and heavier fragments. Turbulent convecting clouds of fine elutriated ash, called **co-ignimbrite ash plumes** may ascend kilometers above moving PDCs. Fallout from these plumes commonly produces the thick, well-sorted ash beds seen capping many ignimbrites. Similarly formed ash beds atop pyroclastic flows having a high initial concentration of coarse ejecta, such as block-and-ash avalanches are typically thin. In other instances, the volumes of co-ignimbrite ash may equal or even exceed the volumes of associated ignimbrites in other instances, indicating that the original ash flows must have been especially gas-rich (Freundt and Bursik 1998). In some cases, as at Montserrat, PDC-related ash plumes become strongly convective, and may even dwarf the heights of source eruptive columns.

Eyewitness accounts show that co-ignimbrite ash plumes rise directly from the turbulent front of advancing flow masses where air is easily entrained, and from places where the pyroclastic flow encounters changes in slope (Figure 8.23). At Merapi volcano [131], pyroclastic flows in November 1994 reached the base of the 30° slope of the cone where their dense basal masses continued moving downhill while the turbulent ash clouds rising from their surfaces detached to form plumes as much as several kilometers high (Bourdier and Abdurachman 2001). Likewise, where ash flows across hills or ridges, deposition of coarser ejecta in the pockets of relatively calm air on the downwind sides of each obstacle enhances the buoyancy of the remaining gas-particle mass, forcing the development of conspicuous ash plumes (Figure 8.24). To distinguish these from the ordinary eruption clouds generated directly above a vent, some volcanologists have called them **phoenix plumes** (Dobran et al. 1993), alluding to the mythological Greco-Phoenician bird whose body "rose from the ashes."

Ignimbrite Aspect Ratios and Criticalities

Walker (1983) distinguished two categories of ash flows. He termed the deposits of conventional ash flows **high-aspect-ratio ignimbrites** (HARIs), and characterized them as being emplaced "quietly and passively in valleys," typically by *en masse* dumping of pyroclastic loads, in contrast to the **low-aspect-ratio ignimbrites** (LARIs) representing ash flows with the "remarkable ability to

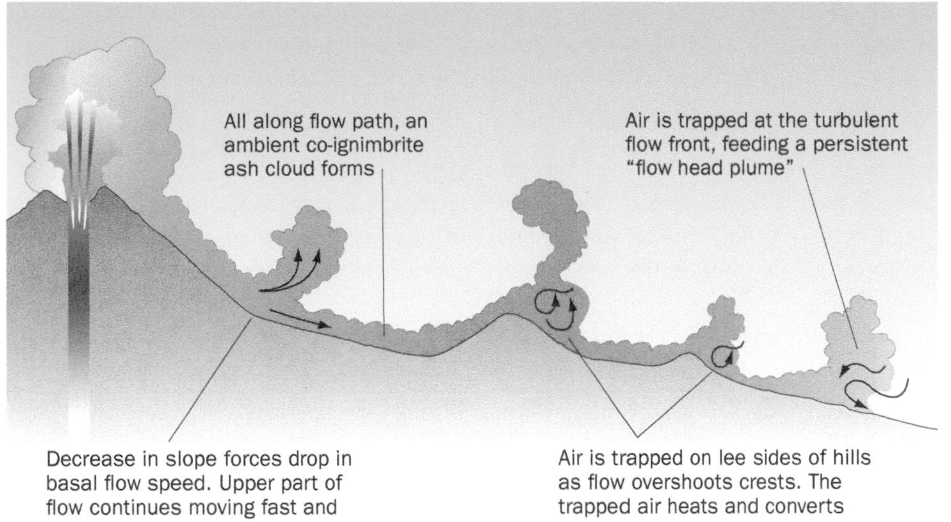
All along flow path, an ambient co-ignimbrite ash cloud forms
Air is trapped at the turbulent flow front, feeding a persistent "flow head plume"
Decrease in slope forces drop in basal flow speed. Upper part of flow continues moving fast and detaches as independent ash cloud
Air is trapped on lee sides of hills as flow overshoots crests. The trapped air heats and converts into plumes as it escapes

FIGURE 8.24 Secondary, or "phoenix" plumes from pyroclastic flows. Source: Based on Dobran et al. (1993).

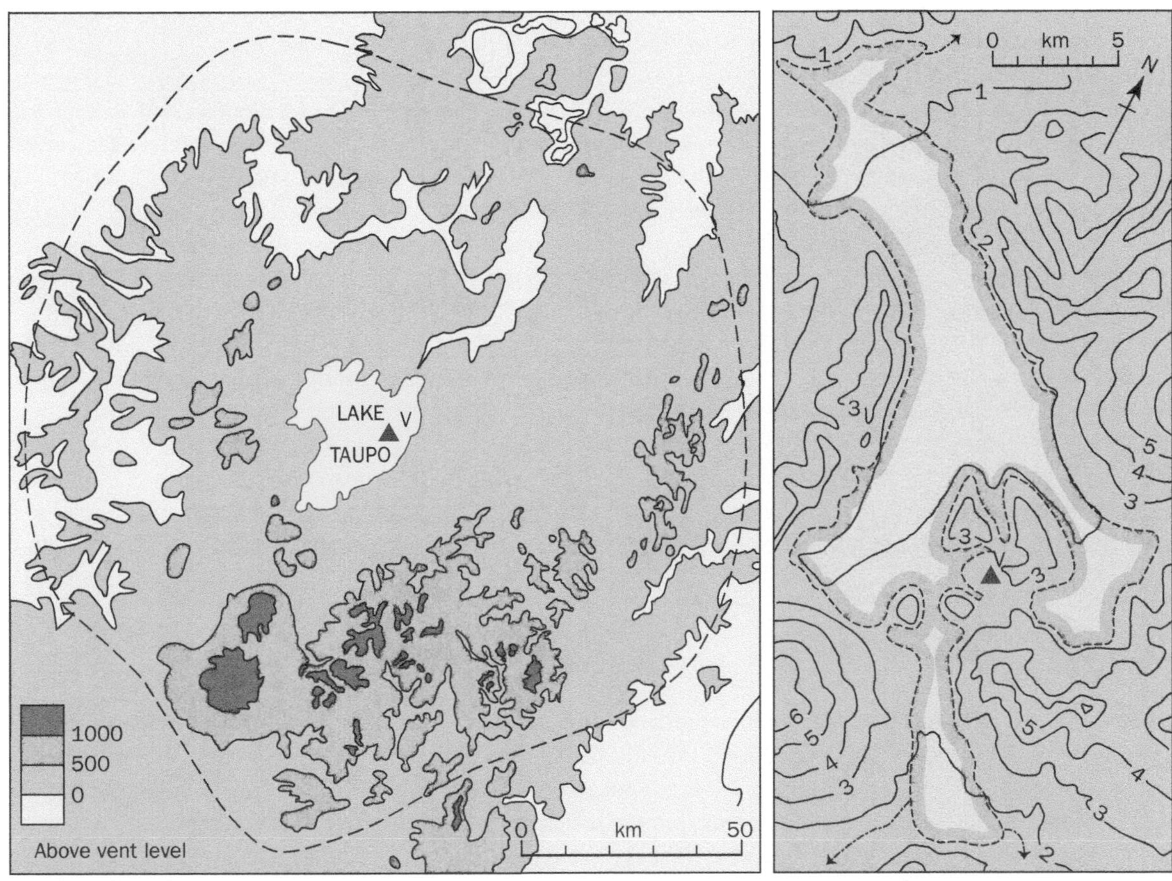
LAKE
TAUPO
V
1000
500
0
Above vent level
0 km 50
0 km 5
N
Taupo ignimbrite, New Zealand, (coverage within the dashed line)
The Valley of Ten Thousand Smokes ignimbrite (orange) is high-aspect ratio. (Contours in 100m intervals)

FIGURE 8.25 LARI and HARI examples. Source vents shown by red triangles. Source: Walker (1983). Reproduced with permission of Elsevier.

scale mountains and cross open water." An example of the former is the 1912 Novarupta [19] deposit in the Valley of Ten Thousand Smokes (VTTS) (Figure 7.34). Examples of the latter include the above-mentioned Aniakchak Tuff [10], the Taupo [180] and Oruanui [181] ignimbrites in New Zealand, and indeed virtually all phreato-Plinian eruption deposits (Figure 8.25). **Aspect ratios** of ignimbrites (or lava flows) are measured by dividing the average thickness of the flow by the area encompassed by a circle with a radius equal to the farthest reach of a flow from its source – that is, its maximum thickness divided by the travel (**run-out**) distance. Walker measured an aspect ratio range of from 1/400 (VTTS) to 1/100,000 (Taupo).

The main current of the Taupo ash flow was indeed fast moving. Wilson (1985) found patches of veneer facies on the slopes of Ruapehu volcano [178], 60 km south of the source vent and 1500 m higher than the surrounding hummock plain. To climb this high at such a distance from the source, the current must have converted all of its available kinetic into potential energy according to:

$$U=(2gh)^{1/2} \tag{8.2}$$

where $h = 1500$ m, and U the speed of the flow – an astonishing 170 m/s in this case. It is on the basis of calculations such as this that Wilson believes emplacement of the Taupo ignimbrite took only a matter of minutes, and estimates a speed of 170 m/s 60 km from the vent, approaching 500 m/s near the Lake Taupo caldera source, from an eruptive column on the order of 10 km in diameter – nearly 70% faster than the speed of sound.

Based on theoretical considerations, Bursik and Woods (1996) categorized ash flows into two categories somewhat differently than Walker's HARI and LARI characterization. They define **subcritical ash flows** as relatively slow (10–100 m/s) and relatively thick, with turbulent suspension loads puffing up to more than 1–3 km above the surface. **Supercritical flows** are faster (100–200 m/s) and thinner (500–1000 m). In general, subcritical ash flows decrease their volumes at a constant, linear rate as they spread away from vents, whereas supercritical ash flows entrain more air and become progressively more voluminous. The entrainment of large amounts of air also means that supercritical ash flows tend to produce larger and more numerous co-ignimbrite plumes, some of which may climb as high as 20–40 km, dwarfing-related Plinian columns. As much as 20–40 percent of the ejecta in a supercritical ash flow, primarily in the form of fine ash, may be removed through plume activity (Sparks and Walker 1973), and flow run-out may be significantly reduced due to loss of kinetic energy during plume generation. Because subcritical ash flows show less plume activity they tend to travel farther than their supercritical counterparts, even though slower moving.

Ash flow speeds rise with increasing eruption intensities, hence larger Plinian eruptions (10^{9}–10^{10} kg/s) tend to generate supercritical ash flows in which air entrainment dominates the development of the flows (Bursik and Woods 1996). For subcritical flows, air entrainment is less important than sedimentation rate in terms of governing how the flows move, and how far they travel.

Single ash flows may behave supercritically at some points along their paths, and subcritically at others. Woods (1988) investigated topographic effects on moving ash flows, suggesting that supercritical ash flows approaching opposing ridge crests would expel air and gases, perhaps through co-ignimbrite plume generation, and become subcritical. But after crossing the ridge crest, the ash flow could entrain large amounts of air as it launches into the atmosphere for a short distance, as off a ski jump, and becomes supercritical again. The distance an ash flow can travel depends significantly upon the degree to which it is in a subcritical or supercritical mode. For instance, an ash flow generated in a high-intensity eruption with an initial gas mass fraction of 0.1 wt. percent and mean particle size of 1 mm can travel 20–25 km down a valley 2 km wide in a supercritical state before stopping, whereas a subcritical flow will travel much further, up to 50–60 km. But on an open plain, given the same pyroclastic current conditions, a subcritical flow will only travel 15–20 km (Bursik and Woods 1996). In general, low-aspect ratio ash flows behave as supercritical flows. However, the local impacts of topography on current dynamics, as introduced above, prevent a strict equivalence between the concepts of ash flow aspect ratio and criticality. (Likewise, the concept of "criticality" as applied to ash flows must not be equated with Reynolds and Froude number "criticalities" to be introduced in different contexts later, though each applies to conditions of flow.)

The Origin of Pyroclastic Surges

Cold War era atmospheric testing of nuclear weapons produced an unexpected benefit for volcanology – revealing a new phenomenon associated with near-surface explosions. During the July, 1946 underwater test of the 20 kt atomic bomb "Baker" at Bikini Atoll (southwestern Pacific), a ring of steam and debris raced out across the lagoon at the base of an ascending mushroom cloud in all directions, at an average speed of 33 m/s. Observers termed this cloud a **base surge**. Nuclear testing at the Nevada Test site in the 1950s demonstrated that base surges formed whenever explosions were focused at or near ground level (Figure 8.26).

In 1952–1953, an eruption of Barcena volcano [46], on an island off the west coast of Mexico produced a similar eruption column and base surge, leading geologists to suggest that the physics of both were the same (Richards 1959). This was confirmed in 1965, when Taal volcano [136] in the Philippines erupted explosively through a submerged vent along the shore of a caldera-filling lake. The blast effects of the Taal base surge closely resembled those of some nuclear tests. The 100°C expanding surge cloud initially spread at 50 m/s, totally destroying trees within a kilometer of the vent, and plastering objects with hot ash out to a distance of 8 km. Moore (1967) described the effects of the eruption and popularized the term *base surge* to describe both the eruption and its deposits.

Base surge eruptions are most powerful when abundant water is involved, and are characteristic of many hydrovolcanic eruptions – a confirmation of Walker's (1973) assertion that "steam means power" when it comes to volcanic explosions. Their origins are also remarkably shallow: Rohrer (1965) related base surge run-out to explosive power, noting that the distance traveled varies as one-third the power of the explosion yield, given an "optimal" explosion depth of only 10–15 m for a one-kiloton charge. Such an explosion would produce a gas-expansion-type base surge traveling 2.5 km. Moore (1967) applied these calculations to derive

FIGURE 8.26 "Priscilla" – a 37-kt atomic blast test at the Nevada Test Center, Nevada, 24 June, 1957. Note well-developed base surge, expanding from the base of the ascending "mushroom cloud." Source: Photo courtesy US Department of Energy.

the explosive yield of the Lake Taal base surge, finding that a surge moving 6 km would require a 16 kt blast originating at a depth of "several meters." Mastin and Witter (2000) went further in using energy equations to constrain the size of the Taal explosive body. Working with fuel-coolant ratios of 1:3 to 1:5, the range observed in hydrovolcanic eruptions (Wohletz 1986), and assuming an efficiency of 10 percent in converting thermal to mechanical energy, they found that on the order of $1–2 \times 10^6$ tons of explosively interacting magma and water would be required to generate the Taal surge, equivalent to a vent 50 m in diameter volatilizing to a depth of 500–1000 m.

By the early 1970s comparison of recent base surge deposits with older features in the geological record showed that surge-like bedding could result from a range of processes other than simply powerful Taal-style hydroexplosions. All shared in common the aspects of powerful, water, or gas-rich ground-hugging energy dispersal. Sparks and Wilson (1976) introduced the term **pyroclastic surge** for this family of related phenomena. Pyroclastic surges tend to be cooler than pyroclastic flows, and many involve condensed water, as evidenced by the presence of accretionary lapilli in many deposits. Although many surges are emplaced "dry" (above condensation temperature of water), there is never evidence of welding in their deposits. Surge currents, although highly dilute, are characterized by extremely turbulent flow, and commonly cause mechanical erosion of earlier layers. In addition to base surges, **ground surges** and **ash-cloud surges** are also recognized (Wohletz 1998). **Directed blasts** have also been considered as a type of surge by some volcanologists, though their deposits imply more complicated emplacement mechanics, and they will also be discussed in the following sections.

Most base surges are relatively local in their effects, rarely extending more than 1 km from hydrovolcanic source vents. Surge processes can be integral components of larger sub-Plinian and Plinian eruptions though, and dry surge phenomena are contemporaneous with pyroclastic flows in most large explosive eruptions. The El Chichón [57] eruption in March and April of 1982 (Chapter 14) illustrates the characteristics of dry surging, where hot air and dominantly magmatic gases rather than a dominance of steam drove PDC generation.

El Chichón is a heavily eroded dome complex in the jungle of southern Mexico (Chapter 1), which had erupted explosively in the eighth, fourteenth, and mid-nineteenth centuries, but despite this record had largely been ignored as a potentially dangerous volcano, because the gaps between its eruptions were so lengthy. The onset of the 1982 eruption was sudden, and a major loss of life took place. Pyroclastic currents, some moving as rapidly as 100 m/s, destroyed nine villages that could not be evacuated on time, with over 2000 fatalities. The six-day-long eruption fluctuated between steam-rich Vulcanian, phreatomagmatic and Plinian phases (Macías et al. 1997). Sigurdsson et al. (1987) describe the initiation of two of the surges:

> *[Activity] began with an explosion which produced a low, incandescent eruption cloud around the summit of the volcano . . . From this cloud issued the first pyroclastic surge, which devastated the flanks of the volcano and spread in all directions, but mainly to the south up to 8 km from the crater. Minor pyroclastic flows and debris flows from the disintegrating dome also accompanied the first stage. Immediately following the surge a Plinian eruption column developed, from which [a] fallout layer. . .was deposited. From the distribution of lithic [fragments] in the fall deposit, it is clear that this eruption column was the most energetic of the El Chichón eruptions, with an average column height of 24 km. The. . . Plinian event also had the highest mass eruption rate, at 6×10^7 kg/s, producing a total of 0.39 km^3 [of ejecta]. . . The Plinian stage was followed by a second pyroclastic surge. . . which spread laterally from the crater and covered 104 km^2. Because of the sequence of events from a Plinian column to a surge, we propose that the second surge originated during column collapse. It was accompanied by pyroclastic flows.*

The depositional fabric of the El Chichón surge beds provides further detailed information about surge dynamics. In the lower halves of surge beds, materials coarsen upward – reversely graded – and the lee sides of sand waves face the distal ends of the deposits. Such a fabric is to be expected as long as the speed of the current is increasing throughout deposition. Heavier material is being plucked up by the strengthening current, and dune crests are migrating vigorously down slope. Just the opposite pattern occurs in the upper half of surge deposits, however, with material fining upward and lee sides of sand waves facing back toward the vent. The late-stage vent-ward migration ("regression") of sand waves indicated by the reversal of slopes implies that a ground-level counterflow of hot gas and air back took place back toward the vent to equalize atmospheric pressure.

Late-stage vent-ward winds are not the only explanation for regressive shifting of sand waves. This can also happen if an eruption suddenly becomes much wetter, with cohesive wet ash accreting to the stoss sides of dunes, or if current speed rapidly accelerates to a high level (Schmincke et al. 1973; Allen 1982; Barwis and Hayes 1985; Cas and Wright 1987). Fisher and Waters (1970) termed the latter category of dunes **antidunes**. In a general way, what happens when antidunes form is that high-speed currents abrasively erode the proximal surfaces over which they flow, plastering sand against the opposing sides of obstacles in the bed further downstream, which grow into mounds continuously shifting their crests backward into the face of the oncoming current. The antidunes themselves will set up internal waves within the current that help shape and maintain additional, regularly spaced antidunes in a smoothly flowing, fully developed system.

Interpreting the dynamics of emplacement based upon stratigraphic structure alone can be difficult. Understanding the origin of particular waveforms in surge deposits requires taking into consideration outcrop location relative to source vent(s), local topographic changes and slopes during eruption, upper flow surface and basal flow influences, eruption history, and other factors. Douillet et al. (2013) identified four distinctive dune types associated with pyroclastic surging during the 2006 eruption of Tungurahua volcano [67], Ecuador. These included elongate, transverse, lunate, and "two-dimensional" (wide, short) dune trains. While largely regressive, these researchers saw no reason to classify any of these structures as antidunes, suggesting that such features could be easily misinterpreted in other surge deposits elsewhere.

In the vicinity of vents, fresh pyroclastic surges roaring down slopes can be highly erosive. In the base surges of Taal and Barcena volcanoes, and even in the 79 CE surge deposits of Vesuvius, proximal erosion channels with U-shaped cross-sectional profiles are common. At Koko Head, a Surtseyan tuff cone on the southeastern coast of Oahu in the Hawaiian Islands, these U-shaped channels range in width from 0.4 m to 5.5 m, with depths of 1–3 m. They occur low in the surge deposit, suggesting that they were scoured during the initial, highest-power portion of the eruption, only to be filled by subsequent surge layers as the speed and energy of surging waned. Fisher (1977) believes that some of these channels, in fact, began as normal fluvial runnels cut into soft ash, which repeated surging later sculpted into deeper and wider troughs. Elsewhere the connection between U-shaped channel cutting and pre-existing drainage is not evident. Fisher also suggested that cutting of furrows on slopes might be specifically correlated with the formation of particle-rich lobes frequently seen extending from the advancing fronts of base surges. The dynamic origin of such lobes is unknown, but at high speeds, the segregation of particles into narrow funnels and channels is also seen in some experimental turbulent suspensions.

Ground and Ash Cloud Surges

Many ignimbrites are immediately underlain by patchy to widespread surge beds, which because of their basal position constitute **ground surge** deposits. Ash-cloud surge beds immediately overly many ignimbrites too, and in many cases are much more extensive and thicker than related ground surge features. Ground and ash-cloud surge layers tend to be fines-depleted, with a very high concentration of crystal fragments. Lacking much glass, they are largely impervious to the welding that may take place in adjacent ash flow deposits (Wohletz 1998).

Several hypotheses have been discussed regarding the origin of ground surges. The simplest one asserts that ground surges initially form when an eruption column collapses, since the falling debris incorporates large amounts of air (Fisher 1979). The surging precedes in advance of a denser, flow-forming ash cloud, and hence lays down a ground pad for it to bury. Others believe that the scattered and localized occurrences of ground surge deposits and their highly fines-depleted compositions require a different mode of formation. Possibly they result from pockets of turbulence, trapped air masses, at the base of advancing ash flows that create locally dilute conditions (Valentine and Fisher 1986). Or perhaps, as in the case of the forest that was overwhelmed by the Taupo ignimbrite, they represent elutriations of dilute, turbulent ejecta from the swiftly advancing turbulent fronts of ash flows. If this latter interpretation is correct, the surges must spasmodically squirt forward, perhaps energized by blast waves, and disperse mere seconds before being covered by the slightly slower pyroclastic flows (Wohletz et al. 1984).

PDCs and Fluid Mechanics

As already discussed, particle transport in pyroclastic flows is regarded as being dominantly laminar, while in pyroclastic surges, turbulence dominates. **Laminar flow** is a sheet-like movement of material in which separate particles maintain parallel directions of motion, though not necessarily at the same speeds. Roughly speaking, the ice crystals in a moving glacier shift down slope this way, with those closest to the base and the margins of the glacier moving more slowly than those at the surface, near the central axis of the ice stream. **Turbulent flow** involves a chaotic movement of particles, all moving in different directions, and in some instances at different speeds. Flow paths are mostly curvilinear. Water molecules in a swiftly flowing stream or river show turbulence. A fluid may flow in a laminar manner at one time, and then become turbulent later. What determines mode of flow in an open channel is summarized by the **Reynolds equation**:

$$\mathrm{Re} = \left(\rho R U_f\right)/\mu \tag{8.3}$$

where Re is the **Reynolds number**; ρ is the fluid density; R is the ratio A/P, with A equaling the cross-sectional area of the channel through which the flow is pouring and P the length of contact between the flow and channel floor and banks ("wetted perimeter"); U_f is the speed of the fluid, and μ is the fluid viscosity. Reynolds numbers of magmas and gases moving through conduits are calculated somewhat differently, but that should not concern us here. The Reynolds number is *directly proportional to the speed and inversely proportional to the viscosity of a moving fluid*, with viscosity greatly increasing at higher particle concentrations. If the Reynolds number in an open channel is greater than about 2000, flow will be turbulent. If less than about 500, it is laminar, and if it is in the range 500–2000, a transitional motion occurs, which is described as **turbulolaminar**.

Pyroclastic surges behave as **inertial flows** in which the moving masses may initially act like a homogeneous Newtonian fluid, then quickly split into two layers – rather as the engineers have observed in gas-particle experiments. One of these is fast and turbulently agitated, with widely spaced particles buoyantly overriding the pull of gravity (see Kunii and Levenspiel 1991). The other is slower, with concentrated grains that shift elastically as a basal drag or traction load that may also be present, as we have seen, in ash flows (Lowe 1982). As a result of this dynamic division, Valentine (1987) refers to pyroclastic surges as **stratified currents**. The traction load forms the surge deposits we eventually see after everything comes to rest. In general, it grows at the expense of the upper, gas-rich stratum as a surge blasts along, although the gas-rich part of the surge widely erodes and redistributes basal materials locally, creating the wide variety of structures and textures that so distinctively characterize surge deposits.

The greater role of turbulence and interstitial gases in pyroclastic surges provides for better sorting than in grain-supported pyroclastic flows. Ash, lithic fragments, and pumice will sort out with maximum grain sizes at different ϕ levels in different parts of the surge cloud. For example, coarser fragments may move almost exclusively within the traction load; intermediate-sized particles may move both by being embedded within the load, or by skipping and jumping along its surface. And finer particles may settle out largely after turbulence dissipates according to their terminal velocities in free air, a process Lowe (1982) refers to as **suspension deposition**.

The model of *sequential fragmentation/transport*, discussed by Mackaman-Lofland et al. (2014), seeks to improve understanding of PDC flow dynamics based upon particle mass considerations. But care must be taken with any sort of modeling to distinguish fragments formed primarily during vent explosions from those that grow in size *during* transport and emplacement. Films of water from stream, and sulfur sublimating from gas, for instance, can act as bonding agents aggregating fine ash particles during surge transport (Scolamacchia and Dingwell 2014).

Directed Blasts

When a shallow magma conduit suddenly depressurizes, molten rock and gases will rapidly escape by taking the path of least resistance to the surface. That path may be sideways through the flank of a volcano, rather than vertically through its summit. Lateral blasting in many, if not most cases involves pre-weakening of the flank of a cone as hydrothermal activity reduces its internal cohesion (e.g.—Andronico et al. 2018). But other structural factors, such as the emplacement of shallow conduit plugs, faulting related to unstable collapse, etc., could also be implicated.

The effects of this type of blast were first observed after the 1956 eruption of Bezymianny volcano [170] by Georgii Gorshkov, who visited this remote wilderness location following the eruption and coined the term "directed blast" to describe the phenomena involved (Gorshkov 1959, 1963). The first opportunity to document a directed blast by eyewitness accounts and film did not come until 1980, however, during the initial phases of the 18 May, 1980 eruption of Mount St Helens. Since the documentation of this eruption and its effects (Lipman and Mullineaux 1981) has had revolutionary impacts on the study of explosive volcanism, a review of this eruption and events that preceded the directed blast is in order.

Case History of A Directed Blast – Mount St Helens, 1980

The first signs of unrest at Mount St Helens (MSH) [31] began with onset of an earthquake swarm on 20 March, 1980, alerting geophysicists to the possibility of an eruption at this volcano for the first time in over a century. Sure enough, small Vulcanian ash bursts commenced on 27 March. A shallow intrusion of dacitic magma was working its way into the northern flank of the volcanic cone, but the earliest minor explosive activity ejected only old rocks as magma ascended closer to the surface. The volcano's northern flank bulged outward over a period of two months as the magma pressure built up within, moving an area of over 2 km^2 as much as 40 m northward – at rates of up to 2 m/day (Lipman et al. 1981). Deformation extended far down the north flank, and tilt changes of up to 50 microradians/h were noted at Timberline, 3 km from the MSH summit, causing visible displacement of level bubbles in survey equipment. The intrusion had formed a classic subsurface *cryptodome* (Chapter 9) and was primed with mounting gas pressure to produce a directed blast. Although similar swelling had preceded the 1956 directed-blast eruptions of Bezymianny volcano [170], Kamchatka (Gorshkov 1959), this example provided only one of several possible outcome scenarios, and most volcanologists on site thought a dome might extrude from the side of MSH without seriously destabilizing the edifice. Most believed that associated ash eruptions probably would not exceed in violence those which had occurred during the volcano's lightly documented Goat Rocks period of activity a century and a half earlier. Although the potential hazards were not fully understood, a risk to visitors *was* recognized, and the area north of MSH was closed to public access – resulting in the saving of hundreds or perhaps thousands of lives on the morning of 18 May.

Early on that morning, a magnitude-5 earthquake triggered a collapse of the volcano's northern flank, producing a massive debris avalanche that cascaded northward. Dorothy Stoffel, observing from a light plane coincidentally flying above the mountain at the time, described the northern side of the mile-high cone as quivering like jelly in the moment leading up to sliding. Depressurized magmatic gases burst from the evolving slide scarp almost at once, jetting horizontally, it appears, owing to the fact that the avalanche had opened a vent to one side, rather than at the top of the volcano's magma column. Within about 15 s, the blast cloud, moving at a speed of 90–110 m/s overtook the advancing front of the avalanche and then killed a promising young U.S. Geological Survey volcanologist, David Johnston, who was monitoring the volcano from a high ridge directly in the blast path, 6 km to the north-northwest (Voight 1990). Blast deposits overtopped this ridge, and David barely had the time to send a radio warning before being overwhelmed. His body was never found. Lateral expansion of the blast cloud had increased its speed to 325 m/s as it descended Mount St Helens toward David, and it retained enough momentum to overtop this 380-m-high ridge, (now known as "Johnston Ridge") at an estimated 235 m/s (Moore and Rice 1984). Blast materials reached supersonic speeds as they accelerated during flow (Kieffer 1981a, 1981b) and quickly overtook and outran the initial debris avalanche.

Rosenbaum and Waitt (1981) record the testimony of some campers caught near the edge of the directed blast surge in forested highlands, an extraordinary 20 km to the west-northwest:

> *A very strong wind, which blew flames from the campfire flat along the ground and held braids of hair out horizontally, preceded the blast cloud by about 10–15 s. The witnesses were able to move about in the wind with little trouble, and no trees were toppled by it. No noise was associated with the approach of the cloud, and no concussion or loud noise was noted prior to the cloud's arrival. When the cloud arrived, it became totally black and all the trees seemed to come down at once. The witnesses were instantaneously buried in a combination of timber and "ash" and probably fell into the root ball of a blown-over tree. They could talk to each other but could see nothing. After perhaps 10 s it got very hot. At this time they could hear their hair "start to sizzle" as it was singed. One witness, who is a baker, estimated the heat to be "like a 300°F oven." Pitch boiled out of trees and remained hot enough to cause minor burns several minutes later. The sky cleared suddenly after several minutes and remained clear for a few more minutes. Then a dense ash fall began.*

Local topography played an increasingly important role influencing deposition with distance from the vent. In many watersheds, small, locally generated, unusually fine-grained pyroclastic flows modified and buried initially emplaced surge layers. These weak PDCs lacked the energy to topple trees, in contrast to the original directed blast, but did transfer large masses of material into low-lying areas, greatly thickening the deposits already accumulated there. The directed blast completely devastated an area

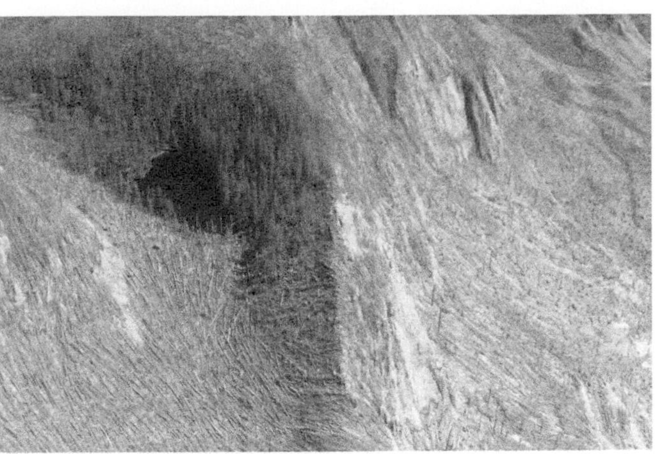

FIGURE 8.27 Effects of the directed-blast eruption at Mount St Helens [31] on 18 May, 1980. (a) Shattered tree stump 8 km north of the blast source. Stumps like this are all that remains of a dense forest that once covered the west edge of Spirit Lake (in the background). Note abrasion on the blast source side. (b) Blown-down trees about 20 km northwest of the MSH blast source. Note the trees that were protected by a ridge and survived blow-down as the surge passed overhead. A few people survived the blast in such locations. Source: USGS photos by J.P. Lockwood.

of about 600 km^2 (Figure 8.27). Trees were uprooted, sheared off, charred, and carried away inside the blast cloud within a broad proximal zone (Figure 8.27a), where temperatures were in excess of 350°C (Moore and Sisson 1981). Further out trees were merely blown down in place (Figure 8.27b), and temperatures were much lower. In the outermost zones, trees were not toppled at all but only lightly burned by hot gases. As the blast cloud reached the limit of its flow, the gas-particle mixture in the upper part of the cloud abruptly became buoyant, separating from a dense, hot "undercurrent" that settled as a poorly sorted, thin ash bed within a wide zone of singed forest. The remaining energy of the surge quickly dissipated as its rising terminal plume mixed into the atmosphere (Gardner et al. 2017). This behavior recalls the stratified surging model of Valentine (1987).

Hoblitt et al. (1981) studied the deposits left behind after the passage of the 18 May directed blast, and found them extremely complex, involving features attributed to pyroclastic flows, pyroclastic surges, and airfalls. Kieffer (1981b) studied the dynamics of the 18 May directed blast and concluded that most of the blast gases and debris were discharged in the first 10–20 s from the exploding magma reservoir at around 100 m/s, but that lateral speeds rapidly increased owing to expanding gases as the directed blast moved radially outward. Total thermal energy release during this short-lived event is estimated at 24 megatons of TNT equivalent.

A comparison of directed blast deposits from three different volcanoes in three very different regions has demonstrated that their pyroclastic stratigraphy is quite similar, implying that very similar processes of transport and deposition operate for all such events (Belousov et al. 2007). Directed blasts are clearly the most violent phenomena associated with smaller, noncaldera-forming eruptions, and we are indeed fortunate that these occur so infrequently! Violent as they are, however, they are much less destructive than the much larger and more catastrophic "super-eruptions" to which we now turn our attention.

"Super-Eruptions"

"Super-volcano" is not a well-defined scientific term proposed by volcanologists; it is a term coined in 2000 by the writers of a popular BBC science program (*Horizon*) to describe volcanoes capable of "super-eruptions." In the words of the BBC narrator of the program, **super-eruptions** were cataclysms that could "plunge the world into a catastrophe and push humanity to the brink of extinction." The new term *super-volcano* and the popularization of the older term *super-eruption* caught the public imagination, and volcanologists were suddenly being asked to discuss the potential for such catastrophic events. Volcanologists quickly adopted these terms to describe such potentially catastrophic, yet extraordinarily rare explosive eruptions, and these words are now well-established in the geologic literature. Miller and Wark (2008) discuss the origins of these terms and the nature of "super-eruptions" in more detail than will be possible here.

There is no standardized definition of what constitutes a super-eruption, although any eruption with a VEI greater than 8 has been widely considered to qualify. Concerned about problems with eruptive column and volumes estimates, however, Mason et al. (2004) rejected the VEI system, proposing to classify super-eruptions strictly upon the basis of total *mass* ejected. Using a logarithmic scale approach like the VEI, they distinguish super-eruptions as those producing log eruptive masses ("M") greater than 10^{15} kg (M > 8). Thus defined, they identified 42 super-eruptions that have occurred over the past 36 million years. The largest super-eruption known to have ever taken place, and the only one with M greater than 9.0 (erupted mass more than 10^{16} kg), was the Oligocene La Garita caldera [48] eruption of the Fish Canyon Tuff in southwestern Colorado (Chapter 11). More recently, Self (2006) defines super-eruptions as having produced erupted magma volumes >450 km^3 (ejecta volume >1000 km^3).

No super-eruption has ever been witnessed by modern humans – none have occurred within Holocene time. The most recent one took place in New Zealand about 26,500 years ago (see discussion below), and only five have happened in the past million years according to Mason et al. (2004). One of the largest Quaternary super-eruptions, from the Toba caldera [127] of Sumatra, produced over 2800 km^3 of pyroclastic material and had major climatic impact on the entire Earth about 75,000 years ago (Chesner et al. 1991; Figure 8.28). This VEI = 8.8 eruption possibly caused a major decline in human populations (Chapter 13), and may have also stimulated an onset of continental glaciation. It covered about 1 percent of Earth's surface with tephra to a depth of 10 cm or more – enough to collapse roofs and sink ships in modern terms (Mason et al. 2004).

Many individual calderas, such as Yellowstone [47] (Christiansen et al. 2007), have produced multiple (2–4) super-eruptions in the past, and are known to be underlain by active magmatic systems capable of large eruptions in the future. Some of these calderas remain geologically restless today, producing volcanic earthquakes, ground deformation, and geothermal changes (Newhall and Dzurisin 1988; Lowenstern et al. 2006). Because super-eruptions occur so infrequently, however, and because so little is known about their precursory activity, volcanologists have little ability to issue credible warnings in time for mass evacuations to take place. Lest anyone "lose any sleep" over the prospect of a super-eruption occurring in their lifetime; however, we can take solace in the remote odds of such an event. Although the infrequency and probable non-random time distribution of such eruptions in the past precludes precise statistical calculations, Mason et al. (2004) suggest the probability of a M greater than 8 super-eruptions occurring in the next million years is about 75 percent, with only a 1 percent chance of such an event taking place in the next 7200 years! Nonetheless, it is important to reflect on the words of Sparks et al. (2005):

> *There may be several super-eruptions large enough to cause a global disaster every 100,000 years. This means super-eruptions are a significant global humanitarian hazard. They occur more frequently than impacts of asteroids and comets of comparable potential for damage. Several of the largest volcanic eruptions of the last few hundred years, such as Tambora (1815), Krakatoa (1883), and Pinatubo (1991) have caused major climatic anomalies in the two to three years after the eruption by creating a cloud of sulphuric acid droplets in the upper atmosphere. . . However, super-eruptions are up to hundreds of times larger than these, and their global effects are likely to be much more severe. An area the size of North America can be devastated, and pronounced deterioration of global climate would be expected for a few years following the eruption. They could result in the devastation of world agriculture, severe disruption of food supplies, and mass starvation. These effects could be sufficiently severe to threaten the fabric of civilization.*

Super-eruption Case Study: The Oruanui [181], New Zealand Phreatomagmatic Eruption

The Taupo Volcanic Zone is a northeast-southwest linear corridor of stretching, sinking crust in the central part of New Zealand's North Island, underlain by a supply of subduction-related magmas that have been continuously active for at least the past 2 million years. The central part of the Taupo Volcanic Zone, an area measuring 115 × 60 km, is one of Earth's most active silicic volcanic fields

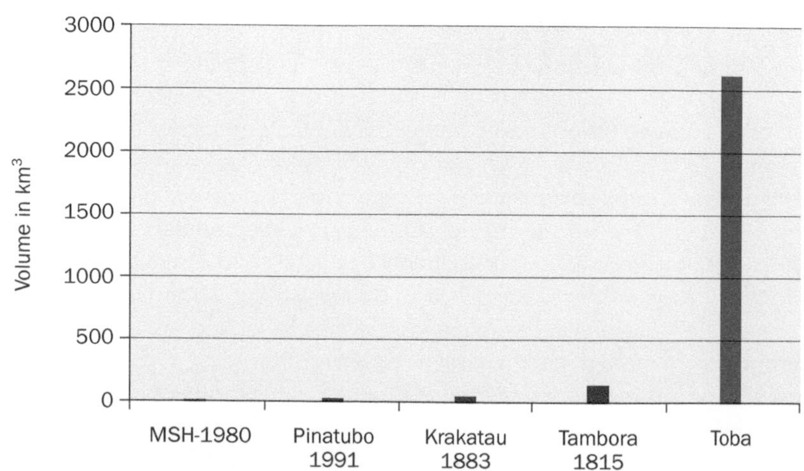

FIGURE 8.28 Comparison of total ejecta volumes (km^3) produced by historical eruptions with volumes produced by the "super-eruption" of Toba volcano [127], 75,000 years ago. Source: Based on Chesner et al. (1991).

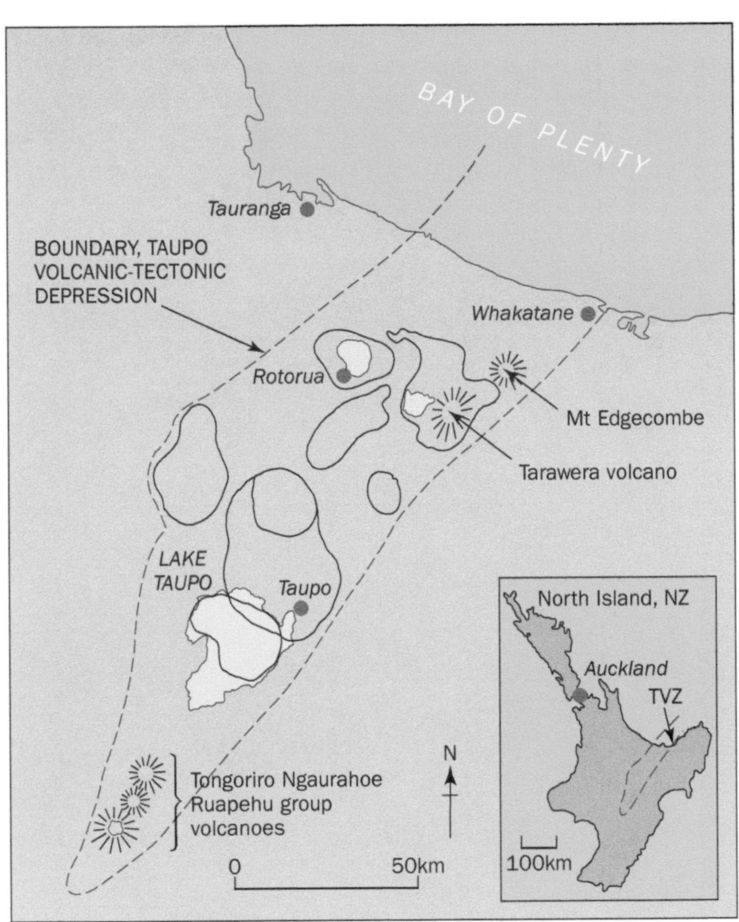

FIGURE 8.29 Taupo Volcanic Zone (TVZ) [180] on the North Island of New Zealand. Borders of major calderas (overlapping in places) indicated by solid lines; major volcanoes shown by hachures circles. Source: Modified from Gravley (2004).

(Wilson et al. 1995). Eight calderas ranging from 10 to 40 km in diameter lie within the zone, each related to past Plinian eruptions (Figure 8.29). The southernmost caldera, occupying 140 km^2, formed 26,500 years ago during eruption of the Oruanui Tuff (Wilson 2001; Wilson et al. 2006). It has since been the location of about two dozen eruptions, mostly small, with the exception of the previously described Taupo [180] eruption of 186 CE. As impressive as the Taupo eruption was, the Oruanui [181] event dwarfed it in comparison, being perhaps 15–20 times larger. The Oruanui eruptive episode was one of the two largest volcanic events in the past 250,000 years, the other being the eruption accompanying the Toba [127], Sumatra eruption 75,000 years ago, which produced about twice as much volcanic material.

In contrast to the Taupo eruption, whose culminating phase probably lasted less than 10 min, the Oruanui episode was protracted, involving 10 independent eruptions, mostly phreatomagmatic, perhaps spanning several years in time. The episode began mildly enough, simply as a sub-Plinian pumice eruption followed by an ash fall. The hiatus that followed was sufficient for soil development on the early deposits. Several additional short-lived outbursts followed by pauses took place, and then the main part of the episode began, perhaps from several simultaneously erupting vents. As it developed, it grew in intensity and magnitude, ultimately incinerating and burying most of the North Island – fortunately long before Polynesian explorers populated New Zealand.

The Oruanui deposits and the proximal portion of the later Taupo ignimbrite (for example, the Hatepe and Rotongaio Ashes) share many characteristics of ordinary wet surge blasts. Waveform facies development, accretionary lapilli, and other evidence of steam–magma involvement occur at several stratigraphic levels. The overall emplacement temperature of the Oruanui Tuff, despite thicknesses approaching 200 m in some sections, was no greater than 150–200°C, suggesting intensive cooling by trapped water, probably by eruption through a deep lake. To cool an ordinary Plinian eruption sheet of this size down to such a temperature would require boiling off as much as 50–100 km^3 of water (Wilson 2001).

The sorting of the Oruanui Tuff differs distinctively from that of drier PDC deposits, which commonly show little distinctive change in sorting with increasing distance from source, apart from localized topographic influences and partitioning of lithic fragments and pumices described earlier. In the Oruanui Tuff, however, overall sorting *decreases* with distance from the source. A similar deterioration of sorting occurs in the deposits of the 1870 Askja [102], Iceland, eruption, which also began at the bottom of a lake. This shows up most strikingly at distal margins, where only coarse fragments are well-sorted while matrices are packed with particles of many shapes and sizes. Such peculiar arrangement is called **coarse-tailing**. Something filtered out the finer particles in an irregular way to create such a dispersal pattern, and that could well have been dense, highly turbulent steam. Booth and Walker (1973) observed that flocculated ash aggregates blown out by steam blasts issuing from a new vent on Mount Etna [112] in 1971 fell to the surface far sooner than they would have in dry air. As a result, ash accumulated in well-sorted beds impacted by ballistic fragments close to the vent. Possibly a similar mechanism operated on a much larger scale during the Oruanui and Askja eruptions, selectively and irregularly removing fine ash, which otherwise would have contributed to more uniform sorting throughout their deposits (Self and Sparks 1978).

Questions for Thought, Study, and Discussion

1. How do we measure the sizes of Plinian eruptions?
2. Describe the dynamic structure of an active eruptive column.
3. What is implied if the prehistoric tephra fall deposit of a particular eruption is skewed to one side or another of a source vent?
4. What explains the fact that many Plinian pumice fall deposits are reversely graded? And what might normal grading in the upper part of a pumice deposit indicate?
5. How and why do pyroclastic flows and surges form?
6. How do pyroclastic flows and surges move?
7. Why are deposits of pyroclastic flows and surges so different?
8. Describe the ways that local topography can affect the movement and deposition of PDCs.
9. What is the evidence for ash elutriation, both during eruptions and within the deposits that these eruptions leave behind?
10. What is ignimbrite "aspect ratio" and why might this be a significant parameter?
11. How do directed blast deposits differ from those of ordinary PDC eruptions?
12. What are super-eruptions, and should we be worried about them? Explain.

FURTHER READING

Freundt, A. (2001). *From Magma to Tephra: Modelling Physical Processes of Explosive Volcanic Eruptions.* North Holland: Elsevier Publishing, 334 p.

Gilbert, J.S. and Sparks, R.S.J. (eds.) (1998). *The Physics of Explosive Volcanic Eruptions.* London, The Geological Society of London Special Publication vol. 145, 186 p.

Miller, C.F. and Wark, D.A. (2008). Supervolcanoes and their explosive supereruptions. *Elements,* 4, 11–15.

Olson, S. (2016). *The Untold Story of Mount St Helens.* New York: W.W. Norton, 336 pp.

Sparks, R.S.J., Bursik, M.I., Carey, S.N. et al. (1997). *Volcanic Plumes.* New York: John Wiley & Sons, 590 p.

PART IV

VOLCANIC LANDFORMS AND SETTINGS

Volcanoes come in all sizes and shapes, not all of which look like the "typical" volcanoes of cartoons and tourist postcards. This part discusses the many landforms created by volcanoes, from towering symmetrical cones, vast lava sheets, craters large and small, to the remote volcanoes beneath the sea and on distant planets. **Chapter 9** describes the **positive** landforms created by volcanoes, including vast lava plains, gigantic lava shields, and the "pointy volcanoes" characterized by the scenic conical structures most people associate with volcanoes. **Chapter 10** describes the perhaps less dramatic **negative** volcanoes, those marked for the most part by large depressions in the landscape – including the all-important, gigantic craters known as **calderas**, sites of Earth's most powerful eruptions. **Chapter 11** describes the processes that can suddenly tear volcanic edifices apart – massive landslides and volcanic mudflows. **Chapter 12** describes the unseen submarine volcanoes of Earth and those of extraterrestrial worlds.

Volcanoes: Global Perspectives, Second Edition. John P. Lockwood, Richard W. Hazlett, and Servando De La Cruz-Reyna.

Companion website: www.wiley.com/go/lockwood/volcanoes2

CHAPTER 9

Constructional (Positive) Volcanic Landforms

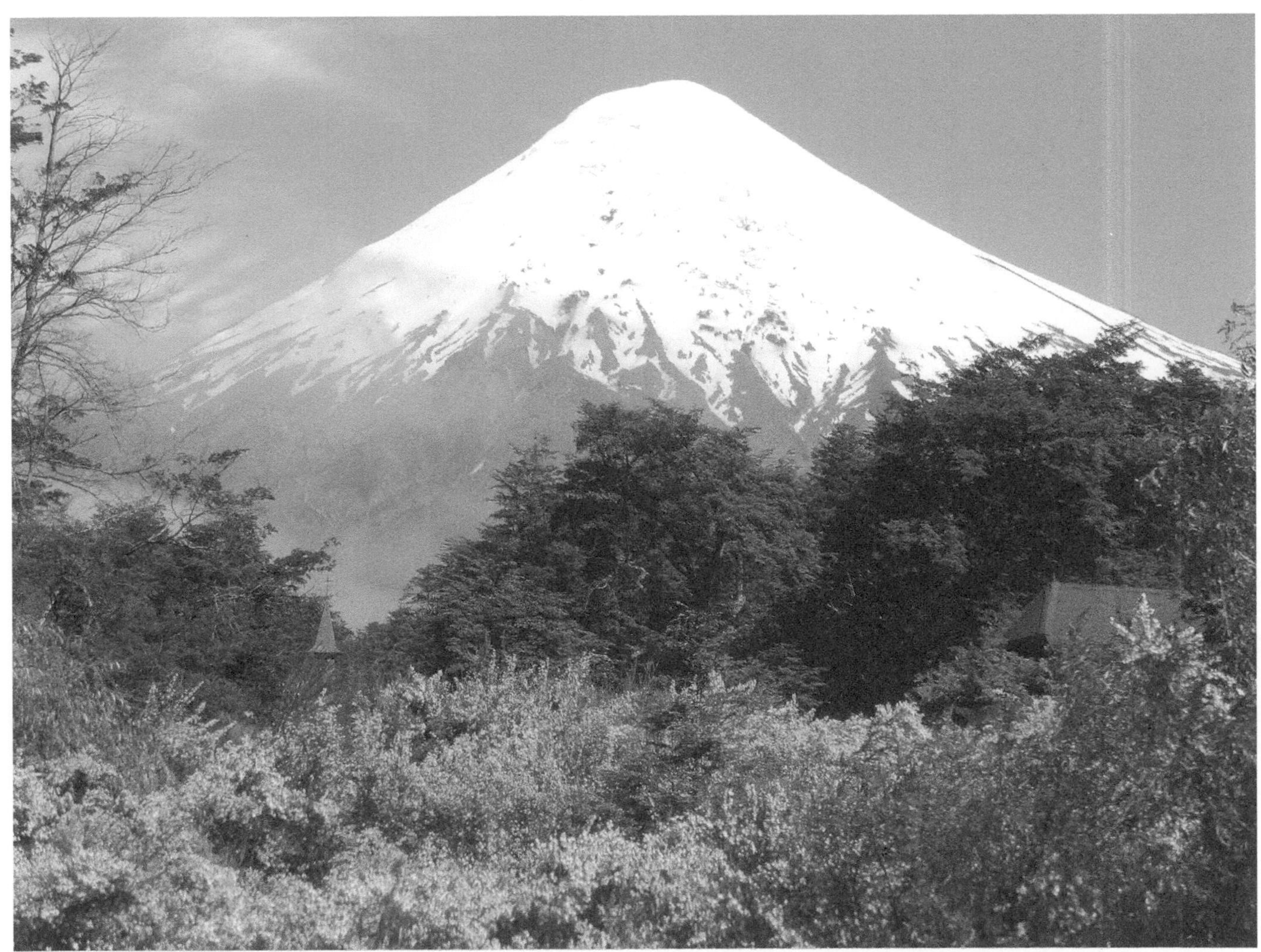

Source: Mary-Ann del Marmol

Volcanoes: Global Perspectives, Second Edition. John P. Lockwood, Richard W. Hazlett, and Servando De La Cruz-Reyna.

Companion website: www.wiley.com/go/lockwood/volcanoes2

It is at once evident that repeated eruptions cannot fail to load the surface of the earth around their source with a mountainous excrescence of a magnitude proportional to the quantity of matter thrown up.

(George Poulett Scrope 1862)

Volcanoes erupt vast quantities of lava and/or ash at Earth's surface, and most construct "positive" (elevated) topographic features. These range from lava plains, plateaus, and shields to majestic composite volcanoes, and to smaller domes and pyroclastic cones. Some volcanoes are largely "holes in the ground," however – we'll discuss those "negative" landforms in Chapter 10. We begin our descriptions of these positive landforms with a look at the most voluminous volcanoes on Earth – the large igneous provinces (LIPs).

Large Igneous Provinces

Large Igneous Provinces (LIPs) are regions in which especially voluminous amounts of volcanic material (more than 100,000 km^3) have erupted in long-lived eruptions that mantle large areas of Earth's surface. They may be sub-classified by the compositions of their dominant products [e.g., Large Basaltic-Rhyolitic Provinces (LBRPs), and Large Andesitic Provinces (LAPs)], or by their tectonic environments, including continental interior, back-arc, rift or plate margin, and oceanic settings (Bryan et al. 2002; Sheth 2007; Ernst 2014). LIPs include the vast flood-basalt tablelands, such as the Columbia River Plain (Chapter 6), caldera-studded ignimbrite plateaus such as the Sierra Madre Occidental in Mexico, and the Great Basin in the western United States. Some 35 LIPs have been erupted in the past 250 million years, and at least 240 others are known from Earth's older geologic record (Ernst et al. 2008; Table 9.1). A general view among earth scientists has been that many, if not all, LIPs represent the focused upwelling of hot mantle, in some instances plainly related to plume hot spot activity (Table 9.1; Chapter 2). The initial impingement of an ascending mantle plume on the underside of the lithosphere may introduce especially large amounts of heat and partial melting to the crust. Intensive, highly explosive silicic volcanism can be superseded by vast outpourings of unusually hot basaltic lava as rifting sets in later on. This appears to be the case, for example, in the Yellowstone–Snake River Plain volcanic region.

Of the LIPs, by far the largest occur on the sea floor. This is no accident. Oceanic lithosphere is thinner than continental, and rifts more easily, permitting the eruption of vast amounts of mantle partial melts that otherwise might underplate or intrude thicker continental crust without erupting. The submarine Ontong Java LIP is the largest flood-basalt province in the world, having an area about the size of western Europe. It rises some 2000 m above the adjacent ocean floor, showing much greater relief than its terrestrial counterparts. In addition to lava flows erupted from deep sea vents, parts of the plateau include deposits of phreatomagmatic pyroclastic material, possibly indicating occurrence of eruptions in shallow water or above sea level as development of the plateau culminated. Most basalt apparently was fed by widespread submarine fissure eruptions (Fitton et al. 2004).

TABLE 9.1 **Some well-preserved, geologically young LIPs. These provinces consist largely if not entirely of flood basalts. note the relationship to hot spots.**

Province	Location	Area (millions of km^2)	Volume (millions of km^3)	Age (millions of years bp)	Associated melting source
Columbia River	Pacific Northwest, USA	0.164	0.175	16.5–14.5	Yellowstone hot spot
Afar	NE Africa, Red Sea area	0.6	n/a	29–31	East African Rift Zone
North Atlantic	Greenland, Iceland, NW Europe continental shelf	1.3	n/a	58–62	Mid-Atlantic Ridge
Deccan	India	1.8	8.6	66	Reunion hot spot?
Madagascar	SW Indian Ocean	1.6	4.4	84–80	Marion hot spot
Caribbean-Colombian	Caribbean Basin	1.1	4.5	90–87	Galapagos hot spot?
Ontong Java	SW Pacific	1.9	4.4	86–94	Louisville hot spot?
Shatsky Rise	NE Pacific	0.5	4.3	115–140	Spreading center
Hess Rise	North Pacific	0.6	5	95–116	Spreading center

Source of estimates: Based on Ernst et al. (2008).

Continental flood basalts

In contrast to the submarine realm, most, if not all continental flood-basalt provinces originate as lavas spread out across broadly subsiding basins. For example, the Columbia River Plain is built of approximately 300 individual lava flows increasing in cumulative thickness toward the center of the province, with the deepest flows now buried as much as 3500 m. This suggests that the underlying continental crust subsided as flows accumulated, or perhaps that the crust was rifting open as eruptions took place. While the base of the volcanic pile sank, the top of the flows formed a low-level plain or basin floor that was frequently resurfaced by freshly erupted lavas.

One of the reasons the word "flood" appears in "flood-basalt province" is because of the way rapid and voluminous effusions of lava cover preexisting topography (Figure 9.1; Chapter 6). The Columbia River lavas buried mountainous landscapes with a relief of over a kilometer. The output of lava is not regular in most flood-basalt provinces, but rather spikes dramatically at one point or another during volcanism. In the Columbia River Plain, eruptions of lava took place over an interval of 11–12 million years. But more than 90 percent of the total volume erupted in only about a million years – that is, less than 10 percent of the province's lifespan. The crescendo in activity came early on – within the first two million years. After that, volcanism in the province gradually tapered off. Six great groups of flows erupted, each representing numerous individual eruptions (Swanson et al. 1979). The flows within each group share similar chemical makeup but have mineralogical differences, suggesting that each had a distinctive mantle source (Tolan et al. 1989; Reidel et al. 2013). Despite the impressive sizes of individual flows and total volume of lava erupted, soils, watercourses, and wildlife re-established themselves in the periods between some eruptions, as indicated by fossiliferous sedimentary deposits interbedded with lava flows in many places. Centuries often passed before another flood of lava covered the landscape.

In most areas of flood basalts, the flows are mainly sheet-type pāhoehoe, though minor amounts of ʻaʻā may also be present. Most flows are less than 10 m thick, but in places, individual flows have thicknesses of many tens of meters – even in excess of 100 m – and are commonly underlain by beds of shattered, glassy volcanic debris called **hyaloclastite** (broken glass rock). These features represent the filling of river canyons with lava, or so-called **intracanyon flows.** Many intracanyon flows show especially well-developed entablature jointing, though entablatures are not exclusive to this environment (Chapter 6).

The relationship of intracanyon flows to their basal hyaloclastite deposits seems quite straightforward. As lava pours into a river channel, sudden quenching by cold water causes the perimeter of the active flow to shed countless flakes of sand and silt-sized

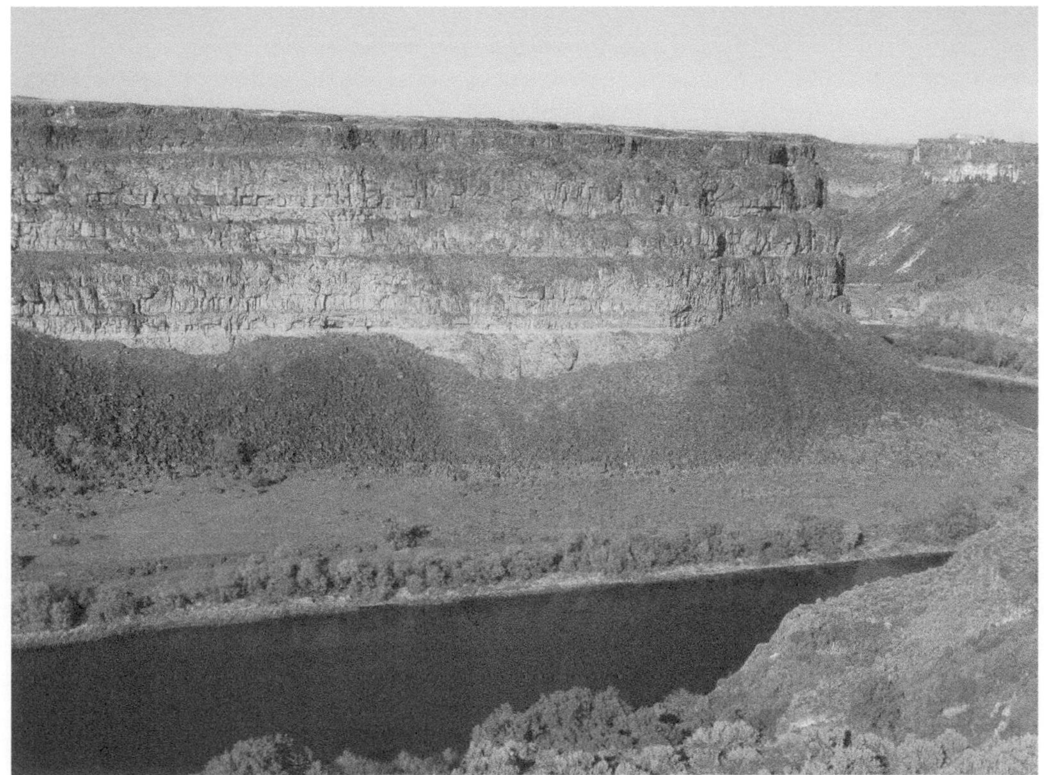

FIGURE 9.1 Snake River flood basalts exposed on north side of Snake River at Twin Falls, Idaho, USA. These thick flows were erupted between 4 and 9.2 million years ago. See Figure 6.43 for a description of flow internal structures. The yellow band at the cliff base are lacustrine sediments that overlie older rhyolite. The flood-basalt flows are overlain by a thin basalt that is less than 100,000 years old that poured over these cliffs (Willsey 2017). Cliff is about 160 m high. Source: Photo by J.P. Lockwood.

grit. This matter piles up at the foot of the lava and then the flow overrides it. Steam explosions may wrench loose larger angular fragments that also mix in with the fine debris. Where the lava enters deep pools, hyaloclastite beds may grow to many meters thickness, and may include detached lava pillows or interfingers of long, contorted pillows extending from the flow. Many hyaloclastite beds become oxidized to a reddish or red-brown color or alter to yellow-brown palagonite as the lava travels across them and boils away pore water.

In some places, intracanyon flows fill dry river gorges, or follow watercourses with very little water. In such cases, little or no hyaloclastite will form, but a bed of rounded stream boulders and cobbles preserved directly beneath the flow indicates the presence of the watercourse. Along some very well-exposed reaches of buried streams, the presence of pools and riffles may be preserved at the flow base as alternating pods of hyaloclastite and stream sediment.

What a Young Flood-Basalt Province Looks Like

The eastern portion of the Snake River Plain is underlain by a relatively small (5×10^4 km^2) flood-basalt province in southern Idaho, in parts so young (a few thousand years) that its original, nearly featureless topography is intact. I.C. Russell (1902, pp. 102–103) described its surface: "On the plain the lava spread out and formed what may be termed a lake of liquid rock. . . the margin of the lake is approximately a contour line. . . No eye can observe that it is not a perfect plain."

This surface is not so flat in detail; it was actually formed by many flows from numerous, widely separated vents (Willsey 2017). Around each vent, the outpouring of highly fluid lava built broad, nearly flat cones or **lava shields** with slopes for the most part less than 1°. These individual shields are typically 30–60 m high and several kilometers across at the base. At the summit of any particular shield a small spatter cone, a row of cones, or a spatter rampart may be found – features easily eroded away from older flood basalts. From the evidence of effusive eruptions seen elsewhere (e.g., in Hawai'i), it seems likely that each shield vent grew along an early formed fissure covered by lava later in the eruption. At Craters of The Moon National Monument, near Arco, Idaho, a well-preserved fissure set and spatter ramparts support this interpretation.

On the Ethiopian and Harar plateaus, flood-basalt flows radiate from some 40 overlapping shield volcanoes, much larger in scale than the modest lava shields of the Snake River Plain. Some edifices rise as much as 4000 m above surrounding lowlands, and the Semain shield volcano alone contains 1×10^4 km^3 of lava. Deep erosion indicates that, even at this scale, dikes whose orientations were controlled by regional tectonic stresses fed the shield-forming eruptions (Mohr and Zanettin 1988). This mode of shield volcano development does not extend throughout the whole of the Ethiopian flood-basalt province, however, perhaps because the shield lavas are alkalic in composition, and were probably somewhat viscous as they flowed away from their vents. Other flood basalts in the region are tholeiitic, and no doubt were more fluid. The most voluminous subaerial flood basalts, also tholeiites, are found in the Deccan Volcanic Province of southwestern India, where more than 500,000 km^2 are covered with basalt flows that were emplaced 64–65 million years ago. The total volume and locations of major vents for these basalts is not known as much of the Province is down-faulted beneath the Indian Ocean, but the volume could be more than a million km^3 (Bondre et al. 2004). Perhaps K.G. Cox (1988, p. 239), a well-known Oxford petrologist, is right: "Every continental flood-basalt province has its own peculiar flavor."

Erosional dissection of the basaltic plains commonly results in the formation of a step-like topography on the sides of the valleys, caused by the variable erosional resistance of different parts of the lava flows. These steps gave rise to an old name for these basalts, which were once widely known as **traps** or **trap rocks**. In the United States, the term is now almost obsolete, but it is widely used elsewhere. Where the beds have been tilted, the treads of the steps may be inclined, or a series of sharp ridges marking the eroded edges of resistant beds, termed **hogbacks** or **cuestas**, may be formed.

Shield Volcanoes

Shield volcanoes are distinctly largerfeatures than lava shields – by one or two orders of magnitude. They include the largest volcanoes on Earth. The biggest are polygenetic and form above long-lasting magma reservoirs. Like lava shields, they are gently sloping structures with broadly rounded cross-sectional profiles (Figure 9.2). The term "shield" derives from a fancied resemblance, in Iceland, to an early Germanic warrior's shield laid concave-side down on the ground. About 15 percent of the subaerial volcanoes on Earth are shield volcanoes (Suzuki 1977).

To illustrate the great variety of shield volcanoes and their fine-scale features, let us explore some examples from around the world.

Shield Volcanoes of Hawai'i

The world's largest and most famous shield volcanoes are those of the Hawaiian Islands, whose eruptions we have already described in Chapters 1 and 6. We shall take a closer look at their construction here.

FIGURE 9.2 Shield shape of Mauna Loa volcano [15], Hawai'i, viewed from the southeast. Only the upper 3 km of this shield is visible in this photo – another kilometer of subaerial shield lies below the clouds, and another 5 km lies beneath sea level. Source: Photo by J.P. Lockwood.

The floor of the central Pacific has a mean depth of about 4200 m. Hawaiian volcanoes originate in this stygian deep. The youngest Hawaiian volcano is a 3000-m-tall seamount, named Kama'ehuakanaloa [17]. It has almost 1000 m to grow before breaching the waves 30 km south of the island of Hawai'i (Malahoff 1987). The slopes of Kama'ehuakanaloa [18] are as much as 10–15° steeper than those of the nearby shields above sea level, where the land is inclined at angles of only 2–12° (Figure 9.2). This is partly due to landslide modification and also because small submarine lava flows, quenched in seawater, do not usually travel as far as subaerial flows. Slopes become more gradual toward the summit. A broadly arching crest marks the top, where a small caldera has formed. From the caldera, two rift zones radiate, one stretching northward about 8 km, the other to the south 25 km.

Volcanic rift zones (not to be mistaken for the "rift valleys" and grabens associated with divergent plate boundaries) are among the outstanding features of Hawaiian volcanoes. They contribute to make the shields more expansive and are marked by dike-fed fissure eruptions, some of which evolve into localized lava lake, cone, and lava shield-building activity. Since we have already discussed their dynamic origin and development (Chapter 4), we'll focus on two particular aspects here related to the overall growth of Hawaiian shields; how rift zones relate to overall shield volcano shapes and the compositions of their lavas.

At their fullest development, there may be as many as three rift zones radiating from the summit of a Hawaiian volcano, with angles of about 120° between them. Usually, however, two of the rift zones are more active than the third. Rift-zone development depends upon the topography and slope of the deep seabed on which the host shield begins growing, and on which directions its flanks consequently spread and split under the influence of gravity (e.g. De Matteo et al. 2018). Where a shield volcano grows from essentially horizontal seafloor, isolated relative to its closest neighbors, flank fissure eruptions still occur, but dikes radiate all points of the compass from the summit – no well-defined rift zones tend to form. Because the basal outline or map plan of each Hawaiian shield is the direct result of its rift-zone orientations and degree of development, shields lacking rift zones tend to have circular basal outlines; while in contrast, volcanoes with two vigorously active rift zones develop highly elongate, blade-like outlines (e.g. Kīlauea [17]).

Magma from older intrusions may lie stored within rift zones for decades, even centuries, gradually evolving through cooling and fractional crystallization into more silicic compositions that can later erupt if a fresh impulse of magma forces it upward (e.g. initial phases of the 2018 Kīlauea eruption – Chapter 1). Lava can drain from summit magma chambers into rift zones triggering summit collapse (e.g. Chapters 1 and 10). Lavas erupted in Mauna Loa's [15] lower southwest rift zone, at elevations a few kilometers lower than the summit, typically include great quantities of olivine phenocrysts that probably derive from the settling and accumulation of crystal grains within lower regions of the volcano's draining magma reservoir. At one shoreline location, Pu'u Mahana, olivine-rich pyroclastic beds have eroded out sufficiently to create one of the world's few-known *green sand beaches.*

Building up to within a few hundred feet of sea level, a young Hawaiian volcano enters a zone of water shallow enough for explosive phreatomagmatic activity to become the dominant eruptive style. Eruptions generate vast amounts of hyaloclastite, which build a platform for an island to grow once summit vents build above sea level (Figure 9.3). Hyaloclastite production

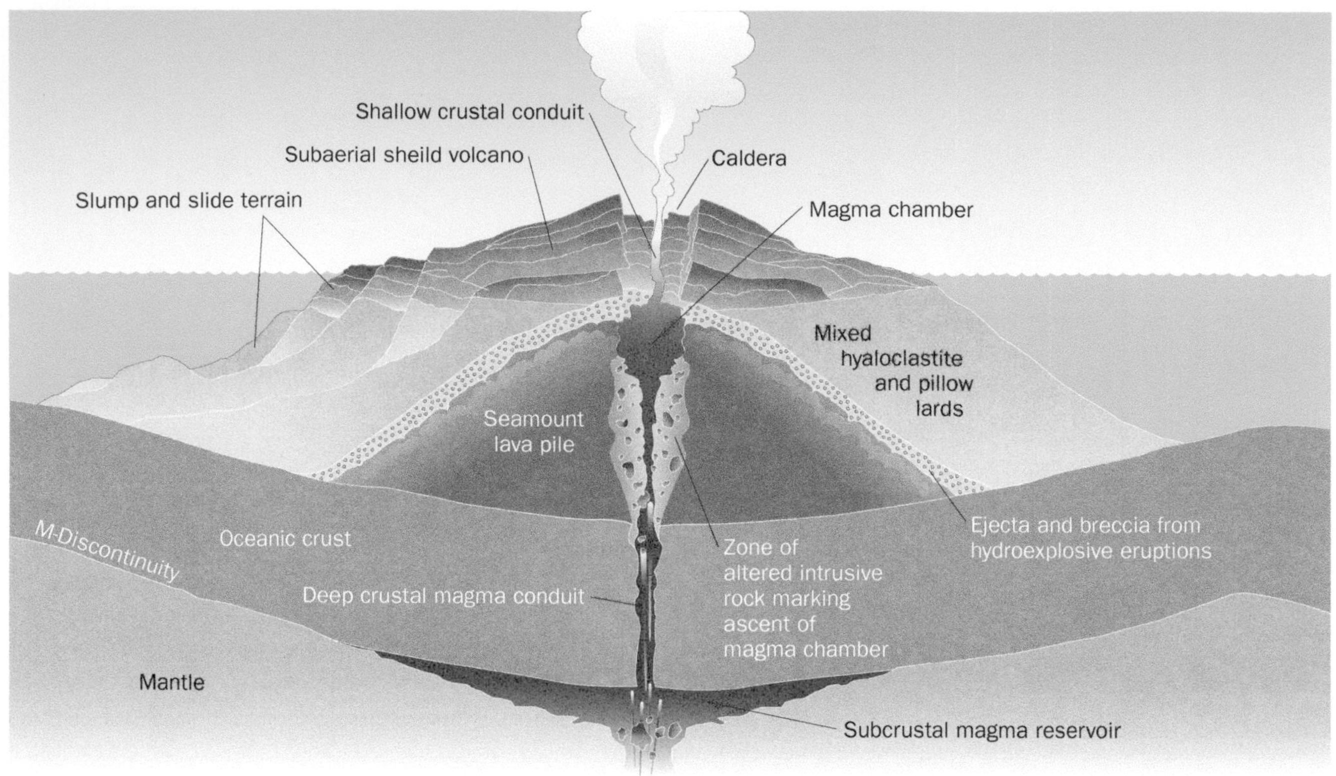

FIGURE 9.3 Schematic cross-section of a typical Hawaiian shield volcano in the main (pre-alkalic) stage of development.

continues all along the coast of the new island, wherever lava flows pour into the ocean, so that, in time, this debris envelops the older seamount formed during the initial stage of shield growth, creating a substantial (not altogether stable) foundation for the subaerial portion of the volcano.

Eventually, Hawaiian shields become so massive that they isostatically depress the underlying oceanic lithosphere to form a moat around the islands as much as 800 m deep – the Hawaiian Deep. At present, the island of Hawai'i is sinking on average 2.5 cm every 10 years, and flooding exacerbated by global sea-level rise (SLR) threatens coastal real estate and infrastructure in low-lying areas. The mean rate of young shield growth is about double this, however, meaning that the younger part of the island is building up faster than its bottom is subsiding (Moore and Thomas 1988). The smallest of the Hawaiian Islands rise nearly 5 km above their bases; and Mauna Loa [15] and Mauna Kea [16] on the island of Hawai'i have absolute heights of about 10 km. Measured from the sunken bases of their volcanic piles, these mountains are even bigger, possibly representing accumulations of lava over 15 km thick! The volume of Mauna Loa alone is at least a hundred times greater than that of typical composite volcanoes such as Mount St Helens [31] or Vesuvius [109]. It is unlikely that mountains much bigger than these can grow on oceanic crust, given the weakness of Earth's lithosphere.

Young Hawaiian shields grow from repeated eruptions perhaps averaging 20–30 per century for many hundreds of thousands of years. Individual eruptions may last for years at a time, and growth rates are quite rapid, geologically speaking. For example, about 95 percent of the 30 × 65 km land area of Kīlauea has been resurfaced by fresh lava within the past 1500 years. To the northwest, Kīlauea's giant neighbor Mauna Loa has had 90 percent of its slopes surfaced with lava flows less than 4000 years old. Within the past 170 years alone, 13 percent of Mauna Loa's landscapes has been buried by fresh lava (Figure 9.4; Lockwood and Lipman 1987; Trusdell and Lockwood 2017). This activity contrasts with the very short lifespans and eruptive brevity of the much smaller Icelandic and Cascadian shield volcanoes.

Eruptions occur more frequently at the summit of a Hawaiian shield volcano than on the flanks, although they tend to be smaller. Eruptions within a summit caldera take place from fissures that cross the caldera floor and often extend up the caldera walls and beyond onto the surrounding slopes. From time to time, long-lasting lava lake activity may also occur within a caldera – the "unroofed," recirculating tops of their underlying magma reservoirs (Chapter 6).

The flattened profile of the upper part of Hawaiian shields is due to a combination of summit caldera collapse and the fact that the most copious discharges of lava are from flank vents. If it were otherwise, the slope would remain constant to the edge of the caldera, or even become steeper because the same number of flows of the same width and thickness would cover a greater proportion of the circumference of the shield near the summit than farther downslope. The summit region would consequently build up faster than the middle flanks.

The smooth profiles of most Hawaiian shields undergo considerable modification, owing to gravitational collapse of flanks in massive slumps and slides, most of which are rooted far underwater. In some instances, these voluminous land failures have head-scarps bordered by rift zones and calderas – natural planes of weakness in the slowly spreading volcano. They are by no means unique to Hawaiian volcanoes and are common on other oceanic shields too (e.g. Delcamp et al. 2012), they are the largest, if not most impressive landslides on Earth (Chapter 11). SONAR mapping reveals that as much as one-third of the landmass of some islands has disappeared into the ocean as volcanic flanks failed, playing a significant role in the development of Hawai'i's spectacular cliff-fringed shorelines. The Nu'uanu slide, which broke from the flank of Ko'olau volcano [11] on the island of O'ahu almost 2 million years ago, had enough momentum to spread 160 km across the seabed, most of the way climbing a gentle slope! The debris field from this single slide is over 35 km in width, and contains as much as 35-km-long and 1.5-km-thick coherent pieces of volcanic shield – as large as many stand-alone seamounts. This catastrophe must have created a gigantic tsunami all around the Pacific Rim. A smaller slide from the flank of Mauna Loa about 105,000 years ago generated tsunami that washed up over 600 m on the coast of Lana'i, an island 150 km to the northwest.

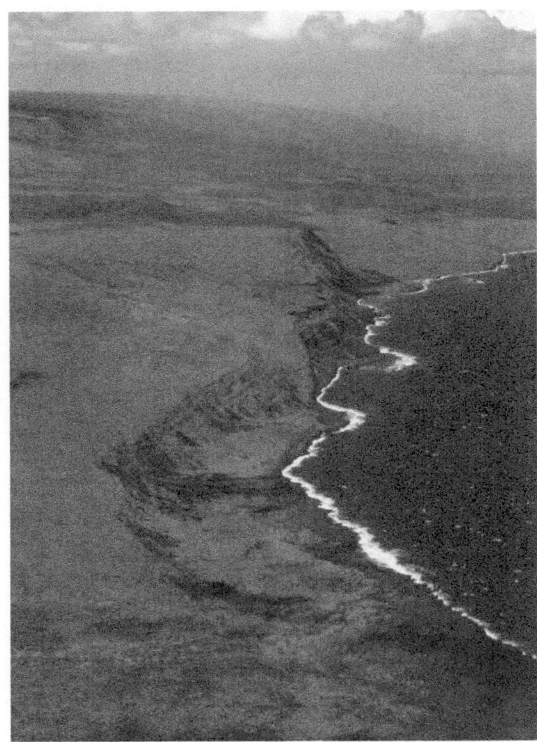

FIGURE 9.4 View east along the southeast coast of Kīlauea volcano [17], Hawai'i, showing the stepped faulting that has developed along this unsupported flank of the Kīlauea shield. Such faulting is common wherever oceanic island volcanoes border deep oceans. Source: USGS photo by J.G.Griggs.

Some Hawaiian volcanoes do not collapse all at once, as happened during the Nu'uanu slide. Instead, their flanks gradually slump into the ocean. This sort of piecemeal flank failure is presently active across most of the southern flank of Kīlauea (Figure 9.4). It is not clear whether slumps such as these are precursory to catastrophic landslides. But some geologists believe that the great mass-wasting events on Hawaiian volcanoes indeed, on most oceanic shields – take place *during the time of volcanic primacy*, when poorly supported slopes are under the greatest strain from intrusions, and subject to intense loading from the accumulation of frequently erupted lava flows. Lowering of sea level during times of continental glaciations may also be an important factor. Fortunately, these voluminous collapse events are infrequent; in Hawai'i, they average about one every 300,000–400,000 years. In any case, shield volcanic islands are in constant existential battle with the sea. Slumping by flank collapses, coastal erosion, and gravitational subsidence into the underlying oceanic lithosphere reduce their sizes. Concurrently, ongoing eruptions work to increase their sizes with the dilation of rift zones and the growth of new lava deltas at the coast (Figure 9.5). The example of Hawai'i shows that the sea always wins, though it may take millions of years for an extinct volcanic island finally to disappear.

Icelandic Shield Volcanoes

In Iceland, where geologists first studied shield volcanoes, the volcanic shields are all quite small relative to those in Hawai'i, ranging in basal diameter from only a few hundred meters across to about 15 km and in height up to 1000 m above their surroundings. Each is built by effusions of thin, fluid basalt pāhoehoe flows from a pipe vent or a very short segment of a fissure, and consequently each is nearly circular in map plan. In the crater atop most shields, lava accumulates as a molten pond from which overflows build up gentle surrounding slopes. Some flows extend well beyond the bases of the shields, a record being a lava flow from the Trolladyngja shield that is 100 km long (Williams and McBirney 1979). Few of these flows exceed a meter in thickness because the lava is extremely hot and fluid as it erupts.

Perhaps the best known of Icelandic shield volcanoes is Skjaldbreiður [91] near Reykjavik – 500 m high and 10 km in diameter, with slopes averaging 7–8°. Other shields nearby have slopes of only 2–3°, similar in profile to the very low lava shields around the vents of some flood eruptions, as in the Snake River Plain. Most of the Icelandic shields formed quickly during single eruptions. It is probable that even, Skjaldbreiður, grew up within a period of less than 10 years (Cas and Wright 1988). This is implied by the rapid growth of Surtsey [90], a new volcanic island that formed off the southern coast of Iceland in 1963–1967. After the vent of Surtsey sealed itself off from contact with seawater, fluid flows poured out, building up a low shield on the new island. In all, about two cubic kilometers of lava issued, comparable in volume to smaller shield volcanoes on the Icelandic mainland. Judging by the numerous extinct, eroding volcanoes that form rocky islands in the vicinity of Surtsey, it seems unlikely this young shield will ever return to life. In the past 11,000 years, about two dozen shields have formed in Iceland.

FIGURE 9.5 View of a lava delta forming on the southwest coast of La Palma, Canary Islands on 04 October, 2021. This delta from the eruption of Cumbre Vieja volcano [97] eventually extended about 700 m into the sea, adding over 75 ha^2 to the island area. Source: Instituto Volcanologico de Canarias (INVOLCAN).

Shields in Volcanic Arcs

Composite cones, domes, and cinder cones are the predominant types of volcanic edifices in volcanic arc settings, but shield volcanoes can be found as well. As you might expect, these shields have significantly different character and composition from those in Hawai‘i and Iceland.

Ambrym [174] is the largest active volcano in the Vanuatu (New Hebrides) island arc in the southwestern Pacific. It forms an island 30 × 50 km long, rising 1270 m above sea level. Atop the volcano is an 11-km-wide caldera rimmed with inner slopes in places as much as 400 m tall. Ambrym typically erupts once every 5–10 years, with lava lakes episodically active in two of its caldera pits, Benbow and Marum. Overflows from the lakes have covered much of the caldera floor with fresh lava. Ambrym has an unusual history. The lower portion of the edifice is a broad shield volcano built up of basaltic pāhoehoe flows; its slopes measure only 2–3°. About 2,000 years ago, a sub-Plinian eruption broke out at the summit, triggered by the intersection of magma with a shallow aquifer. About 60–80 km^3 of ash, lapilli, and other ejecta erupted, forming a thick cap of pyroclastic debris, ranging in composition from dacite at the bottom to basalt at the top. The outer slopes of the explosive deposits rimming the summit caldera are inclined at 10–20°. The caldera itself formed during the eruption of this ejecta, but its mode of formation was not primarily collapse, as is the case with calderas on most volcanoes. Its walls are not vertical, but are made up almost entirely of pyroclastic material lying at the angle of repose, suggesting that powerful explosions played the most important role in shaping the basin. The structure of Ambrym's summit resembles that of a gigantic tuff cone (Robin et al. 1993).

Some shield volcanoes occur adjacent to volcanic arcs, in slight back-arc positions. Mayor Island volcano [182], near the northeastern coast of New Zealand is one example. Unlike most shields, Mayor Island is built almost entirely of pantellerite (highly fluid, alkali-rich rhyolite), an uncommon lava type that characteristically forms in back-arc or intraplate tectonic settings. Eruptions of normal calc-alkaline rhyolite tend to be much more voluminous than their peralkaline counterparts, and typically erupt as ignimbrites, domes, or thick dome-fed flows. The 15-km-wide shield of Mayor Island began to form by the eruption of lava flows

FIGURE 9.6 Thick shield-building pantellerite flows on the northwest coast of Mayor Island, New Zealand [182] (cliff is 110 m high). Flows are separated by soil zones that indicate substantial time elapsed between eruptions. Their bases are marked by quenched obsidian zones that were prized as weapon materials by Maori warriors. Source: USGS photo by J.P. Lockwood.

(Figure 9.6). Emerging above the sea, these flows issued in Hawaiian-style fissure eruptions and from more explosive cinder cones. Then, about 36,000 years ago, the summit of the volcano began collapsing piecemeal, developing a caldera. The latest major subsidence occurred about 6000 years ago. Eruptions showed especially great variety throughout the period of collapse. On typical volcanoes, one might expect these differences in eruptive style to be related to changes in magma chemistry, or to changing volumes of erupted material. This is not the case on Mayor Island volcano – chemical variations are minimal, and no correlation can be drawn between eruption volume and eruption type. Instead, the manner in which molten rock rose and degassed inside the volcano appears to have varied greatly, for unknown reasons, over a geologically short interval of time (Houghton et al. 1992).

Two shield volcanoes much larger than Mayor Island (Newberry volcano [41] and the Medicine Lake Highland [39]) also lie along the back-arc edge of the High Cascades in northern California and Oregon (Chapter 2). Though these volcanoes do not lie directly within the arc, they erupt essentially the same lavas as the main calc-alkaline arc volcanoes, including basaltic andesite and andesite. Structurally the shields lie within a region of extensional faulting called the Basin and Range, which covers much of the interior western United States. Basin and Range faults guided the shallow intrusion and eruption of magma at both shields.

Medicine Lake volcano [39] began growing about a million years ago, atop a flood-basalt plain called the Modoc Plateau (Figure 9.7). The shield grew to a diameter of 25–30 km, and a volume of about 500 km^3. Eruptions were dominantly Hawaiian with some Strombolian activity. Vents opened in widely scattered positions along roughly north-south trends parallel to underlying bedrock faults. A 6 × 10 km caldera had formed by 10,000 years ago. Around the caldera perimeter, at least 10 rhyolite obsidian domes extruded (Donnely-Nolan 1988). The latest eruption took place only a thousand years ago to form the largest dome, Glass Mountain, showing that the volcano remains potentially active. A complex multi-chambered magma system underlies the Highland, fed by melts from both mantle and crustal sources (Chapter 4).

Much smaller shields are more typical of the Cascades, and these, like the shields in Iceland, are the result of single eruptions occurring from single-point sources (Chapter 2). On most shields, a prominent, steep-sided cinder cone marks the crest, giving them a shape that is distinctive even from a great distance. Good examples are Table and Badger Mountains on the northern boundary of Lassen Volcanic National Park [40], Goosenest not far from Medicine Lake volcano [39], and the Whaleback near Mount Shasta [30]. The largest shields are no more than 600–1000 m high and 10–15 km across at the base, comparable in scale to Skjaldbreiður [91]. But unlike the Iceland shields, which are made up of fluid basaltic flows, these small Cascades volcanoes are composed of more

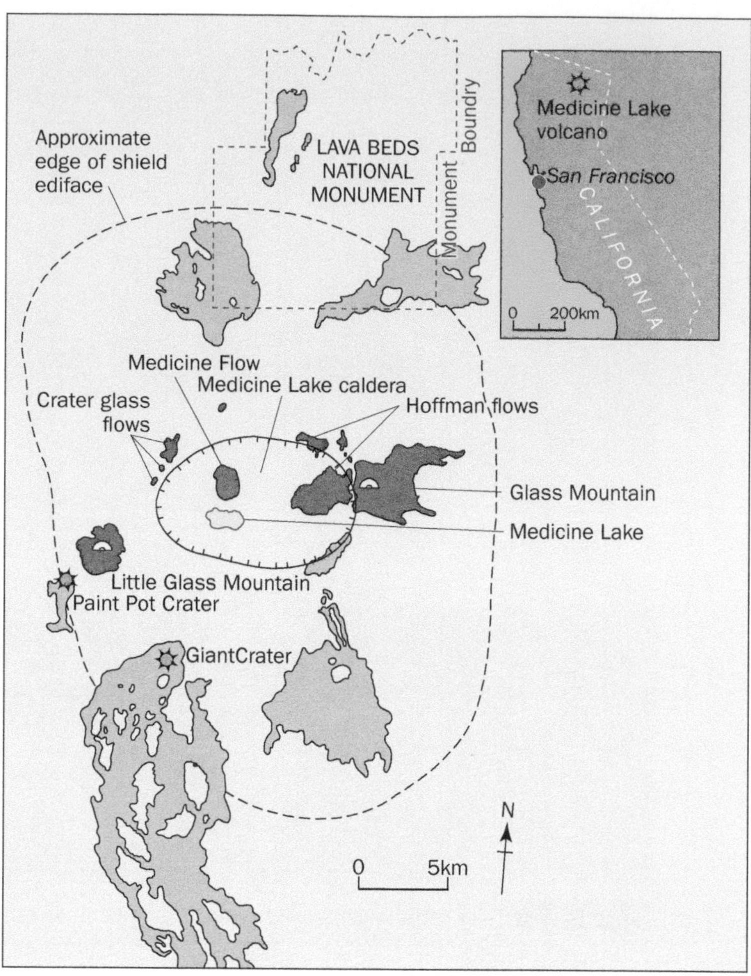

FIGURE 9.7 Medicine Lake shield volcano [39], California. light brown young basaltic and andesitic lava flows; red-brown young glassy rhyolite and dacite lava flows. Remainder of the area is older forest and grassland. Source: Modified from Donnelly-Nolan (1988).

viscous basaltic andesite and andesite lava spread around a pyroclastic core. Within any given shield, the blocky ʻaʻā flows are strikingly homogeneous, though the compositions of adjacent shields may be distinctly different. Lack of significant chemical variation within shields suggests they each must have grown very rapidly, possibly during single eruptions lasting no more than a few years.

Intraplate Shield Volcanoes in Eastern Australia and New Zealand

Though lacking any historical eruptions, one of the world's largest and still potentially active volcanic regions extends 4200 km along the Great Dividing Range, the eastern edge of Australia (Figure 9.8). In places, the volcanic zone is 550 km wide. The nearest active plate boundaries now lie 2000–2800 km away. Much of the volcanic activity here began 80 million years ago, around the time continental rifting led to the opening of the Tasman Sea and the separation of Lord Howe Rise from Australia, and later with the opening of the Coral Sea. Although rifting has ended, volcanic activity has continued, with an eruption at Mount Gambier [155] occurring only 4600 years ago (Johnson and Taylor 1989; van Otterloo et al. 2013).

Large, polygenetic volcanoes, including many shields similar to the Medicine Lake Highland described above, occur in a chain along the Range. Numerous lava fields fed by small, monogenetic lava shields and cinder cones also occur along this line. The ages of volcanism become younger toward the south. This suggests that that the chain formed as Australia drifted northward over a melting source presently positioned in Bass Strait not far from Tasmania (Duncan and McDougall 1989), but slow relaxation of the lithosphere in response to Paleogene rifting has also been suggested to explain this volcanism (Wellman 1989).

The monogenetic lava shields of Australia resemble those found in Iceland and the Snake River Plain of Idaho. They are gently sloping central vent shields made up of piled up thin flows. They are extremely abundant; of the 400 volcanoes found in the Newer Volcanics field of South Australia, fully one-half are shields or related lava cones (Joyce 1975). The largest probably did not exceed 16 km in diameter prior to erosion. In contrast, the bigger polygenetic shields of Australia have extremely complex magmatic histories, reflecting the fact that their source melts derive from continental crust as well as the mantle.

The Australian hot spot shields resemble shield volcanoes found in the northern part of the Kenya Rift Valley (Middlemost 1985). Webb and Weaver (1975) refer to these as **Turkana-type shields**. The African shields are generally trachytic in composition, with slopes of only about 5°. Basal diameters may reach 40 km, with smooth slopes surmounted by a rugged summit region of plugs and cones arrayed around a caldera. A similar set of alkaline shield volcanoes, the largest about 1800 km^3 in volume, marks the

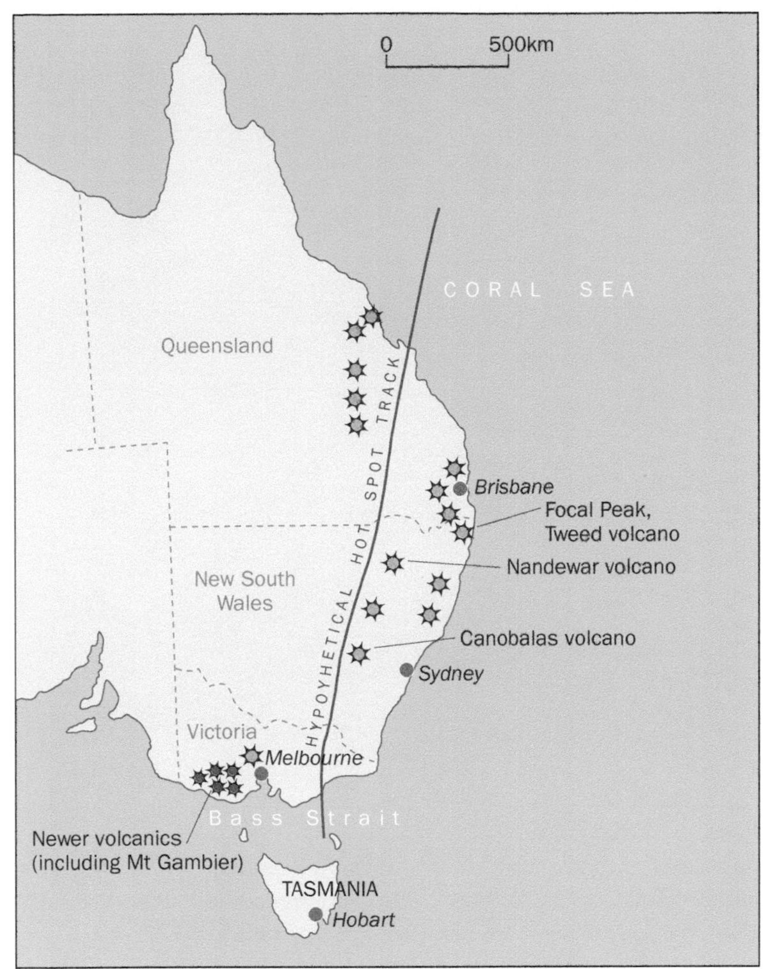

FIGURE 9.8 Locations of eastern Australian shield volcanoes mentioned in the text. Large volcanic edifices are shown by starred circles. Volcanic ages generally decrease to the south with the most recent eruptions occurring about 4600 years ago in the Newer Volcanics [155] region of Victoria. Source: Modified from Johnson (1980).

continental rift system of Marie Byrd Land, West Antarctica (LeMasurier 2013). Their Australian counterparts are comparable in size, composition, and structure, but most are less than 1000 km^3 in volume.

Rangitoto [177], the best-known intraplate shield volcano in New Zealand, forms a small island in the harbor of the country's largest city, Auckland. The volcano began erupting around 1200 CE (Brothers and Golson 1959), heralded by phreatomagmatic explosions as it built up to sea level. A small fragmental cone, not unlike those found on the monogenetic Cascades shields, grew at the summit as activity drew to a close.

Farther south in New Zealand, a group of large intraplate shield volcanoes form peninsulas along the east coast of the South Island. One of the largest is the Dunedin shield, formed 10–13 million years ago. This volcano stretches 25 km across at the base and rises nearly 800 m. It has a polygenetic history, resembling that of the Australian shields. Erupted lavas include basalt, trachyte, and phonolite. Not only flows but extensive pyroclastic deposits, including base surge beds make up the flanks. Trachyte domes roughened the smooth profile of Dunedin's shield prior to erosion (Cas 1989).

Shield Volcanoes of the Galapagos Islands

The basaltic shield volcanoes of the Galapagos Islands, also in an intraplate setting, differ in both general form and structure from those of the Hawaiian Islands to which they might be expected to show close resemblance (McBirney and Williams 1969). In map plan, they are generally circular to somewhat elongate, ranging from 40 to over 70 km across at sea level, and rising as high as 3750 m above the ocean. The volcanoes are positioned at the intersections of tectonic fracture or fault systems on the sea floor (Figure 9.9). In general, the Galapagos lavas are more alkaline than those of the Hawaiian Islands, even during the stage of vigorous shield growth. Tholeiitic and alkali basalt flows are intermixed, reflecting fundamental differences in the degree of melting or composition of the mantle beneath the two regions. Vicenzi et al. (1990) believe that the lavas develop by mixture of magma derived in part from the Galapagos hot spot and from the nearby Galapagos Rise, which drifted across the hot spot 8 million years ago.

Unlike the Hawaiian shields, the Galapagos volcanoes lack distinct rift zones, although radial fissure systems do occur, with linear arrays of cinder cones indicating sites of small Strombolian eruptions on lower flanks. In profile, these mountains are much steeper than young Hawaiian volcanoes, having middle slopes averaging about 25°. Otherwise, there are superficial similarities; lower slopes flatten out to meet the coast, and summit areas are broad and flat-topped (Simkin 1972). The calderas are much larger than those in

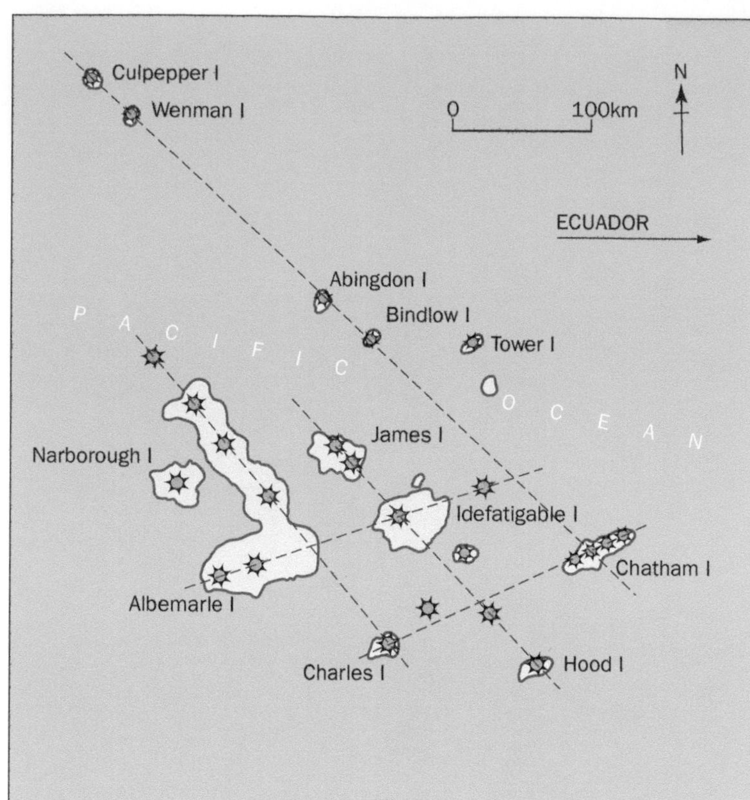

FIGURE 9.9 The Galapagos Islands. Starred circles mark volcanic summits (some of which are submerged). Dashed lines show alignments of volcanoes along some sort of crustal fracture or fault grid. Source: Modified from Ollier (1969).

Hawai'i. They are distinctively enclosed with concentric fracture systems, many of which have served as conduits for erupting lava. Spatter cones and ramparts mark sites of eruptions along most long fractures. Flows from these vents spread both down the steep outside flanks of the volcanoes and into the calderas themselves, partially filling them between repeated episodes of collapse. Williams and McBirney (1979) speculate that the volcanoes began growth with eruptions occurring primarily at centralized summit vents. Later, with the formation of large calderas, the main site of effusive activity shifted to the circumferential ring fractures. In contrast to Hawaiian shields, the proportion of intrusions to effusive lavas may be considerably greater in typical Galapagos volcanoes.

Another reason the Galapagos shields differ so strikingly from their Hawaiian cousins may be that these volcanoes are more widely spaced, proportionate to their volumes. Gravitational instabilities are less likely to arise from the growth of one volcano up against the flank of another. There is no evidence of detachment faults underlying the flanks of the volcanoes, as at Kīlauea [17]. Because a typical Galapagos volcano is structurally stable, it may be that magma reservoirs can more easily develop concentric dikes dipping in toward centralized summit chambers rather than swarms of dikes underlying rift zones. Circumferential vents enclosing Hawaiian calderas are rare, but do exist.

Similar to Hawaiian volcanoes, the Galapagos shields experience a change in the frequency and character of eruptions as they age. Outbursts become more explosive and less numerous in time, gradually building a steep cap of pyroclastic debris and short flows, which eventually inundate calderas and mantle upper slopes. The surfaces of some shields, such as Floreana (Charles Island) [62], are dominantly composed of ejecta (Bow and Geist 1992).

Shield Volcanoes on La Reunion Island

The 50 × 65-km-wide Island of Reunion, in the Indian Ocean 1100 km from the east coast of Africa, represents still another variation on the theme of shield volcanism. It is the youngest of the Mascarene Islands, which rise atop a hot spot melting source less active but almost as old as the one beneath Hawai'i (Bonneville et al. 1988). The island grew from the coalescing of two shield volcanoes (Figure 9.10). The dormant shield of Piton des Neiges [118] forms the northwestern half of the island, rising over 3000 m. The history of this volcano may have begun about five million years ago with initial eruptions on the deep-sea floor, but the subaerial portion exposes rock no more than 2.1 million years in age. Radial fissure vents as well as summit fissure eruptions were numerous throughout the vigorous shield-building stage of activity, which ended about 430,000 years ago. No well-developed rift zones formed. As in the Galapagos Islands and Hawai'i, a brief spasm of late-stage volcanic activity took place, creating a small cluster of alkali basalt and trachyte summit structures 350,000–12,000 years ago (Benard and Krafft 1986; Deniel et al. 1992). Piecemeal caldera collapse occurred throughout this interval of time.

The smaller southeastern neighbor of Piton des Neiges is the 2650 m tall Piton de la Fournaise volcano [125]. Lying directly over the Reunion hot spot, this is the most active volcano in the Indian Ocean basin, and one of the most active in the world. Over 150 eruptions, 15 since the first edition of this book was published 12 years ago!. There are no well-formed rift zones, but radial vent

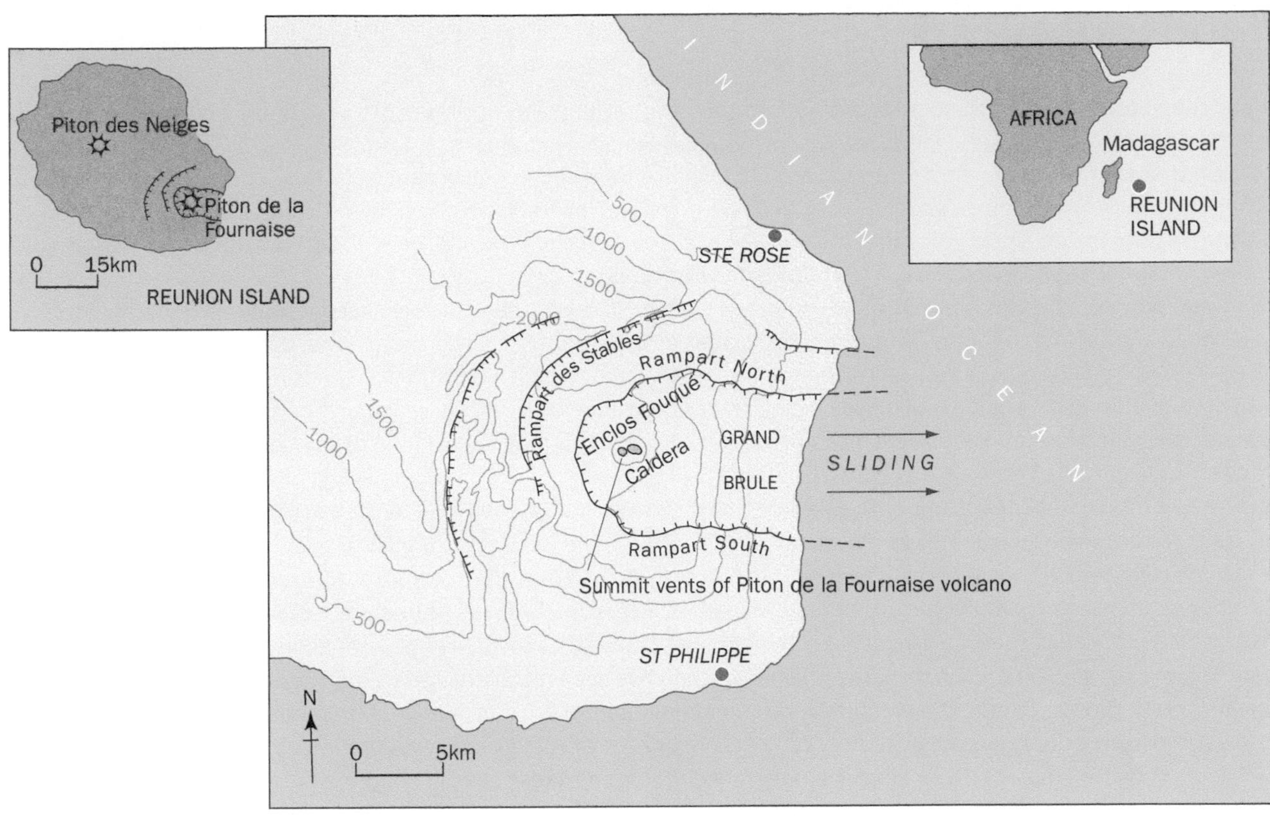

FIGURE 9.10 Map of Piton de la Fournaise volcano [125], Reunion Island, southwest Indian Ocean. Major faults, with teeth toward downthrown blocks, enclose complex area of collapse in the volcano's eastern flank. Like Stromboli [113], this collapse includes the summit vent area. Source: Modified from Duffield et al. (1982).

eruptions are frequent, with some breaking out below sea level. Lavas are extremely fluid, producing spatter cones and ramparts and a few cinder cones at vents.

La Fournaise is actually a relatively thin, younger shield built atop the mass of an older volcano called Les Alizés. The focal point of Fournaise volcanic activity has migrated 5–6 km eastward across this platform over time. Unlike in Hawai'i, the lithosphere beneath La Fournaise does not appear to have subsided significantly (Lénat et al. 2013). Nevertheless, enormous Hawaiian-scale slumping is taking place in the volcano's steep, seaward flank, facing the southeast. Multiple east-facing arcuate scarps have developed indicating a long history of episodic land failure across the mountain, extending as far as 70 km downslope and 4400 m below sea level (Duffield et al. 1982). Enclos Fouqué, the still active scarp that closely encompasses the summit vent area of La Fournaise on three sides, may have become active only 5000 years ago (Zlotnicki et al. 1990). Two huge normal faults border the seaward extension of the Enclos, called the Grand Brule (Figure 9.10). Effusive flank eruptions are concentrated in this strip of sinking land (Bachèlery et al. 1983). Perhaps eastward collapse of the La Fournaise shield relates to the geologically recent eastward migration of its activity, illustrating an important shallow-crustal connection between gravity, landform, and magma intrusion.

The Enclos Foque and Grande Brule are excellent examples of sector collapse; the sliding away of large, wedge-like slices of active volcanoes, including (or nearly including) the summit. Sector collapses are not unique to shield volcanoes. They are even more common on composite volcanoes (see the next section), and will be discussed further in Chapter 11.

Summary

Shield volcanoes form in a variety of geological settings, including intraplate, mid-ocean ridges and continental volcanic arcs. Shield volcanoes share the common aspect of having gentle slopes made up almost entirely of lava flows erupted from a centralized vent area overlying a magma chamber or conduit. They range in composition from low-viscosity alkali-rich rhyolites and trachytes to basalt, though most are basaltic. Flows tend to be fluid enough to spread over considerable areas.

Most shield volcanoes are moderate-sized structures created during single eruptions. However, the biggest mountains on Earth are also shield volcanoes that develop at mid-ocean hot spots, where they form groups of islands. Large shield volcanoes commonly have summit calderas and radial fissure vents extending down their flanks. Cinder or spatter cones, lava shields, domes, and coulees may grow up around some vents. A large, gravitationally unstable shield volcano typically has rift zones, and may experience gigantic landslides. On other more stable shields, there is a tendency for circumferential vent systems to develop around summit calderas instead, and rift zones do not form. These shields tend to be steeper-sided and more circular in shape than the elongate rift-zone shields.

Composite Volcanoes

These are the volcanoes most people regard as "typical," and include the world's most beautifully photogenic, snow-capped cones – the ones like Mount Fuji (Figure 9.11) [146], Kliuchevskoi [171] in Russia (Figure 9.12), and Volcán Osorno [75] in Chile (Figure 9.13) that are featured on postcards and travel brochures from circum-Pacific areas. They are the great scenic volcanoes of Earth (the "gray volcanoes"), made up of interbedded lava flows and layers of ash and cinder (Figure 9.14a). They are also referred to as **stratovolcanoes**, but shield volcanoes also consist of strata (beds) of lava, and we prefer to call those steep-sided volcanoes that consist of both lava and extensive ejecta **composite volcanoes**.

Large composite volcanoes typically have complex eruptive histories, and their rocks record multiple periods of *construction* by Strombolian and Vulcanian summit eruptions, associated with outpourings of short lava flows, dome construction, and *destruction* by mass wasting and sector collapse, often accompanied by major Plinian eruptions. Shapes may be further modified significantly by erosion during long periods of repose. Many of these volcanoes have highly irregular, lumpy profiles that hardly resemble the popular image of "classic" composite volcanoes, and because of this may not even be recognized as potential sources of peril until they erupt (Figure 9.14b).

An example of a composite volcano with a threatening history is Popocatépetl [56], located in a densely populated region in the central Trans-Mexican Volcanic Belt 65 km southeast of Mexico City, with well over 20 million people living nearby. Rising 5,454 m above sea level (Figure 9.15) the volcano is topped by a currently active 600 × 800 m elliptical summit crater. Like all composite volcanoes, Popocatépetl has a complex history, and the present edifice is not the first to form here. A major eruption about 23,000 years ago destroyed the previous cone, causing a massive debris avalanche that covered an area of nearly 1200 km^2 to the southwest of the volcano (Siebe et al. 2017). Since then, the present volcano has been reconstructed by many eruptions. At least seven of these have been major explosive eruptions ejecting several cubic kilometers of ash and pumice. Archaeological remains indicate that the last three of those eruptions, around 3000 BCE, 200 BCE, and 800 CE affected human settlements (Siebe and Macías 2006). In the past 500 years of written records, 16 small and 3 moderate eruptive episodes have been reported – some involving decades-lasting episodes of successive lava dome growth-and-destruction cycles. Popocatépetl began similar activity in 1996 that continues to the

FIGURE 9.11 Mt Fuji [146], Japan (3776 m), viewed from the south. The 1707 eruption ejected huge volumes of ash from the southeast flank Hoei crater (visible just above snow line), and the volcano poses significant risks to central Japan (Yamamoto and Nakada 2015). Source: Photo courtesy of Numazu City. NORIKAZU SATOMI/123 RF.

FIGURE 9.12 Klyuchevskoi volcano [171] in typical eruption, 1987 – view from west. Klyuchevskoi, in central Kamchatka, towers to over 4800 m elevation, and is the largest active volcano in Eurasia. Source: Photo by Aleksei Ozerov.

FIGURE 9.13 Osorno [75], a 2650-m-tall composite volcano in southern Chile, here viewed from south, has not erupted since 1869, but historically erupted frequently before then. Its 1835 eruption was witnessed by Charles Darwin from HMS *Beagle* on his second Pacific voyage. Osorno rivals Mount Fuji in its symmetrical beauty! Source: Photo by J.P. Lockwood.

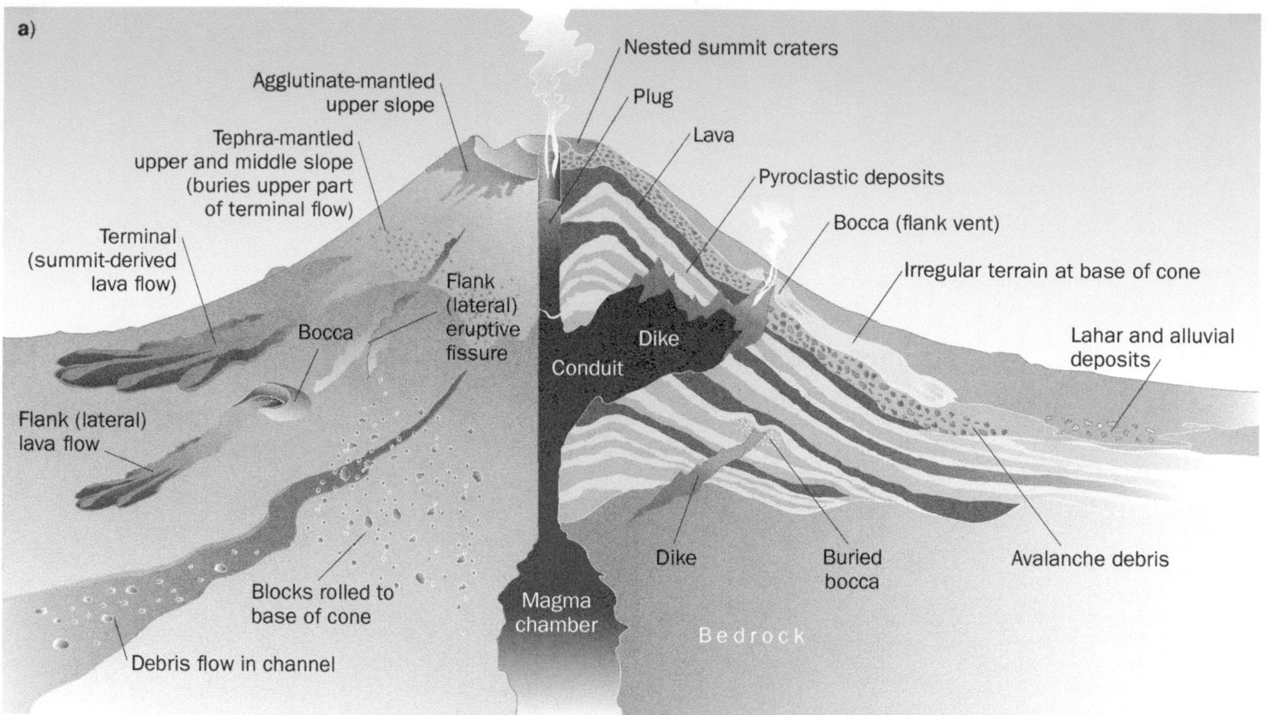

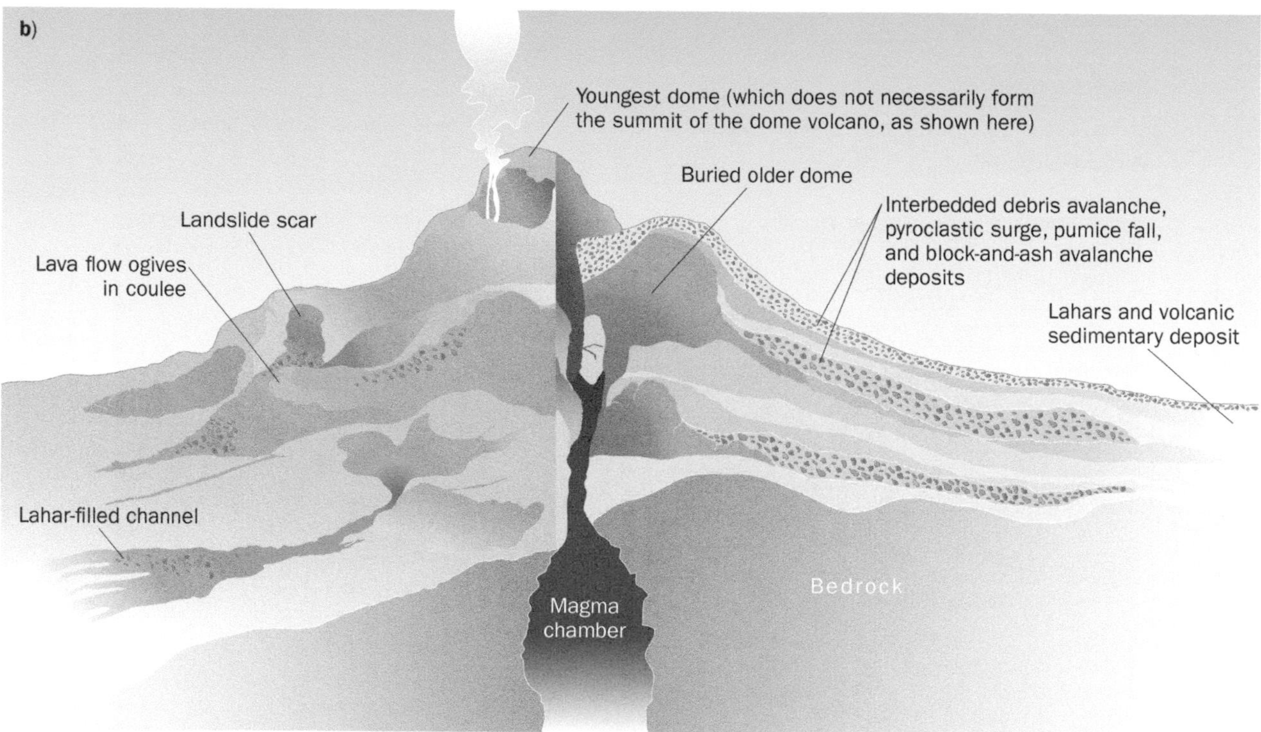

FIGURE 9.14 (a) Anatomy of a simple composite volcano. (This example does not include siliceous domes.) Many composite volcanoes show aspects of both this structure, and that shown in b). (b) A composite volcano made up of high-silica domes, pyroclastics, and related debris is a structurally very complex mountain, lacking graceful symmetry.

present day. Ogburn et al. (2015) report that globally, large explosive eruptions have occurred during or after lava dome eruptions in about 20% of the cases. Considering that the last Plinian phase of Popocatépetl took place around 1100 BCE, one may wonder how long the volcano will continue in the current fashion of lava dome emplacements destroyed by moderate explosions, or if this activity foretells a more dangerous future. Current activity to educate threatened populations about the volcanic risk is most critical (De la Cruz-Reyna and Tilling 2008).

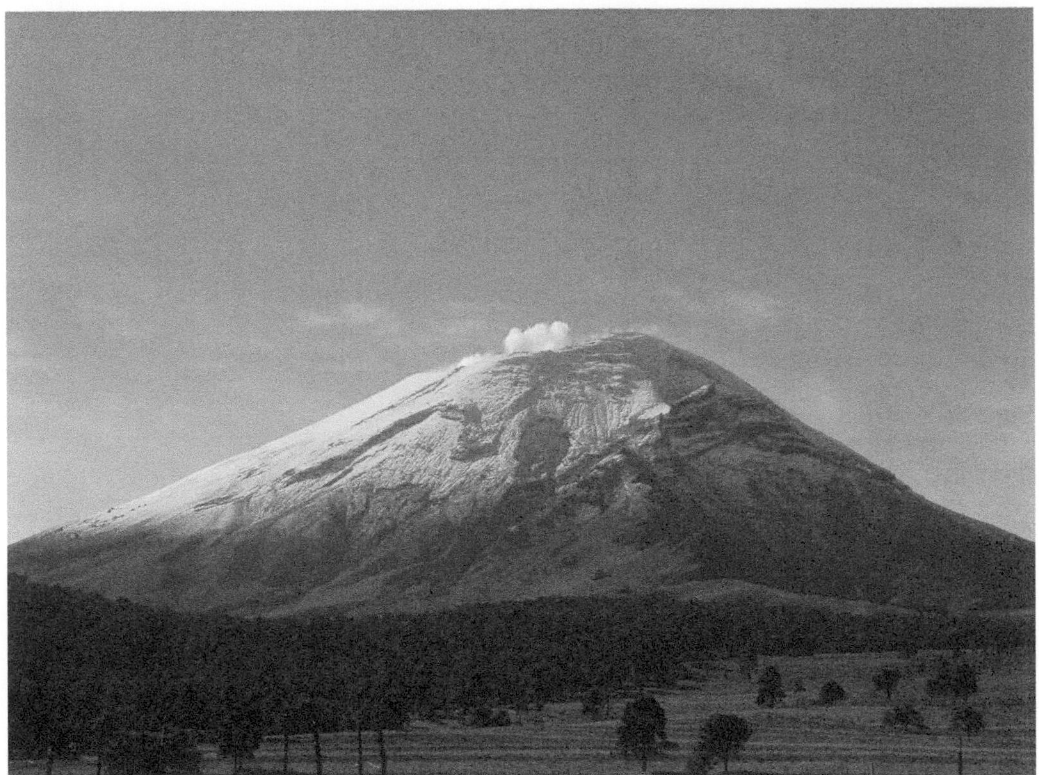

FIGURE 9.15 View of Popocatépetl volcano [56] (5,452 m) central Mexico from the north. The large scarp on the NW slope of the cone is a remnant of an earlier Popocatépetl edifice (Ventorrillo), destroyed by a major eruption about 14,000 years ago (Sosa-Ceballos et al. 2015). Source: Photo by Servando de la Cruz-Reyna.

Mount Fuji [146], 70 km west of Tokyo, one of the world's most scenic composite volcanoes (Figure 9.11), is considered sacred by the Japanese, and is climbed by hundreds of thousands of people every year. The last eruption in 1707 (VEI=5) was preceded by major earthquakes beneath the volcano, and ejected about 800×10^6 m^3 of tephra. Statistically, Fuji is overdue for an eruption, and is being closely monitored because of the great risk posed to visitors and local residents as well as the potential for devastating ash fall in metropolitan areas downwind (Tsuya 1955).

Taranaki volcano (Mount Egmont) [168] is a classic composite cone on New Zealand's North Island. The present edifice is entirely composed of eruptive products less than 10,000 years old, but the volcano is surrounded by an immense ring of volcanic debris that records a volcanic history of growth and destruction that goes back well over 100,000 years. The volume of debris may exceed by an order of magnitude the volume of the present cone of Taranaki (Neall 2001). Similarly, Mount St Helens was a classically symmetrical cone prior to the eruption of 18 May, 1980, before it was decapitated in a few minutes by sector collapse and magmatic evisceration, spreading debris far beyond the foot of the mountain. (Chapter 8; Figure 9.16).

Sub-Plinian and Plinian eruptions often accompany such infrequent and self-destructive flank failures (Chapter 11). Ejecta may make up as much as 70% or 80% of some composite edifices, but in others, an equally large percentage of lava flows may be present (Cas and Wright 1988). The lava flows are most commonly blocky ʻaʻā, but they may rarely be spiny ʻaʻā, or even pāhoehoe. The ejecta is usually cinder or ash, but small amounts of agglutinate may also be present. Depending on the viscosity of the magma erupted, the lava flows may be thin and long or thick and short. Glassy domes are also common. Many large, dangerous composite volcanoes, such as Mount Pinatubo [135] in the Philippines and Montserrat [84] in the Lesser Antilles, consist of an aggregation of domes, block-and-ash flow deposits, and ash and cinder fall beds. Dome growth characterizes the late stages of volcanism at many composite volcanoes, but at others like Pinatubo and Montserrat, dome-related Vulcanian and even Plinian blasts take place through a large part of the history of the volcano. The growth of a dome is not sufficient justification to regard the volcano as approaching extinction.

Survival Tips for Field Volcanologists

Climbing **up** steep composite volcanoes is a lot more work than coming **down**, but descending them is much more dangerous than ascending. "Angle of repose" slopes require special caution – especially if partially mantled by poorly consolidated pyroclastic material. Several geologists (especially students!) have been killed while descending composite volcanoes when their exuberance exceeded their caution. Be aware too that loose rocks dislodged by climbing steep slopes can seriously endanger others hiking down below.

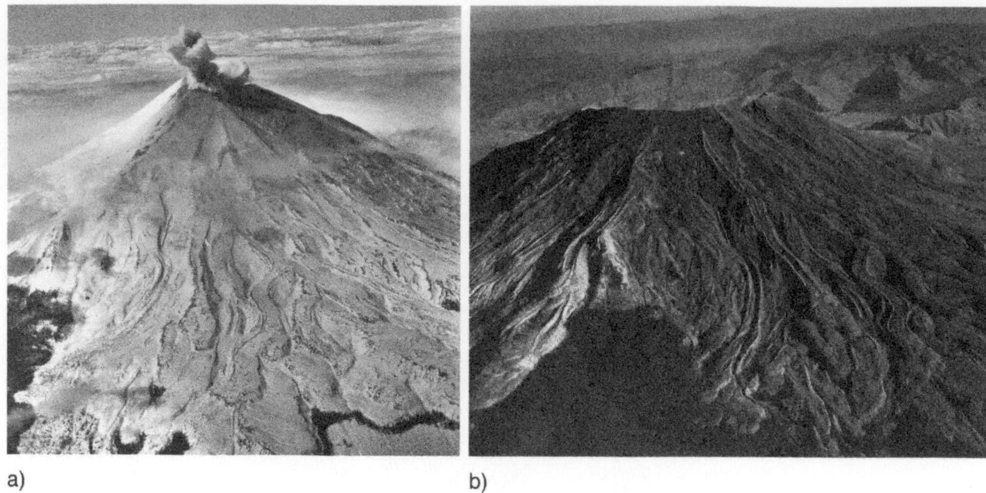

FIGURE 9.16 Mount St Helens [31], before and after 18 May, 1980. (a) View to the northwest, 10 April, 1980 as an awakening, snow-covered MSH was emitting steam and ash from the summit. (b) View to the north-northwest, 10 September, 1980 after this classic composite volcano literally "blew its top" and lost 400 m of its summit. Note the 'a'ā flows with their well-developed lava channels that illustrate how prehistoric summit eruptions formed the volcano. Source: USGS photos by Austin Post (a) and Robert Krimmel (b).

Composite volcanoes range in height from a few hundred to several thousand meters, and in basal diameter up to about 30 km. Most are built by eruptions that come principally from a single central pipe conduit, and consequently their ground plan is roughly circular. This may be greatly modified, however, where considerable proportions of their eruptions have been from lateral vents, termed **boccas** (Italian for "mouths"). "Double volcanoes" such as Mt Shasta [30] and Shastina, or even a line of several quasi-independent cones may form from the same magma reservoir when the main vent shifts its position. The departure from circularity in ground plan is even greater in the rare cases where composite volcanoes are built around lengthy fissure vents. One of the outstanding examples of the latter is Mount Hekla [92] in Iceland. Hekla is a composite volcano built around a fissure 5–7 km long, and the basal plan of the mountain is an oval about 10 km long and only 5 km wide. Viewed in a direction parallel to the fissure, Mount Hekla has the general form of a typical composite volcano, but viewed at right angles to the fissure, it appears as a broadly rounded dome resembling a shield.

On many composite cones, the upper parts have uniform steep slopes near angles of repose (about 35°) and climbing is difficult (Figure 9.17). The upward steepening of composite volcano slopes reflects the fact that most eruptions are relatively small (Figure 9.17), and most pyroclastic debris falls near the central vent. At their bases, slopes typically flatten gradually to near horizontality. The less steep flanks are due to several factors: flank eruptions may have occurred at lower elevations, loose debris high on slopes will be eroded and deposited at bases by rainwash, streams, and lahars, and lava flows that emanate from the summit become thicker as they become more viscous further from vents. Lower flanks of composite cones commonly consist largely of lava flows and lahar deposits, whereas upper parts are mantled by pyroclastic debris from frequent small explosions. In humid regions, particularly in the tropics, the amount of loose material brought to lower volcano flanks by lahars is typically immense. At Mayon volcano [137] (Figure 9.18), nearly the entire broad, gently sloping skirt of the cone is composed of lahar deposits with only minor amounts of ash fall beds.

Large masses of agglutinated spatter and even remnants of lava flows may accumulate on upper cone flanks near crater rims. In some cases, craters do not exist at all, but instead the summit merely tapers to a gently rounded, hydrothermally altered crest mantled with ejecta. In other instances, one or more domes completely fill the crater, giving the cone a steep, craggy top. The vents of some composite volcanoes remain fixed in position throughout the growth of the volcano, but more commonly vent positions shift somewhat during growth, in large part due to the plugging of older vents, or the change in position and configuration of the underlying magma reservoir – or both. In many cases, the multiple vents marked by craters line up, reflecting the control of a basement fault or fracture on magma ascent. On other volcanoes, multiple summit craters of different ages are distributed in a random cluster across the mountaintop.

Gradual rebuilding of composite cones that have experienced sector or caldera collapse may proceed until only the highest rim of the former stump of the mountain remains uncovered by younger volcanic material. This collar-like relict of the older volcano is termed a **somma ridge** (or simply a **somma**) after Monte Somma, Italy, which formed when a former edifice of Vesuvius collapsed 17,000 years ago, leaving a cliffy, dike-laced rampart to the north of the modern cone. Eventually, a renewed edifice may grow large enough even to bury its somma, or sommas, completely, leaving no clue in its shape to a history of former catastrophe(s).

While the lifespans of individual shield volcanoes range from just a few years in Iceland, to as much as a one or two million years for large mountains such as Mauna Loa [15], lifespan estimates of composite volcanoes are less wide-ranging. Some (e.g., Mazama

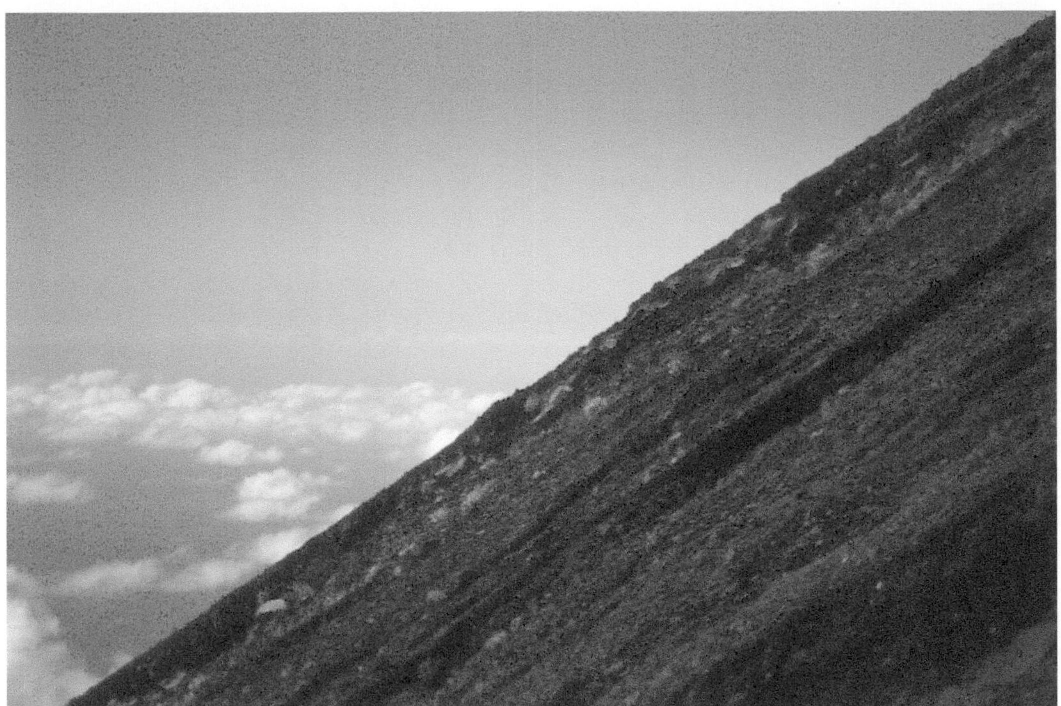

FIGURE 9.17 Upper slope of Mayon Volcano, Philippines [137]. The steep slope is at the angle of repose (35°) and little material can accumulate during eruptions. The lava flow in the background is about 40 cm thick; most erupted lava tumbles downslope to accumulate at lower elevations. Source: Photo by J.P. Lockwood.

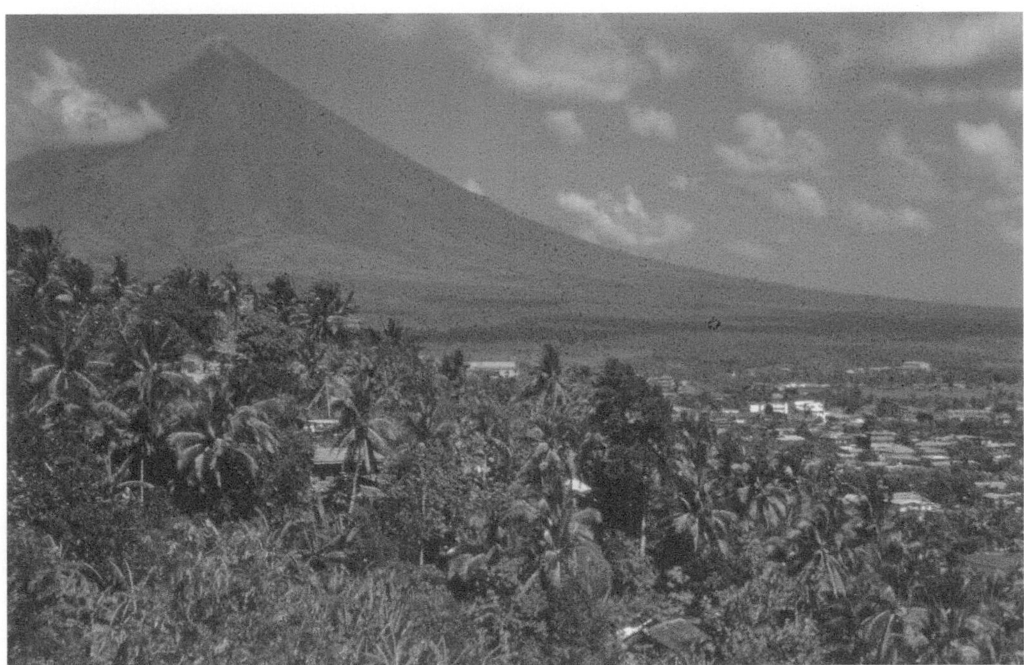

FIGURE 9.18 Mayon volcano [137] (2462 m) rises about 2 km above the surrounding plain of Albay Province, Philippines. This is one of the world's most scenic, yet dangerous volcanoes. More than 2000 people have been killed by lahars and pyroclastic flows on the flanks of Mayon in the past few hundred years. Source: Photo by J.P. Lockwood.

[32], Oregon; Hakone [147], Japan; Mount Pelée [87], Martinique) have eruptive histories extending back 400,000 years or more. Santorini volcano [108], Greece, may date back over a million years. Most composite volcanoes appear to be less than 100,000 years old, however; and many of these are probably in the waning stages of their activity (Wadge 1982). Quite a few composite volcanoes occur in overlapping proximity to one another (**volcanic complexes**), in which one volcano falls extinct while a neighbor just a few kilometers away starts growing. In this manner, individual volcanic complexes may remain active for millions of years. One example

is the island of Martinique, a volcanic complex composed of at least six discrete, coalescing volcanoes that has existed for about 20 million years (Cas and Wright 1988).

Unlike small cinder cones, lava shields and shield volcanoes, all composite volcanoes are polygenetic. Some, such as Stromboli, may be in nearly continuous eruption for thousands of years. Others experience as much as several thousand years between their eruptions. Some exhibit regular intervals between volcanic outbursts. Others appear to have random timing in their activity. Wadge (1982) estimated that on average the duration of quiescence on active composite volcanoes is more than double the duration of their eruptive activity.

A few composite volcanoes pursue essentially a single type of eruptive behavior throughout their lifetimes. Many others alternate or change their behaviors drastically. For example, beneath the edifice of Mount Mazama, a very explosive composite volcano, lies a basaltic shield, which certainly represents an early stage of effusive volcanism in the very same area. Volcanologist Robert Decker likened volcanoes to people; each has its own "personality," and each may experience its own "life changes," some of them traumatic.

Mount St Helens, the most active and explosive volcano in the Cascade Range, is a good example of a young composite volcano with a highly varied eruptive history. Eruptions began in the area of Mount St Helens about 40,000 years ago. These were not mild effusions. From the start, PDCs, Plinian or sub-Plinian eruptions, and lahars spread debris widely across the region. Three early cycles of this kind of activity took place: The earliest, called the *Ape Canyon Period*, 40,000–35,000 years ago, was followed by an apparent hiatus of 15,000 years. While there could have been some eruptions during this long break, they must have been small as they left no traces. From 20,000 to 18,000 years ago, during the *Cougar Period*, renewed pyroclastic outbursts took place. This was followed by the similarly explosive *Swift Creek* (11,000–6000 BCE) and *Smith Creek* (2000–1000 BCE) eruptive periods. Then, beginning about 2500 years ago, the compositions of eruptive products abruptly became more diverse. Initially, the volcano had erupted only andesite and dacite, almost entirely as ejecta. Now basalt also appeared, and fluid lava flows together with more silicic domes began playing a major role in building the composite cone. Lava flows and related radial dikes reinforced the cone, allowing it to grow taller than it could have had it been built solely of pyroclastic debris and dome rubble. Vigorous volcanic activity occurred between 2500–1600 BCE, 1200–800 BCE, and 400 BCE–300 CE. The current period of activity began 520 years ago, culminating in the great blast of 18 May, 1980. It is tempting to regard each of these discrete volcanic spasms as the expression of the rise of a distinct batch of basaltic magma into the base of the reservoir beneath the volcano, though this is not certain. It is difficult to know what this geologically recent introduction of basaltic melt into the Mount St Helens conduit means for the future of the volcano.

Summary

There is a continuous morphological spectrum between broad shield volcanoes and "pointy" composite ones. The primary difference between these end members relates to magma composition: "red" volcanoes almost always erupt low-viscosity lava flows – primarily basaltic – that travel far, forming shields; "gray" volcanoes typically erupt higher viscosity intermediate to silica-rich lavas that do not travel far or explosively disintegrate on eruption to form ash-covered composite cones. Composite volcanoes, like shields, show considerable variation in their features. In general, however, there is a range between those composed of relatively thin lava flows and tephra, with little or no history of self-destructive behavior, and those composed largely of aggregated domes and stubby, thick lava flows with an extremely violent history of construction alternating with self-destruction. Like shield volcanoes, composite volcanoes may experience changes in magma compositions and eruptive behaviors as they age.

Minor Volcanic Landforms

Although massive shield volcanoes like Mauna Loa [15] or large, scenic composite volcanoes like Mayon [136] in the Philippines, Mount Fuji [146] in Japan, or Mount Shasta [30] in California are the ones most commonly envisioned by people as "ordinary" volcanoes, the reality is that most of the world's volcanoes are quite small – typically less than a kilometer in diameter. These "miniature volcanoes" are primarily located out-of-sight on the ocean floor (Chapter 12), but tens of thousands of others are found on land too – not only as satellite structures on larger volcanoes, but more commonly as independent edifices, resulting from single eruptions that may have lasted no more than a few months or years As one example, we have already introduced the lava shields associated with flood basalt eruptions. Let's now consider some additional examples:

Cinder Cones

One of the most familiar of all volcanic structures are the **cinder cones** – cone-shaped small monogenic volcanoes typically truncated with bowl-shaped summit craters (Figures 9.19 and 9.20), built almost entirely of lapilli-sized cinder and scoria with lesser amounts of volcanic bombs and blocks ejected around vents during Strombolian eruptions. Many cinder cones form as satellite structures on large volcanoes, but most are typically found as scattered **monogenic** volcanoes, formed during single eruptions

FIGURE 9.19 Pu'u ka Pele cinder cone in the "saddle" between Mauna Kea [16] and Mauna Loa [15] volcanoes, Hawai'i. This 95-m-high, late Quaternary cone was formed by a flank eruption of Mauna Kea, but is being surrounded and will be eventually buried by younger Mauna Loa flows. The summit crater is 400 m in diameter. Hualalai volcano [14] in background. Source: Photo by J.P. Lockwood.

FIGURE 9.20 SP Mountain is a classic late Holocene cinder cone associated with lava flow in the northern part of the San Francisco volcanic field, Arizona. The cone is 250 m high and 1200 m in diameter; the basaltic andesite flow erupted from its base extends 7 km to the north and is up to 50 m thick. View to north. Source: Photo by Wendell Duffield.

over a relatively short period of time - a decade or less. Cinder cones are mostly of basaltic composition, and so tend to be dark-colored when young, though oxidation when escaping heat mixes with air trapped in the ejecta pile over a period of hours or days may turn their slopes and summits brick red.

Many cinder cones are nearly circular in ground plan. Wood (1980) found that they typically range between 0.25 and 2.5 km in diameter, with a mean width of 0.9 km. The heights of the cones are usually about one-fifth of the width, given the ordinary angle of the slopes. Because crater rims are narrow, this also means that craters widths are typically about 40% that of the total cone diameter. Their slope angles are the result of loose fragments rolling or sliding downward during eruptions until they attain equilibrium, and are initially close to the angle of rest for piled up irregular fragments (the angle of repose – around 30–35°). As crater walls are eroded with time, however, their slopes will become gentler and commonly rilled or gullied. Despite efforts to use these changes as a scale for approximating age, however, doing so has proven elusive for a variety of reasons (Hooper and Sheridan 1998; Kereszturi et al. 2013; Zarazúa-Carbajal . and de la Cruz-Reyna 2020). Bemis et al. (2011), established that in Central America, at least, very little change may be observed in the morphologies of cinder cones less than 500–1000 years old.

The craters at the tops of cinder cones are typically smoothly bowl-shaped. In some cases, slopes meet in the center to form pointed funnels, but on most, the craters have rather flat floors formed by late-stage lava effusion or by material washing or blowing in to form level, clay-rich deposits. In high-rainfall climates, ponds may form in older craters.

The crater may consist of a single depression, indicating a single vent active at the end of the eruption, or it may consist of several depressions, each surrounding a former vent. Multiple craters are commonly aligned, as on some composite volcanoes, reflecting arrangement of the vents along a fissure. Where multiple craters are present during the early stages of the eruption, they may be buried and replaced during the later stages by a single crater when all but one of the vents become inactive. The crater is largely the result of the construction of the surrounding rims, the area of the vent being kept relatively clear of falling ejecta by the force of the escaping gas. To some extent, however, it may also partly result from collapse at the end of an eruption, the lowering of the magma level in the conduit removing some support and allowing the overlying material to subside and the loose material of the crater walls to slide to a new position of rest. Sometimes the resulting collapse is very extensive and greatly alters the shape of the crater and even of the whole cone, but often it is comparatively minor, and the original constructional form of the cone is a little modified. Young cinder cones consist almost wholly of loose material, but with the passage of time, groundwater moving through them may deposit calcium carbonate or some other type of cement that binds the fragments together.

Cones are formed by successive showers of fragments tossed out by successive explosions, and typically show layering that reflects changes in eruption style. The first layer forms a low mound-like rim on the ground around the vent, and each succeeding layer forms a mantle draped over the one below, sloping from the crest of the growing crater rim both outward away from the vent and inward toward it. On the outer slope of the cone, the layers commonly are quite regular, and individual layers extend over large segments of the cone (Figure 9.21). Within the crater, the layers are generally very irregular, distorted, and discontinuous, owing to the truncation of their lower edges by succeeding explosions and resultant slumping of material on the side of the crater downward toward the vent.

Within individual cinder layers, the sizes of the fragments typically decrease upwards, because in general, during any one explosion the larger fragments are thrown less high, fall faster, and strike the ground sooner than the smaller ones. The size of the fragments (from less than a centimeter to as much as 30 cm across) depends in part on the strength of the explosion, more violent expansion of the gas tending to blast the magma into smaller shreds. Commonly, individual cinder cones are characterized by a range of more or less uniform size of fragments – mostly fairly coarse, or mostly fairly fine throughout – resulting from nearly uniform explosiveness of the entire eruption. However, it is also common to find a systematic increase in the size of fragments in the uppermost layers resulting from decreasing explosiveness toward the end of the eruption. It is also common to find occasional bombs embedded in a haphazard manner in finer cinder.

Fusiform and spindle bombs (Chapter 7) are often associated with the cinder, though generally in very minor proportion, and in some cones, they are lacking. Typically, they are most abundant in the outer portion of the cone, and this also seems to result from a decrease in the gas content of the erupting magma and, consequently, in the explosiveness of the last stages of the eruption. The impacts of the larger bombs often form distinct pits, in places as much as a half-meter deep or more, in the surface of the cone, which may rupture the underlying layers or bend them downward as bomb sags. Sometimes, though comparatively rarely, the explosiveness decreases to such a degree that the ejecta changes from cinder to spatter, and a layer of welded spatter may form over the crater rim or even over a large part of the outer slope of the cone. Spatter accumulating on the crater rim may build a nearly vertical wall crowning the cone. Such cones have been called **ruffed cones**, because of the fancied resemblance to a seventeenth- or eighteenth-century ruff, or collar.

Seldom do cones get much higher than 500 m, apparently because even if the eruption continues, the instability of the piled up loose material causes it to slide and slump instead of building vertically, and also because of a common change of eruptive style. Basaltic cinder cone-forming eruptions that begin with high fountaining commonly are the sources of lava flows that break out from cone bases (Figure 4.10). As the eruptive style changes from gas-rich fountaining to lava production, the more dense molten lava may extrude through the enclosing low-density cinders to form boccas at the cone base. First, wisps of fume start to rise from a small area on the side of the cone, then the area starts to glow and sometimes bulges slightly, and finally the molten lava stream emerges. The extrusion deforms the cone, and may carry away large fragments of collapsed cone flank on the flow surface – far

FIGURE 9.21 Quarry wall cross-section of a late Holocene, forest-covered cinder cone in northern California, southern Cascade arc, USA. This cone, near Medicine Lake volcano [39], is one of many prehistoric monogenetic vents in the region. The line separating the steeply dipping layers on the right from the gently sloping strata on the left is called an angular unconformity, which in this case resulted from a shift in eruptive focus during growth of the cone. Note the large number of volcanic bombs scattered randomly in the beds of cinder. Source: Photo by R.W. Hazlett.

downstream of the cone itself (Figure 9.22). Macdonald (1978) described portions of a cinder cone within Haleakala Crater, Hawai'i [13] that were rafted more than 2 km from the source cone. Pieces as large as 30 m in diameter survived lava rafting virtually intact.

Some cones are almost perfectly symmetrical (Figures 9.20; 9.21), although some are lower on one side than on the other. Such asymmetrical cones may result from non-uniform building because of inclined explosive jets that deposit more material on one side of the vent than on the other, or because of strong winds during the eruption that blew the majority of the ejecta in one direction. At other times, molten lava shoves its way through cone walls, as mentioned above, taking away a whole cone sector (Figure 4.10). More commonly cones have a breached appearance simply because a stream of lava poured out from one side of the cone throughout much or all of its formation; cinders or spatter falling on the surfaces of the lava streams are immediately swept away, maintaining openings in horseshoe-shaped edifices. Similar structures develop in large, long-lasting spatter cones (Figure 9.23).

Paricutín [53], in the State of Michoacán, Mexico, is one of the most famous cases of a cinder cone growing in historical time. Foshag and Gonzalez-Reyna (1956) interviewed Dionisio Pulido, the Mexican farmer who witnessed the birth of the cinder cone in his cornfield, 320 km west of Mexico City, on 20 February, 1943. The eruption began as Dionisio was preparing to plow his field for spring sowing. His tale is re-quoted from Luhr and Simkin (1993, p. 56):

> *In the afternoon I joined my wife and son, who were watching the sheep, and inquired if anything new had occurred, since for two weeks we had felt strong tremors in the region. Paula [my wife] replied, yes, that she had heard noise and thunder underground. Scarcely had she finished speaking when I, myself, heard a noise, like thunder during a rainstorm, but I could not explain it, for the sky above was clear and the day was so peaceful, as it is in February. At 4 p.m. I left my wife when I noticed that a [large hole] had... opened on one of the knolls of my farm, and I noticed that [a] fissure... passed through the hole... and continued in the direction of Canicjuata, [about 1 km due west]. Here is something new and strange, thought I, and I searched on the ground for marks to see whether or not it had opened in the night, but could find none; and I saw that it was a kind of fissure that had a depth of only half meter. I felt a thunder, the trees trembled, and I turned to speak to Paula; and it*

FIGURE 9.22 Fragment of a 10 m-high mixed cinder/spatter cone in the Craters of the Moon National Monument, Idaho (USA) that was rafted almost a kilometer downstream from its source cone by a pahoehoe flow during an eruption 2100 years ago. Craters of the Moon is an active volcanic field that may well erupt again within the next few hundred years. Source: Photo by J. P. Lockwood.

FIGURE 9.23 Paricutín Volcano erupting sometime in 1946-48. The new cinder cone destroyed much rich agricultural land and killed thousands of cattle as well as hundreds of horses in just the first few months of eruption. Source: U.S. Geological Survey photo by Ray Wilcox.

> *was then I saw how, in the hole, the ground swelled and raised itself 2 or 2.5 meters high, and a kind of smoke or fine dust – gray, like ashes – began to rise up in a portion of the crack that I had not previously seen near the [hole]. Immediately more smoke began to rise, with a hiss or whistle, loud and continuous, and there was the smell of sulfur. I then became greatly frightened and tried to help unyoke one of the ox teams. I hardly knew what to do, so stunned was I before this, not knowing what to think or what to do and not able to find my wife or my son or my animals. Finally, my wits returned and I recalled the sacred Señor de los Milagros, which was in the church in San Juan Parangaricutiro and in a loud voice I cried, "Santo Señor de los Milagros, you brought me into this world – now save me from the dangers in which I am about to die," and I looked toward the fissure whence rose the smoke; and my fear for the first time disappeared. I ran to see if I could save my family and my companions and my oxen, but I did not see them and thought that they had taken the oxen to the spring for water. I saw that there was no longer any water in the spring, for it was near the fissure, and I thought that the water was lost because of the fissure. Then, very frightened, I mounted my mare and galloped to [the village of] Paricutín, where I found my wife and son and friends awaiting, fearing that I might be dead and they would never see me again.*

The initial vent of Paricutín was quite small, no more than about 30 cm in size, but as the ground distended and lava continued to push up from below, the sides of the vent collapsed and widened. Six or seven hours after Dionisio fled his farm, the eruption became markedly more violent, with incandescent cinder now being ejected in great quantity for the first time, and lightning playing in the ash-laden Vulcanian eruption column. In a few days, activity became Strombolian, and by the end of the first week, the cone was 140 m high. Within two months it grew to 310 m high, nearly its full size. It added only a few tens of meters more to its height during the following 8 years of activity as ʻaʻā lava flows developed (Figure 9.24).

Jorullo cinder cone [52], some 160 km farther southeast in Mexico, commenced its eruption in 1759 with phreatic or phreatomagmatic explosions that continued for about 10 days before the activity became more purely magmatic. In six weeks, the cone grew to a height of 250 m.

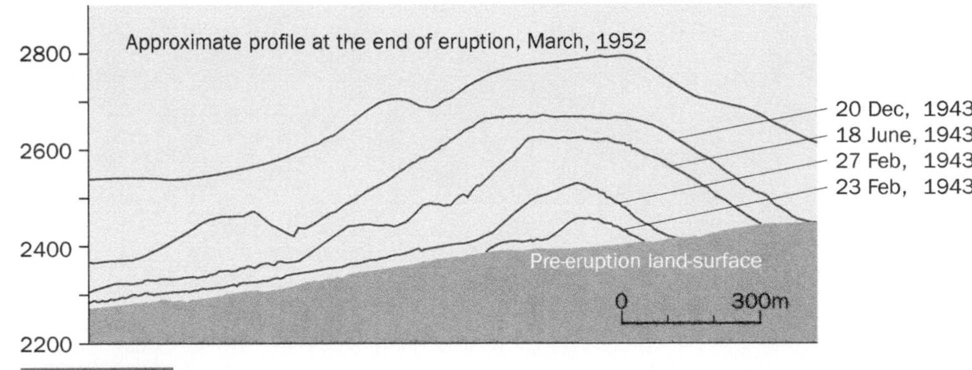

FIGURE 9.24 Growth of Paricutín cinder cone, Mexico [53], 1943–1952. Source: Modified from Luhr and Simkin (1993).

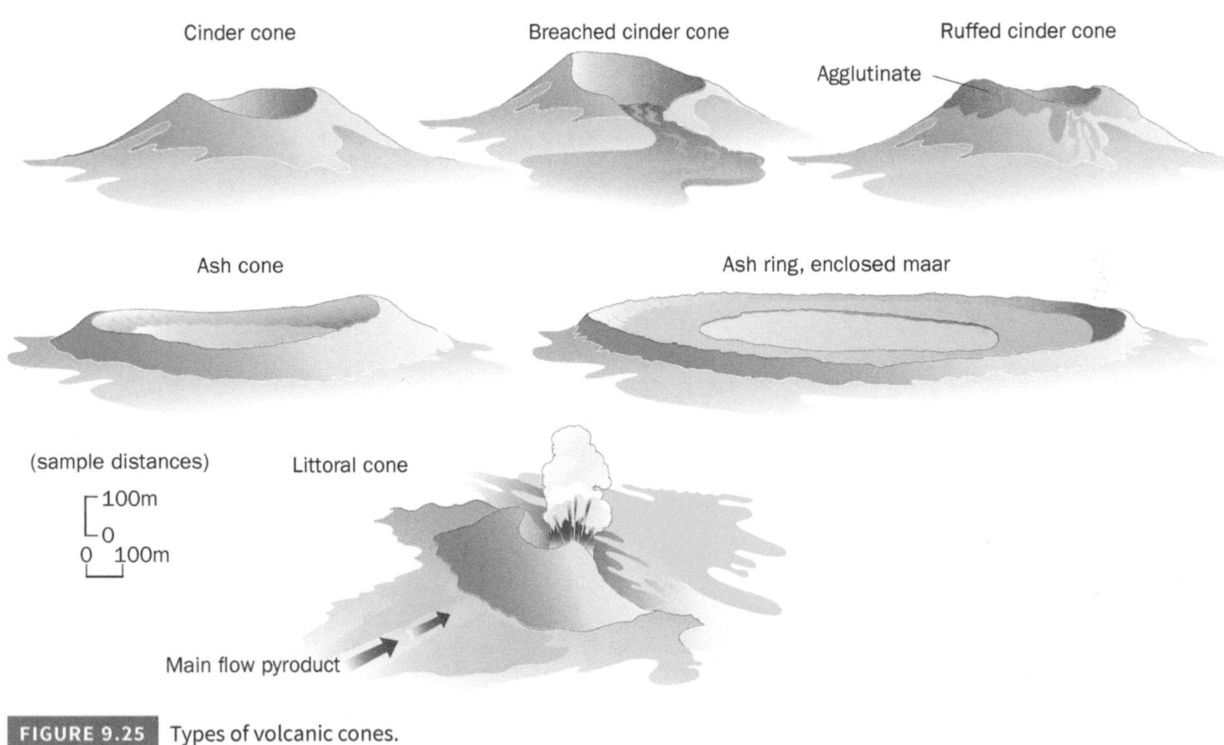

FIGURE 9.25 Types of volcanic cones.

The first recorded "birth of a volcano," in contrast to the opening of boccas on the side of larger composite or shield volcanoes, took place in the Phlegraean Fields, not far from the ancient villa of Pliny near Vesuvius [109], in 1538. The cinder cone, known as Monte Nuovo (New Mountain), is only 135 m high, but was wholly built within 7 or 8 days.

Ash and Tuff Cones

Phreatic and phreatomagmatic eruptions often produce another group of small monogenetic edifices, characterized by the accumulation of ash around vents, usually lying about at sea level or in low country with shallow water tables. These are called **ash cones**, or, if well-cemented and mineralized by palagonite into hard rock, **tuff cones**. Where moderately to very abundant blocks are imbedded in the ash, they are called **block-and-ash cones**. They are especially common around the margins of oceanic islands, where ascending magma encounters seawater near coasts, resulting in explosive eruptions that form landmarks such as the well-known Diamond Head (Leahi), O'ahu Island, Hawai'i. As ascending magma quenches along conduits and is better and better isolated from seawater, eruptions will become less explosive, and lava is commonly erupted in a final eruptive phase (Figure 9.25).

Ash and tuff cones differ from cinder cones not only by the fineness of their constituent pyroclastic material, but by their broader and lower profiles (Figure 9.26). In the case of basaltic eruptions, explosion foci are so shallow that ejecta is blown out at a low angle rather than pitched high into the air, and the fragments accumulate at a considerable distance from the vent. As a result,

FIGURE 9.26 Unnamed prehistoric tuff cone on the southeast coast of Anatahan island, Northern Marianas. This prehistoric tuff cone, about 500 m in diameter (a) has been eroded by the sea, revealing the feeder dike (arrow) that fed a late-stage lava flow that covered the crater floor (b). Source: USGS photos by J.P. Lockwood.

ash and tuff cones commonly have much lower and broader rims than cinder cones, and their craters are more like saucers than bowls or funnels. Typical profiles are shown in Figure 9.25. Very broad, steep-sided ash (tuff) cones are sometimes referred to as **ash (tuff) rings**. Many result from especially powerful, short-lived phreatic explosions (Chapter 7). Many enclose steep-sided, shallow maar craters, typically filled with water. Most maars lack ash beds dipping inward toward their centers. Instead, each is delineated by a circular vertical scarp that may reveal the interior structure of the surrounding ash ring. Maar craters range in diameter from mere meters to several kilometers.

Ash cones may begin forming either under shallow water or on land. In the former case, the cone commonly grows above water level, the base of the cone forming subaqueously and the upper part subaerially. The subaerial part of the cone consists partly or wholly of ash fall deposits, with regular bedding on the flanks, mantle bedding arching over the crater rim, and both normal and reverse-graded bedding. Sorting ranges from poor to good. The thin beds that result from the fall of ash from small to moderate explosions are generally moderately to well-sorted, the degree of sorting increasing with distance from the vent. Thick, massive beds formed close to the foci of strong explosions may have very poor sorting. Slumping constantly distorts the bedding on the crater side of the rim, and less commonly on the outer slope. Falling bombs and blocks cause compaction and distortion of the beds, forming bomb and block-sags. Many ash layers, especially toward the base of the cone, show structures indicative of pyroclastic surging, including base surges (Chapter 8).

In the subaqueous portion of the cone, the beds may show typical sedimentary features, and are apt to be better sorted than those of the subaerial part. Normal graded bedding, with the grains in each bed becoming finer upward, is often present and may be conspicuous. Underwater avalanches of fine sediment, termed **turbidity currents**, may disturb some layers, while redepositing others (Fisher and Waters 1970).

Littoral Cones

Unlike other kinds of volcanic edifices, littoral cones do not form around volcanic conduits but may develop where lava flows enter large bodies of water. The water, reacting with lava, will quickly generate steam explosions, blasting out showers of solid and semi-solid fragments derived from the outer part of the flow, and liquid clots and droplets derived from the still-molten center. Most littoral explosions are caused by 'a'ā flows, because their advance usually is so slow that the amount of very hot lava newly exposed to the water within a small area at any one time is insufficient to generate a large, concentrated, and continuous volume of steam. More rarely, they are caused by pāhoehoe flows, which are encased within a fairly continuous skin that serves to insulate most of the very hot portion of the flow from the water. But small littoral cones can form where the main feeder channels of pāhoehoe flows advance across turbulent surf zones. If the channels are openly exposed or have fragile crusts easily broken by the surf, strong steam blasts can take place, even generating cypressoid eruption clouds.

The larger clots of molten material expelled by littoral explosions solidify into bombs and lapilli. Generally, the bombs are irregular and rather dense, and characteristically, their surfaces show a rather finely meshed shallow breadcrust-like cracking. At Pu'u Hou Hawai'i, a large littoral cone formed during the 1868 Mauna Loa eruption [15], some of the bombs were still sufficiently fluid to flatten out into cow-dung morphology on impact. The smaller drops of liquid spray freeze to little rounded pellets and irregular fragments of glass. Sand-size material of this sort may be washed along the shore to accumulate as beaches of black glass sand, with only minor amounts of stony debris or coral bits.

Where littoral explosions continue for more than a few days at fixed locations, they build cones that may reach 60 m in height and more than a kilometer in breadth at the base. Pu'u Hou reached a height of over 80 m during the 5 days that lava flowed into the ocean. Such cones are commonly built on the surfaces of the lava flows at the edges of the main feeding channels, where the contact of hot lava with the water is most continuous. A large percentage of the ejecta commonly falls into the sea and is washed away, so the cone is semicircular in plan with only the landward portion remaining. It may be quite regular in form with a single rim, or, as in the case of Pu'u Hou, explosions at a series of different centers may build a complex cone with several rims. Often, two separate cones are built at the two sides of the lava channel; or if more than one lava channel entered the water in the same general area, three or more closely spaced cones may grow. Mantle bedding arches over the rims of cones, but bedding and sorting of the material is typically quite poor. Some beds are nearly pure ash, but others contain numerous lapilli and bombs. Where the lava was very fluid, the ejecta may be partly agglutinated in portions of the cone close to the site of the explosions. Local unconformities and slump structures are common in the cones.

It may be difficult or impossible to be sure whether a given prehistoric cone at or near the shoreline was formed above a primary vent or by a littoral explosion. The ash grains in a littoral cone are usually denser than ordinary vitric ash formed by juvenile magmatic explosions and show fewer arcuate forms resulting from the rupturing of vesicles. The difference results from the fact that the ordinary ash is formed by disruption of the magma from within by expansion of its own contained gas bubbles, whereas the littoral ash is formed by the blasting apart of the magma by gas (steam) that originates outside it. The degree of contrast depends on the amount of gas that is still contained in the lava when it reaches the shore. Where the lava has flowed for a long way and lost most of its gas, the fragments produced by the littoral explosion are quite dense. On the other hand, where gas is still exsolving from the lava and forming still-expanding vesicles, as it is in pāhoehoe flows, the fragments may be moderately vesicular.

Volcanic Domes

Where highly viscous lava erupts to the surface, it has great yield strength (Chapter 4) and flows with such difficulty that it tends to pile up forming a steep-sided hill directly over and around a vent (Figure 9.27). Such hills are known as **domes**, although the term "dome" is also used for other types of geological structures, especially tectonically domed strata. Domes that grow by extrusion of new lava erupted onto the surface are termed **exogenous**. Domes that grow primarily by internal intrusion of magma within swelling lava crusts are termed **endogenous**. Some domes, termed **cryptodomes**, are formed by the shallow subsurface intrusion of magma that uplifts overlying rocks into domical shapes – no lava reaches the surface. The term **spine**, which is applied in a couple

FIGURE 9.27 Novarupta dome [19], on the Alaskan Peninsula, viewed from southwest. This 65 m high, 400 m wide lava dome was formed during the major Katmai [20] eruption of 1912. Source: Photo by R.W. Hazlett.

of ways, may refer to any jagged prominence of lava projecting from a dome, or to a rare class of sharp, towering dome (also known as a "plug dome") heaved up from a vent conduit in essentially solid state, like a cork through the neck of a bottle.

When deeply exposed by later erosion, domes may show well-developed columnar jointing formed in their cores as they cooled. Some domes form by extrusion of viscous lava through an opening near their crest, the growth taking place by the piling up of one short flow over another. More commonly, however, new lava being squeezed up through the vent simply distends the mass above it, so that the growth is like that of an expanding balloon. But the amount of stretching that the cooling skin of the dome can sustain without rupturing is limited. Most domes grow by a combination of internal and external accessions of lava with internal development predominant. The rims of growing domes are generally unstable, and are usually marked by borders of talus (Figure 9.27) and large blocks (Figures 9.28 and 9.29).

The degree to which the edges of a growing dome spread out from the margin of the vent depends on the viscosity of the liquid. Some spread very little. Others spread out to several times their height, and they grade into the short, thick coulee flows described in Chapter 6. Occasionally, part of a growing dome breaks away and forms a short lava flow. Most domes are of silicic composition, with rhyolite, dacite, and trachyte most common. Domes of andesite are far less common, and domes of basalt are rare.

Ferdinand Fouqué (1879) provided among the first – and most eloquent – detailed descriptions of a growing volcanic dome while studying an effusive silicic eruption at Santorini volcano [114] Greece in 1866. Subsequent episodes of dome growth on Nea Kameni, the island at the center of Santorini's mostly submerged caldera, occurred in 1925–1928, 1939–1941, and in 1950. The probability of future similar extrusions at Santorini is high, gradually filling in the island's volcanic anchorage (Pyle and Elliott 2006).

Other historically well-documented dome eruptions took place at the summit of Mount Pelée [87] on the island of Martinique early in the twentieth century. T.A. Jaggar (1904) observed the growth of a new dome that began to grow about a week and a half after the deadly May 1902 eruption:

> *On the summit of the cone was seen a most extraordinary monolith, shaped like the dorsal fin of a shark, with a steep and almost overhanging escarpment on the east, while the western aspect of the spine was curved and smooth in profile. The field glass showed jagged surfaces on the steeper eastern side, and long smooth striated slopes on the western. Other horn-like projections from the cone could be discerned with difficulty on the slopes lower down.*

This spine grew to almost 300 m in height after Jaggar left (Figure 7.22), but crumbled to rubble in 1903. Lacroix, who also documented this feature, estimated that without the loss of material from constant collapses, it might have grown to an ultimate height of 800 m. Plug domes such as this are highly ephemeral phenomena.

Localized spines only a few meters or tens of meters tall are very common on volcanic domes, and many domes bristle with them. They are formed by the toothpaste-like extrusion of rapidly solidifying viscous magma through ruptures in the solid to

FIGURE 9.28 Growing dome on the floor of Mount St Helens [31] crater, ringed by its own talus, 1982.
Source: USGS photo by Lyn Topinka.

semi-solid shell of the dome. Such small spines continue to be extruded for only a few hours, but the domes to which they are attached may grow for many months. Internal fracturing as they cool causes them to collapse into spectacularly massive breccias, sometimes accompanied by block-and-ash flows.

The best studied volcanic domes are those that have grown in recent decades within the craters of Mount St Helens [31] and Soufrière Hills [84] volcanoes (Figure 9.30). After the 18 May, 1980 eruption of Mount St Helens, relatively small domes repeatedly formed then were blown apart by explosions, but the volcano remained generally inactive for 18 years after dome growth ceased in 1986. Then, in October 2004, dome growth resumed and remained vigorous until January 2008, when activity again paused. The 2004–2008 resurgence of dome building included the extrusion of giant fin-shaped spines that disintegrated within a few months (Figures 9.31–9.33). (Comparisons with similar dome growth at Bezymianny volcano [170], Kamchatka, active for over 50 years, suggest that dome growth at Mount St Helens is far from over.)

The 1975 dome at La Soufrière volcano [86] on St Vincent island took a little less than a half-year for initial growth, beginning its formation exogenously. The dome developed rapidly during its first 20 days of life, then declined gradually through a phase of mostly endogenous growth over the next 130 days (Huppert et al. 1982). Likewise, the initial Mount St Helens [31] dome, which began forming in the avalanche caldera shortly after the 1980 blast, grew at first by piling one exogenous lobe atop another, but then showing both modes of development as the magma entering the dome grew cooler and more viscous (Swanson and Holland 1990).

Possibly the most common dome-related explosions occur around the base of the dome and from spines protruding from its crust. The surfaces of separation between the dome and the surrounding rocks, and between the spine and the crust of the dome, constitute zones of weakness that allow gases to escape from below and within the dome. In studying the dome that formed on Santorini in 1925 (named "Fouque Kameni" after the pioneer student of domes), H.S. Washington observed jets of gas issuing around the base of the dome like the spikes of a crown, and to these he gave the name **coronet explosions**. Explosions of this sort can cause the collapse of a spine above them and undermine the side of a dome, allowing it to collapse and add to the mass of crumble breccia. Some 350–650 years ago, explosions at the base of the Chaos Crags domes just north of

FIGURE 9.29 The "Federal Building", a 13 m diameter block that rolled off the Mount St Helens [31] dome in 1982. Source: USGS photo – photographer unknown.

FIGURE 9.30 Fast-growing dome at the summit of Soufrière Hills volcano, Montserrat [84] on 27 April, 2007. Note the dust trails left by blocks tumbling down flanks of dome. Source: Photo © NERC/ Government of Montserrat by Graham Ryan.

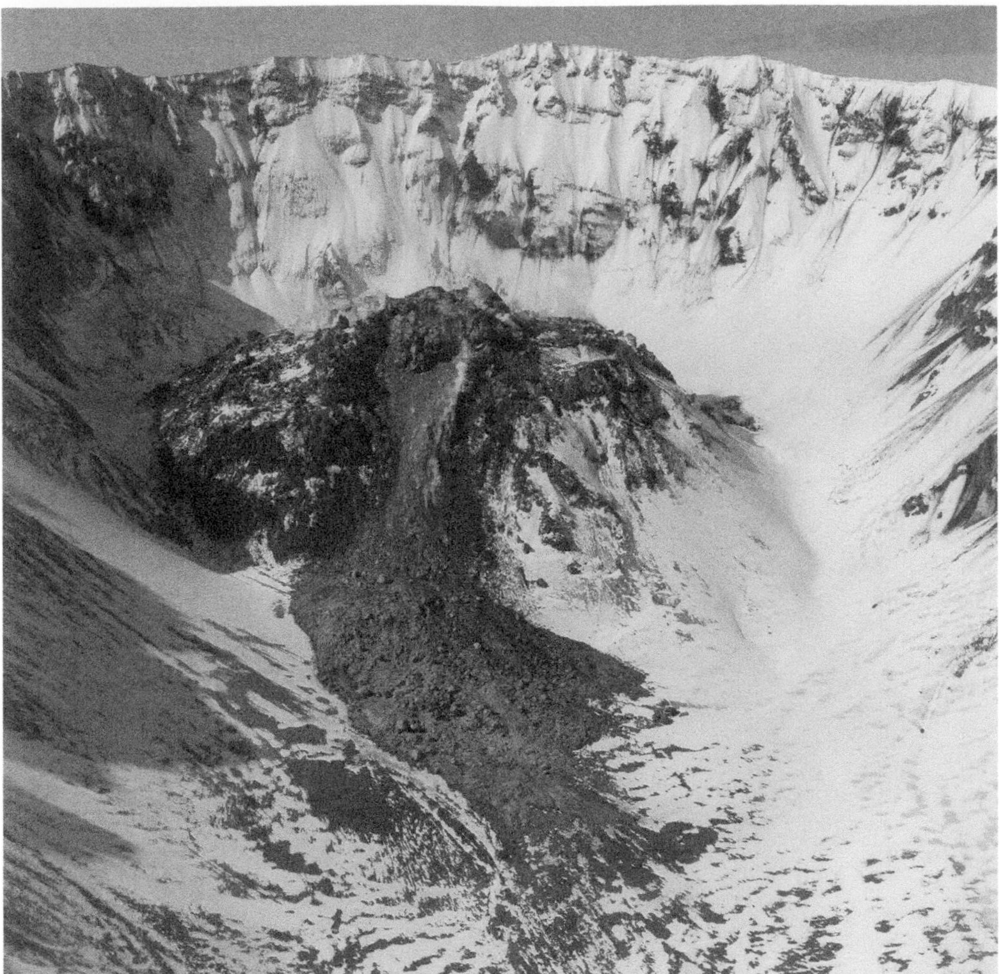

FIGURE 9.31 Mount St Helens [31] dome at the temporary cessation of dome growth in 1986. The dome was unstable, and produced large landslides like the one in the foreground, but except for minor seismic activity and occasional steam venting, MSH showed no activity for the next 18 years. Source: USGS photo by Lyn Topinka.

FIGURE 9.32 Aerial view of the Mount St Helens [31] dome complex from the northwest on 22 February, 2005, after renewal of dome growth in 2004. Note the spectacular growing spine, called the "Whaleback" during its brief life. This spine is nearly 400 m long and rises about 200 m above the crater floor. Source: USGS photo by Steve Schilling.

FIGURE 9.33 The fast-changing morphology of growing domes within the Mount St Helens [31] crater is shown by this view from the northwest on 28 September, 2005. A new dome is growing to the west of the remnants of the Whaleback spine – which has completely disintegrated. Source: USGS photo by John Pallister.

Lassen Peak [40] resulted in collapse of part of the domes, forming great rock avalanches that rushed 5 km down the valley and as much as 100 m up the opposite mountain slope. The blocky deposit left by the avalanches is known as the Chaos Jumbles. Growing domes are extremely hazardous, as they can explode with no warning. Since domes form directly over shallow magma chambers, they are incredibly unstable, and subject to repeated growth and explosive destruction. Many last for years, and the one that formed in the Kelud volcanic crater [132] in 2002–2008 lasted until 2014, when it was destroyed in a massive Plinian eruption in 2014 (McCausland et al. 2019). Others, like the Noverupta [19] dome (Figure 9.27), last indefinitely once underlying magmatic activity ceases.

Some domes have an internal structure consisting of a series of concentric layers like the shells of an onion. These appear to result from gradual expansion from within of a mass of somewhat non-homogeneous magma–outer parts glassier, perhaps, and inner portions more microlite rich. Much more commonly, however, domes are either structureless except for the gradual inward gradation from a brecciated outer part to a massive interior, or show divergent structures, fan-like in cross-section with the ribs of the fan radiating upward from the vent. Surfaces of domes are often marked by a series of concentric ridges resembling the pressure ridges on coulees and blocky ʿaʿā flows. The internal flow structures may be nearly horizontal around dome bases, gradually changing to verticality near centers. It appears easier for a growing dome to expand upward than sideways, because of the confining action of its own shell and the increasing mass of crumbled breccia.

We conclude our discussion of domes with a case history from Japan where an eruption produced both endogenous and exogenous dome growth between 1944 and 1945 (modified from a summary by Macdonald 1972). This eruption of the Showa Shinzan [153] domes occurred during the Second World War, when a news blackout kept all but local citizens and a few Japanese scientists from knowing about this unrest at Usu volcano [152] on the northern island of Hokkaido, at the southern edge of Lake Toya caldera (Minakami et al. 1951). A large endogenous dome (called a "yaneyama" or "roof mountain") formed here in 1910, but for more than 30 years, there was no activity. Then, in late 1943, earthquakes began to be felt in the area. A village postmaster, Masao Mimatsu, kept a meticulous, hand-illustrated journal of events. The volcano was fortuitously visible from his post office window, and throughout the eruption, he kept track of the changing shape of the mountain by outlining its profile on a piece of paper he attached to the window (Figure 9.34).

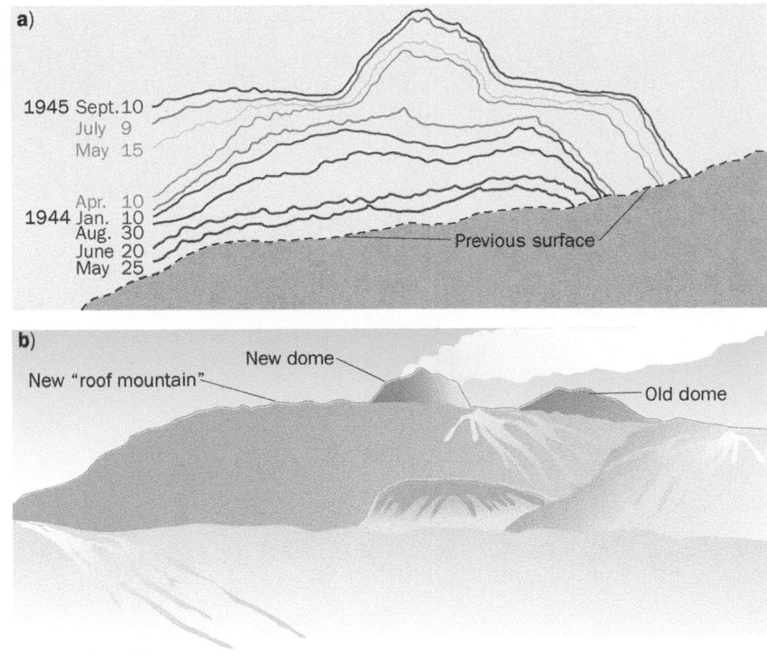

FIGURE 9.34 The growth of Showa Shinzan dome [153], as sketched by Masao Mimatsu in 1944–1945. Source: Courtesy of Saburo Mimatsu.

The first earthquake of the pre-eruption series was felt on 28 December, 1943, and was followed by many more during the next few days. Within a few days, more than 500 earthquakes had been counted at Toya Hot Springs, and quakes continued in increasing number and intensity. At first they were felt most strongly at the northwestern base of the Usu edifice, but after 6 January, they became strongest at the eastern base of the volcano. Past the eastern base of Usu, the Sobetsu River flows southward, and parallel to it ran irrigation canals, roads, and a railroad. Late in January, the ground surface near the southern base of the Usu began to rise. Cracks appeared in the roads and the banks of the canals, and water began to flow less rapidly through the canals. Wells and springs in the rising area dried up, while those in nearby areas flowed more abundantly. By early April, the rising area was roughly circular and about 4 km across, and its central part had risen about 16.5 m. Then the center of the uplift suddenly shifted nearly a kilometer northward, developing a summit very close to the village of Fukaba. Magma was obviously shifting near the surface, but was still underground. In mid-June, more than 100 severe earthquakes were being felt each day, and on 22 June, the number reached 250. The ground surface had by then risen about 45 m. The villagers of Fukaba were literally being taken for a ride!

Then, at about 8:30 on the morning of 23 June, a column of steam was seen rising from a field nearby. Steam discharge increased gradually over the next couple of hours, and at 10 a.m., an explosion hurled out mud, sand, and blocks of rock, creating a crater about 45 m across.

A stream of mud flowed out of the crater and formed a steaming pool in a nearby hollow. Then, after a few hours of quiet, steam again began to rise, followed by another series of explosions. Similar series of explosions continued at short intervals throughout the next 3 months, some of them throwing blocks of rock a 1500 m into the air. Other craters formed near the first one.

On 2 July, a phreatic explosion expelled 2 million tons of debris, and lithic ash resulting from the pulverization of older rocks. The falling ash did much damage to forests and cultivated fields. A similar large explosion occurred on 3 July, and the villagers of Fukaba finally abandoned their homes. By late September, seven separate craters had been formed in a group 600 m west of Fukaba, the earliest craters being partly buried by the material ejected during the formation of the later ones. Avalanches of ash rushed down the mountain slope but were not hot enough to set wooden houses on fire. By late October, the upheaved area formed a flat-topped, dome-shaped hill 140–175 m high and nearly a kilometer across.

The crest and flanks of the hill still consisted of the original soil and rock cover, though largely torn apart. This new dome was named Showa Shinzan – "New Mountain of the Showa Era." The Sobetsu River had been dammed by the rising ground and now had formed a lake nearly a kilometer long. But in spite of the upheaval and numerous phreatic explosions, no new lava had appeared. The uplift could still be characterized as a cryptodome.

At last, in early November, viscous magma began to extrude from the top of the uplift, just to one side of the group of explosion craters. This newly eruptive dome was incandescent, with such high viscosity that it was almost solid. It gradually grew in height and diameter, but it carried on its top a cap of older rock, including stream gravel that had once accumulated on the nearly level surface west of Fukaba. The temperature of the new lava was at least 1000°C, and the heat baked the clay of the old rocks in the cap a natural brick red color. Small pyroclastic flows descended from the north side of the dome, reaching as far as Lake Toya (Okada 2008, pers. comm.) Growth of the 300 m diameter extrusion finally came to an end about September 1945, with its summit

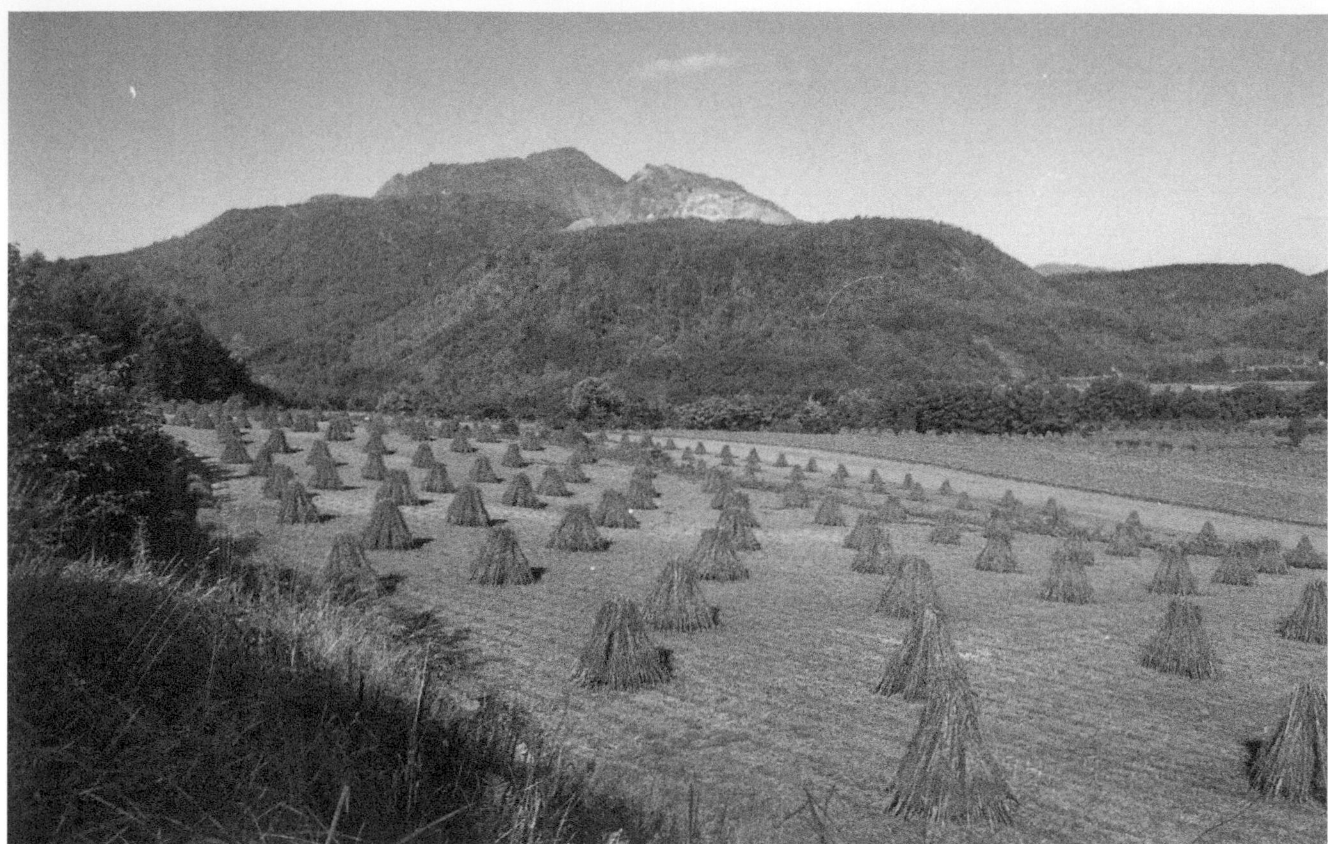

FIGURE 9.35 Showa Shinzan cryptodome viewed from the southeast, forms the brown forested slope in the foreground, with the jagged summit of Usu volcano in the background. The crest of the cryptodome is punctured by a pale-brown pyramidal exogenous dome, itself covered by an uplifted carapace of older sedimentary rocks. Source: Photo by Hiromu Okada.

about 110 m above the top of the earlier cryptodome, and over 300 m above the original level of the ground (Figure 9.35). The dome remains steaming to this day, and has become a popular tourist attraction.

Volcano Old Age and Extinction

No one knows much about how volcanoes die. There is no one path taken to extinction, and volcanoes presumed "extinct" have been known to spring back to life long after being considered dead. Chaitén volcano [73], Chile, for instance, erupted in 2008 after almost 10,000 years of quiescence. Some volcanoes may end their lives in a burst of silicic explosivity, collapsing like Mount Mazama [32] in Oregon, while others may end with minor basalt eruptions. Some appear simply to decline gradually, eruptions tending to pass from Strombolian or Vulcanian rigor to anemic Ultravulcanian or phreatic blasts at the end. Many volcanoes begin their lives effusively and become more explosive with age. Others, like Vesuvius [109] and Mount St Helens [31], show explosive early histories and become more effusive in time. There seems to be no making heads or tails of it at present, and we have to leave it at that.

Hawaiian shield volcanoes pass through a characteristic sequence of changes as plate movement carries them off the hot spot (Chapter 2). After an initial period of producing alkalic lavas in its primacy, a Hawaiian shield begins the voluminous, long-lived production of basalt, erupting dozens of times every century over periods of several hundred thousand years, building its massive bulk almost entirely out of tholeiitic basalt. But moving 100 km or more northwest of the hot spot melting source, the principal magma feeder conduit seals up. Less frequent, more explosive eruptions continue, however, as pockets of deeper, anomalously hot mantle beneath the shield continue partly melting in places. Reflecting increased depths of melting, volcanic products change from tholeiitic to alkalic compositions, with numerous, scattered cinder cones and even domes of highly evolved alkalic lava erupting, including benmoreite and trachyte. The smoothly arching shield surface becomes lumpy and rugged as the new cones develop. Ultimately, however, even this late-stage (or "post-shield") activity wanes within a few hundred thousand years. Carried farther to the northwest, however, the shield volcano may not be altogether dead yet. Persistent flexing of the oceanic plate several hundred kilometers to the northwest, in response to the rapid growth of younger, heavy islands to the southeast, may trigger even deeper decompression melting (Clague et al. 1990), 100–150 km down, which feeds small batches of highly alkalic, silica-undersaturated magma to the surface in locales bearing no relationship to the original shield magmatic plumbing system. This **rejuvenated** (or

FIGURE 9.36 Erosion of a composite volcano in a temperate or tropical latitude.

"post-erosional") **volcanism** may be gradational with late-stage eruptions, or follow a break in activity lasting as long as a couple of million years. No two shields exhibit closely similar patterns of rejuvenation. Some show no signs of it at all. Small, isolated or bunched nepheline-rich cinder and ash cones bearing numerous mantle xenoliths are typical features of this last gasp in volcanic activity, which may take place as long as 5 million years after shield volcano primacy (Clague and Sherrod 2014). Similar, though somewhat distinctive, patterns of volcanic change are also recorded in oceanic shield volcanoes elsewhere (e.g. Moore et al. 2011; Guillou et al. 2014).

Even before most large volcanoes stop erupting, erosion begins to destroy their edifices. How erosion proceeds depends upon the type of volcano and the climate. The primary agents of erosion are flowing water and/or ice. Loose fragments of rock swept downslope by runoff act as an abrasive tool that loosens other fragments. These materials scour the slope along favored paths for runoff, developing channels and gullies that grow wider as they deepen. Many composite volcanoes begin to grow on older, eroded land. As such volcanoes age, water running off slopes is funneled into the preexisting gullies and stream channels at volcano bases (Figure 9.36). These channels gradually erode their way upward into volcano flanks, a process termed **headward erosion**. Within a few thousand years, volcanoes may pass into a stage of aging where the summit of the cone remains youthful and smooth, while lower flanks become deeply channeled. Large strips of uneroded cone flank may continue extending all the way to the base of the cone between the growing channels, but as the channels lengthen, deepen, and widen, these strips grow narrower and finally disappear. At last, the channels work their way to the top of the cone, and the rims of the summit crater, if present, also disappear. At this time, the volcano becomes a roughly radial array of ridges and canyons all meeting at the position of the central conduit. Depending upon the climate, thousands of years may have elapsed since the mountain last erupted. Continuing erosion lowers the ridgelines and channel beds, while also reducing their slopes. The conduit plug and dikes that reinforced the interior of the cone

FIGURE 9.37 Mount Hood, Oregon [37], as viewed from the northeast. Glaciers have deeply sculpted this high volcano (3329 m), but the volcano remains active, and the conflict between fire and ice will continue to modify its shape in the future. Source: Photo by J.P. Lockwood.

emerge as the surrounding, weaker edifice is stripped away, and become topographic prominences in the now rugged terrain of the old volcano. Erosion continues, perhaps for a few million years, until ultimately only the resistant plug and dike rock remain. Even these will weather away and disappear in time, leaving only a bedrock intrusive stock and dikes to mark the former presence of once majestic cones.

For high altitude or high latitude volcanoes, glacial ice may be the primary erosive agent on upper slopes, giving such volcanoes a particularly rugged, jagged appearance (Figure 9.37). Intersection of glacial cirques as they expand summitward may sculpt the mountaintop to a sharp point in time. If late-stage eruptions happen to take place on glaciated volcanoes, the potential for lahars is extremely high, given the presence of enormous volumes of water in the form of ice, the steepness and weakness of the mountainsides, and the presence of unconsolidated glacial debris. This was tragically learned during the 1985 eruption of Nevado del Ruiz volcano [71], Colombia (Chapter 14). Examples of other such potentially active and dangerous glaciated volcanoes include Mount Rainier [36], Washington; Mount Shasta [30], California; Iliamna [22], Alaska; Koryaksky [166], Kamchatka, and Nevado Coropuna [72], the largest of the Peruvian volcanoes.

Shield volcanic islands erode in a manner similar to composite volcanoes, but with some significant differences (Figure 9.38). Headward erosion is triggered as runoff channels develop at the crests of shoreline cliffs, and water collects to flow into small embayments along young island shorelines, initiating channel development at the coast. The drainage system develops in the same manner described above for composite volcanoes. But, as they erode into caldera areas, streams may encounter weak, hydrothermally altered rocks, which wear away much more readily than surrounding, more resistant lava flows. The streams quickly eat out the soft, clay-rich core of the volcano, until the perimeter of the former caldera becomes the high, cliffy rim of an erosional basin, usually with a horseshoe or spoon shape. As the erosion continues, the former caldera may retain its ridge crest outline for as long as several million years. In contrast to a composite volcano, whose summit crater completely disappears, tectonic subsidence may submerge the volcano long before erosion planes it level with the sea.

Cinder cones erode somewhat differently (Figure 9.39). While they will show fine, rill-like channels in their flanks during the early stages of their erosion, the permeability of the loose cinder soaks up all but the heaviest rainfalls. Hence, channel development rarely progresses very deeply. The loose surface material of the cone is subject to mass wasting, and cinder often clogs channels, forcing occasional torrents of water to re-excavate them as erosion progresses. The net effect of this means that aging cinder cones may retain a generally smooth profile. At an advanced stage, the cinder cone may appear as nothing more than a red-brown, gently sloping knoll of oxidized cinder, its crater long wasted away. Studies by Kieffer (1971) of eroding cinder cones in the Massif Central of France indicate that it takes over 10,000 years for this level of decay to occur in that region. In the Big Pine volcanic field in east central California, similar development takes over 100,000 years and perhaps as much as 250,000 years, illustrating the influence of a climate that, for much of that time, has been more arid than in France. At the other end of the climate spectrum, Ollier

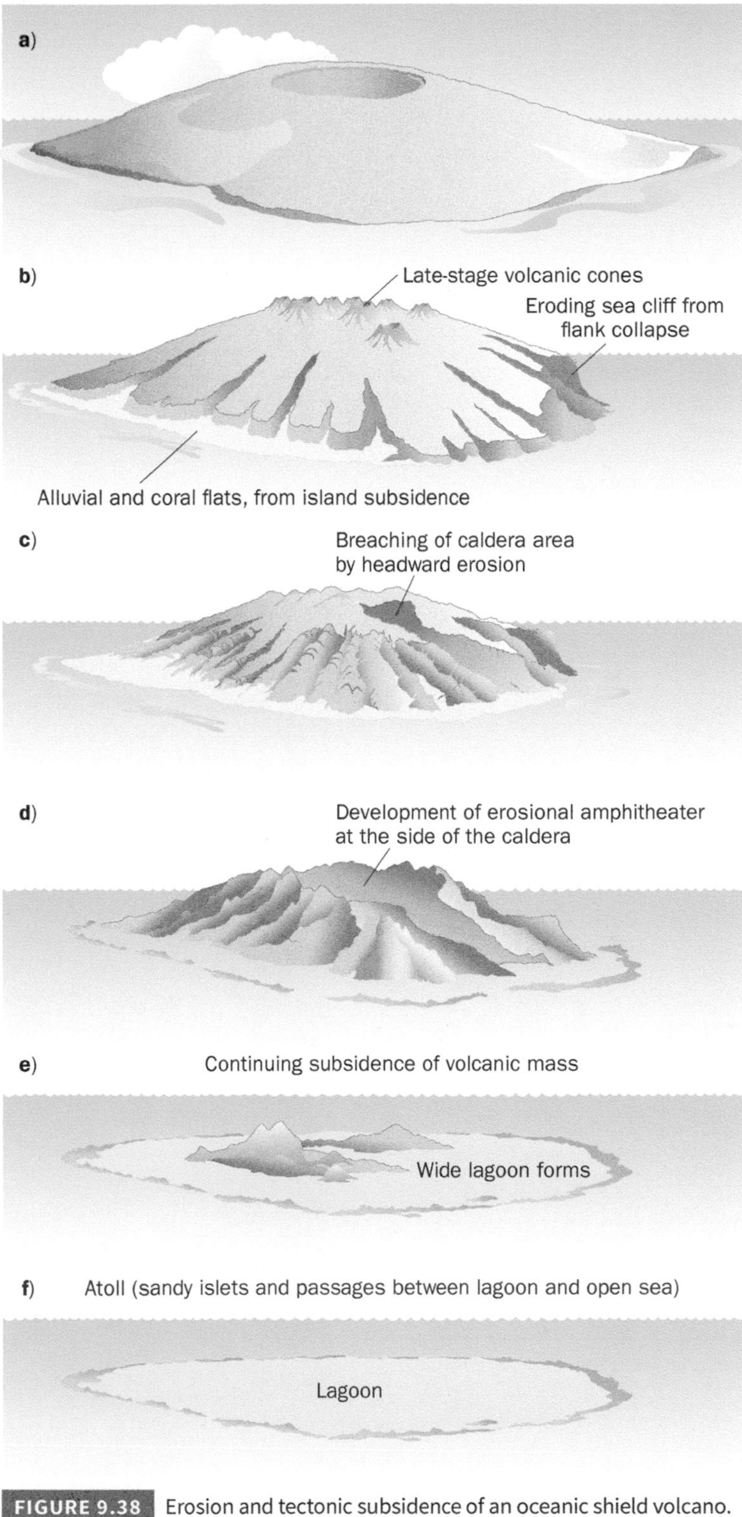

FIGURE 9.38 Erosion and tectonic subsidence of an oceanic shield volcano.

(1969) reports erosion rates on Vulcan, the cinder and ash cone in tropical Rabaul Caldera [160], of as much as a meter of surface lowering over a 30-year interval. In general, the steeper and longer a slope, the faster its surface at any fixed position erodes down. Eventually, a basaltic plug within the cone may be exposed, as in eroding composite volcanoes. The feeder dike beneath the cinder cone will also emerge as a wall of resistant rock as the landscape degrades.

Different types of volcanic deposits respond to weathering and erosion in strikingly different ways. Lava flows tend to be quite resistant to weathering, because of their fine textures and density, providing little opportunity for water penetration. Many lava flows are cut by fine joints, however, and where these joints allow water to penetrate over long periods (millions of years) spheroidal weathering may develop, where hard, unweathered lava cores may be preserved between joints (Figure 9.40). Less resistant rocks surrounding lava flows usually weaken and erode away faster. In many regions, former canyon or valley-filling flows have

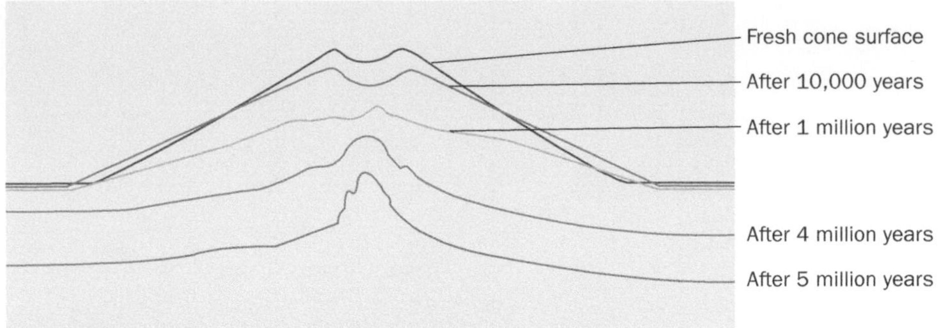

FIGURE 9.39 Changing profile of a cinder cone as it weathers and erodes away. Source: Adapted from Kieffer (1971) and Cas and Wright (1988). Climatic conditions are those of Central France.

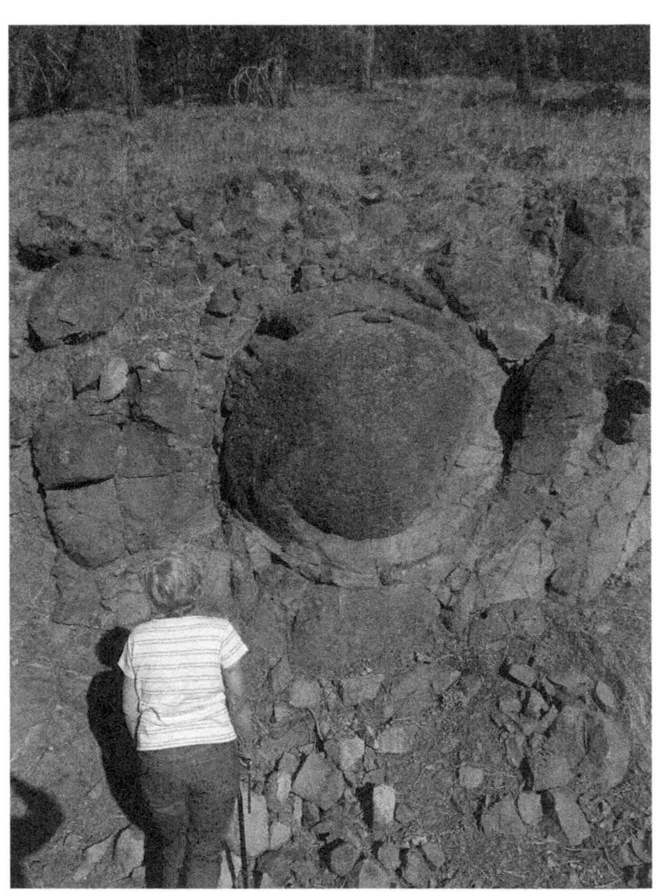

FIGURE 9.40 Spheroidal weathering in 15-MY-old Columbia River basalt east of Yakima, Washington, USA. The dense, spherical lava remnant is enclosed in deeply altered, clay-rich soil. Source: Photo by J.P. Lockwood.

become the caps of long, sinuous ridges after several million years of erosion (Figure 9.41; many of the mesas and buttes of desert lands are capped by thin basalt flows). The flows erode primarily by being undermined by erosion of the weaker rock beneath.

Rocks that weather easily tend to develop gentle slopes, as rainwater washes away detritus quickly – such rocks are commonly weak and cannot support steep faces. Steep faces indicate stronger rocks that weather slowly. Many ignimbrites illustrate this contrast well. The welded centers of the ash flows form dark cliffs, roughly partitioned by widely spaced columnar joints. Their unwelded tops and bases are typically lighter in color, not being oxidized or baked, lack joints, and form gentle slopes. Where fumarolic pipes cemented by mineralization exist, these will rise as vertical pillars or chimneys through the unwelded upper section, since they effectively resist weathering.

Unwelded pyroclastic deposits erode quite easily, becoming laced with steep-walled channels (Figure 9.42). The steepness of the channel walls is not a result of resistance to weathering, but a testimony to the rapidity with which streams can abrade and cut straight down through the deposits. In contrast, lahars, thick with clays and often saturated with water, are not easily eroded, once they come to rest and de-water. Indurated lahar deposits, like lava flows, tend to form resistant, steep-faced outcrops.

a) Young lava flow

b) Eroding landscape begins developing inverted topography (once low area of lava flow becoming high relative to surrounding landscape, owing to resistance to erosion)

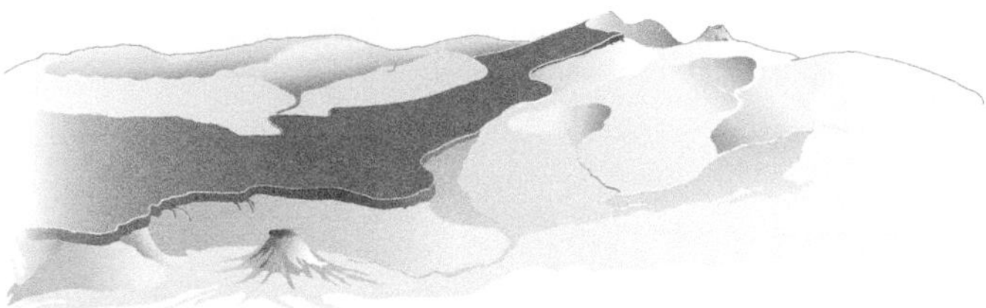

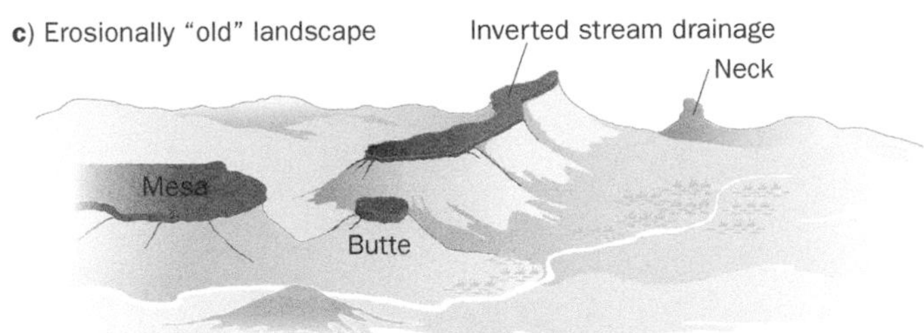

FIGURE 9.41 Erosion of a temperate or desert landscape with a cinder cone and a lava flow.

FIGURE 9.42 Slot canyon formed where New York Creek cuts through the highly resistant, moderately welded portion of the 2050-year-old Okmok ignimbrite [4], Umnak Island, Alaska. Note columnar jointing. The gentler slopes above this consist of the unwelded portion of the deposit and younger volcanic tephra layers. Source: Photo by Richard Hazlett.

Questions for Thought, Study, and Discussion

1 Why do some oceanic shield volcanoes have rift zones in their flanks, while others lack them?
2 Why are some domes endogenous, others exogenous, and some *both* during their construction?
3 If a volcano like Pinatubo appears from a distance merely to be a "non-majestic" group of hills in many places under heavy vegetation, how would you then determine that it is a potentially active volcano? Or would this even be possible?
4 A "perfectly formed" cinder cone is symmetrical, with a circular ground plan. But many cinder cones are asymmetrical or even horseshoe shaped. Why?
5 Consider two composite cones; one with uniformly angled slopes and a small funnel shaped crater on top; the other with slopes that grow much steeper toward its broad summit, which lacks a crater and is approximately level – though quite rough in places. What might explain the differences in the slope aspects of these two cones?
6 Why are some composite volcanoes isolated edifices, while others occur in overlapping clusters (volcanic complexes) or in closely spaced alignments?
7 What factors might be important for determining the pathway a volcano follows toward extinction?
8 Compare the erosional decay of an extinct composite volcano of moderate size built on a landscape near sea level in the tropics with one in the subarctic also built near the sea level, and one in a temperate latitude desert. How and why will these three differ in appearance as erosion progresses?

FURTHER READING

Huggett, R. (2007). *Fundamentals of Geomorphology*. London: Taylor and Francis Publishing, 472 p.
Lu, Z. and Dzurisin, D. (2011). *INSAR Imaging of Aleutian Volcanoes*. Berlin: Springer Praxis, 300 p.
Reidel, S.P., Camp, V.E., Ross, M.E., et al. (eds.) (2013). *The Columbia River Flood Basalt Province*. The Geological Society of America Special Paper 497, 440 p.
Thouret, J-C. and Chester, D.K. (2005). *Volcanic Landforms, Processes and Hazards*. International Association of Geomorphology (IAG) Working Group on Volcanic Geomorphology. Zeitschrift für Geomorphologie, Supplementbände vol. 140, 231 p.
Tilling, R.I., Topinka, L., and Swanson, D. (1990). *Eruptions of Mount St Helens; Past, Present, and Future*. US Geological Survey General Interest Publication, 56 p.

CHAPTER 10

"Negative" Volcanic Landforms – Craters and Calderas

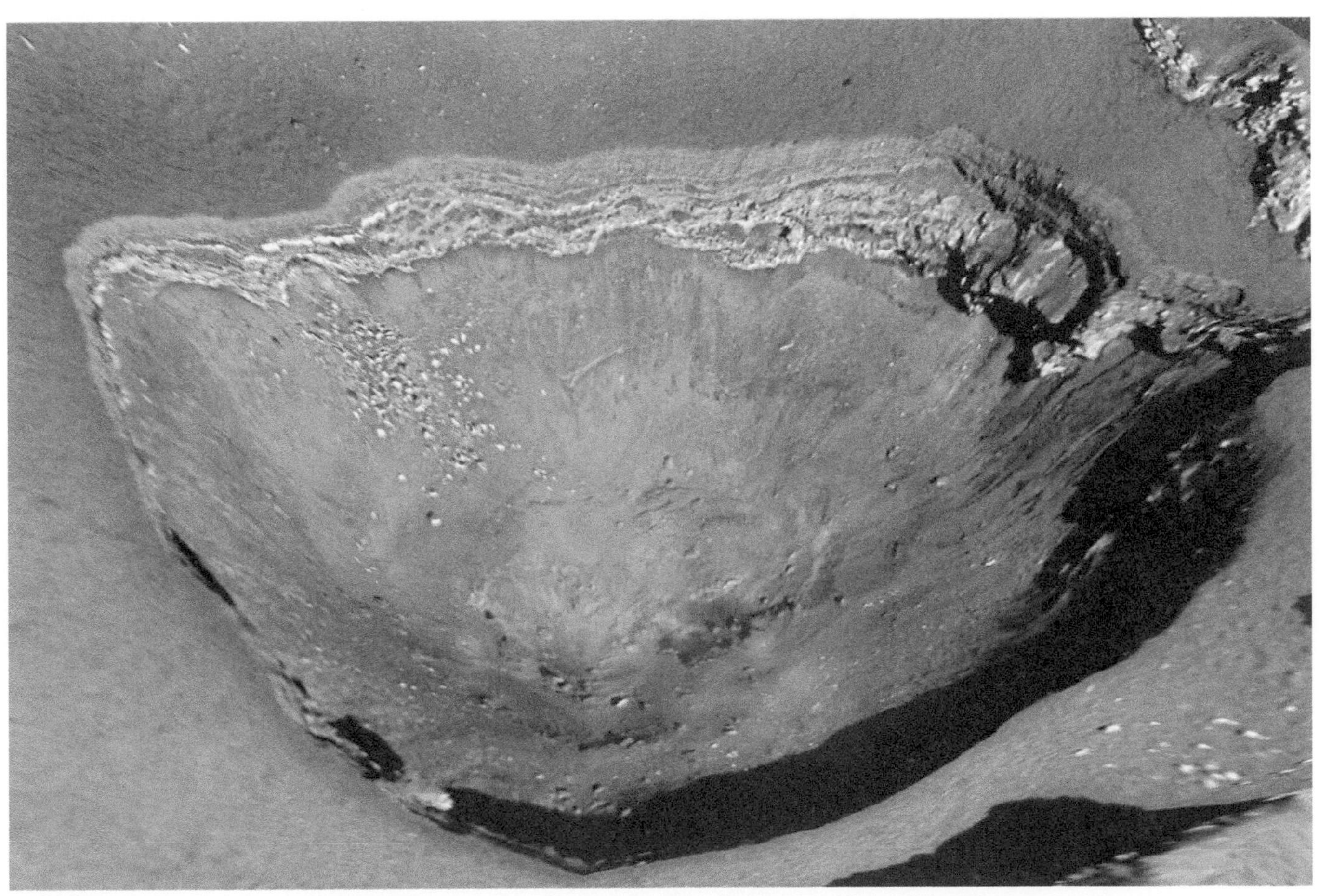

Source: Google Earth.

Volcanoes: Global Perspectives, Second Edition. John P. Lockwood, Richard W. Hazlett, and Servando De La Cruz-Reyna.

Companion website: www.wiley.com/go/lockwood/volcanoes2

Craters of large-scale mountainous volcanoes, instead of marking the site of eternal constructive energy, mark rather the place of sudden withdrawal of lavas which sank back into the depths before their complete solidification.

[Alphons Stübel 1903 (Quoted in Jaggar 1947, p. 360)]

Volcanic craters form in many different ways and come in many different sizes and shapes (Figure 10.1). Some (mostly smaller) craters form by the construction of rims around eruptive vents or by the explosive ejection of rim-forming material. Many involve both construction and collapse mechanisms in their origin. The small cup-shaped craters atop cinder cones and the broader ones of ash cones exemplify such "constructional" craters, and typically span a few tens or hundreds of meters in diameter. Others (including the largest craters on Earth) form primarily by collapse above magma chambers. The largest collapse craters are termed calderas, and will be a major focus of this chapter, since their formation commonly accompanies the most violent and devastating eruptions known. Although geologically active calderas pose a risk to millions of people, they are also beneficial to human society, since younger ones contain vast reserves of geothermal energy, and older ones are the hosts for rich metallic ore deposits (Chapter 15). We will not only discuss calderas from a morphological perspective in this chapter, but also summarize what has been learned about the mechanics of their formation.

Small Craters

Many eruptions vary in intensity, so that one crater rim may be destroyed as blasts widen the crater or as the crater enlarges by collapse. Outer rims may be left to enclose younger crater rims developed by smaller blasts at a later time. Small-diameter crater rims lying within wider ones form **nested craters**. They are common on composite volcanoes, and on some cinder cones. **Multiple craters**, which overlap one another, develop where foci of explosions shift during an eruption or from one eruption to the next. Only the youngest crater in a coalescing set will show complete development of a rim (Figure 10.1).

Survival Tips for Field Volcanologists

Pay attention to sounds when on eruptively active volcanoes – especially along the margins of craters. Rocks commonly begin to fracture slowly at first as they begin to fail – making audible cracking sounds before sudden failure. Major phreatic explosions have been preceded by audible sound changes. Monitor the sounds a volcano makes, and be concerned if those sounds begin to change!

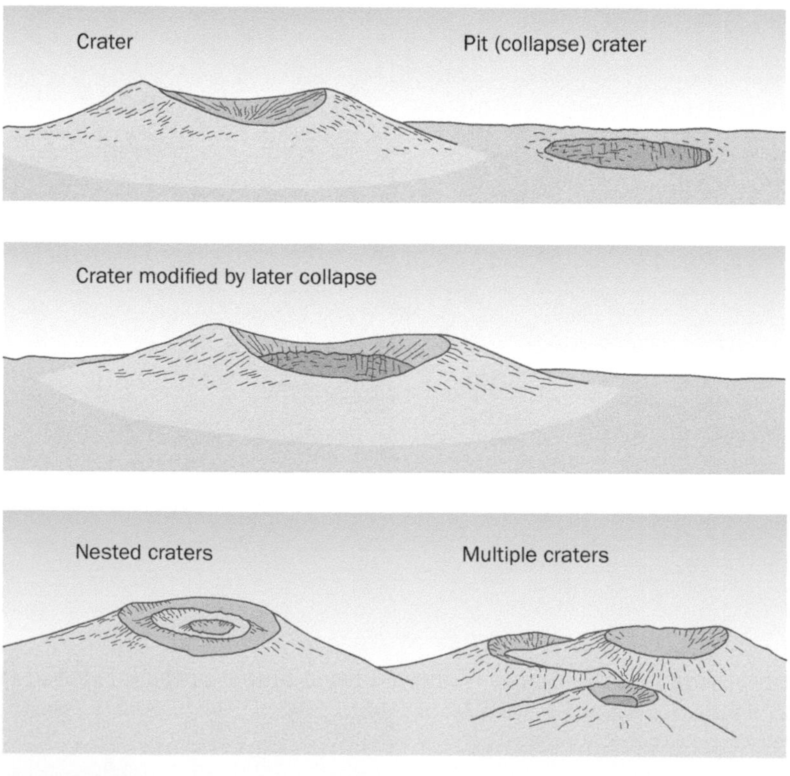

FIGURE 10.1 Types of volcanic craters.

Explosion Craters

Some craters form by purely hydrothermal explosive activity – where no magmatic activity is directly involved. Their rims are formed by the ejection of older country rock blasted out by the action of violent gas explosions. Fluid lava flowing over swampy ground created a vast field of phreatic "pseudo-craters" in the Myvatn [101] area of Iceland. Other good examples are the explosion craters found in many geothermal areas – where violent phreatic blasts excavate the landscape. Numerous examples of such features are found in Yellowstone National Park [47], where prehistoric hydrothermal explosion craters as much as 2 km in diameter have been identified (Chapter 7; Muffler et al. 1971). Some of these explosions evidently resulted from the rapid depressurization of shallow hydrothermal systems when overlying lakes suddenly drained.

Craters repeatedly have formed accompanying explosions of summit domes at the summit of Popocatépetl volcano [56], central Mexico (see the discussion in Chapter 9), but collapse processes are also involved when magma drains from high-level magma storage areas (Figure 10.4).

Explosion craters are typically bounded by low, gently sloping rims that form as the largest fragments and the largest proportion of fragments of all sizes fall closest to the vent. Circular explosion craters with low rims of ejecta may resemble meteorite impact craters, and fieldwork may be necessary to resolve their origins.

Collapse Craters

In contrast to craters formed by the construction of their rims, **collapse craters** lack ejecta rims and tend to have very steep walls, with talus-filled floors that may taper to a point at the center of the crater. The roof rock that drop as such craters form normally shatters. Collapse craters are all formed by the drainage of magma from underlying shallow magma chambers or other intrusive bodies.

Pit crater is the generic name for small collapse craters less than a kilometer or so in diameter (Figure 10.3). Pit craters are common in the rift zones of young Hawaiian shield volcanoes, especially near summits where magma withdrawal to supply flank eruptions at lower elevations can easily remove support from the shallow crust.

In some instances, the subsiding roof block in an opening crater descends as a single coherent piece, perhaps tilting as it does so. This happened, for instance, in Lua Poholo Crater at the summit of Mauna Loa [15], where a level floor that formed when a flow poured into the crater after its initial formation tilted on one side during later subsidence, forming a **trapdoor crater** (Figure 10.3). Roche et al. (2001) did experimental studies of pit crater formation using silicone molds. They determined that if the source cavity is quite shallow – less deep underground than it is wide – the roof will tend to collapse coherently, as in Lua Poholo. If the cavity is somewhat deeper than this, then the roof tends to fall by stoping from beneath – that is, by piecemeal collapse – and will develop a pit with initially overhanging walls and a small opening at the surface.

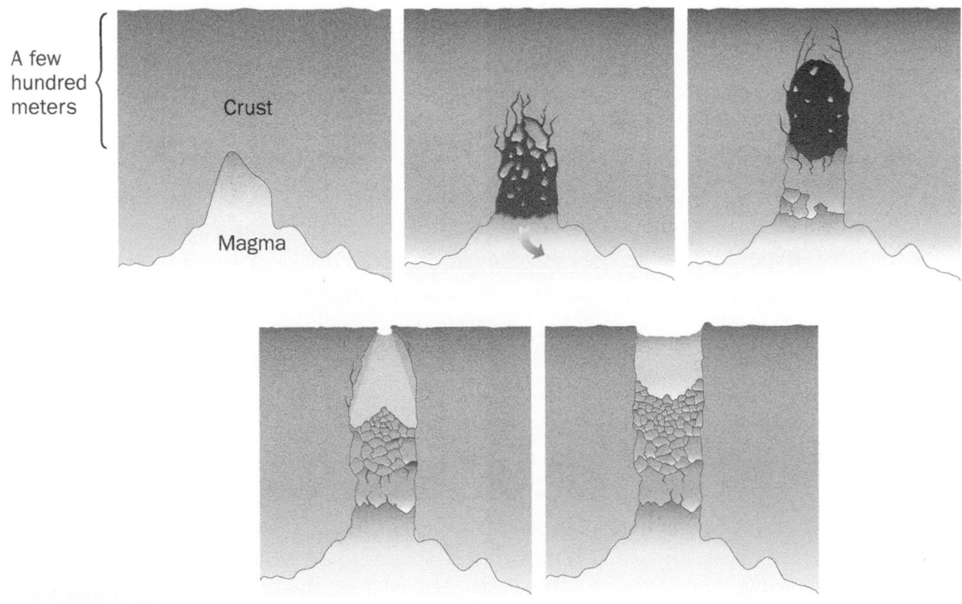

FIGURE 10.2 Pit crater origin, caused by drainage of magma and upward stoping of collapse cavity. Source: Photo © G. Brad Lewis.

FIGURE 10.3 Lua Poholo, a 300-m-diameter, 70-m-deep pit crater on the northeastern floor of Mokuaweoweo caldera, Mauna Loa volcano, Hawai'i [15] is a good example of a "trapdoor" crater. The floor of this crater tilted during collapse as a small underlying magma chamber drained in 1880, feeding an eruption lower on Mauna Loa's flank. Photo by C. Brad Lewis.

On some composite volcanoes, summit craters have deep, pipe-like pits in their centers, indicating that eruptions have a two-stage behavior; an early phase of shallow-focused explosive activity followed by a later stage of magma withdrawal (Figure 10.4). But where does draining magma go in a composite volcano? In part, the decline simply reflects the volume decrease in the underlying magma as exsolution occurs. But as in shield volcanoes, the molten rock also commonly feeds intrusive dikes or sills into the flanks. Radial dike intrusions can lead to large flank eruptions, draining enough melt to stop all but gas emissions at the summit. These emissions can be quite powerful, forming "Vesuvian"-like gas streams that core out the conduit, not only through direct blasting, but by abrading vent conduit walls with loose rubble, like sediment scouring a stream bed. Toward the ends of many explosive eruptions, the volume of gaseous magma lost through pyroclastic ejection exceeds the volume of fresh melt and volatiles ascending from below. Thus, explosion foci and the tops of magma columns drop as eruptions progress. At the end of the activity, a large pit may remain that simply represents the emptied top of the conduit. Dome growth from the remaining, sluggishly ascending magma may later fill this pit to overflowing, leaving no trace of it behind.

Maars (Chapter 9) are another example of craters whose formation commonly combines aspects of both explosions and collapse. Their early phreatic or phreatomagmatic explosive phase, which leads to the development of a tephra ring or ash cone, may be followed by an episode of collapse exposing the wide top of the vent. Most maars are compound structures resulting from shifting of explosion focal points as they form (Graettinger 2018). Deep erosion reveals that many are underlain by diatremes, the wall-rocks of which preserve the evidence of conduit enlargement and collapse during or shortly following eruption (Chapter 4).

Many maars fill partially with water (hence the origin of their name), since their bottoms generally lie below the surrounding water table. On island volcanoes, the intersection of rising magma along radial vents reacts explosively with seawater commonly to form maars in coastal areas. Since most maars in continental settings appear to form where CO_2-rich diatreme magma intersects the surface, they may continue to leak magmatic carbon dioxide through their floors for thousands of years after their formation (Chapter 4). The 10,000-year-old Laacher See maar [105] in northwestern Germany is famous for its springs of carbonated mineral water. Carbon dioxide leaking into waters of the 200-m-deep Lake Nyos maar [107], set the stage for the tragic release of lethal CO_2 in 1986 (Chapter 14).

Fresh maar formation was observed on the Alaska Peninsula in the spring of 1977, when violent phreatomagmatic explosions (Figure 10.5) formed two craters (Ukinrek maars [12]). The initial phreatic explosions sent ash clouds as high as 6 km, and ash drifted as far as 160 km (Kienle et al. 1980). Later magmatic activity constructed a lava dome in the larger eastern crater, but this has now been covered by a lake. The craters formed in an area of previously known CO_2 mofettes, and gas emission continues.

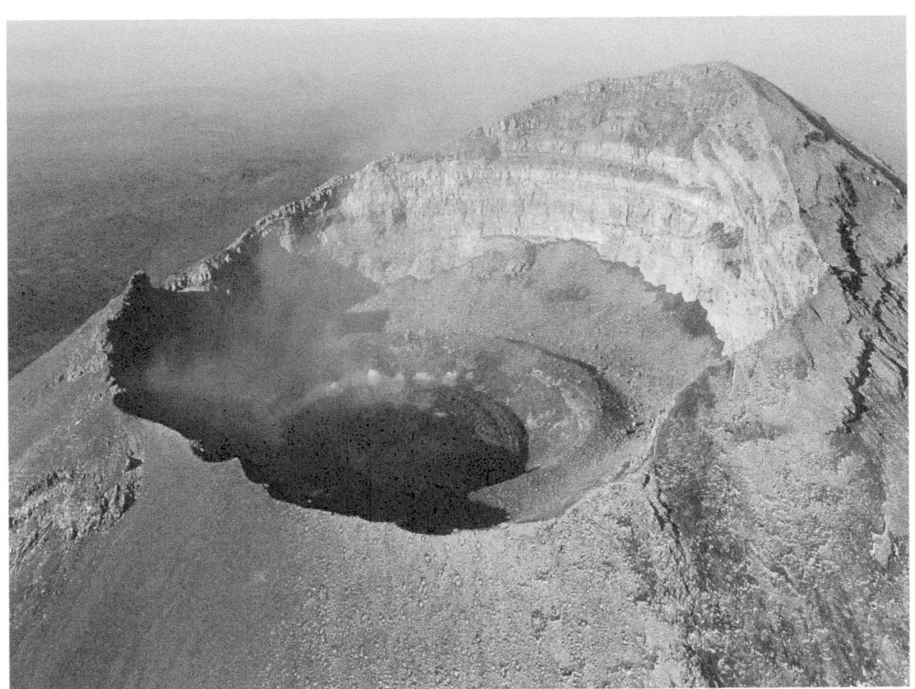

FIGURE 10.4 Aerial view of the summit of Popocatépetl, Mexico [56] showing internal stratigraphy of the cone and a growing lava dome in 2001 shortly before it was destroyed by an explosion. Source: Photo by S. de la Cruz.

Calderas

The term **caldera** has several meanings in volcanology, which has resulted in considerable confusion as to how the term should be used. These are the largest craters of all, by definition exceeding about 1–2 km in diameter. The term was originally proposed to describe any large depression on the summits or flanks of volcanoes – without regard to genetic origins, whether formed by summit collapse or by other processes such as mass wasting, regional tectonic faulting, or the erosion of weak rocks within the hearts of volcanoes. It means "kettle" in Spanish, and was popularized by von Buch after a feature called "La Caldera de Taburiente" on the island of La Palma (Lyell 1855). La Palma is one of the westernmost members of the Canary Island volcanic chain in the Atlantic Ocean off the northwest coast of Africa. The Canarys have a long history of eruptive activity, dating back to 1492, when Christopher Columbus recorded an eruption of Teide volcano as he was preparing to sail west on his voyage of discovery! We now know that La Palma's eponymous "caldera" was mainly formed by erosion at the head of a giant sector-collapse landslide (Chapter 11), and is not a true volcanic crater. [La Palma's Cumbre Vieja volcano [97] has recently been in the news as its first eruption in 50 years caused massive evacuations and destroyed more than 2000 buildings before ending in December, 2021].

Howell Williams, working at Crater Lake [32], Oregon, began our modern understanding of calderas with his classic book *Calderas and their Origin* (Williams 1941). Further insights into the relationships between caldera formation and ignimbrites came in the 1960s and 1970s, largely inspired by the work of R.L. Smith (1960, 1979).

There are various kinds of calderas, but ambiguities about the individual histories of most makes rigid classification systems uncertain. Although some authors have proposed dividing these features into "erosion calderas" and "collapse calderas" (e.g., Karatson et al. 1999), that distinction is difficult as many if not most older features recognized as calderas have been extensively modified by erosion since geologic activity and their initial origins may be debatable. We will here restrict the term caldera to those volcanic features

FIGURE 10.5 Phreatomagmatic eruption during the formation of the eastern Ukinrek maar [12], 6 April, 1977. The formation of these maars on the Alaska Peninsula, 60 km from the 1912 Katmai eruption [20], was the first historical volcanic activity in this area. It took place in an area of previously known CO_2 emissions. Initial explosive activity formed this 300-m-diameter crater, which was later filled by the growth of a basaltic dome. Source: USGS photo by Juergen Kienle.

caused by a collapse of volcanic roofs above underlying magma chambers, although we recognize that their original boundaries may be blurred by subsequent erosion.

Another complexity is that the term caldera sometimes is used to describe volcanic features that no longer have any original topographic expression, but instead consist of the structures and rocks of deep caldera interiors, exposed by erosion in many instances after millions of years. As an example, the classic caldera complex of the San Juan Mountains, Colorado is now largely preserved as a high-standing mountainous terrain, although major caldera centers remain as low-lying areas within the high ranges after nearly 30 million years of erosion (Lipman 1984). In the rugged mountain ranges of the American Southwest, geologists may discern ancient calderas only by the presence of arcuate faults enclosing thick, light-colored tuffs and megabreccias that now stand out in the flanks of mesas and in canyon walls. Detailed geologic mapping and rock compositional analyses are generally required to identify the caldera sources for ancient ash flows in such older terrains, though in many instances, finding a source is simply no longer possible due to post-eruption erosion and burial by later sediments or lava flows.

Most calderas form near the summits of polygenetic volcanoes, directly over or near underlying magma chambers, and are of two general classes. One class, which we call **drainage calderas**, includes those caused by the lateral drainage of underlying magma chambers, typically within basaltic oceanic volcanoes. These calderas form when magma drains to deeper levels, occasionally to erupt at lower elevations on the flanks of volcanoes, causing roofs over their chambers to collapse. The second class, which we refer to as **explosive calderas**, are associated with violent pyroclastic eruptions (Chapters 7 and 8), and are caused by the subsidence of magma chamber roofs as magma is explosively erupted directly above or close to caldera boundaries. This second type of caldera is typically associated with silicic volcanoes of continental or island arc areas. Marti et al. (2009) further divide such calderas on the basis of field relations.

Drainage Calderas

Classic examples of drainage calderas include the summit calderas of Mauna Loa, Hawai'i [15], and Fernandina (Volcán La Cumbre [59]), Galapagos (Simkin and Howard 1970). Mauna Loa's 3 × 5-km-wide caldera, Mokuaweoweo, formed by passive subsidence about 1200 years ago as magma drained to feed a large flank eruption 35 km away and 2500 m lower on the volcano's Northeast Rift Zone (Lockwood and Lipman 1987). It has since enlarged by piecemeal collapse as its floor has undergone repeated periods of partial filling and outflows of lava. At present, the caldera rim looms 180 m above its deepest point (Figure 10.6). The "scalloped" edges of many calderas suggest that they form by coalescence of smaller, discretely formed pit craters, and this is well reflected in the walls of Mokuaweoweo (Figure 10.7). Doubtlessly, many other ancient, long-buried calderas underlie the modern one atop Mauna Loa. These became stacked as the mountain grew kilometers vertically, and much of their structures have likely been obliterated by subsequent hydrothermal activity, faulting, fracturing, and intrusions.

The *smallest* calderas tend to occur on the *largest* volcanic edifices – the intraplate basaltic shields. Although explosive calderas associated with continental volcanoes are much larger, the topographic edifices associated with them may be unassuming. As a basaltic volcano like Mauna Loa inflates with magma and grows through intrusions in its rift zones, the summit is subjected to tremendous tensile stress, which can facilitate summit collapse. Such passive collapse events are typically accompanied by little if any explosive activity, except in cases where high-level water tables allow explosive interaction between groundwater and heated rocks.

One of the best documented recent caldera collapse events occurred at Miyake-jima volcano [148], Japan in 2000, an event that did involve explosive activity. Miyake-jima is a small island volcano along the Izu-Ogasawara island arc system about 200 km south of Tokyo. Seismic activity indicated that magma began to drain from a chamber beneath the volcano's summit in late June along a dike below the west flank, and fed a submarine eruption 2 km off the island coast, prompting an evacuation of all island residents. Miyake-jima's summit began to collapse on 8 July and to form a caldera that continued to enlarge, as phreatomagmatic and pyroclastic surge eruptions devastated the volcano's upper flanks (Figure 10.8). The caldera collapse volume was about 600×10^6 m^3, but only about 9×10^6 m^3 of pyroclastic material erupted, showing that drainage of magma must have been responsible for caldera formation (Nakada et al. 2005).

In April 2007, the summit of Piton de la Fournaise [125] (Réunion Island) collapsed as lava erupted at an unusually high rate of discharge from a flank fissure in the Grand Brûlé, 8 km to the east (Chapter 9). The resulting caldera measured 340 m deep and a kilometer across and had a volume of about 100×10^6 m^3. However, the erupted lava had a volume of 240×10^6 m^3 (Fontaine et al. 2014). This shows that the drained magma at the summit had much less volume than the lower-density material erupted downslope. Unlike at Miyake-jima, caldera collapse was not accompanied by phreatomagmatic activity; the water table lay too deep beneath the summit for critical magma–water interaction to take place. Significantly, however, it occurred in a sequence of steps; 43 incremental drops in the crater floor taking place over a 10-day period, a few meters deeper each time. Following each drop, tiltmeters (Chapter 14) recorded an upward rebound in the ruptured crust around the developing caldera rim, attributed to elastic response much like the relaxation of a rubber band after it has been stretched and snapped (Fontaine et al. 2014).

A very similar sequence of drops-and-rebounds, with net subsidence taking place – occurred at the summit caldera of Kīlauea volcano [17] during the 2018 eruption; 62 episodes or "Type-A" seismic events in all that summer, produced spectacular rockfalls

FIGURE 10.6 View northeast across Mokuaweoweo caldera, Mauna Loa, Hawai'i [15]. The caldera is 4.5 km long and 2.5 km wide, but this foreshortened view across the long axis gives a distorted perspective. Mauna Loa last erupted in 1984; the 1984 eruptive fissure extends across the length of the caldera and 1984 lavas cover the foreground and most of the caldera floor. The present caldera rim is up to 180 m high, but the caldera was estimated to be about 365 m deep when first observed by explorers in 1794. Because lava flows can now flow out of the caldera at either end, it cannot become filled up to higher levels in the short term. Source: Photo by J.P. Lockwood.

FIGURE 10.7 The northwest wall of Mokuaweoweo caldera, Mauna Loa, Hawai'i [15]. The lavas exposed in the 160-m-high vertical wall record the complex caldera history. (A) The youngest (about 1200 year old) caldera overflows from a high-standing lava lake (above thin dashed line); (B) Thick flows that completely filled a pit crater ("Lua Piha" – outlined by thick dashed line) on the flank of an earlier caldera; (C) Earlier caldera overflows. The arrow points to a laccolith intruded into the flanks enclosing an earlier caldera, fed from below by a dike at lower right corner. Source: USGS photo by J.P. Lockwood.

FIGURE 10.8 Formation of Miyake-jima caldera [148], Japan. (a) 9 July, 2000, as caldera collapse began in response to magma chamber drainage. (b) 4 June, 2001, following further caldera formation and accompanying eruptive activity. Source: Photos by S. Nakada and T. Kaneko.

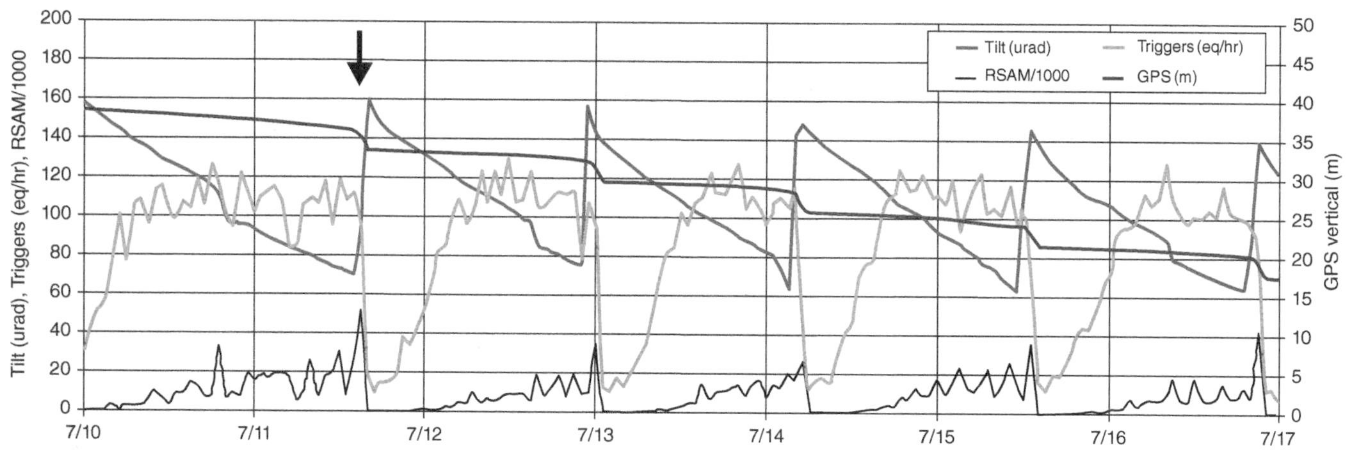

FIGURE 10.9 Plot showing the stepwise collapse in the floor of Kīlauea [17] caldera over a period of a week during the 2018 summit eruption. Red line shows changing elevation in the ground surface at a GPS station mounted on the sinking caldera floor, with sudden drops associated with "Type-A" collapse episodes (black arrow is the first in this week's set). Yellow line is counts of earthquakes per hour (unscaled here). Gray line, labeled "RSAM," is also related to seismicity (Chapter 14). Blue line is tilt, or the angle of slope measured by a nearby tilt recording station. These data illustrate the elastic rebound in the summit crust as stress was released with each collapse episode. Source: Adapted from Brian Shiro and Ingrid Johansen, U.S. Geological Survey.

on the caldera rim (Figure 1.12). These collapse events are the best monitored examples of drainage caldera formation in history (Figure 10.9). Kīlauea caldera has the most detailed historical record of any active caldera on Earth. Photographs and sketches recorded over the past two centuries document the dynamic nature of the relationship between magmatic and deformation activity (Gaddis and Kauahikaua 2021).

Explosive Calderas

Relatively small calderas are typically associated with the summits of collapsing composite volcanoes; larger calderas form from major edifice collapses associated with large magmatic systems that may have originally supplied multiple volcanoes. Irrespective, the explosive silicic eruptions that accompany this category of calderas are typically sub-Plinian or Plinian in character; roughly speaking, the larger the caldera, the greater the magnitude of the eruption. A well-studied example is the collapse of Mount Mazama, Oregon [32] and formation of Crater Lake caldera (Williams 1942). Mount Mazama is a name given to a petrologically complex group of overlapping low to intermediate shields and composite cones that began to form a high mountain in the central Cascades range about a half-million years ago (Bacon 1983). Volcanism continued as late Pleistocene glaciation modified the volcano's shape and deposited extensive moraines on Mazama's flanks. Between about 27,000 and 8000 years ago, rhyodacitic magma erupted north of the summit of Mount Mazama to form a series of flank cones, domes, and coulées as much as 300 m thick. This period of activity remains preserved as Llao Rock, Grouse Hill, Redcloud Cliff, and other prominences. Then, about 7700 years ago, immediately following emplacement of the Cleetwood rhyodacite flow, the climactic, caldera-forming eruption began (Bacon and Lanphere 2006). Bacon calculated that about 30 km^3 of rhyodacitic magma was initially produced as tephra and ash flows from a single vent

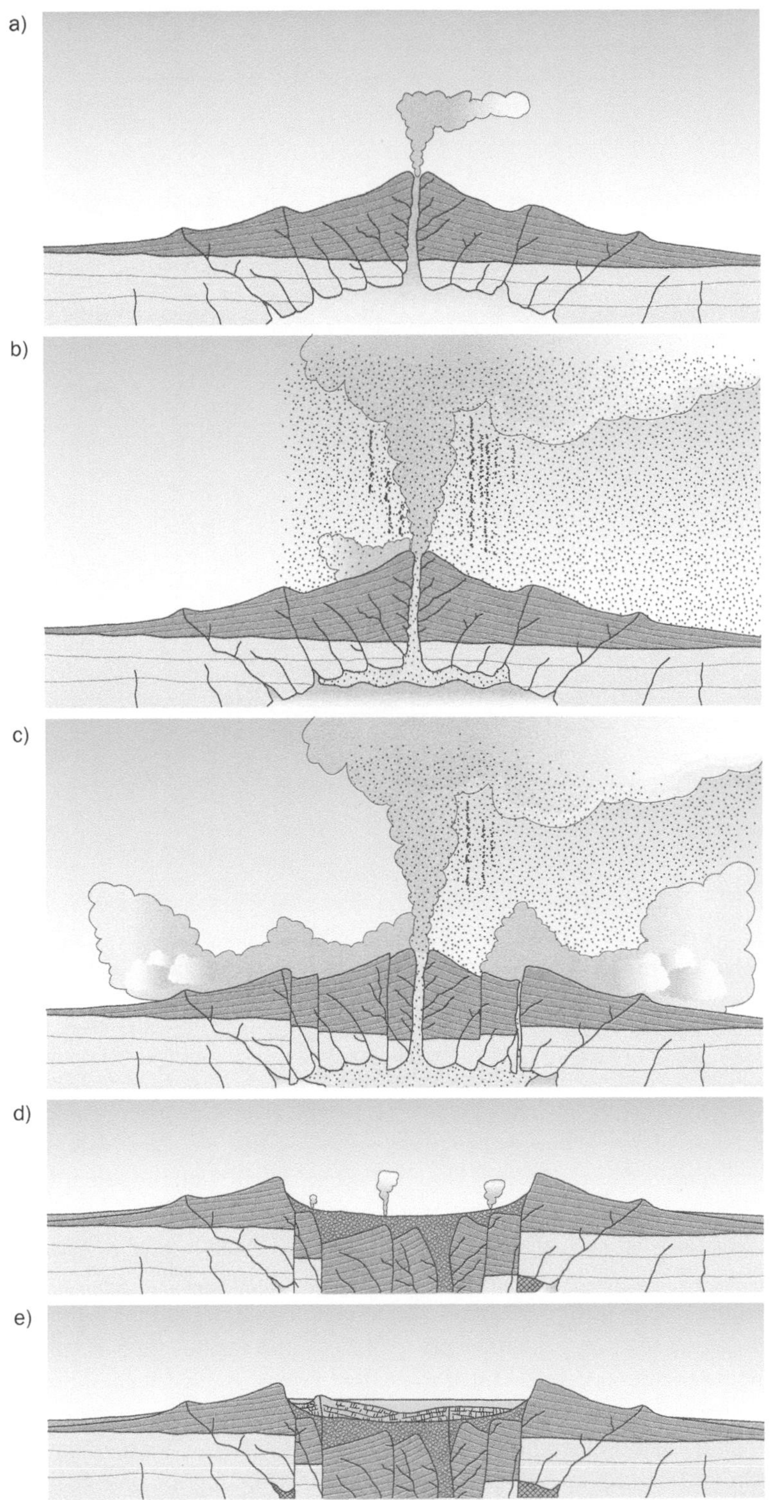

FIGURE 10.10 Cross-sections showing the step-by-step collapse and partial regrowth of Mount Mazama to form Crater Lake [32], Oregon. Source: Modified by C.R. Bacon after Williams (1942).

on the volcano's north flank – before caldera collapse began – with ejecta falling as far away as Alberta, Canada. Another 20–30 km^3 of magma then erupted from ring faults as Mazama's entire summit area – an estimated 40–50 km^3 of solid rock – sank into the underlying, partly evacuated chamber, forming the present-day Crater Lake caldera (Figure 10.10). Pyroclastic flows ejected during collapse rushed down pre-existing canyons burying glacial moraines; and extended as much as 70 km from their vents. While the ash flows continued to erupt, the composition of the magma changed from rhyodacite to andesite and gabbroic crystal mush, while the color of the pumice changed from pale to dark gray. Since its formation, the caldera has partly filled with water from rain and melting snow, and also partly with material from later andesitic eruptions, one of which constructed the cinder cone and blocky `a`ā flows of Wizard Island (Figure 10.11). Another large post-caldera cone is wholly submerged in the lake. The late-stage eruption of low-silica magma into continental calderas that formed during explosive silicic eruptions is a common occurrence, and further evidence for the role of mafic melts in initiating these eruptions.

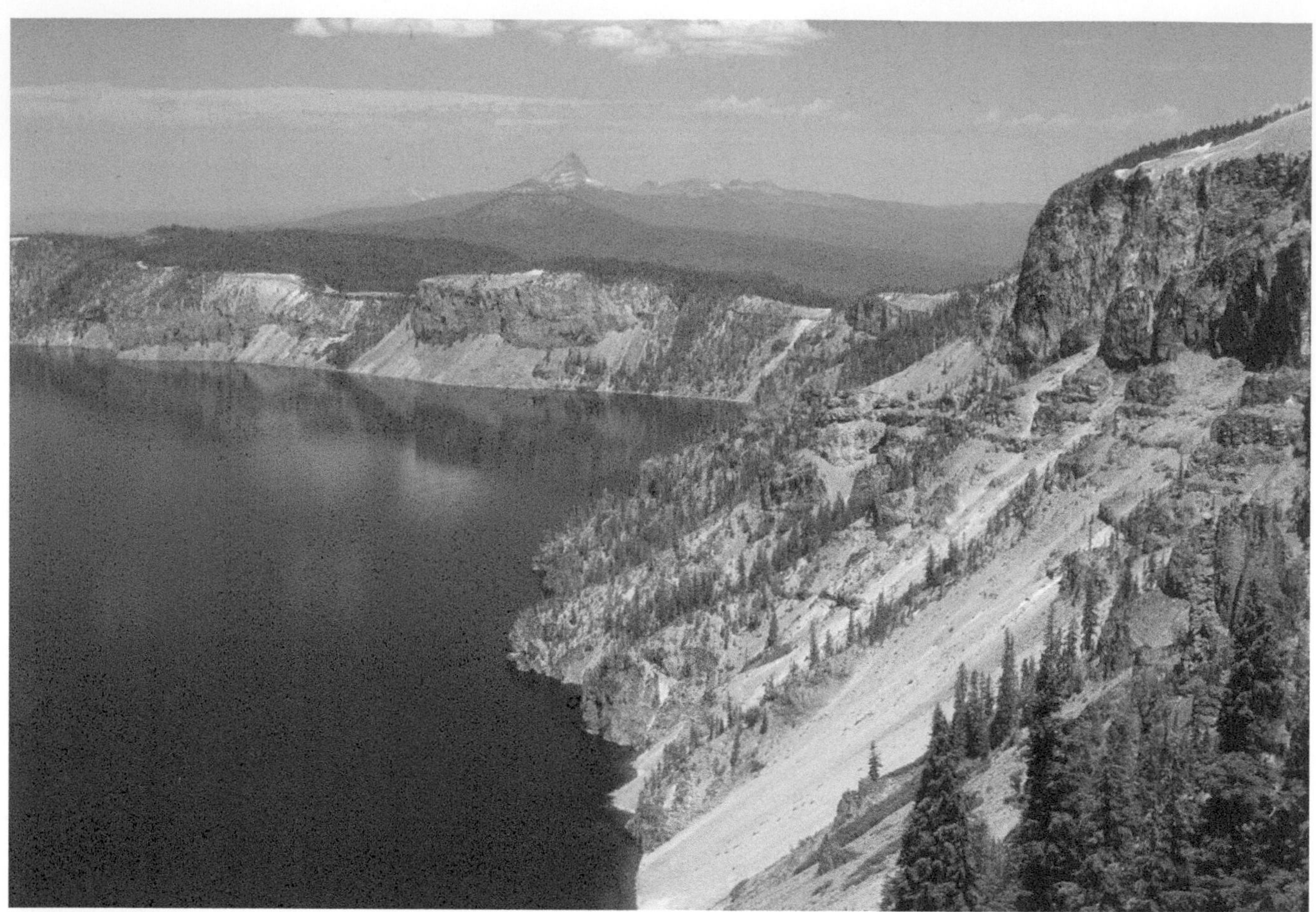

FIGURE 10.11 Eastern rim of Crater Lake caldera [32] – view to north. Relatively thin andesitic and basaltic flows in the lower slope are capped by much thicker glassy dacitic coulées above, which erupted from the side of Mount Mazama shortly before caldera collapse. Their extrusion signified the impending destabilization of the underlying magma reservoir. The eroded volcanic neck of Mount Thielsen [33], known as the "Lightning Rod of the Cascades" by regional mountain climbers, is on the horizon. Thielsen was last active 100,000–250,000 years ago (Harris 2005). Source: Photo by Richard Hazlett.

The frequent occurrence of somma rings that break the slopes of many younger composite cones, for instance, Vesuvius [109] and Merapi [131], indicates that composite volcanoes often collapse in the manner that Mount Mazama did, though the example of Crater Lake is especially large and spectacular. That many composite cones do not experience such collapse may be because the magma chambers underlying them are generally too deep and small for their roofs to fail. A critical ratio, as in the case of pit craters, relates the width of a magma chamber to the thickness of its roof. The analog modeling of Acocella and others (2001) suggests that if this ratio is less than 1:1, the roof will be strong enough to support long-lasting composite volcanoes. If greater than this – that is, if the magma chamber has a very thin roof – then a major volcanic edifice may not develop at all. Composite cones such as Mount Mazama possibly represent volcanoes whose underlying magma chambers come to lie at some depth close to this critical ratio.

As discussed in Chapter 4, depressurization on magma chambers can lead to eruption – and in the case of especially large, shallow magma bodies, caldera collapse. But the development of fracturing required for this to happen does not have to originate right at the ceiling of a magma chamber. In fact, the crust there may be so hot and ductile (viscoelastic) that fracture propagation permitting shallower intrusion is simply impossible. In such instances, uplift of the entire roof and gradual faulting that propagates *downward* from the surface may be sufficient to induce catastrophic eruption – an "external" trigger (Gregg et al. 2012). In some cases, pre-existing faults may influence caldera development, as at Deception Island, Antarctica [87], where a polygonal grid of normal faults related to regional extension guided collapse and associated eruption of 90 km^3 of basaltic-andesitic pumice and ash (Marti et al. 2013). The physical properties and condition of the magma chamber roof, in addition to aspects of size and depth, introduced above, are significant for caldera collapse (Gregg et al. 2015).

Very large, shallow magma chambers develop the largest calderas of all, the Cenozoic record being held by 75,000-year-old Toba caldera [127] in Sumatra, which measures 40 × 105 km (Chesner et al. 1991). Such huge calderas typically form independently of any single large volcano. Many, if not all, appear to develop in association with major volcanic complexes. Caldera collapse destroys these edifice clusters, though volcanic eruptions may resume afterward to build new ones. The area of structural subsidence does not exceed and may be somewhat less than the maximum diameter of the underlying magma chamber.

Five Modes of Caldera Collapse

Anderson (1937) referred to the processes of caldera formation as **cauldron subsidence**, but he could not prove the relationship with surficial volcanism. That relationship is now well-established, and it has been noted that eroded mid-Tertiary calderas in the San Juan volcanic field serve as examples of structures that must underlie major active volcanic fields in the Andes (Lipman and McIntosh 2008). Caldera collapse mechanics and geometries vary widely, however, and it can be said that no two calderas form in exactly the same way. Lipman (1997) described five types of caldera collapse geometries, adding that examples exist that show structural gradation between each:

1. **Plate** or **piston** calderas, first proposed by Anderson (1937), involve floor crust, which remains intact as it drops along enclosing ring faults, like a piston going down a cylinder. This is the "classic" model of caldera collapse. As already discussed, single, high-volume Plinian eruptions take place during this process, commonly breaking out along eruptive fissures rimming the edge of the dropping block, in whole or part. The La Garita caldera in the San Juan Mountains of Colorado, mentioned earlier, appears to belong to this group. The La Garita [48] roof block dropped over 2.5 km, accompanying eruption of the mammoth Fish Canyon Tuff.
2. **Piecemeal calderas** have floors broken into blocks that have subsided to different levels. They resemble the "chaotic-collapse calderas" described by Scandone (1990), but rather than resulting from a single collapse and eruption event, piecemeal calderas are the result of multiple subsidence events and eruptions, resulting in compound or nested caldera features.
3. **Trapdoor calderas** result from eruption and collapse along one side of a ring fracture system, with floors sloping at an angle toward the location of maximum subsidence (Steven and Lipman 1976). Fisher Caldera [7] formed in a trapdoor manner, as did the La Pacana caldera [83], Chile, which has among the world's largest "resurgence" structures (explained in the following text; Lindsay et al. 2001). Trapdoors may display pre-caldera stage vents only on the collapsed margin (Steven and Lipman 1976).
4. **Down-sag calderas**, originally described by Walker (1983), overlie magma bodies that are nearly too small or deep to form calderas. The land simply sinks without rupturing. Down-sag calderas are not as common as the other types discussed above.
5. **Funnel calderas** may be even less common. Funnel calderas are small calderas whose enclosing faults converge downward into wide pipe-like conduits. Lipman (1997) believes that many funnel calderas have been misinterpreted, owing to the retreat of caldera rims by syn-eruptive landsliding to produce inward-sloping boundaries. Most are probably piston calderas. Nevertheless, some true funnel calderas such as Nigorikawa [151] in Japan exist, where exploration drilling has demonstrated the subsidence of roof rocks several kilometers downward into a narrow vent. Funnel calderas are typically small depressions, only a few kilometers across. They lack the related resurgence structures and volcanic fields of larger calderas.

No matter how they form, recent advances in high-precision radiometric dating have revealed that major caldera systems can develop their entire cycles of collapse, multiple eruptions, and resurgence in 50,000–100,000 years (Lipman and McIntosh 2008). Though most silicic calderas form coincidently with the explosive eruption of large amounts of tephra and PDCs, not every such caldera shows this association (e.g., Gilbert et al. 1996; Giannetti 2001). As is well documented by petrologic and field studies, the ultimate cause of the magmatic instability that culminates in silicic caldera-forming eruptions is related to input of heat from basaltic melts ascending into continental crust from the mantle (Bachmann et al. 2002).

Volcanic fields related to large-volume caldera collapse exist in almost every conceivable volcanic setting. They are especially common in some continental volcanic arcs and back-arc environments. Fisher Caldera [7] in the Alaskan-Aleutian volcanic arc includes a typical example of a recently formed caldera-related volcanic field. The caldera lies between two very large, independently formed composite volcanoes, Shishaldin [8] and Westdahl [6], near the terminus of the Alaska Peninsula. Measuring 12 × 18 km, Fisher is the largest of the dozen Holocene calderas within the arc. A ridge largely encloses the caldera, having gentle outer flanks and steep inner slopes. It largely consists of the remains of small volcanoes that existed prior to caldera collapse. Because collapse and subsequent volcanic activity in and around the caldera disrupted stream drainages, two large lakes have pooled on the caldera floor.

The oldest discernable episode of volcanism at the site of Fisher Caldera consisted of widespread basaltic lava effusions between 450,000 and 200,000 years ago, building up the tableland on which the present-day volcanic field partly lies (Stelling et al. 2005). A long erosional hiatus ensued, with renewed activity beginning around 40,000 years ago. Eight small composite cones grew between then and 9000 years ago, each no wider at the base than 2–3 km. The cones grew in a ring defining the rim of the future caldera, with most of them clustered on the west side, suggesting a trapdoor geometry. They represented incipient fracturing in the roof of a young, shallow-level magma body, probably different than the one related to the much older outpourings of tableland lavas. Magma exploited the roof fractures during intervals of over-pressurization to build the cones. Pressure in the chamber may have shoved the roof straight upward, partly decoupling it from the surrounding crust. Evidence of a similar

pre-caldera stage of development also occurs at many other caldera-related volcanic fields, and at composite volcanoes like Mount Mazama [32] (Figures 10.10 and 10.12). The pre-caldera Glass Mountain obsidian domes along the eastern side of California's Long Valley caldera [44], source of the Bishop Tuff (Chapter 4), are one easily visible example. Along the eastern edge of Yosemite National Park, not far to the north of Long Valley, the 400–6000-year-old Mono Domes form an arc that may outline part of the rim of a future caldera.

The pre-caldera phase of volcanism at Fisher Caldera [7] ended abruptly 9100 years, when a Plinian point-source eruption broke out at one of the ring cones, or perhaps from a fresh vent opening in the midst of the pre-caldera volcanic field. In any case, piecemeal collapse across an area measuring 10 × 16 km ensued, and in a period probably not exceeding a few days some 70 km^3 of material spewed out in the form of pumice-rich ash flows that traveled northwest to the Bering Sea. The volume erupted was more than double that of the 1883 Krakatau [128] eruption (Stelling et al. 2005). The spreading pyroclastic currents were powerful enough to overtop Tugamak Ridge, more than 500 m high, 15 km from the vent (Miller and Smith 1977). Finer ash fall drifted hundreds of kilometers farther and covered thousands of square kilometers.

What must such a vast area of total devastation look like soon after a major pyroclastic eruption? One can possibly imagine by reading the account of Robert Griggs after coming upon the still-hot ignimbrite plain of Novarupta [19], Alaska, four years after the 1912 eruption (Griggs 1922, pp. 191–192; Figure 7.28):

> *The whole valley as far as the eye could reach was full of . . . literally tens of thousands . . . of smokes curling up from its fissured floor . . . Some were sending up columns of steam which rose a thousand feet before dissolving . . . After a careful estimate, we judged there must be a thousand whose columns exceeded 500 feet . . . It was as though all the steam engines in the world, assembled together, had popped their safety valves at once and were setting off surplus steam in concert . . . Some of the fumaroles were seen to be closely grouped in lines along common fissures; others stood apart . . . In some cases the steam issued from a large deep hole; in others there was no opening at all – the vapors simply escaped through the interstices of the soil particles . . . There was no relation between the size of the vent and its output. Some of the largest jets had very narrow throats, while from some cavernous holes there issued only faint breathes of steam. In many cases steam poured from the sides of the drainage gullies, where it did not break through the more compact surface layer of (ash) . . . The ground was hot in places where no visible exhalation was being given off, as we found to our dismay when we sat down on a bank seemingly safe enough.*

The development of a caldera by no means ends the life of the underlying magma chamber. Influxes of new melt from below usually continue, re-pressurizing the reservoir and forcing new eruptions during the **post-caldera stage** of development. After a period of tens or several hundreds of thousands of years, other Plinian eruptions associated with renewed chamber collapse may occur from different magmatic sources in the same area, creating new calderas overlapping or lying within the older one. Nested and overlapping (**compound**) calderas show that two or three caldera-forming events may originate from individual

a) Pre-caldera volcanism

b) Caldera development

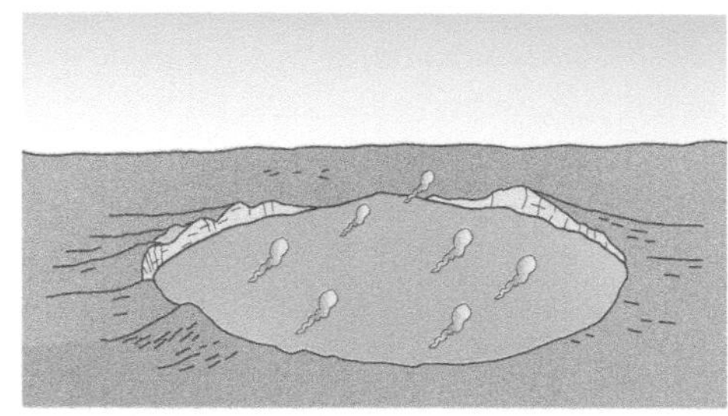

c) Lake filling, hydrological stabilization

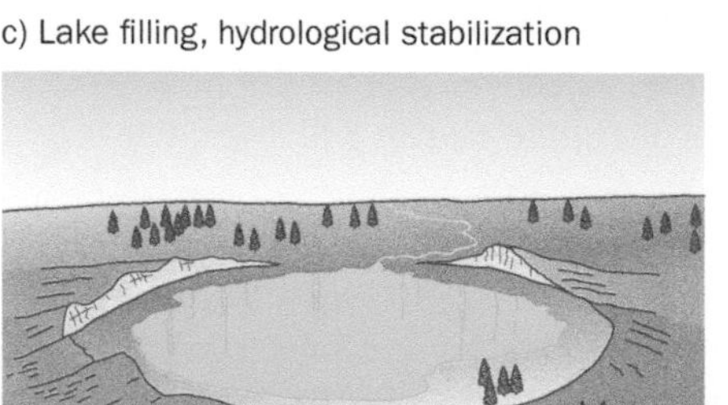

d) Post-caldera volcanism

FIGURE 10.12 Four common stages in the development of a large continental caldera with an associated volcanic field.

magma reservoirs during their lifetimes, as has been well documented in the San Juan caldera cluster of Colorado (Lipman 2006; Figure 10.13). The middle Tertiary San Juan volcanic field is the most active area of the much larger Southern Rocky Mountain Volcanic Field (SRMVF) which was active 37–23 million years ago. The SRMVF originally covered over 100,000 km^2 from northern New Mexico into central Colorado (excluding far-field ash deposits of unknown extent) and generated more than 60,000 km^3 of eruptive products (Lipman et al. 2015). The giant La Garita caldera formed during a single initial eruption pulse about 28 million years ago, but with multiple subsidence segments forming concurrently – perhaps like Fisher Caldera, described earlier. Eruption of the resulting Fish Canyon Tuff, comprising at least 5000 km^3 of ash flows and tephra, is the most voluminous pyroclastic eruption yet documented anywhere on Earth. An additional 2500 km^3 of unaccountable ash and pumice may have accumulated as air fall debris downwind. Over the next 1.1 million years, an additional seven calderas formed within the giant basin of La Garita (Figure 10.13; P. Lipman, pers. comm. 2008). The isolated, older Bonanza Caldera to the northeast is remarkable because the pre-caldera floor on which the 20 × 25 km caldera developed is exposed, and the entire sequence of caldera subsidence, voluminous pyroclastic flow production, and resurgent dome activity has been reconstructed by careful geologic mapping (Lipman 2020).

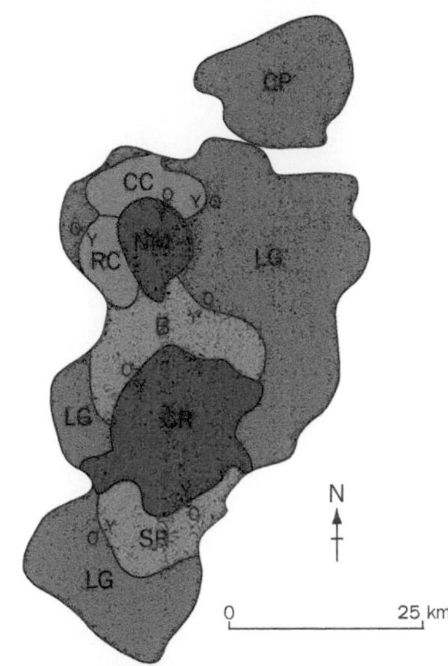

FIGURE 10.13 Geometry of giant nested calderas in the San Juan caldera cluster, southern Rocky Mountains, Colorado. These overlapping calderas formed over a period of less than a million years in Oligocene time (Lipman and McIntosh 2008). LG, La Garita [48] (source of the giant Fish Canyon Tuff), CP, Cochetopa Park, which preceded formation of the later smaller calderas: B, Bachelor [38]; CC, Cebolla Creek; CR, Creede; NM, Nelson Mountain; RC, Rat Creek; SR, South River. Younger calderas in brighter colors; y/o indicates younger vs. older contact relations. Source: Based on Lipman (2006).

Much younger than the San Juan volcanics, the Yellowstone Plateau of Montana–Wyoming encloses a compound caldera some 60 km in diameter formed during three Plinian eruptions, 1.8, 1.3, and 0.6 million years ago (Christiansen 2001). The collapse of the 1.3-million-year-old Mesa Falls caldera, part of this complex, began along a vent system 30 km long, which released silicic magma that had been in storage for only a few thousand years – a very brief interval between intrusion and eruption at this scale (Stelten et al. 2018). Hydrothermal activity, including hot springs, geysers, and travertine deposition, has developed around the margins and within the caldera complex, apparently heated by still restless magma. Calderas make excellent "chimneys" for the circulation of mineralizing fluids following their formation, and many of the world's most important metallic ore deposits are associated with caldera-fill rocks (Chapter 15).

As explosive calderas form, pyroclastic flows and tephra commonly fill the developing depression simultaneously during roof collapse, and typically rush over the rims to bury vast areas. This indicates that Plinian eruption and surface subsidence is a *culminating*, not *causal* event in caldera collapse. By the end of the eruption, so much ejecta may accumulate that no crater may be topographically expressed. Instead, a gently arching plateau of steaming pyroclastic debris may occupy the site of the caldera, although usually some depression remains to trap water – if the climate and water table favor this. The sedimentary deposits of intracaldera lakes often contain valuable clues about the history of post-caldera volcanism, especially caldera resurgence as described below (Murphy et al. 2016; Gooday et al. 2018). It is important to note that the *visible* depth of a caldera is usually much less than the *actual* amount of structural subsidence, in some cases, by an order of magnitude, owing to pyroclastic infilling synchronous with bedrock subsidence. Steep caldera rims commonly disintegrate soon after foundering of a caldera roof commences, shedding huge volumes of country rock that can be preserved as breccia and large, detached slices of bedrock inter-fingering with pyroclastic deposits. The Picayune Megabreccia formed where caldera walls collapsed during the formation of the 28 Ma Uncompahgre-San Juan caldera in the San Juan volcanic field, and contains blocks greater than 500 m in length (Lipman 1976; Figure 10.15). Early during the collapse of the growing diameter of a caldera is controlled by structural processes, whereas mass wasting tends to dominate toward the end (Geshi et al. 2012). Some calderas, however, show little evidence of modification by mass wasting during their formation.

Sub-caldera Intrusions

Since the early twentieth century, structural geologists have been intrigued by a peculiar class of intrusions known as **ring complexes** (e.g., Anderson 1936). Where exposed by deep erosion, these consist of dikes that describe arcs or have circular outcrop patterns many kilometers in diameter. The dikes may be divided into two groups. **Ring dikes** extend vertically into the earth, or may flare outward with depth, away from the centers of the rings that they form at the surface. **Cone sheet** dikes taper downward toward a common point. Both kinds of intrusive structure typically coexist in the same complex, showing orientation around common centers, and both appear to be related to caldera development, the former principally with magma drainage and collapse,

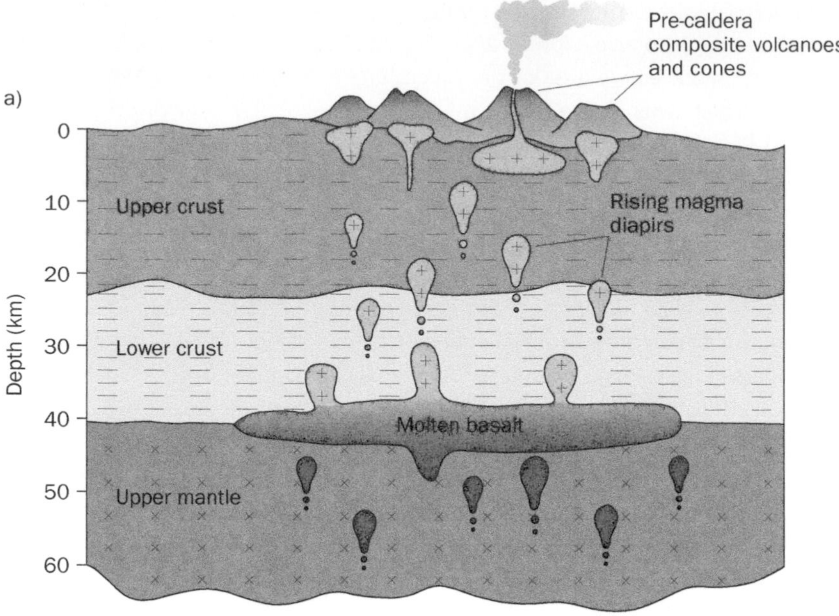

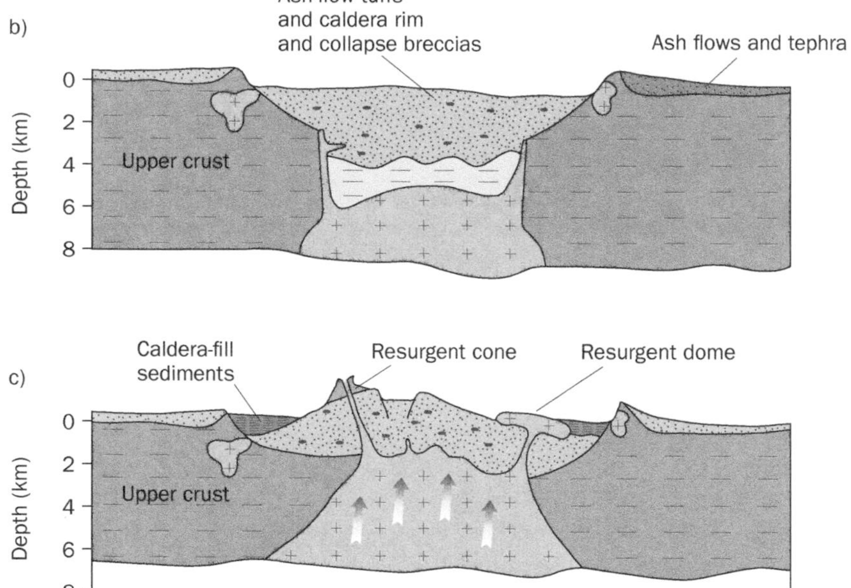

FIGURE 10.14 Model for the origin of silicic ash-flow calderas. Depths shown are schematic approximations. (a) Pre-caldera activity, showing a cross-section of continental crust and upper mantle. A lens of molten basalt from differential melting sources deeper in the mantle is accumulating at the Mohorovicic discontinuity. This hot basalt causes melting and mobilization of overlying crust and the formation of magma diapirs that move upward to supply surface volcanoes and ever-larger shallow magma chambers. (b) The explosive eruption of immense volumes of ash-flow tuffs and tephra has caused the magma chamber roof to collapse along steep marginal faults as the eruption proceeds. The steep walls rimming the caldera collapse, widening the caldera surface diameter as the eruption continues, forming extensive wall-rock breccias that mix with erupting tuffs to partially fill the new caldera. (c) As new magma rising from below continues to inflate the sub-caldera magma chamber, this relatively low-density material causes uplift of the caldera floor and feeds resurgent volcanic activity. Sources: Adapted from Lipman (1984, 1988).

and the latter with build-up of intrusive pressure, and presumably, uplift of roofs. The degree to which ring dikes and cone sheets develop depends in part upon magma chamber geometry (e.g., Gudmundsson 2012), and in part on pre-existing crustal structure (e.g., Schirnick et al. 1999).

Bailey et al. (1924) called the Loch Ba ring dike on the Isle of Mull, Scotland, the "most perfect known example" of a ring dike. It forms a nearly continuous oval 5.8 × 8.5 km in diameter. The maximum width of the dike is about 300 m. Stratigraphic correlation indicates that the bedrock inside the ring complex subsided about 150 m (Lewis 1968), suggesting that the ring dikes intruded along active faults concurrently with downfaulting. Given their great widths in many places and their compositional diversity, Sparks (1988) suggests that ring dikes began forming a long time – perhaps many thousands of years – before any associated caldera collapse.

Research at Rabaul caldera [160] on New Britain Island in Papua New Guinea (Figure 10.14) supports the notion of gradual ring fracture weakening and dike intrusions: a 9 × 14-km-wide Rabaul caldera surmounts a low volcanic shield, and is mostly flooded by the Bismarck Sea across a drowned eastern rim. Matupit Peninsula, next to the regional airport, marks the center of caldera resurgence. Like most major calderas, Rabaul is a compound feature. Its last major collapse 3500 years ago was accompanied by the ejection of 11 km^3 of pyroclastic debris – almost as much material as thrown out by the 1912 Katmai [20], Alaska eruption. Two polygenetic cones, Vulcan and Tavurvur lie 7 km apart along the perimeter of a zone of intense, shallow seismicity that forms

FIGURE 10.15 Caldera-fill breccia associated with rim collapse of the Oligocene Uncompahgre-San Juan caldera, 5 km west of Lake City, northwestern San Juan Mountains, Colorado. Elsewhere, these breccias contain blocks that exceed 500 m in size. Note the angular to subrounded shapes of breccia fragments. Shovel for scale (arrow) is 60 cm long. Source: Photo by J.P. Lockwood.

an epicentral ellipse in the southern floor of the caldera. Hypocenters deeper than about 2 km lie slightly farther from the center of the ellipse than most hypocenters above that level, indicating a zone of crustal failure dipping outward with depth. This is the anticipated pattern of a developing ring fracture system, but with a bit of a twist: Many earthquakes less than about a kilometer deep define inward-dipping fault surfaces, the reverse of the pattern observed for deeper quakes. Saunders (2001) attributes these to **antithetic faulting** (Figure 10.16), a type of normal faulting that essentially cuts off the corners of crustal blocks slipping along larger, opposite-dipping normal faults. Magma and hydrothermal fluids are exploiting this fracture system, as shown by recurrent and occasionally simultaneous eruptions of Tavurvur and Vulcan, most recently in 1994. Continuing eruptive and hydrothermal activity along the Tavurvur (east) side of the seismic zone, with quiescence on the Vulcan (west) side, indicates that deformation is taking place asymmetrically around the caldera.

Under prolonged tension in the shallow crust, ring dikes can form wide, complex networks, ranging in width from a few hundred meters to over a kilometer. They are usually coarse-grained and grade from very mafic gabbros and peridotites to granite. The centers of the ring complexes are commonly occupied by coarse-grained plutonic rocks of the same composition to the ring dikes. In places, ring intrusions extend upward into volcanic rocks in eroded exposures (Macdonald 1972). In contrast, a cone sheet tends to consist of only one or a few dikes at any given point along its ring. Individual dikes may be thin, seldom exceeding 5 m in width, and many are fine-grained and basaltic.

Several hundred cone sheets, most enclosing others, lie in Devonian ring complexes on the Isle of Mull and on the nearby Ardnamurchan peninsula in Scotland. The outermost sheets dip inward at angles of only 35–45°, but those closer toward the centers of the complexes dip more steeply – up to 75°. Projections of the dips of the cone sheets converge toward a focus at a depth of about 5 km beneath the middle of each complex. The structural geologist E.M. Anderson (1937) explained this pattern by suggesting the dikes formed when the upward pressure of a magma body at that focal depth lifted the roof above it at least 300 m, producing upward-diverging fractures that were injected with the melt.

Caldera Roots – Relationships to Plutonic Rocks

The magma chambers once present beneath all calderas are doubtless never completely evacuated, and frozen remnants of these once-molten bodies can be found beneath all calderas – if erosion exposes deep enough levels, and if these plutonic rocks are not made unrecognizable by subsequent magmatic activity or metamorphism. Very large granitic batholiths are widely exposed by erosion on the world's continents (Figure 2.7), and it is probable that most of these once-molten magmatic bodies at one time connected

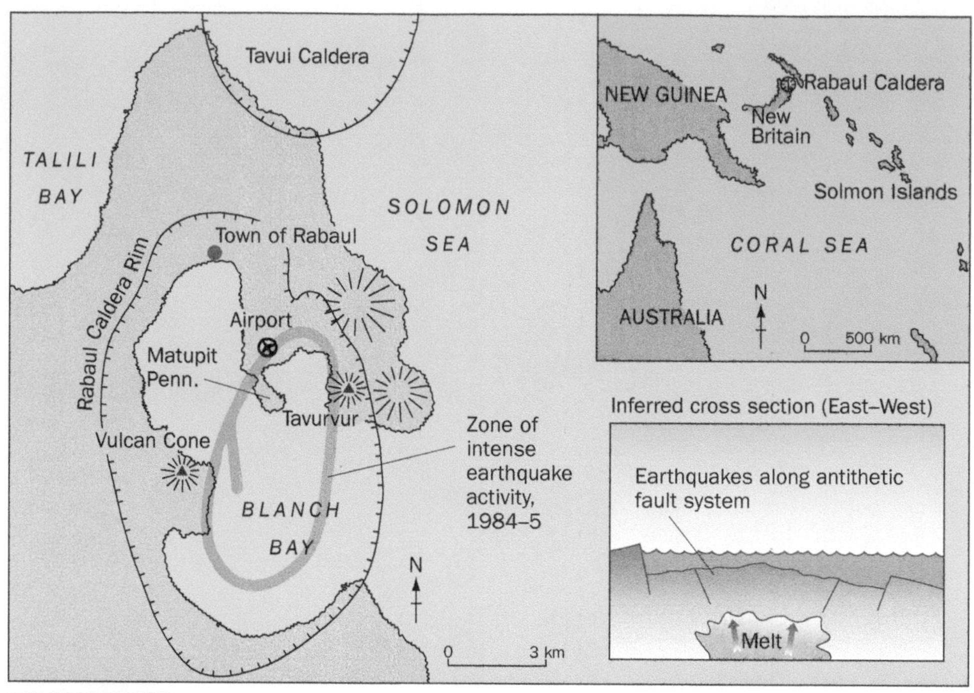

FIGURE 10.16 Rabaul caldera [160], Papua New Guinea. Source: Based on Saunders (2001).

to overlying volcanic systems and in all likelihood were associated with calderas, of which traces rarely remain. The problem in understanding the relationship between caldera formation and plutonism is mostly a matter of identifying areas where identifiable calderas have been eroded to sufficient depths to expose their roots – but not so deeply as to have erased the volcanic–plutonic connections (as in the case of the Scottish ring complexes discussed above). Such areas do exist in many parts of the world, and are especially well-exposed in the South American Andes, and in western North America, as described by Lipman (1984). Bachmann et al. (2002) note that the great, dominantly dacitic ignimbrite sheets that also occur in these regions act as "a link between the plutonic and volcanic realms. Indeed, they resemble erupted batholiths in the sense that they have comparable volumes, occur in the same tectonic settings, and are characterized by high phenocryst contents (~40–50%)" erupted close to crystallization temperatures. Hildreth (1981) refers to these products of continental caldera eruption as the "Monotonous Intermediates," because they lack evidence of compositional zonation, much like the vast bulk of intermediate and silicic plutons found in the same regions. In a classic review of the entire middle Tertiary Southern Rocky Mountain Volcanic Field, Lipman and Bachmann (2015) document the undeniable evidence linking long-lived deep batholith formation to the eruption of voluminous overlying volcanic rocks.

Post-Caldera Resurgence

The post-caldera stage of volcanism is commonly characterized by the uplift of caldera floors, as the underlying magma chamber reinflates. This uplift, which may range from just a few meters to over a thousand meters, is termed **caldera resurgence**, and is typically accompanied by small eruptions that form minor cones and lava flows as illustrated by Okmok [4] (Figure 10.17) and Aso [143] calderas (Figure 10.18). Resurgent activity results mostly from replenishment of gas-depleted magma reservoirs with fresh batches of magma after caldera formation. Marsh (1984) also cites increased water pressures in maturing, caldera-confined aquifers as a causal agent for minor resurgences. Batholith-scale magma accumulations beneath calderas can feed volcanic dike emplacement long after caldera eruptions have ended. Lipman and Zimmerer (2019) document regional dike emplacement activity as much as 9 Ma after cessation of Platoro caldera activity (San Juan volcanic field, Colorado).

Acocella et al. (2001) modeled caldera resurgence using layered sands to represent the crust and silicone to represent magma (Figure 10.19). They found that two modes of resurgence appear to be typical, depending upon the previously mentioned aspect ratio of magma chamber width to crustal roof thickness. At an aspect ratio of around one (1:1), a large domical uplift in the crust takes place, surmounted by a smaller capping mound. Inward-dipping and steepening reverse faults trim the base of the main uplift, while normal faults fringe the summit rise. Crustal layering within the area of uplift is little disturbed, and may remain largely congruent with surrounding flat-lying bedding. Melt may work its way up along the outer fault system, feeding small volcanic eruptions within the so-called caldera **moat** – the area between the central uplift and caldera perimeter. A wide variety of domes, cinder cones and – if a lake is present – maars and tuff cones may develop. Virtually all of these landforms are monogenetic, though they

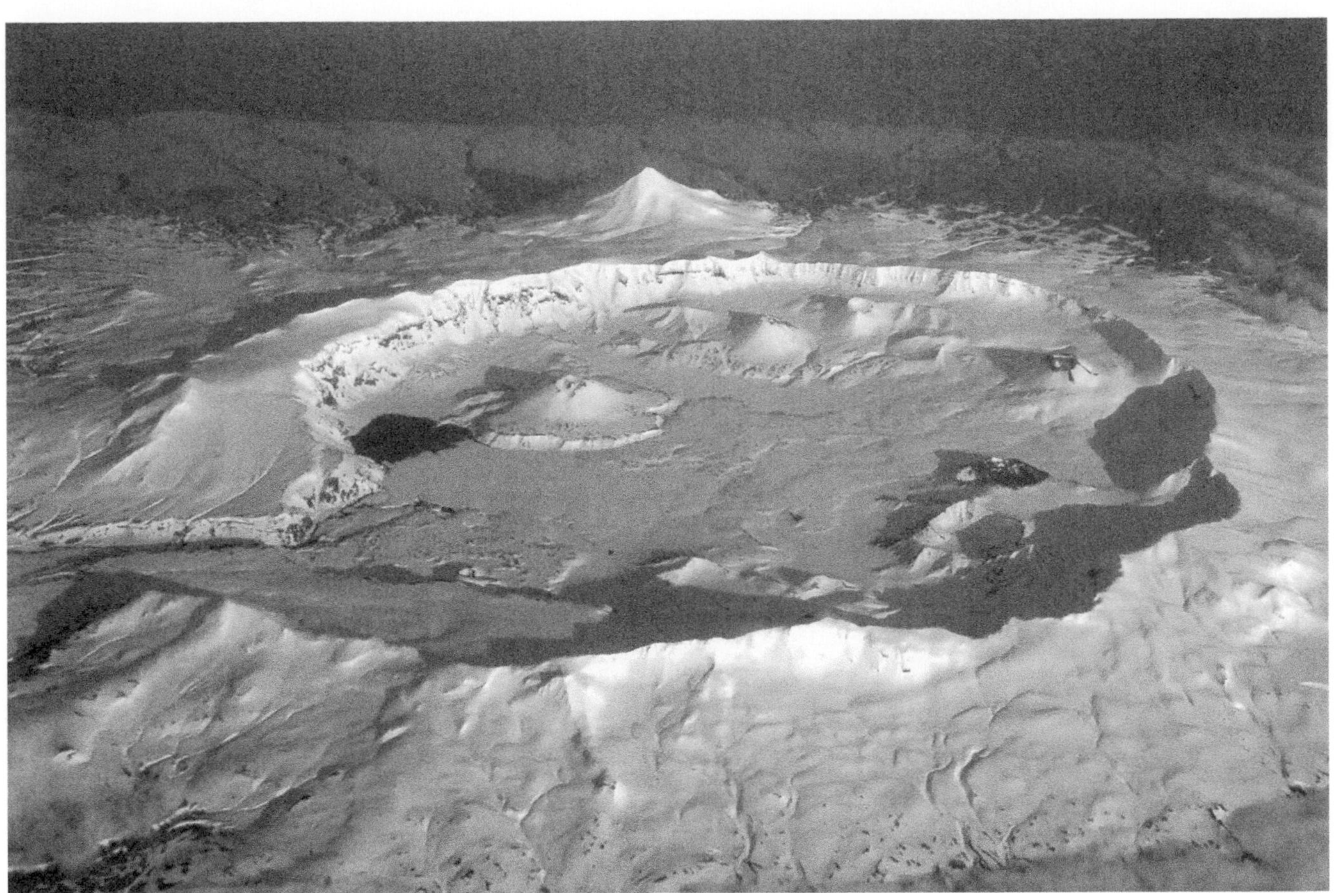

FIGURE 10.17 Okmok [4] summit caldera is a classic collapse caldera on Umnak island, in the Aleutian island arc, 1400 km west of Anchorage, Alaska. The caldera is almost 10 km in diameter, and formed about 2400 years ago. For much of its history the caldera was filled with a deep lake, so that most of the resurgent cones on its floor have formed in the presence of water. A month-long eruption that began on 12 July, 2008, sent ash plumes as high as 12 km. Source: USGS photo by Cyrus Reed.

can nearly grow atop one another in places. The Italian islands of Ischia and Pantelleria are examples of this kind of resurgence lifting the centers of shallow, largely submerged calderas above the sea. An actively intruding laccolith perhaps including as much as 80 km^3 of fresh magma could be responsible for the resurgence at Ischia (Carlino 2012).

Modeling studies show that for even shallower magma bodies, with chamber width more than double the thickness of its roof crust, resurgence takes on a different character. Central uplift again occurs, and the area of resurgence is also fringed with inward-dipping reverse faults. But layering within the uplift tends to dip away toward the base, and the crest of the uplift may subside in time. The area of subsidence can simply be a circular, dimple-like depression, but if regional crustal extension prevails, this may modify the stress imposed by upwelling magma to create a graben whose orientation lies perpendicular to the direction of regional extension. Post-caldera volcanism not only occurs within the moat, but also in the graben or dimpled center of the uplift as well.

Outstanding examples of the latter kind of caldera resurgence include Jemez caldera [49], New Mexico, and Long Valley [44], California. Fisher Caldera [7] (p. 327) probably also belongs within this class. Following its collapse 9100 years ago, a period of quiescence lasting several thousand years allowed a prehistoric lake in Fisher Caldera to accumulate over 6 m of clays. Then, after resurgence began, two small polygenetic cones and at least seven monogenetic vent structures grew within the caldera. Resurgence drained much of the original lake water, with several square kilometers of clay beds and a maar now exposed above water. During this time, one polygenetic cone and a solitary monogenetic edifice also grew outside, though close to, the Fisher Caldera rim. The volcanic field at Fisher Caldera will probably continue to develop over the next few thousand years.

What explains differences in these modes of resurgence? In the case of Ishcia and Pantelleria, the rising melt was able to displace the block of crust upward without causing significant internal deformation. The block was simply too thick to lose its coherency relative to the width of the rising magma body. At Jemez, Fisher, and Long Valley, this apparently was not the case. The thin roof block fractured internally during uplift, permitting the ascent of magma through the center of the caldera floor to the surface. The subsidence or collapse at the center of the uplift simply resulted from stretching of the crustal block beyond its tensile strength. Experimental work by Walter and Troll (2001) suggests that a complete gradation of caldera structures exists between these two styles of resurgence.

FIGURE 10.18 Aerial view of the 25 × 18 km diameter Aso caldera [143], Japan, looking north. The caldera was formed during four major explosive eruptions, the most recent about 90,000 years ago. The resurgent cones in the center of the caldera include Nakadake (fuming), one of the most active volcanoes in Japan. Source: Photo by Yasuo Miyabuchi.

a) Resurgence of magma body about as deep as it is wide

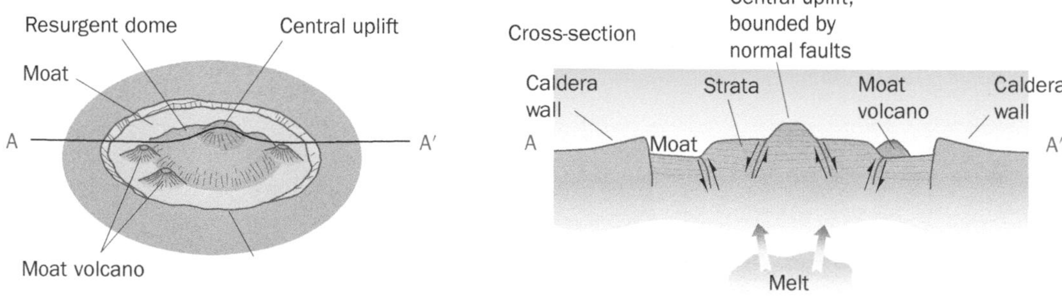

b) Resurgence of a wide, shallow magma body

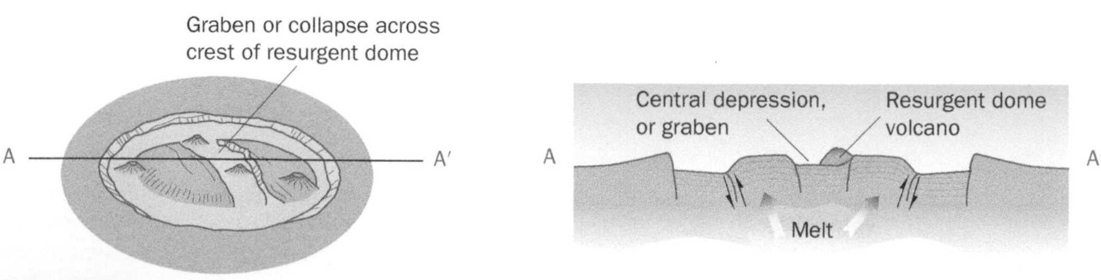

FIGURE 10.19 Types of caldera resurgence. Source: Based on Acocella et al. (2001).

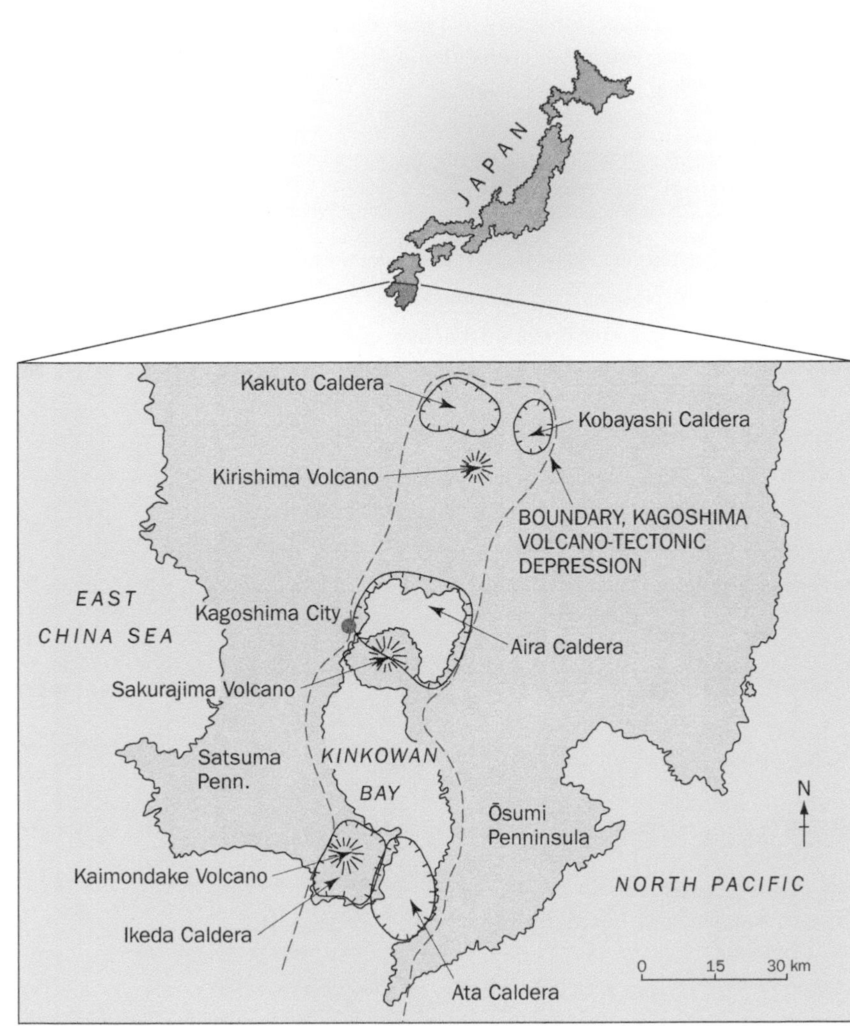

FIGURE 10.20 Map of Kagoshima volcano-tectonic depression, Kyushu, Japan. Source: Modified from Okuno et al. (1998).

During resurgence, eruptions can be highly effusive, often involving lavas of the same or similar composition as the magma that erupted during caldera formation, as for example, at Yellowstone caldera [47]. Here, over the past 640,000 years, numerous chemically evolved rhyolite flows have poured into the area of most recent collapse (Wicks et al. 2006). Post-caldera resurgent activity may also involve eruption of low-silica lavas, as at Crater Lake [32], described above. Within the 760,000-year-old Long Valley caldera, crystal-poor rhyolite lava erupted as resurgence began 700,000–600,000 years ago, but following uplift of the central caldera floor, cooler, crystal-rich silicic flows spilled into the moat at roughly 200,000-year intervals (~500,000, 300,000, and 100,000 years ago). The eruptions took place in clockwise succession around the area of a resurgent uplift. Scattered basaltic eruptions may also occur around the fringe of resurgent silicic calderas, revealing the involvement of mafic magma in the reservoir system. Some erupted lavas may even contain evidence of basalt-rhyolite mingling at depth.

Volcano-Tectonic Depressions

The term *volcano-tectonic depressions* has been used to describe very large grabens or compound grabens, best displayed in continental areas stretching along the axes of volcanic arcs (Chapter 2). Regional stresses associated with plate convergence and heating from deep-seated magma generation and emplacement lead to the development of these major landforms. They enclose multiple volcanic edifices and centers, and allow for the build-up of thick sequences of volcanic and sedimentary material in their floors. Some authors have also associated the term with single very large calderas or volcanic complexes (e.g., Toba [127], Yellowstone [47]), but we believe that this is an erroneous use of terminology. Volcano-tectonic depressions are bigger features than calderas, and as mentioned above, result from broader crustal stress fields than imposed by any single magma reservoir – hence the presence of "tectonic" in the term. Individual volcano-tectonic depressions range up to 800 km in length and may be as much as 150 km wide. Boundary faults generally involve both extensional (normal) and strike-slip movements. Sedimentation and erosion may obscure these faults in places, but elsewhere, they form prominent scarps stretching hundreds of kilometers. Large water bodies such as lakes Managua and Nicaragua in Central America and Kagoshima Bay in Japan result from flooding of depression floors (Figure 10.20). The development of most, if not all, continental volcano-tectonic depressions coincides with intensive ash-flow eruptions, many of the sort described in the previous section (e.g., De Silva et al. 2006).

Questions for Thought, Study, and Discussion

1. What is the fundamental difference between eruptions that create nested craters and those creating multiple craters?
2. In what essential, physical, respects does a crater formed by volcanic explosions differ from one formed by collapse?
3. A volcanic crater consists of an upper rim of loose ash and blocky debris, while the bottom of the crater consists of a cylindrical shaft of vertical walls made up of interlayered pyroclastic material, breccia, and solid, vesicle-poor lava. What kinds of eruptions have taken place at this volcano, and how did this crater form?
4. What physical characteristics distinguish drainage from explosive calderas?
5. How might the "pre-caldera stage" in the history of a typical continental volcanic field differ from the "post-caldera stage," or, in fact, is it even meaningful to speak of these stages?
6. What geologic circumstances favor development of trapdoor calderas as opposed to piston calderas?
7. What evidence exists to suggest that vast areas of ignimbrite sheets along continental margins may be underlain at depth by granitic plutons?

FURTHER READING

Baer, R.B. and Siegel, L.J. (2000). *Windows into the Earth: The Geologic Story of Yellowstone and Grand Teton National Parks.* Oxford: Oxford University Press, 242 p.

Goff, F. (2009). Valles Caldera: A Geologic History, University of New Mexico Press, 130 p.

Gottsmann, J. and Martí, J. (2008). *Caldera Volcanism: Analysis, Modelling, and Response.* North Holland, Elsevier, 516 p.

Van der Pluijm, B. and Marshak, S. (2003). *Earth Structure: An Introduction to Structural Geology and Tectonics.* New York: W.W. Norton, 672 p.

CHAPTER 11

Mass-Wasting Processes and Products

Source: dirkr/Shutterstock.com.

Volcanoes: Global Perspectives, Second Edition. John P. Lockwood, Richard W. Hazlett, and Servando De La Cruz-Reyna.

Companion website: www.wiley.com/go/lockwood/volcanoes2

> *Although volcanoes have collapsed many times in history, it was Mount St Helens [31] in May 1980 that really made volcanologists aware that an entire flank of a volcano could fail.*
>
> *(Peter Francis 1993)*

For all of the awesome power of their eruptions, volcanoes are inherently weak structures, their interiors broken by fractures and poorly consolidated zones of breccia, and their internal cores commonly altered into soft, putty-like clays by hydrothermal activity. Volcanic edifices are continuously stressed by intrusions and the steady influence of gravity, and inherently become less stable as they grow higher. Almost all large volcanoes have significant histories of falling apart from time to time as they grow – ongoing struggles between the eruptions that build them higher and the sometimes-catastrophic gravitational mass-wasting failures that can suddenly tear away their flanks and obliterate their summits. Because the slopes of composite "gray volcanoes" are typically mantled by unconsolidated ash and other loose material, the erosive powers of rainfall and volcanic mudflows (lahars) are also potentially important agents of volcano destruction.

Landslides, Avalanches, and Sector Collapses

The great English boxer Robert Fitzsimmons (1863–1917) popularized the phrase "the bigger they are, the harder they fall," and what is true for boxers is also true for volcanoes! The largest landslides known on Earth, many of them submarine, are derived from the flanks of great volcanoes, and although infrequent, such landslides are an ever-present hazard (Voight 2000). Volcano landslides are of several varieties and sizes. They form at composite volcanoes, at dacitic dome complexes, and as vast submarine deposits on the flanks of oceanic shields (Chapter 9).

The general term **mass wasting** refers to the sudden collapse and sliding away of mountainsides, and can take place in a variety of ways. **Block-glide landslides** tend to be at the small end of the volcano mass-wasting scale, in which a layer of hard, surficial rock breaks free high up on the side of a cone and slides semi-coherently downhill across underlying weaker material. If the slide plane cuts across many layers, however, the detached mass tends to rotate along horizontal, transverse axes during movement. Some mass-wasting features on volcanoes are giant **rotational slides** of this nature. Rotational slides generally move more slowly than another category of even more violent mass movement – avalanches.

FIGURE 11.1 Mass wasting at Ontake volcano [144], Japan. Landslide scar and debris avalanche deposits on the southeast flank. This landslide was triggered by a M=6.8 earthquake in 1984. The slide mass and related lahars traveled 13 km downslope – causing great property damage and loss of life. Source: Photo provided by T. Kobayashi; Ontaki Village Office.

An **avalanche** is a detached rock mass that breaks into many smaller fragments on its way down slope. Avalanches may transition to lahars (discussed later) if sufficient water is present. They are very common on composite cones, and may combine aspects of block-glide and/or rotational slide movements in their development (Figure 11.1). For example, numerous small block-glide avalanches took place during the April 1944 eruption of Vesuvius [109] triggered by the collapse of fresh lava flows deposited atop loose ejecta on the upper slopes of the volcano. Earthquake activity associated with a subsequent explosive phase of the eruption evidently triggered the failure. Each avalanche lobe, though completely fragmentary, may show a general layering of clasts reflecting the stratigraphic structure of its source area upslope. In other words, little relative mixing of fragments took place during transport (Hazlett et al. 1991).

Debris avalanches are especially large, fast-moving landslides that can destroy a substantial portion of a volcanic edifice in a single event (Figures 11.2 and 11.3). Like all smaller avalanche deposits that are localized on volcanic slopes, debris avalanches are disaggregated masses. Their fragments may become so well mixed during transport that they lack any clear reflection of original stratigraphic layering. Sorting is very poor, with pulverized, variably fractured angular to subrounded fragments, in some instances quite large (Figures 11.4 and 11.5). Inter-clast matrices may be ground up into a fine powder or

FIGURE 11.2 Aerial photograph of Unzen volcano [141], Japan, showing sector collapse scar on the east flank of Mayuyama dome, source of a major sector collapse in 1792 that destroyed the city of Shimabara, generated a major tsunami that surged across Shimabara Bay, and killed almost 15,000 people on the opposite shore. This collapse of Mayuyama dome was apparently triggered by deformation within the adjoining Fugendake dome, which had erupted a month earlier. Fugendake in background. Source: Photo courtesy Japanese Ministry of Land, Infrastructure, and Transport.

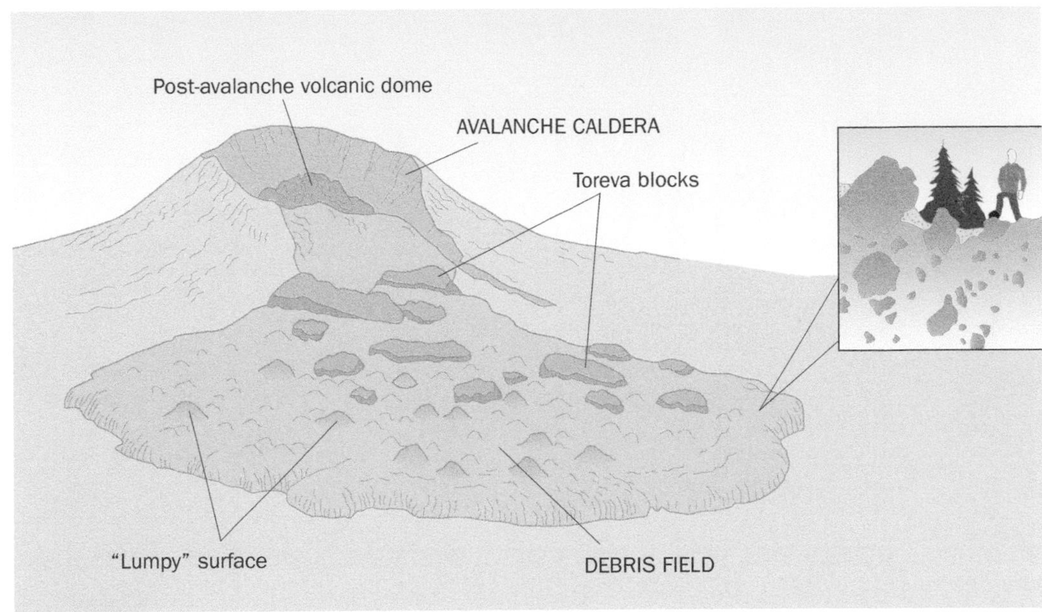

FIGURE 11.3 Common volcanic debris avalanche features.

"rock flour" during shearing and transport. Some older debris avalanche deposits have been mistaken for the ground moraines left by ancient glaciers (Ponomareva et al. 1998), or for other types of brecciated sedimentary rocks, but their distinctive internal characteristics and hummocky surfaces (Figure 11.6) usually make recognition straightforward (Ui 1983).

Whole slices of volcanic edifices, as blocks measuring hundreds of meters – or even kilometers – across may be present within an especially large debris avalanche. Called **toreva blocks** (Reiche 1937), these mega-fragments are large enough to form substantial hills or ridges in the deposit surface (Figure 11.6). **Debris fields** or **debris plains** may extend for tens of kilometers from the source volcano, and be nearly as wide as they are long, depending upon pre-existing topography. Their hummocky landforms, even where mantled by younger deposits, are typically unmistakable for thousands of years. The submarine Hawaiian Alika debris field in Hawaiʻi records the collapse of about 25% of the subaerial Mauna Loa volcano [15] volume about 120,000 years ago. It traveled 100 km along the seafloor and has a volume probably in excess of 500 km^3 (Lipman et al. 1988). Breakup during transport is sustained until the moment the avalanche comes to rest, which takes place quite suddenly since many of the grains in the shifting body interlock, giving the mass a fairly high yield strength once inertial momentum subsides.

FIGURE 11.4 Prehistoric debris avalanche deposit on the southwest flank of El Misti volcano [78], Peru. Note the subangular shapes of boulder clasts and the variety of clast sizes. Source: Photo by J.P. Lockwood.

FIGURE 11.5 Block in the prehistoric "Ten Thousand Hills" debris avalanche deposit, Tasikmalaya, Indonesia. This subrounded fragment shows the internal fracturing typical of large blocks carried by debris avalanches – in this case about 25 km from its source within Galunggung volcano [129]. Source: USGS photo by J.P. Lockwood.

Many debris avalanches are triggered by the collapse of large, growing volcanic domes. Snow-clad Sheveluch volcano [172], Kamchatka, has produced 13 large debris avalanches in the past 4500 years, most recently in 1964. The intervals between these avalanches range from 30 to 340 years. Each collapse has been followed at once by an explosive eruption – a cycle of dome extrusions terminated by large-scale failures of dome flanks (Ponomareva et al. 1998). The 1964 debris avalanche began as a series of ordinary rock falls from an active dome that rapidly accelerated into wholesale collapse. One and a half cubic kilometers of debris roared downslope, covering 98 km^2 out to a distance of 16 km from the vent (Gorshkov and Dubik 1970). This avalanche triggered a modest phreatic dome explosion followed by a powerful Plinian eruption. Shiveluch has been erupting almost continuously since 1980, generating several small debris avalanches, pyroclastic flows, lahars, and Plinian eruptions – in what is fortunately a remote and unpopulated area.

Several major prehistoric debris avalanche deposits on the flanks of volcanoes in Oregon and Washington, including Shasta [30], Mount McLoughlin [28], Rainier [36], and Mount St. Helens [31], involved failure of the north slopes of these volcanoes, possibly

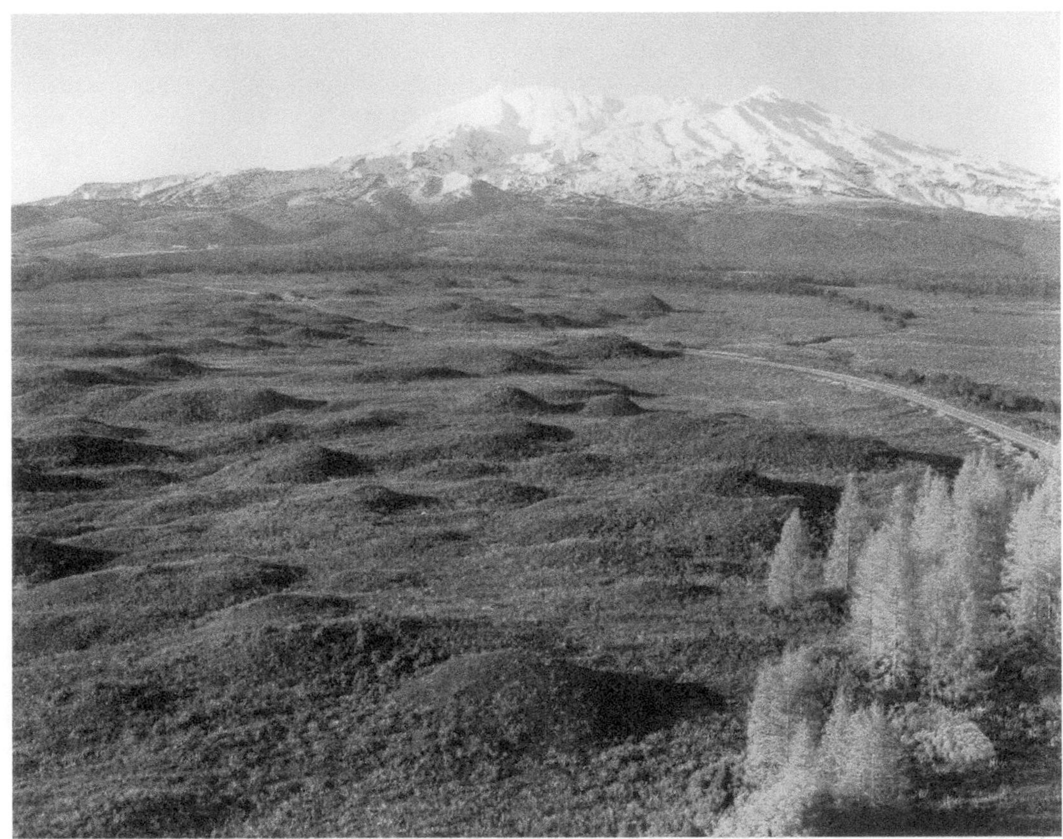

FIGURE 11.6 View of the west flank of Ruapehu volcano [178], North Island, New Zealand. The debris avalanche deposits in the foreground formed during a sector collapse of Ruapehu's north flank 9500 years ago. Such hummocky surfaces are characteristic of debris avalanche deposits all over the world. Source: Photo by J.P. Lockwood.

reflecting the fact that the northern flanks are subject to more sustained snow and ice accumulation, and are more vulnerable to hydrothermal weakening due to erosion and ground water infiltration. Sector collapse of composite volcanoes can also be triggered by earthquakes, by the intrusion of magma (Ellsworth and Voight 1995), and simply by the inherent gravitational instability of growing volcanoes, which become internally unstable as more and more material is added to their edifices by ongoing eruptions.

Dormant or inactive volcanoes can also produce debris avalanches. Forested Casita volcano [64], Nicaragua, for example, with deeply eroded valleys on all flanks, hardly resembles a volcano in profile, and has not erupted since at least the sixteenth century. Hydrothermal activity is widespread across Casita's summit and upper eastern flank, however, and the combination of perched, high-standing ground water and mineral alteration has led to the deformation of the edifice and steepened slopes that frequently send small landslides into adjacent valleys. As Hurricane Mitch blew across the region in 1998, heavy rainfall destabilized the head wall of one very large, prehistoric landslide on the south flank, causing a debris avalanche that covered 12 km^2 of mostly rich agricultural land, devastating two towns and killing 2500 people (Kerle and van Wyk de Vries 2001). At least seven major Quaternary debris avalanche deposits have been identified in nearby Guatemala (Vallance et al. 1995), and hundreds more have been identified worldwide, especially in the past three decades as their features have become better known. Debris avalanches are usually unpredictable, and pose great risks to populations anywhere in the world where steep volcanic slopes stand above populated regions.

Sector collapse

Sector collapses associated with debris avalanches take place where an *entire segment* (sector) of a volcano, commonly including the summit, tears free and slides away, leaving a **sector collapse scar** or **amphitheater caldera** (e.g. Figure 11.2). The largest debris avalanches, those containing giant toreva blocks in debris fields spread across tens of square kilometers, or more, are associated with sector collapses. But not all sector collapses are necessarily associated with debris avalanches: Depending upon whether land failure of a volcanic sector takes place incrementally or near-instantaneously, the mechanics of failure may differ, consist of rotational sliding (slow, as at Piton de la Fournaise [125], mentioned in Chapter 9) or debris avalanching (fast, as at Mount St Helens [31] in 1980). Major sector collapses may be followed by even more devastating directed blasts (Chapter 8), as high-standing magma chambers are depressurized and explode violently (as at Mount St. Helens).

The composite volcano Stromboli [113] rises from the central Mediterranean Sea and is in a state not only of almost constant eruption but also of gradual, stepwise collapse. Gravitational stress on the volcano is greater on its northwestern flank, which slopes steeply down to the deepest part of the surrounding ocean floor, three kilometers lower than the summit. Destabilized by the near continuous intrusion of fresh magma high into the volcanic cone, and by the weight of erupted material atop steep, fractured slopes, this side of the mountain has experienced at least seven episodes of incremental land sliding over the past 100,000 years (Tibaldi 2001). The wedge-shaped sector prone to failure begins underwater and extends on land up to Stromboli's summit. Romagnoli et al. (1993) estimate that 4.7–6.1 km^3 of landslide debris has accumulated at the submarine base of this sector. Glowing ejecta and occasional small flows from the classically Strombolian eruptions are channeled down the channelway of this failing wedge of land, locally called the Sciara del Fuoco ("Ski-slope of Fire").

In contrast to Stromboli, a sector collapse that took place about 7200 years ago at the 6-km-high Socompa volcano [81], high above the Atacama Desert in the Andes of northern Chile (de Silva and Francis 1991), probably took only a few minutes to complete. The 36 km^3 of avalanche deposits spread across 600 km^2 of land. The largest blocks of detached edifice material are several kilometers long and up to 450 m high. The sector collapse, combined with a resulting directed blast, tore a 70° wedge out of the 360° flank of Socompa – 12 km wide at the mouth and over 400 m deep in places. Only about 20 percent of the volume of the avalanche consists of material from the edifice, however; the rest is from underlying ignimbrite, gravel, and salt flat deposits that evidently failed under the weight of the growing volcano, essentially "squirting" from beneath its base all at once and causing collapse of the cone above (van Wyk de Vries et al. 2001). Socompa had the misfortune of trying to build atop a weak and actively deforming substrate. Most of the sector collapse scar has now been filled in by later eruptive deposits.

The triggers for sector collapse include gravitational instability of poorly supported, over-steepened slopes, deformation related to shallow magmatic intrusion (Ellsworth and Voight 1995; Acocella et al. 2013); and weakening of rocks owing to hydrothermal activity – as noted for a major Pleistocene collapse of Nevado del Toluca [54], Mexico. Here, a highly fluid Pleistocene debris avalanche resulted from the collapse of a hydrothermally weakened slope, traveling down a graben and then out across a plain to a distance of 55 km from the mountainside. Blocks in the flow deposit range up to 15 m in diameter, and the deposit itself is 15–40 m thick (Capra and Macias 2000; Valderrama et al. 2016). Repeated inflation–deflation cycles (Kendrick et al. 2013); rapid dome growth (Gorshkov and Dubik 1970); and even tectonic fault movements in the underlying crust (Paguican et al. 2012) are other documented causes. The abrupt impact of sector collapse and de-pressurization of magma system of a volcano can be catastrophic, as was apparent during the 1980 eruption of Mount St Helens (Chapter 8).

Izbekov et al. (2013) noted, based on studies of sector collapse influence on magmatic systems for Bezymiany [170], Mount St Helens [31], and Shiveluch [172] volcanoes:

> *"A sudden drop loss of some 1000 m of the upper conduit [from flank collapse] represents a pressure drop in the magma system of over 20 MPa. This would trigger a rapid transition from slow, partially hidden . . . or completely concealed . . . cryptodome growth to high-discharge-rate tephra eruption. Moreover, if we imagine the magma reservoir as a pump with flow impeded by the head and stickiness of fluid in the exit pipe, then the protracted period of generally more effusive eruptions following substantial removal of the pipe is to be expected as well."*

Lahars

Catastrophic mudflows cascading down the slopes of composite volcanoes and onto the surrounding plains have been responsible for great loss throughout the world, and have probably killed more people than any other volcano-related phenomena worldwide (Blong 1984). They are referred to by the Indonesian word **lahar**, a term restricted to mudflows originating on volcanoes. A 1988 Geological Society of American Penrose Conference broadly defined lahars as "A general term for a rapidly flowing mixture of rock debris and water (other than normal streamflow) from a volcano" (Smith and Fritz 1989). The same phenomena are termed **debris flows** where mudflows originate from other sources. For example, debris flows plague the Los Angeles metropolitan area, far from any active volcano. Most of these are triggered by intense rainfall and rapid erosion of steep, fire-burned slopes of mountains above the city. In many parts of the world, hydraulic mining, overgrazing, clear-cutting of timber, and the construction of roads high on mountainsides also frequently trigger debris flows.

Lahars are major agents of mass wasting, and transport large amounts of massive rock debris down volcano slopes, They commonly transition into more dilute but equally destructive floods as they extend farther from their volcano sources, depositing a variety of sediments that are perhaps best studied using the approaches of sedimentology combined with the perspectives of volcanology. Many phenomena can potentially trigger lahars, including eruptions, earthquakes, heavy rainfall, snow melt, or landslides (Table 11.1; Figure 11.1). By whatever processes that initiate them, large amounts of water must combine with fine-grained clay-size material high on the volcano slopes to generate high-density fluids capable of eroding and transporting vast amounts of mud, water, and loose rock debris downslope. Lahars may contain over 65 volume percent of entrained sediment (80 percent by weight), which greatly raise their bulk densities, allowing them to pick up and transport huge boulders. The increasing erosive

TABLE 11.1 Causes of Lahars.

General category	Specific examples
Volcanic activity	• Hot ejecta, PDC, or dome collapses spread onto snow- or ice-covered slopes (e.g. Pistolesi et al. 2013) • Debris avalanches or sector collapses (e.g. Endo et al. 1989) • Hydrothermal system explosions (e.g. Uesawa 2014) • Failure of crater rims • Surtseyan ejection of crater lake waters onto snow- and ice-covered slopes, or onto slopes laden with fresh, loose volcanic ash
Seismic activity and volcano deformation	• Landslides from shaking or oversteepening of deforming slopes, entering stream drainages or snow- and ice-covered slopes
Heavy rainfall	• Heavy regional rainfall onto volcanic slopes newly mantled with ash
Melting of snow and ice	• Thaw of snow on volcanic slopes covered with fresh mantles of ash or heated from below
Crater lake eruptions	• Volcanic crater lakes contain large volumes of water that can be mobilized by eruptive activity or may be released suddenly if crater rims fail
Deforestation	• Removal of vegetation on volcanic slopes by logging, agriculture, or fires that exposes loose soils and volcanic materials to heavy rainfalls. (Many debris flows in non-volcanic settings such as suburban Los Angeles originate this way.)

FIGURE 11.7 The ruins of Armero, Colombia, after it was destroyed by deadly lahars on the night of 13 November, 1985. The lahar from Ruiz volcano [71] was channeled by the Lagunillas River in a narrow flow up to 35 m deep (inset) but was less than 3 m deep where it widened and flowed through the city. The lahar contained entrained boulders more than a meter in diameter, however, which acted like battering rams, destroying three-story-high concrete buildings, of which only the foundations remain. Source: USGS photos by J.P. Lockwood.

effects of these bulked-up mudflows have a snowball effect as they roar downslope, entraining ever more mud and debris as they increase in volume. Because of their bulk density and the ramming effect of entrained massive boulders, a vigorous lahar can sweep away bridges and obliterate massive buildings with ease (Figure 11.7) – they are far more mechanically destructive than PDCs (Chapter 8) – which at least usually leave the burned shells of buildings in their wake.

Lahars produce runoff that far exceeds the ordinary discharge capacity of streams and rivers draining a volcano's slopes. The term **stream discharge** refers to the amount of water moving per unit time past a particular point along a stream channel. For example, the average annual discharge of the Amazon River, the largest river in the world, is some 2×10^6 m^3 of water per second, while the Mississippi River has a discharge about one-tenth of this value. The discharges of lahars range up to 10^6 m^3/s – comparable with the Amazon – and they move as fast as 5–20 m/s, even faster on steep slopes (Pierson 1995).

To observe an active lahar descending a volcano's flanks is an incredible experience. I (JPL) watched the fast-moving lahars that coursed down rivers below Galunggung [129] volcano's active summit in 1982. They flowed like serpentine rivers of viscous lava – only much faster than lava would on similar slopes. Their surfaces roiled with standing waves of brown colored liquid, and their great density became evident when giant boulders boiled up out of the flow and then re-submerged, repeatedly bobbing back to the surface as the lahars continued downstream. After the lahars had passed, the river channels were difficult to cross because of all the deep sticky mud filling them.

A typical lahar may be part of a sequence of flowing material that begins when muddy water starts overflowing channels near a volcano's summit and bulks up as it roars downslope, incorporating more water, mud, and rock debris through increasing erosive power. When the lahar reaches the lower, more gently-sloping flank of the volcano, it begins to deposit its thick cargo of entrained sediment and spread out – potentially destructively. It transitions to a somewhat more fluid **hyperconcentrated flow**, and eventually just "muddy water" at its distal end, having already dumped its coarser sediment load upstream (Pierson 2005). A key aspect of lahar flows is the large proportion of clay-sized suspended material, which gives them the high density required to lift and carry huge boulders. Indonesian geologists use a "tongue-test" to identify lahar deposits and to distinguish them from ordinary alluvial sediments – one's tongue will stick to samples of clay-rich lahar matrices, but not to the interstitial material of other sedimentary deposits. (This does not work so well for hyperconcentrated flow material, however!)

In outcrop, lahar deposits are seen to be non-bedded, poorly sorted with a high proportion of clay- and silt-size matrix, and most lack the boulder-to-boulder clast-supported internal structure characteristic of conglomerates and landslide breccias (Figure 11.8). Larger fragments, which may become segregated toward the top or margins of a deposit during flow (e.g. Vásquez et al. 2014), show slight to extensive rounding due to abrasion during transport, but finer clasts (sand and gravel size) are typically angular. Because of the typical high proportion of clay in matrices, deposits are commonly well-indurated and may form stable cliffs in outcrop. Pumice is readily pulverized in debris flows, so any pumice clasts in the initial flow may be missing, although fine depositional matrices can have a significant glassy component derived from pumice fragmentation. Lahar deposits commonly show variation in clast composition with distances from source areas, reflecting the incorporation of non-volcanic material eroded from streambeds beyond the

FIGURE 11.8 Lahar deposit on the north flank of Villarica volcano [76], Chile. Note the clay-rich matrix for the boulders, which are "matrix-supported" by mud and not "clast-supported" as would be the case with fluvial deposits. Source: Photo by J.P. Lockwood.

volcanic edifice. The boulders that destroyed Armero, Colombia, in 1985 were of gneiss and granite from basement rocks exposed kilometers from the base of Ruiz volcano; it was difficult to find any volcanic clasts in these lahar deposits at all.

Causes of Lahars

Let's explore some of the most common causes of lahars (Table 11.1).

Ongoing Eruptive Activity and Dome Growth

Explosive eruptive activity can cover the slopes of volcanoes with vast amounts of unconsolidated, unstable tephra that can be re-mobilized under moist conditions to form lahars, as happened for more than a decade after the 1991 eruption of Pinatubo, Philippines [135] (Chapter 1). Dome growth, for example, at Mount St Helens [31] or Bezymianny [170] results in minor collapse events and related explosions that commonly generate lahars.

Seismic Activity and Volcano Deformation

Seismic unrest at volcanoes can trigger landslides that quickly turn into deadly mudflows, as at Ontake volcano [144], Japan (Figure 11.1a; Endo et al. 1989). Lahars triggered by an $M = 6.4$ earthquake near Huila volcano [70], Colombia, in 1994 killed several hundred people. The fact that a major regional earthquake may be "overdue" in the Washington State/British Columbia area poses an indirect threat to populated areas because of the large amounts of snow, ice, and hydrothermally weakened rocks that loom high on the slopes of Mount Rainier [36].

Heavy Rainfall

Following explosive eruptions, tephra deposited on steep slopes provide vast amounts of loose sediment for rainfall to mobilize. Water falling on wooded hillsides is ordinarily captured by vegetation and held within forest litter to slowly percolate into the ground below, evaporate, or evapotranspirate through living plant tissues. A large pyroclastic eruption, however, strips a forest of leaves, even blowing down and sweeping away large trees to render the land bare (Figure 11.9). Covers of loose pumice and ash accumulate, and may develop crusts as much as a centimeter thick. Rainfall cannot percolate through this crust, and immediately runs downslope in rivulets soon laden with abrasive ash and gravel pellets. The rivulets scour the tephra cover into *rills* – narrow, vein-like channels characteristic of rapidly eroding landscapes. As they merge into streams, the potential for generating lahars greatly increases. This origin for lahars is especially common on tropical composite volcanoes covered with loose debris because they are seasonally subjected to heavy rainfall. Recall from Chapter 1 that precipitation from Typhoon Yunya and other monsoonal rains produced devastating lahars in the hours and days following the 15 June, 1991 eruption of Mount Pinatubo [135] (Newhall and Punongbayan 1996). Heavy rainfall can also saturate loose or weakened material on slopes of inactive volcanoes, producing devastating lahars unrelated to eruptive activity, such as those derived from the debris avalanche on Casita volcano [64] (discussed earlier). Rainfall can make accumulations of tephra and loose rock susceptible to mass movements in two ways: it *increases the weight of the mass*, and also can *increase internal pore pressures* – a major factor in the triggering of landslides as well as ash slumping.

Not all of the rain causing lahars is necessarily the product of regional weather systems. Eruption columns, too, can generate their own rainfall (Chapter 6). Large amounts of steam carried to high altitudes in eruptive plumes will condense and may fall near volcanoes as muddy rain, generally in the waning stage of Plinian column development. The shock waves of volcanic blasts, some traveling at supersonic speed, can also force condensation of moisture as they propagate from a vent. Shock-generated vapor clouds formed in advance of, and were overtaken by the 1980 Mount St Helens blast surge, adding moisture to that pyroclastic density current. Lahars were among the final events of the 79 CE Vesuvius eruption that destroyed Pompeii and Herculaneum, and may have been generated by water condensed from the cooling plume above.

Melting of Snow and Ice

The sudden melting of snow and ice, caused by eruptive activity, heavy warm rains, or accelerated geothermal activity can cause small water floods that quickly transition to major lahars as they travel downslope. Though not as common as rainfall-generated lahars, these lahars are among the largest, longest, highest-discharge-rate flows in the world, especially those originating from

FIGURE 11.9 Forest devastation after massive 1982 tephra fall – Galunggung volcano [129], Indonesia. Note the erosive rills beginning to develop – such small sources provided the material that fed massive, destructive lahars downstream. Source: USGS photo by J.P. Lockwood.

subglacial eruptions, considered further in Chapter 13. Major and Newhall (1989) documented more than 40 worldwide volcanoes that have generated lahars and floods through the melting of snow and ice by eruptive activity in historical times. They point out that most of these eruptions are at volcanoes located at latitudes higher than 35° or with summits exceeding 4000 m elevation.

While rainfall-generated lahars may take place months or even years after a volcanic eruption, lahars related to melting snow and ice caps are a more immediate consequence of eruptions, typically forming less than a month following initial activity. Even small eruptions beneath glaciated volcanoes can generate devastating lahars, as was demonstrated by the devastating 1985 Ruiz [71] eruption, which was only a moderate-sized summit pyroclastic outburst.

Mount Rainier, Washington [36] (Figure 11.10), is an example of a snow- and ice-covered volcano that has produced major lahars in the past. Deeply scoured and capped by 23 major glaciers with 13.6 million m^3 of ice, this deceptively beautiful mountain is potentially the most dangerous volcano in the United States (Driedger and Kennard 1986). The volumes of Rainier's lahar deposits exceed by far all other products of the volcano for the past 10,000 years. Rainier was especially active between 6500–4500 years and 2500–2000 years ago, when sector collapses generated major debris avalanches and very large associated lahars. The 5700-year-old Osceola mudflow has a volume of about 8 km^3, traveled over 100 km from Rainier, and blanketed about 300 km^2 of landscape, most of it in the now crowded Puget Sound lowlands. The lahar followed the course of the White River, scooping up stream-side granitic boulders as much as 10 m in diameter. As it burst from the Cascade Range, it spread as widely as 12 km across the surrounding lowlands, burying at least one Native American camp site and probably some local residents as well (Harris 2005). The 500-year-old Electron mudflow is not nearly as voluminous as the Osceola mudflow, but did travel more than 50 km from Rainier down the Puyallup River, almost to the present-day city of Tacoma. It mostly consists of hydrothermally altered material derived from the weak western flank of Mount Rainier (Crandell et al. 1979).

Crater Lake Eruptions

Many volcanic craters on active volcanoes, floored with impermeable hydrothermal clays and other volcanic detritus, quickly fill with water and may become hot, highly acidic pools with submerged active fumaroles. Renewed eruptive activity beneath these lakes can suddenly expel large amounts of water onto upper volcanic slopes, owing to hydrovolcanic explosions, infilling of lake basins with pyroclastic material, overflow and rapid erosion of low points in their rims – or some combination of these events. Mixing with loose ash, soil, rocks, and downed vegetation, such floods may transform quickly into lahars. Lahars from draining crater lakes typically flow out at speeds as high as 2–8 m/s (Pierson 1995). Although not as common as other forms of lahar generation,

FIGURE 11.10 Ice-covered summit of Mount Rainier volcano [36], Washington (4392 m), looking west, with the Emmons glacier just right of center. The last well-documented magmatic eruption of Mount Rainier was about 900 CE. Damaging lahars could be triggered by future eruptions or by major earthquakes in this tectonically active region. Source: USGS photo by Rocky Crandell, courtesy of Tom Sisson.

such eruptions within Kelut [132], Indonesia and Ruapehu [178], New Zealand crater lakes have killed thousands of people. On Christmas Eve, 1953, an explosive eruption within the summit crater lake generated a lahar that rushed down the Whangaehu River and washed out a bridge just as a train was crossing. The train plunged into the lahar and 151 people quickly lost their lives (Figure 11.11).

Lahar Dynamics

Before discussing the ways in which lahars, hyperconcentrated flows and other mixtures of sediment and water move, a brief discussion of fluid behavior is in order. As discussed in Chapter 4, pure water is a Newtonian fluid; that is, it has no yield strength and will continue flowing readily down slopes. In contrast, non-Newtonian fluids exhibit yield strength, and will come to rest with measurable thicknesses. Because pure water has no strength, it cannot suspend dense particles like sand or gravel – these will fall to the bottom of flowing water. When water is mixed with sufficient clay- and silt-size particles, however, its bulk density will increase, it develops yield strength, no longer behaves as a Newtonian fluid, and may suspend large amounts of coarse sand and gravel in the moving flow. When flows contain more than around 70–80 percent by weight of suspended sediments, bulk densities can increase 1.6–2.1 times that of pure water. Particles partially support each other during transport, preventing much turbulence; flows are “cohesive” and flow much like wet concrete (Pierson et al. 1996).

Fluids at rest may be described in terms of their density, viscosity, or yield strength, but when they are moving an additional parameter, **criticality** governs behavior. Fluids, whether water or lahars, behave differently as they flow at contrasting speeds through different types of conduits or channels. Water cascading through rapids, for example, behaves very differently from water flowing placidly down a river. In the former case, the water flows chaotically and somewhat unpredictably, driven mainly by inertial forces, and is said to behave **supercritically**. Slow-moving river waters are responding only to gravitational forces in predictable ways, and are behaving **subcritically**. As mixtures of mud, rock, and water flow down the slopes of volcanoes they may exhibit both super- and subcritical behavior in different places depending on their speeds and channel geometries. For the open-channel

FIGURE 11.11 Scene of train wreckage a short time after the 1953 Tangiwai (Whangaehu River) Ruapehu [178] lahar disaster, New Zealand. Source: Photo courtesy of Archives New Zealand/ Te Rua Mahara o te Kāwanatanga/Wellington Office (Archive ref. AAVK W 3493 D-1022).

conditions that concern us, fluid criticality can be determined by calculating a **Froude Number (*Fr*)**, a dimensionless value that depends only on flow speeds (V) and channel dimensions (Eq. 11.1).

$$Fr = \frac{V}{\sqrt{g.D}} \tag{11.1}$$

where V = fluid speed (m/s); g = acceleration of gravity (m/s/s); and D = "hydraulic depth" of channel (cross-sectional area of channel/channel width).

When *Fr* is greater than 1, moving fluids will exhibit supercritical behavior, tend to flow *around* obstacles in their paths rather than wash over them, and exhibit standing waves and hydraulic jumps (sudden deepening of the current with upstream backwash) wherever flow speeds slow suddenly downslope. Such phenomena help mix particles and contribute significantly to scour and erosion of underlying beds. Many dense lahars (true debris flows) will behave as supercritical fluids throughout much of their progress – particularly on steep slopes – though they may transition in and out of this condition as they sweep along.

When *Fr* is less than 1, moving fluids will exhibit subcritical behavior and tend to flow *over* obstacles in their paths, with little lateral displacement or hydraulic jump behavior. Subcritical lahars tend to be somewhat less erosive than supercritical flows, even though they still transport very high concentrations of particles. More dilute **hyperconcentrated flows** and water floods commonly derive from more dense lahars at distal ends of lahars as these mixtures of water and sediment flow downslope onto gentler terrain. They behave subcritically, except when their speeds are high or their channels are very wide and shallow (i.e. with *Fr* values greater than 1; Eq. 11.1). Hyperconcentrated flows contain lesser amounts of clay-size particles than do the denser, mostly supercritical lahars, are "incohesive" and cannot support their loads when they come to rest (Figure 11.12). Individual particles have more room

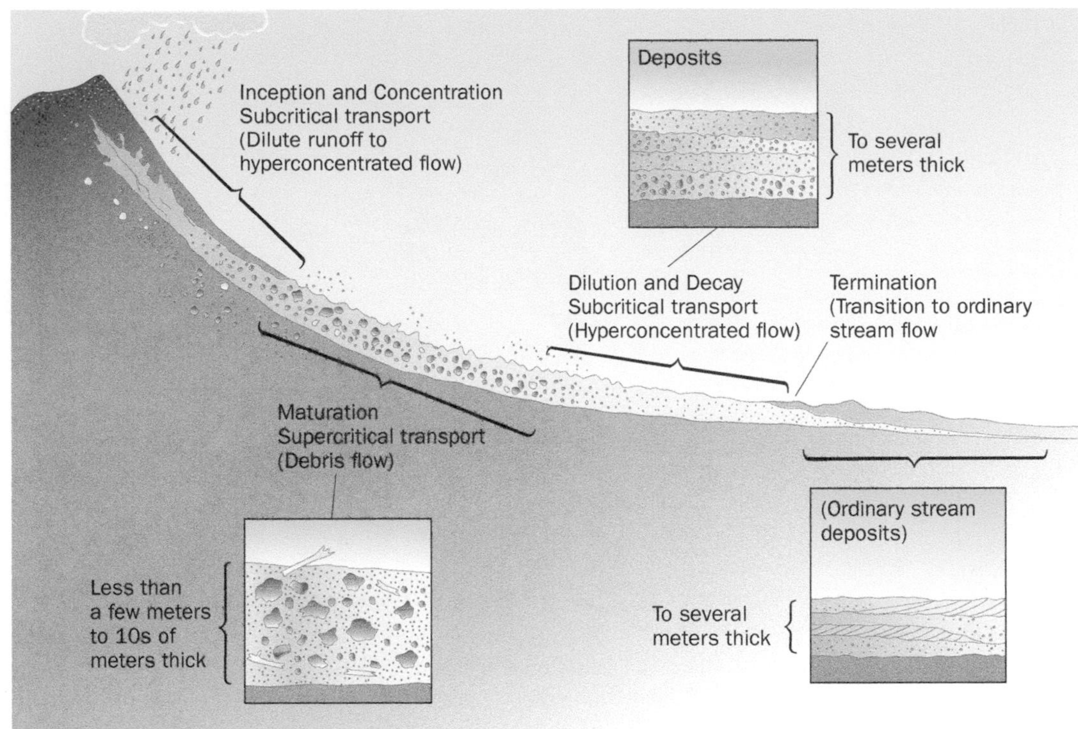

FIGURE 11.12 The relationship of lahars, hyperconcentrated flows, and water floods as a function of sediment concentration and criticality.

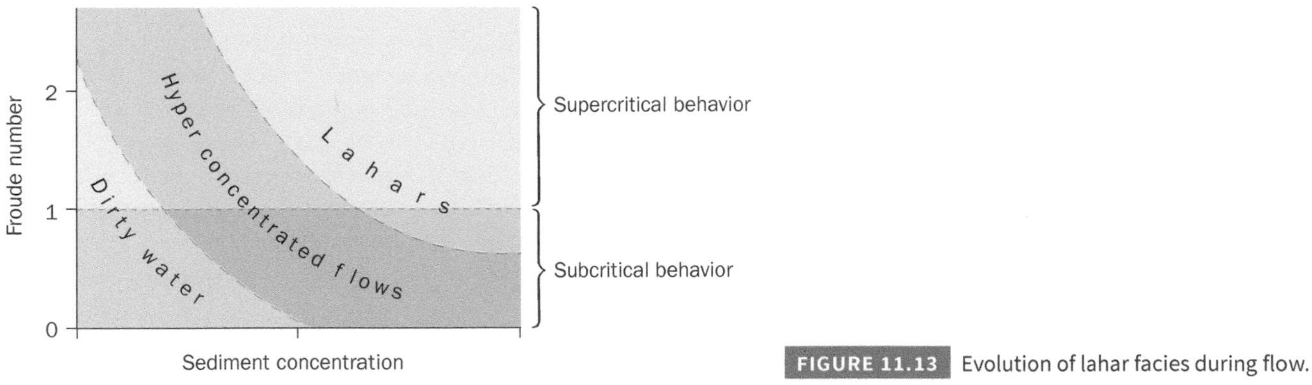

FIGURE 11.13 Evolution of lahar facies during flow.

to move about during transport, and flow tends to be turbulent rather than laminar. Gravitational forces become dominant in guiding the flow, permitting the heaviest particles to settle quickly to the bottom as in an ordinary stream.

Dense, muddy lahars may stop suddenly, with their particle loads "freezing" in place with only minor adjustments in position of individual grains as the deposit de-waters and settles – similar to the *en masse* deposition of block and ash flows that we described in Chapter 8. Such poorly sorted, clay-rich lahars commonly form well-consolidated, cliff-forming deposits around the eroded flanks of many volcanoes. Hyperconcentrated flows are less clay-rich, however, and their moderately sorted deposits are not cliff-forming.

Lahar Destructiveness

Most lahars begin as dilute, subcritical flows high on volcano slopes, but then increase their volumes and may become supercritical as they incorporate sediment along travel paths (Figure 11.13). Lahars cannot deposit sediment on steeper slopes, where they are highly erosive to stream bottoms and banks – streamlining their own channels. The very large boulders and logs that they can pick up are especially effective tools for shattering bridges, walls, and other obstacles in their way downstream (Figure 11.14). The lahars

FIGURE 11.14 Gigantic lahar boulder near Mulaló, Cotopaxi volcano [68], Ecuador. This boulder, which weighs at least 3000 tonnes, was carried about 20 km down the west flank of Cotopaxi by a massive, devastating lahar in 1877. The boulder came to rest when the lahar was no longer able to transport it because of diminishing velocity and thickness. The bus is about 10 m long. Source: Photo by J.P. Lockwood.

that flowed into Armero on 13 November, 1985 carried boulders up to a meter in diameter into the city, which acted like inertial battering rams, destroying major, three-story concrete buildings (Figure 11.7). Geology student Jose Luis Restrepo was in Armero that night on a university geology fieldtrip. Jose and the other students knew Ruiz volcano [71] was erupting, because ash had started falling in the afternoon, but they had no idea that lahars had already begun flowing down the flank of the volcano toward them. Here are translated excerpts from what he told me (JPL) two weeks later:

> *About 11:30 all the lights went out as Armero's generating plant was inundated and the flow arrived a few minutes later. It was only about 2 m deep as it entered town, rushing down streets, and the only lights were from the headlights of cars being tumbled about by the flow. We ran to our hotel and three of us climbed to the third-floor terrace, thinking this would be a safe place since the walls were of concrete. Lower-standing buildings around us were swept away as the flood kept rising. I felt great shocks as rocks slammed into the hotel and suddenly it all crashed down. The screams of people were terrible, and I was sure I was going to die a horrible death, but with four companions we climbed into a concrete water cistern that had come from the hotel roof. We were swept along with debris and screaming people, as burning gas tanks lit up the terrible scene. We saw a high wave of mud coming at us, but somehow our tank stayed afloat. Some mud was warm – but most was cold. We crashed against a standing tree. There were many people around us stuck in the mud, but we were unable to help them. We dragged one man into our tank, but his legs had been amputated and he was in great pain. When morning came, we saw that we were only 50 m from the edge of the mud, but when I saw the terrible devastation behind us, I lost all control and cried – Armero was gone – except for our little island of trees. About 300 people were able to struggle through the mud to a small hill – the Armero cemetery – where we waited among the tombstones for rescue.*

Of the 32 students in Jose's group, only 19 survived.

The destructive potential of lahars changes dramatically after they slow down and transition completely to subcritical flow. Subcritical, hyperconcentrated flows are not harmless, but they do not knock buildings over – they bury them (Figure 11.15). Like any large, fast-moving flood, they are effective at eroding riverbanks, undermining foundations and depositing large amounts of sand-size sediments in a short amount of time. Repeated hyperconcentrated flows following the 1991 eruption of Pinatubo [135] blanketed perhaps 300,000 hectares of rich Luzon farmland with deposits up to 25 m thick, and destroyed the homes of more than 50,000 people (Chapter 1). Efforts to dam or divert these sediment rich flows cost hundreds of millions of dollars, but have been largely ineffective.

In conclusion, volcanic edifices are susceptible to gravitational failure by massive landslides and by rapid erosion; they are relatively ephemeral geomorphic features and few last as significant topographic forms for more than a few million years after they have ceased activity. Their histories typically involve long battles between the eruptive processes that build them higher and the gravitational processes that tear them down. For most extinct volcanoes, their one-time glory is best preserved in surrounding sedimentary deposits – deposits that commonly far exceed the original volumes of the largest volcanic edifices that produced them.

FIGURE 11.15 Burial of homes in 1982 by hyperconcentrated flow deposits alongside Cibanjaran River, 8 km below lahar sources, Galunggung volcano [129], Indonesia. Roof tiles have been salvaged from the abandoned homes. Source: USGS photo by J.P. Lockwood.

Questions for Thought, Study, and Discussion

1. Volcanoes do not always experience sector failures because of inherent weaknesses within their edifices. In some cases, other geologic factors can contribute to that failure. Explain.
2. How would you distinguish a debris avalanche deposit from that of a typical lahar?
3. How and why are many lahar deposits also related to debris avalanches?
4. What gives a lahar energy and increases its capacity to erode as it travels?
5. What happens where a lahar starts, and what happens where a lahar stops?
6. Why are lahars just as destructive (if not more so) than most PDCs?
7. On what bases would you highlight areas of extreme lahar risk if you were making a map of potential danger around a gray volcano to advise local authorities? Also, what criteria would you use to designate evacuation routes for a local population?

FURTHER READING

Committee on Natural Disasters, Division of Natural Hazards, National Research Council (1991). *The Eruption of Nevado del Ruiz Volcano, Colombia, South America, November 13, 1985,* Natural Disaster Studies*: A Series.* National Academies Press, 128 p.

Jakob, M. and Hungr, O. (2005). *Debris-Flow Hazards and Related Phenomena.* New York: Springer Praxis Publishing, 739 p.

McGuire, W.J., Jones, A.P., and Neuberg, J. (1996). Volcano instability on the Earth and other planets. Geological Society of London Special Paper 110, 388 p.

Pierson, T.C., Janda, R.J., Umbal, J.V. et al. (1992). Immediate and long-term hazards from lahars and excess sedimentation in rivers draining Mount Pinatubo, Philippines. US Geological Survey Water Resource Investigations Report 92-4039, pp. 1–35.

CHAPTER 12

Volcanoes Unseen and Far Away

Source: Credit_NASA.

Volcanoes: Global Perspectives, Second Edition. John P. Lockwood, Richard W. Hazlett, and Servando De La Cruz-Reyna.

Companion website: www.wiley.com/go/lockwood/volcanoes2

> *Представляется весьма вероятным, что вулканизм мог играть определяющую роль в формировании внешних оболочек многих космических тел. [It appears highly likely that volcanism might play a defining role in the formation of the outer shells of many cosmic bodies.]*
>
> *(Yevgenii Markhinin 1985)*

Half a century ago, the science of volcanology was for the most part focused only on subaerial volcanoes of planet Earth. Although submarine eruptions had been documented as the causative phenomenon responsible for shallow-water explosive volcanism, volcanologists had no inkling that submarine volcanism was the *dominant* manifestation of volcanic activity on Earth, nor that active volcanism was taking place anywhere else in our solar system. There were theories that volcanic activity had occurred on the moon and Mars, but the extent or complexity of this volcanism was unimagined. **Subglacial** volcanism was widely recognized, principally from observations in Iceland, but much fieldwork was needed to document its extent and to understand the processes involved, and a new term **glaciovolcanism** has been coined to describe these phenomena.

All of this "unseen" volcanism gradually came into view with the advent of new technologies, including specialized cameras, deep-sea manned submersibles, sonar, high resolution telescopes, satellites, and related instrumentation. We are still exploring new volcanic terrains with these tools and continue to gain fresh insights about volcanism in environments that are physically hostile to humans. We have even been fortunate enough to glimpse actual eruptions taking place on the sea floor and other worlds. Far from being restricted to Earth's dry land surface and coastal waters, we now appreciate that volcanoes are truly a cosmic phenomenon. These "unseen volcanoes" are discussed in this chapter in two sections – the first will document the previously unseen volcanoes of Earth, and the second will explore the volcanism that has accompanied the formation of the inner rocky planets and the current cryptovolcanic activity that is ongoing in the far reaches of the solar system.

Submarine and Subglacial Volcanoes – The Meeting of Fire, Water, and Ice

Submarine Eruptions

Although submarine volcanism, which has been estimated to account for over 80 percent of all volcanic activity on Earth (Crisp 1984) poses no direct threats to human life, this underappreciated activity is a manifestation of the principal dynamic forces acting on Earth's crust, and an understanding of these phenomenon is critical to an understanding of many terrestrial mineral deposits (Chapter 15) and perhaps to an understanding of life itself (Chapter 13). Although this section refers to "submarine" eruptions, the same physical processes apply to volcanoes that erupt beneath any water cover, including the relatively uncommon interaction of "fire and water" on the floors of freshwater lakes. We will first discuss *where* submarine volcanism occurs (and is probably occurring somewhere at this moment as you read these words, deep beneath the ocean's surface) and then describe the details of *how* these unseen volcanoes erupt. Rubin and others (2012) document many submarine eruptions that are now well-documented world-wide.

Along the Mid-Ocean Ridge Oceanographers discovered the Atlantic Mid-Ocean Ridge (MOR) in a piecemeal fashion over a century, and for many decades researchers believed that there were multiple discrete ridges on the seafloor (hence the historical pluralization "Mid-Ocean Ridges"), when in fact the MOR is a single, albeit sinuous planetary feature (Chapter 2). The misnomer goes even deeper than that, however, since the MOR only really bisects one ocean basin, the Mid-Atlantic – though small spurs also symmetrically divide the Gulf of California, Red Sea, and other very young seaways. The MOR is the principal locus where oceanic plates are pulled apart by deep convective forces, new crust is born, and young volcanic rocks are directly exposed on the oceans' floors. A great deal of oceanographic work has been carried out on the MOR in the past few decades as petrologists have sampled and characterized the uniquely primitive volcanic rocks erupted at many localities, naming them MOR basalts (MORBs; Chapter 3) (Blatt et al. 2006). We now know that MORBs cover over half of Earth's surface, though in most places they are veneered by a thin layer of deep marine muds.

Mantle upwelling and heating elevates the MOR an average of about 500 m relative to the surrounding deep-sea floor, with gently inclined flanks stretching as much as 25–50 km to either side of the crest (Figure 12.1). Elongate magma chambers as shallow as 1–4 km and 10 km wide underlie the crest, fed from mantle sources as deep as 75 km (Kent et al. 1990; Boudier et al. 2000; Mencke et al. 2002; Singh et al. 2006). The gross morphology of the MOR is greatly influenced by spreading rates – how fast a new crust is formed along its axis and thus how fast the flanks are spreading away from the axial rift zone (Figure 12.2). The MOR can be regarded virtually as a single "mega-volcano" – the largest on Earth, because its crest almost everywhere overlies shallow pockets of magma fed by a single source – Earth's upper mantle, and eruptions can occur practically anywhere along its length (Chapter 2).

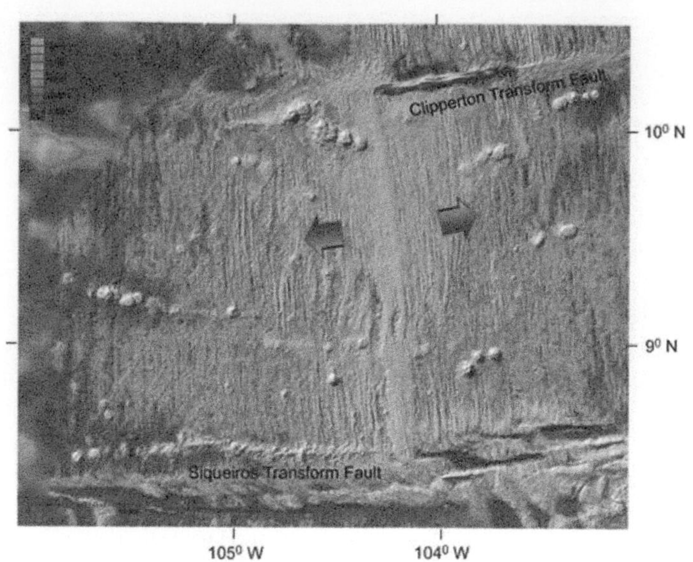

FIGURE 12.1 Detailed bathymetry of a well-studied mid-ocean ridge spreading section, the East Pacific Rise near 10°N, off the coast of Central America. Note the ridge offsets along transform faults and the large number of "off-axis" submarine volcanoes, which indicate the upwelling of magma on the flanks as well as the center of the ridge crest. The figure is 275 km across; scale is in the lower left corner. Source: Compiled by Dan Fornari from the Marine Geoscience Data System database (http://www.marine-geo.org/) using GeoMapApp software.

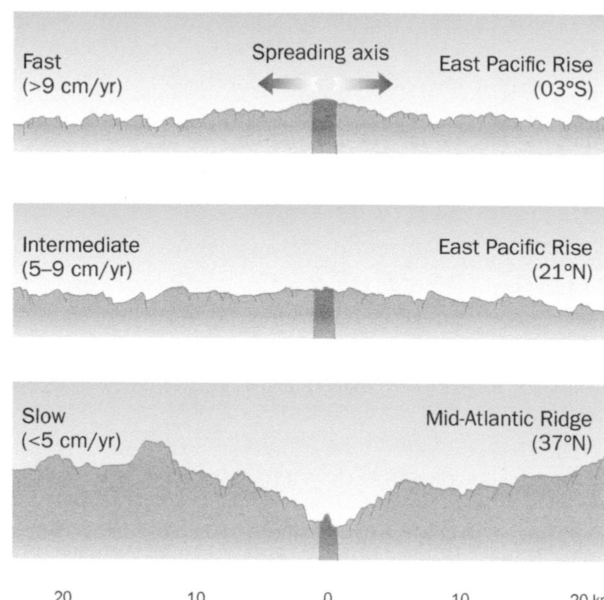

FIGURE 12.2 Profiles across mid-ocean ridge spreading centers, showing characteristic morphologies associated with differing spreading rates. Zones of most active volcanism shown by shaded areas. Source: Based on Macdonald (1982).

Gentle effusive activity from fissures is dominant in the MOR axial rift. Highly fluid lava pours out as pāhoehoe, commonly forming flow fields that look remarkably like their counterparts on land. Close to eruptive fissures where initial eruption rates are high, broad sheet lava fields form. Lava channels and lava caves similar to terrestrial analogs occur in many places, and collapse features associated with fissure drainback or other types of drainage also rupture flow surfaces (Fornari et al. 1980; Fornari et al. 2004; Garry et al. 2006). Unlike terrestrial flows, however, the interiors of deep submarine pāhoehoe flows contain numerous shallow, complex voids resulting from entrapment and boiling of seawater under thin lava crusts. Later collapse of roof crusts reveals that the pockets beneath are supported by pillars which probably grow when large amounts of heated seawater continue to flush through the flow interiors as the lava advances and inflates (Chadwick 2003; Perfit et al. 2003; Figure 12.3).

Pillow lavas are a distinguishing feature of basalt lavas erupted underwater, were first recognized in cross-sectional exposures of ancient, uplifted submarine lavas on dry land (Figure 12.4) but their origins and three-dimensional morphologies were not initially understood. Submersible exploration of the seafloor has shown their true shapes (Figure 12.5). They have been observed forming by divers in shallow water off the island of Hawaiʻi, who report that their formation is a remarkably gentle process: cylindrical lava toes swell into pillows that can break free and tumble downslope into deeper water, accumulating in a bed of glassy lava fragments with other sediments. Some pillows may expand to the point of rupturing, feeding the growth of new offshoot pillows as they continue to grow (Yamagishi 1985). Pillows commonly trap steam and exsolving gases in their cores, and at times highly vesicular, steam-filled pillows forming in shallow water may break off their source feeders and may rise to the surface (Figure 12.6). They do

FIGURE 12.3 Drained pāhoehoe flow formed during the April 2015 eruption of Axial Seamout submarine volcano [25], 475 km off the Oregon Coast. This image shows well the extreme fluidity of ocean floor basalts. In this case, rapidly erupted pāhoehoe formed a small pond, whose upper surface quenched, but was pierced by upwelling steam from seawater trapped beneath the flow. The steam was a little cooler than the surrounding melt and was surrounded by quenched selvages. As the melt drained away downslope, the quenched flow surface and the pillars around the steam (about 1m high) remained. Image courtesy of the Monterey Bay Aquarium Research Institute.

FIGURE 12.4 Glacially polished Precambrian pillow lavas exposed in cross-section, overlain by columnar volcanics in the Flaherty Formation; Belcher Group – on Flaherty Island near Sanikiluaq, Nunavut, Canada. Age about 1900 million years. Such exposures were originally interpreted as the accumulations of "pillow-shaped" masses, and only subsequently was their elongate tubular shapes described. Source: Courtesy of Richard Bell.

FIGURE 12.5 Typical pillow lavas, as seen through the porthole of a submersible (US Navy DSV *Sea Cliff*) diving in volcanic terrain off the west coast of Hawai'i Island. These pillows, at about 800 m depth are lightly dusted with calcareous planktonic ooze, showing that they are prehistoric, but probably no more than 2–300 years old. Source: USGS photo by J.P. Lockwood.

FIGURE 12.6 Discrete pillows floating above a site where pāhoehoe lava was flowing into the ocean on the south flank of Kīlauea volcano, Hawai'i [17] in 1989. The pillows were derived from "buds" breaking off from active submarine pillow lavas flowing 25 m below the surface. Source: USGS photo by J.D. Griggs.

not float for long, however, as they become waterlogged when the interior steam condenses, and sink back into the depths. Such "floating pillows" have been observed forming in the Azores as well as Hawaiʻi, and may be a more common phenomenon than previously appreciated. Rare thin-shelled floating pillows ("balloons") of alkalic compositions were also observed bobbing up-and-down in water near the Island of Pantelleria (Italy) in 1891, during an eruption on the ocean floor 250 m below (Kelly et al. 2014).

The insulating character of quenched flow crusts coupled with the effect of high H_2O heat capacity and the efficient circulation of cold, deep seawater is demonstrated by a recently documented eruption along a portion of the East Pacific Rise near 9° 50′ N latitude, south of Acapulco, Mexico (Soule et al. 2007). This area is one of the most intensely studied MOR systems in the world, and a dozen ocean bottom seismometers (OBSs) had been deployed along the ridge axis in 2005 at 2500 m depth to monitor seismic activity along this fast-spreading ridge segment. The plan was to recover these OBS instruments later for data retrieval, but when a surface crew arrived in 2006, they were only able to recover four instruments – the others had apparently been engulfed by fresh lava! Five of the OBS units were never found and were apparently buried by lava, but three were still able to transmit acoustic signals – although they did not rise to the surface on command – they were apparently "stuck" in fresh submarine flows (Figure 12.7). OBS No. 212, however, was eventually found almost 100 m away from its point of initial deployment – where it had ridden atop a lava flow after being partially engulfed. The OBS had survived its wild ride in good condition, but was thoroughly embedded, and the lava needed to be carefully broken away by the arms of the remotely operated vehicle Jason-2. A delicate plastic flag was still attached to the seismometer casing less than half a meter above the flow top, showing that water temperatures above the moving lava had never been very hot (Figure 12.5). A similar wild ride was taken in 1998 by another ocean floor monitoring instrument on Axial Seamount [25], an active volcano on the Juan de Fuca Ridge 475 km west of Oregon. This instrument contained a pressure sensor, which showed that it was uplifted more than 3 m as new lava flowed beneath it and inflated, but the temperature of surrounding seawater only rose 2°C (W. Chadwick 2008, pers. comm.). Axial Seamount is now well monitored by a remote observatory, and remotely operated vehicle (ROV) imagery obtained after the 2015 eruption documented the extreme fluidity of spreading ridge basalt lavas (Figure 12.7).

As scientists became aware of the critical role of the MOR system in the origin of the oceanic crust, they first assumed that most of the world's numerous undersea volcanic peaks (**seamounts**) developed there too, only to be carried away from their sites of origin by seafloor spreading. Subsequent detailed study of these volcanoes shows that with few exceptions, volcanic seamounts originate where discrete magma sources erupt on the flanks of the MOR, and not along its crest. Large volcanic edifices stand little

FIGURE 12.7 An ocean bottom seismometer (OBS) trapped in a slabby pāhoehoe lava flow during the 2005–2006 East Pacific Rise eruption at around 2540 m depth. The OBS was found to be partly enveloped by lava and had been transported almost 100 m by the flow. The acoustic command components were still functional, which facilitated location and recovery of the OBS. Note the undamaged plastic flag. Source: Photo courtesy of the Woods Hole Oceanographic Institution, National Deep Submergence Facility – ROV Group, the National Science Foundation – Ridge 2000 Program, and the scientific party of cruise AT15-17, E. Klein, chief scientist.

chance of growing in an axial rift valley because of diffuse eruption loci, rifting, and faulting (Chapter 2), though there are some spectacular exceptions, including 10-km-wide Axial Seamount along the Juan de Fuca Ridge, which has erupted over 50 times in the past 1,600 years – most recently in 2011 and 2015 (Clague et al. 2017). Most geographically – offset seamounts begin growing within 100 km of MOR axes, and the occurrence of new volcanism on oceanic crust generally diminishes as the lithosphere grows thicker and older with distance from the ridge (Batiza 1981; Hillier 2007).

These off-ridge volcanoes come in many shapes and sizes – from small isolated monogenic volcanoes, a hundred meters or less in height to true mountains tens of kilometers wide at their bases and 3–4 km high (Gardener et al. 1984). Smaller seamounts typically have conical shapes, but larger ones have more complex forms, with summits either flattened or dimpled with submarine calderas. The lava flows they produce can be far traveling. Individual flows more than 60 km long have been observed off Hawai'i and near the base of seamounts associated with the East Pacific Rise (Fornari et al. 1985; Holcomb et al. 1988).

Large undersea volcanoes can form very rapidly. For example, beginning in 2018, a single six-month-long submarine eruption near the island of Mayotte in the western Indian Ocean built a new seamount 800 m high and 5 km across at the base. The neighboring island sank 13 cm and shifted eastward about 10 cm during the course of the eruption, which seismicity indicated was fed by magma from a depth of 20–50 km – well below the level of reservoirs typically underlying MORs. Sonar detected plumes of bubbles rising from the active volcano, and a single, ultralow-frequency humming related to the unrest spread worldwide through the sea. But otherwise, no expression of this prodigious volcanic activity could be detected on the surface of the water (Pease 2019).

Harry Hess (Chapter 2) discovered a number of flat-topped seamounts in the north-central Pacific during wartime bathymetric surveys, and, as you may recall, named these features **guyots** (Hess 1946). Hess correctly surmised that the flat summits of many guyots represent volcano-fringing coral reefs – atolls – drowned by tectonic submergence. The coral polyps that flourish in warm, sunlit water could not maintain upward reef construction fast enough to compensate for sinking of their volcanic foundations. For a time, this persuasive argument convinced oceanographers that all flat-topped seamounts formed in this way – biogenically as well as geologically, but the distribution and variable depths of their tops, and the fact that some guyots lack coral caps, demonstrated that this cannot be the whole story. As some submarine volcanoes grow large, magma infiltrating their interiors can more easily erupt from the flanks than from the summit, building up extensive flanks so that the summit acquires a more flattened profile (Fornari et al. 1985). Volcanoes may also develop flat tops by growth from circumferential vent systems, as is shown by subaerial volcanoes of the Galapagos Islands (Simkin 1972). Like oceanic volcanoes that project above the surface of the sea, large submarine volcanoes are also inherently unstable; their summits can collapse, and major flank failures may blanket surrounding ocean floors with extensive debris flows.

Oceanic plateaus and hot spot tracks The most voluminous volcanic eruptions ever to occur on Earth formed the extensive basaltic sheets referred to as large igneous provinces (LIPs) (Mahoney and Coffin 1997). Although these include the great subaerial flood basalt plateaus (Chapter 9), we now know that the world's largest LIPs formed on the ocean floor (Taylor 2006). Oceanographic research and deep-sea drilling programs over the past few decades have revealed the enormous size of these high-standing submarine basalt plateaus, which may rise more than a kilometer above the surrounding seafloor. The Ontong Java Plateau, the most voluminous LIP on Earth, underlies the Solomon Islands, covers an area of 2 million km^2 and may have a volume as high as 60 million km^3 (Fitton and Godard 2004). Radiometric dating shows that the massive LIPs in the western Pacific formed in two episodes during Late Cretaceous time. The impact of these voluminous eruptions may have led to major chemical changes in ocean waters and to mass extinctions of marine fauna (Kerr 1998; Chapter 13). The second most voluminous LIP is represented by the Kerguelen Plateau, which developed on the Antarctic plate as the Indian Ocean began to form 130 million years ago. This LIP has a long history, and probably involved subaerial volcanism in its early history. Minor volcanism continues on the Heard and McDonald islands, which are the only subaerial expressions of this LIP.

Many submarine seamount volcanoes are aligned in linear chains that may extend thousands of kilometers across the seafloor, and commonly show no indication that they ever formed subaerial islands. Some of these volcanic chains show age progressions that suggest they formed above intraplate hot spots, (Chapter 2) but others trend at angles to known plate motions and may have formed above linear zones of weakness in the oceanic crust, or may indicate deep-seated linear melting anomalies. Radiometric dating has shown that most of the intraplate submarine volcano chains in the Pacific are of Cretaceous age, and are chronologically related to the widespread magmatic processes responsible for the formation of the afore mentioned oceanic plateaus – spasms of tremendous magma generation and heat release from Earth's interior (Rea and Vallier 1983).

Shallow-water explosive submarine eruptions While effusive eruptions predominate in the deep sea, explosive eruptions are also capable of taking place there. In fact, detailed seafloor exploration by manned submersibles and ROVs have shown that explosive volcanism can occur at *any* water depths, although such eruptions tend to be larger and more violent in shallower water (e.g., Clague et al. 2011).

Eruptive vents that open at water depths of less than a few hundred meters, where hydrostatic confining pressures are low, readily trigger Surtseyan eruptions. Over a period of days or weeks these may lead to the emergence of new volcanic islands

(Chapter 7). But most new islands do not survive the erosive power of tides and waves for long. Three factors are necessary for long-term survival: First, the eruption rate must be fast enough to build a natural dam around the vent, keeping out the surrounding water so that effusive activity may ensue. Then, after effusive activity has begun, lava flows or welded agglutinate must mantle enough of the loose fragmental base of the new island to defend it against erosion. Finally, if the island is to grow large enough to become a substantial platform for life, follow-up eruptions must take place with a short-enough repose interval to increase overall land area or simply to replace land that eroded away in the time since the last volcanic activity.

The historic record contains numerous instances of "wannabe" volcanic islands that never made it. A good example took place in 1831, amid a Mediterranean conflict between Great Britain, France, and the Italian kingdoms. In July of that year, a strange event happened off the southwest coast of Sicily, in waters claimed by all three powers. The surface of the sea boiled, ash and rocks were blasted upward, dead fish were noted, and the smell of sulfur was strong in Sicily and North Africa. By August, a small island had risen above the sea and a British crew landed, claiming the new land for England. They named it Graham Island. The Sicilians, wanting this fresh speck of land for themselves, sent a crew there, pulled down the Union Jack, claimed it for Sicily and renamed it Ferdinandea in honor of their Neopolitan king. Graham Island/Ferdinandea grew to 65 m above sea level and had an area of 4 km^2 at one point – potentially sufficient for a naval outpost (Figure 12.8). The French also attempted to claim the island, but volcanic activity stopped, and the disputed territory quickly disintegrated and washed away. By December 1831, it was no more. Another eruption briefly rebuilt an island in 1863, but the site is now represented only by a shoal – the top of a large submarine volcano. The volcano became seismically active again in 2006, and will no doubt again build new land above the sea one day. (Who will claim it then?)

Shallow submarine eruptions are perhaps more common than generally recognized, and several are recorded each year, especially along island-arc chains. A typical submarine eruption is heralded by large areas of persistently discolored water, caused by the convective rising of ash-laden hot water above an active vent (Chapter 7). Jagged Surtseyan clouds may later burst above the sea, but whether events progress that far or not is unfortunately unpredictable, as shown by the tragic case of Myōjin-Shō [149] in 1952. Myōjin Shō is a submarine volcano rising some 500 m high above the surrounding seabed, located about 450 km south of Tokyo on the Izu-Ogasawara Ridge, a chain of volcanic islands and submarine volcanoes. Numerous submarine eruptions had been recorded from this area for almost a century, and on 17 September, 1952, a fishing vessel reported a new eruption 12 km northeast of the Bayonaisse Rocks. The reawakened volcano was named after the fishing boat (the *Myōjin Maru*). Ships and aircraft rushed to the scene to investigate the eruption. On 24 September, a Japanese research ship, the *Kaiyo-Maru No. 5* had the misfortune to be conducting a marine survey directly over the active vent, some 50 m below, when an extremely violent hydrovolcanic explosion occurred. The 10-m-long vessel was completely destroyed, and the 31 scientists and crew aboard were all killed – the worst disaster in the history of Japanese oceanographic research. Episodic explosions continued for the next year, forming a new island that rose as much as 100 m above sea level, but which also disappeared, only to reappear and disappear again repeatedly (Minakami 1956). After 1953, no reappearances of land occurred, though submarine volcanic activity continued. It was studied using a pioneering unmanned radio operating survey boat, the *Manbou* (Sunfish), which for the first time utilized the SOFAR (Sound Fixing and Ranging) channel to listen safely to the volcano's underwater rumblings. Eruptive activity finally tapered off in 1970.

FIGURE 12.8 Ferdinandea, the "ephemeral island" as sketched in 1831. Source: Engraving from the *Bilderbuch fur Kinder*, Bertuch, 1849.

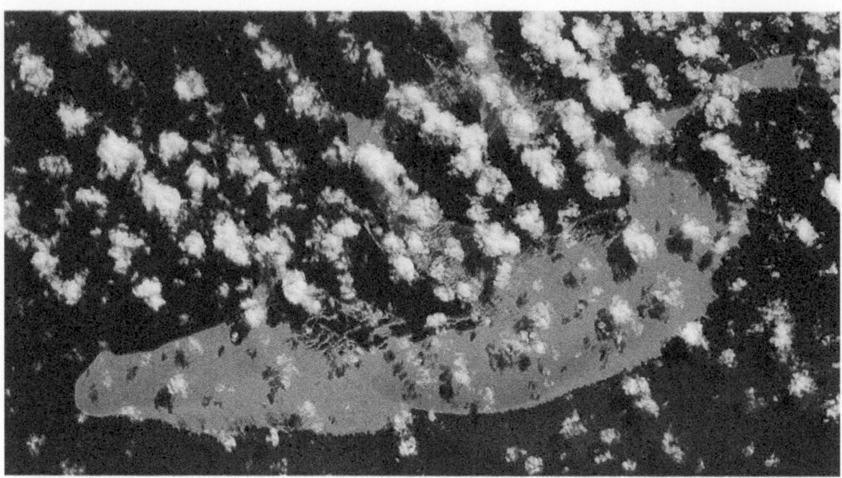

FIGURE 12.9 Pumice raft south of Vava'u island, Tonga, as observed by the Landsat 8 satellite on 13 August, 2019. This is only one of several rafts produced by the 6 August eruption, and covers an area of about 165 km². Source: NASA image Tonga_OLI-2019225.

Subsequent studies have revealed that Myōjin-Shō is a satellite volcano on the northeast rim of a 5.6-km-wide, extensively mineralized submarine crater, now called the Myōjin Knoll caldera, whose floor lies 1100 m below sea level. The post-collapse volcanism within and at the margins of the caldera has been extensive and is characterized by voluminous accumulations of silicic pumice that appear to have been erupted underwater at over 500 m depth (Fiske et al. 2001). Despite its low density, most of this pumice likely never reached the sea surface, but became waterlogged and sank soon after formation.

Pumice-rich deposits related to submarine eruptions well below wave-base and even accretionary lapilli deposits formed entirely underwater have been identified elsewhere in the western Pacific (Cunningham and Beard 2014; Jutzeler et al. 2014). At the 900-m-deep Havre volcano in the Kermedec arc, a 2012 eruption produced blocks of pumice as much as a few meters across that piled up around the vent area. Nevertheless, roughly 75% of the pumice and finer erupted fragments floated toward the surface and drifted away.

Pumice rafts from some silicic eruptions can persist for months (Chapter 7). An underwater eruption at Tonga's Home Reef volcano [2] in the summer of 2006 (Figure 12.9) produced an 8-km-wide raft of floating dacite pumice, parts of which reached northeastern Australian beaches over 4000 km away by the spring of 2007. The frequency of small to moderate-sized silicic submarine eruptions is so great in the Tonga–Kermadec island arc that similar pumice rafts are reported once every 5–15 years. A voluminous five-day eruption took place on an unnamed submarine volcano 80 km northwest of Vava'u island in the Tonga group in August 2019 (Figure 12.9; Brandl et al. 2019). Separate pumice rafts up to 30 cm thick initially spanned an area greater than 200 km². Massive amounts of pumice drifted westward, and after a year, pumice reached the east coast of Australia.

Deep water explosive eruptions The "conventional wisdom" (a dangerous thing!) that was accepted by all volcanologists not so long ago was that the high hydrostatic pressures of the deep oceans would prevent the explosive eruptive activity so characteristic of shallow-water eruptions. The assumption was that phreatic explosions could only take place at very shallow depths (less than 100 m). Submersible studies over the past two decades have shown however, that pyroclastic deposits are widespread in all submarine volcanic environments – *regardless of water depths.* It was also erroneously assumed that high confining pressures on the deep-sea floor would prevent dissolved magmatic gases from exsolving and contributing to pyroclastic activity, but that was quickly disproved by the observation that vesicles are present in lava samples recovered from anywhere on the ocean floor. The realization that deep-sea explosive volcanism was an important phenomenon was given economic impetus by the revelation that many of Earth's largest metallic ore deposits (Kuroko-type, Chapter 15) are associated with pyroclastic rocks that are clearly marine in origin (Burnham 1983). Proof-positive, real-time eye-witness evidence that explosive eruptive activity occurs in the deep sea was recently provided by submarine exploration of active volcanoes in the Northern Marianas arc, when ROV operators in 2004 discovered ongoing explosive activity near the 555-m-deep summit of NW Rota-1 volcano [57], 175 km northwest of Guam (Figure 12.10; Embley et al. 2006). Video of the ongoing eruption showed cauliflower-shaped white clouds pulsating up to 50 m above the small volcanic vent named "Brimstone Pit," roiling plumes that rained down large blocks of highly vesicular basaltic andesite and blobs of molten sulfur. Similar phenomena were documented in 2005, and in 2006, incandescent lava was observed being ejected from the vent – the first-ever direct observation of high-temperature submarine volcanism (W. W. Chadwick 2008, pers. comm.). The flanks of the 16-km-diameter NW Rota-1 volcano are mantled with pyroclastic debris from these ongoing summit eruptions, which appear to be continuous phenomena during the periods of observation.

Submarine eruptive activity in island-arc environments is much more common than previously assumed. In May 2009, an interdisciplinary survey team discovered an active eruption at a water depth of 1200 m (by far the deepest active eruption site ever studied),

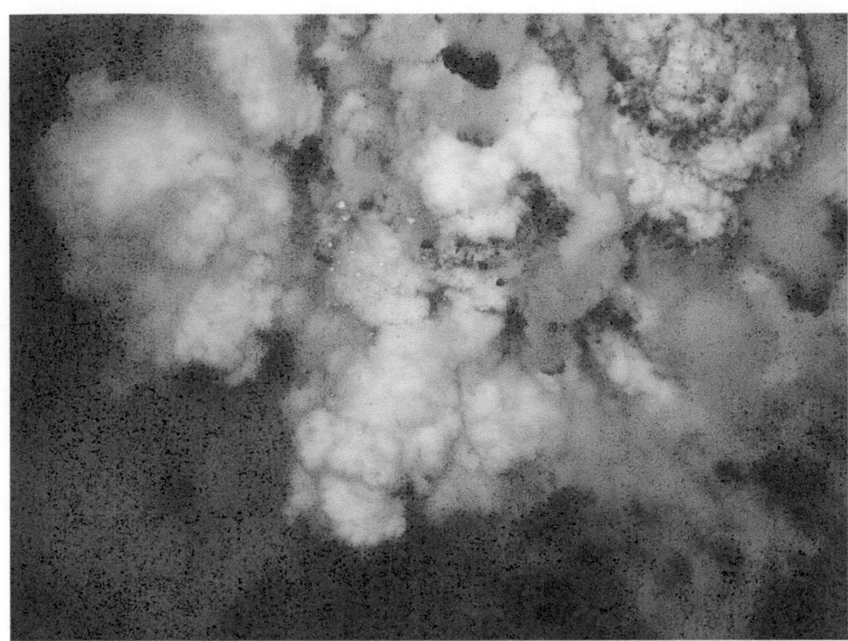

FIGURE 12.10 Eruptive plume above an active vent at 550 m depth during the 2010 eruption of NW Rota-1 submarine volcano [149] in the Mariana volcanic arc, located 100 km north of Guam. Note basaltic lava bombs falling out of plume. ROV photo by the NOAA Ocean Exploration Program, and NOAA Vents Program. Source: Photo courtesy of W.W. Chadwick.

at W. Mata volcano [3] in the Lau basin – a backarc environment 200 km southwest of Samoa. Observations from the tethered ROV Jason-II showed spectacular eruption of gas rich magma from at least two vents, which both produced pulsating explosions from the tops of 10-m-high vent edifices, and vesicular pillow lavas near their bases (K. Rubin 2009, pers. comm., Clague et al. 2011).

Explosive submarine eruptions involve two separate processes that contribute to the fragmentation of lava and to the formation of deep-sea pyroclastic rocks. Simple quenching of hot lava results in differential stresses and fracturing that produce angular fragments of broken rock – hyaloclastite – as is normal when lava flows pour into the sea. But recent studies of submarine pyroclastic deposits also indicate that *actual lava fountains* have operated on the deep seafloor, producing primary vitric ejecta that clearly erupts as globules of molten rock in the form of delicate fragments of glass bubbles and "Pele's tears" (Chapter 6). How can lava fountains operate under the great pressures existing on the deep-sea floor, at depths where "steam" (gaseous water) cannot exist? Explosive eruptions at these depths must be driven by exsolving volatiles contained within the erupting lava (Wohletz 2003; Perfit et al. 2003). The concentration of these gases (in large part CO_2) is not normally adequate to drive explosive activity thousands of meters below sea level, but pockets of gases at sufficient concentration can certainly be formed by magmatic processes before they are erupted (Head and Wilson 2003). Resulting explosions may feed plumes of hot water and entrained pyroclastic debris many hundreds of meters above a vent (Barreyre et al. 2011).

Subglacial Eruptions

The specialized field of **glaciovolcanology** first developed in Iceland, a country where the concentration and frequency of subglacial eruptions is the greatest. Subglacial historical eruptions have occurred in Antarctica, and eruptions have taken place beneath snow-capped volcanoes on all continents except Australia. Where these eruptions occur, molten lava never touches ice directly, or if so not for long, as liquid water and steam form an interface between the lava and ice near instantaneously, and cooling lava may produce ten times its volume in melted water (e.g. Oddsson et al. 2016). Many of the rock structures that form where "fire and ice" meet (pillow lavas and hyaloclastites) are thus similar whether formed beneath glaciers, in lakes, or under the sea. Where lava does contact melted ice water, it typically develops intricate, irregular networks of cooling cracks due to rapid quenching (e.g. Conway et al. 2015). Mee et al. (2006) describe distinctive fracture patterns in lava flows quenched at snow-lava contacts on a Chilean volcano. Prehistoric lava flows near the summit of Mauna Kea volcano [16] show similar fracturing in pillowed basalt formed where they flowed beneath summit glaciers (Figure 12.11).

Distribution Earth's glaciers are divided into two types: the thick "continental" ice sheets, which cover vast areas of land (and sea) in polar regions, and the smaller "montane" glaciers, which form on high mountain at any latitude. Many of Earth's most active volcanoes are covered by snow and ice, and when these volcanoes erupt, unusual processes generate unique features and hazards.

FIGURE 12.11 Subglacial pillow lavas formed at the contact between a prehistoric lava flow and melted ice water at 4050 m elevation on Mauna Kea volcano, Hawai'i [15]. This large, glassy pillow was later abraded by subsequent glacial movement after the flow cooled. Note hammer for scale. Source: Photo by J.P. Lockwood.

The Antarctic ice sheet is Earth's largest, and accounts for about 90 percent of all the world's glacial ice. Like East Africa (Chapter 2), the western part of Antarctica is gradually rifting apart along a 3000-km-long line of faults and related basins. Volcanoes dot this alignment, and geophysical investigations reveal that at least 138 are buried deeply beneath the ice cap (de Vries et al. 2008, 2018). Ice-penetrating radar studies (Corr and Vaughn 2008) have documented a major volcanic eruption (VEI = 3–4) that occurred near the coast of the West Antarctic ice sheet and blanketed an area of over 20,000 km^2 with tephra about 2200 years ago. This tephra layer now lies buried beneath 100–700 m of ice, and was derived from a subglacial volcano that stands about a kilometer above the glacial bedrock. There is evidence that this and nearby volcanoes remain thermally active, and raise concern that subglacial heating may be contributing to the well-documented thinning and accelerated sliding of the adjacent Pine Island glacier.

Elsewhere on Antarctica, volcanoes are only partly ice covered, their summits easily visible at the surface. They form a belt extending from the continuously active Mount Erebus [173], across West Antarctica, and along the length of the Palmer Peninsula (Simkin and Siebert 1994). Erebus is one of the most active volcanoes in the world. A frequently active lava lake churns within its ice-rimmed summit pit, 3800 m above sea level. Deception Island volcano [88], near the other end of the Antarctic rift system, last erupted from its caldera margins in 1967–1970, destroying buildings and forcing evacuation of resident island research personnel (Smellie 2002).

Although not on a continent, "continental" glacier ice fields cover almost 10 percent of Iceland and bury or partially bury several active volcanoes, including the dangerous Katla volcano [94]. The large Vatnajökull glacier (8100 km^2) has long been known to overlie a volcano, Grímsvötn [99], one of the world's most frequently active with over two dozen eruptions recorded over the past thousand years of Icelandic human history (Simkin and Siebert 1994). Grímsvötn is characterized by a large, ice-filled caldera athwart the Icelandic eastern rift zone, and is related to the fissure system that fed the great Laki [96] eruption of 1783–1784 (Chapter 6). Melting of ice above the Grímsvötn magma chamber sustains a subglacial warm-water lake whose water level is constantly monitored because of the threat of **jökulhlaups** (glacial lake outburst floods) which increases as lake levels rise. A subglacial fissure opened just north of Grímsvötn on 30 September, 1996, and provided an excellent opportunity for scientists to make a detailed study of subsurface eruptive phenomena and magma-ice thermal exchange beneath Vatnajökull. The eruption (named Gjálp) eventually produced 0.8 km^3 of basaltic hyaloclastites, but only an estimated 2–4 percent of total erupted material made it through the ice sheet into the atmosphere. The eruption melted more than 2.5 km^3 of ice at the base of the glacier, at an initial rate of about 500,000 m^3/day. This could be calculated by measuring the volume of a large down sag and collapse in the overlying glacial surface (Gudmundsson et al. 2004). Meltwater poured into the adjacent Grímsvötn caldera lake, and was stored there until sufficient hydrostatic pressure at the lake bottom exceeded the ice overburden pressure. At that point, the water began wedging its way downslope along the ice–bedrock interface. It probably also melted new passageways by release of heat. Although the Gjálp eruption stopped on 13 October, the meltwater did not reach the edge of Vatnajökull, 50 km away, until 5 November, when it explosively burst forth, creating a jökulhlaup that destroyed bridges, roads, and powerlines while depositing house-size blocks of ice kilometers from the glacier's ruptured margin. The torrential floods formed lahars that added several square kilometers of new land to the coast of Iceland. The initial discharge rate of jökulhlaup water was over 45,000 m^3/s – much greater than the average flow rate of the Mississippi River!

Grímsvötn returned to life beginning in the summer of 2003 with heightened seismic activity accompanied by a caldera uplift. Icelandic scientists increased their monitoring in preparation for a likely eruption. Earthquake activity suddenly jumped to a more intense level on 25 October, prompting a public warning of impending volcanism. Meanwhile, meltwater input from the surrounding ice increased, and the caldera lake level rose to the point where the base of the surrounding glacier could retain it no longer. On 28 October, the lake began draining again, and a jökulhlaup took place the next day in the nearby Skeidara River watershed. Prior experience had demonstrated that drop in lake level by roughly 10–20 m could trigger an explosive eruption by reduction of pressure on the underlying magma chamber. So alerted, the scientists warned the Volcanic Ash Advisory Centre in London on 29 October of a possible tephra cloud that might threaten air traffic (Chapter 14). Sure enough, a swarm of intense volcanic earthquakes and tremor began on 1 November, 2004, announcing the beginning of the eruption, which soon sent ash to altitudes of more than 10 km (Figure 12.12). All air traffic was safely rerouted by then (Vogfjörd et al. 2005).

Wherever volcanoes rise to sufficient height and precipitation is adequate (even near the Equator), they will be covered with snow and ice. This is true for most of the high Cordilleran volcanoes of North and South America, and for volcanoes of the Aleutians and Kamchatka. Unfortunately, glaciers are shrinking or have disappeared from many of these volcanoes within the past century. Although the volumes of meltwater and resultant floods caused by volcanic melting of these montane glaciers and snowfields are generally much less than that produced by eruptive activity beneath continental glaciers, these volcanoes too can produce disastrous jökulhlaups and lahars. Where lava flows move down steep ice, snow, and water-saturated slopes, they may disintegrate into pyroclastic flows downslope, as has been observed at Etna [112]and Klyuchevskoy [171] volcanoes (Belousov et al. 2011).

A tragic case in point involves the 1985 eruption of Ruiz volcano [71] (Colombia) . The summit of Ruiz (5321 m) was covered by a small glacier only about 30 m thick on the evening of 13 November when hot pyroclastic material blasted up through deeper ice and overlying snow that filled Arenas crater. The initial pyroclastic surges formed deposits that consisted mostly of angular ice clasts with sparse lithic debris (Figure 12.13). Little meltwater was apparently produced by this initial explosive activity, but subsequent

FIGURE 12.12 Eruption of Grímsvötn volcano [99] on 2 November, 2004, triggered by the drainage of a subglacial lake (view from south). The column of ash and steam is about 700 m wide at the base and was rising to over 11 km at this time. This vigorous explosive phase lasted for about 30 hours and was followed by minor explosive activity until the eruption was over on 6 November. Note the black ash ejecta at the plume base – which is black not because of increased ash content, but because its temperature is over 100°C, and steam has not yet condensed. Source: Photo by M. Gudmundsson.

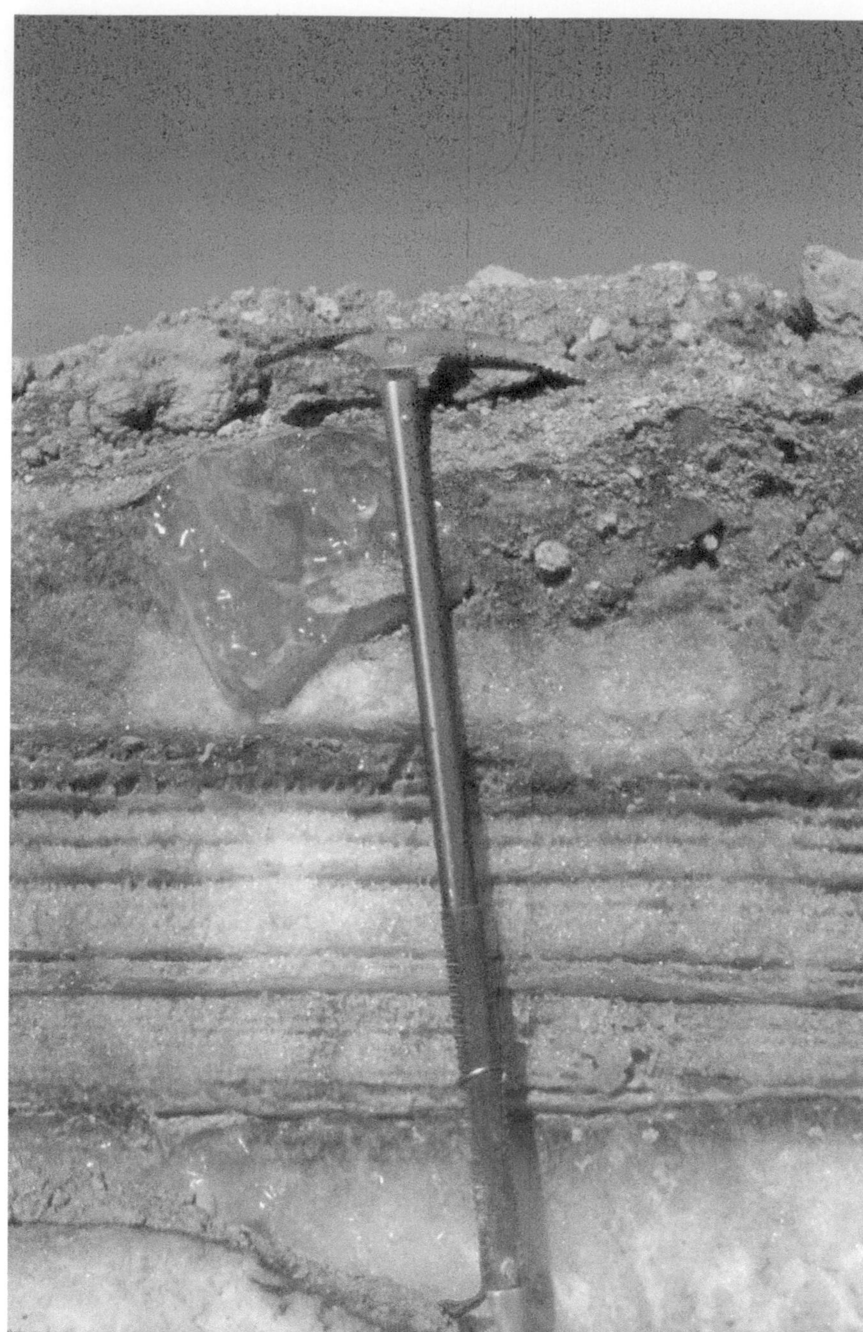

FIGURE 12.13 Pyroclastic surge ice breccias on the rim of Arenas crater, north of the summit of Ruiz volcano [71], Colombia, formed during the eruption of 13 November, 1985. This subglacial eruption produced pyroclastic deposits with similar structures to lithic breccias – but most of the angular clasts were of ice. Later phases of the brief eruption produced hot pyroclastic flows that overrode earlier surge deposits. Source: USGS photo by J.P. Lockwood.

pyroclastic flows melted snow overlying the glacier and triggered small lahars that incorporated unconsolidated pyroclastic and water-saturated sediments to form the massive debris flows, already described, that devastated cities as far as 50 km away (Chapters 11 and 14; Pierson et al. 1990).

The Kamchatka peninsula (Chapter 2) is home to nearly 200 volcanoes, two dozen of which have been erupted in historical time. All of these are covered with snow and ice most of the year, and Kamchatka volcanologists are very accustomed to observing the interactions between fire and ice. When these volcanoes send lava or pyroclastic flows down their flanks, abundant steam hides much of the activity (Figure 12.14), but because lava and PDCs are much denser than snow, flows usually tunnel beneath thick snowbanks, melting large volumes of water as they proceed. Edwards et al. (2015)) reported that fast-moving flows descending the steep upper slopes of Tol'batchik volcano [168] during the 2012–2013 eruption "generally moved on top of snowpacks" before later melting downwards after they came to rest. These quickly melted the underlying or overlying snow and generated small lahars downslope. While mapping a fluid prehistoric lava flow on the slopes of Gorelli volcano [165] I (JPL) encountered extensive pillow lavas at basal contacts where the flow had tunneled under a snowfield. The pillows had been quenched so quickly by the surrounding meltwater that the mat of underlying tundra plants had only been carbonized in a few places! Volcanic eruptions beneath glaciers can also incorporate large blocks of ice in pyroclastic flows, which may generate phreatic explosions, as occurred after the 1980 eruption of Mount St Helens [31] (Figure 12.15).

FIGURE 12.14 Lava flowing down north-west slope of Klyuchevskoi volcano [171], Kamchatka, in April 2007. Flowing lava is denser than snow and commonly flows beneath the short-lived snow covers. Klyuchevskoy (4750 m), is the largest and most active volcano in Eurasia. Source: Photo by Yurii Demyanchuk.

FIGURE 12.15 Steam explosions in pyroclastic flow deposits on the 1980 “Pumice Plain” near the southwest corner of Spirit Lake, Mount St Helens [31], Washington. These phreatic explosions were caused by the melting of large blocks of glacial ice embedded in the hot pyroclastic flow deposits, and continued for months after the 18 May, 1980 eruption. Note the phreatic explosion craters from previous events. Source: USGS photo by Dan Dzurisin.

Landforms As we discussed earlier, subglacial effusive eruptions occurring beneath thick ice fields produce large volumes of meltwater that will enclose the erupting lava in a watery interface. If lava supply rates are very high, and hydrostatic pressures relatively low, extensive fragmentation of quenched lava will occur and extensive subglacial hyaloclastite deposits can be formed; if effusion rates are slower, and the hydrostatic pressure is higher, pillow lavas similar to those formed in submarine environments will be formed instead (Gudmundsson et al. 2004).

The recession of once more-extensive Icelandic ice sheets has revealed distinctive flat-topped mountains, called by Icelanders **table mountains**, which are characterized by pillow lavas at their bases, hyaloclastic deposits on their flanks, and thin, subhorizontal layers of lava capping their summits. This stratigraphy reflects the change in the depth of ice-water cover during edifice growth. The steep flanks of most of these structures exist because of original confinement by ice, while the flat tops represent eruption into the floors of overlying meltwater lakes that may be filled or displaced as eruption continues. However, because of confusion with the more common use of "table mountains" to describe flat-topped mountains elsewhere formed by the erosion of flat-lying resistant sedimentary beds or lava flows (e.g., Cape Town, South Africa), the term **tuya** is now preferred to describe these distinctive glacier-related features, after Tuya Mountain in British Columbia (Figure 12.16).

If an erupting volcanic cone or fissure fails to melt all the way through an icecap to develop a lava-caped tuya over time, the resulting volcanic mound, which is seen from a distance may be smooth in profile and quite elongate, is called a **tindar**. Tindars tend to be much smaller (<0.01 km^3) than tuyas (>0.1 to tens of km^3), and may consist mostly of pillow lava with lesser amounts of hyaloclastite. Both tindars and tuyas tend to be monogenetic in origin and simply represent varying degrees of ice–basaltic lava interaction at point-source or short fissure vents.

In some parts of Iceland during the last ice age, long eruptive fissures such as that of the Eldgjá–Laki [95–96] eruptions (Chapter 6) developed beneath the ice sheet that covered most of the island. These ancient flood basalt eruptions are recorded by the presence today of curvilinear, steep-sided ridges of hyaloclastite, commonly altered to brownish palagonite (Figure 12.17). Termed **hyaloclastite ridges**, these features are now major topographic divides as much as 44 km long. In a well-studied part of Iceland's western Volcanic Zone, there are almost twice as many hyaloclastite ridges as there are tuyas (Zimbelman and Gregg 2000). A new hyaloclastite ridge developed along a 6 km fissure during the 1996 Grímsvötn eruption. Radio echo soundings and gravity

FIGURE 12.16 Tuya Butte, British Columbia, viewed from the south. This is the type locality for **tuyas**, the distinctive flat-topped volcanic features that form beneath thick glaciers. This tuya formed during the Pleistocene within a glacier estimated to be almost 1 km thick, and is about 400 m high, having a volume of 2–3 km^3. Tuyas and other subglacial volcanic features are widespread in British Columbia where subglacial volcanism has been extensively studied. Source: Geological Survey of Canada photo by Cathy Hickson.

FIGURE 12.17 Late Pleistocene hyaloclastite formed beneath glacial ice about 40 km east of Thingvullir, Iceland. The mechanical pencil is 16 cm long. Source: Photo by R.W. Hazlett.

surveying have since helped map the ridge, still covered by Vatnajökull glacier, which at the highest point rises some 450 m above the pre-eruption bedrock, with a maximum width of some 2 km. The original thickness of the ice into which this ridge developed was 550–750 m (Gudmundsson et al. 2002).

Impacts of glaciation cycles on volcanism Gravitational loading of the crust by thick ice sheets can influence the ascent of magma in the crust, where shallow magma chambers are in delicate balance (Schmidt et al. 2013). When a massive ice sheet retreats, as in many parts of the northern hemisphere at the close of the last ice age, the reduction of stress acting on the lithosphere generally takes place faster than the lithosphere is capable of rebounding. This causes relatively rapid decompression of the underlying crust and mantle, which can lead to increased rates of partial melting and increase in eruptive activity throughout volcanic areas. This may explain the increased rates of eruptions in the Canadian Cascades and at Mauna Loa volcano [15] that apparently followed rapid melting of snow and ice above delicately balanced magma reservoirs at the end of the Pleistocene (Lockwood 1995). Changes in the chemistry of erupted lavas may also take occur during the immediate aftermath of de-glaciation, including the eruption of slightly more siliceous compositions (e.g. Eason et al. 2015).

Extraterrestrial Volcanoes

Volcanoes are not unique to Earth. Our neighboring planets in the solar system have spectacular volcanoes, and it may well be a universal rule of thumb that all the inner, rocky planets and satellites more than a few hundred kilometers in diameter have produced effusive volcanoes similar to those on Earth at one time in their histories (Table 12.1). The basic role of all volcanoes, no matter their chemistry, is to help cool worlds (Chapter 2). Most rocky planets and moons have high-temperature origins, and all continue to produce heat through radioactive decay, gravitational adjustments, and tidal stresses. In general, the bigger a rocky planet, the greater its ability to sustain internal heat production through self-generated radiogenic decay, and with that, the continuation of volcanism (Chapter 3). While volcanic activity remains widespread on Earth, for example, it died out on the much smaller Moon hundreds of millions of years ago (Spudis 2000).

Not all volcanoes erupt silicate or carbonate-based magmas: the outer, icy planets also produce volcanoes of very different sorts from those of Earth. Some volcanoes further out in the solar system erupt much cooler matter, including water and nearly freezing gases! We certainly have not seen all of the permutations of volcanic activity possible in nature, but what we have seen expands imagination and understanding far beyond what would have been possible had we remained strictly Earth-bound.

The Moon – Flood Basalts in Space

Our moon has a complex igneous history, and it is probable that the entire satellite was molten at some point in its early, unrecorded history before its now dominant, light-colored plutonic rocks formed more than 3 billion years ago. Subsequent volcanism has covered significant areas of the near side of the moon with flood basalts. Most cultures have names for the pattern of these dark

TABLE 12.1 **Planets and satellites in the solar system known to have volcanic eruptions (listing is from the closest to the farthest from the sun). Question marks (?) indicate existence of these features is not proven.**

Planet or moon	Still volcanically active?	Major source(s) of heating	Major kinds of volcanic landforms	Known or possible types of volcanic eruptions*
Mercury	Potentially yes, but not at present	Fracturing and tearing of crust from large impacts and planetary contraction	Low, broad, pyroclastic vents	"Dry" explosive activity of diverse intensity
Venus	Yes	Radiogenic	Shield volcanoes, lava shields, basalt plains, coronae, paterae, montes, canali	Flood basalt, Hawaiian
Earth	Yes	Radiogenic	Composite cones, shield volcanoes, lava shields, basalt plains, Plinian calderas, cinder cones, domes	Flood basalt, Hawaiian, Strombolian; Vulcanian; Plinian; Surtseyan
Moon	Potentially yes, but not at present	Early radiogenic, impacts	Shield volcanoes(?), lava shields, basalt plains, cinder cones, rilles	Flood basalt, Hawaiian, Strombolian, Vulcanian, Plinian
Mars	Potentially yes, but not at present	Early radiogenic, impacts	Shield volcanoes, lava shields, basalt plains, cinder cones(?), paterae, tholi, montes	Flood basalt, Hawaiian, Strombolian(?), Vulcanian(?), Plinian
Io	Yes	Solid tidal	Shield volcanoes, paterae	Flood basalt(?), Hawaiian, Vulcanian(?), sulfur plume
Enceladus	Yes	Solid tidal	None	H_2O plume
Triton	Yes	Seasonal solar	None	N_2 plume

* Applying terrestrial eruption classifications to other worlds is somewhat misleading, because planetary atmospheres and gravity fields differ greatly. Explosive eruption behavior in particular differs significantly. Take the fifth column of this table somewhat with a grain of salt (or sulfur)!

patches on the white face of the full moon. Consider, for example, "the man on the moon" many of us learned to identify as children, or the giant rabbit pattern familiar to people in eastern Asia. Whatever the image evoked, we are actually looking at tholeiitic flood basalt plains across 400,000 km of space when we view these dark patches. Most of these dark fields of lava accumulated between 3.9 and 3.1 billion years ago. Termed **maria**, these lunar volcanic fields, averaging 1–2 km thickness, cover 17 percent of the otherwise brightly reflective lunar surface (Figure 12.18; Condie 2016). Curiously, they are distributed almost entirely on the hemisphere that permanently faces Earth. Were the "far side" of the moon to face Earth instead, we would hardly have a clue as to their existence, apart from satellite observations made in recent decades.

The wide range of ages in the maria flows suggest that melting in the moon's interior was episodic, possibly triggered by the shock of giant meteorite impacts. When a meteorite of more than a few hundred meters diameter strikes, a pressure pulse shoots through the crust at speeds of kilometers per second. Almost at once this is followed by intensive tensional relaxation as the lithosphere and underlying mantle rebound, much of the superficial material spreading as tendrils of showering ejecta extending for hundreds of kilometers. This **rarefaction** event, accompanied by transient heating to several million degrees, can instantaneously melt a large volume of impacted bedrock. Intensive fracturing extending kilometers beneath and around the new crater can also trigger decompression melting and open up pathways for molten rock to ascend well after the ejecta has settled, in some cases raising and fracturing crater floors while active sills or laccoliths spread out beneath them (Wilson and Head 2017). In many craters, a conspicuous central peak stands above a plain of slightly younger lava, illustrating the connections between impact, rebound, and volcanism (Wilhelms et al. 1987; Wilhelms 1993).

Since the heyday of maria formation, the surface of the moon has gradually stabilized and contracted into its present sterile configuration, much in contrast to neighboring Earth, where radiogenic decay continues to sustain vigorous volcanism and drive plate tectonics (Byrne 2020). The moon may now be too cold to generate renewed volcanic activity without, at least, the "boost" of major meteoritic impacts – increasingly less likely as the solar system ages. The youngest radiometric age for a sample of lunar lava collected by astronauts is about 800 million years, with the vast majority of volcanic products issuing billions of years earlier (Spudis 2000). At about 70 widely scattered locations on the lunar near side, however, small, irregular patches of lightly cratered lava, each averaging no more than 500 m across, suggest that very minor eruptions could have taken place as recently as 50–100 million years ago – not long ago given the great age of the solar system (Braden et al. 2014).

FIGURE 12.18 The Earth-facing surface of the moon, showing the scattering of dark volcanic maria, which for the most part fill large, impact related basins. Source: Photo: NASA.

Eruptive styles include the vast flood basalt activity of the maria plains and low, broad shield construction related to lithospheric flexure around the margins of a few of the larger basin plains (Spudis et al. 2013). Some individual, large basalt flows have thermally eroded their channels to create conspicuous, sinuous canyons, called **rilles**, which stretch in some cases hundreds of kilometers. Smaller, younger vents have produced cones and fragmental deposits, generally aligned to indicate feeding by dikes dozens of kilometers long, showing evidence of Hawaiian, Strombolian, Vulcanian, and even Ultraplinian eruptions. The sudden exsolution of gases in ascending dikes triggered by the lack of an atmosphere and the low-pressure environment of a shattered shallow crust sets off these infrequent explosive eruptions (Wilson and Head 2017).

Mercury – A "Moon-like" Volcanic History

Like the moon, Mercury is a small, rocky body only about 40% the size of Earth, and the innermost planet in our solar system. It, too, went through an early phase of intense, flood basalt volcanism related to major impacts billions of years ago, but has cooled and contracted rigidly since (Byrne 2020). Unlike the moon, Mercury lacks highlands made of compositionally distinctive rocks. Its crust is likely the least evolved of the stony planets.

Nevertheless, there is evidence of volcanic activity continuing long after the end of large-volume effusive eruptions. About 150 explosive volcanic source areas have been identified, active to at least a billion years ago. These produced circular, bright blankets of pyroclastic debris extending as far as 10 km or more from vents, which are low, subdued affairs. Think of flattened ash-cones. Vents are associated with large impact craters and other areas of crustal weakness. The vacuum of space doubtlessly played a role inducing explosivity of shallow intrusions, which formed during the tearing or fracturing of Mercury's ancient, shriveling crust (Rothery 2017; Thomas and Rothery 2019).

Venus – A "Mantle Plume" World

It might be expected that Venus, our closest planetary neighbor, would show many similar geological characteristics to Earth, including plate tectonics, hot spot tracks, continents, and ocean basins, even in the absence of liquid water. But despite the similarity in the sizes and mean densities of Earth and Venus – the diameter of Venus is only 330 km less than that of Earth – the two planets lose their heat in remarkably different ways and have very different physical surfaces. While large, ancient masses of intensely fractured, elevated crust superficially resemble continents set in a global ocean basin, there is no evidence for plate tectonics on Venus, such as subduction zones, alpine-style mountain ranges, and hot spot tracks. In fact, of the approximately 1750 volcanoes (greater than 20 km in diameter) that have so far been identified, few are aligned whatsoever, in striking contrast to the linear belts of volcanoes found on Earth (Crumpler and Aubele 2000). Neither is there any evidence that the crust of Venus has differentiated into "oceanic" and "continental" components. The continent-sized, highly fractured and faulted highland plateaus of Venus, termed **tessarae**, appear to be made up of ordinary basaltic crust, crumpled and uplifted from powerful compressive forces. The tessarae cover about 15 percent of the surface of Venus; the rest is basaltic lowland spotted with volcanic structures and uplifts, in places complexly rifted and fractured (Cattermole 1994; Crumpler 1996; Moore 2002).

The surface conditions of Venus are inhospitable in the extreme, with mean atmospheric temperatures of 430°C, and pressures equivalent to what one would expect at a depth of a kilometer beneath the sea on Earth, roughly one-hundred times Earth's sea-level atmospheric pressure. The "air" in this oven-like environment is mostly carbon dioxide (97 percent), with clouds of sulfuric acid droplets forming a permanent layer 45–60 km above the surface. Carbon dioxide and sulfur dioxide are certainly important volcanic gases, and infrared imaging in 2014 revealed flashes related to a probable volcanic eruption in the Ganiki Chasma rift system, similar to phenomena seen in 2008–2009 (Shalygin et al. 2015). At least three large volcanoes, Maat Mons, Ozza Mons, and Sapas Mons, can be regarded as active based on weathering studies (Filiberto et al. 2020). The relative lack of meteorite impact craters on Venus and the fresh appearance of most volcanic structures indicate that the planet's surface is at most no more than a few hundred million years old (Bougher et al. 1997; Moore 2002).

The distribution of volcanoes on Venus is non-uniform. One hemisphere, including the Beta, Alta, and Themis regions, contains several times the global average concentration of vents (Head et al. 1992). This area of the planet is also somewhat elevated, though not as high as the tessarae, and is deeply cut by faulted rift blocks and fractures indicating that the surface has stretched and ruptured throughout. The lithosphere of this part of the planet is unusually hot and possibly supported by an enormous bulge of upwelling mantle (Bindschalder and Parmentier 1990). Lightly cratered lava covers virtually the entire surface of Venus (Strom et al. 1994; Basilevsky and Head 1996).

Soviet Venera landers sampled the surface of Venus in the 1980s, demonstrating that the crust, at least in the lowlands, consists mostly of tholeiitic basalt, the same composition making up Earth's ocean floor (Barsukov 1992). Hence it should be little surprise that lava shields and small (less than 20 km across) shield volcanoes are common volcanic landforms (Guest et al. 1992). Flood basalt plains and plateaus, many deeply fractured and faulted, abound, and fields of small lava shields are the most common planetary volcanic feature. Larger shield-like volcanoes with compound elements also occur, in many cases exceeding 100 km in diameter; larger than any volcano found on Earth. They stand as tall as 10 km; the highest landforms on Venus other than the tessarae themselves (Head et al. 1992). As in the case of the Hawaiian Islands, they exert tremendous stress and flexural strain in the underlying lithosphere (McGovern et al. 2014).

Not everything may be of basaltic composition. Flattened, circular dome-like masses, termed **farra** (singular: **farrum**) or **pancake domes**, which somewhat resemble the rhyolite and dacite domes found on Earth, occur in rare scattered clusters (Figure 12.19). The largest farrum so far observed is 65 km in diameter, and its steep, rough margin rises nearly a kilometer. Whether these masses are highly siliceous, like most domes on Earth, is unknown. Physical modeling suggests that they are only slightly siliceous – basaltic-andesites – and emplaced rapidly, on the order of 2–16 years each (Quick et al. 2016). In any event, their presence implies that some shallow reservoirs of Venusian magma have undergone a significant degree of chemical differentiation. Many Venusian domes and other volcanoes have large landslide scallops in their sides, and some closely resemble the catastrophic sector collapses observed on Earth (Chapter 11; Basilevsky and Head 2003).

Sinuous channels, termed **canali**, wind torturously across many gradually sloping Venusian surfaces. They are analogous to the lunar rilles, and superficially resemble the unroofed drainage channels of especially large basalt flows seen on Earth. But they are much larger than similar terrestrial features, being as much as several kilometers wide, hundreds of meters deep, and hundreds of kilometers long. (The longest canal stretches 1400 km). Canali remain enigmatic features more closely resembling meandering river beds than huge lava channels, perhaps related to some poorly understood process of thermal or mechanical erosion occurring during large-volume flow under atmospheric conditions very different from Earth's (Gregg and Greeley 1993).

Many Venusian volcanoes occur in dense clusters, localized atop areas of domically uplifted crustal "hot spots." Hot spots are also indicated by the presence of **coronae**, calderas tens of kilometers in diameter bordered by concentric ring fractures (Figure 12.20). In addition to the ring structure, radial fractures commonly extend hundreds of kilometers from each corona. Some fractures mark the courses of enormous dikes fed at shallow depths, showing that the Venusian lithosphere is weak and vulnerable

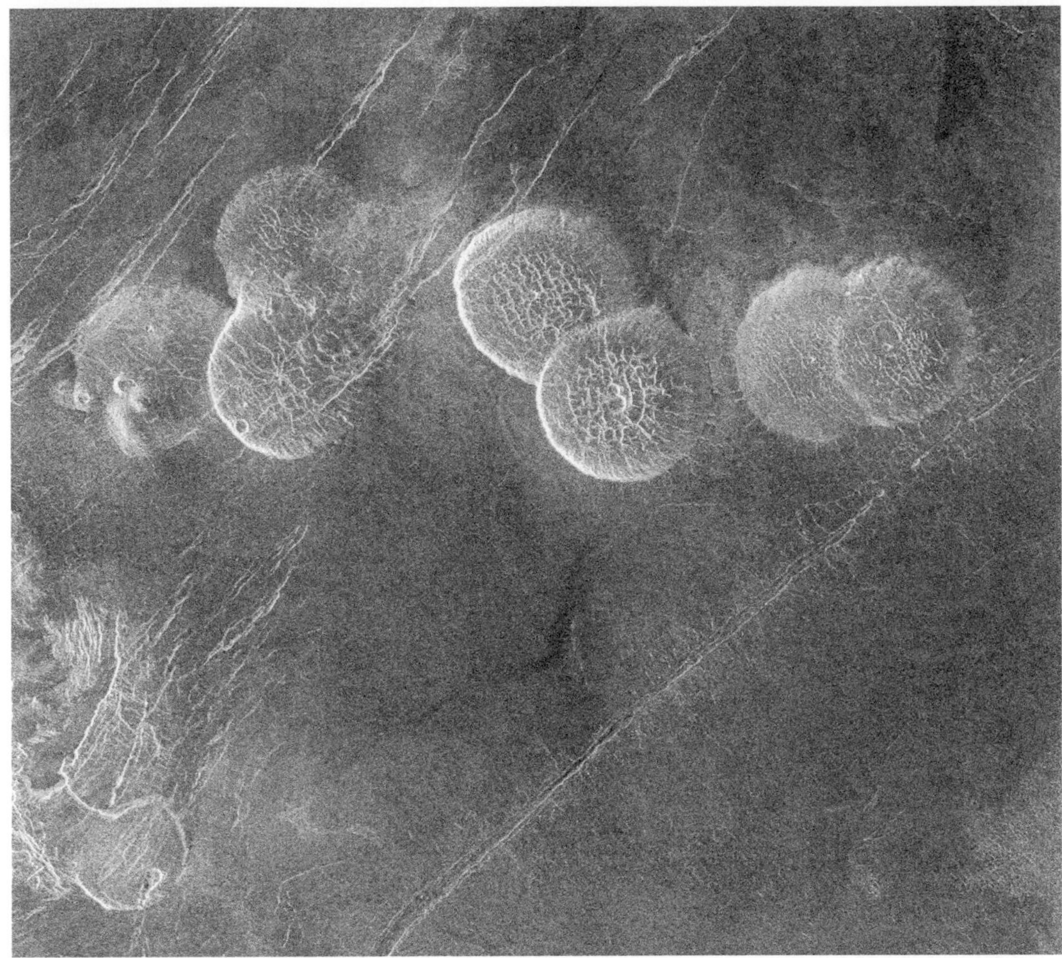

FIGURE 12.19 A chain of farra ("pancake domes") partly cut by younger fractures on a lava plain, Venus, imaged by Magellan Mission radar. The largest farrum is about 25 km in diameter and 700 m thick – much larger than any silicic flow on Earth. Light areas represent rougher radar-reflective surfaces, darker areas are smoother. Source: Photo: NASA, JPL/Caltech. NASA Archive PIA00215[1].

to point-source dike injection beneath areas of tens of thousands of square kilometers (Ernst et al. 1995; Galgana et al. 2013). In many cases, the radiating fractures curve into the trends of regional tectonic fracture sets, giving the overall fracture pattern the appearance of a giant spider with curving legs. Planetary geologists refer to these particular coronae as **arachnoids** (Stofan et al. 1992). Unlike terrestrial calderas, coronae lack associated widespread deposits of pyroclastic ejecta. In fact, the great pressure of the Venusian atmosphere discourages explosive eruptions, including anything approaching a Plinian outburst. Some coronae calderas show little evidence of erupting at all, in fact, and probably form simply from collapse of rising, stretching crust. In other cases, effusive eruption plainly accompanied corona development, in many places through vents on the flanks of coronae-crowned uplifts (Crumpler and Aubele 2000).

Radiating networks of grabens that are likely to overlay shallow dikes, form starburst patterns hundreds of kilometers in diameter in some locations. No accompanying volcanism or uplift accompanies these structural magmatic or radial fracture centers. Each structural magmatic center probably marks the initial impingement of a rising plume of hot, partly molten mantle at the base of the lithosphere, which could in time evolve into a fully developed corona. In fact, all states of gradation are seen between structural magmatic centers and coronae. They tell us that mantle plumes and hot spot volcanism play a primary role in facilitating heat release from Venus (Soloman et al. 1992).

On Earth's Precambrian crustal cratons, remnants of giant radiating dike swarms are preserved in several regions, implying that Earth too once produced some now distinctly Venusian volcanic features (Ernst et al. 1995). If this is so, what accounts for the change in the paths these two planets took? More particularly, why does Venus lack the vigorous plate tectonic dynamic now helping keep Earth cool?

One possible answer to the above question leads us back to the oppressively thick and hot Venusian atmosphere. While a surface temperature of 430°C is well below the melting point of basalt, it does keep the underlying surface significantly warmer than it would be under terrestrial conditions. Ascending magma would not lose heat as rapidly as it does at equivalent depths inside Earth.

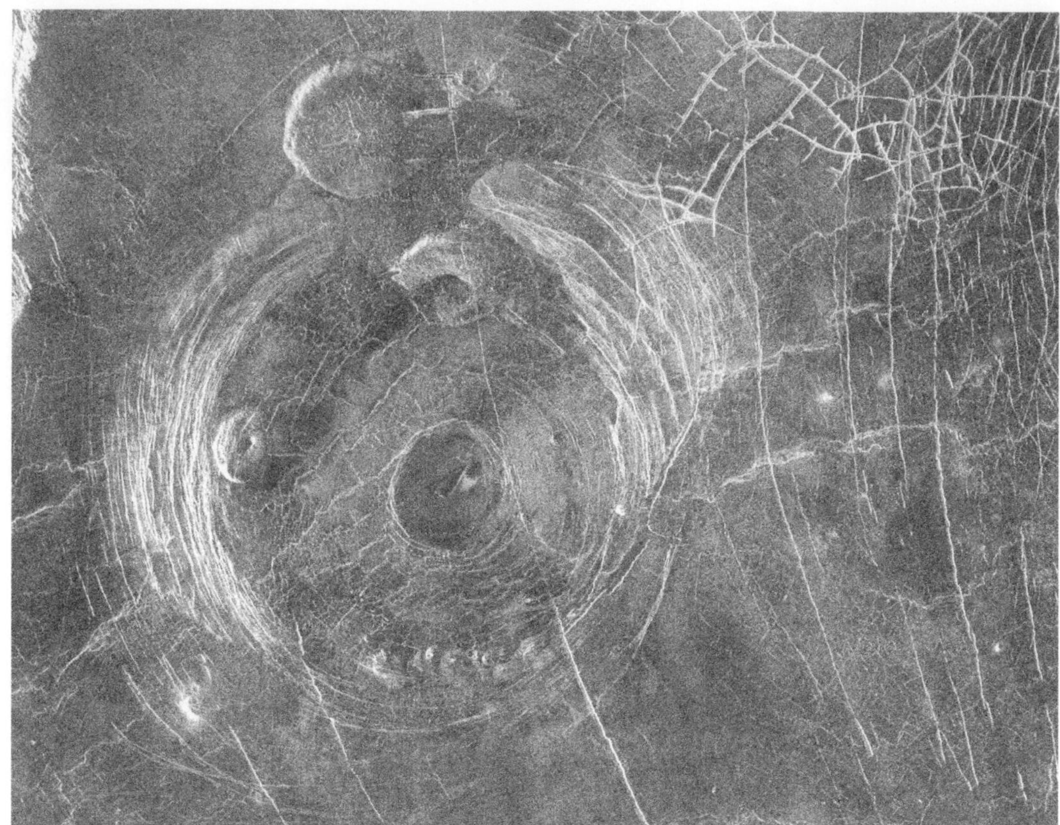

FIGURE 12.20 Aine Corona, a partly collapsed area of magmatic uplift on the surface of Venus displaying related concentric fractures and farra. Source: NASA image, JPL/Caltech. Archive PIA00202[1].

More importantly, the very high air pressure at the surface means that levels of neutral buoyancy are much shallower in the Venusian crust. In fact, it is likely that molten rock penetrating the lithosphere of the Venusian lowlands remains buoyant all the way to the surface, and fails to form substantial, long-lasting magma chambers. This is supported by the fact that lava shield fields dominate lowlands, while larger shields, complex volcanoes, and coronae are more abundant at higher elevations. It is simply easier for magma to escape from the Venusian surface than on Earth – or at least on Earth's continents (Head et al. 1992; Grosfils et al. 1999).

The floor of Earth's oceans, at a mean depth of over 2 km, has a pressure more than double that of the atmosphere of Venus, and it is certainly easier for magma to issue here than elsewhere on Earth's surface. The thousands of seamounts scattered across the ocean floor form what may be regarded as enormous volcanic fields, far larger than any seen on Venus with the arguable exception of the Beta, Alta, and Themis regions. In other words, plate tectonics may largely be a manifestation of our planet's surface environment, just as mantle plume volcanism may be stimulated by the Venusian atmosphere. Earth's hydrological cycle also plays critical roles via weathering of rocks to produce clays and other sediments, and providing the source fluid for substantial volatile flux during subduction, described in Chapter 2 as vital for creating arc volcanism. The circulation of water in Earth's mantle and lithosphere leads to the development of important continent-building suites of rocks that simply have no chance of forming on a world as anhydrous and hot as Venus. Indeed, the whole of Earth's crust is recycled through plate tectonics. On Venus, the crust may be continuously repaved thanks to mantle upwellings, but it is not recycled.

Mars – Large Volcanoes on A Small Planet

Larger than the Moon, but only about a third of the diameter of Earth or Venus, Mars has a volcanic history that blends aspects of volcanism from all three of these neighboring worlds. Being intermediate in size, Mars has substantially fewer volcanoes than either Venus or Earth, but certainly more than have been identified with certainty on the Moon. There are only about two dozen named, notable volcanoes on the Martian surface, but these include some volcanic giants, the biggest known volcanic features in the solar system (Zimbelman 2000).

Like Venus, Mars shows no evidence of plate tectonics. As befits the smallness of this planet, the atmosphere is tenuous and thin. Though water once played across the planet's surface, this world is now drier than anyplace on Earth. Volcanism there appears to have resulted from primarily two forces of nature: the impacts of enormous meteorites, and plume-like upwelling from the deep interior.

Perhaps two or three billion years ago (dating Martian eras is not certain), an enormous impact scoured out the 2100-km-wide Hellas Basin, which quickly flooded with a smooth plain of fluid, homogenously cooling lava. Intense impact fracturing opened up a northeast–southwest line of weakness to either side of the Basin through which molten rock erupted to construct the four Hellenic volcanoes, Amphitrites, Hadriaca, Peneus, and Tyrrhena. These belong to a special morphological group, termed **paterae**, which in fact makes up almost half of the large Martian volcanoes, and a number of volcanoes on Venus and the Jovian moon Io as well. A typical patera consists of a single large caldera that may have collapsed in multiple stages, perched at the nearly level crest of a very gently sloping (up to 1°) volcanic rise. The diameters of the Hellenic paterae range from 120 to 180 km, with summits no more than a kilometer above the surrounding older terrain. Individual calderas range up to 30 km across. Like other ancient volcanoes on this part of Mars, the Hellenic paterae are deeply gullied and furrowed, and quite possibly constructed of easily eroded ejecta generated by massive Plinian-style eruptions.

Paterae continued to form long after the Hellenic impact. The largest, Alba Patera, has a diameter of 450 km, and rises 4 km above an adjoining flood lava plain. The shallow central caldera is a hundred kilometers across – the biggest volcanic crater on Mars. Alba Patera appears to be made up mostly – if not entirely – of lava flows morphologically resembling ordinary ʻaʻā and pāhoehoe. Individual flows are truly enormous, perhaps as much as an order of magnitude greater than the largest flows on Earth.

Alba Patera is but one of a cluster of enormous volcanoes, for the most part shields, in the Tharsis region of Mars (Figures 12.21 and 12.22). The biggest of them all is Olympus Mons, the largest shield volcano in the solar system, with a diameter of 500 km and a surface area equal to that of the State of Arizona! With slopes not exceeding 5° in most places, Olympus Mons rises 27 km above its base. The top of the volcano is so high that atmospheric pressure is only 2 percent of what it is in the nearby lowlands. The summit caldera, a composite of at least five collapse events, measures 90 × 60 km and is as deep as 3 km – comparable in scale to reconstructions of the largest calderas on Earth (Chapter 10). Episodes of lava lake infilling interspersed with renewed collapse have created a complex, terraced caldera floor, well-preserved because erosion is such a very slow process on Mars (Mouginis-Mark and Robinson 1992). One of the most striking aspects of Olympus Mons is the scarp enclosing the foot of the mountain, a

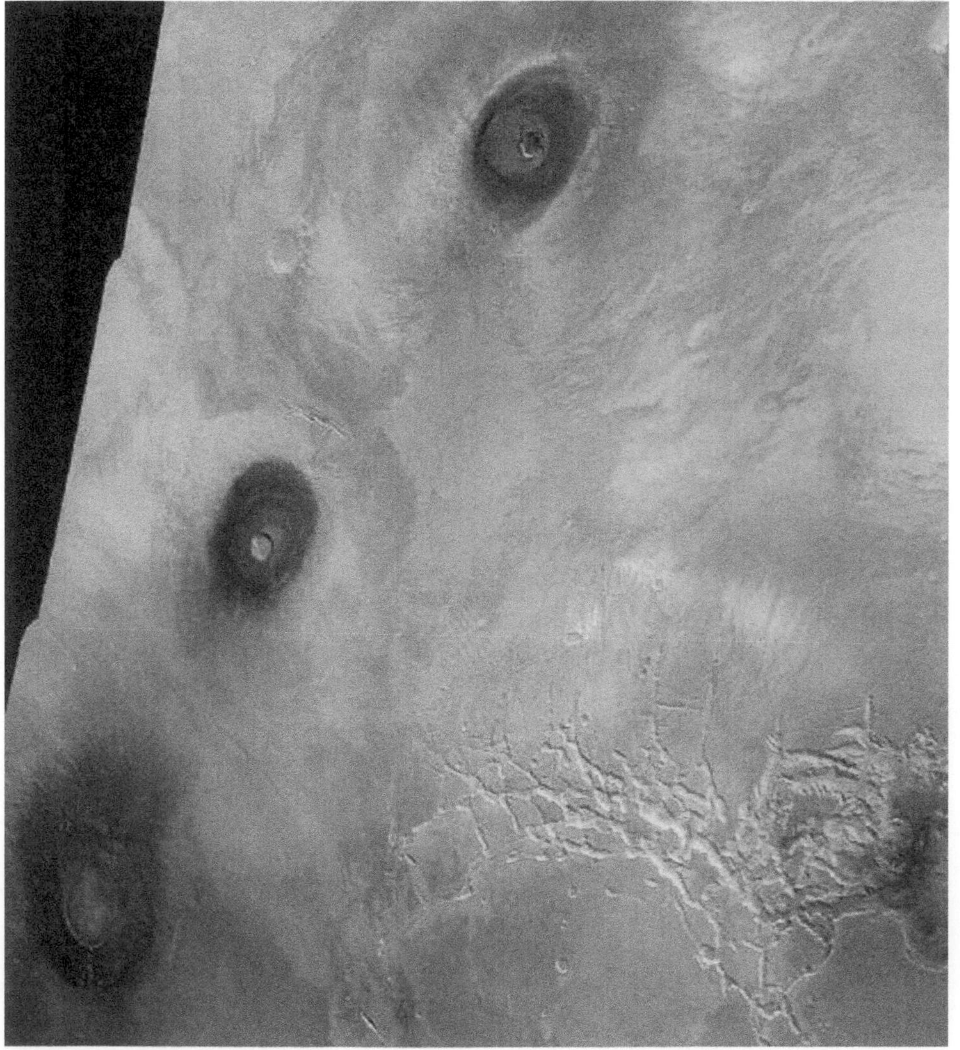

FIGURE 12.21 Based upon cratering density and estimates of growth rates, this trio of shield volcanoes in the Tharsis region of Mars probably formed on the order of 700 million–3 billion years ago: At the lower left is Arsia Mons (oldest of the group), with Pavonis Mons in the center, and Acraeus Mons (the youngest) in the upper right. The spaces between the volcanoes are each about 700 km. The volcanoes lie at the crest of a gently sloping lava-covered uplift some 10 km high – the Tharsis Bulge. In the lower right is the head of the 4000-km-long, 6–7-km-deep Valles Marineras, a tectonic rift valley probably related to early growth of the Bulge. Source: Photo: NASA, JPL/Caltech. NASA Archive PIA02987[1].

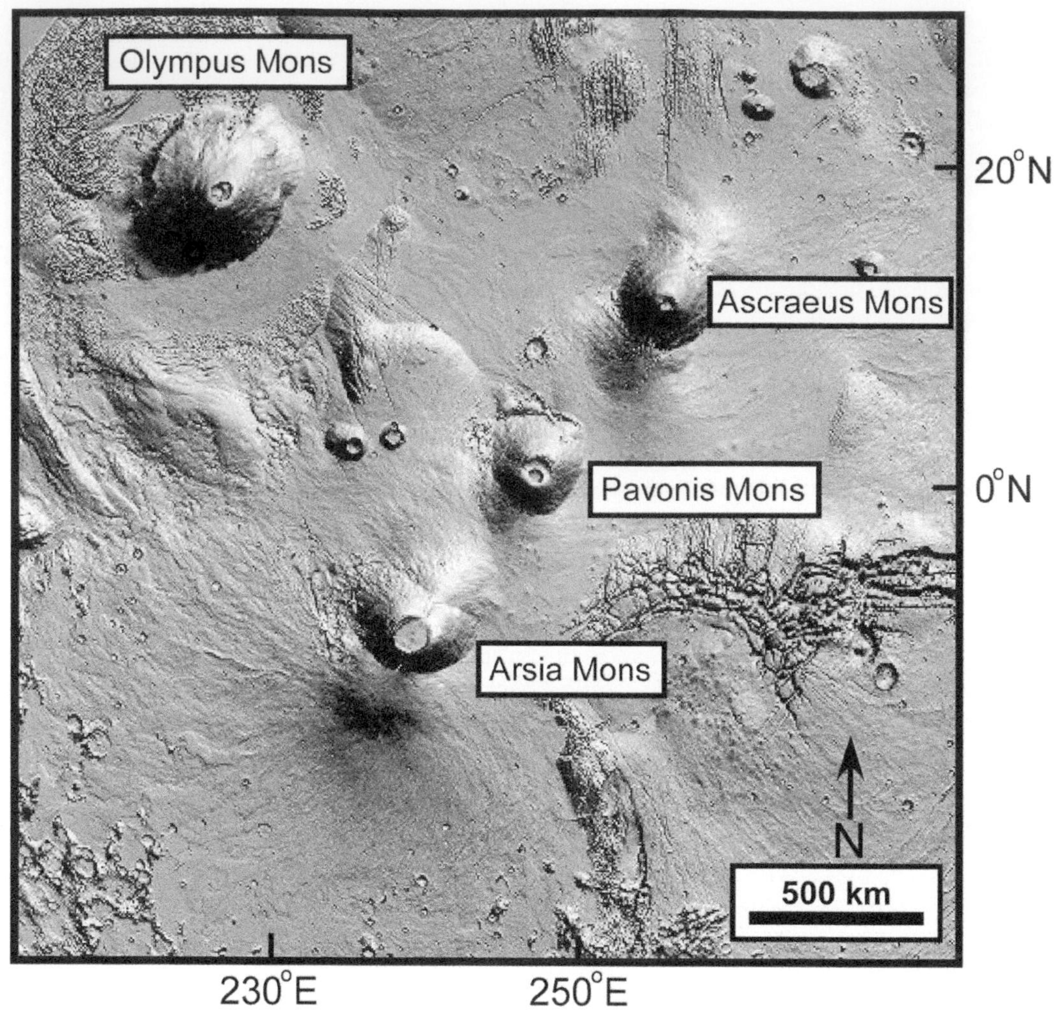

FIGURE 12.22 Shaded relief map showing Olympus Mons and other large shield volcanoes in the Tharsis Region on Mars (vertical exaggeration about 3x) Illumination is from the northeast. Source: From Garry et al. (2007).

fault cliff 3–6 km high, which faces outward, away from the volcano. Large landslides shed from the scarp created debris fields and lobate masses in some cases stretching hundreds of kilometers onto the surrounding plain. Some deposits resemble the sea floor debris fields formed by the catastrophic collapse of Hawaiian shields (Lopes et al. 1980). The scarp clearly developed late during the growth of the volcano, but the presence of numerous unbroken flows draped across its face indicate that eruptions continued for a while after it developed. Edifice spreading and faulting over a downward flexing lithosphere accompanied by collapse in the lower flanks could be implicated, as in young Hawaiian volcanoes; but the basal scarp of Olympus Mons remains one of the mysteries of Martian topography.

Unlike Earth's moon, volcanic activity may be far from over on Mars. Detailed photographic studies by NASA's Global Surveyor starting in 1997, and the Mars Express High Resolution Stereo Camera, which began orbiting Mars in December of 2003, provide evidence of effusive volcanism perhaps as young as 2–2.5 million years, with flows mantling the flanks of Olympus Mons that erupted mostly within the past 200 million years. Recent discovery of a fissure system in the Cerberus Fossae area provides even more startling evidence of a pyroclastic eruption there sometime within the past 53–210 ka (Horvath et al. 2021). NASA's InSight lander seismometers indicate that this area is still seismically active. It cannot be assumed that Mars is now internally cold and volcanically dead, and in fact, volatile efflux may still be in progress.

Io – An "Insanely Active" Volcano World

When Galileo Galilei first turned his new telescope to look at the giant planet Jupiter in 1610, he quickly spotted four large moons circling that body, which have since been named the **Galilean moons**. There are over 60 known satellites orbiting Jupiter, but most are small, irregular bodies, little more than giant, icy rocks trapped in the Jovian gravity field. The Galilean moons are notably

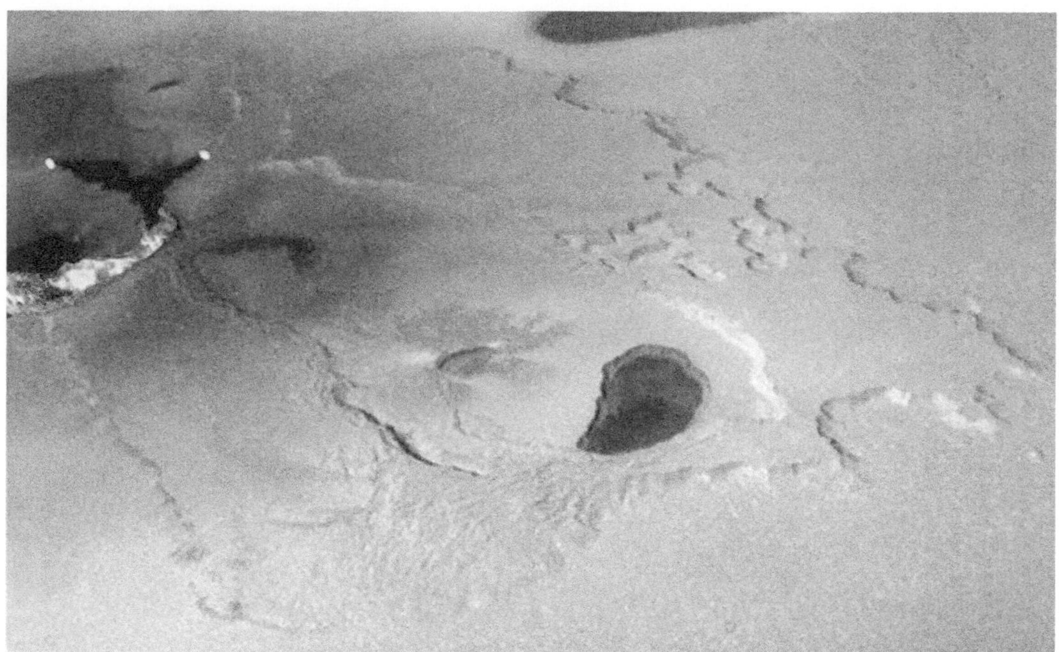

FIGURE 12.23 Scalloped (possibly sapped) volcanic tableland and compound caldera features of Tvashtar Patera, Io, with an ongoing effusive eruption highlighted to the left. The frame is roughly 200 km across, left to right. Volcanic activity in this area has been observed by spacecraft on two other occasions. In 2000, the Galileo spacecraft imaged a 25-km-long eruptive fissure with lava fountaining as high as 1 km, which subsequently cooled to form the dark strip just to the right of the active area. Tvashtar was erupting again at the time of the New Horizons spacecraft Jupiter fly by in 2007, when a massive plume was observed. Source: Photo: NASA Archive PIA02550[1].

larger, however, comparable in diameter to Earth's moon, or the planet Mercury. Listed in order of their distance from Jupiter, the four are named Io, Europa, Ganymede, and Callisto.

The first spacecraft to explore this system, Voyagers 1 and 2, flew past the Galilean moons in the spring and summer of 1979. Observers had expected to see small, dead worlds, but were instead surprised as they collected data showing the presence of interior oceans on at least two of the moons thought to be too cold to sustain liquid water – Europa and Callisto. Even more stunning was Io, the innermost moon, which was volcanically active almost beyond imagination.

The heating of the Galilean moons derives in large part from the tidal forces acting on them as they orbit Jupiter, the largest body in the solar system other than the sun itself (Peale et al. 1979). The continuous flexing of the moons by intense tidal forces generates great frictional heating. The surface of Io rises and falls a hundred meters with each tidal cycle, compared with a mere 0.1–0.4 m on Earth. This is all the more impressive when we consider that Io is only about the size of our moon.

In March 1979, Voyager 1 photographed a powerful volcanic cloud shooting 300 km into space from the volcano Pele, seen on the horizon of Io. In the moon's low-density sulfurous atmosphere (nine orders of magnitude less dense than Earth's atmosphere) the cloud resembled a giant opened umbrella. Its fall ejecta ultimately covered 10,000 km^2 of the moon's surface. To add to the excitement, eight additional volcanic clouds were quickly discovered during the Voyager fly-bys, and spacecraft thermal sensors showed that nine other volcanoes were "hot," perhaps ready to erupt or recently having done so. With the return of the Galileo Mission in 1996 and the later Cassini mission (2000–2001), many changes in the surface could be precisely documented. These included the outpouring of fresh lava flows exceeding 100 km in length, and the apparently continuing eruption at a number of volcanoes, including Prometheus, Loki Patera, and Pele, all active in 1979. The Galileo spacecraft photographed 10 additional eruption clouds, and brought the total number of known active volcanoes on Io up to 74, with dozens of additional hot spots (Figure 12.23; McEwen et al. 1998).

One of Io's most striking and telling aspects is its color. The surface is yellow, red, and orange, mottled with white, black, and dark gray patches (Figure 12.24). This coloration, together with spectrographic analyses of plumes, initially led observers to believe that only sulfur or sulfur-related compounds erupt from Io's volcanoes, and that these materials have coated virtually the whole moon, with the exception of the white patches, which are likely to be sulfate frost condensing out of the frigid atmosphere. (The surface temperature on Io is a chilly −143°C.) Astronomer Carl Sagan pointed out most of the colors, especially the warm ones, could signify the presence of different forms of molecular sulfur known to absorb light at different wavelengths according to their temperature. However, subsequent thermal measurements by infrared detectors on Earth-based telescopes and on the Galileo

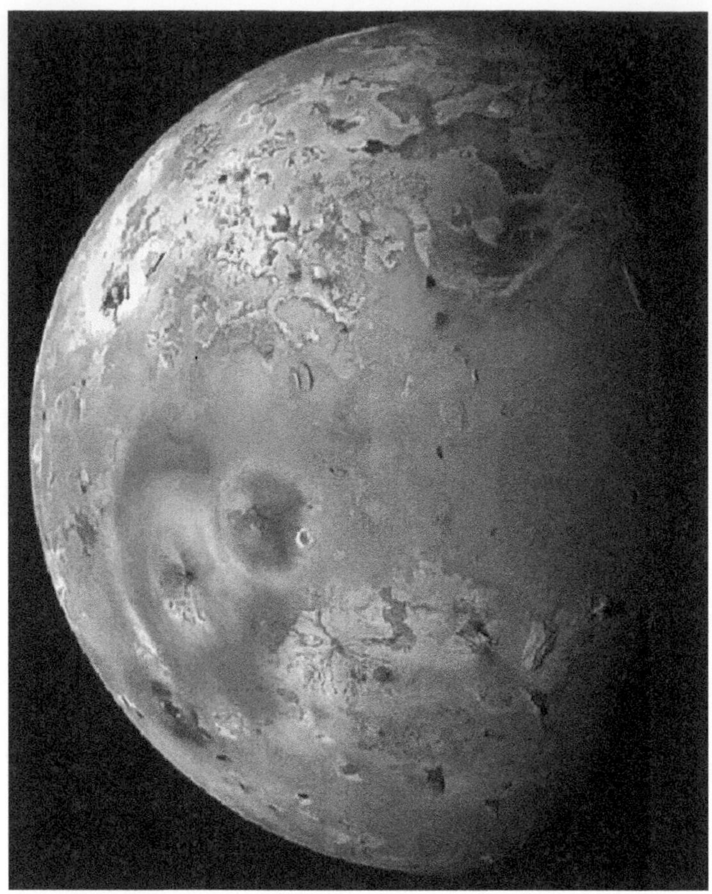

FIGURE 12.24 The surface of Io, the lava flows and other surface features of which are coated with dark smudges of pyroclastic debris – localized around volcanoes – and brightly-colored, globally distributed gas sublimates. The mineralized, nearly complete red ring to the lower left is centered around the volcano Pele. The ring resulted from the explosive eruption and deposition of a sulfurous plume in the very tenuous Ionian atmosphere. When first imaged by the Galileo Orbiter, the ring was complete. But between the 7th and 10th orbit of the spacecraft around Jupiter, the 400-km-wide Pillan Patera grew to cover its upper right section – stark testimony to the vigor of volcanic activity on this strange world. Source: Photo taken in 1999: NASA Archive PIA01667.

spacecraft have shown that the temperatures at active volcanic vents are typically consistent with silicate volcanism (Johnson et al. 1988; Veeder et al. 1994). Values as high as 1500°C have been measured, indicating that substantial volumes of mafic silicate lava must also pour from Io's volcanoes (Keszthelyi et al. 2006; Davies 2009; Figure 12.24). Indeed, some fresh flows retain their dark coloration too long to be composed of sulfur; and larger volcanoes, as high as 18 km, are likely to be composed mostly of silicate rock – stronger natural construction material. We now know that mafic to ultramafic silicate magmas, as on planets closer to the sun, are the primary drivers of Io's volcanoes.

Ionian volcanoes take a wide variety of forms. Large areas of the surface are covered in extensive lava flows, some with visible pyroduct and channel systems. One of the longest known lava flows in the solar system, erupted from Amirani volcano, extends 500 km. Patera calderas occur by the hundreds, set into flat plains and squat mesas, some of which contain active lava lakes. The biggest, most powerful volcano is Loki Patera, the vent of which is marked by a restlessly molten lava lake with an area of over 20,000 km^2 – almost the size of Lake Erie! Io's volcanoes individually erupt vastly larger volumes of lava than their terrestrial counterparts (Ashley Davies, pers. comm. 2009; Davies 2009). Calderas show long histories of producing abundant lava flows (Figure 12.25).

Eruptions commonly issue Hawaiian-style from fissures – often generating massive volcanic clouds whose deposits cover large portions of Io's surface as sulfur-rich "snowfalls." Vent-derived clouds are rich in sulfur (S_2) and may shoot as far as 500 km into space before spreading and settling out. They originate where ascending silicate magma melts deeply buried deposits of sulfur and sulfur dioxide in volcanic conduits at temperatures well above surface boiling points. (The same principal acts to explain water geysers on Earth.) The great heights of these columns may be attributed to the low gravity and very tenuous atmosphere of Io (with about one-billionth the atmospheric pressure at sea level on Earth). Were such eruptions to occur from a terrestrial volcano, their columns would rise only a few kilometers. Additional, less substantial volcanic cloud development takes place where silicate flows snake across the ordinarily frigid landscape, boiling adjacent sulfate-rich ices. Lava flows made of molten sulfur also almost certainly form as a result of thermal interaction with hot silicate lava (Geissler 2003; Davies 2009).

Like Earth's moon, Io keeps one face continuously toward Jupiter, 440,000 km away. While volcanoes are scattered across Io's entire surface, they are concentrated at roughly the 165°W and 352°W meridians, theoretical points of maximum tidal stress on opposite sides of the moon (Schenk et al. 2001).

Io shows no evidence of plate tectonics. The satellite's lithosphere has to be at least 30 km thick to support its highest mountains. Volcanic activity is apparently resurfacing Io faster than any other planetary body in the solar system and its sulfur-rich lithosphere is being continuously recycled. Among the material escaping from Io, a fine mist of sulfur particles escapes into space, cast there by the larger sulfur plumes spraying from Io's surface. Some of it coats the neighboring moon, Amalthea, painting it red and

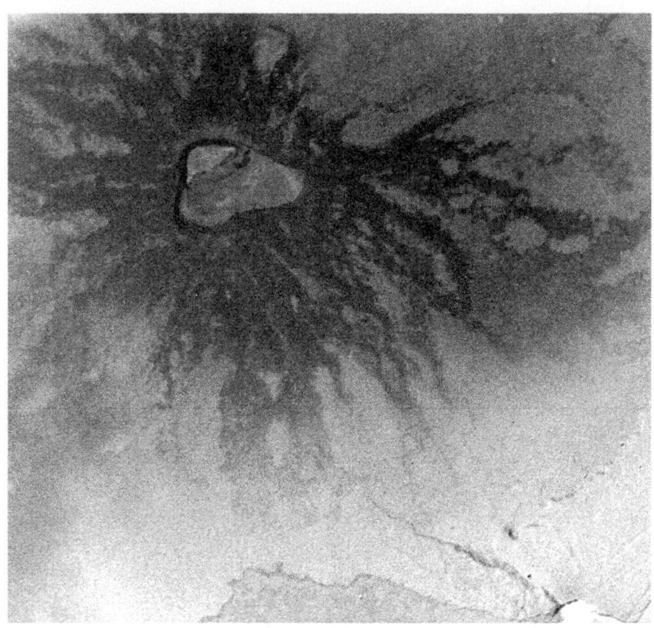

FIGURE 12.25 Volcanic caldera on Io, showing 50-km-wide caldera with radiating lava flows of probable basaltic composition up to 100 km long. The Galileo Jupiter orbiter obtained this image in 1979 at Io 67° south latitude and 328° longitude. Source: Photo: NASA PIA02280.

orange. The rest escapes to form a diffuse ring, or **torus**, orbiting Jupiter. Io is exceptional for bodies its size, thanks to powerful external forces. It is a world seemingly on the verge of tearing itself apart (Figure 12.24).

Deeper Space – Colder Volcanoes

Deeper in space, volcanic activity takes the form of erupting water and gases, rather than silicate-based lavas, a non-magmatic phenomenon called **cryovolcanism**. So far cryovolcanic (or "cryogeyser") eruptions have been directly observed on only two worlds beyond the orbit of Jupiter; Enceladus, a moon of Saturn; and Triton, a moon of Neptune. Other candidate outer solar system worlds where such activity may be common include Europa, Miranda, Pluto plus its partner Charon, and Quaoar – the largest body yet discovered in the remote Kuiper Belt.

Enceladus is a small moon with a diameter of only 500 km – barely the diameter of Olympus Mons on Mars. Planetary bodies of this size barely maintain a spherical shape, and in fact Enceladus is slightly oblong. Moderately cratered water ice covers the northern hemisphere, where crater counts suggest that freezing took place at least 10–100 million years old. Not much has happened since then. In the south, however, surface conditions are much different – and far more dynamic. A fresh veneer of tectonically disrupted ice covers the region, laced with a network of light blue fractures termed "tiger stripes" similar to the leads crisscrossing broken river ice after refreezing. Infrared spectrometer measurements reveal that the ice in the tiger stripes is about 15°C warmer than the surrounding ice "plates," indicating that relatively warm matter, probably liquid water, is active below.

The south pole of Enceladus shows the most extreme tectonic deformation. Fractures and broken platelets, some heaved and thrust aside, and boulders of ice from 10 to 100 m diameter lie strewn across the dim, blue-white landscape (Porco et al. 2006). In 2005, the Cassini spacecraft observed a spectacular geyser of water shooting up 500 km and dissipating into the oxygen-rich E-ring circling Saturn. Enceladus was feeding matter into one of Saturn's rings! A sprawling system of fluid pockets beneath the polar surface, hardly above freezing temperature, may have supplied the high-pressure jet of liquid water (Spencer et al. 2006). But what caused the eruption? Like the Galilean worlds that orbit Jupiter, Enceladus experiences tremendous tidal forces exerted by Saturn, and possibly also by a neighboring satellite, Dione. There is also some evidence to suggest that the orbit of Enceladus is librating (oscillating), causing tidal stress to heat up its interior (Hurford et al. 2009). This pent-up energy may be released as highly pressurized water, gases, and any suspended particles suddenly break icy crust and blast into space. In any event, this world seems to defy the "small planet – no heat" equation.

Triton is a much bigger world, with a diameter of 2700 km, almost matching Io. Being even farther from the sun as it orbits Neptune, frozen gases and ice also cover the surface. Floating high in a tenuous, nitrogen-rich atmosphere, wispy clouds of nitrogen ice crystals provide weak daylight shade. Perhaps these clouds are residues of volcanic plumes. In August 1989, Voyager 2 swept close past Triton, and photographed jets of dark matter shooting from two locations in the southern hemisphere. Each column rose about 8 km above Triton's surface, then flattened out and spread downwind about 150 km before dissipating. Eruptions like this must be commonplace on this Neptunian moon, since Voyager recorded the dark ejecta blankets of more than a hundred other, earlier plumes (Geissler 2000).

Triton's surface temperature is −235°C, only a bit above absolute thermodynamic "zero." On Earth, the physics of matter interactions at this temperature can only be studied in a laboratory. We know that gases such as nitrogen, carbon dioxide, and methane, all found on Triton, freeze when it gets so cold. But nitrogen will sublimate at −225°C, not much above an average day's temperature on that world, so it seems likely that it would take only a little bit of heating to generate violent nitrogen gas eruptions from the frozen crust. One proposal is that carbon particles trapped in the shallow, translucent crust capture weak sunlight seasonally, then re-radiate this energy in thermal infrared wavelengths. This **solid-state greenhouse effect** heats the surrounding nitrogen ice just enough to turn it into gas. Like water flashing to steam, vaporizing nitrogen expands with great force – about a hundredfold per unit volume. Voyager happened to pass Triton's southern hemisphere at a time when that region was nearing summer solstice – the optimal moment for such heating, and rare good fortune indeed given that the length of one year on Triton is equivalent to 165 years on Earth.

From what we have observed on Enceladus and Triton, it would be surprising if other worlds in the outer fringes of our solar system were not also volcanically active. Perhaps gas eruptions take place unseen on Titan, Saturn's largest moon, which is permanently obscured in a thick, smog-like atmosphere. Recent surveying by the Cassini spacecraft has detected glaciers resembling lava flows, ice shields, and even ice calderas on that world (Elachi et al. 2005; Kerr 2005). Maybe Pluto will prove to have nitrogen plumes also, much like those on Triton. Flyby of NASAs New Horizons spacecraft in 2015 suggested the presence of at least two cryovolcanoes on Pluto's icy surface. The common occurrence of volcanic activity throughout our solar system virtually assures that volcanism must be taking place on planets around other suns in our galaxy and beyond.

Earth's terrestrial eruptions "get all the press," especially ones in populated areas and those easy for reporters to access. Although these are the ones we know most about, eruptions from the unseen volcanoes are significantly more numerous and voluminous. We have recently begun to learn about them, and to document those occurring in far more remote reaches of our solar system. Volcanism clearly plays a fundamental role in the redistribution of heat on massive bodies, and widespread volcanic activity must be commonplace throughout the Universe. Even if we cannot be sure that life exists on other worlds, we *can* be sure that volcanoes do!

Questions for Thought, Study, and Discussion

1. Would you expect to find a composite volcano like Mount St Helens on the deep ocean floor? Why or why not?
2. What is the origin of Earth's many submarine seamounts?
3. What factors are most important for building a long-lasting volcanic island?
4. Some shallow submarine eruptions take place without developing Surtseyan eruption clouds. Yet, we can still see evidence of their taking place. How can this happen?
5. How does the flow of pāhoehoe erupted from vents along the MOR differ from that of subaerial pāhoehoe erupted from, say, the flank of Kīlauea?
6. For many years, earth scientists presumed that explosive volcanic eruptions could not take place in water more than a few hundred meters deep. Why were they wrong?
7. How and why do volcanic landforms formed subglacially differ in appearance from volcanoes formed subaerially?
8. Why do many large impact craters on the moon have central peaks, with floors of basaltic lava?
9. How and possibly why do the volcanoes of Venus appear so different from those of Earth, a world that is otherwise very similar in terms of its density and overall size?
10. Intensive volcanic activity is confined (as far as we know) to only smaller planetary satellites in the outer solar system. Why, and how does this volcanism differ from that of the inner solar system?

FURTHER READING

Condie, K.C. (2016). *Earth as an Evolving Planetary System*. Cambridge, MA: Academic Press, 430 p.

Davies, A.G. (2009). *Volcanism on Io: A Comparison with Earth*. Cambridge: Cambridge University Press, 355 p.

Frankel, C. (1996) *Volcanoes of the Solar System*. Cambridge: Cambridge University Press, 232 pp.

Lopes, R.M.C. and Carroll, M.W. (2008). *Alien Volcanoes*. Baltimore: The Johns Hopkins University Press, 176 p.

Ruben et al. (2012). *Volcanic Eruptions in the Deep Sea*. Oceaonography, Rockville, Maryland, 95 p.

Rubin, K.H., Soule, S.A., Chadwick Jr., W.W. et al. (2012). "Volcanic eruptions in the deep sea." *Oceanography* 25(1): 142–157, http://dx.doi.org/10.5670/oceanog.2012.12.

White, J.D.L., Smellie, J.L., and Clague, D.A. (eds.) (2003). *Explosive Subaqueous Volcanism*. Washington, DC: American Geophysical Union, 379 p.

PART V

HUMANISTIC VOLCANOLOGY

This part discusses the interactions of human society and volcanism – the important ways in which volcanoes have affected Earth's life, impacted human safety, and provided society's necessary mineral and energy resources. **Chapter 13** discusses the climatic and biological impacts of volcanism and the role of volcanism in human evolution and history. **Chapter 14** describes the nature of volcanic hazards and risks and the ways in which that risk can be lessened through volcano monitoring and crisis management. **Chapter 15** discusses the role that volcanoes play in the concentration of metallic ores and geothermal resources critical to modern civilization.

Volcanoes: Global Perspectives, Second Edition. John P. Lockwood, Richard W. Hazlett, and Servando De La Cruz-Reyna.

Companion website: www.wiley.com/go/lockwood/volcanoes2

CHAPTER 13

Volcanoes: Life, Climate, and Human History

Source: Credit D. Swanson.

Volcanoes: Global Perspectives, Second Edition. John P. Lockwood, Richard W. Hazlett, and Servando De La Cruz-Reyna.

Companion website: www.wiley.com/go/lockwood/volcanoes2

Le volcanisme contribue au développement de l'humanité; il ya un revers à la médaille: les volcans tuent, provoquent parfois d'effroyables catastrophes et à l'occassion, perturbent le climat. [Volcanism contributes to the development of humanity; but on the other side of the coin: volcanoes kill, sometimes cause horrible catastrophes, and on occasion disrupt the climate.]

(Patrick Barois 2004)

Volcanoes and the Origin of Life

Volcanoes likely played an essential role in the origin of life early in Earth's history. From the time of Darwin's idea of a "warm little pond" as a natural nursery for life, scientists have speculated that organic molecules, life's building blocks, could form under natural conditions, perhaps aided by phenomena such as lightning and volcanic activity. In 1953, Stanley Miller and Harold Urey attempted to replicate Earth's primitive, reducing atmosphere in a pressure flask – a mixture of methane, ammonia, water, and hydrogen instead of carbon dioxide, nitrogen, oxygen, and water. Subjecting this brew to electrical discharges, the two chemists produced glycine and alanine as well as (quite likely) other amino acids (Miller 1953). Yevgennii Markhinin (1980) later coined the term **biovolcanology** to describe the possible interactions of volcanic activity and biology, and discussed the unique chemical environments primordial volcanoes could have provided for the development of life. He proposed that before free oxygen appeared in Earth's atmosphere, lightning from Plinian eruption clouds may have catalyzed the formation of organic molecules from the Precambrian atmospheric soup of CH_4, H_2, H_2O, CO_2, and N_2. In the same vein, A.P. Johnson et al. (2008), using an apparatus also coincidentally developed by Miller in the 1950s, modeled a steam-rich, electrically active eruption column to produce some 22 amino acids and 5 amines. They demonstrated that a reducing global atmosphere was not necessary to build these molecules. Localized prebiotic chemical synthesis could take place even in an atmosphere more like today's, noting (p. 404):

In... volcanic plumes, HCN, aldehydes, and ketones may have been produced, which, after washing out of the atmosphere, could have become involved in the synthesis of organic molecules. Amino acids formed in volcanic island [arc] systems could have accumulated in tidal areas, where they could be polymerized by carbonyl sulfide, a simple volcanic gas that has been shown to form peptides under mild conditions.

Recent discoveries of simple life forms (bacteria) living within hydrothermal vents on undersea volcanoes – so-called **hyperthermophilic environments** – has stimulated exciting new thinking about the possible range of volcanic settings in which life might have arisen (see, e.g., Huber and Wachtershauser 2006). Perhaps instead of Darwin's warm little pond or tidal pool, a "warm little sea" (or at least a portion of that sea) hosted the world's first biological ecosystem instead.

In any event, none of these discoveries tells us with certainty what provided the "spark" that got life going. Whether this magic spark was struck under the guidance of a Higher Power or a natural happening is best left for religious authorities and philosophers to argue, but either way, volcanism is one likely provider of the crucibles in which life as we know it first began. Further important thoughts about how *abiotic* chemistry may have transitioned to *bio*chemistry were given by Miller and Orgel (1974).

Once established, natural processes of selection will favor the most successful life forms, and it has been suggested that volcanoes accelerate evolutionary processes even today, by creating isolated ecological niches where limited numbers of individuals and restricted gene pools magnify the importance of random mutations in forcing evolutionary adaptation (Carson et al. 1990).

Volcanoes, Atmosphere, and Climate

Economist Kenneth Boulding (1968) popularized the concept of "Spaceship Earth," which asserts that our planet's life-support system is based upon natural cycles of matter and energy that are fully interlocking. One component of the system cannot change without all others ultimately being affected. Life developed and evolved through exploiting these cycles. It modified them favorably, if perhaps unwittingly, in the process. Because of the integration of life with natural fluxes, Earth System scientists, who study Spaceship Earth holistically, refer to these fluxes as **biogeochemical cycles**.

A good example of a biogeochemical cycle is the natural flux of water, which geologists call the **hydrological cycle**. This flux is powered by solar energy and gravity: Water evaporates from water bodies, rises over high land, cools and precipitates to feed streams and rivers or groundwater which eventually return the water to places where it can evaporate again. Water is essential for life, being a vital ingredient in the transport of nutrients and construction of cells, tissues, and other organic structures. Irrespective of how the water is ingested, it does not remain long in any given organism. Plant leaves, for example, continuously yield water vapor via evapotranspiration, which contributes significantly to the notable haziness seen above many forests.

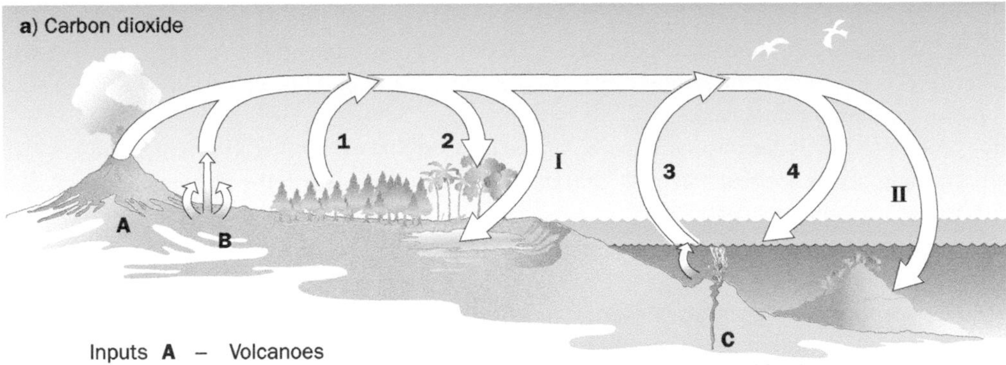

Inputs **A** – Volcanoes
B – Carbonated springs (and related deep seated metamorphism)
C – Submarine fumaroles, volcanic vents

Cycle components
1 – Decomposition of dead organic matter
2 – Incorporation of CO_2 in living tissue
3 – Exhalation of CO_2 from (primarily warm) sea surface
4 – Dissolving of CO_2 from (primarily cold) sea surface

Sequestrations
I – Incorporation of carbon from incomplete decomposition of organic matter in soils, swamps and other bodies
II – Incorporation of carbon dioxide as carbonate minerals by reef-building organisms and by precipitation as lime sands and carbonate rocks

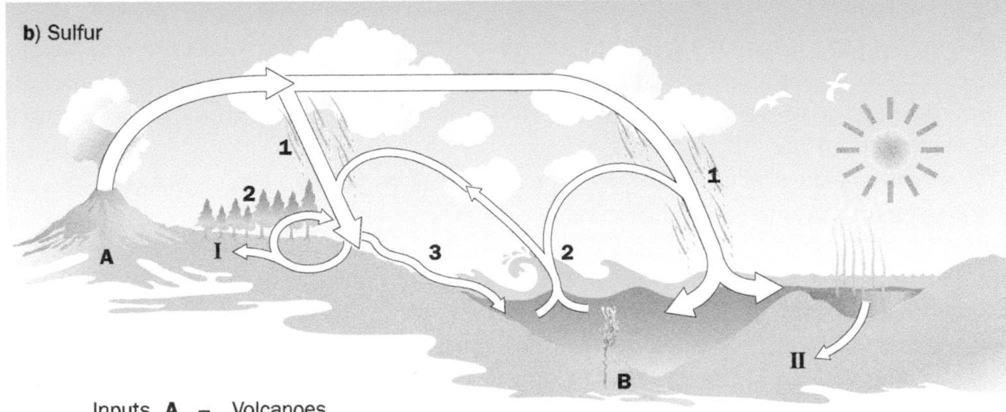

Inputs **A** – Volcanoes
B – Submarine fumaroles, volcanic vents

Cycle components
1 – Acidic rainfall
2 – Microorganic (land) or phytoplanktonic (sea) sulfate emissions
3 – Erosion

Sequestrations
I – Deeply buried soils, swamp depoits
II – Precipitation as sulfate rocks (e.g. gypsum, anhydrite beds)

FIGURE 13.1 The role of volcanoes in the global biogeochemical cycles of carbon dioxide and sulfur.

Volcanoes play a critical role in biogeochemical cycling. It is fair to say, in fact, that Earth's life as we know it might not exist without volcanic activity. It is beyond the scope of this book to explore fully the reasons for this, but it is worth surveying two examples: the **carbon** and **sulfur cycles** to illustrate the point (Figure 13.1).

Carbon Dioxide

Carbon dioxide is a **greenhouse gas**, meaning that it acts to make the atmosphere warmer than it would be if sunlight re-radiated directly back into space after reaching Earth's surface. The ground converts many wavelengths of solar energy to infrared radiation, which we feel as heat. Escaping back into the air, the infrared radiation is absorbed and re-radiated by various gases, so that atmospheric residence time of this energy is prolonged, making conditions warmer. Though present in only a bit over 400 ppm in the atmosphere, but increasing at 2 ppm/year (largely reflecting human combustion of fossil fuels), carbon dioxide is one of the most effective gases in this regard, and one whose CO_2 concentration is greatly sensitive to changes in the **carbon cycle**, which is the continuous exchange of CO_2 between the atmosphere, lithosphere, oceans, and life. Volcanoes are major sources of CO_2, and as such are an important factor in this global balancing.

Carbon dioxide combines with water through metabolic reactions to make organic compounds. Photosynthesis by plants is a very straightforward example of this. Hence, much CO_2 that otherwise would end up in the air is instead stored in organisms, especially vegetation. CO_2 is also absorbed by seawater. The colder the seawater, the greater its capacity for retaining dissolved carbon dioxide. Much of the deep-marine carbon storage may take the form of methyl hydrate crystals, the proposed possible mining of which concerns many environmental scientists.

In epochs when temperatures are high, rates of evapotranspiration and biochemical reactions increase, favoring blossoming of plant life and greater storage of carbon dioxide. Greater biomass mitigates warming by reducing greenhouse retention of heat – a negative feedback process that has been vital throughout the latter part of Earth history. However, warmer atmosphere also means warmer seas, so that oceanic degassing can lead to an enhanced level of CO_2 in the atmosphere despite the increase in biological activity. Ice ages have come and gone at least 17 times over the past 2.4 million years, forced in large part by changes in Earth's orbital revolution, tilt, and albedo. During the most recent Ice Age, temperatures were 4–5°C cooler than at present, terrestrial biomass was considerably reduced, and atmospheric CO_2 levels were less by about 120 parts per million. Ocean–atmosphere interaction has also been important throughout this period, modulating heating of Earth's atmosphere.

Volcanoes are critical as an abiotic source of "new" CO_2 for atmospheric balance. As life respires or dies and decays, CO_2 returns to Earth's atmosphere, completing the carbon cycle. But the cycle is far from perfect owing to the fact that dead organisms require oxygen to decay.

Beneath piles of litter or sediment in deep, stagnant water, organic remains accumulate, ultimately converting to coal, petroleum, gas, or limestone. Along convergent plate margins, much of this material – especially limestone – is subducted. Weathering of silicate minerals also removes CO_2 from the atmosphere. Hence, carbon dioxide is "scrubbed" from the system. In fact, Holland (1978) calculated that without replenishment the atmosphere would lose all of its carbon dioxide in only about 10,000 years – and within the oceans in about 500,000 years, at given current rates of natural removal. Metamorphism, volcanic outgassing, and other important natural feedbacks step in to replenish the supply. Contemplate this for a moment, and one appreciates more deeply the ecological and physical balances of Boulding's "Spaceship Earth."

Of course, the rate of volcanic CO_2 production has not been constant through long spans of geologic time – or even on a decadal basis. The late Cretaceous period, around 65–80 million years ago is a notable contrast to the present. At that time the continents were arrayed in such a way that no Arctic Ocean existed. Warm equatorial waters could mix with polar seas to produce a higher mean ocean temperature than at present. This contributed significantly to the build-up of CO_2 in the atmosphere. Moreover, the super-continent of Pangaea had broken apart as heat trapped beneath this enormous landmass finally weakened the lithosphere, stimulating intensive seafloor spreading. The level of volcanism worldwide was considerably greater than the average for at least the past half-billion years. This activity, too, augmented the large amount of CO_2 in the atmosphere – perhaps four times greater than the current level. Since it is conceivable that human industrial activity could add just as much carbon dioxide to the air we breathe over the next few centuries, it is worth reflecting that during late Cretaceous time, temperatures at the North Pole were not unlike those of Washington State today. Conifers flourished around the pole and tropical flora existed above the Arctic Circle (Flannery 2001).

Sulfur

Like carbon dioxide, the annual global production of volcanic sulfur is something for which we humans may be grateful, since it facilitates protein development in our bodies and that of all other living organisms. The sulfur cycle is considerably more complex than the carbon cycle in large part due to the multivalent nature of sulfur. Volcanic sulfur is released primarily as SO_2, which can adhere to ash particles or dissolve in atmospheric moisture to precipitate later as acid rain or mists containing a mixture of H_2SO_4, H_2S, and sulfate ions (SO_4^{2-}). Ultimately, then, volcanogenic sulfur becomes an important component of soil and soil pore water. Sulfate ions are also the dominant form of sulfur where it is present in water bodies and seas.

The fates of sulfur on land and in the sea differ significantly. On land, nutrient recycling by growing and decaying vegetation maintains a reservoir of sulfur to support ecosystems. Deforestation and erosion can disrupt this supply. Microorganisms are important for breaking down sulfur-bearing compounds from decaying organic matter, making sulfur available for living plants to re-scavenge. Sulfates may be returned to the air as a part of this microorganic activity, so recycling is not 100 percent efficient. In the oceans, too, marine life incorporates sulfur, primarily through processing by phytoplankton, which is then consumed at higher levels in the food chain. Shallow sea temperatures and nutrient supply in turn regulate phytoplankton activity. High surface temperatures dampen nutrient upwellings that nourish marine life. But in counterbalance, phytoplankton releases sulfur back to the air in the form of dimethyl sulfide (DMS), which drifts inland as wave-tossed sea spray to fertilize coastal soils, and plays an even more important role, perhaps, by nucleating water droplets to form clouds. This in turn helps cool the atmosphere by increasing reflection of sunlight back out into space. DMS production, together with the thermal properties of water helps explain the moderating effects of oceans on the global environment.

As in the case of the carbon cycle, natural mechanisms exist to sequester sulfur and prevent the build-up of an intolerably acidic atmosphere through continuing volcanic activity. On land, organic decay in oxygen poor swamps and bogs is incomplete and considerable amounts of sulfur end up locked in plant remains, eventually to form sulfur-rich coal deposits such as those found in southern China. Since sulfate is the second most abundant dissolved ion in the oceans (after carbonate), evaporation in warm, shallow gulfs and lagoons, called *sabkhas*, forces precipitation of sulfate minerals such as gypsum and anhydrite. At certain times in geologic history, truly enormous supplies of these minerals have accumulated. The sulfate beds of the Delaware Basin, formed by evaporation of a shallow Permian sea in western Texas 250–280 million years ago is one example. More recently, repeated evaporation of the Mediterranean at the close of Miocene times (5–7 million years ago) produced gypsum beds as much as a kilometer thick. Volcanism counteracts this sequestration of sulfur, while biological activity acts to regulate the level of sulfur build-up in the atmosphere at any one time, generally at levels that favor the continuation of life.

Vast amounts of sulfate aerosols produced by Plinian or flood-basalt eruptions may penetrate the tropopause at the base of the stratosphere. On the laterally moving air masses at this level in the atmosphere, aerosols from a single great eruption may spread worldwide. Distribution largely depends on the latitude of the eruption. At high latitudes, a volcano might lie pole-ward of the subtropical jet stream, and since mixing of the air masses to either side of a jet stream is sluggish, its aerosols are likely to be restricted mostly to the hemisphere (north or south) in which it lies. Large eruptions near the equator are more likely to spread their aerosols worldwide because of trans-hemispheric circulation. The VEI = 7 eruption of Samalas volcano [133] (Lombok Island, Indonesia) was the Earth's most powerful eruption of the past millennium, depositing distinctive sulfate layers in the Arctic and Antarctica, causing worldwide famines in Europe, and possibly triggering the Little Ice Age (Vidal et al. 2014).

Volcanoes are also major sources of significant quantities of sulfur gases that do not reach stratospheric levels. Most active volcanoes emit some sulfur continuously through fumarolic activity (Chapter 3). Kīlauea volcano [17], for instance, erupted an average of 1250–1500 tons of sulfurous gases daily during a period of continuous lava lake activity in 2009 (Figure 13.2). It is estimated that the world's volcanoes produce between 1.5 and 50 Tg SO_2 per year, varying because of episodic explosive eruptions that insert much larger amounts into the stratosphere all at once (Textor et al. 2004).

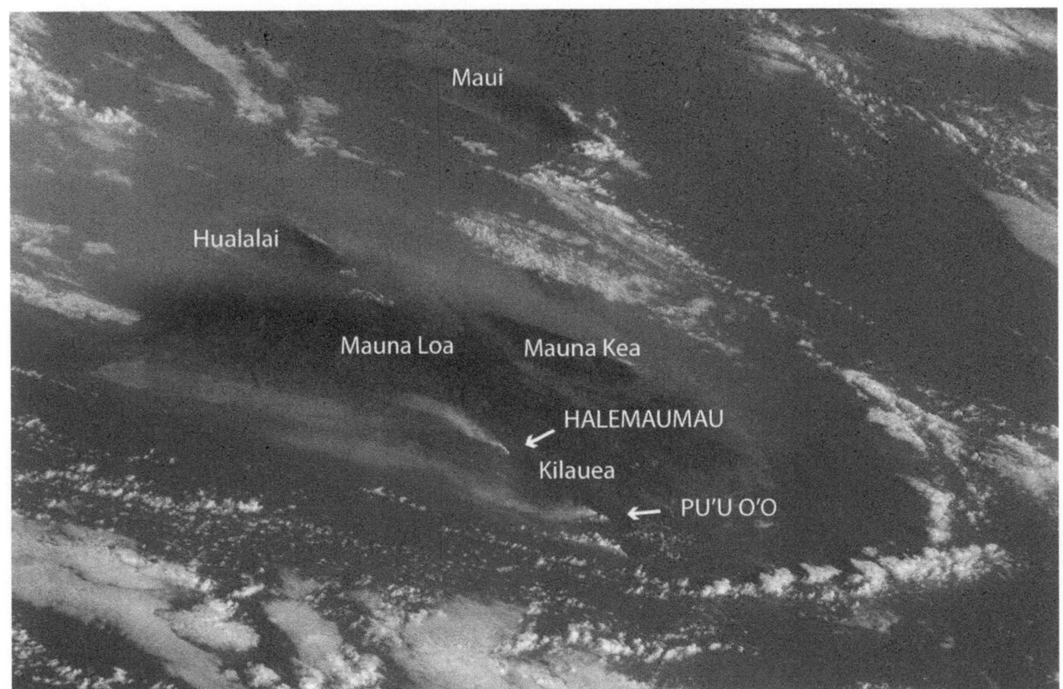

FIGURE 13.2 Volcanic gas emissions from Kīlauea volcano, Hawai'i [17], as viewed by astronauts from the Space Shuttle Atlantis on 13 May, 2009. View to northwest shows Maui island in the background, and conversion of H_2O and SO_2 plume from Halema`uma`u and Pu`u O`o vents to brown particulate-rich "vog" as it collects downwind of Hawai'i Island, and circles around the north side of the island. Note "bow wave" of clouds forming upwind of the island. Source: Photo courtesy of Image Science and Analysis Laboratory, NASA – Johnson Space Center: STS-125-E6569.

The lower stratosphere does not readily mix with the convective air masses closer to Earth's surface. Hence, hazes of volcanic aerosols introduced by great Plinian and flood-basalt eruptions may persist at this high altitude (15–20 km near the equator, 10 km in polar regions) for as long as several years. Like ordinary water clouds, volcanic aerosols increase Earth's albedo (reflectivity), cooling the atmosphere so much that they have significant impacts on global climate (Rampino and Self 1984). Because large volcanic eruptions are not frequent, these impacts are historically noteworthy. As mentioned in Chapter 6, for example, the largest historical flood-basalt eruption took place in south-central Iceland with opening of the 57 km long Eldgjá [95] fissure in 934 CE. More than 19 km^3 of basalt lava poured out episodically over a period of 3–8 years, forming flows as long as 40 km, and as wide as 20 km. Degassing of these flows produced about 35 megatons of SO_2, all of which remained within the lower atmosphere around Iceland. The sulfur output directly from eruptive vents was much greater, and had widespread climatic impact. Thordarson et al. (2001) estimate that the vents dumped 185 megatons of SO_2 into the upper troposphere and stratosphere. Combined with water, this corresponds to a potential yield of 450 megatons of H_2SO_4 aerosol. Reports of moist dust and sulfur particle fogs came from Europe and the Middle East in 934–9 CE as the fume from the eruption spread across the Atlantic. In China, half-way around the world, a string of intensely hot summers simultaneously killed hundreds in Luoyang, capital of the Later Tang Empire, and in the four years following the eruption, severe winters and locust-plagued summer droughts starved many thousands, leading to the collapse of the central government (Fei and Zhou 2006).

Two factors muted the potential impact of the Eldgjá gas release. One is the extreme northerly location of Iceland. Sulfates were apparently swept up in the westerly jet stream and largely confined above 30°N latitude (Lamb 1970). The other is that the Eldgjá eruption took place over several years, allowing atmospheric mixing to reduce sulfate concentrations below the levels that would have occurred had the volcanic activity taken place all at once. Thordarson et al. (2001) calculate that an instantaneous release of 185 megatons of sulfur compounds to the lower stratosphere would cause global atmospheric cooling of about 1.2°C. To put things in perspective, the difference between current mean atmospheric temperature and that near the close of the last Ice Age is only about 4–5°C.

The 1600 CE eruption of Huaynaputina volcano [80] in southern Peru was perhaps the largest eruption of the past 500 years, also had major global impact, which has so far received little scientific attention. It produced almost 20 km^3 of tephra, which blanketed an area of at least 300,000 km^2 (de Silva and Zielinski 1998), depositing ash and sulfuric acid aerosols in Antarctica and probably aerosols in Arctic ice too. Tree ring analyses and historical accounts suggest that this eruption caused the coldest summers in the past several hundred years, compounding Little Ice Age cooling trends.

The more recent eruptions of Tambora [134] (1815), Krakatau [122] (1883), and Pinatubo [135] (1991) each caused major global cooling. These were equatorial, highly explosive eruptions, each injecting the stratosphere with millions of tons of fine ash, aerosols, and sulfate particles. Each eruption was also less energetic, though more powerful than the Laki [96] outbreak. The Krakatau eruption, in particular, impressed European observers with remarkable, smog-like sunsets and silvery midday skies. This inspired a number of paintings, possibly including the lurid sky in Edvard Munch's famous work titled *The Scream*, which he painted in 1893 (Figure 13.3). In Poughkeepsie (New York) and New Haven (Connecticut), fire companies were turned out by local alarms from panicked citizens misconstruing lurid volcanic sunsets for large forest fires (Simkin and Fiske 1983). For over an hour after sunset, the western skies continued to show a bright yellow, orange, or red bands from sunlight reflecting off high-altitude aerosols, a phenomenon known as *noctilucence*. Sunrises were also affected, and over three months after the 20 May eruption, Reverend Sereno Bishop in Honolulu related the following observations from friends in the Caroline Islands (Simkin and Fiske 1983, pp. 155–156):

> *They state that, while they were dressing their children on the morning of September 7, the natives came anxiously asking what was the matter with the sun, which rose over the mountains with a strange aspect. It was cloudless, but pale, so as to be stared at freely. Its colour Dr. Pease called a sickly greenish-blue, as if plague stricken. Mrs. Pease's journal described it as "of a birds-egg blue, softened as this colour would be by a thin gauze. Around the sun the sky was a silvery-gray. At the altitude of 45° the sun appeared of its usual brightness, but resumed its pallid green aspect as it declined in the west."*

Reverend Bishop also made the first and perhaps most literary description of another optical effect of Krakatau's volcanic aerosols, a corona or circular patch of hazy white or bluish-white light enclosing the sun. The coronas Bishop observed extended to an angular distance of as far as 20° from the sun, and were fringed with a faint ring of red, brown, pink, or orange color, now termed **Bishop's ring**. Bishop's rings were most intense a year following the Krakatau eruption, but continued to be seen for as long as three years. The radius of Bishop's rings, and of the corona they enclose, is inversely proportional to the size of the aerosol particles that produce them by means of diffraction and reflection of incident sunlight (Bary and Bullrich 1959). Similar, but much less dramatic visual displays were associated with the 1982 and 1991 eruptions of El Chichón [57] (Rampino and Self 1984) and Pinatubo volcanoes [135] (Figure 13.4).

Given the well-known short-term climatic impacts of these large historical volcanic eruptions, is it possible that much larger prehistoric eruptions had much greater impacts? This question tempts scientists seeking to explain various mass extinctions

FIGURE 13.3 Norwegian painter Edvard Munch's *The Scream* – possibly influenced by the atmospheric effects of the 1883 Krakatau [129] eruption. Source: Edvard Munch/ National Gallery of Norway. Reproduced with permission from Erich Lessing/Art Resources, New York; National Gallery, Oslo, Norway.

FIGURE 13.4 Atmospheric effects of the 4 April, 1982 El Chichón [57] eruption, as observed from Hawai'i two weeks later. This eruption released abnormally large amounts of SO_2 into the stratosphere, causing anomalous conditions that persisted for many months in the northern hemisphere. Source: Photos by J.P. Lockwood.

observed in the fossil record. Of particular interest are the gigantic flood-basalt eruptions that formed the Deccan Traps in central India and the even more voluminous submarine large igneous province (LIP) eruptions of the late Cretaceous (Chapter 6). Much of this intensive volcanism took place within a million years of the extinction of the dinosaurs and most other large animal species living at the end of Mesozoic times. The extinctions created huge niches in the environment ultimately filled by mammals and birds. Was so much sulfur released by this volcanism that the world plunged into an altered climatic regime – a prolonged "volcanic winter"? Only a few years of cooler temperatures imposed on a tropical world could have had devastating impacts through disruption of reproductive cycles, photosynthesis, and food chains. Acidification of rainfall and atmospheric moisture would be especially hard on photosynthesis, while polluted waters could impact terrestrial and even marine biota (Officer and Drake 1983). The correlation between these great volcanic events and the extinction of the dinosaurs is compelling, and certainly reinforced by the fact that as *individual* phenomenon, LIP eruptions show strong correlations with four other major extinction events between 300 and

150 million years ago. The eruption of the Siberian Traps of eastern Russia, gigantic flood-basalt eruptions at the end of the Permian period 250 million years ago, coincides with the largest-known mass extinction in Earth history, when 90–95 percent of all living species became extinct (see, e.g., Rampino and Strother 1988; Morgan et al. 2004). But the case is not watertight. For one thing, younger LIP events *are not apparently* related to large-scale die-backs (Wignall 2005). Eruption of the voluminous Columbia River basalts between 16 and 18 million years ago, for instance, produced no discernable breaks either in the North American or global fossil records. In addition, other factors may be involved to explain extinctions. The Chicxulub meteor impact off Yucatan 60 million years ago, with its well-documented global impact (Alvarez and Zimmer 1997) clearly was catastrophic for the world of the dinosaurs and other living things. Improved paleontological data for the Cretaceous–Tertiary transition reveal that extinctions of both flora and fauna increase in severity toward its 100-km-wide crater, now largely buried by younger limestone (Hildebrand et al. 1991; Flannery 2001). Perhaps the most that may be said with respect to the closely antecedent Deccan and mid-Pacific volcanism is that these eruptive episodes possibly stressed existing biota so that when the Chicxulub strike occurred shortly thereafter (geologically speaking), extinctions were worse than may otherwise have been (Frankel 1999; Arens and West 2008).

Volcanic Influence on Soil Fertility, Agriculture

Fertile soils have attracted human populations since the dawn of agriculture nearly 11,000 years ago. Initial farming and grazing could take place in a wide range of environments, but depletion of soil nutrients by overplanting of crops and erosion has rendered, and continues to render, many of these areas sterile – a problem of mounting global concern that undermines the highly applauded Green Revolution. In only a few geologic environments does natural replenishment of nutrients readily take place, often sustaining very fertile soils. One such environment includes the world's floodplains, such as that of the Huang He River in eastern China, the Euphrates-Tigris breadbasket of Iraq, and the Egyptian Nile. The other regions of renewable fertility are associated with the gray volcanoes of volcanic arcs, where frequent ash falls create fertile soils allowing in some cases for as many as three crop harvests per year. On train rides across the countryside of Java, one observes that the wealth of the towns (as measured by numbers of bicycles or quality of homes and clothing) increases wherever the train approaches a volcano – indicating the prosperity provided by farming in these areas. The most productive tea plantations in Java grow on ash-derived soils, and indeed about one-tenth of the human population lives upon the soil products of volcanoes, especially around the Pacific Rim (Chapter 14). Many early civilizations, too, drew their wealth and power from the fertility of volcanic soils, including the Aztec, Inca, and Highland Maya in the Americas, and the Mataram of central Java (Newhall et al. 2000).

Rock and rock fragments contain almost all of the nutrients that plants and animals need to thrive, but the nutrients are locked up in crystalline minerals or glasses inaccessible for biological exploitation. Residual soils form in the same places as their original source materials through chemical weathering. They characteristically develop three main layers (A-B-C) over time, with the rate of layer development dependent upon the climate and rock type. In the humid temperate climate of northern Japan, the topmost fertile A-horizon may start to develop from a bed of volcanic ash within a century, and the underlying B-horizon, also a source of soil fertility within 100–500 years. The complete sequence, A-B-C is usually in place after about a thousand years, though development of a rich organic component in ash soils (≥ 10 percent by weight carbon) may take 4000–5000 years (Wada and Aomine 1973; Ugolini and Zasoski 1979). Slower rates of weathering and soil development occur in other non-volcanic temperate regions, and it may take as long as 20,000–100,000 years for clays with valuable primary nutrient sources to form fully (Gibbs 1968).

Although rich in mineral nutrients, juvenile volcanic soils contain little nitrogen, but they yield abundant crops given appropriate fertilization. A key factor in their natural fertility may be the high, unsaturated water content of the soil, which helps promote rapid cation exchange. Volcanic ash soils (**andosols**) commonly support a wide range of crops, even within the same region. But the response of agriculture to fresh ash fall is not always positive. If tephra accumulation amounts to less than a few centimeters, the addition of ash will rejuvenate the soil by adding new nutrients without damage. Heavier accumulations bury the soil so deeply, however, that without intensive tilling, farmers must abandon their fields and wait for weathering to restore the land. That can take a long time indeed, possibly well past a lifetime. In the third-century CE, the explosive eruption of Ilopango volcano [63], El Salvador deposited tephra exceeding a meter thickness as far as 77 km from the eruption focus. Sheets (1979) estimated that it may have taken as much as two centuries for the land in the devastated area to recover to its pre-eruption productivity.

Tephra fall not only impacts, for better or worse, the agriculture of a region, but may greatly alter fisheries as well, with harmful implications for populations dependent upon them. Workman (1979, p. 346) notes:

> *Red salmon were beginning their annual run when Katmai [Alaska] erupted in June 1912. Salmon already in the streams of Afognak Island stayed there until they suffocated with their gills filled with liquid mud. About 4000 perished in the Litnik Stream Hatchery. Rains kept the lakes and streams muddy, preventing or delaying salmon from reaching their spawning grounds. A heavy rain in mid-August put so much ash in the water that salmon suffocated. . .. In this later run salmon were observed ascending the polluted streams a short way, going back to sea, then trying to ascend again, repeating erratic movement a number of times.*

In the following year, however, salmon populations exploded, perhaps in response to the input of nutrients from fresh volcanic ash entering their waterways.

Volcanoes have also provided unexpected resources for biologists who study the ecology and rates of forest growth. Dating of individual prehistoric lava flows that cross climatic zones governed by elevation, temperature, and rainfall have provided a boon to biologists who can evaluate details and rates of forest succession processes as a function of differing environmental factors (Vitousek et al. 1995).

Volcanoes and Human History

Volcanoes in the East African Rift system forged the ecological cradles in which primates began to walk upright and human species evolved. Although primates lived throughout much of Eurasia and the Americas throughout the Tertiary Period, it was only in the African East Rift system where hominid species evolved, and where modern humans first appeared. Volcanic activity in the East Rift frequently stressed the resident proto-humans there, isolated small populations, and together with tectonic deformation created new environmental niches, where restricted populations needed to rapidly adapt to survive (WoldeGabriel et al. 2000). Fresh lava flows there also were a source of easily worked glassy rocks where toolmaking could be learned, critical to the evolution of human intelligence and social structures. Volcanism played a vital role in our evolution.

It was also volcanism that almost ended the human evolutionary experiment, when a catastrophic eruption 8000 km east of Africa caused worldwide climactic changes in Late Pleistocene time. The Ultraplinian super-eruption of Toba volcano [127] on Sumatra, 75,000 years ago (Chapter 8) injected vast quantities of ash and volcanic aerosols into the upper atmosphere, causing a volcanic winter that probably lasted for several years (Strothers 1984). Mitochondrial DNA studies have conclusively shown that human populations were drastically reduced about this time, and that perhaps only a few thousand individuals survived as ancestors of the billions of humans who inhabit Earth today (Rampino and Self 1993; Ambrose 1998; Rampino and Ambrose 2000). The exact cause of this population "bottleneck" is not known, but the fact that the timing approximately correlates with the Toba eruption suggests that catastrophic climatic change may have been responsible for a famine that drastically reduced our ancestral populations.

Social Impact of Volcanic Eruptions

Volcanoes impact society in many ways. Over a quarter-million people have been killed directly by volcanic eruptions over the past 500 years (Brown et al. 2017) and hundreds of thousands of others have perished because of "collateral damage" such as starvation and related disease. Their psychological influence on society throughout history is also great, and extends beyond fear or respect for the unknown (which in some traditions motivated human sacrifices and religious worship) but also generates artistic respect for their beauty. Perhaps the first depiction of an erupting volcano is of Hasan Dagi [119], a now-dormant volcano on the Anatolian Plateau of central Turkey – from a fresco recovered in the ruins of 6200-year-old Çatal Höyük, one of the world's oldest known cities. Likewise, classical Japanese artists are renowned for their portrayals of Mount Fuji [146]. Volcanoes remind us that primal forces of nature are still operating in the world to change and create anew.

The physical impacts of volcanic eruptions can have major societal repercussions and displace or kill thousands of people. In 260 CE, a thriving, sophisticated Mayan culture occupied the highlands of what is now El Salvador when the a forementioned Ilopango [63] Plinian eruption took place, forcing perhaps hundreds of thousands of survivors to flee to coastal Guatemala and Belize. The catastrophe shifted native trade routes far to the north, possibly bringing new wealth and power to such Lowland Mayan centers as Tikal in Guatemala and drastically altering the whole course of Mayan history (Sheets 1979).

Similar speculations have been made about the impact of the 1645–1625 BCE caldera-forming eruption of Santorini volcano [114] on the Minoan civilization (McCoy and Heiken 2000). Some archeologists have proposed that this moderately violent (VEI = 6) late Bronze Age eruption, with its attendant tsunami and pyroclastic activity, so devastated Minoan power that a vacuum developed, stimulating the rise of Mycenae and the later Greek city-states (Nixon 1985). Caldera collapse during the eruption probably submerged the center of the strategically central Aegean island of Thera, of which present-day Santorini is but a remnant, and with it buried the thriving Minoan settlement of Akrotiri – the "Minoan Pompeii." The eruption possibly inspired the legend of Atlantis described by Plato, who wrote his dialogs 1300 years later (Friedrich 1999). Consider the following 1871 translation of Professor Benjamin Jowett (University of Oxford) from Plato's *Timaeus*:

> *Now in this island of Atlantis there was a great and wonderful empire which had ruled over the whole island and several others, and over parts of the continent... This vast power, gathered into one, endeavored to subdue at a blow our country... but [then] there occurred violent earthquakes and floods, and in a single day and night of misfortune all*

[the] warlike men [of Atlantis] in a body sank into the earth, and the island of Atlantis in like manner disappeared in the depths of the sea. For this reason the sea in those parts is impassible and impenetrable, because there is a shoal of mud in the way, and this was caused by subsidence of the island.

Likewise, frequent explosive eruptions of Merapi [131] in the eighth–tenth centuries CE apparently forced abandonment of the giant temple and political center of Borobodur and encouraged the decline of the Mataram culture (Newhall et al. 2000). Such impacts have ripple effects through all of subsequent human history.

The competition for volcanic resources has also had impacts on human affairs. Eastern Aleut peoples often fought wars over precious obsidian for spear points and other implements to use in their subarctic island outposts. They mined sulfur from fuming craters to use in starting fires – an important task in that wet, wood-deprived region – and used volcanic hot springs to cook fish, edible roots, and the meat of sea mammals (Workman 1979). The mining of magmagenic silver from Laurion helped provide Athens with the wealth it needed to defeat the Persians at the Battle of Salamis. Likewise, the metals fashioned from subvolcanic ores in southern Spain financed and equipped the Carthaginians in the Second Punic War, which very nearly ended the Roman Republic. Renaissance Spain rose to power on a tidal wave of gold, plundered from American native peoples who had retrieved it from placer deposits related to volcanic and other igneous sources.

Perhaps the most archeologically important eruption in all of human history was the 79 CE outburst of Vesuvius [109], which buried the cities of Pompeii and Herculaneum, the port of Stabiae, and countless villas, farms, and fields within 20 km of the volcano. As though thrown into a time capsule, a portion of Roman society was preserved beneath thick Plinian tephra and PDC (Pyroclastic Density Current – Chapter 8) deposits (Figure 13.5). The story of how these ruins were "found" after over 1500 years of burial is worth recounting. Digging of a canal in 1592 accidentally brought the first artifacts from Pompeii to light, but serious excavation did not begin until 1710 when Austrian Prince D'Elboeuf obtained statuary from a well dug by a peasant on land atop the site of Herculaneum. D'Elboeuf purchased the well site without realizing that a Roman city lay beneath his bucolic countryside property. But with mixed success in finding further statues, and continuing difficulty in digging through hard mudstone and tuff, he soon lost heart. Bullard (1976, pp. 204–205) describes what happened next:

The political fortunes of Italy now changed, and in 1735 Naples and Sicily came under Spanish rule. The eldest son of the King of Spain, 19-year-old Charles of Bourbon, became the absolute monarch of the Two Sicilies. The royal youth, much interested in fishing and hunting, acquired for his pastime the house that had formerly belonged to the Prince d'Elboeuf. Still in the house were many of the statues which d'Elboeuf had recovered from his diggings. In 1738, young King Charles

FIGURE 13.5 Temple of Jupiter, Pompeii, with Mount Vesuvius volcano [109] in the background. Built in the second century BCE, this temple and the entire city of Pompeii was buried in 79 CE. The central areas of Pompeii were excavated during the nineteenth century, but unfortunately, none of the original pyroclastic deposits that buried Pompeii were preserved until later cooperative work between archeologists and volcanologists. Source: Photo by J.P. Lockwood.

married Maria Amalia Christini, whose father, Augustus III of Saxony, was a great patron of the arts. . . When the king brought his young queen to Naples, she was fascinated by the ancient statuary. . . and she begged her husband for more pieces. He organized a digging force, and work was started at the original well which d'Elboeuf had taken over from the peasant. The first discovery consisted of three pieces of a statue of a huge bronze horse. Next they found the torsos of three marble figures in Roman togas and another bronze horse. As work progressed, a flight of stairs was discovered, and on December 11, 1738, a plaque bearing the inscription Theatrum Herculanensem was unearthed. Thus, by sheer luck, it appears that d'Elboeuf had unwittingly hit upon the front of the stage of a theater, on which had collapsed, under the impact of the mudflow, the wall which served as wings and background, with its marble facing and numerous statues. This was one of the few spots, perhaps the only one, where sculpture was literally piled one piece upon the other. Thus was Herculaneum discovered.

These archeological discoveries strongly inspired writers, artists, architects, scholars of history, theologians, and philosophers throughout the Western World. Such finely detailed artifacts as wall frescoes and graffiti, Roman glassware and plumbing, prepared foods, ornamented doors, furniture, tiled bathrooms, medical equipment, and the body molds of many people, and some of their pets, came to light in subsequent excavations (Figure 13.6). Many frescoes are practically as good as photographs in their depiction of Roman dress, manners, hobbies, mythology, taste, religious practice, and industry. As tragic as the 79 CE eruption was, had it not occurred, this "voice" from the past would be missing, and our appreciation of early Roman society would be much less personal and clear. Unfortunately, for the first two centuries of archeological exploration, the pyroclastic deposits that buried these cities were only regarded as "dirt" to be removed, and there was no volcanological effort to understand the nature of the 79 CE eruption (Chapter 7). This has now changed, much of that "dirt" is now being preserved, and studies by Italian volcanologists, working in cooperation with archeologists, have led to much greater understandings of this tragedy and the details of the eruption, which involved precursory earthquakes, early Plinian tephra fall, and later devastating PDCs (Scandone et al. 1993). The area is much more densely populated now than it was in 79 CE, and an eruption of that scale would put millions of people in jeopardy (Figure 13.7).

Dozens of archeological sites buried by volcanic pyroclastic deposits are now known in addition to Akrotiri and Pompeii, including the more recent discovery of a sixth-century Mesoamerican agricultural village known as Joya de Cerén, El Salvador (Sheets 2002), built in an area devastated by the large eruption of Ilopango [63] 200–300 years previously. Cerén was itself destroyed (and preserved) by a small eruption from Loma Caldera volcano [79], only 600 m north of the village. The ongoing archeological investigations at Cerén, in cooperation with volcanologists, sociologists, and agricultural experts, has provided a detailed insight into the daily lives and culture of these rural people, who lived on the edge of the Mayan empire. The village and surrounding fields were buried by about 4 m of tephra, which preserved household belongings and even food supplies. The well-preserved crops showed that the eruption occurred in mid-summer, probably on an August evening as villagers were finishing their meals. The eruption was probably preceded by large earthquakes and explosions that caused the residents to flee to safety, without any known victims. Volcanoes not only change history, but they preserve our past, and will likely continue doing so as future eruptions occur in an ever more populated world. The impact of major explosive eruptions on scales not witnessed in historical times could have devastating impact on global populations. As Payson Sheets writes (2015, p. 1318):

"A very large eruption, on the scale Tambora, or Thera (or for sure Toba) would cause almost unimaginable disruption to worldwide societies [. . .] dense populations in the billions would suffer, and the cost of maintaining mitigating measures such as immense food and water supplies is beyond what any nation is willing to invest."

It is to volcanic hazards and risk that we next turn our attention.

FIGURE 13.6 Human body preserved in pyroclastic surge deposits of the 79 CE eruption of Mount Vesuvius [109] (victim's skull preserved at left). Note that the surge deposits formed a dune structure down-current from the body, which formed an obstruction to the PDC. Source: Photo by J.P. Lockwood.

FIGURE 13.7 Image of Vesuvius [109] and surrounding, densely populated areas obtained by astronaut Chris Hadfield in 2013 as the International Space Station was 400 km above the Bay of Naples area. Source: Image courtesy of the Earth Science and Remote Sensing Unit, NASA – Johnson Space Center. Frame ISS034-E-12262.

Questions for Thought, Study, and Discussion

1. The environmental benefits of atmospheric carbon dioxide and sulfur compounds are very different. Explain.
2. In what ways can volcanic eruptions modify global climate?
3. Suppose that Earth lacked volcanism. How would our planet be different?
4. Look up and describe the biological role of amino acids.
5. Discuss the balance between beneficial and destructive impacts of volcanic eruptions.
6. Why are areas devastated by prehistoric eruptions sought out by archeologists for research excavations?

FURTHER READING

Alvarez, W. (2017). *A Most Improbable Journey – A Big History of Our Planet and Ourselves.* New York: W.W. Norton.

Deamer, D.W. (2019). *Assembling Life: How Can Life Begin on Earth and Otsher Habitable Planets?* Oxford, UK: Oxford University Press, 184 pp.

Fisher, R.V., Heiken, G., and Hulen, J.B. (1997). *Volcanoes: Crucibles of Change.* Princeton, NJ: Princeton University Press.

Frankel, C. (1999) *The End of the Dinosaurs.* Cambridge, UK: Cambridge University Press, 236 pp.

Klingaman, W.K. (2013). *The Year Without Summer: 1816 and the Volcano That Darkened the World and Changed History.* St Martin's Press, 388 p.

McCoy, F. and Heiken, G. (2000). *Volcanic Hazards and Disasters in Human Antiquity.* Boulder: Geological Society of America.

Robock, A. and Oppenheimer, C. (2003) *Volcanism and the Earth's Atmosphere.* Washington: American Geophysical Union.

Rowland, I.D. (2014). *From Pompeii: The Afterlife of a Roman Town.* Cambridge, Massachusetts: Belknap Press, 340 p.

Wood, G.D. (2014). *Tambora: The Eruption That Changed the World.* Princeton, NJ: Princeton University Press, 293 p.

Sheets, P.D. and Grayson, D.K. (1979). *Volcanic Activity and Human Ecology.* New York: Academic Press.

CHAPTER **14**

Volcanic Hazards and Risk – Monitoring and Mitigation

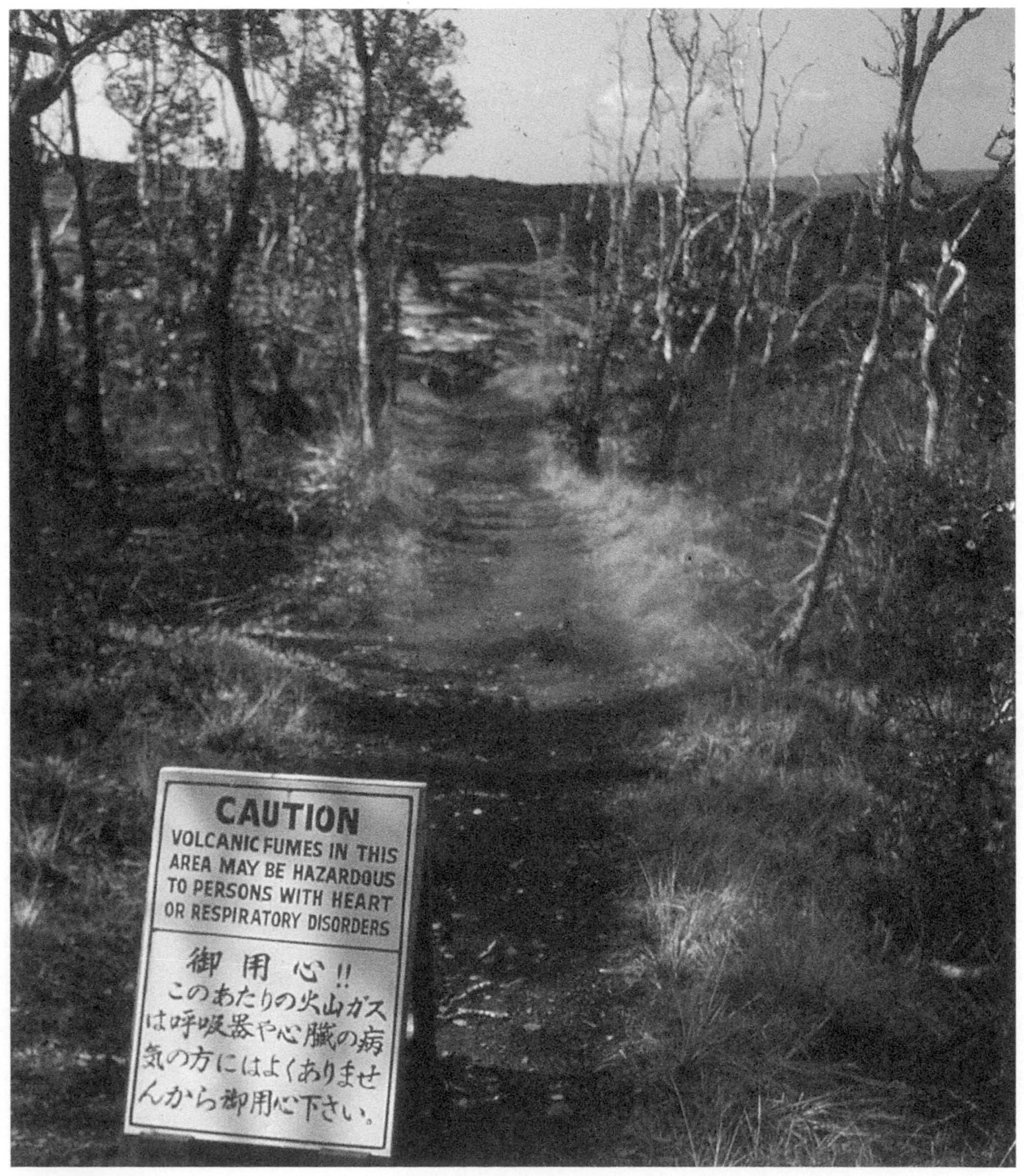

Source: J.P. Lockwood.

Volcanoes: Global Perspectives, Second Edition. John P. Lockwood, Richard W. Hazlett, and Servando De La Cruz-Reyna.

Companion website: www.wiley.com/go/lockwood/volcanoes2

Los desastres naturales no existen.
Se presenta un desastre, no por causa de la Naturaleza,
sino por la falta de preparación por parte de la Sociedad.
[There's no such thing as natural disasters.
Disasters take place, not by Nature's hand,
but by the lack of preparation on the part of Society.]
(Hugo Delgado Granados)

For less complex forms of life on Earth, volcanic activity must be of little long-term concern – eruptions either do or do not destroy them, they may or may not be able to flee their fates, and life, at least elsewhere, goes on. For human beings, however, our greater interconnectedness and awareness makes volcanism a long-term concern, since eruptions can alter our environment and impact our societies in dramatic ways, often over generations. Modern science and technology have revealed that volcanic activity is a manifestation of natural processes and not the work of gods, and as such is subject to rational observation and limited control – in the sense that the detrimental impact of volcanic eruptions on human society can be greatly reduced.

For me (JPL), volcanology was initially a "fun" career, and I regarded eruptive activity as "entertaining" and "beautiful." The scientific papers I was writing were all focused on "better understanding" volcanic activity – important work, but relatively devoid of "applied" aspects that related to the needs of society. Those perspectives changed dramatically one day in late November 1985, when I was dropped by a helicopter with USGS colleagues Dick Janda and Tom Pierson on the outwash lahar deposits about 5 km downstream from the ruins of Armero, Colombia – a city that had been obliterated by lahars two weeks earlier (Chapter 11). Human bodies are less dense than lahar sediments, float to the surface, and we were surrounded by the rotting remains of hundreds of the 23,000 people who had perished here on 13 November. I shall never forget the horrors of that day, nor the sudden realization that, although eruptions may indeed be "beautiful," they can also be deadly, and that "deadly" is not merely an abstraction involving ink on paper. People who have to deal with mass casualties quickly become numb to the realities they must deal with, and we went about our work, digging trenches to evaluate lahar stratigraphy, making size analyses, and calmly writing down our notes – not paying much attention to the grisly scenes around us – as buzzards too full to fly dined on human flesh. That night however, back alone in my Manizales hotel room, I broke down and cried uncontrollably, as hard as I have ever cried in my life. I wept not just for the victims, but I cried because of the realization that very few of these people would have perished if the volcanologists, who had correctly described the risks facing Armero and other cities, had only done more to ensure that their warnings were understood by the people at risk. If the people in Armero had heeded the alarms that sounded as lahars were rushing down at them, they only would have needed to walk short distances to higher ground and survived. I realized that volcanologists who deal with dangerous volcanoes have more than scientific responsibilities to describe hazards and risk – they also have obligations to be sure those warnings reach the decision-makers in charge of the safety of people, and the people at risk.

Real people live on and near volcanoes, and their lives may well depend on the work we do as volcanologists. Volcanology can be "fun," but it is also serious business, and no matter what aspect of volcanology we pursue, we should never forget that the ultimate value of our research and study will be the contributions that allow our colleagues who may be faced with life and death situations to make better decisions, based on clearer understanding of "how volcanoes work."

Peterson (1996) estimated that nearly 10 percent of the world's population lives on or near active volcanoes (Chapter 13). The volcanic risk to most of these people is very low, but risks change as volcanoes become restless, and awareness of risk is important for public well-being. For these reasons, this chapter on volcanic hazards and risks is the most important of our entire book. Communicating hazards and risk to the society we all ultimately serve is not a trivial issue and involves the critical task of explaining volcanological concepts to people with many different backgrounds and perceptions of the natural world. Here in this chapter, we attempt to provide some concepts and semi-quantitative tools that could help volcanologists translate their insights about the threats a volcano may represent into forms that can be better understood by those at risk. First, the difference between hazards and risks.

Hazards and Risks

The terms "hazards" and "risks" are too often used synonymously, but in fact they have very different meanings:

- **Hazards** are natural or sometimes human-triggered processes that have the *potential* to wreak havoc whether people are directly threatened or not;
- **Risk** is the chance, or probability, that any particular hazard or set of related hazards will actually impact people or property within a threatened area.

A large eruption at a remote volcano may thus generate many potentially deadly hazards such as lava flows and explosive blasts, but if the area were unpopulated, there would be no risk to humans. Volcanic hazards are natural phenomena that will occur no matter what humans do, whereas volcanic risks are inevitably the result of people being "in the wrong place at the wrong time." Hazards are usually described in general terms, such as "Hazards at X volcano include pyroclastic flows and lahars," whereas risks are what most concern people, and are best expressed in semi-quantitative terms, such as "The risk of lava flows impacting X area within a period of 1000 years is about 35 percent." Nothing much can be done to control volcanic hazards, but a great deal can be done to lessen volcanic risk to human populations and to their infrastructures through better planning and public education.

Since volcanic phenomena may threaten various sectors of the population dwelling in exposed areas, the risk reduction process requires a coherent response of all involved: the public, the scientists in charge of monitoring volcanic activity and appraising the volcanic hazard, the political decision-makers and civil protection or civil defense bodies. For risk management to be efficient, it is essential that all of those sectors share a common perception of the hazards and risks throughout the exposed region, and this is only possible through an effective communication of the volcanological concepts involved (Jolly and De la Cruz-Reyna 2015; De la Cruz-Reyna et al. 2017). This calls for clear and, to the extent possible, precise characterizations of hazard and risk, discussed in the following sections.

Active, Dormant, and Extinct Volcanoes

Volcanologists use various terms to describe the eruptive state or potential of a volcano. These terms are hardly "standardized," however, and inconsistent use has led to scientific as well as to public confusion. For most volcanologists, an **active volcano** describes a volcano that has erupted in recorded human history and has the potential to do so in the future. The problem with this definition is that the history of human observation varies greatly from place to place – from over 3000 years of written documentation in the Mediterranean area to less than 400 years for the preservation of oral traditions in places like Hawai'i. Understanding this discrepancy, other geologists conservatively regard any volcano that has erupted during the past 10,000 years as "active" (Simkin 1984). For most of the public, however, an "active" volcano refers to a volcano that is *presently erupting*.

Dormant volcanoes are ones that have not erupted in historic times, but are known to have prehistoric eruptive recurrence intervals greater than the elapsed time since their last eruption. Dormant volcanoes may hide their potentials for eruptive activity very well, however, and may not even be recognized as "volcanoes" by local people, especially in tropical regions where heavy rainfall causes deep erosion, and vegetation grows quickly, obscuring signs of past activity. Examples include El Chichón volcano [57] (Figure 1.11) in Chiapas, Mexico, which was not considered threatening by many before its deadly 1982 eruption (Chapter 1), and Mount Lamington [159] in Papua, New Guinea, which no one recognized as having a volcanic origin prior to its catastrophic eruption in 1951 (Taylor 1958).

Extinct volcanoes are those that are not expected to erupt ever again and show no signs of seismic or fumarolic activity. To determine that a volcano is safely extinct, however, requires field mapping to determine past eruptive history. If a volcano can be shown to have not erupted for periods of time *much* longer than its past eruptive recurrence intervals, it is safe to call it extinct. A critical problem, however, is that most of the world's volcanoes have not been mapped or studied in sufficient detail to determine their past geological histories. Simkin and Siebert (1994) noted that "Some of the most calamitous eruptions of recent decades have been from volcanoes with no previously known historical volcanism," and showed that of the 16 largest eruptions of the past 200 years, 12 had no previously known historical activity! As an example of a "lurking" volcano, Chaitén (southern Chile) [73], had not erupted in over 9000 years and was thus considered to be extinct, but came back to life with little warning in May 2008, with a VEI = 3 eruption that caused widespread damage. Connor et al. (2006) warn that "*It is incorrect to assume that volcanoes that have not been active during the Holocene are incapable of future eruptions.*" El Chichón [57] and Mount Pinatubo [135] (described in Chapter 1) are other "surprise" volcanoes that had not erupted within the memories of living humans.

Well beyond the ambiguities of these important formal concepts, however, public perceptions of risk tend to be deeply psychological and socially based – and of serious concern to volcanologists tasked with conveying information during a crisis to the public, since most volcanologists have little or no professional grounding in either psychology or sociology.

Perception of risk directly relates to motivation to take action, so many recent studies have focused on ways of building better understanding in the scientific community of how best to connect with people subject to significant risk during a volcanic eruption (e.g. Bird and Gisladóttir 2012; Ricci et al. 2013; Favereau et al. 2018). Doyle et al. (2014), for instance, note (p. 1):

> *The issuing of forecasts and warnings of natural hazard events, such as volcanic eruptions . . . often involves the use of probabilistic terms, particularly when communicated by scientific advisory groups to key decision-makers, who can differ greatly in relative expertise and function in the decision-making process. Recipients may also differ in their perception of relative importance of political and economic influences on interpretation. Consequently, the interpretation of these probabilistic terms can vary greatly due to the framing of the statements, and whether verbal or numerical terms are used. . . . It is also unclear as to how people rate their perception of an event's likelihood throughout a time frame when a forecast time window is stated. Previous research has identified that, when presented with a 10-year time-window forecast,*

participants viewed the likelihood of an event occurring 'today' as being of less than that in year 10. . . [T]his skew in perception also occurs for short-term time windows (under one week) that are of most relevance for emergency warnings. In addition, unlike the long-time-window statements, the use of the phrasing "within the next. . ." instead of "in the next. . ." does not mitigate this skew, nor do we observe significant differences between the perceived likelihoods of scientists and non-scientists. . . . This finding suggests that effects occurring due to the shorter time window may be "masking" any differences in perception due to wording or career background observed for long-time-window forecasts.

We will return to this personal topic shortly. But first let us explore the physical aspects of volcanic hazards more closely.

Primary and Secondary Volcanic Hazards

Volcanic hazards can be divided into two classes: **primary hazards** – eruptive phenomena directly related to volcanic activity, and **secondary hazards** – those phenomena indirectly related to the eruptions themselves. Lava flows, volcanic gases, and localized airfall of spatter and other tephra are the primary hazards associated with effusive activity of "red volcanoes" (Figure 14.1). Primary hazards associated with explosive "gray volcanoes" include pyroclastic density currents (PDCs; flows and surges), the widespread deposition of airfall tephra, dome explosions, local hydrovolcanic explosions (Figure 14.2), and emissions of caustic gases. Most of those primary eruption hazards have already been explored at length in Chapters 6, 7, and 8, and will not be further described in this section. Such hazards are generally confined to the vicinity (within a few tens of kilometers) of an erupting volcano. Airborne ash clouds are of course an exception. They can spread much farther downwind – even thousands of kilometers – and have serious impacts on travel and communications. The challenges to aviation is especially alarming, and are discussed separately below.

Secondary hazards may be triggered immediately by eruptive activity, or may develop *many years* or *even decades* after eruptive activity has ceased (e.g. see Chapter 1 gray volcanoes case studies). Secondary hazards may be much worse and more widely destructive than the direct actions of the eruptions themselves. The list of secondary hazards is long, and must include cascades of hazardous changes, each one (third order, fourth order, etc.) the result of another acting on a particular environment. For instance, heavy ash falls (a primary hazard) on a watershed may strip it of vegetation and load slopes with loose material, inducing erosion

FIGURE 14.1 View of Goma, eastern Democratic Republic of Congo, after fluid lavas poured through the city in 2002. Nyiragongo volcano [118], source of these lavas, looms over the city in the background. Highly fluid pāhoehoe lavas flowed around and through buildings, but `a`ā flows crushed any structures they encountered. Source: Photo by J.P. Lockwood.

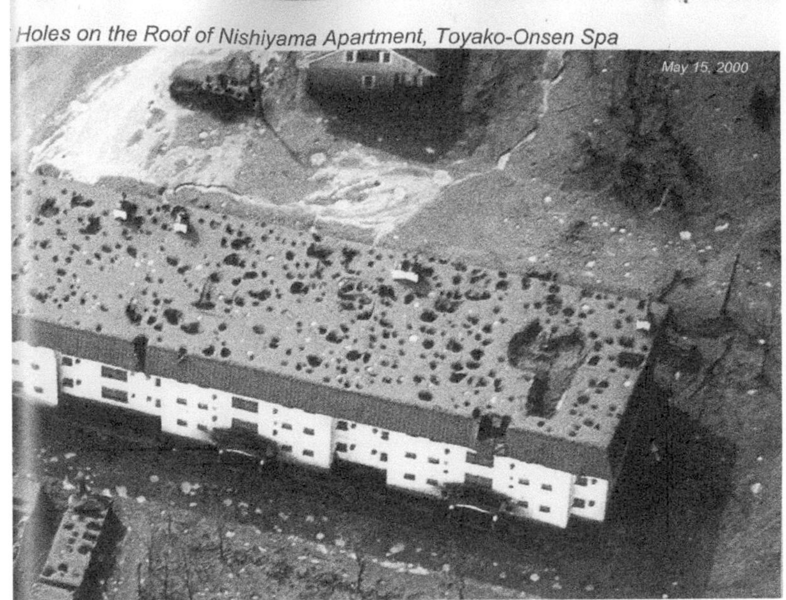

Holes on the Roof of Nishiyama Apartment, Toyako-Onsen Spa
May 15, 2000

FIGURE 14.2 Apartment house at the Toyako-onsen Spa, Toyako Town, Hokkaido, Japan. The building was bombarded by "volcanic mud bombs" blasted out of a hydrothermal vent on the northwest flank of Usu Volcano [152] during a phreato-magmatic eruption in April 2000. Source: Courtesy of Hiromu Okada.

and debris flows (a secondary hazard), which contributes to increased sediment load in rivers and lakes (third-order hazard), which in turn kills off fish populations (for many people possibly the most important consequence). Defoliation of crops by acidic gases during an ash fall destroys harvests, possibly leading to famine with economic or social impacts persisting for years (Delmelle and others, 2002). Serious public health aspects of directly inhaling ash and gases can also be immediate (a primary hazard) or secondarily long-lasting (Buist and Bernstein, 1986; Carapezza et al., 2012). But long-term environmental benefits may also ensue, such as soil nutrient enrichment and better land-use practices (Chapter 13). Ayris and Delmelle (2102, p. 1905) note:

> *Embracing the full complexity of environmental effects caused by tephra fall demands a renewed investigative effort drawing on interdisciplinary and laboratory studies, combined with consideration of the interconnectivity of induced impacts within and between different receiving environments.*

The following box encapsulates the major primary and secondary volcanic hazards and their most common impacts.

MAJOR HAZARDS

Hazard category	Principal source and type	See also ...
PRIMARY	RED VOLCANOES	
	Lava Flows	Chapters 1, 6
	Gases	Chapters 1, 3, 6
	GRAY VOLCANOES	
	Ballistic ejecta	Chapter 7
	Tephra Fall	Chapters 1, 7, 8
	PDCs	Chapters 1, 3, 7, 8
	Gases	
SECONDARY	RED VOLCANOES	
	Methane Bursts	Chapter 6
	Littoral blasts	Chapter 7
	Jökulhaups	Chapters 11, 12
	GRAY VOLCANOES	
	Mass wasting	Chapter 11
	Lahars	Chapters 1, 11
	Flooding	Chapter 1, 11, 12

COMMON IMPACTS

Impact category	Types of impacts	Example eruptions in text
FIRST-ORDER (Immediate)	Destruction of property	Kīlauea, 2018 (Chap. 1)
	Burns, asphyxiation	Mount Pelée, 1902 (Chap. 7)
	Disruption of infrastructure and supply chains	El Chichón, 1982 (Chap. 1)
	Death	Nevado del Ruiz, 1985 (Chap. 11)
SECOND (3rd, 4th, etc.) -ORDER (Longer-term)	Economic and social disruption	Pinatubo, 1991 (Chap. 1)
	Famine	Eldgjá, Iceland, 934 CE (Chap. 13)
	Dislocation of ecosystems	Katmai, 1912 (Chap. 13)
	Climate change	Krakatau, 1883 (Chap. 13)
	Death	Vesuvius, 79 CE (Chapter 7, 8, 13)

Now let's consider a few additional hazards, as case examples with high potential risk that have historically caught people by surprise:

Volcanic ash hazards to aviation The advent of the jet age greatly increased the hazards posed to the aviation industry by major ash-forming eruptions. Large explosive eruptions send vast amounts of fine-grained volcanic ash and aerosols as high as 40 km into the stratosphere, where they may persist for weeks and completely circle the globe (Chapter 7). Volcanic ash was not a major problem for commercial aviation prior to the jet age, since internal combustion engines ingest less air and operate at lower temperatures than do turbine engines, and prop-driven aircraft fly at lower altitudes and speeds. The new risks to commercial aviation were first driven home to me (JPL) in late June 1982, when I inspected a damaged British Airways 747 at Jakarta airport just after it had made an emergency landing there following a direct encounter with an eruption plume from Galunggung volcano [129] 150 km to the southeast. The windshield of the aircraft was completely frosted, and the leading edges of the wings and engine nacelles were stripped of paint. Each of its engines had failed in flight before restarting, were seriously damaged, and each had to be replaced. The experiences of the 263 passengers and crew aboard this flight that night describe what it is like to fly through an ash cloud, reporting static electrical discharges surrounding the jet's wings as cabins fill with fine ash and sulfur gases and engines start to fail, terrorizing the occupants as the aircraft begins dropping (Tootell 1985). This encounter with Galunggung ash was not unique; a second 747, also carrying over 200 passengers, encountered one of Galunggung's ash clouds a few weeks later. The second plane, like the one I inspected, also lost power to all four engines. Engines were restarted only after the aircraft lost several kilometers of altitude, enabling each plane to make emergency landings at Jakarta Airport.

Apart from the externally visible damage of impact abrasion from ash particles, more serious damage occurs within jet engines because they operate at temperatures greatly exceeding the melting point of volcanic ash. Ingested ash turns into a molten spray within engine combustion chambers, which is then deposited on turbine blades and other cooler areas, affecting engine operation. Sulfur gases cause additional metallurgical damage that can result in weakened metals and subsequent failure.

The international aviation community was greatly alarmed by the realization that this previously unrecognized severe hazard could cause expensive damage and future tragedies, but it took a near-disaster in 1989 when another B-747 flew into an ash cloud from Redoubt volcano [23], Alaska, to galvanize a response. The KLM jet lost power in all four engines and descended almost 5 km before restarting engines just before impact in mountainous terrain. That plane made an emergency landing in Anchorage, but suffered over $100 million in damages (Miller and Casadevall 2000). This time, the international aviation community got serious, and realized that scores of highly trafficked international air routes are located downwind of hundreds of potentially explosive volcanoes throughout the world (Casadevall and Thompson 1995). The urgent need for an effective warning system, highlighted by these near-disasters in Indonesia and Alaska, motivated extraordinary international cooperative efforts to warn pilots about inflight ash hazards and to predict the movement of ash plumes. These efforts, largely led by the International Civil Aviation Organization (ICAO) with critical input from volcanologists, has resulted in the establishment of nine Volcanic Ash Advisory Centers (VAACs) around the world for around-the-clock operations to link volcano observatories, meteorologists, satellite monitoring agencies, and air traffic control centers. These VAACs must coordinate input from the different agencies and provide inflight warnings to threatened pilots in near real-time, since ash plumes from large eruptions can reach the 9–11-km-high altitudes at which jets travel in only a few minutes, and can jeopardize hundreds of flights downwind of these volcanoes within a few hours. The VAAC system has worked well, and although expensive encounters between aircraft and ash

FIGURE 14.3 Ash plume drifting downwind from Eyjafjallajökull volcano [93] on 7 May, 2010, aboard NASA's Terra satellite passed just east of the Eyjafjallajökull volcano mid-morning on this date and viewed its ash plume for about 600 km (373 miles) downwind with the multi-angle imaging spectro radiometer (MISR) instrument. Source: NASA/GSFC/LaRC/JPL, Credit: MISR Team.

continue, no new near-disasters like those of 1982 and 1989 have occurred. The threats posed by volcanic ash to aircraft are now well known, but their economic impact on global transportation systems remains severe, as was demonstrated by the 2010 eruption of Eyjafjallajökull volcano [93].

Eyjafjallajökull is a glacier-covered volcano in southern Iceland. It had not erupted in nearly 200 years when seismic unrest was noted in February 2010. Eyjafjallajökull was coming back to life. An eruption began in March, and glacial melting produced damaging jökulhlaups (Chapter 12). Eruption intensity increased on 14 April, and ash columns reached 9 km height, sending dense ash plumes downwind to the east (Figure 14.3). This was a relatively small eruption (VEI=4) but its impact on global aviation over the 15–20 April period was severe. Although the ash-laden eruption plume only reached about 9 km height, ash was dispersed eastward for more than 2,500 km, directly impacting an area of over 7 million km^2 – all of western Europe (Gudmundsson et al. 2012). Over 300 airports in about two dozen European countries, and a correspondingly large airspace, were closed for nearly a week. This resulted in massive impacts on air travel worldwide, as flight cancellations rippled through the global reservation system. Over 100,000 flights were cancelled, affecting 7 million passengers and resulting in US$1.7 billion in lost revenue to airlines (Oxford Economics http://www.oxfordeconomics.com/FREE/PDFS/OEAVIATION09.PDF).

In addition to transportation impacts, accumulating tephra poses a wide range of threats to electrical and communication systems. The most common form of disruption is blackouts from insulator flashover as ash coats transmission lines and equipment (Wardman et al. 2012).

Volcanogenic tsunami

While only 8 percent of the 7900 historically reported eruptions have occurred from submarine volcanic vents, they have caused about 20 percent of all recorded volcano fatalities (Mastin and Witter 2000). **Tsunami** (Japanese for "great harbor wave") are giant sea waves, so large that their wave movements extend downward to the ocean floor. The depth of the sea, in fact, controls their speed. Individual wave sets may travel through the deep open ocean as fast as 800 km/h (compared with a maximum speed of about 100 km/h for ordinary wind-driven waves). The wavelengths of tsunami are as much as 160 km, with amplitudes of only a few tens of centimeters far out at sea (Dudley and Lee 1998). Hence, passengers on ships far from land might not notice the passage of tsunami waves beneath them. Near shore, however, all of this enormous energy is concentrated, causing wavelengths to shorten and wave crests to rise. In inlets and bays, additional wave funneling creates abnormally high waves that may run kilometers up coastal valleys as they expend their energy before receding. Most tsunami result from undersea tectonic earthquakes that cause vertical displacements of the seafloor or generate submarine landslides. These great waves will be generated only when the morphology of the seafloor below changes suddenly.

Volcanogenic tsunami are generated when large volumes of seawater are suddenly displaced by submarine or littoral eruptions or by collapse of volcanic edifices beneath or adjacent to the sea (Figure 14.4). History records over 90 volcanogenic tsunami beginning with Pliny the Younger in his account of the 79 CE Vesuvius [109] eruption, who reported that water "withdrew" from the shore of the Bay of Naples, stranding fish and other marine creatures before a resurgence of the sea. During the 8 May, 1902, eruption of Mount Pelée [87], Martinique, the pyroclastic flows that obliterated the city of St Pierre also entered an adjoining bay, generating small tsunami that entered the harbor of Fort de France, 25 km away.

Of the approximately 60,000 tsunami fatalities related to historical volcanic activity, most were caused by just two incidents: In May 1792, after nearby eruptive activity, a steep dome making up part of the Unzen volcano complex in southern Japan partially collapsed (Chapter 11), forming a voluminous debris avalanche that reached the sea, setting up a series of giant tsunami waves.

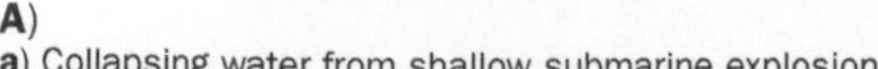

A)

a) Collapsing water from shallow submarine explosions

b) Land slides from volcanic edifices into, or beneath the sea

c) Submarine volcanic earthquakes and caldera collapse

d) Pyroclastic density currents moving out across the sea

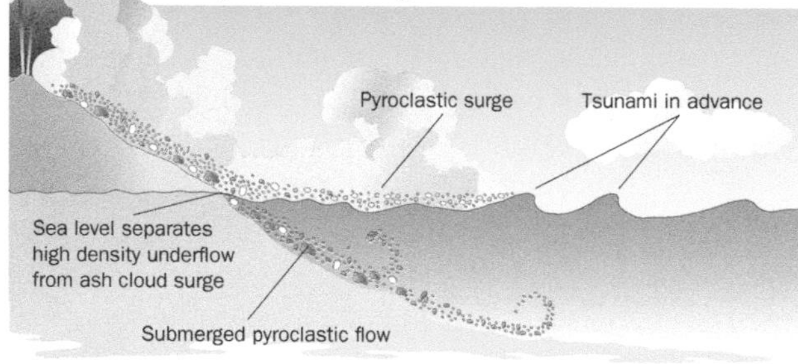

B)

FIGURE 14.4 Some causes of volcanogenic tsunami. (A) Four common tsunami generation mechanisms. (B) 15 January, 2022 eruption plume above the Hunga Tonga -Hunga Ha'apai [1] submarine volcano as recorded by the Japan Meteorological Agency Himawari-8 satellite, courtesy of the Cooperative Institute for Meteorological Studies, University of Wisconsin – Madison. This image, obtained one hour after the initial hydrovolcanic explosion, shows a 200 km diameter plume umbrella forming about 20 km above the ocean surface. High pressure gravity waves have already begun global transmission and are generating "Volcano-tectonic" tsunami on the ocean surface below. A separate set of local tsunami is impacting the islands of Tonga at the same time.

The waves destroyed 17 seaside villages, mostly across a bay from the destroyed town of Shimabara. Damage was worsened by the fact that the disaster coincided with a spring high tide. 15,000 people died, and 80 km of coastline was devastated. In some inlets, wave heights reached nearly 60 m above mean sea level, a record for known historic volcanic tsunami (Dudley and Lee 1998). Mass-wasting of volcanoes near the shorelines of volcanic islands such as at Unzen is a rare, but extremely dangerous volcanic hazard. Ferrer et al. (2021) have documented five major prehistoric events on the Canary Islands that have sent tsunami hundreds of meters above coeval sea levels. Ward and Day (2001) speculated that the next eruption on the Island of La Palma could trigger a major island collapse that would cause catastrophic tsunami damage throughout the northern Atlantic Ocean (extending to the coast of Florida!), but the 2021 eruption has not caused any tsunami yet, and the magnitude of the threat was challenged by Gisler et al. (2006).

An even more disastrous volcanic tsunami occurred during the 1883 Krakatau [128] eruption in Indonesia. On 26–27 August, when the culminating phase of the eruption took place, gigantic waves pounded the shorelines of the Sunda Strait, killing more than 36,000 people, all of them living near or close to the coast. One of the loveliest ports-of-call on the marine route from Europe to Australia was Anjer, a colorful trading center with a dominantly Sundanese, Chinese, and Arab population on the western shore of Java, about 40 km from the volcano. The town was built at the end of a bay constrained by two ranges of hills, with a low agricultural plain stretching inland; a fine tsunami funnel in other words. On a Sunday afternoon, Plinian activity began on Krakatau island across the water, causing loud concussions and ground shaking throughout Anjer and other districts rimming the Strait. A black eruption column could be seen rising off the distant island, which soon spread to darken the sky directly above the town. One resident on the Java coast reported what happened a few hours later (quoted from Simkin and Fiske 1983, p. 73):

> *Before daybreak on Monday, on going out of doors, I found the shower of ashes had commenced, and this gradually increased in force until at length large pieces of pumice-stone kept falling around. About 6 a.m., I was walking along the beach. There was no sign of the sun, as usual, and the sky had a dull, depressing look. Some of the darkness of the previous day had cleared off, but it was not very light even then. Looking out to sea I noticed a dark object through the gloom, traveling toward the shore. At first sight it seemed like a low range of hills rising out of the water, but I knew that there was nothing of the kind in that part of the Sunda Strait. A second glance – and a very hurried one it was – convinced me that it was a lofty ridge of water many feet high, and worse still, that it would soon break upon the coast near the town. There was no time to give any warning, and so I turned and ran for my life. . .. In a few minutes I heard the water with a loud roar break upon the shore. Everything was engulfed. Another glance around showed the houses being swept away and the trees thrown down on every side. . .. Struggling on, a few yards more brought me to some rising ground, and here the torrent of water overtook me. I gave up all for lost, as I saw with dismay how high the wave still was. I was. . . taken off my feet and borne inland [by the current]. . .. I remember nothing more until a violent blow aroused me. Some hard firm substance seemed within my reach, and clutching it. . .. I found myself clinging to a coconut palm-tree. Most of the trees near the town were uprooted and thrown down for miles, but this one. . . had escaped and myself with it. . .. The huge wave rolled on, gradually decreasing in height and strength until the mountain slopes at the back of Anjer were reached, and then, its fury spent, the waters gradually receded and flowed back into the sea. The sight of those receding waters haunts me still. . .. As I clung to the palm-tree, wet and exhausted, there floated past the dead bodies of many a friend and neighbor. Unless you go yourself to see the ruin you will never believe how completely the place has been swept away.*

This tsunami disturbed tide gauges as far away as La Havre, France, 16,000 km distant. Its debris, including trees packed with pumice and human remains, washed up on shorelines up to 4000 km from Krakatau for months following. But what caused this calamity? Volcanologists have been divided about the matter. There are several possibilities. One obviously is the fact that over half of the island of Krakatau vanished at this time, engulfed by caldera collapse that created a large crater on the seafloor. Perhaps huge landslides off the scarp of Rakata volcano, the truncated remnant of the island, or submerged vertical fault rupture and mass wasting also played a role, as did the impact of PDCs on the sea surface (Latter 1981). At 15 localities around the Strait, including Anjer, the initial shoreline water movement was toward land, not the withdrawal that would be expected from seawater suddenly flooding a cavity opening in the seafloor. This suggests to some that the outward displacement of the ocean by submarine explosions caused the big waves. But not all localities showed initial tsunami insurgence (Dudley and Lee 1998). In fact, the Dutch engineer Verbeek (1885) described shoreline water as initially receding in many places, with currents flowing toward the volcano in the open Strait. One modern view suggests that the eruption's pyroclastic activity, as at Mount Pelée [87], was the main trigger (Yokoyama 1981). Because of density differences, the dense basal ash flow components may sink and travel submerged along the sea bed, displacing large amounts of water like a submarine landslide, while the lighter ash cloud surge races out across the surface, also shoving water ahead of it like a powerful wind fan (Figure 14.4 d), while generating large amounts of steam. How far can a pyroclastic surge race across open water? It is certain that a huge amount of ash and pumice traveled at least 10 km laterally from the vent, building up two temporary islets in the sea, originally 35–40 m deep (Verbeek 1885). At Katimbang on the southeast cape of Sumatra, 40 km from Krakatau, a Dutch official and his family barely escaped with their lives while the ordeal in Anjer was unfolding. With a 15-m-high tsunami striking the shore, they fled inland and took refuge in a mountain hut, but were seriously burned by scalding ash gushing through openings in the floor of the shelter. Given their account, it seems certain that an energetic surge traveled across the ocean the 40 km between the vent and Katimbang, ultimately ending on the flanks of Raja Bassa Mountain where the Dutch official and his party had taken shelter (Self 1992; Simkin and Fiske 1983; Yokoyama 1981). Despite questions about the exact origin of the Krakatau tsunami, it is certain that underwater volcanic explosions, in freshwater lakes as well as the sea, have the potential to generate tsunami (Mastin and Witter 2000). Tsunami created by Krakatau continue tragically to impact residents of nearby shores. A sector collapse of Anak Krakatau in 2018 killed 426 people (Ye et al. 2020). With these examples in mind, the populated west coast of Italy has been declared vulnerable to a potential sector collapse of the active submarine volcano Marsili just offshore (Gallotti et al. 2021).

Another, less common type of volcanogenic tsunami, a "**Volcano-meteorological tsunami**" as defined by Lowe and de Lange (2000) was generated by a 15 January, 2022 eruption in Tonga. Before 2014, the islands of Hunga Tonga and Hunga Ha'apai were subaerial rim segments of a massive submarine volcanic caldera. Moderate eruptive activity in 2015 built a new landmass between them, named by using the combined names of the older islands. A new episode of intermittent eruptive activity in and near a new crater began in mid-December 2021. The intensity of the Surtseyan explosions increased by mid-January 2022, and on 14 January, a major subaerial explosion removed the newly-built land, forming a 20 km high plume column with a 240 km diameter mushroom-shaped cap. The volcanic vent, source of the eruptions, continued underwater activity. The next day, a major submarine eruption, the largest anywhere on Earth since 1883, produced extremely powerful hydrovolcanic explosions that generated a Plinian plume, probably exceeding 35 km diameter in its central part and capped by a nearly 600 km umbrella (Figure 14.4B). This eruption caused damaging ashfalls on nearby islands, soundwaves that were measured worldwide, and tsunami that caused unexpected damage around much of the Pacific rim. The unforeseen maximum amplitudes of the tsunami affecting coastlines around the Pacific and other oceans were not much smaller than those affecting nearby islands, an anomalous situation that was likely caused by the influence of large-amplitude gravity waves propagating around the planet through much of the troposphere, transferring energy to most of the Pacific Ocean surface.

Carbon dioxide hazards This section on CO_2 hazards may seem over-emphasized, in light of the fact that other volcanic processes have killed far more people. More than 150 people were killed by CO_2 on the Dieng Plateau, Indonesia [130] in 1979 (LeGuern et al. 1982), and at least 1800 people died in 1984 and 1986 downslope from Lakes Monoun and Nyos [107], Cameroon, when deep volcanic lakes overturned and released vast amounts of exsolving CO_2 (see the following section). The dangers of CO_2 emissions are, however, greatly underappreciated by most volcanologists and others who work around active volcanoes, and we feel the risks involved are worth discussing. Dozens of people are poisoned by CO_2 around volcanoes each year, but the causes of their mysterious deaths are commonly unrecognized, or are incorrectly attributed to "asphyxiation."

Although CO_2 is an essential minor component of Earth's normal atmosphere, critical to regulate pulmonary functions in animals, and is a food source for plants, it is poisonous to plants and animals when concentrated. It normally comprises about 0.4 percent of the atmosphere, although its concentration is slowly increasing owing to human activities. At total atmospheric concentrations above 3–4 percent, serious physiological effects are noted in humans, and the immediately dangerous to life or health (IDLH) concentration has been set at 40,000 ppm (NIOSH 1994). The anesthetic properties of carbon dioxide have long been noted by medical science, and humans will lapse into unconsciousness when CO_2 exceeds about 6 percent of the air they breathe. At CO_2 levels above 8–10 percent (varying with individuals), neurological reactions will cause respiratory failure and death. People who unwittingly venture into areas of concentrated CO_2 and die are thus not "asphyxiated," since there is adequate oxygen available – they are poisoned.

CO_2 is one of the three most abundant volatile constituents of magmas – along with H_2O and SO_2 (Chapter 3). It is a highly mobile gas, and normally diffuses harmlessly to the atmosphere over the surfaces of volcanoes either by dispersed migration through rocks and soil, or as a fume component at eruptive vents and fumaroles. So long as the CO_2 is dissipated over a wide area, no hazard is created, but when emissions are concentrated or the gas is dissolved in the waters of deep lakes, hazardous concentrations may result. Concentrations of CO_2 in fumaroles or solfataras are usually low enough not to form significant hazards. CO_2 is the dominant gas in the less common fume vents called mofettes (Chapter 3; Pfantz 1999), however, where it may constitute more than 90 percent of the gases released. Mofette gases are typically cool, usually as cool as the surrounding air. When no wind or breezes are present, CO_2 may accumulate in hollows or small valleys, since it is a heavier-than-air gas, 1.5 times as dense as normal atmosphere. Because CO_2 is a colorless, odorless gas, it may not be recognized as a hazard until too late.

Animals wandering into such pools of CO_2 may quickly lose consciousness and die (Figure 14.5). A modern "death gulch" of this sort was found by early geological explorers in the Absaroka Range near Yellowstone National Park [47], and described by Thomas Jaggar (Chapter 1) in one of his early publications. Although wind circulation cleared out the CO_2 before the men entered the gulch, the bodies of several grizzly bears indicated the deadly potential of the place. More recently, the shallow rise of magma under Mammoth Mountain Volcano [43] in east-central California, next to a popular ski resort, increased CO_2 content in the soils from an

FIGURE 14.5 Skeleton of dog in mazuku on the slopes of Nyamuragira volcano [117], Democratic Republic of Congo. Skeletons of vultures and other carrion-eating animals were also abundant. Source: USGS photo by J.P. Lockwood.

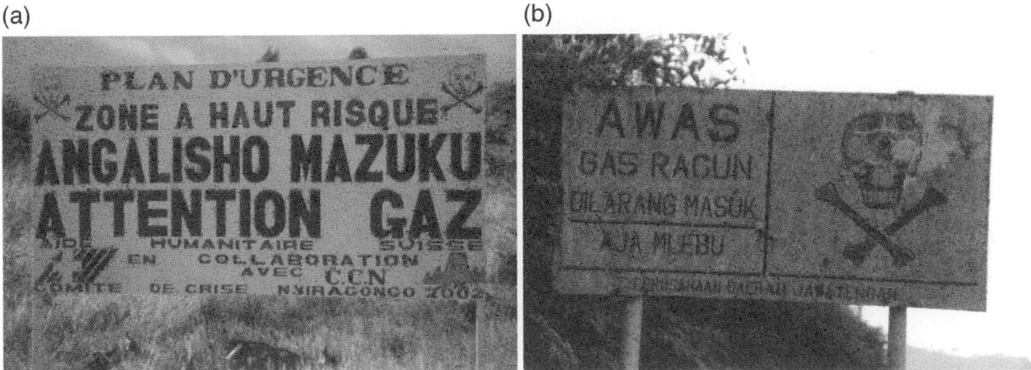

FIGURE 14.6 Carbon dioxide mofettes can be deadly places. Despite warning signs, many people die in these low-lying places every year. (a) Warning sign above a Nyiragongo volcano [118] "mazuku" on the outskirts of Goma, Democratic Republic of Congo. (b) Warning sign below mofettes on the Dieng Plateau [130], Indonesia, in an area where 149 people were killed in 1979. Source: Photos by J.P. Lockwood.

ambient level of 1.5 volume percent of soil gas to 30–90 percent, killing the forest in places, and leading to nausea and dizziness for people staying in local mountain cabins (Chapter 3). Three skiers were killed on Mammoth Mountain in 2006 when they fell through snow into a concealed mofette. Excavations near or beneath volcanoes are especially dangerous, and many people are killed in such places by accumulations of CO_2, most recently beneath Teide volcano [99] in the Canary Islands, where six persons perished while exploring water tunnels beneath that volcano. Volcanic craters are especially hazardous places on windless days because they are closed depressions where CO_2 can concentrate, and dozens of people have died exploring craters within volcanic cones in Japan and at Rabaul volcano [160].

Carbon dioxide mofettes are widespread along the East African Rift Zone, and are called **mazuku** in Swahili. The mazuku are places known for "evil winds" by local people, and are low-lying areas where dense CO_2 accumulates on windless days. Mazuku are frequently associated with sources of water, and in East Africa, the myth of "Elephant Graveyards" is based on the fact that bones of elephants and other large mammals have been found in large concentrations in areas where mofettes are associated with volcanism. Mazuku are common around Nyiragongo [118] and Nyamuragira [117] volcanoes in the eastern Democratic Republic of Congo, and similar mofettes are also deadly places in Indonesia and Kamchatka (Figure 14.6). Where not mixed by breezes, the contact between the denser CO_2 and overlying air is commonly sharp, and one can see such boundaries because of the change in refractive index, and feel the higher density and viscosity of the gas. Because of its density, only places near the ground may be affected. Adults were able to safely walk across the ground with flowing CO_2 currents at Dieng [130] in 1979, while most children collapsed and died, as did parents who stooped to save them. Intentionally breathing to test the potency of a Kamchatka mofette (Dolina Smerti "Valley of Death") in 1990 (while a friend stood by to retrieve me!) I (JPL) quickly became dizzy and almost lost consciousness, but breathing normal air quickly revived me. Volcanologist Tom Miller had his encounter with CO_2 poisoning in the summer of 1977 a few months after the Ukinrek maar [12] eruption, at a nearby place called "Gas Rocks," known for CO_2 mofettes. While squatting down next to a CO_2 "soda spring," he suddenly lost consciousness, but was saved by colleagues standing nearby. He writes: "What surprised me in particular was how fast and with no warning this all took place!" (Miller 2008, pers. comm.).

Survival Tips for Field Volcanologists

Craters or other closed depressions on young volcanoes can become "death traps" on windless days if substantial CO_2 has accumulated. CO_2 is a colorless and odorless gas and gives few warnings of toxicity. Whenever descending into craters or closed depressions on windless days – be careful!

Local residents learn where mazuku are located and how to avoid them, but during the Rwandan refugee crisis of 1994, many passing travelers lost their lives in mazuku when looking for a place to sleep. Normal breezes will quickly dissipate the CO_2 and make the mazuku temporarily safe, but if the wind dies down, these places become deadly. Basements can also accumulate hazardous levels of CO_2 in volcanic areas, and one enterprising factory owner I (JPL) met near Goma controlled rats by throwing food scraps into a deep hole excavated into soil beneath his basement!

Hazards associated with volcanic lakes There are three principal types of volcanic lakes: those formed by the damming of streams or rivers by lava flows, debris flows, or PDCs; those occupying maars or other old, inactive craters; and those occupying the summit craters of active volcanoes. The first type form on the flanks of volcanoes, and pose hazards of flooding and lahars as vast amounts of impounded waters may be suddenly released when these dams are overtopped by rising waters. Pyroclastic surge deposits that dammed a pre-existing stream in Cameroon impound 50×10^6 m^3 of water on the edge of Lake Nyos maar [107], and pose a significant hazard to populated areas downstream (Lockwood et al. 1988). Maar lakes are especially problematic because they are commonly associated with sources of magmatic carbon dioxide and may entrap and dissolve large amounts of CO_2 in their deep waters. This is not a problem in temperate climates, since seasonal temperature changes cause annual lake overturning that will release stored CO_2 before it builds to dangerous levels. In tropical climates, however, especially near the equator, the lack of seasons may preclude natural lake overturning, and quasi-stable lakes may allow dissolved gas to build up to dangerous levels, posing the threat of a catastrophic release of lethal amounts of poisonous CO_2. The 1986 gas release from Lake Nyos maar, Cameroon (Figure 14.7) is a tragic example that killed more than 1800 people (Kling et al. 1987). Lakes that occupy the summit craters of active volcanoes are the most hazardous volcanic lakes of all. Summit crater lakes are particularly important to understand because they serve as chemical traps above magmatic systems, and may accumulate the heat and magmatic volatiles that would normally be released to the atmosphere. Such lakes are commonly extremely acidic, because of the ingress of SO_2 and HCl gases from the underlying magmatic systems. Eruptive activity that might result in minor explosions and degassing activity at dry craters may result in the expulsion of large volumes of lake water and the generation of massive lahars – of the sort that have killed thousands of people on the flanks of Kelud volcano [132] (Indonesia) and at Ruapehu [178], (New Zealand). Means to mitigate the risks associated with each of these three lake types are discussed later in this chapter.

Non-eruptive volcanic hazards Volcanoes can cause widespread destruction long after eruptive activity has ceased. This includes the threat of volcano sector collapse and lahar generation discussed in Chapter 11, but also the long-term hazard of rockfalls and slope collapse events unique to eroded volcanic terrains. Volcanic deposits are inherently gravitationally unstable, owing both to their mechanically heterogeneous nature, and also to the fact that even homogeneous lava flows are typically riven by cooling fractures and are susceptible to mechanical failure. Methods to evaluate these hazards and risk are discussed by Gonzalez de Vallejo et al. (2020) based on the analysis of rockfall events on the Island of Tenerife.

Risk

Since 1700 CE, more than a quarter of a million people have perished in volcanic eruptions, almost a third of them during the nineteenth century (Simkin et al. 2001). However, volcanic eruptions, despite their violence, are not at the top of statistical lists of hazards. Traffic accidents in the United States alone usually kill many more people every year than volcanoes do worldwide. Although the risk to human life posed by volcanic eruptions appears to be less than that of other hazards such as earthquakes, tsunami, and floods, that perception is based on the human experience of only the past few hundred years – the period of modern communications. No "Super-eruptions" (Chapter 8) have occurred near large population centers during this time, yet such highly infrequent

FIGURE 14.7 Lake Nyos, Cameroon [107], eight days after the tragic overturn of 21 August, 1986. More than 1800 people were killed by massive amounts of carbon dioxide gas that flowed downslope into populated areas after being released by overturning of the lake waters. The brown color of the water is caused by the oxidation of the iron-rich deep waters that mantle the lake surface. Note the bare cliffs in the background, scoured of vegetation by water fountains and seiches that accompanied the overturn event. Source: USGS photo by J.P. Lockwood.

events will still occur in the future, and have the potential to claim millions of lives while altering the earth's climate in disastrous ways (Chapter 13). Over 100 million people now live within areas that have been devastated by prehistoric PDCs adjacent to large calderas. If not forewarned and evacuated, there will be no survivors when the next large caldera-forming eruption takes place in a populated region. An uncomfortable reality for volcanologists is the fact that pre-eruptive evacuation of large populations at risk would likely prove infeasible, owing to the imprecision of probable eruption forecasts (Donovan and Oppenheimer 2016). General forecasts such as "There is a high probability that [your city] might be destroyed within the next few months" cannot be used to justify the great economic costs involved, as our present prediction capabilities do not allow 100 percent certainty about the timings or magnitudes of such caldera-forming eruptions.

Volcanic risk to humans and their property is expressed by most people in relatively subjective terms – for example: "People shouldn't live there – that volcano could erupt at any time," or "I think it's safe to build a town there – the volcano hasn't erupted in hundreds of years." But, volcanologists must describe risk in more quantitative terms for decision-making officials and emergency planners, and methods for more rigorous evaluation of risk is the subject of this section.

Fournier d'Albe (1979) suggested that economic aspects of volcanic risk could be qualitatively evaluated according to three general parameters:

$$\text{Volcanic risk} = \textit{Hazard} * \textit{Value} * \textit{Vulnerability} \tag{14.1}$$

"**Hazards**" are the various volcanic phenomena, already introduced, that could possibly impact a given area (lava flows, tephra fall, lahars, etc.), and cannot be quantitatively defined in numerical terms. An objective, quantitative evaluation of risk nonetheless requires measurable parameters. Assuming volcanologists have been able to describe the volcano's past eruptive history, the hazard factor in the above equation is the numerical value of the probability that a "HAZARD" (a particular volcanic phenomenon) occurs and can be statistically estimated in terms of the odds of the hazard(s) impacting a particular area in a specified interval of time.

Values are relatively straightforward for property in the area of interest, but also include the subjective value of human lives at risk and of their social stability. For quantitative assessments, this parameter may be measured as the number of human lives at stake, or the area's total capital value and productive capacity (Fournier d'Albe 1979; Peterson 1988).

Vulnerability is a composite parameter related to the ability of the area in question to withstand the impact of likely volcanic hazards. The lower the ability, the greater the damage caused by a hazardous phenomenon. In their analyses, New Zealanders call this **fragility**. Are the buildings strong enough to survive ash fall loads? Are the residents well informed about volcanic hazards and capable of evacuating the area if needed? Has there been good land-use planning to restrict development in the most vulnerable areas? Consistently with this conceptual definition, we may quantify vulnerability as a measure of the proportion of the value likely to be lost as a result of a given volcanic event.

Risk, as defined in Eq. 14.1, thus establishes a relationship between nature and society. Concerning the natural component, we can measure the hazard factor assuming that essential information of the future behavior of a volcano or volcanic region is contained in its past behavior. However, we cannot change it; no foreseeable technology hints at the possibility of controlling volcanoes. On the other hand, human intervention may reduce or even preclude the loss of value through the reduction of vulnerability. **Risk mitigation** is then feasible, and should not be all that difficult to accomplish once people realize that hazardous phenomena may affect all of the components of society in different ways. The management of risk and the reduction of vulnerability thus requires an organized response of all sectors of an endangered population with common perceptions of each component of risk. Objective and quantitative appraisal of risk factors is essential for building such perceptions. Risk mitigation may be quantified as a "Preparation" parameter, referring to the investment in a series of measures to reduce the vulnerability. It makes sense that the overall risk is reduced as proper measures to reduce vulnerability are implemented.

To depict the role of "preparation," the above Fournier d'Albe equation may be rewritten as follows (De la Cruz-Reyna and Tilling 2008):

$$\text{Volcanic risk} = Hazard * Value * (Vulnerability - Preparation) \tag{14.2}$$

Statistics and Calculating Probabilities

In the previous section, we mentioned that "hazards" are quantifiably measured in terms of probabilities. This assumes that we can envision some general features of the future behavior of a volcano or volcanic region by carefully looking at its past behavior. Indeed, it is common sense that geologic studies of a volcano's eruptive history are needed for evaluating volcanic risk. The challenge is how to justifiably foresee future volcanic behavior given such information. The idea that the volcano will reproduce *exactly* what it did in the past defies experience after all. And the fact is that no eruptions can be precisely predicted yet, making it more challenging to foretell the future.

You may think that if we could only measure all the physical, chemical, and thermodynamic parameters within a magmatic system, we could construct a framework of useful equations describing the dynamics of the system, making it possible for us to predict the time and size of the next hazardous eruption. This would be a **deterministic approach** providing "hard" data for risk managers. Unfortunately, the present scientific and technological state-of-the-art has not yet reached this capability (although we may not be far from that), mostly because we do not have access to any active magmatic system to measure in real-time all of the parameters necessary to understand it fully. We surely can measure some parameters indirectly from the surface, such as depth and approximate volume of an active magma body, but the number of measurable parameters is generally less than the number of equations required to make a deterministic prediction of its behavior, and there is the additional problem of the non-uniqueness of **inverse problem solutions** – put simply, the signals we can detect above a magma reservoir may have several possible causes and be subject to different interpretations. Hence, a deterministic mathematical model for an evolving magmatic system is usually "underdetermined," providing at most *a wide spectrum of possible solutions*. What is left to us is a different, probabilistic approach in which some educated guesswork is ordinarily necessary. One of the main goals of statistical volcanology is how to refine that guessing, to make it valuable in a practical way for communities concerned about future risks.

The first step in formalizing the "guesswork" that goes into risk determinations is looking for a good way mathematically to describe a volcano's hazardous volcanic activity given *available information* – which is rarely complete. A simple way to do that is to represent eruptions (or any of their hazardous manifestations) as "events" occurring along an historical timeline. Optimally, the durations of the events should be short compared with the time between them. Then, they may be represented as a succession of points along a time axis. An observer looking back from the present sees a succession of events taking place as time unfolds. The time gaps (repose intervals) between successive events are a random aspect, since if we witnessed any given event, we could not know for sure when the next one would take place. Because of this, the entire sequence of events may be characterized as a **random** (or **stochastic**) **process**. Or alternatively, because many magma reservoirs tend to replenish themselves, we could call this a **renewal process,** since after each event, the system "renews" its state in preparation for the next event. For such a process, we can average the intervals between past events and assume that this pattern will be maintained in the future. The longer our record of past events, the more confident we can feel statistically about the future pattern. In other words, we develop a **time-averaged basis** for calculating the probabilities of future hazardous events, as we'll shortly illustrate with an example from Mexico.

The widely applied time-averaged approach should be regarded as providing "soft" data about future hazards, since there is no absolute certainty (100% probability) in them. It is important here to emphasize, however, that there is no conceptual conflict between the deterministic and the probabilistic approaches. The "hard" or deterministic approach, if achievable, would tell us when the next volcanic activity is with a high degree of certainty, while a probabilistic approach provides a likelihood of ensuing events with a degree of certainty that depends on the size and quality of the historical sample we are analyzing. Either way, we should be able to calculate the same likelihood of having one or more eruptions in a given time interval, or even the likelihood that *no eruptions* will occur at all within a time frame of interest, given what we know of "average" behavior in the past. Such information may be useful on a long-term sense, for example for insurance companies which have a great need to quantify risk to establish insurance rates; or short-term, as during an actual volcanic emergency, when volcanologists may be called on by government agencies to provide reliable forecasts of changing eruptive behavior (Donovan and Oppenheimer 2012).

During an individual volcanic crisis, there are usually much-increased available data from monitoring, and a deterministic element could be added to a stochastic approach based on the analysis of similar previous eruptions taking place at the volcano. A team of volcanologists can develop a series of possible eruptive scenarios as the crisis begins, each with a somewhat different probability of unfolding, based upon a deterministic understanding of the volcanic system. This is illustrated by Probabilistic Event Trees (Figure 14.12).

Some volcanologists may prefer to leave the preparation of statistics to trained statisticians, but we feel that *it is much less dangerous for volcanologists to learn a little statistics than it is for statisticians to learn a little volcanology*! Important insights on the behavior of a volcano or a volcanic area may be gained through simple statistical tools, as those discussed here. Although this may be insufficient to provide the level of detail required for commercial risk analyses or for decision-making, they can certainly help to develop a more objective view of what can be expected from a volcano in the future. Given that any risk analysis of a volcano's future behavior must be based on knowledge of its past activity, analytical reliability depends upon:

1. How long a history of eruptive activity is known for the volcano in question.
2. How many eruptions have occurred over this period.
3. What we know about the "types" of eruptions (Plinian phases, Vulcanian explosions, etc.) and "sizes" of those eruptions (VEI, emitted mass, column height, affected area, or any other measure of the magnitude or intensity).
4. The dates (based on historical accounts or radiocarbon studies) at which the eruptions occurred.
5. A reasonable certainty that such "eruptive history" is complete over the period under consideration.

A volcano with a well-documented, long history of many eruptive events will yield much better statistical results than will a volcano with a short history and few documented eruptions, or with missing data – ever a critical concern.

Consider the challenge of trying to measure the average height of the students in a school. How successful will you be if many students happen to be absent the day you take your measurements? A relative concentration of shorter or taller individuals who are attending that day will certainly bias the result. In the case of eruptions, it is more likely that smaller events may be missing in a long record; while major events, which tend to be much less frequent, may be excluded in a short record.

To address these problems one can designate a "cutoff" level for the parameter used to measure the eruption (VEI, magnitude, etc.), taking analytical account only of those eruptions that have equal or larger eruption indices. For most practical purposes, this deliberate disregard of "small-scale" volcanic events is acceptable. Since smaller eruptions with unknown size are assigned a default VEI=2 in several catalogs, and considering that such types of eruptions may not represent a major threat, it may be wise to use the eruption record for all the events with VEI>2 (which is the same as saying VEI≥3) for a statistical analysis.

Several catalogs contain valuable compilations of the eruptive histories of most active volcanoes of the world (e.g. Smithsonian Institution Global Volcanism program, https://volcano.si.edu; LaMEVE, www.bgs.ac.uk/vogripa). However, elaborating a complete chronology of a given volcano may require additional fieldwork leading to a revision of the available historical and geological records. These data then can be assembled in the way discussed above to produce a timeline for a particular volcano or volcanic area. We present here an example of a simple approach to characterize the recent eruptive activity of Volcán de Colima [51], Mexico (Figure 14.8), which has an enviably complete historical record compiled from multiple sources, in Table 14.1. In this example, we use a "counting of events" approach for the description of the random process.

Depending on the time length of the record, it should be divided into a sufficiently large number of equal intervals. Here, we divide the 460-year historical record into 46 decades, and show the number of eruptions with VEI equal to or greater than 3 occurring in each:

These data may also be represented graphically in a way that can provide a clear overview of a volcano's past behavior. The dashed black line in Figure 14.8 shows the cumulative number of eruptions with VEI≥3 over time. The blue triangles are reference indicators for each of the individual decades (1–46 in Table 14.1), related to the "years" axis at the bottom.

Figure 14.8 shows that Colima activity consists of periods of (high rate) frequent eruptions, alternating with relatively (low rate) quiescent times. These fluctuate around a "mean" with a long-term eruption rate $\lambda = 18/46 = 0.39$ eruptions per decade represented

TABLE 14.1 Eruptive history of Volcán de Colima [51].

Years (CE)	Interval number	Number of events VEI≥3
1560–1569	1	0
1570–1579	2	1
1580–1589	3	1
1590–1599	4	1
1600–1609	5	1
1610–1619	6	1
1620–1629	7	1
1630–1639	8	0
1640–1649	9	0
1650–1659	10	0
1660–1669	11	0
1670–1679	12	0
1680–1689	13	0
1690–1699	14	1
1700–1709	15	0
1710–1719	16	0
1720–1729	17	0
1730–1739	18	0
1740–1749	19	0
1750–1759	20	0
1760–1769	21	0
1770–1779	22	1
1780–1789	23	0
1790–1799	24	0
1800–1809	25	0
1810–1819	26	1
1820–1829	27	0
1830–1839	28	0
1840–1849	29	0
1850–1859	30	0
1860–1869	31	1
1870–1879	32	1
1880–1889	33	2
1890–1899	34	1
1900–1909	35	2
1910–1919	36	1
1920–1929	37	0

TABLE 14.1 (Continued)

Years (CE)	Interval number	Number of events VEI≥3
1930–1939	38	0
1940–1949	39	0
1950–1959	40	0
1960–1969	41	0
1970–1979	42	0
1980–1989	43	0
1990–1999	44	0
2000–2009	45	0
2010–2020	46	1
Total number		**18**

Sources: Adapted from Medina-Martínez (1983); De la Cruz-Reyna (1993); De la Cruz-Reyna et al. (2019).

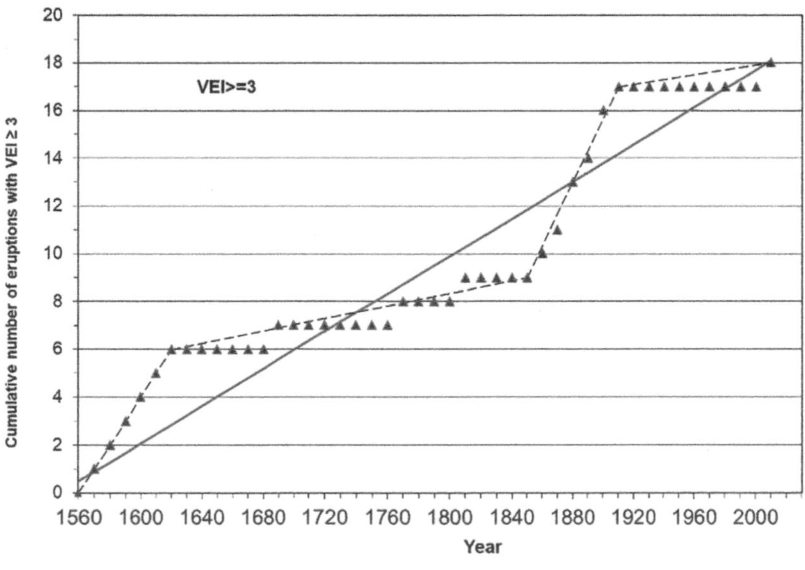

FIGURE 14.8 Graphical representation of the eruptive history of Volcán de Colima [51].

by the slope of the blue line. The inverse of this number is the **mean recurrence time *(T)*** of eruptions with VEI≥3, which is ***T*** = 25.6 years (or 2.56 decades) in this example. This value of course should never be interpreted as fixed, regular periodicity; over the long term, the repose times between Colima eruptions are definitely variable. But neither is ***T*** meaningless; this statistic is useful to long-term planners for emergency response – the "Preparation" term in Eq. 14.2. (How such information is accepted and used politically, socially, and psychologically is another story.) More details about the risks posed by Colima (Figure 14.9) and advanced statistical analysis are given by De la Cruz-Reyna et al. (2019).

Stationary and Independent Volcano Behaviors

Calculating a reliable value of eruption probability depends on the answers to several questions, in addition to the above-mentioned concerns about past eruption history. Among these, two are especially important: Is the eruption sequence process described by the time series "stationary," and are the eruptions themselves "independent" of one another?

A volcano showing **stationary behavior** is one that generates eruptions in a continuous, unchanging sequence, keeping the probability of eruptions unchanged during any specified time. We can assume that it has a stable magmatic plumbing system that operates in a continuous long-term fashion. Of course, the eruption rate may show some fluctuations over periods of variable duration, as shown in the Colima example, but *on average*, it can still be said to show stationary behavior, indicated by the straight blue line in Figure 14.8. Colima is a good example – from all we have seen so far – of a volcano showing stationary behavior, at least over its recent eruptive history. Future eruptions large enough to alter the magmatic plumbing system may change the eruption rate,

FIGURE 14.9 Volcán de Colima [51] (foreground – 3820 m) and Nevado de Colima (background – 4160 m) as viewed from the outskirts of Comala, a small town NW of Colima city, about 26 km SW of the volcano. About 350,000 people live on massive young debris flow deposits from the destruction of an older volcanic edifice, which formed a somma structure around the modern cone. Currently, about 6000 people live in small villages scattered within a radius of 15 km of Colima's summit, likely within the reach of major pyroclastic flows. These people depend on volcano monitoring, risk evaluations, and reliable warning protocols for their safety. Source: Photo by Servando De la Cruz-Reyna.

making the process-non-stationary, which would thus invalidate the use of simple statistics, like the Poisson distribution. Other unexpected future events may also have the same effects.

If the long-term cumulative curve of the eruptive sequence shows a clear increasing (or decreasing) slope, it may be a signal that the volcano is not in a stationary eruptive regime, but evolving into a higher (or lower) level of activity. However, to support this type of conclusion, it is important to determine if the observed rate fluctuations along a historical mean are just "normal" statistical variations or a real acceleration (or deceleration) of the rate and thus the eruptive level of a volcano. A sufficiently long record must be acquired to test for stationary conditions in any volcanic system.

A volcano showing **independent behavior** is said to "lack a memory," meaning that any given VEI-sized eruption does not trigger, nor prevent, future eruptions, and is in no way related to the next one taking place. The notion of "lacking memory" may defy intuition in some cases, since one may assume that each eruption at such a volcano fully taps the eruptive capacity of the source magma reservoir, in terms of energy released and material erupted. This also assumes that a new magmatic system must be established before the next eruption can take place. This is not always the case, however, even for many large eruptions, which only release a minor proportion of the energy stored in a magmatic system, keeping the volcano ready for another important event in any timescale. Think of the sequence of dome-related blasts following Mount St Helens catastrophic 1980 eruption (or Peléean eruptions beginning in 1902) clearly fed from the same batch of magma.

There are many approaches to analyze the stability and independence of past events for random behavior (e.g. Klein 1982; Davis 1986; Mader et al. 2006). However, if a random succession of eruptions does not pass the stationarity and independence-between events tests, it means that the volcanic process is too complex to be described in terms of independent, uniformly distributed events. These conditions are generally categorized as "non-homogeneous" volcanic activity. It is beyond the scope of this book to describe and discuss non-homogeneous processes, but that does not mean that the problem cannot be solved. Examples of the methods used in such cases may be found in Turner et al. (2008) and in Bebbington (2013).

Poisson analyses If the past eruptions of a volcano appear to be randomly distributed in time in a stationary fashion, and the events in a sufficiently large sample of the eruptive history seem to be independent, then the eruptive sequence may be statistically treated as a **Poisson process**. In a Poisson analysis, designated time intervals (or "trials"), as in the case of the "decades" used in Table 14.1, can possibly show only one of two things: no eruption occurs, or at least one eruption takes place. The probabilities of having 0, 1, 2, or any number of eruptions in each trial period are described by the **Poisson distribution** (De la Cruz-Reyna 1996). Constructing a Poisson distribution reveals a challenge however: As Figure 14.9 shows, the repose intervals between eruptions vary significantly. There are more short-term intervals causing clustering of eruptions (e.g. 1570–1629) than long-term ones. The actual probability of an eruption occurring during any given interval of time depends in part on this variability.

To deal with that challenge, we switch the random process approach from one counting the events *to one that analyzes the distribution of the repose intervals* between eruptions. If the number of events in a random process is describable by Poisson distribution, we can expect that shorter repose times between events will be more common than longer repose times. In fact, the proportion between shorter and longer repose time is characterized by an exponential function. The proof of this may be found in most books on statistics (e.g. Papoulis and Pillai 2002).

Hence, in a volcano showing a long-term stable mean recurrence time $\boldsymbol{T}$, calculated from a reasonably complete database, the Poisson probability of a repose time of duration t (which is equivalent to no eruptions occurring in that interval t) is:

$$P(0)=e^{-t/T} \tag{14.3}$$

Likewise, the probability that one eruption does occur within an interval t is given by the equation:

$$P(t)=1-e^{-t/T} \tag{14.4}$$

These relations render values between 0 and 1, but expressing the probability as a percentage multiplying the result by 100 is usually better understood when communicating hazards to the public.

Although it is important to calculate the probabilities that an eruption of a given type will occur on a particular volcano in the future, it is also useful for stakeholders to know the probability of serious damage happening in a particular area close to the volcano; perhaps an area being evaluated for development. A professional volcanologist may be asked the following question by a developer or insurance company in need of evaluating volcanic risk: "What are the chances that my project area could be covered by a lava flow (or impacted by some other hazard) in the future?" To provide the needed answers, a geologic map of the area in question must be made if none is available, and the ages of lava flows or timing of other hazardous events in that area determined. Selection of the area to be studied is not simple – the area must be large enough to reveal the typical flux of lava flows into areas near the property of interest, but not so large as to generalize regional hazard probabilities that do not reflect the specific risks to the area in question.

In the following box, we consider how Poisson probabilities can be calculated to evaluate the risk of lava flows that could impact one landscape – in this case, on the flank of a "red volcano" on the island of Hawai'i.

As noted earlier, eruption probabilities based on the past eruptive activity of a volcano provide a useful tool for communicating risks. But they should be used with great care to avoid misunderstandings, particularly when a hazard analysis is presented to decision-makers or to the people living near a volcano.

To illustrate the conflicting perspectives that may arise in such a situation, let us consider a hypothetical situation. Suppose in 1980 someone had polled Mexican citizens about which volcano is more "dangerous": Colima [51] or El Chichón [57]? Both volcanoes showed random past behavior and could be taken as appropriate for Poisson analyses. At that time, Colima would have been considered more dangerous – by a long shot – given its continuous fuming and recent, well-viewed eruptions. In contrast, El Chichón volcano was hardly known to be an active volcano. In Chapter 1, we describe what happened at this volcano in March–April 1982. Assuming that the eruptive histories of both volcanoes were available at that time, the calculated mean recurrence times for significant eruptions (VEI≥3) at Colima would have been 25.6 years, as established above. For El Chichón, based on the dating of eight major prehistoric explosions over the past 3070 years, it would have been 384 years (Mendoza-Rosas and De la Cruz-Reyna 2010). Simply using these quotients, we would have said that the probability of a new eruption at Colima within the 1980–1989 (or any other decade) decade was 32.7 percent while for El Chichón it was only 2.6 percent. Strictly speaking, these are correct values.

In this case, it was El Chichón that erupted catastrophically in that decade. The low eruption rate of El Chichón was one of the reasons it was underrated and ignored. Consider this, however: The most recent major eruption of El Chichón (pre-1982) took place about 700 years ago. Equation 14.3 would have told us that the probability of an eruption occurring at this volcano over a 630-year span of time was only 20 percent. Equation 14.4 tells us that the probability of El Chichón erupting *after* a 630-year-long repose interval was nearly 80 percent. Each calculation presents a wholly different risk perspective. One describes the probability of an eruption occurring in any given decade, while the other measures the probability of having a too long repose interval. Statistics may not "lie" mathematically, but we have to be careful about which we elect to use. The corollary is that how a probability is determined must be clearly explained and its limitations discussed with decision-makers.

Lava Flow Risk on Hualālai Volcano

Hualālai [14], one of three active volcanoes on the island of Hawai'i, has erupted more than 75 lava flows over the past 5000 years (Moore and Clague 1991). No area on its slopes is completely devoid of volcanic risk. In one recent case, a developer on the western flank of the mountain needed to answer a general question for financial purposes: "What are the odds that my proposed project will be impacted by a lava flow in the future?" To answer this question, it was necessary to prepare a geologic map of all the lava flows that had come close to this facility. "Close" was arbitrarily defined as any lava flow that came within about 5 km of the site (Figure 14.10). Six radiocarbon-dated lava flows were later determined to have entered the area over the past 4700 years in a temporally random fashion (Figure 14.11). The "recurrence interval" for lava flow impact (***T***) is thus 783 years. The probabilities (*P*) that flows will enter this area again in the future for various future time intervals (*t*) may be calculated from Eq. 14.4 above, and are given in Table 14.2.

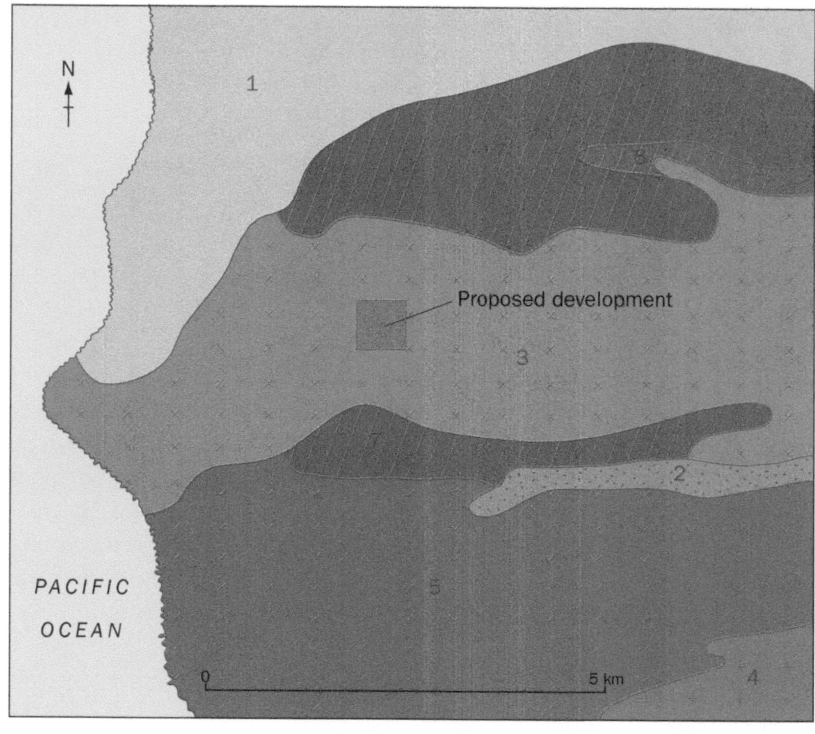

FIGURE 14.10 Geologic sketch map of lava flows surrounding proposed development.

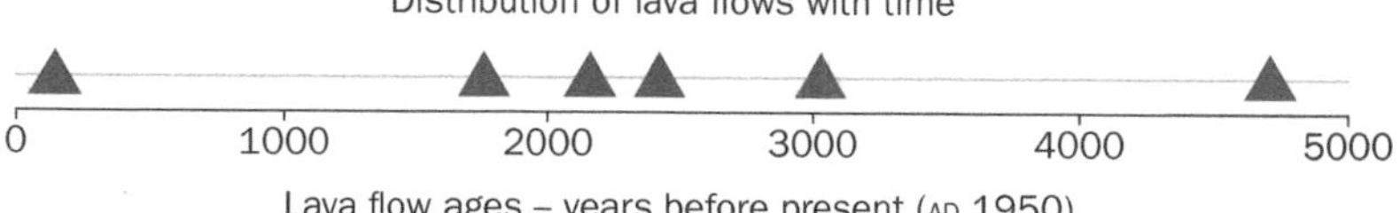

FIGURE 14.11 Age distribution of lava flows entering Hualālai [17] test area.

TABLE 14.2 **Simple Poisson probabilities that the Hualālai [14] test area will be impacted by future lava flows over various future time intervals.**

Future time interval *t* (years)	10	50	100	250	500	1000
Probability *P* (%)	1.3	6.2	12.0	27.3	47.2	72.1

Volcanoes with high eruptive rates may be more likely to erupt in any given time interval, and, for that very reason, people and authorities around them are more aware of the hazards and are generally better prepared. However, a volcano that has a mean recurrence time much longer than the human life or community social memory demands the attention of volcanologists since it becomes their responsibility to estimate and communicate the threat that such volcano may pose to society.

Probabilistic Event Trees

In the previous section, we have discussed the "static" or long-term probabilities of future eruptions at volcanoes. However, when a volcano shows any signal of unrest, the nature of the probabilities that people seek to understand changes rapidly. They also immediately become short-term (Donovan and Oppenheimer 2012). During the unrest leading up to an eruption and as an eruption itself develops, volcanologists on the scene will be asked "What do you think the volcano is going to do next?" Lacking 100 percent certainty, no volcanologist can answer this question with 100 percent confidence. ("Volcano gods" are capricious and always have the potential to deliver surprises!) But the probability tools outlined above, together with field experience, understanding of data revealed through volcano monitoring (described toward the end of this chapter), and skill working with stakeholders, help frame the future in useful ways for all concerned during an emergency.

One of the most powerful tools for understanding a dynamic situation like a volcanic eruption is **Bayesian statistics**. This approach, first formulated by Thomas Bayes in 1763, allows the probabilities of events to change in response to feeding of fresh information into a growing data set (e.g. new field discoveries added to a record of past volcanic activity, or the chance occurrence of a particular event during an ongoing eruption, etc.). Bayes Theorem holds that the probability of one outcome taking place *depends on the probabilities of other related outcomes* in a probability system, as well illustrated in the form of probabilistic event trees (Garcia-Aristizabal et al. 2013).

Probabilistic (event) trees for individual volcanoes diagrammatically assign changing probabilities of future events as volcanic unrest takes place. They are easily grasped tools for assessing risks during a dynamic, unfolding crisis (Figure 14.12; Newhall and Hoblitt 2002; Marti et al. 2008). Generalized answers to specific questions – for example, "What are the probabilities that the volcano will explosively erupt?"; "Will the eruption be large or small?"; "What are the probabilities that lahars, lava flows, or tephra fall will be involved?"; etc. – can be given by event trees, which may be modified as eruptions proceed. In this manner, volcanologists on the scene can readily reach consensus about a situation so they can better advise authorities and to communicate more credibly to public groups about what is most *likely* to happen (Newhall and Pallister 2015; Woo 2008).

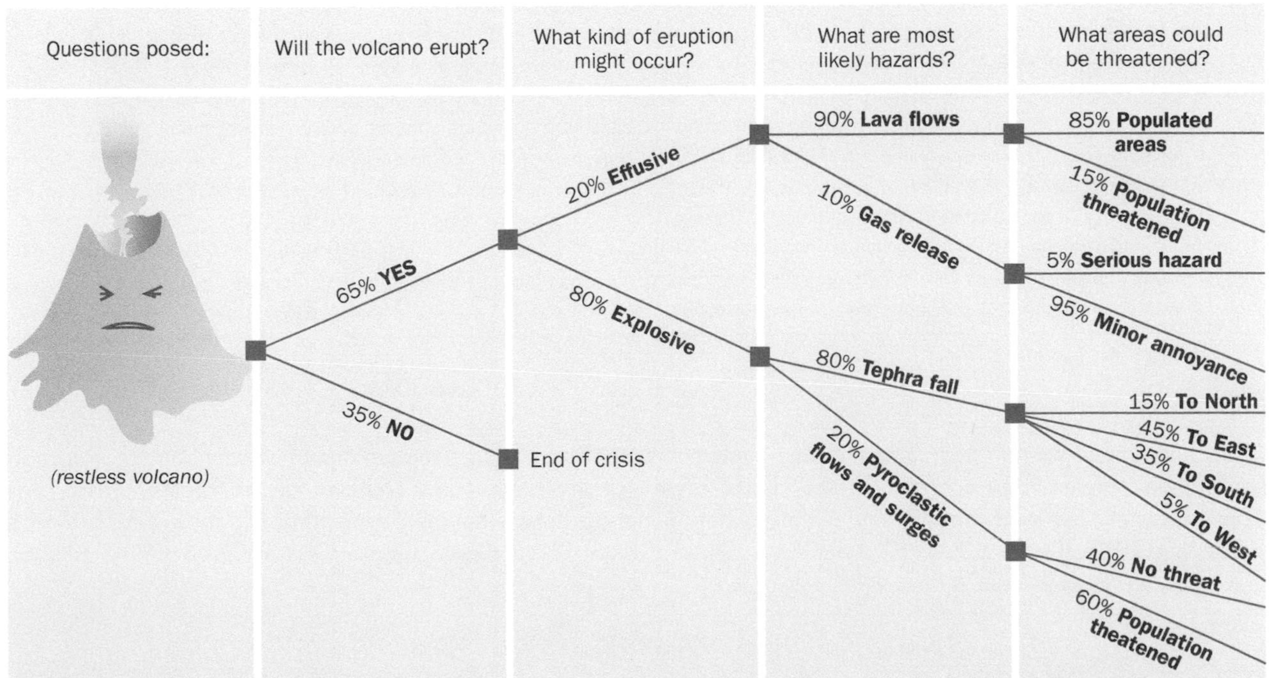

FIGURE 14.12 Probabilistic event trees are important tools to represent schematically the "odds" of possible scenarios developing during volcanic unrest, and are easily understood by the public. As a volcanic crisis unfolds, the "odds" of future developments continually change as new information becomes available, and branches of the tree become outdated by actual events. Source: Adapted from Newhall and Hoblitt (2002).

Acceptable Risk – The Balance Between Public Safety and Personal Freedoms

Despite all the forewarnings and guidance volcanologists can provide, we must always accept not only that volcanoes have their own unpredictable plans (Poland and Anderson 2019), but that people themselves may be willing to take risks authorities regard as highly problematic. People may choose to live in hazardous areas and visitors to visit threatening sites for the rewards of unique experiences. In these cases, culturally defined "Freedom of Choice" rights meet risk-management concerns head-on. What is the correct balance between personal freedoms and the desire for public safety?

Consider tourism: All active volcanoes exhibit hazardous behaviors at times that present risks to visitors who might stray "too close" to the action. How should "too close" be determined? Most active volcanoes lie at least in part on public lands, and governments have the conflicting responsibilities of keeping people safe while allowing them the freedom to enjoy some of Earth's most spectacular displays of nature's forces. There is a fine line to be drawn here, legally and practically, and it is not always easy to determine.

White Island, New Zealand [185] is a very active volcano off the northeast coast of New Zealand. Although privately owned, monitoring of the volcano and warnings about hazardous situations is the responsibility of GNS Science, a quasi-governmental descendant of the New Zealand Geological Survey. Spectacular hydrothermal activity within White Island's central crater is a popular tourist attraction, and some 10,000 tourists visited the island by helicopter and cruise boats every year up until December 2019 when a phreatic eruption there killed 22 visitors, injuring an additional 20. Although GNS had issued warnings about increasing volcanic instability in the weeks before the eruption, no efforts to limit tour visits were made. Several dozen lawsuits against New Zealand government agencies for failure to protect visitors are in progress [at the time of this writing in 2022]. Ground tours have not resumed, but pressure to open White Island to tour operations has intensified.

Kīlauea [17], is perhaps the world's most visited active volcano, in Hawai'i Volcanoes National Park with about two million visitors/year before infrastructure damage associated with the 2018 eruption (Chapter 1) or the more recent Covid-19 pandemic. Keeping those visitors safe while still allowing them to witness eruptive activity whenever possible has always been a park priority. Prior to 1996, park authorities allowed visitors unrestricted freedom to view lava entering the sea from the ongoing Pu'u 'O'o eruption. But one day that year, a 13-hectare slab of lava bench suddenly broke loose and toppled into the ocean – a phenomenon never before observed. Subsequent "bench collapses" killed without warning several visitors who ventured too close to lava entries, and the park hastened to include this newly recognized problem with risk-warning signs and protective barriers. Similarly, dangers from breathing volcanic fumes are now better appreciated, leading to restrictions on access to some active vents that would previously have been left open to the public.

In the early 1970s, an active lava lake at the summit Kīlauea's Mauna Ulu satellite shield was a major visitor attraction. The rim of the crater frequently collapsed; sudden bursts of molten spatter and overflows of the lake were obvious hazards, yet park resources were inadequate to prevent all visitation. Bryan Harry, then park superintendent, made a brave decision in 1972: For safer viewing, he authorized the construction of a wooden overlook platform at the crater's edge, which gave visitors willing to hike over difficult terrain a direct look into boiling lava a few meters below them (Figure 6.6). Hawaiian Volcano Observatory (HVO) staff began a program of monitoring crater edge cracks to evaluate platform safety. This allowed safer (but still not completely "safe") access. The platform was charred at times by the heat of the lava below, and eventually collapsed into the lake one night when no visitors were present. Superintendent Harry confided in me (JPL) that if that platform had collapsed and anyone had died, it would have been the end of his career with the Park Service. But it did survive for a while, and he was proud that his action allowed hundreds of thousands of visitors to have perhaps the most memorable experiences with nature in their lives.

The pendulum between personal freedom and public safety has swung greatly toward restricting freedoms over the past 50 years – and it is doubtful in this litigious, increasingly bureaucratic age that present-day authorities would approve such a viewing area. Ultimately, the question still remains, however: How responsible should individuals be for their own safety and personal rescue on active volcanoes? How much "freedom" should citizens have if their decisions might cause deaths or rescue operations that could jeopardize other lives? How accessible should volcanic activity be for visitors? What should the balance be?

Another related question regards the balance between economic costs of risk reduction efforts and the derived social benefits. Woo (2015) explores the moral complexities involved: When might the economic costs of risk reduction (e.g., mass evacuations) exceed the actual benefits to society? Aspinall and Blong (2015) discuss the various metrics by which volcanic risk assessment can be quantified.

When major economic and political issues are involved in responses to long-lived periods of volcanic unrest, volcanologists may be called on to defend rigorously their assessments before investigative panels or courts of law as expert witnesses. For such forensic purposes, the assumptions that lead to these assessments require a defendable evaluation of all evidence involved by the techniques Willy Aspinall et al. (2003) call "evidence-based volcanology." The development and application of probabilistic event trees may play an important role in these circumstances.

Mitigation of Volcanic Risk

"Potentially destructive phenomena such as volcanic eruptions occur independently of any human action. Volcanic disasters occur when a social group fails to respond to a threatening situation resulting from volcanic activity."

Jolly and De la Cruz-Reyna (2015, p. 1118)

For volcanic risks to be mitigated, those risks for a particular volcano must be well understood as a pre-condition for mitigation. Geologic mapping and the associated radiometric dating studies (Chapter 6) are absolutely essential to decipher a volcano's past eruptive history and to better understand the potential for future eruptions. Geologists conduct their field studies in order to recognize patterns of past volcanic behavior that are critical for interpreting the present behavior, and for looking into the future. James Hutton's assertion that "The present is the key to the past" is true, but must be inverted when it comes to volcanology: "*The past is the key to the present – and to the future*"!

Mitigation (reduction) of volcanic risk for vulnerable populations should not be difficult to accomplish, but how much that risk can be reduced depends on how much is known about the hazards involved, and how great an effort societies will undertake to reduce their exposure (Tilling 1989). The first steps might appear relatively easy – people residing near active volcanoes must become aware of the risks they face, and must learn what they can do to reduce them. In the real world, however, the realities involved in public education are often very complex, especially in rural communities where general education levels may be relatively low and religious beliefs and cultural traditions may be very strong. Sociological studies to identify the degree of awareness about volcanic hazards and the willingness to heed warnings are an important way to mitigate risk (Gregg et al. 2004). Although public understanding of ways to reduce volcanic risk to people and their property seems like an obvious virtue (especially to outside "experts"), there are often intertwined local economic and political vested interests fearful that awareness of risk may reduce property values or somehow threaten economic stability of the involved communities.

Public education Despite these challenges, much has been learned about the most effective ways to disseminate reliable volcanic risk information to vulnerable communities, and thousands of lives have been saved by timely dissemination of warnings (e.g., the 1991 Mount Pinatubo [135] eruption). Failures to deliver effectively credible warnings (e.g., the 1985 Ruiz [71] eruption, where tens of thousands of lives were needlessly lost), weigh heavily on the minds of volcanologists, however, and efforts to improve communications between volcanologists, government authorities, and the public remain one of our most important obligations (e.g. Marzocchi et al. 2012). Schools are important channels for disseminating information about volcanic hazards and risks. Young children are especially fascinated by volcanoes – especially if there are active ones nearby, and the information they learn at school can be shared with others at home. Children are wonderful artists, and a young Chilean student's depiction of volcanic alarm states demonstrates the importance of art (e.g. Figure 14.13).

The most critical need for conveying hazards and risk information to the public at risk is to make the information available in terms that are readily understood by ordinary citizens. Color-code systems have long been used to indicate risk levels and to alert

FIGURE 14.13 A third grader's depiction of volcanic activity conditions for restless volcanoes near her school, the Escuela Epson, Ensenada village, near Puerto Varas, Chile. This school was heavily damaged by tephra fall from the 2015 eruption of Calbuco volcano [74].

the public as to the level of danger. It is very important to convey unambiguous information. **Green-yellow-red** schemes are most easily understood, as volcanic threats increase. Mexican agencies prepared an excellent "traffic light" scheme used around Popocatepetl volcano [56], based on universal familiarity with traffic light colors (Figure 14.14). This poster tells people what the different colors mean and how to respond to different alert levels.

One critical problem with public education that looms over all efforts to keep a citizenry informed and ready to respond to accelerating volcanic unrest is associated with long-lived, low-level eruptions that have the potential to become deadly: "warning fatigue." The longer pre-eruptive or low-level activity continues, the greater the possibility that the public will become bored with warnings, and will resent restrictive decrees that limit their access to dangerous area. The problem was well-discussed for the long-lived Popocatepetl activity by De la Cruz-Reyna et al. (2017).

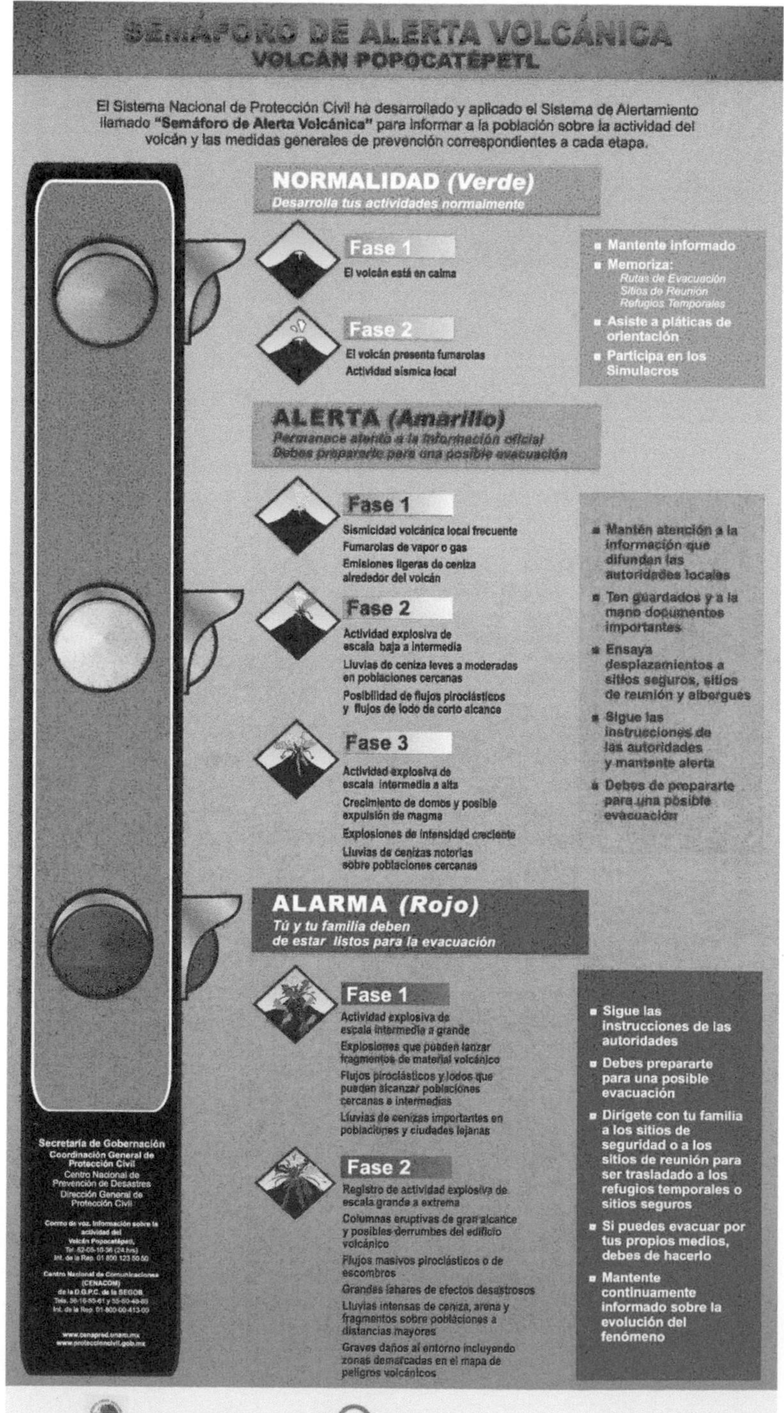

FIGURE 14.14 Semáforo de Alerta Volcanica (Volcanic Traffic Light Alert System) poster as distributed to populations at risk from future major eruptions of Popocatepetl, Mexico [56] (De la Cruz-Reyna and Tilling 2008). This poster explains possible alert levels, what risks to expect, and what actions to take. Source: De la Cruz-Reyna and Tilling (2008). Reproduced with permission of Elsevier.

Volcanic risk maps Geologic maps are critical for understanding a volcano's history and for determining the recurrence intervals for hazardous eruptive events, but by themselves are of relatively little use to the emergency managers and land-use planners who must make decisions about volcanic risk. For these officials, and for the public at risk, **volcanic risk maps** are essential (Figure 14.15) and are best prepared by the geologists who have conducted the geologic mapping of the particular volcano in question. Such field maps can be refined by using tools for carefully mapping land surfaces, such as LIDAR (a portmanteau of "light" plus "radar"), and geographical information systems (GIS) to develop inundation models, for example LAHARZ (Darnell et al. 2013), FLOWGO, and Q-LAHVA (Mossoux et al. 2016). These computer-enabled tools enable mappers to determine potential flow paths for lava flows, lahars, and PDCs, including their speeds, travel times, and even thicknesses.

The 1985 Ruiz [71] disaster (see Chapter 11) caused increased awareness of volcanic hazards throughout Latin America, especially in Ecuador, Colombia's southern neighbor. Cotopaxi [68], south of Quito, is one of the most dangerous active volcanoes in the Andes, and has claimed many thousands of lives over the past centuries. Volcanologists prepared a volcanic risk map to document volcanic hazards for local government authorities (Miller et al. 1978). Ecuadorian emergency agencies, in cooperation with local geologists then prepared simplified posters to inform vulnerable residents in plain language about those risks (Figure 14.15). Similar posters have now been prepared for distribution around many hazardous volcanoes throughout the world by local civil defense agencies, and have become invaluable public education tools.

FIGURE 14.15 Lahar hazards poster, Cotopaxi volcano [68], Ecuador. This widely distributed poster depicts the areas and municipalities that could be threatened by future lahars and provides easily understood information about the nature and dangers of lahars, the areas at risk, the locations of evacuation centers, and basic information on where to obtain warnings and what to do in case alarms are sounded. Source: Poster created by Patricia Mothes of the Ecuadorian Geophysical Institute, National Polytechnic School in cooperation with the Ecuadorian National Office of Civil Defense.

Land-use planning Appropriate zoning regulations can be designed and enforced to prevent populations and critical facilities from being established in high-risk areas. This depends ultimately in the credibility of governments, which in turn depends largely on the reliability of volcanological information that is provided to authorities. In some cases, governments have supported active resettlement programs to remove threatened populations from high-risk volcanic zones, as around Mayon Volcano [137] in the Philippines. This clearly requires close interaction with impacted communities, and assessments that factor in local livelihoods, culture, and other social circumstances (e.g. Usamah and Haynes 2012).

Citizen science One of the most effective tools for building public awareness of, and respect for potential volcanic hazards, is **citizen science**, enlistment of members of the public for the routine collection of field data important for monitoring the state of a restless volcano. Done in collaboration with a scientific facility or government agency, this also helps build trust and cooperation between public and authorities that can prove invaluable in a crisis. Joseph et al. (2019) describe the successful implementation of citizen science for tracking potentially unhealthful volcanic emissions at Sulphur Springs on the Island of St Lucia in the Lesser Antilles. Volcanologists provided volunteers training in personal protection, use of field instrumentation and basic analytical techniques for routine sampling. They note (p. 50) that "This participatory approach was an effective option for scientists to engage local stakeholders and the wider community as partners in geoscience hazards education."

Direct Intervention

Humans may directly intervene in volcanic processes to lessen the impact of associated hazards through engineering means or individual efforts in various ways – depending on the hazard involved. The three principal volcanic hazards that are amenable to human modification to reduce risk are tephra falls (Chapter 7), lahars (Chapter 11), and lava flows (Chapter 6). We discuss each separately below. Pyroclastic surges and flows (PDCs) overwhelm most structures in their paths, however, and no direct intervention means to protect lives or property from these hazards have been proposed or seem feasible at this time.

Tephra fall risk reduction Falling volcanic ash can be a rather beautiful sight at first, as it is commonly a gentle process, sounds are deadened, and the soft light that may soon turn to total darkness casts an eerie, peaceful ambience over the land as verdant landscapes are quickly transformed into uniform grayness (Chapter 1). Any admiration for the aesthetic aspects vanishes quickly however, as the fine ash penetrates all manner of mechanisms (including the engines of automobiles and aircraft), and breathing may become difficult. The first steps are obvious as people scurry to cover their heads and faces with umbrellas or newspapers, and will soon seek particle masks or damp clothes to facilitate breathing and to limit lung damage. The next steps are not so obvious, and rushing inside homes to avoid the ash may not be a wise move if the ashfall is heavy and prolonged. Many people who sought shelter inside the homes of Olongapo, Philippines during the 1991 eruption of Mount Pinatubo perished when roofs collapsed under the weight of water-soaked ash. Major buildings at Clark Air Force Base at the foot of Pinatubo also collapsed (Figure 14.16), but aircraft and personnel had been evacuated before the climactic 15 June eruption. Unless roof pitches are steep – not common in the tropics (!), roofs may collapse under as little as 15 cm of ash – especially if wet (Spence et al. 1997). Individuals can greatly reduce damage to their homes and business buildings by removing ash before it builds up to dangerous levels – a sometimes difficult process when ash is still falling, particularly when day may have suddenly turned to night, and lightning may be crackling overhead (Figure 14.17). During the 1982 eruption of Galunggung [129] volcano, Java, many villagers returned daily to their homes to remove accumulated ash, but where ash was not removed, roofs collapsed, completely destroying homes (Figure 14.18). Ash removal from city streets and airport runways is also critical, and must be done repeatedly if the tephra fall is ongoing. Keeping ash from clogging sewer systems is important, as is ensuring that the removed ash does not block drainage channels, as this can cause subsequent flooding.

Lahar Risk Reduction Lahar risks can best be reduced through public education and prevention of construction activities in potential lahar channels, but also by major engineering efforts in advance of eruptive activity or before the rainy seasons that frequently trigger them. Techniques used to lessen lahar damage (known as **sabo engineering** in Japan) are focused on three principal strategies:

1. channelization of lahars by reinforcing and augmentation of natural or man-made pathways;
2. impounding of lahar debris by the construction of dams and basins; and
3. removal of large boulders from flowing lahars by massive grates that trap the largest boulders (Figure 14.19).

This last strategy is to eliminate the ramming effect of entrained boulders that can destroy even the strongest buildings at risk downslope.

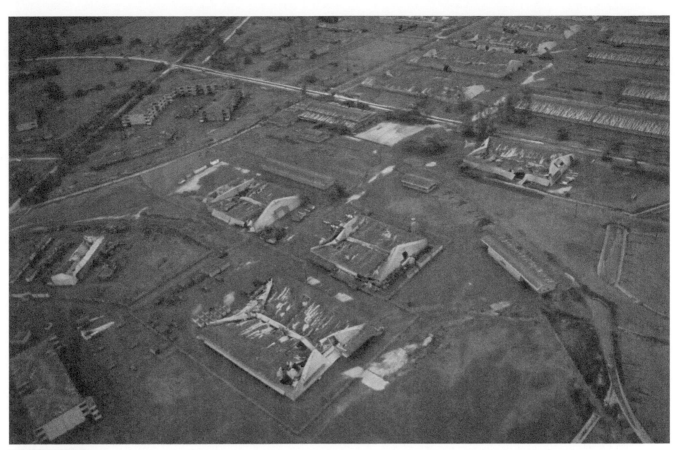

FIGURE 14.16 View of Clark Air Force Base, 25 km east of Mount Pinatubo [135], Philippines, after the eruptions of June 1991. The hangars in the foreground were destroyed by the weight of accumulated airfall tephra, and hundreds of millions of dollars in damage were caused to this important facility. Fortunately, owing to timely warnings from volcanologists, all aircraft were removed from these hangars and flown to safety before the paroxysmal eruption of 15 June. After major Plinian ashfall events, everything is gray – this is a color photograph. Source: USGS photo by Rick Hoblitt.

FIGURE 14.17 Villager removing accumulated ash from his roof in Cikasasah between eruptive episodes of Galunggung volcano [129] in 1982. People who returned to their homes after eruptive episodes to remove ash generally were able to save their homes from destruction. Source: USGS photo by J.P. Lockwood.

FIGURE 14.18 Ruins of home destroyed by the weight of accumulated ash near Cipanas, Galunggung volcano [129], Indonesia. Note trees in background – defoliated and killed by falling ash. Source: USGS photo by J.P. Lockwood.

FIGURE 14.19 Sabo dam for retarding lahars across the Mizunashi River system below Unzen [141] volcanic dome, Japan (in background). Dome growth in 1991 caused pyroclastic flows to move repeatedly down this river and set the stage for lahars and debris avalanches that remain a hazard at present. The sabo dam, 290 m wide and 10 m high, is designed to impound lahar deposits and to remove large lahar boulders by "filtering" them from lahars that may overtop the sabo. The lahar catchment basin above the sabo must be continuously excavated of debris to accommodate future lahar material. Source: Photo courtesy of Etushi Sawada, Unzen Restoration Work Office, Nagasaki.

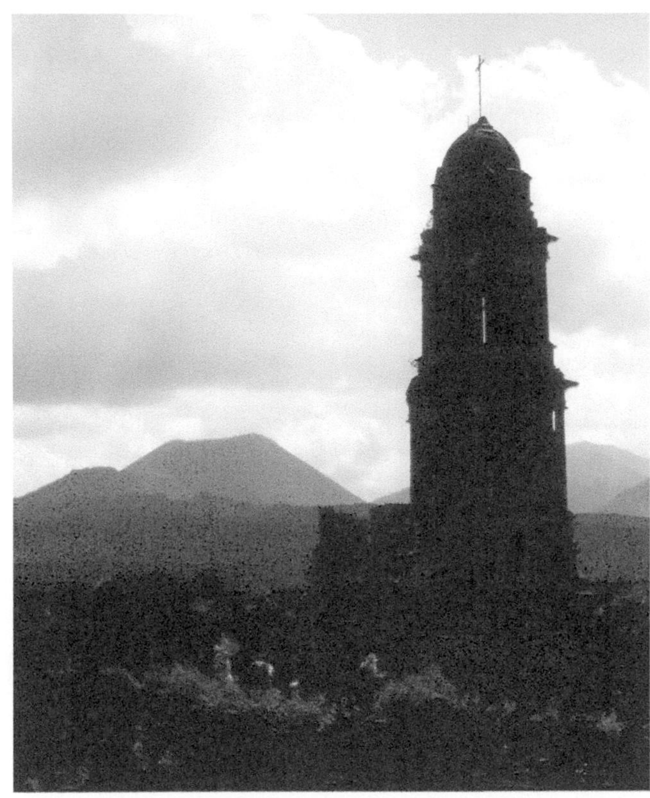

FIGURE 14.20 San Juan Parangaricutiro Church, projecting through lavas erupted in 1944 from Paricutín [53] volcano (seen on skyline), Michoacan State, Mexico. The Paricutín eruption began in a farmer's cornfield on 20 February, 1943 and lasted for 9 years. Church steeples seem to commonly survive when lower-standing structures are buried by lava flows and lahars in volcanic areas. Source: Photo by J.P. Lockwood.

A problem with lahar mitigation efforts is that they are mostly temporary, as repeated lahars can overwhelm any human-built structures with time. Lahar mitigation efforts are most sophisticated and extensive in Japan, although the maintenance of these extensive sabo systems involves huge costs, as accumulation basins have to be repeatedly excavated, and lahar diversion structures rebuilt.

Lava flow risk reduction When lava flows through towns, destruction is usually complete, although it is of interest to note that in some cases, only churches may remain standing (Figure 14.20). Although taken by some to demonstrate the religious sensitivities of lava, this actually reflects the fact that those churches are usually the most imposing and strongest built structures in rural towns, and also are frequently constructed on hills!

Like lahar risk, the risks of lava flow impact can be best reduced or eliminated by careful advance planning, so that cities and vulnerable structures are built away from or above likely lava flow paths. But for many reasons, notably the attractiveness of flat land close to water, many communities in volcanic areas are sited in drainages subject to lava inundation. People rarely plan for low-probability future events, so that wherever effusive eruptions occur in populated areas, human-built structures are commonly threatened.

Lava flows are fluids, albeit much more viscous than lahars, and by and large flow downslope following the easiest paths. Under certain circumstances human intervention may be employed to alter those "easiest paths" and divert them away from destructive courses. Efforts to change those directions or geometry of active lava flows are known as **lava diversion** and have been employed with varying success for several centuries. The earliest recorded attempt to change the path of a lava flow occurred during the 1669 eruption of Etna volcano [112], when citizens attempted to divert a lava flow from inundating the important port city of Catania. Macdonald (1972, pp. 419–420) described the attempt:

> *Under leadership of a man named Diego Pappalardo, 50 or so men from Catania covered themselves with wet cowhides for protection against the heat and dug a channel through the wall of hot lava at one edge of the flow. At first the operation was successful. Molten lava escaped through the gap thus created and flowed away at an angle to the path of the original flow, reducing the amount of lava moving toward Catania. However, the new stream was headed toward the town of Paterno, and 500 indignant citizens of that town armed themselves and drove the Catania men away. The channelway in the main flow wall [levee] soon clogged up and the flow continued toward Catania, where it came up against the feudal city wall. For several days the wall withstood the flow and diverted it around the city toward the sea, but eventually the lava broke through a weak place in the wall and flooded part of the city. Thus, the 1699 eruption provides examples of lava diversion both by destroying the flow wall and by turning the flow with an artificial barrier.*

Although the 1669 efforts were ultimately unsuccessful, evolving technology (the invention of bulldozers, airplanes, and powerful water pumps) has given society powerful new tools to divert lava flows, and humans have had considerable success in changing the direction or limiting the spread of advancing flows during the twentieth century. Diversion of flowing lava is only appropriate to consider in special, uncommon circumstances, where terrain is appropriate, sufficient time is available, economic benefits outweigh potential costs, and the threatened populations and governments are willing to consider "messing with nature." Techniques that have been successfully utilized to alter the courses of lava flows or to impede their progress include: (1) construction of earthen barriers; (2) application of large volumes of water to solidify advancing flows; and (3) use of explosives to disrupt active pyroducts and flow channels.

Earthen barriers Relatively small lava diversion barriers were built to attempt lava flow diversion during the 1906 eruption of Vesuvius [109] and the 1955 and 1960 eruptions of Kīlauea [17]. None of these barriers worked, because they were viewed as dams to stop the advance of flows and were far too small to achieve their objectives. The barriers built to protect the town of Kapoho in 1960 (Macdonald 1962) were constructed of highly vesicular, low density pāhoehoe, and were simply floated away by lava flows of greater density. The most successful lava diversion barriers to date were built during the 1983 and 1992 eruptions of Etna volcano [112]. During the 1983 eruption, massive berms up to 14 m high were built and successfully diverted flows around a valuable resort and threatened astronomic facilities (Figure 14.21) (Colombrita 1984; Lockwood and Romano 1985). Lava barriers were also built directly across flow paths in 1992 not to divert flows, but to temporarily delay flow advance and to protect the town of Zafferana, while efforts to explosively disrupt pyroducts upslope were being carried out (Barberi et al. 1992). Experience has shown that lava diversion barriers can never "stop" lava flows except temporarily as they will eventually be overrun if an eruption continues long enough. They can only be successful if designed to change the direction of lava flow paths, but only where sloping terrain allows diverted lava to readily flow downslope away from the barrier. Lava diversion barriers can also be constructed in advance of eruptive activity to protect valuable facilities from anticipated future lava flows, as has been done in Iceland, Japan, and Hawai'i (Figure 14.22). Even small, hastily constructed diversion structures can be effective (Figure 14.23).

Water cooling During the 1960 Kīlauea eruption, firefighters noted that small amounts of water sprayed on advancing pāhoehoe flows could temporarily stop thin flows by freezing immobile crusts over liquid interiors. The amounts of available water

FIGURE 14.21 Diversion of an active 'a'ā flow above the Sapienza resort and tramway complex, Etna Volcano [112], Sicily in May 1983. A 25-m-wide 'a'ā flow is being diverted by a 10-m-high lava barrier, which is being built higher as this photo was taken. The barrier, built from 100,000 m^3 of transported rocks in 10 days as lava piled up against it, crosses an earlier flow that had nearly overwhelmed the Sapienza buildings a week before the diversion effort began. The barrier successfully diverted the flows about 100 m laterally from their natural terrain path and saved the vital Sapienza structures. Source: USGS photo by J.P. Lockwood.

FIGURE 14.22 Lava diversion barrier, constructed to divert future lava flows around the NOAA Mauna Loa Observatory on the north slope of Mauna Loa volcano, Hawai'i [15]. The longest arm of the 5–7-m-high barrier is about 700 m long. This NOAA facility is of global importance, as it maintains the longest complete record of atmospheric CO_2 monitoring in the world, and protection from future lava flows is essential. The ʿaʿā flows surrounding the Observatory were erupted in 1843. Source: Photo © G. Brad Lewis.

FIGURE 14.23 **A** 3-m-high lava diversion structure built by property owner above Pahoa during the 2014 eruption of Kīlauea volcano [17]. This quickly built barrier successfully diverted this pāhoehoe flow, and the owner was fortunate that (1) the eruption stopped before this flow and the barrier was overtopped by a subsequent flow, and (2) because the diverted flow caused no damage to properties downslope! Source: Photo by J. P. Lockwood.

were too small to have any lasting effect in this case, and it was not until 1973 that water application was shown to be a viable lava diversion technique. In that year, a new volcano erupted on the island of Heimaey, Iceland, just upslope from the island's only city. Icelandic emergency officials ordered enormous volumes of seawater to be pumped and sprayed on advancing *'a'ā* flows to quench molten lava and to cause those flows to thicken and slow their advance (Williams and Moore 1983). The heroic efforts of Icelanders at Heimay, utilizing massive water pumps and diversion barriers, saved much of the city and spared the economically invaluable harbor from closure by advancing lavas (Figure 14.24). About six million cubic meters of seawater were pumped onto the advancing flow over a five-month period, and it was found that roughly a cubic meter of water was required to solidify and immobilize a cubic meter of flowing lava. Water application is most feasible where advancing lava flows are close to large bodies of water, although small amounts of water may suffice to freeze small pāhoehoe flows that threaten to overtop barriers.

Explosives Explosives have been used several times to attempt lava flow diversion in Hawai'i (Mauna Loa volcano [15]) and Italy (Etna volcano [112]). The Mauna Loa efforts in 1935 and 1942, conducted by US Army Air Corps bombers, used obsolete munitions and aircraft, and were not successful, although in 1942 levee walls were broken and a flow diverted for a short distance before it rejoined the original channel. Field experiments with modern delivery systems have shown that aerial bombing does have the potential to disrupt pyroducts and active lava channels (Lockwood and Torgerson 1980), but these modern techniques have never been employed on active flows.

Large quantities of explosives were successfully hand-emplaced in the walls of lava channels and pyroducts to disrupt supply conduits during 1983 and 1992 eruptions of Etna volcano (Lockwood and Romano 1985; Barberi et al. 1992), although the lava diversion effects were short-lived. The use of explosives to divert lava flows has proven to be culturally controversial in Hawai'i, and is most appropriate to employ where target areas are remote, and where the negative impacts of unsuccessful efforts or unintended consequences are acceptable.

Volcanic Lake Risk Reduction Hazards from the three types of volcanic lakes mentioned earlier in this chapter are each amenable to different means of mitigation. Lakes formed by the blockages of pre-existing streams pose risk to downstream populations if they are overtopped by rising lake waters. To mitigate these risks, drainage tunnels or canals can be excavated in erosion-resistant rocks below the dam level, as was done successfully to keep the levels of a rapidly rising Spirit Lake stable at Mount St Helens [31] in 1981–1982. The second category of volcanic lakes, those known to contain dangerous and rising concentrations of CO_2 gas can be made safer by controlled degassing efforts, as is being done at "killer lakes" in Cameroon (Kling et al. 2005). Degassing is

FIGURE 14.24 Pumping seawater to solidify and stop the advance of lava down a Vestmannaeyjar street, Heimaey, Iceland, in March 1973. The flow, here about 8 m thick, was slowed and greatly thickened by the cooling operation, and was successfully blocked. By July 1974, the lava had been completely removed, and the street had been returned to use. Source: Photo courtesy of Sveinn Eiriksson.

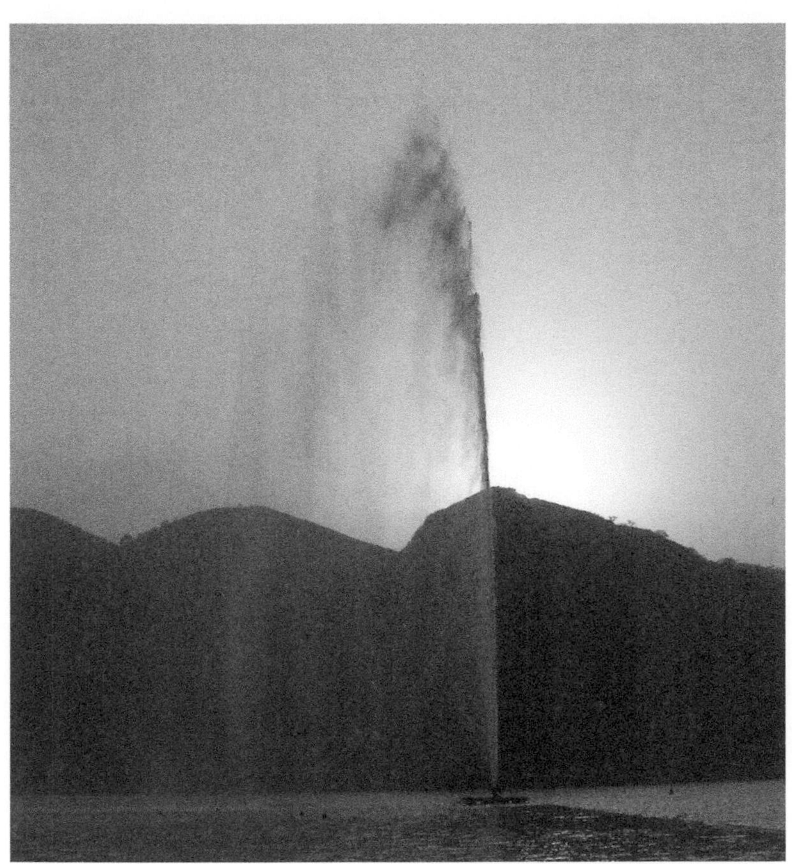

FIGURE 14.25 Self-sustaining degassing of Lake Nyos, 2006. This fountain of water and CO_2 gas is being propelled upward about 40 m by the expansion of carbon dioxide gas in a 200-m-long pipe whose intake is in CO_2-saturated waters near the bottom of Lake Nyos. Source: USGS photo by W.C. Evans.

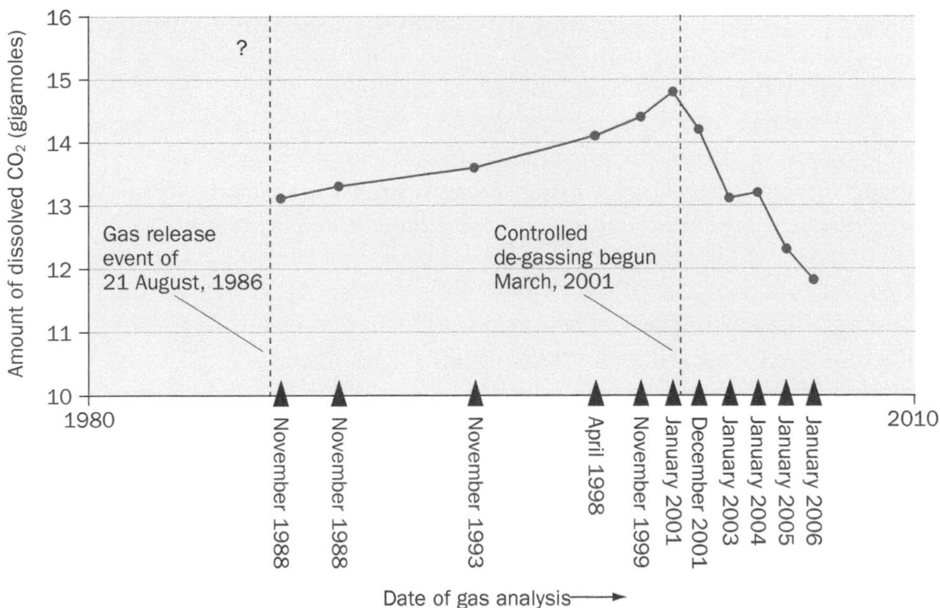

FIGURE 14.26 Amount of carbon dioxide dissolved in the waters of Lake Nyos [107], Cameroon after the tragic lake overturn event of 21 August, 1986, showing the effect of controlled degassing efforts. CO_2 content in gigamoles (1 gmole $CO_2 = 44 \times 10^6$ kg). CO_2 content before the 21 August event has been estimated at about 27 gmoles. Controlled degassing efforts must continue to offset gas recharge rate and to make the lake safer. Source: Data from Kusakabe et al. (2008).

achieved by lowering a large diameter plastic pipe to CO_2-saturated lower levels of these lakes, initiating upward flow by lowering pressure in the pipe until the buoyant rise of bubble-filled water becomes a self-sustaining process and fountains of water and gas are released harmlessly to the atmosphere (Figure 14.25). The controlled degassing of Cameroonian volcanic lakes has now been carried out since 2001, and has become an important example of successful human efforts to mitigate volcanic risk (Figure 14.26).

The third category of volcanic lakes, those at volcano summits, are the most dangerous of all because of the interaction of water and eruptive activity, and have generated lahars that have killed thousands of people on the flanks of volcanoes (Chapter 11). Keled volcano [132], in central Java, is one of Indonesia's most active volcanoes. Eruptions within its large central crater lake have ejected waters onto the outer slopes of the volcano many times in the past, forming lethal lahars. After a lahar-producing eruption in 1919 that killed over 5000 people, major efforts were initiated by Dutch engineers to lower lake levels by a series of tunnels through crater walls. These and subsequent tunneling efforts reduced the volume of the summit lake dramatically, and subsequent eruptions have produced much less damaging lahars.[1]

Volcano Monitoring

There are two purposes for volcano monitoring: (1) to understand better "how volcanoes work"; and (2) to provide public warnings of potentially harmful future activity. Prognostications about the future involve defining the terms **predictions** and **forecasts**, which are used to describe prospects for future events. *Predictions* are highly specific about the time, place, and nature of future events, whereas *forecasts* are much more general. Wise volcanologists should never make predictions, however, for reasons well stated by Bob Decker in his important 1973 paper "State-of-the-art in Volcano Forecasting" (p. 372):

> *Forecasting the time and place of volcanic eruptions is one of the major goals in volcanology. I could have used the title "Prediction of Volcanic Eruptions" [in my paper], but the word "prediction" sounds precise and deterministic, and the state-of-the-art in this field is far from precise. The practice of weather forecasting has introduced all of us to a more probabilistic and less precise notion of scientific prediction that is still most useful even if not perfectly precise. Choice of the term "forecasting" is an attempt to convey this same sense of useful though uncertain prediction of events which lie in the future."*

Volcanoes always seem to give some warnings before they erupt (at least in hindsight), and in an ideal world, there would be some means to monitor all of Earth's potentially active volcanoes – to provide early warning of impending eruptive activity that could threaten nearby populations, or could pose threats to aircraft that fly above or downwind of them. Unfortunately, there are far too many potentially dangerous volcanoes in the world to monitor on a routine basis. Of the 538 volcanoes known to have erupted during historic times (Simkin and Siebert 1994), less than half are under regular surveillance (McGuire 1995). As noted earlier, many active volcanoes have never even been mapped, and most have geologic histories that are poorly known. Even where routine monitoring of a particular volcano is taking place, the type and amount of instrumentation will likely be insufficient to provide a complete picture of volcanic activity. When a potentially hazardous volcano does become "restless," equipment and personnel may need to be brought from thousands of kilometers away. At times, such equipment may arrive after an eruption has already begun, too late to warn people in advance of impending trouble. Fortunately, the time between initially observed volcanic unrest and subsequent eruptive activity is commonly long enough for emergency monitoring and successful risk mitigation (e.g., Mount St Helens [31], 1980; and Mt Pinatubo [135], 1991). Unfortunately, the tragedy of the 1985 Ruiz [71] eruption (Chapter 11) demonstrates that a long warning period by itself will not prevent disaster, if the links between scientists, emergency officials, and the endangered public are not forged well (see later section titled "Volcano crisis management").

Although new technical developments are constantly improving our ability to understand volcanoes as time goes on, the increases of populations living in volcanically hazardous areas constantly boost the need for better monitoring. While volcanologists are now much more capable of providing warnings about volcanic risk, the overall threat of volcanic eruptions to populated areas has ironically never been higher, especially in lesser developed countries where competition for agricultural lands has increased, leading to ever more farming on the fertile slopes of potentially destructive volcanoes (McGuire 1995).

Nonetheless, two of the most extensive volcano observatory systems in the world are located in less-wealthy countries (Indonesia and the Philippines). More lives have been saved by volcanologists in these countries than by volcanologists in all of the rest of the world combined! An appreciation of the magnitude of the volcano-monitoring efforts in these countries is best obtained by being in the modern volcano observatories in Bandung or Manila and listening to the morning radio reports coming in from dozens of field posts located in remote locations all over these volcanic lands – some accessible only by footpaths. The observers at many of these observatories may not be highly trained, and they may only have records from a single, smoke-drum seismograph to add to visual reports, but their great dedication to monitoring "their volcano" to protect the lives of villagers below is admirable and should never be forgotten. Volcano observatories, no matter their level of sophistication, are critical to understanding volcanic behavior. Some history about them follows.

[1] A lava dome that began to grow in Kelud crater in November 2007 completely displaced the preexisting lake, thus eliminating the lake-lahar hazard – but this dome was later destroyed by a large Plinian eruption in 2014 (McCausland et al. 2019). A new lake has begun to form and will become a hazard again!

Volcano Observatories

The Vesuvius Volcano Observatory (Osservatorio Vesuviano – OV) is the world's oldest volcano observatory, and was established by the King of Naples in 1841 on a heavily forested hill overlooking the port of Naples, halfway up the western side of Vesuvius. This location was close to the volcano, hence providing easy access to important field localities, while remaining topographically protected from the volcano's frequent lava flows. It was also upwind of the heaviest historical ash falls from the volcano, which the builders knew could readily collapse roofs. The new institute was initially christened the Osservatorio Meteorologico Vesuviano, because the state of scientific knowledge about volcanoes at that time was so slim that learned people, following the suggestion of Aristotle (!) felt it appropriate to lump volcanic eruptions together with weather phenomena. Earth magnetism was also considered appropriate to study at the OV; perhaps, for all anyone knew, it might turn out that volcanic eruptions were somehow related to changes in Earth's magnetic field, or vice versa. (In fact, shallow magma intrusions *can* have striking effects on local magnetic fields, but their eruptions are not triggered by changes in the strength of planetary magnetism.) The OV has been a beacon for volcanologists because of its long history and proximity to the great variety of both effusive and explosive eruptive action characteristic of Vesuvius. Frank Perret (Chapter 1) made his first detailed volcanological observations there (Perret 1924), and witnessed the devastating 1906 eruption – which also nearly destroyed the observatory. The OV had another close call during WWII, when it was alternatively occupied by German and Allied soldiers and reduced to the operation of a single seismograph – which Director Guiseppe Imbo kept in his home! Imbo was able to give early warnings about the eruption of March 1944, the most recent eruption of this dangerous volcano.

The OV was the world's only volcano observatory for over 70 years, as for the most part, volcanoes elsewhere in the world were studied only *after* they erupted. This began to change after the widely publicized Mount Pelée [87]disaster of 1902, when tens of thousands of people perished. After viewing the carnage in St Pierre, Thomas Jaggar (Chapter 1) realized that continuous monitoring of volcanoes would be required to understand them and to warn vulnerable populations of impending eruptions (Dvorak 2017). He founded the HVO (Chapter 1), now operated by the US Geological Survey, on the rim of Kīlauea caldera [17] in 1912. HVO set an example for many more observatories around the world, and many of the most important tools and techniques for volcano monitoring were developed there. The World Organization of Volcano Observatories (WOVO) now lists 79 observatories, located in 32 countries. These facilities come in many varieties, depending on available funding and the perceived threat of the volcano being monitored. Some observatories in developing countries consist of one or two technical observers manning a single seismometer. Others are much more sophisticated, centralized facilities, typically operated by governments or in some cases by universities, which bring together teams of multi-disciplinary monitoring specialists. The pioneering Vesuvius and Hawaiian observatories each monitor only a few local volcanoes. In contrast, the Alaska and Cascade Volcano Observatories in the United States; the Institute of Volcanology in Petropavlovsk, Kamchatka; the Japan Meteorological Agency; the Center for Volcanology and Geological Hazard Mitigation in Indonesia; and the Philippine Institute of Volcanology and Seismology (Chapter 1) each monitor several dozen **potentially active** volcanoes and volcanic fields, many in remote locations. Indonesia has more active volcanoes (129) than any other country on Earth, and maintains observatories or observation posts on 66 of them! Seventy-nine volcanoes have erupted in Indonesia in the past 400 years, and each of these has the potential to cause widespread death and destruction in this densely populated land, so that the work of the dedicated volcanologists and technicians at these facilities is essential for public safety. New facilities continue to develop, but long-term funding support for volcanological studies waxes and wanes in response to short-term volcanic crises. It is no surprise that satellite remote sensing and telecommunications have become so important for monitoring volcanoes in remote locations and for providing early warnings about volcanoes that are becoming restless. All major observatories maintain portable equipment that can be deployed quickly to unmonitored volcanoes when needed ("expeditionary monitoring"). The US Geological Survey has established a special unit, the Volcano Disaster Assistance Program to send specialized equipment and volcanologists to any area in the world where emergency monitoring is requested. Several volcano observatories have been established as temporary responses to particular eruptions and have evolved into first-class observatories with wide regional responsibilities. An example is the Montserrat Volcano Observatory, established in response to the 1995 eruption of the Soufrière Hills volcano [84] (Aspinall et al. 2002). The Martinique Volcano Observatory (Figure 14.27) founded by Alfred Lacroix after the disastrous Mount Pelée [87] eruption, has now moved to beautiful new facilities and coordinates work of several other French observatories.

The newest volcano observatory on Earth sits directly on top of an active volcano – Axial Seamount – 1400 m below sea level off the coast of Oregon. This new observatory, linked to mainland facilities by a cable network, was established by the National Science Foundation (NSF) Ocean Observatory Initiative in 2014 as one of several dozen ocean floor observatories worldwide – mostly primarily dedicated to real-time monitoring of biological processes and ocean chemistry. The Axial Seamount [25] observatory is one of only two situated on an area of submarine active volcanism (the other is located along the Juan de Fuca Ridge west of Vancouver Island, Canada). The observatory proved its worth in 2015 when it detected premonitory activity and allowed a forecast of the April 2015 eruption.

FIGURE 14.27 L'Observatoire Volcanologique et Sismologique de Martinique has a long history, but its new building that overlooks the adjacent Mount Pelée volcano [87] (in the background), is the most futuristic volcano observatory in the world. Source: Courtesy of Anne-Marie Lejeune.

Only a thousand or so practicing scientists and technicians staff the world's volcano observatories, but millions of people living near active volcanoes depend on their monitoring efforts to warn them of impending danger. Research carried on at these observatories is improving our understanding of active volcanoes, and is thus contributing to public safety.

Direct monitoring methods

Volcano monitoring involves visual observations, eruption documentation, and instrumental surveillance. The instrumental monitoring efforts are each based on the need to detect changes in the underlying magma chambers and the migration of magma within the volcano. Most of these instruments either *listen* to the volcano (seismic monitoring, infrasound), or *measure changes* in volcano shape, gas output, or thermal or magneto-electrical characteristics. Volcanologists also study the chemistry and mineral compositions of erupted rocks for hints about changing magma properties, and collect samples for archiving. No single monitoring method can adequately determine a volcano's behavior. Successful monitoring requires an appreciation of all the various techniques that can be brought to bear, and even then will depend heavily on individual human talents of interpretation and intuition. Here are brief descriptions of some of the most important monitoring techniques.

Event family, VT (Volcano-tectonic), and LP (Long-period) earthquake monitoring Seismic monitoring is the single most critical activity at any volcano observatory and commonly provides the earliest warnings of pending eruptive activity. When the strength of rocks is exceeded by strain rate in the rocks, caused by magma movement, they break. This generates acoustic waves, which we can detect instrumentally or feel as an earthquake (Figure 14.28). Volcanic earthquakes are much smaller than the largest tectonic ones, with greatest magnitudes (M) usually less than 5.0. Most volcanic earthquakes have M less than 2.0 and are only detected instrumentally. Hundreds of such tiny earthquakes may be recorded each day by sensitive instruments, though most have M less than 0 and cannot be located precisely at all.

In contrast to the gradually weakening main shock–aftershock sequences typical of non-volcanic regions undergoing tectonic strain, volcanic earthquakes tend to occur in swarms, called **event families**. Individual event families are caused as just one or a few individual sources within a volcano repeatedly break. Bezymianny [170], Kīlauea [17], Redoubt [23], and Mount St Helens [31] are all examples of volcanoes that characteristically exhibit seismic event families, with each showing its own unique earthquake type or signature. At Mount St Helens, for instance, hundreds of similar earthquakes took place at a focal area atop the magma reservoir, 6 km deep, in 2018. More mysteriously, tens of tiny earthquakes a day have originated in a small volume of the upper mantle 30 km beneath the summit of the dormant volcano Mauna Kea [16], for over three decades. Do these also represent the top of some deep-seated, magma body? Slow cooling could induce the fracturing, leading to such persistent earthquake activity. An

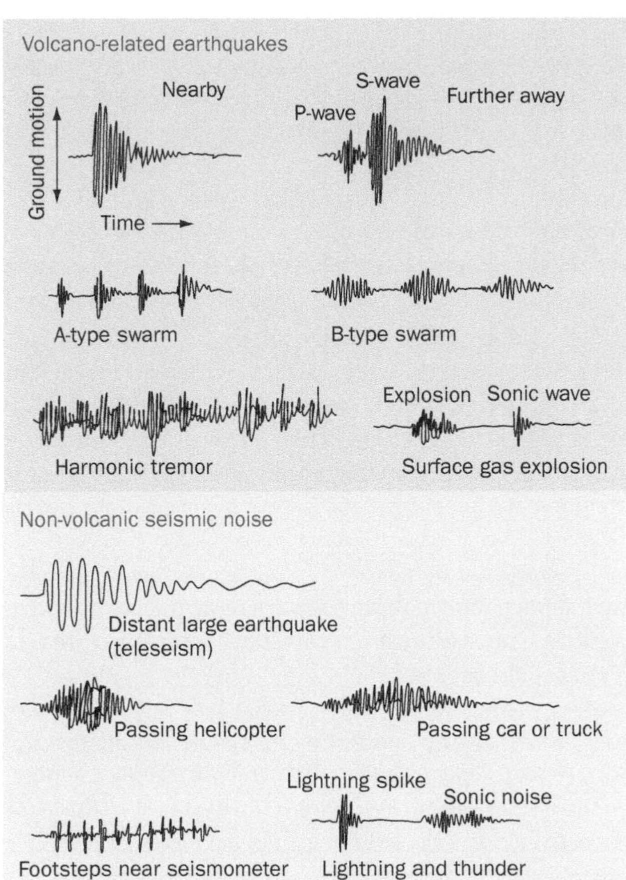

FIGURE 14.28 Typical seismic signals that might be recorded on a volcano observatory seismograph. These are sketches – not actual seismic signal traces. Source: R.Y. Koyanagi (1974, personal communication) and other sources.

event family linked to a single trigger space is generally less of a concern than multiple sources of unrest firing off all at once. In that case, eruptive activity could be pending.

Earthquakes within a volcanic edifice may be set off by processes other than the movement of magma culminating in an eruption. Large-scale, non-volcanic landslide or fault movements under the great weight of a growing volcanic cone can cause them. The accumulation or magma in a fixed reservoir or active intrusion can also destabilize the surrounding volcanic flanks as they swell and experience high levels of strain. These are all termed volcano-tectonic, or **VT earthquakes**.

Distal VT earthquakes can be triggered in the crust many kilometers away from restless volcanoes as a result of swelling magma reservoirs (White and McCausland 2016). For example, as many as four clusters of distal VT earthquakes active in the summer of 2018 occurred 8–10 km from the summit of Mount St Helens during a period of ongoing inflation. Intrusions of dikes into the rift zones of Kīlauea have frequently generated VT earthquakes in that volcano's southern flank within a matter of a few weeks (Bell and Kilburn 2013).

Some very large tectonic earthquakes that may occur hundreds or even thousands of kilometers away can trigger swarms of VT earthquakes within a volcano. The M7.9 Denali, Alaska earthquake in 2002 caused seismic unrest beneath Mount Rainier, Washington [35], and Long Valley caldera [44], California. The type of earthquake and the orientation of its fault plane are critical correlation factors. The Denali earthquake took place on a strike-slip fault, and seismic energy concentrated along the fault plane spread directly toward the Cascade volcanic arc. Similarly powerful earthquakes on normal or oblique-slip faults, or faults with orientations that channel energy in directions other than toward active volcanoes, may have no impact on local volcanic seismicity. Likewise, each volcano seems to show a different level of sensitivity to distant earthquakes. This reflects the unique structures of individual volcanic conduits and the surrounding crust (Weston Thelen, pers. commun. 2018).

In contrast to VT earthquakes, **LP**, or **long-period** (also known as low-frequency) **earthquakes** can relate directly to magma on the move, though they may also result from the slow tectonic rupture of fault planes or the interaction of seismic energy from very shallow (<1 km deep) earthquakes with the overlying surface. A series of LP quakes scattered inside a volcano is generally a solid indication, in any case, of possibly impending eruption. Such earthquakes may be triggered at narrow points or bends in a volcanic conduit through which magma is flowing. Surrounding brittle wall-rock, weakened by heat and pressure, may snap and shear. Collapsing gas bubbles in rising magma or the sudden flashing of steam underground could also produce this type of shaking (Chouet 1996). **Hybrid earthquakes** may start as ordinary events, resembling typical VTs, but then evolve into LPs as they take place.

As magma rises within 1–3 km of the surface, hours to a few months in advance of an eruption, the character of seismicity commonly changes. Newhall and Endo (1987) note that in 25 percent of the 192 eruptions they considered, VT quakes decline significantly before eruptive activity – an important "calm before the storm" signal. This might be due to intense shallow fracturing in the crust (McNutt 1996). LP earthquakes predominate in these immediate pre-eruptive periods, often accompanied by volcanic tremor, discussed later.

Adding to the interpretive challenge, seismic epicenters do not always correspond to the locations of ensuing eruptions. For example, the precursory earthquake swarm at Pinatubo 1991 eruption began almost two weeks before the eruption – 5 km from the eventual outbreak site. Seismic activity declined at the original swarm site, then resumed directly beneath the impending eruption location, reflecting magma migration (Harlow et al. 1997).

Volcanic tremor If a series of overlapping or near continuous LP earthquakes takes place owing to sustained movement of magma, the result is **volcanic tremor**. This, at least, is one commonly accepted explanation. **Harmonic** (or "monochromatic") **tremor** is shuddering of the earth at a single dominant frequency, rather like the plucking of a particular piano string. **Broadband tremor** shows a wider range of frequencies, with strong peaks following one after the other in the range of 1–20 Hertz. Spattering of lava in a vent can generate both forms of tremor, though broadband tremor more often develops in explosively erupting craters (Brian Shiro, pers. commun. 2018).

At Kīlauea [17], an interval of intense summit deflation accompanied by strong harmonic tremor often begins tens of minutes to several hours before opening of the first eruptive fissures, especially when activity breaks out along one of the volcano's rift zones. This enables volcanologists to warn local authorities in time to safely evacuate visitors from potentially threatened areas in Hawai'i Volcanoes National Park (Chapter 1) and residents along Kīlauea's East Rift Zone.

Seismologists can average the amplitude of tremor-related seismic energy waves for particular frequencies and intervals of time to generate **RSAM** – "rectified" or "real-time" seismic amplitude measurements (Endo and Murray 1991). The seismic spectral amplitude measurement (SSAM) approach of Stephens et al. (1994) is closely related. The typical interval of time is usually a minute, but may be as long as 10 minutes. RSAM are a measure of the strength of the tremor seismicity, perhaps the best tool we have for evaluating the up-to-the-moment magmatic activity of a volcano. RSAM need not correlate entirely to magma moving up toward the surface. Strong winds or even waves crashing on a nearby shore can also cause tremor-like vibrations in the shallow crust of a volcano. Observers must consider the circumstances in which the tremor signal is received. A closed magmatic system, for instance a dike that is propagating underground, creates stronger RSAM than an open magmatic system – an actively erupting vent – from which seismic energy dissipates to the atmosphere. The wedge-like tip of an opening dike as it passes a seismic station often generates the strongest RSAM signals. Later, once fluid flow is well established inside the dike, RSAM levels drop off.

McNutt (1994) identified a useful way of comparing seismic energy release with volcanic activity, studying the codas of volcanic tremors recorded at various seismic stations around erupting volcanoes. He showed that the energy released as volcanic tremor (primarily harmonic) is directly proportional to the intensity of eruptions, and to such eruption parameters as the amount of ejecta released and the heights of ash columns. By this method, were a strong eruption to take place at night with no visual observations or radar facilities nearby, tremor measurement alone could provide a general idea of potential aviation threat from rising ash plumes with a certainty of about 95 percent.

Once an eruption has begun, all types of volcanic earthquakes may be generated, not simply tremor alone. Their relative abundances, frequency distribution, and times of onset provide important clues about eruption mechanics. At explosively erupting volcanoes where eruptions occur in a series, precursory seismicity may be strong for the initial outburst, but once the vent is cleared of obstructions, little or no seismic precursors may precede further explosions. McNutt (1996) recommends careful and cautious monitoring around such volcanoes at least 3–6 months following an explosive event to make sure people are not caught by surprise. Following eruptions, episodes of renewed deep (5–15 km) VT earthquakes commonly occur beneath many volcanoes, which Abe (1992) relates to readjustment in the crust as a result of the withdrawal of magma. These earthquakes may signify a return to volcanic quiescence, the true end of an eruption.

Infrasound Complimenting seismic monitoring, **infrasound measurements** can provide specific information about what is happening unseen somewhere nearby at the surface; especially important for volcano watchers with limited lines of sight or wishing to track activity well after nightfall (Garces et al. 1999, 2003). Ordinarily, humans generally hear sound in the range of 20–20,000 Hertz. Sound is essentially pressure pulses in the atmosphere akin to the P-waves of earthquakes, though much slower (300 m/s vs. 3 km/s). "Infrasound" means sound waves vibrating at frequencies below what the human ear can detect. An infrasound network can pick up the occurrence of distant volcanic explosions that may not be detectable by routine seismic monitoring, a meaningful application for important but remote regions such as the Alaska Peninsula and Aleutian Islands (e.g. Smith et al. 2016). Infrasound can also distinguish the opening of effusive fissures otherwise seismically invisible. Once located, hazard managers can use maps to track the most likely path lava will follow from the new fissures, providing timely warning to people downslope. It may be possible to use infrasound to detect lahars as well, especially from large volcanoes such as Mount Rainier.

Deformation monitoring During their active lives when they overlie or contain dynamically evolving magma chambers, volcanoes change shapes almost constantly, and monitoring their changing shapes is one of the most critical methods to evaluate internal processes. We pity those poor geologists who have to study stable structures in continental heartlands where deformation proceeds very slowly if at all !! As has been discussed elsewhere in our text, volcanic edifices are rather weak structures, mostly comprised of relatively brittle, easily fractured lava flows and loose accumulations of volcanic ash and breccia. Conceptually, it is useful to think of active volcanoes as balloon-like coverings over their underlying magma chambers. Whenever internal pressures within magma chambers change, or magma migrates to a different part of the volcano, volcano surfaces deform accordingly. The many geodetic techniques for monitoring these changes will be discussed briefly here, but are treated in much greater detail by Dzurisin (2007).

Deformation changes on an active volcano are usually very small and physically imperceptible to humans, and can only be detected by sensitive instruments. At times, however, volcanic edifices deform rapidly due to the internal movement of shallow magma, such as at Showa Shinzan [153] in the early 1940s (Chapter 9), or at Mount St Helens [31] in the early spring of 1980 (Chapter 7). Sakurajima [142], is a composite volcano rising 1100 m above the mostly submerged Aira Caldera in southern Japan (Chapter 10). The influential Japanese seismologist F. Omori studied its powerful 1914 eruption, and noticed that benchmarks around the mountain rose in elevation as magma moved toward the surface before the eruption, inflating the landscape over many square kilometers (Omori 1914). His observations were the first showing that changes in the level and slope of the surface could be used to discern subterranean magma movement. Ground deformation studies have since been applied to many active and erupting volcanoes worldwide, notably achieving refinement at Kīlauea in Hawai'i, where eruptions are usually nonviolent, access is easy, and the climate mild.

There are three basic parameters of volcano deformation that need to be monitored: horizontal distance, vertical elevation, and surface tilt. These three parameters are interrelated, and normally all three variables are involved as any volcano changes its shape. Until recently, each of these geometric variables were measured separately with different instruments and techniques, but increasingly capable, newly developed instruments, many involving earth satellite technology, have revolutionized geodetic studies and can simultaneously measure multiple parameters over large areas. The new instruments are relatively expensive, however, and time-tested traditional monitoring tools are still useful – especially for surveys of smaller areas.

Old-fashioned optical theodolites and stadia rods have been used to measure horizontal distances and elevations for over 200 years, and are inexpensive and useful tools for volcano deformation studies. They are still used for **leveling surveys**, which are the most versatile methods for precise measurement of elevation changes around the flanks of deforming volcanoes. Leveling depends upon a network of benchmarks that must be surveyed repeatedly to measure volcanic deformation over time. For reference to other surveys, these networks must be tied into locations that surveyors assume represents "stable" ground. Such a reference benchmark ordinarily is not far from the base or summit of a volcano, since the ground heave associated with volcano inflation tends to be largely confined to the area of the edifice itself. On island volcanoes mean sea level (MSL) is a handy reference point.

Although horizontal distances can be measured with theodolites, they have low precision compared with various types of electronic distance meter (EDM) devices that have been developed over the past 40 years for precise measurements along lines of sight. Differences in vertical angle (elevation) as well as distance between benchmarks may be combined using modified EDM systems called **total field stations**. Total field instrumentation, as well as EDM equipment is somewhat expensive, but less sophisticated means of measuring tilt and leveling certainly lie within the means of low-budget volcano-monitoring efforts. During the Mount St Helens activity in late 1980–1982, ordinary steel tapes served to measure rapid ground deformation, permitting successful prediction of 13 dome eruptions (Swanson et al. 1982).

Tiltmeters record the slope of the ground. As the interior of a volcanic edifice inflates and fills with new magma, the flanks of the mountain ever so slightly steepen. **Wet tilt measurements**, largely superseded by electronic tilt measurement devices, utilize three water pots connected by hoses arrayed in an equilateral triangle with sides approximately 15 m long. The level of water in each pot, achieving equilibrium, changes as the slope of the ground changes, with water flowing into the lowermost pot. **Dry-tilt** measurements are a simpler alternative to wet tilt, involving theodolites and stadia rods. Tilt measurements are remarkably precise; a well-maintained tiltmeter can detect angular changes of less than a **microradian**, equivalent to the angular change of a kilometer-long board lifted a mere millimeter at one end (Decker and Decker 1997).

The advent of satellite-based surveying using the Global Positioning Satellite **(GPS)** and other systems has revolutionized volcano deformation studies and largely replaced older conventional surveying methods at most modern observatories. GPS surveys allow real-time horizontal and vertical data to be remotely collected in hard-to-access places, and do not require lines of sight between survey points. Portable GPS stations can also be installed on short notice to monitor ongoing eruptions. Space-based **remote-sensing** technology is providing even more capable tools for volcano deformation studies – these are described in a later section.

K. Mogi (1958) first provided a mathematical model relating ground deformation around active volcanoes to volume changes of underlying magma chambers. His work permitted volcanologists to calculate depths to the sources of deformation. Mogi applied his new model to Omori's 1914 Sakurajima data, which showed that between 1895, when local benchmarks were emplaced, and 1914, during the eruption, a 60-km-wide region centered around the middle of Aira Caldera subsided by as much as a meter, while the immediate vicinity of Sakurajima rose by several meters. Mogi determined that the source of deformation lay 8–10 km beneath the center of Kagoshima Bay, about 10–15 km away from the volcano. Subsequent uplift in this broad region beginning in 1919 indicated replenishment of the caldera magma system, which culminated in another Sakurajima eruption and regional subsidence in

1946 (Murray et al. 1995). Several refinements of Mogi's equations to allow for analyses of more complex magma body geometries have subsequently improved the utility of ground deformation studies to interpret subsurface processes.

Gravity and Electrical methods Geophysical monitoring techniques of many varieties complement seismic and ground deformation monitoring. Microgravity studies, for example, seek to detect changes in mass within volcanoes arising from shallow intrusions, although they must be combined with deformation monitoring to filter out elevation changes that also affect gravity. Gravity surveys are also useful for evaluating internal density structures of volcanoes, which can define caldera infill deposits, shallow conduits, and hydrothermally altered bedrock areas. Different kinds of volcanoes have different gravitational responses to intrusions. Changes in gravity are strongest, and in some respects most easily interpreted, around composite volcanoes. Eggers (1987) argued that this was due to displacement of high-density rock by vesiculating magma. An integrated approach to monitoring magma movement through simultaneous measurements of gravity and deformation changes shows promise for evaluating pre-eruptive behavior and may be useful for eruption forecasting (Rymer and Williams-Jones 2000; Williams-Jones and Rymer 2002).

A great variety of electrical methods may be utilized for volcano monitoring, and include measurement of the electromagnetic (EM) fields generated by volcanic activity, induced EM fields generated by large surface transmitters, surface measurement of electrical self-potentials, surface or subsurface studies of heat flow, and perturbation of very-low-frequency (VLF) radio signals by shallow magma conduits. VLF-EM studies at Kīlauea volcano have recently used routinely to evaluate the flux rates of shallow magma movement in pyroducts (Kauahikaua et al. 2003).

Gas monitoring As was discussed in Chapter 3, all magmas contain dissolved gases as important components, and as these magma bodies rise toward the surface or undergo cooling and crystallization, large amounts of these gases are released and rise to the surface. Monitoring of these gases is an important component of overall monitoring efforts, and many techniques are in use. Those methods can be divided into three categories:

1. direct sampling of gases emitted from known fumaroles or soils – which can be done through field sampling for later laboratory analysis or continuous instrumental monitoring;
2. direct air sampling to monitor volcanic pollutants, either through continuously operated air quality monitors or from airborne sampling devices; and
3. remote-sensing techniques – either from satellite-based monitors that categorize and track volcanic plumes as they migrate downwind of active volcanoes or by looking up at volcanic plumes from ground-based instruments. These techniques are discussed in the following section.

As has been mentioned earlier (Chapter 3), the three most important volcanic gases emanating from volcanoes are water (H_2O), sulfur dioxide (SO_2), and carbon dioxide (CO_2), although many more volatile species are usually present in minor amounts. H_2O is the most visible gas, since it readily condenses to the steam seen as "smoke" above volcanoes and volcanic fumaroles. It is the least important gas to monitor, however, because it largely comes from secondary meteoritic sources, and does not reveal much about underlying magmatic systems. SO_2 is much more important, as it is almost entirely derived from magmatic sources and is an atmospheric pollutant of concern (Chapter 13). Interpretations of SO_2 flux variations are made difficult by the fact that it is highly soluble in water, however, and its release from solfataras and fumaroles is greatly affected by rainfall and soil moisture. Nonetheless, it is easy to monitor semi-quantitatively in volcanic plumes by remote-sensing techniques (discussed later), and gives an important measure of magmatic production rates at active volcanoes. CO_2 is increasingly viewed as an important volcanic gas for study, since it is highly mobile, rises quickly to the surface from magmatic sources, is not greatly affected by surface water, and may be one of the best indicators of the emplacement of new magma beneath a volcano (Chapter 3). New analytical tools have made CO_2 flux monitoring much easier and applicable for volcano monitoring (Gerlach et al. 2002). Large declines in CO_2 emissions preceding each of three Plinian eruptions at Mount St Helens [31] in 1980, detected by direct plume sampling, helped to forecast the eruptions of 7 August and 16–18 October (Harris et al. 1981).

Volcano Remote Sensing

In its broadest sense, *remote sensing* refers to any technique that allows observers to monitor volcanic activity from distant vantage points, and could include visual and photographic documentation. Since the development of Earth-orbiting satellites in the past half-century, however, the term has been increasingly associated with observations made from space (Mouginis-Mark et al. 2000). These new methods allow for global monitoring of volcanoes that are either too remote, or at times too dangerous for conventional ground-based study. Quantitative remote-sensing methodologies can be deployed from three platform levels: ground, aircraft, and satellites.

Ground-based techniques The earliest remote-sensing studies involved the analyses of SO_2 emissions in volcanic fume clouds using a correlation gas spectrophotometer (COSPEC). Observers could breathe fresh air, well away from eruptive perils, from fixed points or by driving COSPEC-mounted vehicles back-and-forth beneath plumes to quantify daily emissions of sulfur. Newer techniques and instrumentation, including miniaturized UV-spectrometry (FLYSPEC), Fourier Transform InfraRed Spectroscopy (FTIR), and mini-Differential Optical Absorption Spectroscopy (DOAS) analyze the ratios and absolute amounts of a wide variety of gas concentrations issuing from volcanoes (Platt and Stutz 2008). Since such ratios can systematically change before, during, and after eruptions, these new techniques are powerful and relatively inexpensive tools for gas geochemists. Ground-based radar in support of civil aviation has proven remarkably effective in tracking the appearance and expansion of large eruption ash clouds when eruptions occur near major airports, as during the Mount St Helens eruptions (Chapter 8).

Airborne techniques Airborne COSPEC monitoring is useful, though it is far more expensive than ground-supported studies, and cannot be repeated following the same precise routes. Thermal infrared multispectral scanning (TIMS) and the related multispectral infrared and visible imaging spectrometer (MIVIS) system hold more promise for airborne applications, permitting researchers to map volcanoes in reconnaissance detail. TIMS work is especially useful in discriminating lava flows of different ages based upon weathering and other aging characteristics (Kahle et al. 1988). Highly detailed and accurate mapping is also facilitated by topographic synthetic aperture airborne radar (TOPSAR), which creates digital elevation models (DEMs) of volcanoes. Francis et al. (1996) cite the example of the DEM mapping of a lava flow on Hekla [92] having a variable thickness of 5–33 m, with relief of up to 15 m on its surface. Ridge lines as small as 2 m high could be mapped remotely, with minutes of data collection and a few days of processing in the place of months or years of arduous groundwork.

Satellite techniques Satellite observation has proven to be a low-cost and effective alternative to ground monitoring for many remote volcanoes. The technology has yet to achieve its full potential, but is already showing exciting possibilities. The earliest non-military detection of volcanic heat from space took place over Surtsey volcano [90], Iceland, in 1967 (Williams and Friedman 1970). Since then, volcanic reconnaissance has become an important service of many different satellites. Five kinds of space platforms have demonstrated different kinds of utility (Francis et al. 1996): High-resolution satellite imagery has great potential for volcano monitoring and evaluation of hazards and risk when available to civilian scientists. In one example, satellite monitoring of eruptive activity during the 2010 eruption of Merapi [131] volcano (Indonesia) supported emergency evacuation efforts that saved an estimated 10,000 to 20,000 lives (Surono et al. 2012; Pallister et al. 2012).

1. High-spatial-resolution sensors, such as the LANDSAT thematic mapper (TM) and the French Systeme Probatoire d'Observation de la Terre (SPOT) sense visible and infrared energy reflected or radiated from Earth's surface back into space. They can resolve the details of landforms down to the scale of 10–30 m, which can assist in a large-scale study of volcanic landforms. De Silva and Francis (1991) scrutinized TM images of the central Andes to estimate the number of potentially active volcanoes in that region, based upon degrees of post-ice age erosion and volcanic deposition. The same locality may be observed by TM or SPOT cameras from once a week to once a month, depending upon the satellite and its orbit. This makes these sensors of little use in examining ongoing eruptions, but they can provide the first information we get about volcanic change in little-visited localities. Rowland and Munro (1992) document a large intra-caldera avalanche and eruption within Fernandina caldera [59] in the Galapagos, which came to the attention of volcanologists, thanks to SPOT reconnaissance. TM observations, together with those of the advanced very-high-resolution radiometer (AVHRR) have also been the primary means of monitoring eruptions at Láscar volcano [82], Chile (Francis and Rothery 1987; Wooster and Rothery 1997). Despite its name, the AVHRR lacks the fine resolution of TM and SPOT surveys. But this system can provide data at least four times a day for any volcano in the world, making it potentially valuable as a means of monitoring eruptions. It provides data changes in the heat emission of large bodies, including lava lakes (Wiesnet and D'Aguanno 1982), cooling lava flows (Oppenheimer 1991), and pyroclastic deposits (Harris et al. 1997). AVHRR images are wide-ranging – to 3000 km, and may be downloaded for study within 10 minutes.
2. Environmental satellites, like high-resolution sensors, detect visible light or infrared radiation, and can obtain multispectral images. Their spatial resolution is low (hundreds of meters), as they image large areas and can survey the entire globe. This makes it possible to spot volcanic plumes that might be missed by other, more narrowly focused satellite scans. Such satellites travel in polar orbits, providing frequent repeat viewing of single locations. Hence it is easy to track the progress of eruption plumes to warn people downwind, including aviators. The total ozone mapping spectrometer (TOMS), aboard the Nimbus-7 and Meteor-3 satellites, is intended to monitor the condition of the ozone layer, but has also proven invaluable in monitoring sulfate aerosol injection of the stratosphere during big explosive eruptions. The Moderate Resolution Imaging Spectroradiometer (MODIS) instruments carried aboard the Terra (EOS AM) and Aqua (EOS PM) satellites have replaced TOMS for volcanic plume monitoring.

3. Space Shuttle and related space vehicle photography has produced spectacular images of volcanoes and eruptions from space, and are readily available from government facilities or private vendors. Unfortunately, such platforms are not in continuous operation, like most other satellites, and are only of use when subject volcanoes are cloud-free.

   Synthetic aperture radar (SAR) methodologies, first developed for terrain mapping of remote or inaccessible areas from aircraft, can now be employed on satellites, which cover vast areas of Earth. Rowland et al. (1993) used SAR data to examine Aleutian and Alaskan volcanoes, which, for other approaches, are often very difficult to study owing to inclement weather and shortage of daylight. Radar analysis enabled them to map volcanic deposits around Aniakchak caldera [10], Westdahl [6], and Spurr [24] volcanoes. More impressively, perhaps, radar showed the presence of dimples in the ice cap of Veniaminoff [9] volcano, which may have marked the occurrence of recent subglacial eruptions. The Phased Array L-band SAR (PALSAR) instruments carried on Japanese satellites are able to provide detailed terrain images through dense vegetation – very useful for geologic mapping of tropical volcanoes.

4. SAR **radar interferometry** has proven to be extremely useful for remote detection and analysis of volcanic deformation – for volcanoes not covered with deep vegetation, and also for volcanoes covered by ice (Scharrer et al. 2008). The basic methodology was proven following the 1992 M7.3 Landers, southern California earthquake, when SAR images taken from almost 800 kilometers above Earth were compared with images taken by happy chance not long before the seismic displacements. Wavelength analysis provided a precise measure of how distances had changed between satellite and ground as a result of the earthquake, leading to the construction of an **interferogram (InSAR)**, or ground displacement map (Massonet et al. 1993). In just a few moments, SAR had gathered data that would have taken years to collect using standard methods, and revealed patterns of strain that might never have become evident otherwise. Scientists quickly recognized that this technique also holds great promise for volcanic ground deformation monitoring. For example, interferometry showed uplift from renewed dike injection beneath the floor of remote Okmok caldera [4] in the Aleutian Islands following a 1997 eruption, with shifting loci of uplift tracking the migration of the shallow melt. Since no other means of closely monitoring the volcano existed at the time, SAR satellites became instant, low-budget "volcano observatories." An example of an interferogram that demonstrates volcanic deformation on the island of Hawai'i over a four-year period (Figure 14.29) demonstrates the utility of interferograms in monitoring volcano deformation over large areas. Interferograms have even been made using ground-based monitoring of particular active volcanoes (Di Traglia et al. 2014). But in any case, they are notoriously difficult to analyze, and their interpretations may be ambiguous and dependent on the experience of the investigator involved. Conventional methods have the advantage of greater precision and simpler interpretation, and can be conducted at low cost with traditional survey equipment over shorter – often critical – intervals of time, but they only reveal local changes at specific survey points.

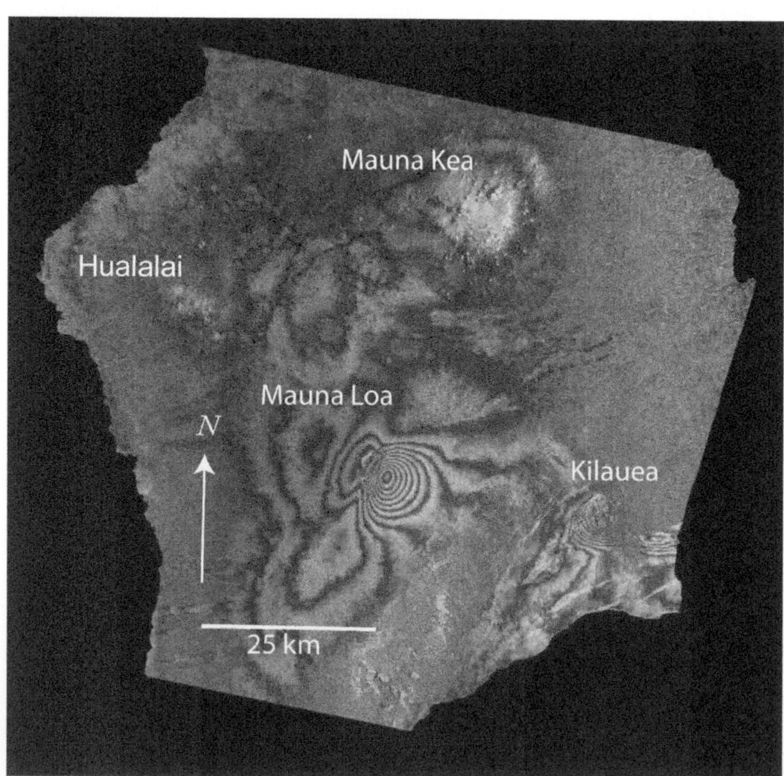

FIGURE 14.29 SAR interferogram of part of Hawai'i Island, showing deformation of Mauna Loa [15] and Kīlauea [17] volcanoes over a four-year period (27 January, 2003 to 1 January, 2007). Surface deformation between those two time periods is represented by colored fringes. Each set of fringes represents 2.8 cm of deformation along the radar's line of sight, but because the radar's look angle is inclined from vertical (about 22° in the image), the observed deformation is a mix of vertical and horizontal motion. As a result, many deforming regions (for example, southeast of Mauna Loa's caldera) have an asymmetric appearance. A progression of fringes from pink to yellow to blue indicates decreasing range (distance) between the satellite and ground. The opposite fringe progression indicates lengthening distance between the satellite and ground. Fringes in the summit region of Mauna Loa indicate inflation, with the double-lobed pattern suggesting both a spherical magma reservoir at depth just southeast of the caldera plus a tabular magma body running the length of the caldera. Inflation is also occurring at Kīlauea's summit, although the East Rift Zone is deflating. The fringes around Mauna Kea and northwest of Mauna Loa are related to atmospheric distortion and not to deformation. Source: Interferogram and caption by Mike Poland, USGS.

Volcano crisis management

Volcanic disasters are usually preceded by volcanic crises or premonitory activity, and this pre-eruptive period is the most crucial time period for scientists, emergency managers, and political authorities to coordinate their activities to avert preventable losses. Pre-eruptive crises involve complex interplays among the scientists themselves, and between scientists, emergency managers, and the political authorities who must ultimately make the difficult fiscal and humanitarian decisions required to protect lives. Successful management of volcanic crises and disasters is best achieved when there is a prior awareness of potential hazards by *all* of the above, and the roles of these key players are well defined in advance. Unfortunately, this is all too rarely the case, and in practice, these relationships usually need to be established or refined after a volcano becomes restless or a disaster strikes. The most important factor in successful crisis management is the establishment and maintenance of effective channels of communication between each of these groups and with the media representatives who will ultimately convey most of the information to the affected public. As noted earlier, effective, ongoing public education about volcanic risk and hazards can facilitate this communication and enable timely citizen responses to hazards alerts as they are issued. Scientists have the most critical roles to play during volcanic crises, not only because of their knowledge but because they initially have great public credibility. Maintaining that credibility is important, and depends on their professional behavior (IAVCEI 1999).

Fiske (1984) presents a particularly telling example of the interaction of volcanologists, government, and the media in the case of two nearly simultaneous, close-by volcanic eruptions in the Caribbean. La Soufrière-Guadeloupe[2] [85] is an eroded, 12-km-wide composite volcano surmounted by a younger cone at the southern end of Guadeloupe Island, about 500 km southeast of Puerto Rico. About 72,000 people lived on the flanks of La Soufrière in 1976, most of them in the district capital of Basse Terre on the west coast of the volcano. Earlier seismic monitoring indicated that the base-level earthquake count for La Soufrière is 1–10 shallow tremors a month. Between July 1975 and July 1976, however, this number steadily rose to over 200 events a month. Plainly magma was on the rise somewhere in the volcano, though little scientific attention was given to the matter. On 8 July, 1976, the first of many explosions suddenly burst forth as dense clouds of ash and steam roiled from summit fissures. This triggered panic as local residents fled the mountainside. Over the next month, intermittent ash outbursts continued while steam emissions increased, and earthquake numbers remained high. On 12 August, scientists declared that falling ash now included fresh glassy material, not the pulverized older rock fragments that had characterized the initial Ultravulcanian phase of the eruption. This suggested that magma had ascended to a shallow level beneath the summit and had begun discharging directly to the surface. To some of the scientists, this indicated that dome growth and violent explosions were possibly imminent. Remembering the Mount Pelée [87] volcanic disaster on nearby Martinique 74 years earlier (Chapter 1), these scientists urged the local governor to begin immediate evacuation of the tens of thousands of residents remaining in the area.

It later came as a shock for the governor and members of the public to realize that the volcanologists urging evacuation had spoken in haste. In fact, the fine glassy ash had been misinterpreted. Later study indicated that no juvenile material had been present. But the damage had been done. Organized evacuation of any large population is not only a logistical nightmare, made worse when pressured by immediate dangers, but an economic and social catastrophe as well. The population of Guadeloupe was not wealthy, depending heavily upon agriculture, fishing, and tourism for sustenance. These slim sources of income disappeared as people rapidly streamed from La Soufrière to safer quarters of the island. The government evacuated nearly a quarter of the endangered population within the first two days.

Meanwhile, more than a dozen other scientists converged on the island, mostly from France, and began work out of a hastily assembled volcano observatory at Fort St Charles. There was little coordination between these scientific groups, and no single spokesperson who could speak for all of them. Instead, the observers haphazardly worked in separate groups, in two rival teams. Members of the press wandered freely in the field, including areas that were closed to the public. Ever eager for a story, media personnel hurriedly interviewed members of the different teams, amplifying scientific differences and speculations publicly. This deepened personal rivalries among some of the scientists, exacerbated public fears, and created needless tension for authorities who were trying to constrain the impact of the crisis. One group of scientists concluded that the eruption was not so dangerous as originally feared. The other held fast to the alarmist scenario. During the next few months, as volcanic activity gradually waned, public disagreements that should have been privately discussed in the setting of an organized team raged in public. In the end, the French government convened a special committee, including foreign volcanologists, to develop a scientific "consensus" on the situation, based upon all available information. The committee concluded that the doomsayers had erred in their judgment. Careers were needlessly savaged, the economy of Guadeloupe wastefully disrupted, and the credibility of the volcanological community was greatly diminished.

[2] There are two volcanoes named "La Soufrière" in the lesser Antilles – this one, more specifically referred to as "La Soufrière-Guadeloupe," and another 300 km to the south on the island of St Vincent.

Less than 3 years later another crisis struck the Caribbean when a different volcano, also coincidentally named Soufrière [86], began erupting on St Vincent island, about 300 kilometers south of Guadeloupe. This eruption occurred only 77 years after a tragic eruption in 1902 had killed almost 1700 people on the island, so there was great potential for panic. In this case, however, a well-established seismic monitoring and tiltmeter array operated by the University of the West Indies provided advance warning of impending activity and prevented panic as about 20,000 people evacuated from threatened areas. Only 5–7 scientists converged on the island during the eruption, organized under a single aegis at a volcano field station that was kept strictly off-limits to the press. A single scientific consensus presented by telephone or in writing reached the media through appointed representatives, while differences in scientific opinion were resolved privately. Although the explosive eruption caused great agricultural damage owing to tephra fall and formed a new dome, no lives were lost (Fiske and Sigurdsson 1982). The lessons from these eruptions have been learned, and have been effectively applied in many (though not all) volcanic crisis situations ever since.

Whims of fortune and judgment also play an important role in the "management" of every volcanic crisis. Had the volcanic ash on La Soufrière not been misidentified, and a major eruption had taken place, how would the situation have played out? Had a general in command of Clark Air Force Base in the Philippines not taken seriously the timely warning of USGS and PHIVOLC scientists to evacuate Clark Air Force Base before the climactic 1991 Pinatubo eruption (Chapter 1), how many more people might have died there?

When life and death decisions must be made as to whether people should be evacuated from threatened areas, how far they should be moved for safety, and how long they should stay away, the most important thing volcanologists can do is to first reach observation-based "best-judgment" consensus among themselves. They must then convey consistent recommendations to the government authorities and agencies that will ultimately make the difficult decisions required to save lives. Those decisions depend not only on scientific forecasting, but also on critical economic, political, and social factors that must ultimately be considered.

This does not mean, however, that volcanologists can simply provide their factual observations and recommendations to government authorities and then walk away from all responsibility as to how (or even if) their recommendations are acted upon. The tragic 1985 eruption of Nevado del Ruiz [71] (Colombia) was a type example of what can happen when knowledgeable volcanologists fail to concern themselves with the "end-use" of their findings. Ruiz gave ample warning of an impending eruption when small earthquakes and increased fume emissions were noted from the ice-clad, 5300-m-high summit crater in November 1984 – a year before the climactic eruption. Local geologists, who had no previous experience dealing with restless volcanoes, realized the potential hazards involved, and alerted national Colombian officials about the potential risks to populated areas. John Tomblin of the United Nations Disaster Relief Organization (UNDRO) was invited by Colombian officials to assess the situation in March 1985, and afterward issued an international appeal for technical assistance (Mileti et al. 1991). Unfortunately, there was little response until a large phreatic eruption occurred on 11 September, 1985, deposited ash around the volcano and sent a lahar almost 30 km downslope. UN-supported technical experts came to Colombia from Costa Rica, Ecuador, Italy, and the United States. They provided advice, some seismic monitoring equipment, and recognized the extreme risk to cities located along river valleys downslope, but made no efforts to see that their findings reached the people at risk. The Colombia Geological Survey (INGEOMINAS) took the lead in assessing the hazards, but had little success in convincing others about the gravity of the situation. INGEOMINAS published an excellent risk map on 7 October, but made no major effort to contact threatened populations directly. The Colombian Civil Defense Agency and local authorities made uncoordinated efforts to educate people living in high-risk areas, but were thwarted by sensational stories in Colombian newspapers, which in some cases accused government scientists of attempting to lower property values for personal gain. The vital risk map was ultimately distributed to 17 different government agencies, but insufficient efforts were made to present the map to affected citizens, although it was published by a national newspaper on 9 October (Herd and Comite de Estudios Vulcanologicos 1986). In the end, 23,000 people died on the night of 13 November (Voight 1990). This could easily have been avoided.

"Top-down" volcanic crisis management works fine to assess risks and to coordinate agency responses, but to save lives, successful crisis management requires "bottom-up" input and participation from the people whose lives are in jeopardy. Unless local citizens and their leaders are directly informed about their peril by credible scientists and trusted higher-level authorities, and become personally involved in the means of their salvation, disasters like that of Ruiz are bound to be repeated in the future. When effective communication channels are never established (ideally before disaster strikes), confusion and tragedy are likely to result (Figure 14.30).

Public education about volcanic hazards helps pave the way for such communication channels to open quickly and effectively as need demands. Consider, for example, that community-based education carried out by dedicated volcanologists from the Rabaul Volcano Observatory facilitated the timely self-evacuation of more than 10,000 people when Tavurvur volcano [160] erupted in the middle of the night in 1994. Governments that tend to be paternalistic in nature assume that bureaucratic efforts alone can protect their citizens, but a well-informed population can take measures to save itself and avoid mass tragedy as soon as warnings pertaining to volcanic eruptions are issued.

Another very important aspect of volcanic crisis management concerns the relationships between volcanologists and the media. The media (newspapers, radio, and television) are the most important intermediary between scientists and citizens who are

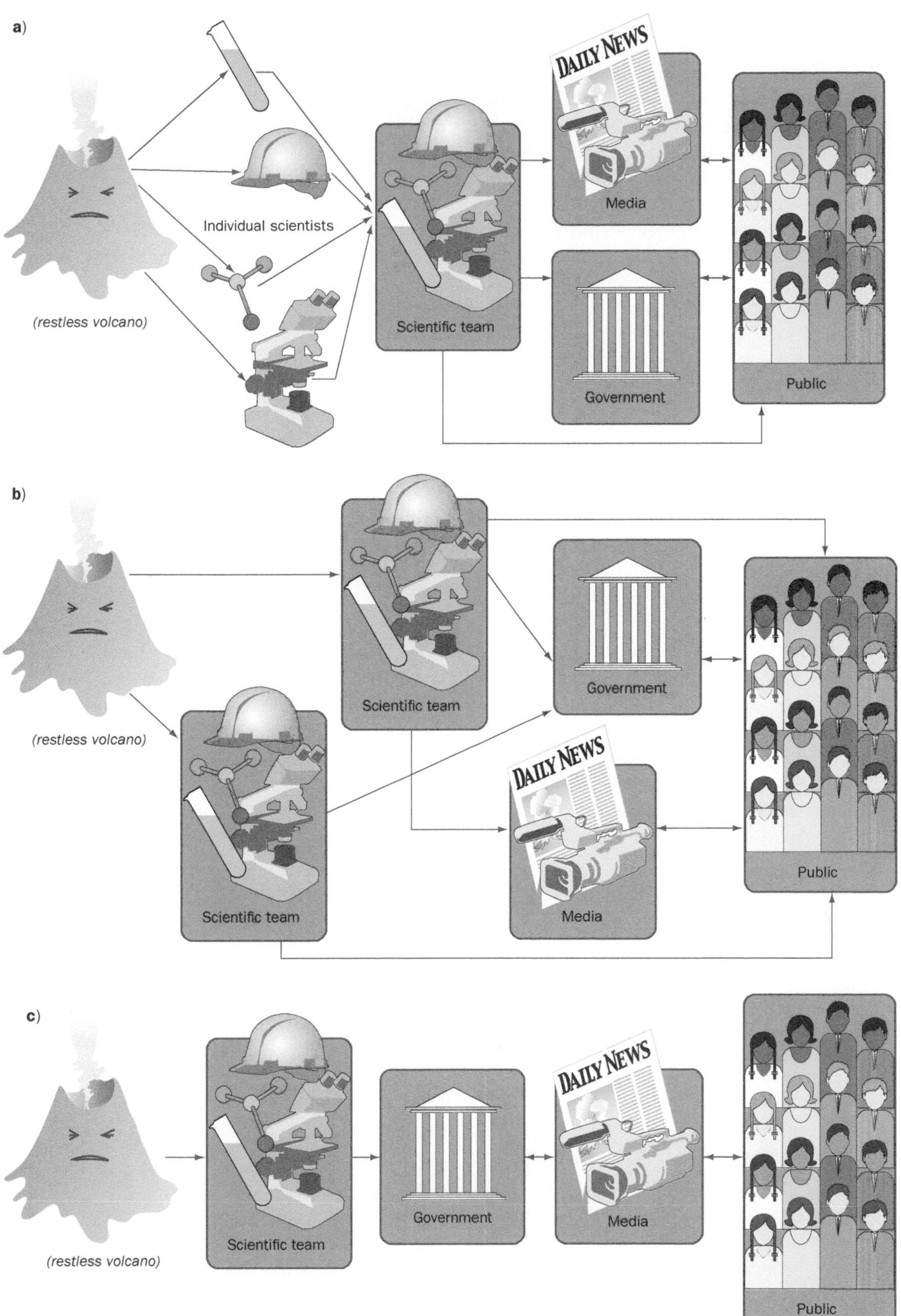

FIGURE 14.30 Volcanic crisis-management relationships flow sheet. (a) Ideal relationships: (1) The volcano provides data to individual scientists. (2) The scientists form one team and speak with one voice. (3) The scientists first provide information to government authorities, then via press conferences to the media, then directly to the affected public via public meetings. (b) The La Soufrière-Guadeloupe [85] experience (1976): (1) The volcano provides data to independent scientific teams. (2) The individual scientific teams contact the media, then the authorities and public with their individual opinions. (3) Mass confusion and misunderstandings result. (c) The Ruiz [71] experience (1985): (1) The volcano speaks to scientists, who form one team. (2) The scientists speak mostly to government authorities. (3) The media provides confusing reports to the public. (4) The public receives mixed messages that are mostly ignored – a mass tragedy results.

at risk. This is because the media are able to reach a much broader audience of potentially vulnerable people than feasibly could be reached by scientists through public meetings, and in many cases, the media will have already established great credibility with their audiences. It is therefore critical that scientists know how to communicate effectively with media personnel and that a scientifically credible representative be chosen to act as a single spokesperson for scientific teams during a crisis. This person needs to understand and respect the special needs of reporters who can become either important allies – or harsh adversaries if good relationships are not carefully fostered.

There are generally two types of media representatives who attend press conferences and/or seek interviews with scientists during major volcanic crises or disasters; the approaches and needs of these two groups tend to differ greatly. **National and international reporters** are often interested in sensational headline-grabbing stories and may be prone to exaggeration to attract an audience. Depending on circumstances, they also face clear-cut deadlines that may vary significantly from those of the more laid-back local media. Such persons should be treated more formally, with a certain degree of caution to avoid misunderstandings and damaging misquotes. Reporters representing **local media**, however, fall into a second distinct category and are the most important messengers of critical information that local citizens need to understand for personal survival. These reporters usually have a vested interest in "getting the story right" and are the ones whose friendship and trust are best cultivated *before* a volcanic crisis develops. In an ideal situation, local volcanologists will have established relationships with these reporters long before the "outside" media descend during a major crisis. One can speak frankly with them about details that might be impossible or dangerous to discuss at formal press conferences when video cameras are rolling and tape recorders are turned on!

Another factor in crisis management involves the length of the pre-eruption crisis. Eruptions preceded by short, intense periods of forewarning activity (e.g., Kīlauea 2018) are more easily dealt with by public officials than are those preceded by several months or years of pre-eruptive activity. People can too easily become used to dangerous pre-eruptive events if they occur over a long enough period, and the warnings of volcanologists may become less and less credible. The catastrophic May 1980 eruption of Mount St Helens (Chapter 7) was preceded by less than two months of relatively low-level activity. What terrible loss of life might have occurred had "nothing much" continued to happen until the peak of the summer tourist season? Would public officials have continued to heed the warnings of scientists as political and economic pressures to open the area to the public increased?

Volcanic eruptions will always be accompanied by hazardous activity, and the risks to society posed by that activity are constantly growing, as populations increase near active volcanoes. To lessen those risks, new generations of volcanologists must continue to develop enhanced techniques to understand volcano behavior better and for volcano monitoring. The odds that a major caldera-forming eruption will occur in a densely populated area in the near future are very low, but such an eruption will occur one day, and one can hope that by then future volcanologists will have developed reliable eruption prediction tools that will allow credible warnings to be given before disaster strikes!

Questions for Thought, Study, and Discussion

1. What is the most important work a volcanologist can do – volcanic *hazards* evaluations, or volcanic *risk* mitigation? Defend your answer!
2. Does a volcano have actually to be in eruption to be considered active? Explain.
3. Why are dormant volcanoes especially dangerous?
4. What is the difference between "primary" and "secondary" volcanic hazards? Provide an example of each category.
5. Why are tropical maar lakes apt to be more dangerous than their counterparts in more temperate climates?
6. Under what circumstances are Poisson analyses appropriate for the evaluation of volcanic risks?
7. What are probability event trees, and how are they useful?
8. What is the value, and what are the limitations of volcanic hazards maps?
9. How can the hazards posed by lava flows, lahars, and volcanic lakes be mitigated through human intervention? Is "messing with Mother Nature" appropriate?
10. What are some easy-to-make mistakes in volcano crises, and what are some successful strategies to be considered?
11. No human activity can ever be made completely "safe" – as some degree of risk is involved in crossing streets. In the case of living on or near active volcanoes, how much risk should society tolerate for those wishing to live or work in hazardous areas? What are the boundaries between personal freedoms and societal responsibilities to "keep people safe"?
12. In the case of settlements in highly exposed areas on the slopes of Volcán de Colima [51], people have refused to be relocated, or even evacuated during volcanic crises, even when they were fully informed of risks. At Popocatepetl [56] during an eruptive crisis in 2000, residents of endangered villages refused to evacuate, saying *"If we are going to die, we prefer to die here."* This becomes a very complex ethical issue: Do civil authorities have the right to force a relocation or an evacuation of such people?

FURTHER READING

Blong, R.J. (1984). *Volcanic Hazards: A Sourcebook on the Effects of Eruptions.* Orlando: Academic Press.

Dzurisin, D. (2007). *Volcano Deformation: Geodetic Monitoring Techniques.* Chichester, UK: Springer-Praxis.

Fearnley C.J., Bird, D.K., Haynes, K., McGuire, W.J., and Jolly, G. (eds.) (2019). *Observing the Volcano World – Volcanic Crisis Communication.* New York: Springer, 320 pp. https://doi.org/10.1007/11157_2017_28; https://doi.org/10.1007/11157_2016_37

Kieffer, S.W. (2013). *The Dynamics of Disaster.* W. W. Norton& Co.: New York, 315 pp.

Later, J.H. (ed.) (1989). *Volcanic Hazards: Assessment and Monitoring.* Berlin: Springer- Verlag.

Lowe, D.J. and de Lange, W.P. (2000). "Volcano-meteorological tsunamis, the CE 200 Taupo eruption (New Zealand) and the possibility of a global tsunami." The Holocene, 10, 401–407 pp.

Mader, H.M., Coles, S.G., Connor, C.B. et al. (2006). *Statistics in Volcanology.* London: Geological Society of London.

Mouginis-Mark, P.J., Crisp, J.A., and Fink, J.H. (eds.) (2000). *Remote Sensing of Active Volcanism.* Washington, DC: American Geophysical Union.

Papale, P. (ed.) (2015). *Volcanic Hazards, Risks, and Disasters.* Amsterdam: Elsevier. 505 pp.

Scarpa, R. and Tilling, R.I. (eds.) (1996). *Monitoring and Mitigation of Volcano Hazards.* Heidelberg: Springer-Verlag.

Varley, N., Connor, C.B. and Komorowski, J.C. (eds.) (2019). *Volcán de Colima: Portrait of a Persistently Hazardous Volcano.* Berlin: Springer. ISBN 978-3-642-25911-1 (eBook). https://doi.org/10.1007/978-3-642-25911-1

Economic Volcanology

Source: New Zealand Free Photos, https://en.wikipedia.org/wiki/Wairakei_Power_Station#/media/File:Wairakei_Geothermal_Power_Station-5834.jpg. Licensed under CC BY-SA 3.0 nz.

Volcanoes: Global Perspectives, Second Edition. John P. Lockwood, Richard W. Hazlett, and Servando De La Cruz-Reyna.

Companion website: www.wiley.com/go/lockwood/volcanoes2

Volcanoes are nature's forges and stills where the elements of Earth, both rare and common, are moved and sorted. Some elements are diluted and some pass through concentrated into those precious lodes that people seek for fortune or industry.

(Decker and Decker 1979)

The few key, basic activities separating the human species dramatically from the rest of the living world include widespread agriculture, the ability to control fire (energy), and the utilization of metals (mineral ores). As we saw in Chapter 13, volcanoes are economically important because they provide nutrients for the rich growing soils of volcanic areas, but they also transport great amounts of exploitable heat energy to Earth's upper crust, and are responsible for the formation of many mineral deposits. In this chapter, we will focus on the role of volcanic energy in its many forms, and on the ways volcanic processes concentrate metallic elements and form other deposits of economic use to society.

Earth Energy Relationships

Measurements of increasing temperature with depth in drill holes and deep mines show that huge amounts of thermal energy are being transferred upward to the surface of Earth by conduction ("heat flow" – Chapter 3). Araki et al. (2005) estimate that the total amount of heat released from Earth's surface by heat flow from below (to the atmosphere and oceans) is around 3×10^{13} watts (9.8 $\times 10^{20}$ joules/year). The input of energy to Earth's surface from absorbed solar radiation is much greater yet – estimated at 5×10^{24} joules/year (extrapolated from data of Li et al. 1997). The energy released by volcanic eruptions turns out to be miniscule relative to the overall Earth energy budget, and is estimated by Verhoogen (1980) at less than 8×10^{11} watts (2.6×10^{19} joules/year), or only about 2% of the conductive heat flow production! 8×10^{11} watts is still a staggering value, given the tiny relative areas of the vents and fissures through which all that energy escapes relative to the whole surface of the planet. Long-term heat release from the active Icelandic volcano Grímsvötn [99] alone is on the order of 2–4 gigawatts, enough energy to provide the needs of 3–4 million homes (Reynolds et al. 2018).

Volcano Energy

Volcanic eruptions release vast amounts of energy in several different forms, many not normally included in a review of this subject, but each important in its own way. We find it useful to separate volcano energy into two categories for the sake of discussion: (1) **eruptive volcano energy**, which reaches Earth's surface directly during eruptions; and (2) **stored volcano energy**, trapped within and beneath volcanoes as a result of intrusive and seismic activity.

Eruptive Energy

Most eruptive volcano energy during effusive eruptions is released as thermal energy by cooling lava or ash brought to the surface. Kinetic energy is also released through physical impact as lavas rise and flow downhill, but this is rarely considered in calculating the total energy budget of an effusive eruption. Thermal energy is almost entirely radiative as lava flow surfaces cool, but flows may also continue transferring energy conductively to percolating rainwater and to adjacent rocks – sometimes for many years in those rare instances where deep lava lakes are formed (Chapter 6).

The energy released during explosive eruptions at "gray" volcanoes is also primarily thermal in nature (Hedervari 1963; Sparks 1986; Pyle 1995: Yokoyama 1956; Yokoyama 1957). Other forms of energy can also be released, though they are ordinarily of lesser magnitude. These include the energy of shock waves transferred through the atmosphere (sonic waves) or water (tsunami waves), and the kinetic energy represented by movement of volcanic ejecta or mass wasting of the volcanic edifice. In explosive eruptions, most of the released thermal energy is either transferred quickly to the atmosphere by buoyant clouds of ash and gas (Sparks et al. 1986), or more slowly by the cooling of PDCs (Pyroclastic Density Flows – Chapter 8). Phreatic eruptions are a special case – where the principal energy release may be kinetic rather than thermal (Shimozuru 1968).

Comparison of Eruptive Energy with Other Forms of Natural Energy Release

To witness a major explosive eruption, where huge amounts of hot ash and gases are ejected from Earth, or a large effusive eruption where millions of cubic meters of molten rock flow across the land, one would naturally assume that volcanoes are one of the most important energy producers among all natural surface phenomena. But – is this true? The energy of explosive eruptions is usually

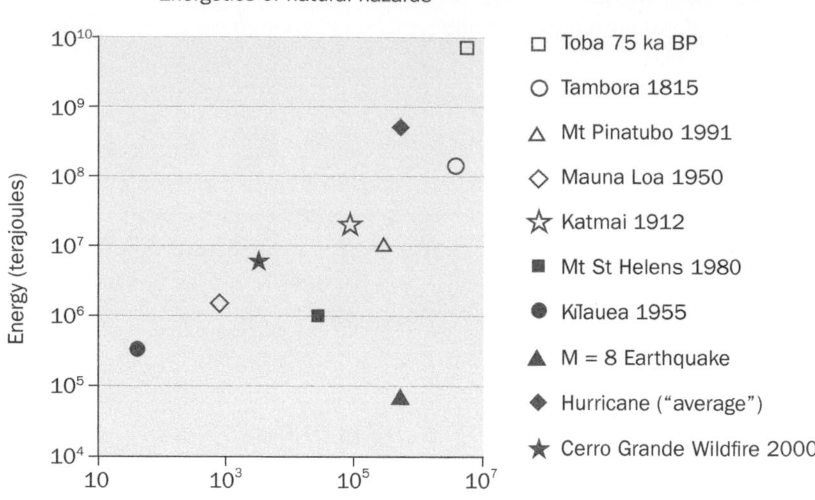

FIGURE 15.1 Comparison of the energy and power of volcanic eruptions and other natural hazards. Data obtained from various sources, including Emanuel (1999), Kasahara (1981), and Pyle (1995). For reference, 1 terajoule is equivalent to the explosion of about 240 tons of TNT. Sources: Adapted from Emanuel (1999); Kasahara (1981); Pyle (1995).

more impressive and destructive than that of longer-lived effusive eruptions since it is released in a much shorter period of time. In other words, explosive eruptions show much greater **power** (where the term "power" signifies *rate* of energy expenditure). While a magnitude 8 earthquake pales in comparison with the energy of larger volcanic eruptions, it is far more powerful than most eruptions in general, and in terms of kinetic energy release, potentially far more destructive. Even then, most eruptions are dwarfed in terms of total energy release by other hazardous natural phenomena, which occur somewhere on Earth every year (including wildfires and cyclonic storms). A plot of total energy versus power (Figure 15.1) shows that, though the infrequent great eruptions of VEI (Volcanic Explosivity Index – Chapter 5) greater than 7 (Chapter 8) can "compete" with other natural phenomena, the more frequent large earthquakes, hurricanes, typhoons, and wildfires exceed most eruptions in these respects; and even average hurricanes are stronger than all but the largest eruptions in terms of energetics and power (Emanuel 1999).

Stored Energy: Geothermal Power

Although the amount of energy transferred directly to Earth's surface by volcanic eruptions is large (see Verhoogan's estimate earlier), most volcanic energy is stored within and beneath volcanoes, and reaches the surface only by slow conduction or by upward migration of heated groundwater. Geothermal resources (in the form of hot springs) have long been used by humans for bathing and for cooking food, but the exploration for deeper, higher temperature resources that could be used to produce electricity only began in 1904, with pioneering efforts by Prince Piero Conti in the Larderello area of Italy (DiPippo 1988). Geothermal production of electricity was insignificant until after the middle of the twentieth century, when increasing petroleum prices and evolving technologies allowed geothermal electric power production to accelerate rapidly (Lund 2000, Figure 15.2). The generation of electricity from geothermal resources has increased an annual average of 5 percent in recent years, with geothermal power plant capacity doubling worldwide from 2005 to 2015 (GEA 2015) – in large part owing to the expected rise in economic and environmental costs of fossil fuel. Nearly 700 geothermal power stations presently operate around the globe, with the United States leading the world in geothermal electricity production. Still, geothermal power remains only a bit player on the world energy scene; less than 0.7% of total energy use. Bear in mind though, that some regions (e.g. the country of Iceland and the San Francisco metro area) depend quite heavily indeed on geothermal resources, and that perhaps only 6–7% of the total worldwide potential for generating electricity geothermally has so far been tapped (Matek 2015). "Earth-heat" appears to have a bright energy future.

Most polygenetic volcanoes are underlain by magma chambers at some point during their active lives (Chapter 3), and direct energy recovery from these molten bodies has been frequently proposed (principally in the popular media). Tapping energy directly from a magma chamber might seem to be a promising proposition, but it appears such efforts are futile, since experimental AQ1
placement of heat exchangers directly into magma (as into the Kīlauea Iki lava lake – Chapter 6) shows they become immediately encased in an insulating glassy selvage, making extraction of heat inefficient. Magma chambers do transfer vast amounts of thermal energy to adjacent rocks over long periods of time, together with dikes and other intrusive bodies (Chapter 4). Volcanic earthquakes and tremor that typically accompany or precede eruptive activity also provide heat energy (Cristofolini et al. 1987). The shallow crust beneath and within the volcanoes stores this heat effectively because of the low thermal conductivity of volcanic

rocks. Individual volcanoes and their surrounding rocks can remain at elevated temperatures for tens of thousands of years after all eruptive activity has ceased.

Although Earth's temperature increases with depth everywhere, only areas located near volcanic fields are characterized by the high-temperature rocks necessary to support commercially feasible, large-scale geothermal electrical energy production. Unfortunately, hot rock alone is not sufficient for energy production, as no technology exists for the direct conversion of hot rock or magma heat to power. To exploit volcanic heat from hot rocks, water must be present (either as hot water or steam) as an intermediate medium to transport heat to surface generation facilities, and the rocks themselves must be sufficiently permeable to allow this water to circulate. Three types of volcanic geothermal systems (Figure 15.3) have proven to be commercially exploitable for the generation of electrical power: **Vapor-dominated systems**, where water is present as steam that can be used to directly power turbines; **water-dominated systems**, where hot water (usually above about 200°C) can be brought to the surface, partially flashed to steam to drive turbines; and **moderate-temperature water systems**, where water temperatures are too cool to directly drive turbines, but where another fluid with a lower boiling point (e.g. butane, pentane) can be vaporized to power turbines in binary fluid

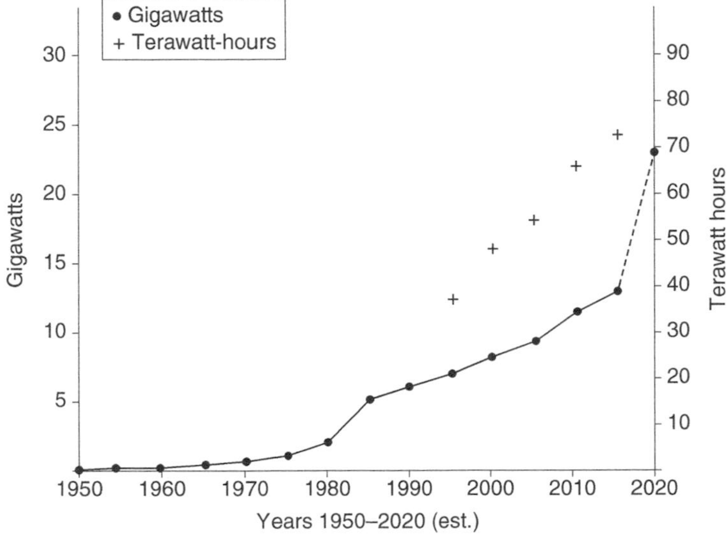

FIGURE 15.2 History of worldwide geothermal power generation capacity (gigawatts) and production (terawatt-hours). Data from various Geothermal Energy Association informal reports.

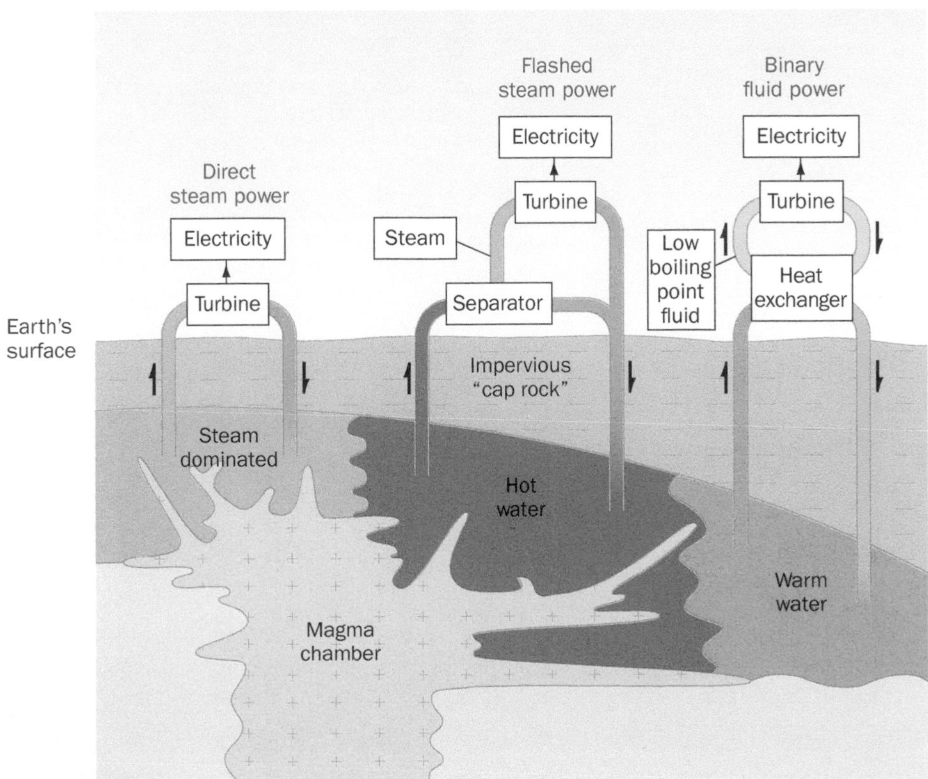

FIGURE 15.3 Schematic diagram of three geothermal reservoir types and power generation facilities above a magmatic system.

systems. The **vapor-dominated** systems are the most efficient and profitable geothermal fields, but are also the least common. An example is the Geysers field north of San Francisco, the world's largest, which has generated as much as 2000 megawatts (MW) of power in the past. Poor steam reservoir management has now lowered current production to about 1000 MW, still enough to supply the energy needs of San Francisco. Experience at the Geysers has revealed the risks for non-sustainable overproduction of geothermal energy, but has also shown how to maintain long-term power production by careful production management and water reinjection – an essential technique at any geothermal field. Water-dominated systems are the most commonly developed geothermal fields in the world, producing over 9000 MW of electrical power worldwide (Bertani 2006). An example of such a field is the Coso geothermal area, being developed in a volcanic field 250 km north of Los Angeles, California (Adams et al. 2000). Moderate-temperature geothermal systems (100–200°C) are most commonly developed to supply binary fluid power plants. A good example is in the Mammoth Lakes [43] area of California, where three binary fluid power plants produce around 40 MW of power. Development of the Wairakei moderate-temperature geothermal field in the Taupo Volcanic Zone [180], New Zealand, resulted in the second-oldest large-scale geothermal power plant in the world (after Lardarello, Italy). Wairakei went online in 1958, expanded its resources to include exploitation of lower temperature fluids with a binary power plant in 2005, and now produces more than 180 MW annually, helping to make New Zealand one of the world's leaders in using renewable resources for power generation (Bixley et al. 2009).

Geothermal plants situated on the flanks of active volcanoes exploit some of the highest-temperature geothermal fluids, but they potentially have to deal with volcanic hazards. A 30-MW geothermal facility on Kīlauea volcano [17] was partially inundated by lava flows during the summer, 2018 eruption (Chapter 1; Figure 15.4), but after a two-year hiatus to repair damages and recover lava-buried wells, is now back online, producing about 5% of Hawai'i Island's electrical power.

Direct-Use Applications

Large volumes of low-temperature warm water (less than 100°C) are a common byproduct of geothermal power generation in volcanic areas, and can also be obtained from non-volcanic areas of high heat flow or tectonic activity. These resources are also of great economic value, as they can be used for multiple agricultural, residential, and industrial uses – drying crops, heating buildings, running refrigeration systems, and supplying warm water to public baths. Iceland is a pioneer in the development of geothermal resources for electrical energy production and direct-use applications, and now supplies almost all of its energy needs with sustainable geothermal and hydroelectric projects. The more recent development of geothermal heat pump systems, where heat

FIGURE 15.4 Aerial view of the Puna Geothermal Venture powerplant on June 10, 2018, as a 150-m-wide lava river from Kīlauea volcano [17] sends lava flows over the channel edge toward the facility. The dark flows were emplaced a few weeks earlier and partially inundated the plant and some production wells. Source: Photo copyright Bruce Omori.

exchanger loops are buried within a few feet of the ground surface, are proving to be an economic alternative to conventional heating (and cooling) of residential buildings over much broader areas of North America and Europe.

Volcanoes and Ore Deposits

Modern technology depends on two natural resource pillars: fossil fuels and metals. Metals, in turn, come from a multitude of minerals that have formed in many different ways, and are divided into three classes: **iron** (modern society's most economically important metal), **base metals** (Cu, Pb, Zn, Sn, etc.) and **precious metals** (Au, Ag, Pt, etc.). We will not discuss the common metals iron or aluminum, as their deposits are not generally related to igneous activity. Almost all of the world's deposits of base and precious metals are, however, directly related to volcanism or to intrusive processes, and will be the focus of this section.

Metallic **ore deposits** are defined as economically exploitable concentrations of metal-bearing minerals. Such deposits are not commonplace, nor can their occurrences be easily prospected, although new techniques in remote sensing, airborne geophysics, and geochemical prospecting provide powerful new tools for identifying and prioritizing exploration targets that would never have been suspected a few decades ago. Without metallic ore deposits, civilization would hardly have moved past its beginning. The terms "Iron Age," "Bronze Age," and "Industrial Age" record the incremental progress of technology and manufacturing enabled by metallic ore deposits throughout history.

Ore mineralization around a subvolcanic intrusion can take place in numerous settings and at different times as a magma body cools and crystallizes. Some mineralization may occur within the magma body itself. Much may occur within the volcanic cover overlying an intrusion (Table 15.1). Most can take place more broadly in the enclosing host bedrock facilitated by fluids set in motion by volcanic fluids. Volcanogenic massive sulfide (VMS) deposits form directly on seafloors (see section on Seafloor Mineralization) Grateful commodities brokers might well view magmas as the agents by which they earn their living! Weathering and erosion in turn can develop additional deposits of considerable value, for example, supergene copper deposits and placer gold.

The different ways that metallic ore deposits can develop may be divided into two general classes: those associated directly with high-temperature **magmatic processes**, and those associated with **hydrothermal activity** at or near Earth's surface. An example of a purely magmatic ore deposit is given by the two-billion-year-old Bushveld Complex in the Transvaal region of South Africa. This immense saucer-shaped body might have originally included over a million cubic kilometers of molten rock, emplaced by multiple replenishments within a period of less than a hundred thousand years (Cawthorn and Walraven 1998). Most of the complex is mafic in composition, suggesting that it may represent the feeding system for a large igneous province that has long since eroded away (Hatton 1995). As repeated injections of mantle-derived melt entered the Bushveld magma reservoir and cooled, fractional

TABLE 15.1 Plutonic associations.

Deposit Type	Host Rock	Examples (reference)	Average Grade
Porphyry Cu/Mo/Au	Granite, granodiorite, and related breccias/stockworks	Chuquicamata, Chile (1)	0.55% Cu
	Dacite porphyry-intermediate sill complex	El Teniente, Chile (2) Grasberg, Indonesia (3)	1.31%Cu 1.04% Cu; 0.9 ppm Au
	Basalt/monzo-diorite dikes	Oyu Tolgoi, Mongolia (4)	1.25% Cu 0.24 ppm Au
Epithermal Au/Ag	a. Low sulfidation Andesite/rhyodacite	Hishikari, Japan (5)	40 ppm Au
	b. Intermediate sulfidation Andesite	Fruta del Norte, Ecuador (6)	20 ppm Au
	c. High sulfidation Basalt-Rhyolite	Pueblo Viejo, Dominican Rep (7)	3.22 ppm Au
VMS Cu/Zn/Ag	Archean felsic meta-volcanics	Kidd Creek, Canada (8)	1.82% Cu, 5.61% Zn 54 pm Ag

1: Camus (2002, pp. 5–21).
2: Cannell et al. (2005, pp. 979–1003).
3: Underlying Value. Freeport McMoRan Copper & Gold Inc., Annual Report. 88 pp http://www.fcx.com/inrl/annlrpt/2006/FCX%20AR%202006.pdf
4: Kirwin et al. (2005).
5: Izawa et al. (1990, pp. 1–56).
6: Aurelian (private report).
7: Barrick Gold Corp web page. http://www.barrick.com/GlobalOperations/NorthAmerica/PuebloViejoProject/default.aspx
8: http://www.falconbridge.com/our_business/copper/operations/kidd_creek.htm
Source: Compiled by Karl Roa (2008).

crystallization laid down beds of platinum, gold, and chromium-rich ore. These extremely valuable layers, termed **reefs** by local miners, may be individually traced across the rolling landscape as far as 70 km. These reefs contain a very large percentage of the world's mineable platinum and chromium, with the Merensky Reef alone (Viljoen 1999) estimated to contain 17,000 tons of platinum in one widespread sub-layer, commonly only about 25 cm thick.

Ores related to hydrothermal activity may consist of metals contributed directly by magma itself, or leached and concentrated from the enclosing host rock as fluids cool farther away from their heat sources. The fluids commonly travel through cracks and fissures in the rock, filling them with mineral deposits to form **ore veins**. Many vein-filled cracks may be related directly to intrusions and volcanic activity. A dense network of veins called a **stockwork** in many cases develops directly over or close to related intrusions. Commonly minerals such as milky quartz and pyrite ("fool's gold") precipitate with ore minerals. In and of themselves, quartz and pyrite are virtually worthless, but as indicators of possible ore enrichment, they are useful prospecting indicators and are called **gangue minerals** to distinguish them. The host rock in the vicinity of veins may also break down and oxidize at shallow depths, transforming into masses rich in yellow clays, red to black hematite, and golden-brown limonite. Different metals will be deposited in different areas, depending on the temperatures, oxygenation, acidities, and chemical compositions of the transporting fluids, and on the chemistry and temperatures of the rocks these fluids encounter (Figure 15.5). Valuable ore deposits formed by hydrothermal activity include gold and copper such as those found at Bingham Canyon (Utah) and Chuquicamata (Chile).

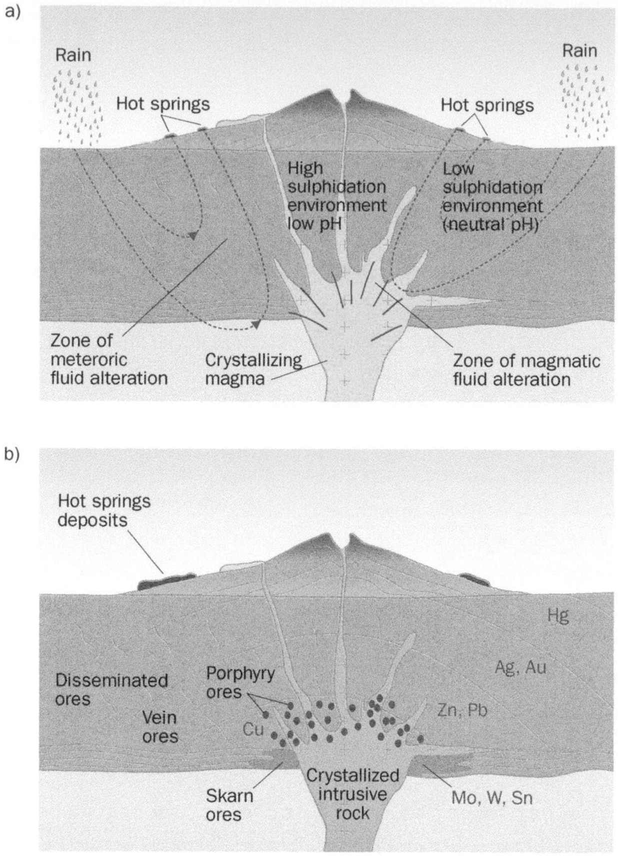

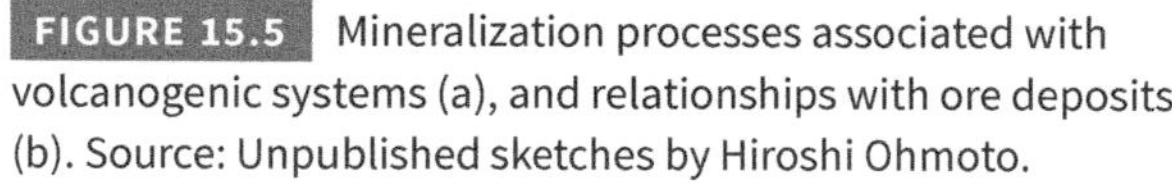

FIGURE 15.5 Mineralization processes associated with volcanogenic systems (a), and relationships with ore deposits (b). Source: Unpublished sketches by Hiroshi Ohmoto.

Researchers have long debated the question of whether the heated groundwater and associated ore metals come from magma itself or from surrounding country rocks, but isotopic data suggest that in the early stages of magma emplacement there can be a significant juvenile metal component. Some ore and gangue minerals trap tiny bubbles of the fluids from which they precipitate as they crystallize. Such fluid inclusions provide valuable information about processes that remove metals from magma. Gold, for example, appears to bond ionically with volatile chlorine, forming a compound that is soluble in high-temperature solutions, but breaks down to precipitate native gold upon cooling. Many gold particles are rimmed or embedded in pyrite in a later stage of sulfide precipitation, to be followed finally by milky quartz (Bodner et al. 2014). Chaplygin et al. (2015) observed that gold was transported in extremely small amounts by high-temperature volcanic gases and deposited as euhedral, submicroscopic gold crystals in gas sublimates during the 2012–2013 Tol'batchik [168] eruption.

Evidence for potential ore mineralization around the shallow conduits of some volcanoes may be seen in their accidental ejecta. Consider, for example, the active conduit of Mount Vesuvius [109], which passes up through beds of limestone and calcareous marl that provide a foundation for the Campanian Plain in Italy. Xenoliths of these formations embedded in Vesuvian lavas and ejecta show that hot, acidic solutions emanating from the conduit have penetrated the surrounding fractured wall-rock, dissolving some of the minerals encountered there while simultaneously precipitating others that are more stable at high-temperature, including minor ore deposits. Ore petrologists call this process **metasomatism**. The resulting calcium-rich metamorphic rock is a **skarn**. The envelope of skarn enclosing the Vesuvian conduit may be several hundred meters wide, given that fractured carbonate rocks are readily permeable (Fulignati et al. 2004). Two common products of skarn metasomatism are the minerals scheelite and wolframite, both calcium-tungstates. One of the largest tungsten mines in the world (now closed) lies along Pine Creek on the eastern side of the Sierra Nevada Range, California (Bateman 1965). The metasomatism there occurred sometime during the Mesozoic Era, when numerous granitic plutons doubtless fed an active chain of volcanoes above the levels where glaciated mountain crests rise today. Its operation was of strategic importance during WWII.

Aqueous solutions around subvolcanic magma bodies can easily permeate bedrock beneath volcanoes for many kilometers beyond their bases. As they alter the crust through various mineral reactions and mix with the groundwater, they lose their acidity and volatiles. Where these fluids reach Earth's surface, hot springs, geysers, fumaroles, and related features develop. There are two general kinds of hydrothermal fields that permeate many active volcanoes. Those lying close to or upon volcanic edifices are characterized by highly acidic hot springs (pH 0.5–1.5), sulfur deposits, and high-temperature fumaroles. Hydrothermal fields farther away are characterized by hot springs whose waters have become neutral in acidity, and show little or no sulfur deposition. Geysers commonly form in this latter environment, partly because groundwater supplies are more stable than closer to active volcanoes. Geysers recycle water of entirely meteoric origin, though ultimately derive their heat from cooling magma. The famous geyser basins of Yellowstone National Park are an example of this latter kind of hydrothermal system.

Ore deposits may form within a few hundred meters of the surface in hydrothermal areas, though occurrences are notoriously spotty. Mining geologists call such ore bodies **epithermal** ("shallow-heated") **ores**. There are a few noteworthy examples of epithermal gold deposits in still active hydrothermal systems, including Mina Limón in Nicaragua, and Ladolam in Papua New Guinea. Miners of epithermal ores in active hydrothermal fields must deal with potential flooding and sweltering conditions underground, however, with mine walls greatly altered to hot, sticky mud by hydrothermal circulation – a veritable Hades for miners.

Porphyry Coppers and Related Alteration

Copper is one of the most basic and oldest metals used by humans, and can form in various volcanic environments. For reasons not yet understood, veinlets and amygdules of pure copper formed late in the cooling of some Precambrian basalt flows on the Upper Peninsula of Michigan. It could be picked directly out of outcroppings with stone tools and patience, and stimulated metalworking and widespread trade among Native Americans throughout the region. In the Old World, copper deposits in the ancient basaltic seafloor crust of the island of Cyprus sustained an important bronze industry throughout the Eastern Mediterranean in Classical times, once smelting technology was developed. The name Cyprus, in fact, stems from *aes cyprium*, the Roman name for Cyprus, meaning "Island of Copper."

Deep erosion of low-grade, high-tonnage porphyry copper deposits in some parts of the world shows evidence for the passage of hydrothermal fluids in and around certain granitic plutons and in places their overlying volcanic rocks. Each pluton was formerly a silicic magma chamber that stewed and underwent chemical transformation in its own volatile-rich residual "juice" as it cooled and crystallized, producing concentric alteration shells that extend into the surrounding country rock for kilometers (Banks and Page, 1980). The alteration often is centered in an upper portion of the pluton known as the **potassium-silicate alteration zone** that is enriched in reddish potassium feldspar and biotite. Farther out, the surrounding **sericite zone** includes rocks containing abundant hydrothermal quartz, calcite, sericite, and pyrite; while even further away the most far-reaching alteration effects are displayed by rocks of the greenish **propylitic zone**, which contain abundant chlorite, epidote, and albite. The most productive occurrences of copper ores tend to occur in the potassium-silicate alteration zone, taking the form of scattered crystals of chalcopyrite and other cupriferous minerals embedded in granite, or as vein-filling minerals. The crystallized host rock also ordinarily contains especially large crystals of potassium feldspar, called **megacrysts**. The resulting texture, termed **porphyritic**, lends its name to this class of ores; **porphyry coppers**. As a bonus for miners, molybdenum, gold, tungsten, and tin may also be abundant in association with the copper. The American Southwest is noteworthy for its abundance of porphyry copper deposits. About 60% of the world's total copper production comes from such features. Their associated volcanoes and probable epithermal ore deposits have long since been destroyed, however, by the erosion of several kilometers of overburden (Cooke et al. 2005; Miller and Groves 2019).

Nickel

Nickel plays a vital role in the manufacture of steel to make it harder. Like copper and gold, nickel ores are ultimately related to shallow magma chambers, but for the most part in far more unusual ways. One of the greatest nickel deposits in the world is the Sudbury intrusive complex in Ontario, Canada. This large body evidently formed in response to the impact of a **bolide** (large meteorite) during Precambrian times (Faggart et al. 1985). The meteorite blew an enormous hole in Earth's crust, perhaps as deep as several kilometers and from 200–250 km in diameter! As matter blasted out of the crater, decompression of the underlying mantle triggered almost instantaneous melting, resulting in a gigantic body of basaltic magma that welled up into the shallow crust and possibly even into the crater itself to produce the terrestrial equivalent of a small lunar mare (Chapter 12). During ascent of the magma, large pockets of immiscible sulfur-rich liquid developed within the dominantly silicate melt. Nickel concentrated within these pockets, bonding with the sulfur to form nickel-sulfide minerals such as pentlandite as the melt cooled. Subsequent erosion has exposed the ore, making the Sudbury complex one of only about a half-dozen major nickel mines in the world. Another great nickeliferous body of similar age, the Stillwater complex in Montana may be of similar origin.

Another very different kind of nickel ore occurs on the island of New Caledonia. Basalt lava flows there contain unusually nickel-rich phenocrysts of olivine. Although the fresh flows cannot be mined directly, since the phenocrysts are too few and far between, tropical weathering concentrates the nickel-bearing crystals in the lateritic soils forming from these flows. New Caledonia, like Sudbury, has become a major world supplier of nickel. Weathered serpentinites of Cuba were also once a major nickel source, but production has been halted owing to trade embargoes on necessary processing chemicals.

Lithium

Lithium is an increasingly valuable metal as the world coverts to non-organic energy sources. Most lithium is produced from brines in volcanic areas of South America, but the 16 my old McDermitt caldera in northern Nevada, at the western end of the Yellowstone "Hotspot Chain", USA is also a major producer. Here caldera-filling sediments have been intensely mineralized with magmatically-derived Li-rich brines (Castor and Henry, 2020).

Seafloor Mineralization

Ore mineralization is not restricted to terrestrial geological environments. The ocean floor proves to be a potent environment for ore development too. In many island arcs and at mid-oceanic ridge, submarine volcanic eruptions and black smoker fumaroles release dark clouds of flocculating mineral precipitates laden with copper, zinc, gold, lead, and other metals into the sea (Karson et al. 2015). Research suggests that some of these metals come directly from ascending, degassing magma, whereas others are concentrated from the surrounding seafloor crust aided by associated hydrothermal activity. There are usually no strong currents in the deep ocean to disperse the metal-bearing precipitates far, hence metal-bearing minerals rain down onto the surrounding seabed to accumulate as white, yellow, and dark-gray beds. Most of the metals bond with

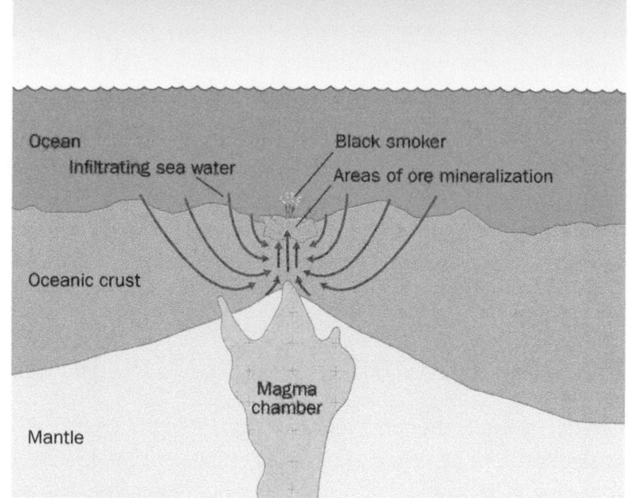

FIGURE 15.6 Formation environment for volcanogenic massive sulfide (VMS) deposits on the seafloor, where rising fluids from magmatic bodies create "heat engines" that drive the circulation of sea water through oceanic crust. Although this cross-section depicts a mid-ocean ridge location, the same processes apply anywhere oceanic crust is undergoing extension, including active submarine calderas. Red arrows: ascending hot fluids; blue arrows: descending cool fluids. Source: Based on Mercier-Langenin et al. (2014).

FIGURE 15.7 Open pit Kidd Mine, Ontario, Canada. This is one of the world's largest VMS deposits, and produces over 100,000 tons annually of Cu, Zn, Pb, plus other metals. The pit is about 800 m long, 220 m deep, but was mined out in 1977. The mine's production is now entirely from underground sources, which extend to almost 3 km below the surface – the world's deepest sulfide ore mine. Source: Photo courtesy Xstrata Copper.

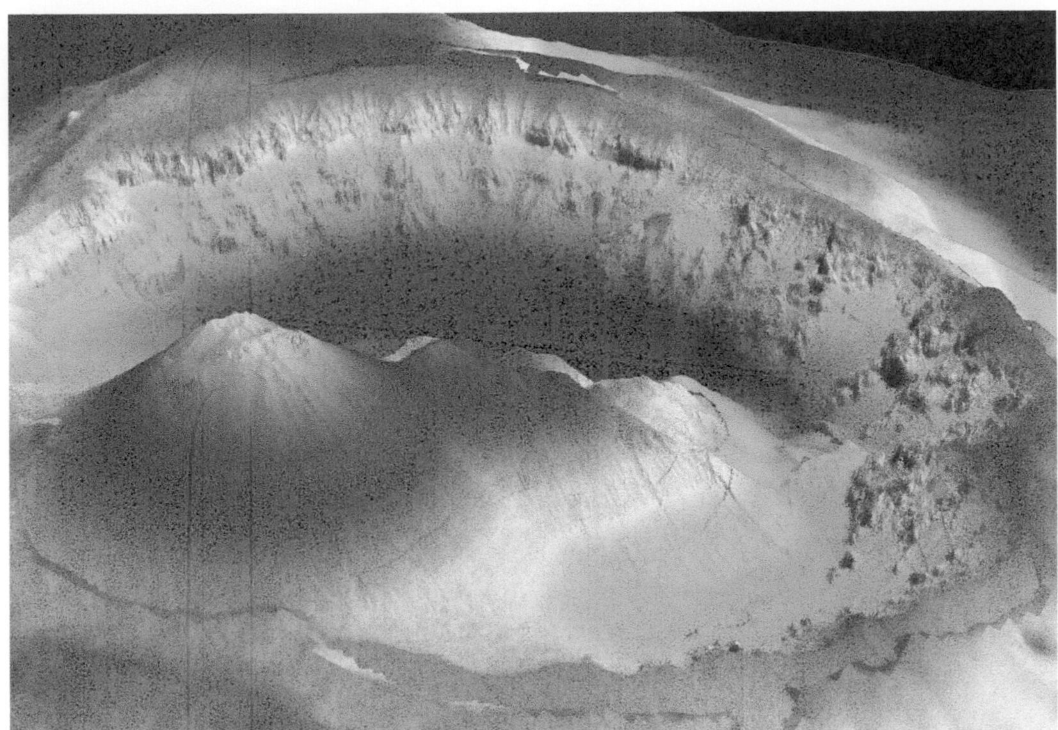

FIGURE 15.8 Side-scan sonar bathymetric view of the Brothers Volcano [186], Southern Kermadec Arc, New Zealand. This view to the north shows the 500-m-high rim bounding an 8 × 13 km caldera. The young post-caldera cone rises to within 1100 m of the sea surface, and is marked by active hydrothermal venting. This vent, along with vents on the NW caldera rim which reach 300°C temperature, is associated with the deposition of Cu, Zn, Pb sulfides, and indicates that subsurface VMS ore deposits may be forming today. Source: Image courtesy of New Zealand American Submarine Ring of Fire 2007 Exploration, NOAA Vents Program, the Institute of Geological & Nuclear Sciences and NOAA-OE.

sulfur, forming various sulfide and some sulfate minerals. The influx of heat, acids, and chemicals from the submarine vents may force precipitation of minerals directly out of the sea water as well, including gypsum and anhydrite. The resulting **stratiform** volcanogenic massive sulfide (VMS) deposits, are typically rich in copper, zinc, and lead, and are economically important sources of gold, silver, and lesser amounts of arsenic, cobalt, and tin (Mercier-Langenin et al. 2014; Figures 15.6 and 15.7). VMS ore deposits have been found on all continents (except Antarctica – so far!). Because of their great economic importance, a great deal of geologic research has been devoted to interpreting their origins. Two types of VMS deposits exist: one is associated with ocean floor basalts and associated oceanic crustal rocks formed along divergent plate boundaries (Chapter 12); the other with felsic volcanic rocks erupted in shallow near-continental back-arc basins or along volcanic arcs (Chapter 2; Figure 15.8). Japan, an island-arc nation, features some of the best examples of the latter type of ore deposit in the world. These are the classic **Kuroko-type deposits** (Ohmoto and Skinner 1983), where rich ores formed directly on the seafloor or in shallow dacite-thyolite domes (Figure 15.9). Marine investigations have shown that similar VMS deposits are presently forming beneath submarine calderas along the Bonin Arc (Fiske et al. 2001; Glasby et al. 2008).

Other Useful Volcanic Materials

Volcanoes produce many non-metallic resources of great economic importance. Not so glamorous as gold, silver, or diamonds, but more important to the infrastructure of society are the vast amounts of volcanic materials that are quarried for construction purposes, either as cut blocks or as aggregate and crushed rock. Fine-grained basalt is especially important for construction because of its strength and resistance to weathering, and quarries in the Eifel region of Germany have long provided building stones for cathedrals, as well as the finest millstones all over central Europe. Early Romans discovered that vitric tephra from Campi Flegrei volcano [108] on the Bay of Naples had unique properties when mixed with calcinated lime [$Ca(OH_2)$], as well as the finest millstones. The lime reacts with volcanic glass to form calcium silicate, which forms a durable concrete when mixed with sand. This material, called **pozzolan** after the nearby city of Pozzuoli, was used as mortar and concrete in classic structures like the Pantheon and Colosseum of Rome, which are

FIGURE 15.9 Large polished outcrop sample (40 cm high) of classic Kuroko-type ore from the Fukazawa Mine, Akita Prefecture, Japan, showing original sedimentary structure indicative of seafloor deposition. The black ore layers consist of fragmental, fine-grained aggregates of sphalerite-galena with subordinate tetrahedrite and barite. Although primarily mined for Cu, Pb, and Zn, these ores also are rich in Au and Ag. Source: Photo courtesy of Ryoichi Yamada, DOWA Holding Co.

still standing after two millennia. Pozzolanic volcanic tephra is a valuable commodity, which can enhance the qualities and lower the costs of concrete, and is being used to build durable roads and runway surfaces in many volcanic areas of the world.

Elemental sulfur deposits are found on most volcanoes of the world, and have been exploited by peoples for medicinal purposes since the dawn of civilization. The Chinese invention of gunpowder over a thousand years ago increased the demand for sulfur, and the European conquerors of the Americas were doubtlessly delighted to find so many volcanoes in Central and South America. These volcanoes enabled the conquistadores to manufacture gunpowder along their routes, and thus played an important role in the bloody conquest of Native Americans.

Questions for Thought, Study, and Discussion

1 Where does most of Earth's energy come from?
2 What are the sources of most of the energy utilized by society for power generation?
3 How can volcano energy be best utilized?
4 Why are most of the world's metallic ore resources associated with igneous bodies and the roots of volcanoes?
5 Why have the interfaces between submarine volcanoes and seawater been places for ore deposition throughout Earth's history?
6 Why are Earth's geothermal energy production fields almost all associated with past volcanism?
7 What are the major types of geothermal resources, and how are they exploited to produce electrical power?
8 What geological field evidence would you seek to discover a metallic ore deposit?

FURTHER READING

Arnorsson, S., Thorhallsson, S. and Stefansson, A. (2015) Utilization of geothermal resources (Ch 71, pp. 1235–1252). In: *The Encyclopedia of Volcanoes, Second Edition* (eds. Sigurdsson, H. et al.). Amsterdam: Elsevier Press, 1421 pp.

Boden, D.B. (2017). *Geologic Fundamentals of Geothermal Energy.* London and New York: CRC Press, 425 p.

Gibson, H.L. (2005). "Volcano-hosted ore deposits." In *Volcanoes and the Environment* (eds. J. Martí and G.G.J. Ernst), pp. 333–386. Cambridge: Cambridge University Press.

Heiken, G. (2005). "Industrial uses of volcanic materials." In *Volcanoes and the Environment* (eds. J. Martí and G.G.J. Ernst), pp. 387–403. Cambridge: Cambridge University Press.

McPhie, J. and Cas, R. (2015) Volcanic successions associated with ore deposits – facies characteristics and ore-host relationships (Ch. 49, pp. 865–879). In *The Encyclopedia of Volcanoes, Second Edition* (eds. Sigurdsson, H. et al.). Amsterdam: Elsevier Press, 1421 pp.

Wohletz, K.H. and Heiken, G. (1992). *Volcanology and Geothermal Energy.* Berkeley: University of California Press, 432 p.

Epilogue: The Future of Volcanology, Second Edition

Revising this second edition has been "fun" for us, but it has also been humbling, as we are forced to confront the fact that the abundance of important new volcanological research conducted and published over the past decade is impossible to summarize in any single book. This voluminous new work can now be published in several new print journals, as well as in digital-only, online journals that have made it possible to publish research findings quickly and to distribute them globally. Increasingly, volcanologists are becoming more specialized in their expertise, and find it impossible to keep up with all advances. For this reason, praise be to authors who write general papers that summarize major developments important to all of us! A wonderful reference book that describes advances in more than 80 specialized research fields was compiled by Haraldur Sigurdsson and co-editors (2015). This massive volume is an invaluable source for information about aspects of volcanology that we only "touched lightly" in our book.

Most of our readers will not need to keep up with ongoing research, since, for many students, this may be the only "rock course" they will ever take. We do hope our book gives all readers the basic tools for understanding news stories about future eruptions, and, most importantly, for appreciating the volcanoes you will encounter in your lives. There are over 150 national and local parks throughout the world that are located on and around scenic volcanoes, and we hope you will be able to visit many of them!

As we wrote in the epilogue of our first edition, *The future of volcanology is bright!* The challenges ahead are great, but the increased capabilities of the next generations of more diverse volcanologists give great hope that they will enable our science to serve society better! Twelve years ago, when the first edition of this book was published, women were graduating in record numbers from graduate schools. Many of these women have now risen to leadership positions in academia and as directors of volcano observatories around the world! We make a special plea to these young volcanologists not to specialize too much in their future careers, and not to forget the importance of basic fieldwork. Less than 10 percent of the Earth's active volcanoes have been mapped in sufficient detail to understand their past behavior and evaluate their potential for future activity – you can choose your own volcano to honor with a published geologic map!

Rick, Servando, and I (JPL) now pass on to readers our best wishes for your further learning in the field from volcanoes – *the best teachers you will ever have*!

Volcanoes: Global Perspectives, Second Edition. John P. Lockwood, Richard W. Hazlett, and Servando De La Cruz-Reyna.

Companion website: www.wiley.com/go/lockwood/volcanoes2

Bibliography

Abe, K. (1992). "Seismicity of the caldera-making eruption of Mt. Katmai, Alaska in 1912." *Bulletin of the Seismological Society of America*, 82, 175–191.

Acocella, V., Cifelli, F., Funiciello, R. et al. (2001). "The control of overburden thickness on resurgent domes: Insights from analogue models." *Journal of Volcanology and Geothermal Research*, 111 (1–4), 137–153.

Acocella, V., Puglisi, G., and Anmelung, F. (2013). "Flank instability at Mt. Etna." *Journal of Volcanology and Geothermal Research*, 251, 1–4.

Adams, M.C., Moore, J.N., Björnstad, S. et al. (2000). "Geologic history of the Coso geothermal system." *Proceedings of the World Geothermal Congress 2000*, Kyushu-Tohoku, Japan.

Aiuppa, A., Baker, D.R., and Webster, J.D. (2009). "Halogens in volcanic systems." *Chemical Geology*, 263 (1–4), 1–18.

Aki, K. and Koyanagi, R.Y. (1981). "Deep volcanic tremor and magma ascent mechanism under Kilauea, Hawaii." *Journal of Geophysical Research*, 86, 7095–7109.

Allen, J.R.L. (1982). *Sedimentary Structures: Their Character and Physical Basis*. New York: Elsevier.

Allen, S.R. (2001). "Reconstruction of a major caldera-forming eruption from pyroclastic deposit characteristics: Kos Plateau Tuff, eastern Aegean Sea." *Journal of Volcanology and Geothermal Research*, 105, 141–162.

Allison, C.M, Roggensack, K., and. Clarke, A.B. (2021). Highly explosive basaltic eruptions driven by CO2 exsolution. *Nature Communications*, 11 January 2021. doi: 10.1038/s41467-020-20354-2.

Alvarez, W. (2017). *A Most Improbable Journey – A Big History of Our Planet and Ourselves*. New York: W.W. Norton, 246 pp.

Alvarez, W. and Zimmer, C. (1997). *T. Rex and the Crater of Doom*. Princeton: Princeton University Press, 185p.

Ambrose, S.H. (1998). "Late Pleistocene human population bottlenecks, volcanic winter, and differentiation of modern humans." *Journal of Human Evolution*, 34 (6), 623–651.

Ancochea, E., Huertas, M., Hernán, F., and Brändle, J. (2014). "A new felsic cone-sheet swarm in the Central Atlantic Islands: The cone-sheet swarm of Boa Vista (Cape Verde)." *Journal of Volcanology and Geothermal Research*, 274, 1–15.

Anderegg, C.R. (2000). *Ash Warriors*. USAF Office of PACAF History, Hickam AFB, Honolulu, pp. 135.

Anderson, C.A. (1941). "Volcanoes of the Medicine Lake highland." California: University of California Publications, *Bulletin of the Department of Geological Sciences*, 25, 347–422.

Anderson, D.J. and Lindsley, D.H. (1988). "Internally consistent solution models for Fe-Mg-Mn-Ti oxides: Fe-Ti oxides." *American Mineralogist*, 73, 714–726.

Anderson, E.M. (1936). "Ring dykes and cauldron subsidence: The dynamics of formation of cone sheets." *Proceedings of the Royal Society of Edinburgh*, 56, 128–157.

Anderson, E.M. (1937). "Cone-sheets and ring dykes: The dynamical explanation." *Bulletin Volcanologique*, 2 (1), 35–40.

Anderson, J.L. and Smith, D.R. (1995). "The effect of temperature and oxygen fugacity on Al-in-hornblende barometry." *American Mineralogist*, 80, 549–559.

Anderson, J.L., Barth, A.P., Wooden, J.L. et al. (2008). "Thermometers and thermobarometers in granitic systems." *Reviews in Mineralogy and Geochemistry*, 69, 121–142.

Anderson, L.W. and Hawkins, F.F. (1984). "Recurrent Holocene strike-slip faulting, Pyramid Lake fault zone, Western Nevada." *Geology*, 12, 681–684.

Anderson, S., Smrekar, S., and Stofan, E. (2012). "Tumulus development on lava flows: insights from observations of active tumuli and analysis of formation models." *Bulletin of Volcanology*, 74, 931–946.

Andrew, R.G. (2022). *Super Volcanoes: What They Reveal about Earth and Worlds Beyond*. New York and London: W.W. Norton, 372.

Andronico, D., Di Roberto, A., De Beni, E. et al. (2018). Pyroclastic density currents at Etna volcano, Italy: The 11 February 2014 case study. *Journal of Volcanology and Geothermal Research*, 357, 92–105.

Annen, C. and Zellmer, G.F. (2008). *Dynamics of Crustal Magma Transfer, Storage, and Differentiation*. Geological Society of London Special Publication 304, 288 p.

Arakami, S. and Ui, T. (1982). "Japan." In: *Andesites: Orogenic Andesites and Related Rocks*. (ed. R.S. Thorpe), 259–292. Chichester: John Wiley.

Araki, T., Enomoto, S., Furano, K. et al. (2005). "Experimental investigation of geologically produced antineutrinos with KamLAND." *Nature*, 436, 499–503.

Arens, N.C. and West, I.D. (2008). "Press-pulse: A general theory of mass extinction." *Paleobiology*, 34, 456–471.

Armienta M.A., De la Cruz-Reyna, S., and Macías J.L. (2000). "Chemical characteristics of the crater lakes of Popocatépetl, El Chichón and Nevado de Toluca volcanoes, México." *Journal of Volcanology and Geothermal Research*, 97, 105–125.

Armienta M.A., De la Cruz-Reyna, S., Ramos S. et al. (2014). "Hydrogeochemical surveillance at El Chichón volcano crater lake, Chiapas, México." *Journal of Volcanology and Geothermal Research*, 285, 118–128. http://dx.doi.org/10.1016/j.jvolgeores.2014.08.011.

Arno, V., Bakashwin, M.A., Baker, A.Y. et al. (1980). "Geodynamic evolution of the Afro-Arabian rift system." *Atti dei Conveigni Lincei Academia Nazionale*, pp. 629–643.

Arnorsson, S., Thorhallsson, S., and Stefansson, A. (2015). Utilization of geothermal resources: Chapter 71 in *The Encyclopedia of Volcanoes* (eds. H. Sigurdsson et al.), 2nd edition, pp. 1235–1252. Amsterdam: Elsevier Press, 1421 pp.

Arpa, M.C., Zellemr, G.F., Christenson, B. et al. (2017). "Variable magma reservoir depths for Tongaririo Volcanic Complex eruptive deposits from 10,000 years to present." *Bulletin of Volcanology*, 79 (56). doi: 10.1007/s004445-017-1137-5.

Aspinall, W., Loughlin, S.C., Michael, F.V. et al. (2002) "The Montserrat Volcano Observatory: Its evolution, organization, role and activities." *Geological Society of London, Memoirs*, 21, 71–91.

Aspinall, W.P. and Blong, R.J. (2015). "Volcanic risk assessment." Chapter 70 in *Encyclopedia of Volcanoes*, 2nd edition (ed. H. Sigurdsson), pp. 1215–1231. Amsterdam: Elsevier Press, 1423 pp.

Aspinall, W.P., Woo, G., Voight, B. et al. (2003). "Evidence-based volcanology: Application to eruption crises." *Journal of Volcanology and Geothermal Research*, 128 (1–3), 273–285.

Atkinson, A., Griffin, T.J., and Stephenson, P.J. (1975). "A major lava tube system from Undara Volcano, North Queensland." *Bulletin Volcanologique*, 39 (2), 266–293.

Avard, G. and Whittington, A. (2012). "Rheology of arc dacite lavas: experimental determination at low strain rates." *Bulletin of Volcanology*, 74, 1039–1056.

Avdeiko, G.P., Palueva, A.A., and Khlebrodova, O.A. (2006). "Geodynamic conditions of volcanism and magma formation in the Kurile-Kamchatka Island Arc System." *Petrology*, 14, 248–265.

Ayris, P.M. and Delmelle, P. (2012). "The immediate environmental effects of tephra emission." *Bulletin of Volcanology*, 74, 1905–1936.

Volcanoes: Global Perspectives, Second Edition. John P. Lockwood, Richard W. Hazlett, and Servando De La Cruz-Reyna.

Companion website: www.wiley.com/go/lockwood/volcanoes2

Bachèlery, P., Chevallier, L., and Gratier, J.P. (1983). "Caracteres structuraux des eruptions historiques du Piton de la Fournaise." *Comptes Rendus de l'Académie de Science Paris*, II, 296, 1345–1350.

Bachmann, O. and Bergantz, G.W. (2008). "The magma reservoirs that feed supereruptions." *Elements*, 4, 17–21.

Bachmann, O., Dungan, M.A., and Lipman, P. (2002). "The Fish Canyon magma body, San Juan volcanic field, Colorado: Rejuvenation and eruption of an upper-crustal batholith." *Journal of Petrology*, 43, 1469–1503.

Bacon, C.R. (1983). "Eruptive history of Mount Mazama and Crater Lake caldera, Cascade Range, USA." *Journal of Volcanology and Geothermal Research*, 18 (1), 57–115.

Bacon, C.R. and Lanphere, M.A. (2006). "Eruptive history and geochronology of Mount Mazama and the Crater Lake region, Oregon." *Geological Society of America Bulletin*, 118, 1331–1358.

Baer, R.B. and Siegel, L.J. (2000). *Windows into the Earth: The Geologic Story of Yellowstone and Grand Teton National Parks*. Oxford: Oxford University Press, 242p.

Bailey, E.B., Clough, C.T., Wright, W.B. et al. (1924). "Tertiary and pre-Tertiary geology of Mull, Loch Aline and Oban." *Memoirs of the Geological Survey of Great Britain*. London, HMSO, 445p.

Balagizi, C.M., Kies, A., Kasereka, M.M. et al. (2018). "Natural hazards in Goma and the surrounding villages, East African Rift System." *Natural Hazards*, 93 (1), 31–66.

Banks, N.G. and Page, N.J. (1980). "Some observations that bear on the genesis of porphyry-copper deposits." In *Proc 5th Q IAGOD (International Association on the genesis of ore deposits) Symposium* (ed. J.D. Ridge), vol. I. Stuttgart, pp. 49–73.

Barberi, F., Carpezza, M.L., Valenza, M. et al. (1992). *L'eruzione 1991–1992 dell'Etna e gli interventi per fermare o ritardare l'avanzata della lava*. Rome: Giardini.

Bardintzeff, J.-M. and McBirney, A.R. (2000). *Volcanology*. Sudbury, MA: Jones and Bartlet.

Barois, P. (2004). *Guide Encyclopédique des Volcans*. Paris: Délachaux et Niestlé, 416p.

Barreyre, T., Soule, S.A., and Sohn, R.A. (2011). "Dispersal of volcaniclasts during deep-sea eruptions: Settling velocities and entrainment in buoyant seawater plumes." *Journal of Volcanology and Geothermal Research*, 205, 84–93.

Barsukov, V.L. (1992). "Venusian igneous rocks." In: *Venus Geology, Geochemistry, and Geophysics: Research Results from the Soviet Union* (eds. V.L. Barsukov, V.P. Volkov, and V.N. Zharkov), pp. 165–176. Tuscon, AZ: University of Arizona Press.

Barth, T.W. (1952). *Theoretical Petrology*. New York: John Wiley, 387 pp.

Barwis, J.H. and Hayes, M.O. (1985). "Antidunes on modern and ancient washover fans." *Journal of Sedimentology*, 55, 907–916.

Bary, E. de and Bullrich, K. (1959). "Zur theorie des Bishopringes." *Meteorolgische Rundschau*, 29, 89.

Basilevsky, A.T. and Head, J.W. III (1996). "Evidence for rapid and widespread emplacement of volcanic plains on Venus: Stratigraphic studies in the Baltis Vallis region." *Geothermal Research Letters*, 23, 1497–1500.

Basilevsky, A.T. and Head, J.W. III (2003). "The surface of Venus." *Reports on Progress in Physics*, 66, 1699–1734.

Bateman, P.C. (1965). "Geology and tungsten mineralization of the Bishop district, California." Washington, DC, US Geological Survey, Professional Paper 470, 208p.

Bateman, P.C. and Eaton, J.P. (1967). "Sierra Nevada Batholith, California." *Science*, 158, 1407–1417.

Batiza, R. (1981). "Lithospheric age dependence of off-ridge volcano production in the North Pacific." *Geophysical Research Letters*, 8 (8), 853–856.

Baudouin, C., Parat, F., and Michel, T. (2018). "CO2-rich phonolitic melt and carbonatite immiscibility in early stage of rifting: Melt inclusions from Hanang volcano (Tanzania)." *Journal of Volcanology and Geothermal Research*, 358, 261–272.

Baxter, P. J., Ing, R., Falk, H., and Plikaytis, B. (1983). "Mount St. Helens eruptions: The acute respiratory effects of volcanic ash in a North American community." *Archives of Environmental Health*, 38, pp. 138–143.

Beaulieu, S.E., Baker, E.T., and German, C.R. (2015). "Where are the undiscovered hydrothermal vents on oceanic spreading ridges?" *Deep-Sea Research II*, 121, 201–212.

Bebbington, M. (2013). "Models for temporal volcanic hazard." *Statistics in Volcanology*, 1, 1–24. https://doi.org/10.5038/2163-338X.1.1.

Behnke, S. and McNutt, S. (2014). "Using lightning observations as a volcanic eruption monitoring tool." *Bulletin of Volcanology*, 76, 847. doi 10.1007/s00445-014-0847-1.

Behrens, H. and Jantos, N. (2001). "The effect of anhydrous composition on water solubility in granitic melts." *American Mineralogist*, 86, 14–20.

Bell, A.F. and Kilburn, C.R.J. (2013). "Trends in the aggregated rate of pre-eruptive volcano-tectonic seismicity at Kilauea volcano, Hawaii." *Bulletin of Volcanology*, 75. doi 10.1007/s00445-012-0677-y.

Bell, K. and Simonetti, A. (1996). "Carbonate magmatism and plume activity: Implications from the Nd, Pb, and Sr systematics of Oldoinyo Lengai." *Journal of Petrology*, 37, 1321–1339.

Belousov, A., Behncke, B., and Belousova, M. (2011). "Generation of pyroclastic flows by explosive interaction of lava flows with ice/water-saturated substrate." *Journal of Volcanology and Geothermal Research*, 202, 60–72.

Belousov, A.B., Voight, B., and Belousova, M. (2007). "Directed blasts and blast-generated pyroclastic density currents: A comparison of the Bezymianny 1956, Mount St Helens 1980, and Soufrière Hills, Montserrat 1997 eruptions and deposits." *Bulletin of Volcanology*, 69 (7), 701–740.

Bemis, K., Walker, J., Borgia, A. et al. (2011). "The growth and erosion of cinder cones in Guatemala and El Salvador: Models and statistics." *Journal of Volcanology and Geothermal Research*, 201, 39–52.

Benard, R. and Krafft, M. (1986). *Au Coeur de la Fournaise*. Nourault-Benard, Saint-Denis. France, La Reunion, 220p.

Beranek, L.P., McClelland, W.C., Stael, C.R. et al. (2017). "Late Jurassic flare-up of the Coast Mountains arc system, NW Canada, and dynamic linkages across the northern Cordilleran margin." *Tectonics*, 36, 877–901.

Bergantz, G.W. and Dawes, R. (1994). "Aspects of magma generation and ascent in continental lithosphere." In: *Magmatic Systems* (ed. M.P. Ryan), pp. 291–217. San Diego: Academic Press.

Bertani, R. (2005). "Geothermal power generation in the world 2010–2014 (update report)." *Proceedings World Geothermal Congress 2015*, Melbourne, Australia, 19–25 April 2015.

Bierwirth, P.N. (1982). "Experimental welding of volcanic ash." Master's thesis, Monash University, Australia.

Bindschalder, D.L. and Parmentier, E.M. (1990). "Mantle flow tectonics: The influence of a ductile lower crust and implications for the formation of topographic uplands on Venus." *Journal of Geophysical Research*, 95, 21, 329–344.

Bird, D.K. and Gisladóttir, G. (2012). "Residents' attitudes and behavior before and after the 2010 Eyjafjallajökull eruptions: A case study from southern Iceland." *Bulletin of Volcanology*, 74, 1263–1279.

Bixley, A.W., Clothworth, A.W., and Mannington, W. (2009). "Evolution of the Wairakei geothermal reservoir during 50 years of production." *Geothermics*, 38, 145–154.

Blatt, H., Tracy, J., and Owens, B.E. (2006). *Petrology: Igneous, Sedimentary, and Metamorphic*. New York: Macmillan Publishing.

Blong, R.J. (1982). *The Time of Darkness: Local Legends and Volcanic Reality in Papua New Guinea*. Canberra: Australian National University Press.

Blong, R.J. (1984). *Volcanic Hazards: A Sourcebook on the Effects of Eruptions*. Orlando: Academic Press.

Blong, R.J., Fallon, S., Wood, R. et al. (2018). "Significance and timing of the mid-17th-century eruption of Long Island, Papua New Guinea." *The Holocene*, 28 (4), 529–544. https://doi.org/10.1177/0959683617735589.

Blower, J.D. (2001). "Factors controlling permeability-porosity relationships in magma." *Bulletin of Volcanology*, 63, 497–504.

Blundy, J. and Cashman, K. (2006). "Reconstructing the 1980–86 Mount St. Helens magma reservoir using melt inclusions." American Geophysical Union Fall Meeting, San Francisco.

Bobrowski, N., Hönninger, G., Galle, B. et al. (2003). "Detection of bromine monoxide in a volcanic plume." *Nature*, 423, 273–276.

Boden, D.B. (2017). *Geologic Fundamentals of Geothermal Energy*. New York and London: CRC Press, 425 pp.

Bodner, R.J., Moncada, D., Lecumberri-Sanchez, P., and Steele-MacInnis, M. (2014). "Fluid inclusions in hydrothermal ore deposits." Chapter 13.5 in *Treatise in Geochemistry* (eds. K.K. Turekian and H.D. Holland), pp. 119–142. Oxford: Elsevier Publishing.

Bohrson, W.A. and Spera, F.J. (2001). "Energy-constrained open-system magmatic processes II: Application of energy-constrained assimilation fractional crystallization (EC-AFC) model to magmatic systems." *Journal of Petrology*, 42, 1019–1041.

Bonadonna, C., Cioni, R., Costa, A. et al. (2016). (Forum Contribution), "MeMoVolc report on classification and dynamics of volcanic explosive eruptions." *Bulletin of Volcanology*, 78, 84. doi 10.1007/s00445-016-1071-y.

Bond, D.P.G. and Sun, Y. (2021). "Global warming and mass extinctions associated with large igneous province volcanism." Chapter 3 in *Large Igneous Provinces: A Driver of Global Environmental and Biotic Changes* (eds. R.E. Ernst, A.J. Dickson, and Bekker, Audry), pp. 83–102. American Geophysical Union Monograph 255, 508 pp.

Bondre, N.R., Duraiswami, R.A., and Dole, G. (2004). "Morphology and emplacement of flows from the Deccan Volcanic Province, India." *Bulletin of Volcanology*, 66, 29–45.

Bonneville, A., Barriot, J.P., and Bayer, R. (1988). "Evidence from geoid data of a hotspot origin for the Southern Mascarene Plateau and Mascarene Islands (Indian Ocean)." *Journal of Geophysical Research*, 93 (B5), 4199–4212.

Bonney, T.G. (1899). *Volcanoes: Their Structure and Significance*. New York: G. P. Putnam and Sons.

Bonnichsen, B. and Kaufmann, D.F. (1987). "Physical features of rhyolite lava flows in the Snake River Plain volcanic province, southwestern Idaho." In: *The Emplacement of Silicic Domes and Lava Flows* (ed. J.H. Fink), pp. 119–145. Geological Society of America Bulletin, 212.

Booth, B. and Walker, G.P.L. (1973). "Ash deposits from the new explosion crater, Etna, 1971." *Philosophical Transactions of the Royal Society of London*, 274, 141–151.

Boreham, F., Cashman, K., Rust, A., and Höskuldsson, Á. (2018). "Linking lava flow morphology, water availability and rootless cone formation on the Younger Láxa Lava, NE Iceland." *Journal of Volcanology and Geothermal Research*, 364, 1–19.

Boudier, F., Godard, M., and Armbruster, C. (2000). "Significance of gabbronorite occurrence in the crustal section of the Semail ophiolite." *Marine Geophysical Researches*, 21, 307–326.

Bougher, S.W., Hunten, D.M., and Phillips, R.J. (1997). *Filiberto Venus II: Geology, Geophysics, Atmosphere, and Solar Wind Environment*. Tucson: University of Arizona Press.

Boulding, K. (1968). *Beyond Economics: Essays on Society, Religion, and Ethics*. Ann Arbor: University of Michigan Press, 281p.

Bourdier, J.L. and Abdurachman, E.K. (2001). "Decoupling of small-volume pyroclastic flows and related hazards at Merapi Volcano, Indonesia." *Bulletin of Volcanology*, 63, 309–325.

Bove, D. (2001). "Geochronology and geology of Late Oligocene through Miocene volcanism and mineralization in the Western San Juan mountains, Colorado." US Geological Survey Professional Paper 1642, 30p.

Bow, C.S. and Geist, D.J. (1992). "Geology and petrology of Floreana Island, Galapagos Archipelago, Ecuador." *Journal of Volcanology and Geothermal Research*, 52, 83–105.

Bowen, N.L. (1913). "The melting phenomena of the plagioclase feldspars." *American Journal of Science*, 35, 577–599.

Bowen, N.L. (1928). *The Evolution of The Igneous Rocks*. Princeton: Princeton University Press.

Boyd, F.R. (1961). "Welded tuffs and flows in the rhyolite plateau of Yellowstone Park, Wyoming." *Geological Society of America Bulletin*, 30, 41–50.

Braden, S.E., Stolpar, J.D., Robinson, M.S., and Lawrence, S. (2014). "Evidence for basaltic volcanism on the Moon within the past 100 million years." *Nature Geoscience*, 7, 782–791.

Brandl, P.A., Schmid, F., Augustin, N. et al. (2019). "The 6-8 Aug 2019 eruption of 'Volcano F' in the Tofua Arc, Tonga." *Journal of Volcanology and Geothermal Research*, https://doi.org/10.1016/j.jvolgeores.2019.106695.

Branney, M.J. and Kokelaar, B.P. (1994). "Volcanotectonic faulting, soft-state deformation and rheomorphism of tuffs during development of a piecemeal caldera." *Geological Society of America Bulletin*, 106, 507–537.

Branney, M.J. and Kokelaar, B.P. (1997). "Giant bed from a sustained catastrophic density current flowing over topography: Acatlan ignimbrite, Mexico." *Geology*, 25, 115–118.

Branney, M.J. and Kokelaar, B.P. (2002). *Pyroclastic Density Currents and the Sedimentation of Ignimbrites*. London: Geological Society of London.

Breitkreuz, C. and Petford, N. (2004). *Physical Geology of High-Level Magmatic Systems*. Geological Society of London Special Publication 234, 253 p.

Briffa, K.R., Jones, P.D., Schweingruber, F.H., and Osborn, T.J. (1998). "Influence of volcanic eruptions on Northern Hemisphere summer temperature over the past 600 years." *Nature*, 393, 450–455.

Brigham, W.T. (1909). "The volcanoes of Kilauea and Mauna Loa on the island of Hawaii." *Memoir* 2, no. 4, Bernice Pauahi Bishop Museum, Honolulu, 222p.

Brothers, R.N. and Golson, J. (1959). "Geological interpretations of a section of Rangitoto ash on Motutapu Island, Auckland, New Zealand." *Journal of Geology and Geophysics*, 2, 569–577.

Brown, G.C., Rymer, H., Dowden, J. et al. (1989). "Energy budget analysis for Poas crater lake: Implications for predicting volcanic activity." *Nature*, 339, 370–373.

Brown, R. and Branney, M. (2004). "Bypassing and diachronous deposition from density currents: Evidence from the giant regressive bed form in the Poris ignimbrite, Tenerife, Canary Islands." *Geology*, 32, 445–448.

Brown, R.J. and Andrews, G.D.M. (2015). "Deposits of pyroclastic density currents." Chapter 36 in *The Encyclopedia of Volcanoes*, 2nd edition (eds. H. Sigurdsson et al.), pp. 631–648. Amsterdam: Elsevier, 1421 pp.

Brown, R.J. and Branney, M. (2013). "Internal flow variations and diachronous sedimentation within extensive, sustained, density-stratified pyroclastic density currents flowing down gentle slopes, as revealed by the internal architectures of ignimbrites on Tenerife." *Bulletin of Volcanology*, 75, 727. doi 10.1007/s00445-013-0727-0.

Brown, S.K., Jenkins, S.F., Sparks, R.S.J. et al. (2017). Volcanic fatalities database: analysis of volcanic threat with distance and victim classification. *Journal of Applied Volcanology*, 6, 15. https://doi.org/10.1186/s13617-017-0067-4.

Browne, B. and Saramek, L. (2018). "Rates of magma ascent and storage (Chapter 9)." In *The Encyclopedia of Volcanoes* (eds. H. Sigurdsson, B. Houghton, and H. Rymer), pp. 203–214. Cambridge, Massachusetts: Academic Press, Elsevier Publishing.

Browne, P.R.L. and Lawless, J. (2001). "Characteristics of hydrothermal explosions, with examples from New Zealand and elsewhere." *Earth-Science Reviews*, 54, 299–331.

Browning, J., and Gudmundsson, A. (2015). "Caldera faults capture and deflect inclined sheets: An alternative mechanism for ring dike formation." *Bulletin of Volcanology*, 77. doi: 10.1007/s00445-014-0889-4.

Bryan, S.E., Cook, A., Evans, J.P. et al. (2004). "Pumice rafting and faunal dispersion during 2001–2002 in the Southwest Pacific: Record of a dacitic submarine explosive eruption from Tonga." *Earth and Planetary Science Letters*, 227, 135–154.

Bryan, S.E., Riley, T.R., Jerram, D.A. et al. (2002). "Silicic volcanism; an undervalued component of large igneous provinces and volcanic rifted margins." *Geological Society of America Special Papers*, 362, 97–118.

Buddington, A.L. and Lindsley, D.H. (1964). "Iron-titanium oxide minerals and synthetic equivalents." *Journal of Petrology*, 5, 310–357.

Buist, A.S. and Bernstein, R.S. (eds.) (1986). "Health effects of volcanoes: An approach to evaluating the health effects of an environmental hazard." *American Journal of Public Health*, 76, 1–2.

Bullard, F. (1976). *Volcanoes of the Earth*. Austin: University of Texas Press, 579p.

Buller, A.T. and McManus, J. (1973). "Distinction among pyroclastic deposits from their grain size frequency distributions." *Journal of Geology*, 81, 97–106.

Burke, K. (2001). "Origin of the Cameroon Line of volcano-capped swells." *Journal of Geology*, 109, 349–362.

Burnham, C.W. (1983). "Deep submarine pyroclastic eruptions." In: *The Kuroko and Related Volcanogenic Massive Sulfide Deposits* (eds. H. Omoto and B. J. Skinner), pp. 142–148. Littleton: The Economic Geology Publishing Company.

Bursik, M., Gilbert, J.S., and Carey, S. (1992). "I. Theory and its comparison with a study of the Fogo A Plinian deposit, Sao Miguel (Azores)." *Bulletin of Volcanology*, 48, 109–125.

Bursik, M.I. and Woods, A.W. (1996). "The dynamics and thermodynamics of large ash flows." *Bulletin of Volcanology*, 58, 175–193.

Bursik, M.I., Sparks, R.S.J., Gilbert, J.S. et al. (1992). "Sedimentation of tephra by volcanic plumes. I. Theory and its comparison with a study of the Fogo A plinian deposit, Sao Miguel (Azores)." *Bulletin of Volcanology*, 48, 109–125.

Byrne, D.K. (2020). "A comparison of inner Solar System volcanism." *Journal of Nature Astronomy*, 4, 321–327.

Calder, E.S., Sparks, R.S.J., and Gardeweg, M.C. (2000). "Erosion, transport and segregation of pumice and lithic clasts in pyroclastic flows inferred from ignimbrite at Lascar Volcano, Chile." *Journal of Volcanology and Geothermal Research*, 104 (1–4), 201–235.

Calvari, S., Inguaggiato, S., Puglisi, G. et al. (2008). *The Stromboli Volcano: An Integrated Study of the 2002–2003 Eruption*. American Geophysical Union Geophysical Monograph 182, 397 pp.

Camp, V.E. (2013). "Origin of the Columbia River Basalt: Passive rise of shallow mantle, or active upwelling of a deep-mantle plume?" In: *The Columba River Flood Basalt Province* (eds. S.P. Reidel, V.E. Camp, M.E. Ross et al.), pp. 181–199. Geological Society of America Special Paper 497.

Camus, F. (2002). "The Andean porphyry systems." In: *Giant Ore Deposits: Characteristics, Genesis and Exploration* (eds. D.R. Cooke and J. Pongratz), pp. 5–21. Hobart: University of Tasmania.

Cannell, J., Cooke, D.R., Walshe, J.L. et al. (2005). "Geology, mineralization, alteration, and structural evolution of the El Teniente Porphyry Cu-Mo deposit." *Economic Geology*, 100 (5), 979–1003.

Canul, R.F. and Rocha, V.S. (1981). "Informe geológico de la zona geotérmica de 'El Chichónal,' Chiapas" Informe 32–81, unpublished report of the Geothermal Department of the Comisión Federal de Electricidad, Morelia, Michoacán, México, completed September 1981, 30 p., 5 figs., and 9 plates.

Capponi, A., Taddeucci, J., Scarlato, P., Palladino, D. (2016). "Recycled ejecta modulating Strombolian eruptions." *Bulletin of Volcanology*, 78, article 13. doi: 10.1007/s00445-016-1001-z.

Capra, L. and Macías, J.L. (2000). "Pleistocene cohesive debris flows at Nevado de Toluca Volcano, central Mexico." *Journal of Volcanology and Geothermal Research*, 102, 149–168.

Carapezza, G., Tait, S., Kaminski, E., Gardner, J. (2012). "The recent Plinian explosive activity of Mt. Pelée volcano (Lesser Antilles): The P1 AD 1300 eruption." *Bulletin of Volcanology*, 74, 2187–2203.

Carey S., Sigurdsson, H. (1986). "The 1982 eruptions of El Chichon volcano, Mexico (2): Observations and numerical modelling of tephra-fall distribution" *Bulletin of Volcanology*, 48 (2–3), 127–141.

Carey, R., Soule, S.A., Manga, M. et al. (2018). "The largest deep-sea oceanic silicic volcanic eruption of the past century." *Science Advances*, 4. doi:10.1126/sciadv.1701121.

Carey, S.N. (1991). "Transport and deposition of tephra by pyroclastic flows and surges." *Society for Sedimentary Geology, SEPM Special Publication*, 45, 39–57.

Carey, S.N. and Sigurdsson, H. (1987). "Temporal variations in column height and magma discharge rate during the 79 AD eruption of Vesuvius." *Geological Society of American Bulletin*, 99 (2), 303–314.

Carey, S.N. and Sigurdsson, S. (1989). "The intensity of Plinian eruptions." *Bulletin of Volcanology*, 51, 28–40.

Carey, S.N. and Sparks, R.S.J. (1986). "Quantitative models of the fallout and dispersal of tephra from volcanic eruption columns." *Bulletin of Volcanology*, 48, 109–125.

Carey, S.N., Sigurdsson, H., and Sparks, R.S.J. (1988). "Experimental studies of particle-laden plumes." *Journal of Geophysical Research*, 93, 15,314–15,328.

Carlino, S. (2012). "The process of resurgence for Ischia Island (southern Italy) since 55 ka: The laccolith model and implications for eruption forecasting." *Bulletin of Volcanology*, 74. doi 10.1007/s00445-012-0578-0.

Carr, M.J., Feigenson, M.D., Patino, L.C. et al. (2003). "Volcanism and geochemistry in Central America: Progress and problems." In: *Inside the Subduction Zone Factory* (ed. J. Eiler), 153–174. Washington, DC: American Geophysical Union, Geophysics Monograph Series, 138.

Carroll, M.R. and Holloway, J.R. (1994). "Volatiles in magmas." *Reviews in Mineralogy*, 30, Chantilly, Mineralogical Society of America, 517p.

Carson, H.L., Lockwood, J., and Craddock, E.M. (1990). "Extinction and colonization of local populations on a growing shield volcano." *Proceedings of the National Academy of Sciences, USA*, 87, 7055–7057.

Cas, R.A.F. (1989). "Eruptive centres." In: *Intraplate Volcanism in Eastern Australia and New Zealand* (ed. R.W. Johnson), 60–63. Cambridge: Cambridge University Press.

Cas, R.A.F. and Wright, J.V. (1987). *Volcanic Successions: Modern and Ancient*. London: Chapman and Hall.

Cas, R.A.F. and Wright, J.V. (1988). *Volcanic Successions: Modern and Ancient*. London: Unwin Hyman.

Casadevall, T.J. and Thompson, T.B. (1995). World Map of Volcanoes and Principal Aeronautical Features. US Geological Survey, Geophysical Investigations Map GP-1011.

Casadevall, T.J., De la Cruz-Reyna, S., Rose, W.I., Jr. et al. (1984). "Crater lake and post-eruption hydrothermal activity, El Chichón Volcano, Mexico." *Journal of Volcanology and Geothermal Research*, 23 (1–2), 169–191.

Casadevall, T.J., Tormey, D., and Roberts, J. (2019). *World Heritage Volcanoes: Classification, Gap Analysis, and Recommendations for Future Listings*. Gland, Switzerland: IUCN. viii + 68pp.

Cashman, K.V. and Giordano, G. (2014). Calderas and magma reservoirs. *Journal of Volcanology and Geothermal Research*, 288, 28–45. https://doi.org/10.1016/j.jvolgeores.2014.09.007

Cashman, K.V. and Cronin, S. J. (2008). "Welcoming a monster to the world: myths, oral tradition, and modern societal response to volcanic disasters." *Journal of Volcanology and Geothermal Research*, 176, 407–418.

Cashman, K.V., Thornber, C.R., and Pallister, J.S. (2008). "From dome to dust: shallow crystallization and fragmentation of conduit magma during the 2004–2006 dome effusion of Mount St Helens, Washington." In: *A Volcano Rekindled: The Renewed Eruption of Mount St. Helens, 2004–2006* (eds. D. Sherrod, W. Scott, and P.H. Stauffer), pp. 387–414. US Geological Survey Professional Paper 1250.

Castor, S.B. and Henry, C.D. (2020). "Lithium-rich claystone in the McDermitt Caldera, Nevada, USA: Geologic, mineralogical, and geochemical characteristics and possible origin." *Minerals*, 10 (1), 68–106. https://doi.org/10.3390/min10010068

Cattermole, P. (1994). *Venus: The Geological Story*. Baltimore: Johns Hopkins University Press.

Cawthorn, R.G. and Walraven, F. (1998). "Emplacement and crystallization time for the Bushveld complex." *Journal of Petrology*, 39, 1669–1687.

Chadwick, W.W. (2003). "Quantitative constraints on the growth of submarine lava pillars from a monitoring instrument that was caught in a lava flow." *Journal of Geophysical Research*, 108 (B11), 2534.

Chang, W.-L., Smith, R.B., and Wicks, C. (2007). "Accelerated uplift and magmatic intrusion of the Yellowstone Caldera, 2004 to 2006." *Science*, 318, 952–956.

Chapin, C.E. and Elston, W.E. (eds.) (1979). *Ash-flow Tuffs*. Geological Society of America Special Paper 180, 29–42.

Chaplygin, I., Yudovskaya, M., Vergasova, L., and Mkhov, A. (2015). "Native gold from volcanic gases at Tolbatchik 1975–1976 and 2012–2013 fissure eruptions, Kamchatka." *Journal of Volcanology and Geothermal Research*, 307, 200–209.

Chappell, B.V. and White, A.J.R. (2001). "Two contrasting granite types: 25 years later." *Australian Journal of Earth Science*, 48, 489–499.

Chappell, W.M., Durham, J.W., and Savage, G.E. (1961). "Mold of a rhinoceros in basalt, lower Grand Coulee, Washington." *Geological Society America Bulletin*, 62, 907–918.

Chédville, C., and Roche, O. (2018). "Autofluidization of collapsing bed of fine particles: Implications for the emplacement of pyroclastic flows." *Journal of Volcanology and Geothermal Research*, 368, 91–99.

Chesner, C.A., Rose, W.I., Deino, R. et al. (1991). "Eruptive history of Earth's largest Quaternary caldera (Toba, Indonesia) clarified." *Geology*, 19, 200–203.

Chevrel, M.O., Harris, A., James, M.R., and Calabro, L. (2018). The viscosity of pāhoehoe lava: In situ syn-eruptive measurements from Kilauea, Hawaii. *Earth and Planetary Science Letters*, 493 (1–2): 161–171. doi:10.1016/j.epsl.2018.04.028

Chouet, B.A. (1996). "New methods and future trends in seismological volcano monitoring." In *Monitoring and Mitigation of Volcanic Hazards. Bulletin of Volcanology*, v. 69 (eds. R. Scarpa and R. Tilling). Springer: New York.

Choux, C.M. and Druitt, T.H. (2002). "Analogue study of particle segregation in pyroclastic density currents, with implications for the emplacement mechanisms of large ignimbrites." *Sedimentology*, 49, 907–928.

Chretien, S. and Brousse, R. (1988). *La Montagne Pelee se Réveille (Mt. Pelée Revealed)*. Societe Nouvelle des Edicions Boubee, Paris, 243 pp.

Chrétien, S. and Brousse, R. (1989). "Events preceding the great eruption of 8 May, 1902 at Mount Pelée, Martinique." *Journal of Volcanology and Geothermal Research*, 38, 67–75.

Christiansen, R.L. (1979). "Cooling units and composite sheets in relation to caldera structure." In: *Ash-flow Tuffs* (eds. C.E. Chapin and W.E. Elston), pp. 29–42. Geological Society of America Special Paper 180.

Christiansen, R.L. (1984). "Yellowstone magmatic evolution: Its bearing on understanding large-volume explosive volcanism." In: *Explosive Volcanism: Inception, Evolution, and Hazards*. Geophysics Research Forum. Washington, DC: National Academy Press, pp. 84–95.

Christiansen, R.L. (1987). "Rhyolite-baslt volcanism of the Yellowstone Plateau and hydrothermal activity of Yellowstone National Park, Wyoming." *Proceedings of the Rocky Mountain Sectional Meeting*, Geological Society of America.

Christiansen, R.L. (2001). "The Quaternary and Pliocene Yellowstone Plateau volcanic field of Wyoming, Idaho, and Montana." US Geological Survey Professional Paper 729-G: 145.

Christiansen, R.L. and Lipman, P.W. (1966). "Emplacement and thermal history of a rhyolite lava flow near Fortymile canyon, southern Nevada." *Geological Society of America Bulletin*, 7, 671–684.

Christiansen, R.L., Foulger, G.R., and Evans, J.R. (2002). "Upper mantle origin of the Yellowstone hot spot." *Geological Society of America Bulletin*, 114, 1245–1256.

Christiansen, R.L., Lowenstern, J.B. Smith, R.B. et al. (2007). Preliminary Assessment of Volcanic and Hydrothermal Hazards in Yellowstone National Park and Vicinity: U.S. Geological Survey Open-file Report 2007-1071, 4 pp.

Church, A.A. and Jones, A.P. (1995). "Silicate-carbonate immiscibility at Oldoinyo Lengai." *Journal of Petrology*, 36, 869–890.

Cioni, R., Pistolesi, M., and Rosi, M. (2015). "Plinian and subplinian eruptions." Chapter 29 In: *The Encyclopedia of Volcanoes* (eds. H. Sigurdsson et al.), 2nd edition, pp. 519–535. London: Academic Press, 1421 pp.

Clague, D.A. (1987). "Hawaiian xenolith populations, magma supply rates and development of magma chambers." *Bulletin of Volcanology*, 49, 577–587.

Clague, D.A. and Dalrymple, G.B. (1987). "Hawaiian-Emperor volcanic chain." In: *Volcanism in Hawai'i: U.S. Geol. Survey Prof. Paper 1350* (eds. R.W. Decker, T. L. Wright and P. H. Stauffer), US Geological Survey, pp. 5–54.

Clague, D.A. and Sherrod, D.R. (2014). "Growth and degradation of Hawaiian Volcanoes." In: *Characteristics of Hawaiian Volcanoes* (eds. M.P. Poland, T.J. Takahashi, and C.M. Landowski), pp. 97–146. United States Geological Survey Professional Paper 1801.

Clague, D.A., Holcomb, R.T., Sinton, J.M. et al. (1990). "Pliocene and Pleistocene alkalic flood basalts on the seafloor north of the Hawaiian Islands" *Earth and Planetary Science Letters*, 98 (2), 175–191.

Clague, D.A., Paduan, J.B., Caress, D.W. et al. (2017). "High-resolution AUV mapping and targeted ROV observations of three historical lava flows at Axial Seamount." *Oceanography*, 30 (4), 82–99. doi.org/10.5670/oceanog.2017.426.

Clague, D.A., Paduan, J.B., Caress, D.W., and Thomas, H. (2011). "Volcanic morphology of West Mata Volcano, NE Lau Basin, based on high-resolution bathymetry and depth changes." *Journal of Geochemistry, Geophysics, and Geosystems*, 12 (11). doi: 10.1029/2011GC003791.

Clarke, A.B., Voight B., Neri, A. et al. (2002). "Transient dynamics of Vulcanian explosions and column collapse." *Nature*, 415, 897–901.

Clemens, J. (1998). "Observations on the origins and ascent mechanisms of granitic magmas." *Journal of the Geological Society of London*, 155, 843–851.

Cloos, H. (1941). "Bau und Tätigeit von Tuffschloten." *Geologische Rundschau*, 30, 405–527.

Coan, T. (1844). "Journey to Mauna Loa." *Missionary Herald*, 40, February, 44–47.

Coan, T. (1857). "Volcanic action on Hawaii." *American Journal of Science*, 73, 435–437.

Coan, T. (1882). *Life in Hawaii: An Autobiographical Sketch*. New York: Anson D. F. Randolph & Company.

Coats, R.R. (1951). "Volcanic activity in the Aleutian arc." US Geological Survey, pp. 35–47.

Cohen, B., Vascancelos, P.M.D., and Knesel, K.M. (2004). "Tertiary magmatism in southeast Queensland." In: *Past, Present, and Future* (eds. J. McPhie and J. McGoldrick), 256. Sydney: Geological Society of Australia.

Cole, P., Guest, J., Duncan, A. et al. (2001). "Capelinhos 1957–1958, Faial, Azores: Deposits formed by an emergent Surtseyan eruption." *Bulletin of Volcanology*, 63, 204–220.

Colombrita, R. (1984). "Methodology for the construction of earth barriers to divert lava flows: The Mt. Etna 1983 eruption." *Bulletin of Volcanology*, 74 (4), 1009–1038.

Committee on Natural Disasters, Division of Natural Hazards, National Research Council (1991). *The Eruption of Nevado del Ruiz Volcano, Columbia, South America, November 13, 1985*. Natural Disaster Studies: A Series. National Academies Press, 128p.

Condie, K.C. (1982). *Plate Tectonics and Crustal Evolution*. New York: Pergamon Press.

Condie, K.C. (2021). *Earth as an Evolving Planetary System*, 4th edn. Amsterdam: Elsevier Press, 430 p.

Condie, K.C. and Pease, V. (2008). *When Did Plate Tectonics Begin on Earth?* Boulder: Geological Society of America.

Connor, C.B., McBirney, A.R., and Furlan, C. (2006). "What is the probability of explosive eruption at a long-dormant volcano?" In: *Statistics in Volcanology* (eds. H.M. Mader, S.G. Coles, C.B. Connor, et al.), 39–46. London: Geological Society.

Conway, C., Townsend, D., Leonard, G. et al. (2015). "Lava-ice interaction on a large composite volcano: A case study from Ruapehu, New Zealand." *Bulletin of Volcanology*, 77. doi 10.1007/s00445-015-0906-2.

Conway, S., Wauthier, C., Fukushima, Y., and Poland, M. (2018). "A retrospective look at the February 1993 east rift zone intrusion at Kīlauea volcano, Hawaii." *Journal of Volcanology and Geothermal Research*, 358, 241–251.

Cooke, D.R., Hollings, P., and Walshe, J.L. (2005). "Porphyry copper deposits: Characteristics, distribution, and tectonic controls." *Economic Geology*, 100, 801–818.

Cooke, R.S.S., McKee, C.O., Dent, V.F. et al. (1976). "Striking sequence of volcanic eruptions in the Bismarck volcanic arc, Papua NG, in 1972–1975." In: *Volcanism in Australasia* (ed. R.W. Johnson), 149–172. Amsterdam: Elsevier.

Coppola, D., Campion, R., Laiolo, M. et al. (2016). "Birth of a lava lake: Nyamulagira volcano 2011–2015." *Bulletin of Volcanology*, 78, article 20. doi: 10.1007/s00445-016-1014-7.

Corner, J.B. (1974). "Genesis of Jamaican bauxite." *Economic Geology*, 69, 1251–1264.

Corr, H.F.J. and Vaughn, D.G. (2008). "A recent volcanic eruption beneath the West Antarctic ice sheet." *Nature Geosciences*, 1, 122–125.

Cox, K.G. (1988). "The Karoo Province." In: *Continental Flood Basalts* (ed. J.D. Macdougal), pp. 239–271. Dordrecht: Kluwer Academic.

Craig, H. and Lupton, J.E. (1981). "Helium-3 and mantle volatiles in the ocean and oceanic crust." In: *The Sea: Oceanic Lithosphere, Vol. 7* (ed. C. Emiliani). Harvard University Press, pp. 391–428.

Crandell, D.R., Mullineaux, D.R., Miller, C.D. et al. (1979). "Volcanic-hazards studies in the Cascade Range of the Western United States." In: *Volcanic Activity and Human Ecology* (eds. P.D. Sheets and D.K. Grayson), pp. 195–219. New York: Academic Press.

Crisp, J.A. (1984). "Rates of magma emplacement and volcanic output." *Journal of Volcanology and Geothermal Research*, 20, 177–211.

Cristofolini, R., Gresta, S., Imposa, S. et al. (1987). "An approach to problems on energy sources at Mount Etna based on seismological and volcanological data." *Bulletin of Volcanology*, 49, 729–736.

Crosweller, H.S., Arora, B., Brown, S.K. et al. (2012). "Global database on large magnitude explosive volcanic eruptions (LaMEVE)." *Journal of Applied Volcanology*, 1, 4. https://doi.org/10.1186/2191-5040-1-4.

Crowe, B.M. and Fisher, R.V. (1973). "Sedimentary structures in base-surge deposits with special reference to cross-bedding, Ubehebe Craters, Death Valley, California." *Geological Society of America Bulletin*, 84, 663–682.

Crumpler, L.S. (1996). "Venus." In: *Macmillan Encyclopedia of Earth Sciences* (ed. J. Dasch), pp. 1129–1135. New York: Simon and Schuster.

Crumpler, L.S. and Aubele, J.C. (2000). "Volcanism on Venus." In: *Encyclopedia of Volcanoes* (ed. H. Sigurdsson), pp. 727–769. San Diego, Academic Press.

Cuaresma, J.C., Hlouskova, J., and Obersteiner, M. (2008). "Natural disasters as creative destruction? Evidence from developing countries." *Economic Inquiry*, 46 (3), 214–226.

Cullen-Tanaka, J. (1980). *Fire Mountain*. New York: Kensington Publishing, 352 pp.

Cunningham, J.K. and Beard, A.D. (2014). "An unusual occurrence of mafic accretionary lapilli in deep-marine volcaniclastics on 'Eua, Tonga: Paleoenvironment and process." *Journal of Volcanology and Geothermal Research*, 274, 139–151.

Cuoco, E., Spagnvolo, A., Balagizi, C., and De Francesco, S. (2013). Impact of volcanic emissions on rainwater chemistry: The case of Mt. Nyiragongo in the Virunga Volcanic Region (ROC). *Journal of Geochemical Exploration*, 125, 69–79.

Currier, R., Forsythe, P., Grossmeier, C. et al. (2017). "Experiments on the evolution of laccolith morphology in plan-view." *Journal of Volcanology and Geothermal Research*, 336, 155–167.

Damon, P. and Montesinos, E. (1978). "Late Cenozoic volcanism and metallogenesis over an active Benioff Zone in Chiapas, Mexico." *Arizona Geological Society Digest*, 11, 155–168.

Dana, J.D. (1852). "Note on the eruption of Mauna Loa." *American Journal of Science*, 2nd Series, 14, 1–4, 254–329.

Darnell, A.R., Phillips, J.C., Barclay, J. et al. (2013). "Developing a simplified geographical information system approach to dilute lahar modelling for rapid hazard assessment" *Bulletin of Volcanology*, 75. doi 10.1007/s00445-013-0713-6.

Davies, A.G. (2009). *Volcanism on Io: A Comparison with Earth*. Cambridge: Cambridge University Press.

Davies, R.J., Brumm, M., Manga, M. et al. (2008). "The East Java mud volcano (2006 to present): An earthquake or drilling trigger?" *Earth and Planetary Science Letters*, 272 (3–4), 627–638.

Davis, J.C. (1986). *Statistics and Data Analysis in Geology*. New York: John Wiley & Sons.

Dawson, J.B. (1998). "Peralkaline nephelinite-natrocarbonatite relationships at Oldoinyo Lengai, Tanzania." *Journal of Petrology*, 39, 2077–2094.

Dawson, P.B. and Chouet, B.A. (1999). "Three-dimensional velocity structure of the Kilauea caldera, Hawaii." *Geophysical Research Letters*, 26 (18), 2805–2808.

Day, S.J. (2015). Volcanic tsunamis: Chapter 58 in *The Encyclopedia of Volcanoes* (eds. H. Sigurdsson et al.), 2nd edition, pp. 993–1009. Elsevier Press, Amsterdam, 1421 pp.

Day, S.J. (1993). "The structural evolution and mechanics of volcanoes and subvolcanic intrusions." *Journal of the Geological Society of London*, 150, 207–208.

De Carolis, E. and Patricelli, G. (2003). *Vesuvius AD 79: The Destruction of Pompeii and Herculaneum*. Rome: Roberto Marcucci.

De Gregorio, S., Camarda, M., and Gurrieri, S. (2014). "Change in magma supply dynamics identified in observations of soil CO2 emissions in the summit area of Mt. Etna." *Bulletin of Volcanology*, 76, article 846. doi: 10.1007/s00445-014-0846-2.

De Guzman, E.M. (2005). Eruption of Mount Pinatubo in the Philippines in June, 1991: Asian Disaster Reduction Center, https://www.adrc.asia/publications/recovery_reports/, 18 pp.

De la Cruz-Reyna, S. (1993). "Random patterns of activity of Colima Volcano, Mexico." *Journal of Volcanology and Geothermal Research*, 55, 51–68.

De la Cruz-Reyna, S. (1996). "Long-Term probabilistic analysis of future explosive eruptions." In: *Monitoring and Mitigation of Volcanic Hazards* (eds. R. Scarpa and R.I. Tilling, R.I.), pp. 599–629. Berlín-Heidelberg: Springer-Verlag.

De la Cruz-Reyna, S. and Martin Del Pozzo, A.L (2009). "The 1982 eruption of El Chichón volcano, Mexico: Eyewitness of the disaster." *Geofísica Internacional*, 48 (1), 21–31.

De la Cruz-Reyna, S. and Tilling, R.I. (2008). "Scientific and public responses to the ongoing volcanic crisis at Popocatepetl volcano, Mexico: Importance of an effective hazards-warning system." *Journal of Volcanology and Geothermal Research*, 170, 121–134.

De la Cruz-Reyna, S. Mendoza-Rosas, A.T., Borselli L., and Sarocchi D. (2019). "Volcanic hazard estimations for Volcán de Colima." In: *Volcán de Colima, Active Volcanoes of the World* (eds. N. Varley, C.B. Connor, and J.-C. Komorowski), 267–289. Berlín-Heidelberg: Springer-Verlag. https://doi.org/10.1007/978-3-642-25911-1_6.

De la Cruz-Reyna, S., Tilling, R.I., and Valdés González, C. (2017). "Challenges in responding to a sustained, continuing volcanic crisis: The case of Popocatépetl Volcano, México, 1994-Present." In: *Observing the Volcano World: Volcano Crisis Communication* (eds. C.J. Fearnley, B. McGuire, G. Jolly et al.). Advances in Volcanology IAVCEI Barcelona series, Springer Verlag. https://doi.org/10.1007/11157_2016_37.

De Matteo, A., Corti, G., van Wyk de Vries, B. et al. (2018). "Fault-volcano interactions with broadly distributed stretching in rifts." *Journal of Volcanology and Geothermal Research*, 362, 64–75.

De Rita, D., Di Filippo, M., and Rosa, C. (1996). "Structural evolution of the Bracciano volcanotectonic depresión, Sabatini Volcanic District, Italy." In: *Volcano Instability on the Earth and Other Planets* (eds. A.W.J. McQuire, P. Jones, and J. Neuberg), 110. London: Geological Society of London.

de Silva, S.L. and Francis, P.W. (1991). *Volcanoes of the Central Andes*. New York: Springer-Verlag.

de Silva, S.L. and Zielinski, G.A. (1998). "Global influence of the ad 1600 eruption of Huaynaputina, Peru." *Nature*, 393, 455–458.

de Silva, S., Zandt, G., Trumbull, R. et al. (2006). "Large ignimbrite eruptions and volcano-tectonic depressions in the Central Andes: A thermomechanical perspective." *Geological Society of London, Special Publications*, 269, 47–63.

De Vries, B.V.M. Bingham, R.G., and Hein, A.S. (2018). A new volcanic province – an inventory of subglacial volcanoes in West Antarctica. In: *Exploration of Subsurface Antarctica: Uncovering Past Changes and Modern Processes* (eds. M.J. Siegert, S.S.R. Jamieson, and D.A. White), pp. 231–248. Geological Society, London, Special Publications 461.

De Vries, B.V.W., Self, S., and Francis, P.W. (2001). "A gravitational spreading origin for the Socompa debris avalanche." *Journal of Volcanology and Geothermal Research*, 105, 225–247.

Deamer, D.W. (2019). *Assembling Life: How Can Life Begin on Earth and Other Habitable Planets?* Oxford: Oxford University Press, 184 pp.

Decker, R.W. (1973). "State-of-the-art in volcano forecasting." *Bulletin Volcanologique*, 37 (3), 372–393.

Decker, R.W. (1990). "How often does a Minoan eruption occur?" In: *Thera and the Aegean World III* (eds. D.A. Hardy, S. Keller, P. V. Galanpoulos et al.), Vol. 2 (Earth Sciences), pp. 444–454. London: The Thera Foundation.

Decker, R.W. and Decker, B. (1979). *Volcanoes*. New York: W. H. Freeman, 244p.

Decker, R.W. and Decker, B. (1997). *Volcanoes: Third Edition*. New York: W. H. Freeman.

Decker, R.W. and Decker, B.B. (1991). *Mountains of Fire*. Cambridge: Cambridge University Press.

Decker, R.W. and Kinoshita, W. (1971). *Geodetic Measurements: The Surveillance and Prediction of Volcanic Activity*. Paris: UNESCO.

Decker, R.W., Wright, T.L., and Stauffer, P.H. (eds.) (1987). "Volcanism in Hawaii." US Geological Survey Professional Paper 1350, 1667p.

Delaney, P.T. and Pollard, D.D. (1981). "Deformation of host rocks and flow of magma during growth of minette dikes and breccia-bearing intrusions near Shiprock, New Mexico." US Geological Survey, 61.

Delcamp, A., Troll, V.R., van Wyk de Vries, B. et al. (2012). "Dykes and structures of the NE rift of Tenerife, Canry Islands: A record of stabilization and destabilization of ocean island rift zones." *Bulletin of Volcanology*, 74, 963–980.

Dellino, P., Dioguardi, F., Mele, D. et al. (2014). "Volcanic jets, plumes, and collapsing fountains: Evidence from large-scale experiments, with particular emphasis on the entrainment rate." *Bulletin of Volcanology*, 76, 834. doi 10.10071500445-014-0834-6.

Delmelle, P., Stix, J., Baxter, P.J. et al. (2004). "Atmospheric dispersion, environmental effects and potential health hazard associated with the low-altitude gas plume of Masaya volcano, Nicaragua." *Bulletin of Volcanology*, 64 (6), 423–434.

Deniel, C., Kieffer, G., and Lecointre, J. (1992). "New 230Th-238U and 14C age determinations from Piton des Neiges Volcano, Reunion: A revised chronology for the differentiated series." *Journal of Volcanology and Geothermal Research*, 51, 253–267.

Deniel, C., Vincent, P.M., Beauvilain, A., and Gourgaud, A. (2015). "The Cenozoic volcanic province of Tibestic (Sahara of Chad): Major units, chronology, and structural features." *Bulletin of Volcanology*, 77, article 74. doi:10.1007/s00445-015-0955-6.

Deruelle, B., Moreau, C., Nkoumboum C. et al. (1991). "The Cameroon line: A review." In: *Magmatism in extension structure settings; The Phanerozoic African Plate* (eds. A.B. Kampunzo and R.T. Lubala), pp. 275–327. Berlin: Springer Verlag Press.

Devine, J.D. (1995). "Petrogenesis of the basalt-andesite-decite association of Grenada, Lesser Antilles Island arc." *Journal of Volcanology and Geothermal Research*, 69, 1–33.

Di Piazza, A., Del Bello E., Mollo, S. et al. (2017). "Like a cannonball: Origin of dense spherical basaltic ejecta." *Bulletin of Volcanology*, 79, 37. doi: 10.1007/s00445-017-1121-0.

Di Traglia, F., Intrieri, E., Nolesini, T. et al. (2014). "The ground-based InSAR monitoring system at Stromboli volcano: Linking changes in displacement rate and intensity of persistent volcanic activity." *Bulletin of Volcanology*, 76. doi: 10.1007/s00445-013-0786-2.

Dieterich, J.H. and Decker, R.W. (1975). "Finite element modeling of surface deformation associated with volcanism." *Journal of Geophysical Research*, 80, 4094–4102.

Dilek, Y. and Robinson, P.T. (eds.) (2003). *Ophiolites in Earth History*. London: Geological Society of London.

DiPippo, R. (1988). "International developments in geothermal power production." *Geothermal Resources Council Bulletin*, 8–19.

DN-III-E. (1983). "El Plan DN-III-E y su aplicación en el área del volcán Chichonal." In: *El Volcan Chichonal*. Instituto de Geologia, Universidad Nacional Autonoma de Mexico. Mexico D.F., pp. 90–99.

Dobran, F., Neri, A., and Macedonio, G. (1993). "Numerical simulation of collapsing volcanic columns." *Journal of Geophysical Research*, 98, 4231–4259.

Dogliani, C., Green, D., and Mongelli, F. (2005). "On the shallow origin of hot spots and the westward drift of the lithosphere." In: *Plates, Plumes, and Paradigms* (eds. G.R. Foulger, J.H. Natland, D.C. Presnall, and D.L. Anderson), pp. 735–749. Geological Society of America Special Paper 388.

Doillet, G.A., Pacheco, D.A., Kueppers, U. et al. (2013). "Dune bedforms produced by dilute pyroclastic density currents from the August 2006 eruption of Tungurahua volcano, Ecuador." *Bulletin of Volcanology*, 75, 762, doi: 10.1007/s00445-013-0762-x.

Donnelly-Nolan, J.M. (1988). "A magmatic model of Medicine Lake Volcano, California." *Journal of Geophysical Research*, 93 (B5), 4412–4420.

Donovan, A. and Oppenheimer, C. (2016). Imagining the unimaginable: Communicating extreme volcanic risk. In: *Observing the Volcano World* (eds. C.J. Fearnley, D.K. Bird, K. Haynes et al.), Advances in Volcanology (An Official Book Series of the International Association of Volcanology and Chemistry of the Earth's Interior – IAVCEI, Barcelona, Spain). Springer, Cham. https://doi.org/10.1007/11157_2015_16.

Donovan, A., and Oppenheimer, C. (2012). The aviation sagas: Geographies of volcanic risk, The Geographical Journal, Royal Geographical Society, doi:10.1111/j.1475-4959.2011.00458x

Dorn, R.I. (2009). *The Role of Climatic Change in Alluvial Fan Development*. New York: Springer-Verlag.

Doronzo, D.M. (2013). "Aeromechanic analysis of pyroclastic density currents past a building." *Bulletin of Volcanology*, 75, 684. doi: 10.1007/s00445-012-0684-z.

Douillet, G.A. (2014). "Saltation threshold for pyroclasts at various bedslopes: Wind tunnel experiments." *Journal of Volcanology and Geothermal Research*, 278–279, 14–24.

Douillet, G.A., Rasmussen, K., and Kueppers, U. (2014). Saltation threshold for pyroclasts at various bedslopes, wind tunnel measurements. *Journal of Volcanology and Geothermal Research*, 278, 14–24.

Douillet, G.A., Tsang-Hin-Sun, E., and Kueppers, U. (2013). Sedimentology and geomorphology of the deposits from the August 2006 pyroclastic density currents at Tungurahua Volcano, Ecuador. *Bulletin of Volcanology*, 75, 1–21.

Doyle, E., McClure, J., Johnston, D., and Paton, D. (2014). "Communicating likelihoods and probabilities in forecasts of volcanic eruptions." *Journal of Volcanology and Geothermal Research*, 272, 1–15.

Driedger, C.L. and Kennard, P.M. (1986). "Ice volumes on Cascade volcanoes: Mount Rainier, Mount Hood, Three Sisters, and Mount Shasta." U.S. Geological Survey, 28.

Druitt, T.H. (1992). "Emplacement of the 18 May 1980 lateral blast deposit ENE of Mount St. Helens, Washington." *Bulletin of Volcanology*, 54, 554–572.

Druitt, T.H. (1996). "Turbulent times at Taupo." *Nature*, 381, 476–477.

Druitt, T.H. and Kokelaar, B.P. (2002). *The Eruption of Soufriere Volcano, Montserrat, from 1995 to 1999*. London: Geological Society of London.

Druitt, T.H. and Sparks, R. (1982). "A proximal ignimbrite breccia facies on Santorini, Greece." *Journal of Volcanology and Geothermal Research*, 13, 147–171.

Duba, A.G., Durham, W.B., Handin, J.W. et al. (1990). The brittle-ductile transition in rocks: The heard volume. *American Geophysical Union Geophysical Monograph Series*, 56, 243 p.

Dubosclard, G., Donnadieu, F., Allard, P. et al. (2004). "Doppler radar sounding of volcanic eruption dynamics at Mount Etna." *Bulletin of Volcanology*, 66, 443–456.

Ducea, M.N., Saleeby, J., Morrison, J., and Valencia, V.A. (2005). "Subducted carbonates, metasomatism of mantle wedges and possible connections to diamond formation: An example from California." *American Mineralogist*, 90, 864–870.

Dudley, W. and Lee, M.H. (1998). *Tsunami!* Manoa: University of Hawaii Press.

Duffield, W.A. (1972). "A naturally occurring model of global plate tectonics." *Journal of Geophysical Research*, 77 (14), 2543–2555.

Duffield, W.A. (2003). *Pele Stirs – A Volcanic Tale of Hawaii, Hemp, and High-Jinks*. New York: Universe Press, 279 pp.

Duffield, W.A. (2007). *Yucca Mountain Dirty Bomb*. New York: Universe Press, 184 pp.

Duffield, W.A., Stieltjes, L., and Varet, J. (1982). "Huge landslide blocks in the growth of Piton de la Fournaise, La Reunion, and Kilauea volcano, Hawaii." *Journal of Volcanology and Geothermology Research*, 12 (1), 147–160.

Duffield, W.A., Tilling, R.I., and Canul, R. (1984). "Geology of El Chichón Volcano, Chiapas, Mexico." *Journal of Volcanology and Geothermal Research*, 20, 117–132.

Duncan, R.A. and Clague, D.E. (1985). "Pacific plate motion recorded by linear volcanic chains." In: *The Ocean Basins and Margins: The Pacific Ocean* (eds. A.E.M. Nairn, F.G. Stehli, and S. Uyeda), vol. 7b, pp. 89–121. New York: Plenum Press.

Duncan, R.A. and MacDougall, I. (1989). "Volcanic time-space relationships." In: *Intraplate Volcanism in Eastern Australia and New Zealand* (ed. R.W. Johnson), pp. 43–53. Cambridge: Cambridge University Press.

Dutton, C.E. (1884). "Hawaiian volcanoes." Fourth Annual report of the US Geological Survey, 1882–83, Government Printing Office, Washington, DC, 75–219.

Dvorak, J. (2017). *The Last Volcano: A Man, A Romance, and The Quest to Understand Nature's Most Magnificent Fury*. Pegasus Books, 344 p.

Dwiggins, T. (2020). *Volcano Watch*. Cupertino, California: Milbank Press, 386 pp.

Dzurisin, D. (1980). "Influence of fortnightly earth tides at Kilauea volcano, Hawaii." *Geophysical Research Letters*, 7 (11), 925–928.

Dzurisin, D. (2007). *Volcano Deformation: Geodetic Monitoring Techniques*. Chichester: Springer-Praxis.

Eason, D.E., Sinton, J.M., Grönwald, K., and Kurz, M.D. (2015). "Effects on the petrology and eruptive history of the Western Volcanic Zone, Iceland." *Bulletin of Volcanology*, 77. doi 10.1007/s00445-015-0916-0.

Easton, R.M. and Lockwood, J.P. (1983). "'Surface-fed dikes'–the origin of some unusual dikes along the Hilina Fault Zone, Kilauea Volcano, Hawaii." *Bulletin Volcanologique*, 46 (1), 45–53.

Eaton, G.P. (1979). "A plate-tectonic model for late Cenozoic crustal spreading in the western United States." In: *Rio Grande Rift: Tectonics and Magmatism* (ed. R.E. Riecker), pp. 7–32. Washington, DC: American. Geophysical Union.

Ebinger, C.J., Baker, J., Menzie, M.A. et al. (2002). "Volcanic rifted margins." Geological Society of America Special paper 362, 236p.

Edmonds, M., Pyle, D.M., Oppenheimer, C.M. et al. (2003). "Trends in SO2 fluxes 1995–2001 at Soufriére Hills Volcano, Montserrat, West Indies and their implications for changes in conduit permeability, hydrothermal interaction and degassing regime." *Journal of Volcanology and Geothermal Research*, 124, 23–43.

Edmunds, M. and Woods, A. (2018). "Exsolved volatiles in magma reservoirs." *Journal of Volcanology and Geothermal Research*, 368, 13–30.

Edwards, B.R., Belousov, A., Belousova, M., and Melnikov, D. (2015). "Observations on lava, snowpack, and their interactions during the 2012–2013 Tolbatchik eruption, Klyuchevskoy Group, Kamchatka, Russia." *Journal of Volcanology and Geothermal Research*, 307, 107–119.

Eggers, A.A. (1987). "Residual gravity changes and eruption magnitudes." *Journal of Volcanology and Geothermal Research*, 33, 201–216.

Eichelberger, J. (1978). "Andesites in island arcs and continental margins: Relationship to crustal evolution." *Bulletin of Volcanology*, 41, 481–500.

Eichelberger, J., Gordeev, E., Izbekov, P. et al. (2007). *Volcanism and Subduction: The Kamchatka Region*. American Geophysical Union Geophysical Monograph 172, 369 pp.

Elachi, C., Wall, S., Anderson, R. et al. (2005). "Cassini view the surface of Titan." *Science*, 308, 970–974.

Ellsworth, W.L. and Voight, B. (1995). "Dike intrusion as a trigger for large earthquakes and the failure of volcano flanks." Journal of Geophysical Research, 100, 6005–6024.

Emanuel, K.A. (1999). "The power of a hurricane: An example of reckless driving on the information superhighway." *Weather*, 54, 107–108.

Embley, R.W., Chadwick, W.W., Baker, E.T. et al. (2006). "Long-term volcanic activity at a submarine arc volcano." *Nature*, 44, 25 May, 494–497.

Endo, E.T. and Murray, T. (1991). "Real-time seismic amplitude measurement (RSAM): A volcano monitoring and prediction tool." *Bulletin Volcanologique*, 53, 533–545.

Endo, K., Sumita, M., Machida, M. et al. (1989). "The 1984 collapse and debris avalanches of Ontake volcano, Central Japan." In: *Volcanic Hazards Assessment and Monitoring* (ed. J.H. Latter), pp. 210–229. Berlin: Springer-Verlag.

Enlows, H.E. (1955). "Welded tuffs of Chiricahua National Monument, Arizona." *Geological Society of America Bulletin*, 66, 1215–1246.

Epp, D. (1979). "Age and tectonic relationships among volcanic chains on the Pacific Plate, Hawaii Symposium on Intraplate Volcanism and Submarine Volcanism," July 16–22, University of Hawaii at Hilo, 121.

Erlich, E.N., Melekestsev, I.V., Tarakanovsky, A.A. et al. (1973). "Quaternary Calderas of Kamchatka." *Bulletin of Volcanology*, 36 (1), 222–237.

Ernst, R.E., Dickson, A.J., and Bekker, A. (eds.) (2021). *Large Igneous Provinces: A Driver of Global Environmental and Biotic Changes*. American Geophysical Union Monograph 255, 508 pp.

Ernst, R.E. (ed.) (2014). *Large Igneous Provinces*. Cambridge: Cambridge University Press, 653 p.

Ernst, R.E., Buchan, K.L., Aspler, L.B. et al. (2008). Large Igneous Provinces Commission, LIP Record. International Association of Volcanology and Chemistry of the Earth's Interior. available at: http://www.largeigneousprovinces.org/record.html.

Ernst, R.E., Head, W.J., Parfitt, E.A. et al. (1995). "Giant radiating dike swarms on Earth and Venus." *Earth Science Reviews*, 39, 1–58.

Espíndola J.M., Macías, J.L., Tilling, R.I., and Sheridan, M.F. (2000). "Volcanic history of El Chichón Volcano (Chiapas, Mexico) during the Holocene, and its impacts on human activity." *Bulletin of Volcanology*, 62 (2), 90–104.

Fagents, S., Gregg, T.K., and Lopes, M.C. (eds.) (2013). *Modeling Volcanic Processes: The Physics and Mathematics of Volcanism*. Cambridge, UK: Cambridge University Press, 431 p.

Faggart, B.E., Basu, A.R., and Tatsumoto, M. (1985). "Origin of the Sudbury Complex by meteoritic impact: Neodymium isotopic evidence." *Science*, 230 (4724), 436–439.

Favereau, M., Robledo, L., and Bull, M. (2018). "Analysis of risk assessment factors of individuals in volcanic hazards: Review of the last decade." *Journal of Volcanology and Geothermal Research*, 357, 254–260.

Faybishenko, B., Witherspoon, P.A., and Benson, S.M. (2000). Dynamics of fluids in fractured rocks. *American Geophysical Union Monograph Series*, vol. 112, 400 p.

Fearnley, C.J., Bird, D.K., Haynes, K. et al. (eds.) (2019). *Observing the Volcano World – Volcano Crisis Communication*. New York: Springer, 320 pp. https://doi.org/10.1007/11157_2017_28; https://doi.org/10.1007/11157_2016_37

Fedotov, S.A. (1985). "Estimates of heat and pyroclast discharge by volcanic eruptions based upon the eruption cloud and steady plume observations." *Journal of Geodynamics*, 3, 275–302.

Fedotov, S.A. and Masurenkov, Y.P. (1991). *Active Volcanoes of Kamchatka*. Moscow: Nauka Publishers.

Fei, J. and Zhou, J. (2006). "The possible climatic impact in China of Iceland's Eldgjá eruption inferred from historical sources." *Climate Change*, 76, 443–457.

Ferrer, M., de Vallejo, L.G., Madeira, J. et al. (2021) "Megatsunamis Induced by volcanic landslides in the Canary Islands – age of the Volcanic Landslides in the Canary – Islands: Age of the Tsunami deposits and source landslides." *GeoHazards*, 2, 228–256, https://doi.org/10.3390/geohazards2030013

Ferry, J.M. and Watson, E.B. (2007). "New thermodynamic models and revised calibrations on the Ti-in-rutile thermometers." *Contributions to Mineralogy and Petrology*, 154, 429–437.

Fierstein, J. and Hildreth, W. (1992). "The Plinian eruptions of 1912 at Novarupta Volcano, Katmai National Park, Alaska." *Bulletin of Volcanology*, 54, 646–684.

Fierstein, J. and Wilson, C.N. (2005). "Assembling an ignimbrite: Compositionally defined eruptive packages in the 1912 Valley of Ten Thousand Smokes ignimbrite, Alaska." *Geological Society of America Bulletin*, 115, 1094–1107.

Figueroa, J. (1973). "Sismicidad en Chiapas. Series del Instituto de Ingeniería." *Sismología e Instrumentación Sísmica,* 316. 50 pp.

Filiberto, J., Trang, J., Treiman, A.H., and Gilmore, M.S. (2020). "Present-day volcanism on Venus as evidenced from weathering rates of olivine." *Science Advances*, 6. doi:1126/sciadv.aax7445.

Finlaysen, D.M., Gudmundsson, O., Ikitarai, I. et al. (2003). "Rabaul Volcano Papua New Guinea; Seismic tomography imaging of an active caldera." *Journal of Volcanology and Geothermal Research*, 124, 153–171.

Fisher, R.L. and Hess, H.H. (1963). "Trenches." In: *The Sea* (ed. M.N. Hill), 411–436. New York: John Wiley & Sons.

Fisher, R.V. (1961). "Proposed classification of volcaniclastic sediments and rocks." *Geological Society American Bulletin*, 72, 1402–1414.

Fisher, R.V. (1964). "Maximum size, mean diameter, and sorting of tephra." *Journal of Geophysical Research*, 69, 341–355.

Fisher, R.V. (1977). "Erosion by volcanic base-surge density currents: U-shaped channels." *Geological Society of America Bulletin*, 88, 1287–1297.

Fisher, R.V. (1979). "Models for pyroclastic surges and pyroclastic flows." *Journal of Volcanology and Geothermal Research*, 6, 305–318.

Fisher, R.V. and Schmincke, H.U. (1984). *Pyroclastic Rocks*. New York: Springer-Verlag.

Fisher, R.V. and Waters, A.C. (1970). "Base surge bedforms in maar volcanoes." *American Journal of Science*, 268, 157–180.

Fisher, R.V., Heiken, G., and Hulen, J.B. (1997). *Volcanoes: Crucibles of Change*. Princeton: Princeton University Press.

Fisher, R.V., Smith, A.I., and Roobol, M.J. (1980). Destruction of St. Pierre, Martinique, by ash-cloud surges, May 8 and 20, 1902. *Geology,* 8, 472–476. https://doi.org/10.1130/0091-7613(1980).

Fisher, R.V. (2000). *Out of the Crater – Chronicles of a Volcanologist*. Princeton University Press, 180 pp.

Fisher, T.P., Burnard, P., Marty, B. et al. (2009). "Upper-mantle volatile chemistry at Oldoinyo Lengai volcano and the origin of carbonatites." *Nature*, 459, 77–80.

Fiske, R.S. (1984). "Volcanologists, journalists, and the concerned local public: A tale of two crises in the eastern Caribbean." In: *Explosive Volcanism: Inception, Evolution, and Hazards*, N. R. C. Geophysics Study Committee. Washington DC: National Academy Press, pp. 170–176.

Fiske, R.S. and Sigurdsson, H. (1982). "Soufrière volcano, St. Vincent: Observations of its 1979 eruption from the ground, aircraft, and satellites." *Science*, 216 (4550), 1105–1106.

Fiske, R.S., Naka, J., Lizasa, K. et al. (2001). "Submarine silicic caldera at the front of the Izu-Bonin arc, Japan: Voluminous seafloor eruptions of rhyolite pumice." *Geological Society of America Bulletin*, 113 (7), 813–824.

Fitton, J.G. (1980). "The Benue Trough and Cameroon Line: A migrating rift system in West Africa." *Earth and Planetary Science Letters*, 51, 132–138.

Fitton, J.G. and Godard, M. (2004). "Origin and evolution of magmas on the Ontong Java Plateau." In: *Origin and Evolution of the Ontong Java Plateau* (eds. J.G. Fitton, J. Mahoney, P.J. Wallace, et al.), 151–178. London: Geological Society of London.

Fitton, J.G., Mahoney, J.J., Wallace, P.J. et al. (2004). *Origin and Evolution of the Ontong Java Plateau*. London: Geological Society of London.

Flannery, T. (2001). *The Eternal Frontier: An Ecological History of North America and Its Peoples*. Washington, DC: Atlantic Monthly Press, 368.

Foit, F.F., Gavin, D.G., and Hu, F.S. (2004). "The tephra stratigraphy of two lakes in south-central British Columbia, Canada and its implications for mid-late Holocene volcanic activity at Glacier Peak and Mount St. Helens, Washington, USA." *Canadian Journal of Earth Sciences*, 41, 1401–1410.

Fontaine, F.R., Roult, G., Michon, L. et al. (2014). "The 2007 eruption and caldera collapse of the Piton de la Fournaise volcano (La Réunion Island) from tilt analysis at a single very broadband seismic station." *Geophysical Research Letters*, 41 (8), 2803–2811. doi 10.1002/2014GL059691.

Forbes, A., Blake, S., and Tuffen, H. (2014). "Entablature: Fracture types and mechanisms." *Bulletin of Volcanology*, 76, article 820, doi: 10.1007/s00445-014-0820-z.

Forbes, A., Blake, S., McGarvie, D., and Tuffen, H., (2012). "Pseudopillow fracture systems in lavas: Insights into cooling mechanisms and environments from lava flow fractures." *Journal of Volcanology and Geothermal Research*, 245–246, 68–80.

Fornari, D.J., Batiza, R., and Allan, J.F. (1987). "Irregularly shaped seamounts near the East Pacific Rise: Implications for seamount origin and Rise axis processes." In: *Seamounts, Islands, Atolls* (eds. B.H. Keating, P. Fryer, R. Batiza, et al.), pp. 35–47. American Geophysical Union Monograph 43.

Fornari, D.J., Lockwood, J.P., Lipward, P.W. et al. (1980). "Submarine volcanic features west of Kealakekua Bay, Hawaii." *Journal of Volcanology and Geothermal Research*, 7 (2), 323–337.

Fornari, D.J., Ryan, W.B.F., and Fox, P.J. (1985). "Sea-floor lava fields on the East Pacific Rise." *Geology*, 13 (6), 413.

Fornari, D.J., Tivey, M.D., Schouten, H. et al. (2004). "Submarine lava flow emplacement at the East Pacific Rise 9° 50´N–Implications for uppermost ocean crust stratigraphy and hydrothermal fluid circulation." In: *The Thermal Structure of the Ocean Crust and the Dynamics of Hydrothermal Circulation.* Geophysical Monograph 148, Washington, DC, American Geophysical Union, pp. 187–218.

Foshag, W.F. and Gonzalez-Reyna, J. (1956). U.S. Geological Survey Bulletin: Birth and Development of Paricutín Volcano, Mexico. US Geological Survey, pp. 355–489.

Foulger, G.R. and Jurdy, D.M. (2007). *Plates, Plumes and Planetary Processes.* Boulder: Geological Society of America.

Fouqué, F.A. (1879). *Santorin et ses éruptions.* Paris: G. Masson.

Fournier d'Albe, E.M. (1979). "Objectives of volcano monitoring and prediction." *Journal of the Geological Society of London*, 136, 321–326.

Francis, E.H. (1982). "Magma and sediment-I: Emplacement mechanism of late Carboniferous tholeiite sills in northern Britain." *Journal of the Geological Society of London*, 139, 1–20.

Francis, P. (1993). *Volcanoes: A Planetary Perspective.* Oxford: Oxford University Press, 443p.

Francis, P.W. and Rothery, D.A. (1987). "Using the Landsat Thematic Mapper to detect and monitor active volcanoes." *Geology*, 15, 614–617.

Francis, P.W., Wadge, G., and Mouginis-Mark, P.J. (1996). "Satellite monitoring of volcanoes." In: *Monitoring and Mitigation of Volcanic Hazards* (eds. R. Scarpa and R. I. Tilling). Berlin: Springer, 841p.

Frankel, C. (1996). *Volcanoes of the Solar System.* Cambridge University Press, 232 pp.

Frankel, C. (1999). *The End of the Dinosaurs.* Cambridge University Press, 236 pp.

Freundt, A. and Rosie, M. (eds.) (2001). *From Magma to Tephra: Modelling Physical Processes of Explosive Volcanic Eruptions.* North Holland: Elsevier, pp. 334.

Freundt, A. and Bursik, M.I. (1998). "Pyroclastic flow transport mechanisms." In: *From Magma to Tephra: Modelling Physical Processes of Explosive Volcanic Eruptions, Developments in Volcanology 4* (eds. A. Freundt and M. Rosi), pp. 173–245. New York: Elsevier.

Freundt, A. and Schmincke, H.-U. (1986). "Emplacement of small-volume pyroclastic flows at Laacher See (East-Eifel, Germany)." *Bulletin of Volcanology*, 48, 39–59.

Fridleifsson, I.B. (2003). "Status of geothermal energy amongst the world's energy sources." *Geothermics*, 32, 379–388.

Friedman, I., Lipman, P.W., Obrudovich, J.D. et al. (1974). "Meteoric water in magmas." *Economic Geology*, 184, 1069–1072.

Friedrich, W.L. (1999). *Fire in the Sea: The Santorini Volcano: Natural History and the Legend of Atlantis.* Cambridge: Cambridge University Press.

Fryer, P., Lockwood, J.P., Becker, N. et al. (2000). "Significance of serpentine mud volcanism in convergent margins." In: *Ophiolites and Oceanic Crust: New Insights from Field Studies and the Ocean Drilling Program* (eds. Y. Dilek, E.M. Moores, D. Elthon, et al.), pp. 35–51. Colorado, Geological Society of America. Special Paper 349.

Fulignati, P., Marianelli, P., Santacroce, R., and Sbrana, A. (2004). Probing the Vesuvius magma chamber-host rock interface through xenoliths. *Geological Magazine*, 141, 417–428.

Furman, T., Meyer, P.S., Frey, F.A. et al. (1992). "Evolution of Icelandic central volcanoes: Evidence from the Austurhorn Intrusion, southeastern Iceland." *Bulletin of Volcanology*, 55, 45–62.

Gaddis, B. and Kauahikaua, J. (2021). "Views of a century of activity at Kīlauea Caldera: A visual essay." chap. B of Patrick, M., Orr, T., Swanson, D., and Houghton, B., eds., *The 2008–2018 summit lava lake at Kīlauea Volcano*. Hawai'i: U.S. Geological Survey Professional Paper 1867, 23 p. https://doi.org/10.3133/pp1867B.

Galgana, G.A., Grosfils, E.B., and McGovern, P.J. (2013). "Radial dike formation on Venus: Insights from models of uplift, flexure and magmatism." *Icarus*, 225, 538–547.

Gallotti, G., Zaniboni, E., Pagnoni, G. et al. (2021). "Tsunamis from prospected mass failure on the Marsili submarine volcano flanks, and hints for tsunami hazard evaluation." *Bulletin of Volcanology*, 83, 1–15, https://doi.org/10.1007/s00445-020-01425-0

Gamble, J.A., Price, R.C., Smith, I.E.M. et al. (2003). "40Ar/39Ar geochronology of magmatic activity, magma flux, and hazards at Ruapehu Volcano, Taupo volcanic Zone, New Zealand." *Journal of Volcanology and Geothermal Research*, 120, 271–287.

Garces, M.A., Harris, A.J.L., Hetzer, C. et al. (2003). "Infrasonic tremor observed at Kilauea Volcano, Hawaii." *Geophysical Research Letters*, 30, 20, No. 2023.

Garces, M.A., Iguchi, M., Ishihara, K. et al. (1999). "Infrasonic precursors to a Vulcanian eruption at Sakurajima Volcano, Japan." *Geophysical Research Letters*, 26 (16), 2537–2540.

Garcia, M.O., Tree, J.P., Wessel, P., and Smith, J.R. (2020). "Pūhāhonu: Earth's biggest and hottest shield volcano." *Earth and Planetary Science Letters*, 542, 15 July 2020, 116296.

Garcia-Aristizabal, A., Selva, J., and Fujita, E. (2013). "Integration of stochastic models for long-term eruption forecasting into a Bayesian event tree scheme: A basis method to estimate the probability of volcanic unrest." *Bulletin of Volcanology*, 75. doi 10.1007/s00445-013-0689-2.

Gardener, J.V., Dean, W.E., and Blakely, R.J. (1984). "Shimada seamount: An example of recent midplaste volcanism." *Geological Society of America Bulletin*, 95, 855–862.

Gardner, J.E., Andrews, B.J., and Dennen, R. (2017). "Liftoff of the 18 May 1980 surge of Mount St. Helens (USA) and the deposits left behind." *Bulletin of Volcanology*, 79, 8. doi 10.1007/s00445-016-1095-3.

Gardner, J.E., Thomas, R.M.E., Jaupart, C. et al. (1996). "Fragmentation of magma during Plinian volcanic eruptions." *Bulletin Volcanologique*, 58, 144–162.

Garry, W.B., Gregg, T.K.P., Soule, S.A. et al. (2006). "Formation of submarine lava channel textures: Insights from laboratory simulations." *American Geophysical Union*, 111, B3104, 1–21.

GEA (2015). The International Geothermal Market at a Glance, May 2015, the Geothermal Energy Association, www.geo-energy.org/reports.aspx.

Geissler, P. (2000). "Cryptovolcanism in the outer solar system." In: *Encyclopedia of Volcanoes* (ed. H. Sigurdsson), 785–800. San Diego, Academic Press.

Geissler, P.E. (2003). "Volcanic activity on Io during the Galileo Era." *Annual Review of Earth and Planetary Sciences*, 31, 175–211.

Gerlach, T. (2011). "Volcanic versus anthropogenic carbon dioxide, EOS Transactions." American Geophysical Union, 14 June. doi:org/10.1029/2011EO240001.

Gerlach, T.M. (1991). "Present-day CO2 emissions from volcanoes." *Transactions of the American Geophysical Union [EOS]*, 72, 249, 254–255.

Gerlach, T.M., McGee, K.A., Elias, T. et al. (2002). "Carbon dioxide emission rate of Kilauea volcano: Implications for primary magma and the summit reservoir." *Journal of Geophysical Research*, 107 (B9), 2189.

Geshi, N., Acocella, A., and Ruch, J. (2012). "From structure- to erosion-controlled subsiding calderas: evidence thresholds and mechanics." *Bulletin of Volcanology*, 74, 1553–1567.

Geze, B. (1964)."Sur la classification des dynamisms volcaniques." *Bulletin Volcanologique*, 27, 237–257.

Ghiorso, M.S. and Sack, R.O (1991). "Fe-Ti oxide thermometry, thermodynamic formulation and estimation of intensive variables in silicic magmas." *Contributions to Mineralogy and Petrology*, 108, 485–510.

Giannetti, B. (2001). "Origin of the calderas and evolution of Roccamonfina volcano (Roman Region, Italy)." *Journal of Geothermal Research*, 106 (3–4), 301–319.

Gibbs, S.H. (1968). *Volcanic Ash Soils in New Zealand*. Department of Science and Industrial Research, New Zealand, 39.

Gibson, H.L. (2005). "Volcano-hosted ore deposits." In: *Volcanoes and the Environment* (eds. J. Martí and G.G.J. Ernst), pp. 333–386. Cambridge: Cambridge University Press.

Giggenbach, W.F. (1987). "Redox processes governing the chemistry of fumarolic gas discharges from White Island, New Zealand." *Applied Geochemistry*, 2, 143–161.

Gilbert, J.S. and Lane, S.J. (1994). "The origin of accretionary lapilli." *Bulletin of Volcanology*, 56, 398–411.

Gilbert, J.S. and Sparks, R.S.J. (eds.) (1998). *The Physics of Explosive Volcanic Eruptions*. London, The Geological Society of London, Special Publication No. 145.

Gilbert, J.S., Lane, S.J., Sparks, R.S.J. et al. (1991). "Charge measurements on a particle fallout from a volcanic plume." *Nature*, 349, 598–600.

Gilbert, J.S., Stasiuk, M.V., Lane, S.J. et al. (1996). "Non-explosive, constructional evolution of the ice-filled caldera at Volcan Sollipulli, Chile." *Bulletin of Volcanology*, 58 (1), 67–83.

Gill, J.B. (1981). *Orogenic Andesites and Plate Tectonics*. Berlin: Springer-Verlag.

Gisler, G., Weaver, R., and Gittingsm, M.I. (2006). SAGE calculations of the tsunami threat from La Palma. *Science of Tsunami Hazards*, 24, 288–391.

Glasby, G.P., Iizasa, K., Hannington, M. et al. (2008). "Mineralogy and composition of Kuroko deposits from northeastern Honshu and their possible modern analogues from the Izu-Ogasawara (Bonin) Arc south of Japan: Implications for mode of formation." *Ore Geology Reviews*, 34 (4), 547–560.

Godchaux, M.M., Bonnichesen, B., and Jenks, M.D. (1992). "Types of phreatomagmatic volcanoes in the Western Snake River Plain, Idaho, USA." *Journal of Volcanology and Geothermal Research*, 52, 1–25.

Goethe, J.W. (1827). Zahme Xenien VI. Reprinted in: Richter et al. (eds.) (2006), *Sämtliche Werke nach Epochen seines Schaffens (Münchner Ausgabe)*, Vol. 13.1. (the years 1820-1826, eds. G. Henckmann and I. Schneider), pp. 221–228. München: Carl Hanser Verlag.

Goff, F. (2009). *Valles Caldera: A Geologic History*. Albuquerquem: University of New Mexico Press, 130 p.

Gonnermann, H.M. and Manga, M. (2007). "The fluid mechanics inside a volcano." *Annual Review of Fluid Mechanics*, 39, 321–356.

Gooday, R.J., Brown, D.J., Goodenough, K.M., and Kerr, A.C. (2018). "A proximal record of caldera-forming eruptions: the stratigraphy, eruptive history and collapse of the Palaeogene Arran caldera, western Scotland." *Bulletin of Volcanology*, 80. doi 10.1007/s00445-018-1243-z.

Gorshkov, G.S. (1959). "Gigantic eruption of the volcano Bezymianny." *Bulletin of Volcanology*, 20, 77–109.

Gorshkov, G.S. (1963). "Directed volcanic blasts." *Bulletin of Volcanology*, 26, 83–88.

Gorshkov, G.S. and Dubik, Y.M. (1970). "Gigantic directed blast at Shiveluch Volcano (Kamchatka)." *Bulletin Volcanologique*, 34, 261–288.

Gottman, J. and Martí, J. (2008). *Caldera Volcanism: Analysis, Modelling, and Response*. North Holland: Elsevier, 516p.

Gourgaud, A., Camus, G., Gerbe, M.-C. et al., (1989). "The 1982–83 eruption of Galunggung (Indonesia): A case study of volcanic hazards with particular relevance to air navigation." In: *Volcanic Hazards – Assessment and Monitoring* (ed. J.H. Latter), pp. 151–162. Berlin: Springer-Verlag.

Graettinger, A.H. (2018). "Trends in maar size and shape using the global Maar Volcano Location and Shape (MaarVIS) database." *Journal of Volcanology and Geothermal Research*, 357, 1–13.

Graettinger, A.H. and Valentine, G.A. (2017). "Evidence for the relative depths and energies of phreatomagmatic explosions recorded in tephra rings." *Bulletin of Volcanology*, 79, 88. doi:10.1007/s00445-017-1177-x (open access).

Granier, C., Artaxo P., and Reeves, C.E. (eds.) (2012). *Emissions of Atmospheric Trace Compounds* (Advances in Global Change Research, 18), 4th edn. New York City: Springer Publishing.

Gravley, D. (2004). "Evolution and eruptive products of the Whangapoa Basin, Taupo Volcanic Zone." Doctoral dissertation, University of Canterbury.

Greeley, R. (1971). "Observations of actively forming lava tubes and associated structures, Hawaii." *Modern Geology*, 2, 207–223.

Gregg, C.E., Houghton, B.F., Johnston, D.M. et al. (2004). "The perception of volcanic risk in Kona communities from Mauna Loa and Hualalai volcanoes, Hawaii." *Journal of Volcanology and Geothermal Research*, 130, 179–196.

Gregg, P.M., de Silva, S.L., Grosfils, E.B., and Parmigiani, J.P. (2012). "Catastrophic caldera-forming eruptions: Thermomechanics and implications for eruption triggering and maximum caldera dimensions on Earth." *Journal of Volcanology and Geothermal Research*, 241–242, 1–12.

Gregg, P.M., Grosfils, E.B., and de Silva, S.L. (2015). "Catastrophic caldera-forming eruptions II: The subordinate role of magma buoyancy as an eruption trigger." *Journal of Volcanology and Geothermal Research*, 305, 100–113.

Gregg, T.K.P. and Greeley, R. (1993). "Formation of Venusian canali: Considerations of lava types and their thermal evolution." *Journal of Geophysical Research*, 98, 10,873–10,882.

Gregg, T.K.P. and Keszthelyi, L.P. (2004). "The emplacement of pahoehoe toes: field observations and comparison to laboratory simulations." *Bulletin of Volcanology*, 66, 381–391.

Griggs, R.F. (1922). *The Valley of Ten Thousand Smokes [Alaska]*. Washington, DC: National Geographic Society.

Grosfils, E., Aubele, J., Crumpler, L. et al. (1999). "Volcanism on Venus and Earth's Seafloor." In: *Environmental Effects on Volcanic Eruptions* (eds. J. Zimbleman and T.K.P. Gregg), pp. 113–142. New York: Plenum Press.

Gudmundson, A. (1998). "Magma chambers modeled as cavities explain the formation of rift zone central volcanoes and their eruption and intrusion statistics." *Journal of Geological Research*, 103, 7401–7412.

Gudmundsson, A. (2012). "Magma chambers: Formation, local stresses, excess pressures, and compartments." *Journal of Volcanology and Geothermal Research*, 237–238, 19–41.

Gudmundsson, M., Thordarson, T., Höskuldsson, Á. et al. (2012). Ash generation and distribution from the April-May 2010 eruption of Eyjafjallajökull, Iceland. *Scientific Reports*, 2, 572. https://doi.org/10.1038/srep00572.

Gudmundsson, M.T., Jónsdóttir, K., Hooper, A. et al. (2016). "Gradual caldera collapse at Bárdarbunga volcano, Iceland, regulated by lateral magma outflow." *Science*, 15 Jul 2016, 353 (6296), aaf8988. doi: 10.1126/science.aaf8988.

Gudmundsson, M.T., Pálsson, F., Björnsson, H. et al. (2002). "The hyaloclastite ridge formed in the subglacial 1996 eruption of Gjálp, Vatnajökull, Iceland: Present day shape and future preservation." In: *Remote Sensing of Terrestrial and Martian Subglacial Features*. Geological Society of London Special Publications, 202, 319–335.

Gudmundsson, M.T., Sigmundsson, F., Björnsson, H. et al. (2004). "The 1996 eruption at Gjálp, Vatnajökull ice cap, Iceland: Efficiency of heat transfer, ice deformation and subglacial water pressure." *Bulletin of Volcanology*, 66, 46–65.

Gudmundsson, S.L. (2006). "Magma chambers, dyke injections and surface deformation in composite volcanoes," Geological Research Abstracts, European Geosciences Union. 8.

Guest, J.E., Bulmer, J.L., Aubele, J. et al. (1992). "Small volcanic edifices and volcanism on the plains of Venus." *Journal of Geophysical Research*, 97, 15,949–15,996.

Guffanti, M. and Weaver, C.S. (1988). "Distribution of late Cenozoic volcanic vent in the Cascade Range: Volcanic arc segmentation and regional tectonic considerations." *Journal of Geophysical Research*, 93, 6513–6529.

Guillou, H., Maury, R.C., Guille, G. et al. (2014). "Volcanic successions in Marquesas eruptive centers: A departure from the Hawaiian model." *Journal of Volcanology and Geothermal Research*, 276, 173–188.

Gunn, L.S., Blake, S., Jones, M.C., and Rymer, H. (2014). "Forecasting the duration of volcanic eruptions: an empirical probabilistic model." *Bulletin of Volcanology*, 76. doi 10.1007/s00445-013-0780-8.

Gunnarsson, B., Marsh, B.D., and Taylor, H.P. Jr. (1998). "Geology and petrology of postglacial silicic lavas from the southwest part of the Torfajokull Central Volcano, Iceland." *Journal of Volcanology and Geothermal Research*, 83, 1–45.

Gurioli, L., Pareschi, M.T., Zanella, E. et al. (2005). "Interaction of pyroclastic density currents with human settlements: Evidence from ancient Pompeii." *Geology*, 33, 441–444.

Haase, K.M., Struncik, N., Garbe-Schnoberg, D. et al. (2006). "Formation of island arc dacite magmas by extreme crystal fractionation: An example from Brothers Seamount, Kermadec island arc (SW Pacific)." *Journal of Volcanology and Geothermal Research*, 152, 316–330.

Halliday, W.R. (1998). "Hollow volcanic tumulus caves of Kilauea caldera, Hawaii County, Hawaii." *International Journal of Speleology*, 27B, 95–105.

Hamilton, D.L., Burnham, C.W., and Osborn, E.F. (1964). "Solubility of water and effects of oxygen fugacity and water content of crystallization in mafic magmas." *Journal of Petrology*, 5, 21–39.

Hamilton, W. (2013). *Observations on Mount Vesuvius, Mount Etna, and Other Volcanoes: In a Series of Letters Addressed to the Royal Society from the Honourable Sir W. Hamilton* [Reprint of late 18th century letters], Hardpress Publishing, 212 p.

Hamilton, W.L. (1973). "Tidal cycles of volcanic eruptions: Fortnightly to 19 year periods." *Journal of Geophysical Research*, 78 (17), 3363–3375.

Hamilton, Sir William. (1768). An account of the eruption of Mount Vesuvius, in 1767. *Philosophical Transactions of the Royal Society of London* 58: 1–14.

Hammarstrom, J.M. and Zen, E. (1986). "Aluminum in hornblende: An empirical igneous geobarometervol." *American Mineralogist*, 71, 1297–1313.

Harlow, D.H., Power, J.A., and Laguerta, E.P. (1996). "Precursory seismicity and forecasting of the June 15, 1991 eruption of Mount Pinatubo." In: *Fire and Mud: Eruptions and Lahars of Mount Pinatubo, Philippines* (eds. C.G. Newhall and R.S. Punongbayan), pp. 285–305. Philippine Institute of Volcanology and Seismology and University of Washington Press: Quezon City.

Harper, B.E., Miller, C.F., Koteas, C.C. et al. (2004). "Granites, dynamic magma chamber processes and pluton construction: The Aztec Wash pluton, Eldorado Mountains, Nevada, USA." *Transactions of the Royal Society of Edinburgh, Earth Sciences*, 95, 277–295.

Harris, A., Rowland, S., Villeneuve, N., and Thordarson, T. (2017). "Pāhoehoe, ʻaʻā, and block lava: an illustrated history of the nomenclature." *Bulletin of Volcanology*, 79, article 7. doi: 10.1007s/00445-016-1075-7.

Harris, A.J.L., Butterworth, A.L., Carton, R.W. et al. (1997). "Low-cost surveillance from space: Case studies from Etna, Krafla, Cerro Negro, Fogo, Lascar, and Erebus." *Bulletin Volcanologique*, 59, 49–64.

Harris, D.M., Sato, M., and Casadevall, T.J. (1981). "Emission rates of CO2 from plume measurements." In: *The 1980 eruption of Mount St. Helens, Washington* (eds. P.W. Lipman and D.R. Mullineaux), 201–207. Washington, DC: US Geological Survey. Professional Paper 1250.

Harris, S.L. (2005). *Fire Mountains of the West: The Cascade and Mono Lake Volcanoes.* Mountain Press, Missoula, MT, 454p.

Harrison, T.N., Brown, P.E., Dempster, T.J. et al. (1990). "Granite magmatism and extensional tectonics in southern Greenland." *Geological Journal*, 25, 287–293.

Hartley, M.E. and Thordarson, T. (2012). "Formation of Öskjuvatn caldera at Askja, North Iceland: Mechanism of caldera collapse and implications for the lateral flow hypothesis." *Journal of Volcanology and Geothermal Research*, 227–228, 85–101.

Hatton, C.J. (1995). "Mantle plume origin for the Bushveld and Ventersdorp magmatic provinces." *Journal of African Earth Sciences*, 21, 571–577.

Haughton, D.R., Roeder, P.L., and Skinner, B.J. (1974). "Solubility of sulfur in mafic magmas." *Economic Geology*, 69, 451–467.

Havskov, J., De la Cruz-Reyna, S., Singh, S.K. et al. (1983). "Seismic activity related to the March-April, 1982 eruptions of El Chichón Volcano, Chiapas, Mexico." *Geophysical Research Letters*, 10 (4), 293–296.

Hawkins, D. and Wiebe, B. (2007). "Construction of the subvolcanic Vinalhaven Intrusive Complex, coastal Maine, Abstracts with Program Volume, Twentieth Annual Keck Symposium in Geology." Twentieth Annual Keck Symposium in Geology.

Hazlett, R., Buesch, D., Anderson, J. et al. (1991). "Geology, failure conditions, and seismogenic avalanches of the 1944 eruption at Vesuvius, Italy." *Journal of Volcanology and Geothermal Research*, 47, 249–264.

Hazlett, R.W. (1990). "Extension-related volcanism in the Mopah Range volcanic field, southeastern California." *Geological Society of America*, 174, 133–145.

Head, J.W. and Wilson, L. (1992). "Magma reservoirs and neutral buoyancy zones on Venus: Implications for the formation and evolution of volcanic landforms." *Journal of Geophysical Research*, 97 (E3), 3877–3903.

Head, J.W. and Wilson, L. (2003). "Deep submarine pyroclastic eruptions: Theory and predicted land forms and deposits." *Journal of Volcanology and Geothermal Research*, 151, 155–193.

Head, J.W.I., Crumpler, L.S., Aubele, J.C. et al. (1992). "Venus volcanism: Classification of volcanic features and structures, associations, and global distribution from Magellan data." *Journal of Geophysical Research*, 97, 13,153–13,197.

Heald, E.F., Naughton, J.J., and Lynus Barnes, I. (1963). "The chemistry of volcanic gases, 2, Use of equilibrium calculations in the interpretation of volcanic gas samples." *Journal of Geophysical Research*, 68 (2), 545–557.

Hedenquist, J.W. and Aoki, M. (1991). "Meteoric interaction and magmatic discharges in Japan, and its significance for mineralization." *Geology*, 19, 1041–1044.

Hedervari, P. (1963). "On the energy and magnitude of volcanic eruptions." *Bulletin Volcanologique*, 25, 373–386.

Heiken, G. (2005). "Industrial uses of volcanic materials." In: *Volcanoes and the Environment* (eds. J. Martí and G.G.J. Ernst), 387–403. Cambridge, Cambridge University Press.

Heiken, G. (2006). *Tuffs: Their Uses, Hydrology, and Resources.* New York: Geological Society of America.

Heiken, G. (2013). *Dangerous Neighbors – Volcanoes and Cities*. Cambridge University Press, 185 pp.

Heiken, G. and Bierwith, P.N. (1982). "Experimental welding of volcanic ash." Master's thesis, Monash University, Australia.

Heiken, G. and Wohletz, K. (1992). *Volcanic Ash.* Ann Arbor: University of California Press.

Heirtzler, J.R. (1968). "Sea-floor spreading." *Scientific American*, 219 (6), 60–70.

Helz, R. and Thornber, C.R. (1987). "Geothermometry of Kilauea Iki lava lake, Hawaii." *Bulletin of Volcanology*, 49, 651, 668.

Helz, R.T. (1980). "Crystallization history of Kilauea Iki lava lake as seen in drill core recovered in 1967–1979." *Bulletin of Volcanology*, 43 (3), 675–701.

Henry, C.D. and Wolff, J.A. (1992). "Distinguishing strongly rheomorphic tuffs from extensive silicic lavas." *Bulletin of Volcanology*, 54, 171–186.

Herd, D.G. and Comite de Estudios Vulcanologicos (1986). "The 1985 Ruiz volcano disaster." *EOS (American Geophysical Union Transactions)*, 67 (19), 457–460.

Hess, H.H. (1946). "Drowned ancient islands of the Pacific basin." *American Journal of Science*, 244, 772–791.

Hess, H.H. (1960). *The Evolution of Ocean Basins.* Princeton: Princeton University Department of Geology, Special Paper.

Hibbard, M.J. and Hibbard, M. (2002). *Mineralogy: A Geologist's Point of View.* McGraw Hill, 576 p.

Higgins, M.D. (2009). "The Cascadia megathrust earthquake of 1700 may have rejuvenated an isolated basalt volcano in western Canada: Age and petrographic evidence." *Journal Of Volcanology and Geothermal Research*, 149–156.

Hildebrand, A.R., Penfield, G.T., Kring, D.A. et al. (1991). "Chicxulub Crater: A possible Cretaceous-Tertiary boundary impact crater on the Yucatan Peninsula, Mexico." *Geology*, 19, 867–871.

Hildreth, W. (1981). "Gradients in silicic magma chambers: Implications for lithospheric magmatism." *Journal of Geophysical Research*, 86, 10,153–10,192.

Hildreth, W. (2007). *Quaternary Magmatism in the Cascades: Geologic Perspectives.* Washington, DC: US Geological Survey.

Hildreth, W. and Mahood, G. (1986). "Ring fracture eruption of the Bishop Tuff." *Geological Society of America Bulletin*, 97, 396–403.

Hill, M. (2006). *Geology of the Sierra Nevada.* Berkeley: University of California Press, 468 p.

Hill, R.L. (2004). *Volcanoes of the Cascades: Their Rise and Their Risks.* Guilford: Falcon.

Hillier, J.K. (2007). "Pacific seamount volcanism in space and time." *Geophysical Journal International*, 168 (2), 877–889.

Hobiger, M., Sonder, I., Büttner, R., and Zimanowski, B. (2011). "Viscosity characteristics of selected volcanic rock melts." *Journal of Volcanology and Geothermal Research*, 200, 27–34.

Hoblitt, R. (1994). "An experiment to detect and locate lightning associated with eruptions of Redoubt Volcano." *Journal of Volcanology and Geothermal Research*, 62, 499–517.

Hoblitt, R.P., Miller, C.D., and Vallance, J.W. (1981). "Origin and stratigraphy of the deposit produced by the May 18 directed blast." In: *The 1980 eruptions of Mount St. Helens, Washington* (eds. P.W.E. Lipman and D.R. Mullineaux), pp. 401–419. US Geological Survey, Professional Paper 1250.

Holcomb, R.T., Moore, J.G., Lipman, P.W. et al. (1988). "Voluminous submarine lava flows from Hawaiian volcanoes." *Geology*, 16, 400–404.

Holland, H.D. (1978). *The Chemistry of the Atmosphere and Oceans.* New York: Wiley Inter-Science, 352p.

Hon, K., Gansecki, C., and Johnson, J. (2000). *Lava Flows and Pyroducts: What They Are, How They Form.* DVD, Volcano Video Productions, Volcano Hawai'i, 75 minutes.

Hon, K., Kauahikaua, J., Denlinger, R. et al. (1994). "Emplacement and inflation of pahoehoe sheet flows: Observations and measurements of active lava flows on Kiluaea Volcano, Hawaii." *Geological Society of America Bulletin*, 106, 351–370.

Hon, K.H. and Gansecki, C.A. (2003). "The transition from 0a0a to pahoehoe crust on flows emplaced during the Pu0u O0o-Kupaianaha eruption." In: *The Pu/u O/o-Kupaianaha Eruption of Kilauea Volcano, Hawaii* (eds. C. Helliker, D.A. Swanson, and J.T. Takahashi), pp. 89–103. New York: US Geological Survey, 1676.

Hooper, D.M. and Sheriden, M.F. (1998). "Computer-simulation models of scoria cone degradation." *Journal of Volcanology and Geothermal Research*, 83 (3–4), 241–267.

Horvath, D.G., Moitra, P., Hamilton, C.W. et al. (2021). "Evidence of geologically recent explosive volcanism in Elysium Planitia, Mars." *Icarus*, v.365, 1 September 2021, 114499. https://doi.org/10.1016/j.icarus.2021.114499.

Horwell, C.J. and Baxter, P.J. (2006). "The respiratory health hazards of volcanic ash: A review for volcanic risk mitigation." *Bulletin of Volcanology*, 69, 1–24.

Houghton, B. (1993). "Wet explosive eruptions." *Explosive Volcanism: Processes and Products, Commission on Explosive Volcanism Short Course*, J. McPhie. Canberra, Australia, 3.1–3.8.

Houghton, B., White, J.D.L., and van Eaton, A.R. (2015). "Phreatomagmatic and related eruption styles." Chapter 30 in *Encyclopedia of Volcanoes* (ed. H. Sigurdsson), 2nd edition, pp. 537–569, 1421 pp. https://doi.org/10.1016/B978-0-12-385938-9.00030-4.

Houghton, B.F. (1993). "Dry explosive eruptions: Processes and products." *Explosive Volcanism: Processes and Products, Commission on Explosive Volcanism (IAVCEI) Short Course.* J. McPhie, Canberra, Australia.

Houghton, B.F. and Wilson, C.J.N. (1989). "A vesicularity index for pyroclastic deposits." *Bulletin Volcanologique*, 51, 451–462.

Houghton, B.F., Weaver, S.D., Wilson, C.J. et al. (1992). "Evolution of a Quaternary peralkaline volcano: Mayor Island, New Zealand." *Journal of Volcanology and Geothermal Research*, 51, 217–236.

Huber, C. and Wachtershauser, G. (2006). "Alpha-hydroxy and alpha-amino acids under possible Hadean, volcanic origin-of-life conditions." *Science*, 314, 630–632.

Huckfeldt, M., Courtier, A.M., and Leahy, G. (2013). "Implications for the origin of Hawaiian volcanism from a converted wave analysis of the mantle transition zone." *Earth and Planetary Sciences Letters*, 373, 194–204.

Huff, W.D., Bergström, S.M., and Kolata, D.R. (1992). "Gigantic Ordovician volcanic ash fall in North America and Europe: Biological, tectonomagmatic, and event-stratigraphic significance." *Geology*, 20 (10), 875–878.

Huggett, R. (2007). *Fundamentals of Geomorphology.* London: Taylor & Francis Publishing, 472p.

Hulme, G. (1974). "The interpretation of lava flow morphology." *Geophysical Journal of the Royal Astrological Society*, 39, 361–383.

Huntingdon, A. (1973). "The collection and analysis of volcanic gases from Mount Etna." *Philosophical Transactions of the Royal Society of London, Series A. Mathematical and Physical Sciences*, 174, 119–127.

Huppert, H.E., Shepherd, J.B., Sugurdsson, H. et al. (1982). "On lava dome growth, with application to the 1979 lava eruption of the Soufrière of St. Vincent." *Journal of Volcanology and Geothermal Research*, 14 (3–4), 199–222.

Hurford, T.A., Bills, B.G., Helfenstein, P. et al. (2009). "Geological implications of a physical libration on Enceladus." *Icarus*, 203, 541–552.

IAV-CEI Subcommittee for Crisis Protocols (1999). "Professional conduct of scientists during volcanic crises." *Bulletin of Volcanology*, 60, 323–324.

Imai, A., Listancio, E.L., and Fujii, T. (1993). "Petrologic and sulfur isotopic significance of highly oxidized and sulfur-rich magma of Mt. Pinatubo, Philippines." *Geology*, 21, 699–702.

INGEOMINAS (1985). "Mapa Preliminar de Riesgos Volcanicos Potenciales del Nevado del Ruiz, Republica de Colombia." *Ministerio de Minas y Energia*, 7 October.

Ishimine, Y. (2006). "Sensitivity of the dynamics of volcanic eruption columns to their shape." *Bulletin of Volcanology*, 68, 516–537.

Izawa, E., Urashima, Y., Ibaraki, K. et al. (1990). "The Hishikari gold deposit: High-grade epithermal veins in Quaternary volcanics of southern Kyushu, Japan." *Journal of Geochemical Exploration*, 36, 1–56.

Izbekov, P., Gordeev, E., Eichelberger, J., and West, M. (2013). "Magma system response to edifice collapse." *Journal of Volcanology and Geothermal Research*, 263, 1–2. doi: 10.1016/j.jvolgeores.2013.09.001.

Izett, G.A., Wilcox, R.E., Powers, H.A. et al. (1970). "The Bishop ash bed, a Pleistocene marker bed in the western United States." *Quaternary Research*, 1, 121–132.

Jaggar, T.A. (1904). "The initial stages of the spine on Pele." *American Journal of Science*, 17 (97), 34–40.

Jaggar, T.A. (1909). "The Messina earthquake: Prediction and protection." *The Nation*, 88, 2271, 22–23.

Jaggar, T.A. (1917). "On the terms aphrolith and dermolith." *Washington Academy of Science*, 7, 10, 277–281.

Jaggar, T.A. (1919). *Monthly Bulletin of the Hawaiian Volcano Observatory*, 7, 5 (May).

Jaggar, T.A. (1945). *Volcanoes Declare War: Logistics and Strategy of Pacific Volcano Science*. Honolulu: Paradise of the Pacific Press, 166 pp.

Jaggar, T.A. (1945). *Volcanoes Declare War: Paradise of the Pacific.* Honolulu, Ltd.

Jaggar, T.A. (1947). *Origin and Development of Craters.* Geological Society of America Memoir 21, 508p.

Jaggar, T.A. (1956) *My Experiments with Volcanoes.* Hawaiian Volcano Research Association, Advertising Publishing Company, Honolulu, 198p.

Jakob, M. and Hungr, O. (2005). *Debris-Flow Hazards and Related Phenomena.* New York: Springer-Praxis Publishing, 739p.

James, M.R., Lane, S.J., Wilson, L. et al. (2008). "Degassing at low magma viscosity volcanoes: Quantifying the transition between passive bubble burst and Strombolian eruption." *Journal of Volcanology and Geothermal Research*, 180, 81–88.

Jean-Baptiste, P., Allard, P., Fourré, E. et al. (2016). "Spatial distribution of helium isotopes in volcanic gases and thermal waters along the Vanuatu (New Hebrides) volcanic arc." *Journal of Volcanology and Geothermal Research*, 322, 20–29.

Jellinek, A.M. and DePaolo, P. (2003). "A model for the origin of large silicic magma chambers: Precursors of caldera-forming eruptions." *Bulletin of Volcanology*, 65, 313–381.

Jellinek, A.M. and Ross, K. (2001). "Magma dynamics, crystallization, and chemical differentiation of the 1959 Kilauea Iki lava lake, Hawaii, revisited." *Journal of Volcanology and Geothermal Research*, 110, 235–263.

Jenkins, S.F., Spence, R.J.S., Fonseca, J.F.B.D. et al. (2014). "Volcanic risk assessment: Quantifying physical vulnerability in the built environment." *Journal of Volcanology and Geothermal Research*, 276, 105–120.

Jicha, B.R., Scholl, D.W., Singer, B.S. et al. (2006). "Revised age of Aleutian Island Arc formation implies high rate of magma production." *Geology*, 34, 661–664.

Jiménez, Z., Espíndola, V.H., and Espíndola, J.M. (1999). "Evolution of the seismic activity from the 1982 eruption of El Chichon Volcano, Chiapas, Mexico." *Bulletin of Volcanology*, 61, 411–422.

Johnson, A.P., Cleaves, H.J., Dworkin, J.P. et al. (2008). "The Miller volcanic spark discharge experiments." *Science*, 322, 404.

Johnson, J.B., Aster, R., Jones, K.R. et al. (2008). "Acoustic source characterization of impulsive Strombolian eruptions from the Mt. Erebus lava lake." *Journal of Volcanology and Geothermal Research*, 177, 673–686.

Johnson, R.W. (ed.) (1980). *Intraplate Volcanism in Eastern Australia and New Zealand.* Cambridge, Cambridge University Press, 408p.

Johnson, R.W. (ed.) (2009). *Intraplate Volcanism in Eastern Australia and New Zealand.* Cambridge: Cambridge University Press, pp. 29–37.

Johnson, R.W. and Taylor, S.R. (1989). "Preview: Introduction to intraplate volcanism." In: *Intraplate Volcanism in Eastern Australia and New Zealand* (ed. R.W. Johnson), Cambridge: Cambridge University Press, 408p.

Johnson, S.E. and Schmidt, K.L. (2002). "Ring complexes in the Peninsular Range Batholith, Mexico and the USA: Magma plumbing systems in the Middle and Upper Crust." *Lithos*, 61, 187–208.

Johnson, T.V., Veeder, G.J., Matson, D.L. et al. (1988). "Io: Evidence for silicate volcanism in 1986." *Science*, 242, 1280–1283.

Johnston-Lavis, H.J. (1885). "Some speculations on the phenomena suggested by the geological study of Vesuvius and Monte Somma." *Geological Magazine*, 2, 302–307.

Jolly, G. and De la Cruz-Reyna, S. (2015). "Volcanic crisis management." Chapter 68 in *The Encyclopedia of Volcanoes* (eds. H. Sigurdsson, B. Houghton, S. McNutt et al.), 2nd edition, pp. 11/87–1202. Amsterdam: Academic Press. doi:10.1016/B978-0-12-385938-9.01001-4.

Jones, A.E. (1943). "Classification of lava surfaces." *AGU Transcripts of 1943*, 1, 265–268.

Joseph, E.P., Jacksonb, V.B., Beckles, D.M. et al. (2019). "A citizen science approach for monitoring volcanic emissions and promoting volcanic hazards awareness at Sulphur Springs, Saint Lucia in the Lesser Antilles arc." *Journal of Volcanology and Geothermal Research*, 369, 50–63.

Joyce, E.B. (1975). "Quaternary volcanism and tectonics in southeastern Australia." *Quaternary Studies*. Wellington: Royal Society of New Zealand, pp. 169–176.

Jurado-Chichay, Z. and Rowland, S.K. (1995). "Channel overflows of the Pohue Bay flow, Mauna Loa, Hawaii: Examples of the contrast between surface and interior lava." *Bulletin of Volcanology*, 57, 117–126.

Jurado-Chichay, Z. and Walker, G.P.L. (2001). "Variability of Plinian fall deposits: Examples from Okataina Volcanic Center, New Zealand." *Journal of Volcanology and Geothermal Research*, 111, 239–263.

Jutzeler, M., McPhie, J., and Allen, S.R. (2014). "Submarine-fed and resedimented pumice-ruch facies: The Dogashima Formation (Izu Peninsula, Japan)." *Bulletin of Volcanology*, 76. doi 10.1007/s00445-014-0867-x.

Kadereit, J.W. and Beyschlag, W. (eds.) (2004). *Progress in Botany*, 65. Berlin/Heidelberg: Springer.

Kahle, A.B., Gillespie, E.A., Abbott, M.J. et al. (1988). "Relative dating of Hawaiian lava flows using multispectral thermal images: A new tool for mapping of young volcanic terrains." *Journal of Geology and Geophysical Research*, 93, 15329–15351.

Kaminski, E. and Jaupert, C. (2001). "Marginal stability of atmospheric eruption columns and pyroclastic flow generation." *Journal of Geophysical Research*, 106, 21, 785–798.

Kant, E. (1803). *Physische Geographie*, V. 2. Mainz, Gottfried Vollmer, 242p.

Karatson, D., Thouret, J.C., Moriya, I. et al. (1999). "Erosion calderas: Origins, processes, structural and climatic control." *Bulletin of Volcanology*, 61 (3), 174–193.

Karson, J.A., Fornari, D.J., Kelley, D.S. et al. (2015). *Discovering the Deep: A Photographic Atlas of the Seafloor and Oceanic Crust*. Cambridge: Cambridge University Press, 430 p.

Kasahara, J. (2002). "Geophysics, tides, earthquakes, and volcanoes." *Science*, 297, 348.

Kasahara, K. (1981). *Earthquake Mechanics*. Cambridge: Cambridge University Press.

Kauahikaua, J., Cashman, K., Mattox, T. et al. (1998). "Observations on basaltic lava streams in tubes from Kīlauea Volcano, Hawaii." *Journal of Geophysical Research, Solid Earth*, 102. doi.org/10.1029/97JB03576.

Kauahikaua, J., Sherrod, D.R., Cashman, K.V. et al. (2003). "Hawaiian Lava-flow dynamics during the Pu'u 'Ō'ō-Kupaianaha eruption: A tale of two decades." In: *The Pu'u O'o-Kupaianaha Eruption of Kilauea Volcano, Hawai'i: The First 20 Years* (eds. C. Heliker, D.A. Swanson and J. Takahashi), US Geological Survey, Professional Paper, 1676, pp. 63–88.

Kavanagh, J.L., Menand, T., and Sparks, R.S.J. (2006). "An experimental investigation of sill formation and propagation in layered elastic media." *Earth and Planetary Sciences Letters*, 245, 799–813.

Keating, B.H., Fryer, P., Batiza, R. et al. (eds.) (1987). *Seamounts, Islands, Atolls*. American Geophysical Union Monograph 43, 35–47.

Kelly, J.T., Carey, S., Pistolesi, M. et al. (2014). "Exploration of the 1891 Foerstner submarine vent site (Pantelleria, Italy): insights into the formation of basaltic balloons." *Bulletin of Volcanology*, 76. doi 10.1007/s00445-014-0844-4.

Kempe, S. (2002). "Lavaröhren (Pyroducts) auf Hawaii und ihre Genese." *Angewandte Geowissenschaften in Darmstadt–Schriftenreihe der deutschen Geologischen Gesellschaft*. Darmstadt, W. H. Rosendahl, 15, 109–127.

Kempe, S. (2008). "Immanuel Kant's remark on lava cave formation in 1803 and his possible sources." *Proceedings of the 13th International. Symposium on Volcanospeleology*, Jeju Island, Korea, Sept. 1–5, 35–37.

Kempe, S. (2012). Volcanic rock caves. In: *Encyclopedia of Caves* (eds. W. White and D.C. Culver), 2nd edition, pp. 865–873. Elsevier: Academic Press.

Kempe, S. (2019). "Volcanic rock caves (revised)." Chapter 131 in *Encyclopedia of Caves* (eds. W. White, D.C. Culver, and T. Pipan), 3rd edition, pp. 1118–1127. London: Academic Press/Elsevier.

Kempe, S., and Kempe C. (2016). "Pyroducts, the third most common lava type on Earth." *17th Symposium of Vulcanospeleology Proceedings*, Hawaii, p. 1.

Kempe, S., Middleton, G., Addison, A. et al. (2021). "New insights into the genesis of pyroducts of the Galápagos Islands, Ecuador." *Acta Carsologica*, 50, 55–74.

Kendrick, J.E., Smithy, R., Sammonds, P. et al. (2013). "The influence of thermal and cyclic stressing on the strength of rocks from Mount St. Helens, Washington." *Bulletin of Volcanology*, 75. doi 10.1007/s00445-013-0728-z.

Kent, G.M., Harding, A.J., and Orcutt, J.A. (1990). "Evidence for a smaller magma chamber beneath the East Pacific Rise at 9°30N." *Nature*, 344, 650–653.

Kereszturi, G., Geyer, A., Marti, J. et al. (2013). "Evaluation of morphometry-based dating of monogenetic volcanoes: A case study from Bandas del Sur, Tenerife (Canary Islands)." *Bulletin of Volcanology*, 75. doi 10.1007/s00445-013-0734-1.

Kerle, N. and van Wyk de Vries, B. (2001). "The 1998 debris avalanche at Casita Volcano, Nicaragua – Investigation of structural deformation as the cause of slope instability using remote sensing." *Journal Volcanology and Geothermal Research*, 105, 49–63.

Kerr, A.C. (1998). "Oceanic plateau formation: A cause of mass extinction and black shale deposition around the Cenomanian-Turonian boundary." *Journal of the Geological Society of London*, 155, 619–626.

Kerr, R.A. (2005). "Icy volcanism has rejuvenated Titan." *Science*, 308, 193.

Kerr, R.A. (2013). "The Deep Earth Machine is coming together." *Science*, 340, 22–24.

Keszthelyi, L., Self, S., and Thorvaldur, T. (2006). "Flood lavas on Earth, Io, and Mars." *Journal of the Geological Society*, 163, 253–264.

Kieffer, G. (1971). "Aperçu sur la morphologie des régions volcaniques de Massif Centrale." Symposium Jean Jung: Géologie, Géomorphologie et Structure Profonde de Massif Central Français. Clermont-Ferrand, pp. 479–510.

Kieffer, S.W. (1981a). "Blast dynamics of Mount St. Helens on 18 May 1980." *Nature*, 291, 568–570.

Kieffer, S.W. (1981b). "Fluid dynamics of the May 18 blast at Mount St. Helens." In: *The 1980 Eruption of Mount St. Helens, Washington* (eds. P.W. Lipman and D.R. Mullineaux). Washington, DC, US Geological Survey. Professional Paper 1250.

Kieffer, S.W. (2013). *The Dynamics of Disaster*. New York: W. W. Norton & Co., 315 pp.

Kienle, J., Kyle, P.R., Self, S. et al. (1980). "Unirek maars, Alaska I. April 1977 eruption sequence, petrology and tectonic setting." *Journal of Volcanology and Geothermal Research*, 7, 11–37.

Kirwin, D.J., Forster, C.N., Kavalieris, I. et al. (2005). "The Oyu Tolgoi copper-gold porphyry deposits, South Gobi, Mongolia." In: *Geodynamics and Metallogeny of Mongolia with a Special Emphasis on Copper and Gold Deposit* (eds. R. Seltmann, O. Gerel, and D.J. Kirwin), pp. 1–14. London: CERCAMS.

Klein, F.W. (1982). "Patterns of historical eruptions at Hawaiian volcanoes." *Journal of Volcanology and Geothermal Research*, 12, 1–35.

Klein, F.W., Koyanagi, R.Y., Nakata, J.S. et al. (1987). "The seismicity of Kilauea's magma system." In: *Hawaiian Volcanism* (eds. R.W. Decker, T.L. Wright, and P.H. Staufer), pp. 1019–1185. Geological Survey Professional Paper 1350.

Klein, J., Mueller, S.P., Helo, C. et al. (2018). "An expanded model and application of the combined effect of crystal-size distribution and crystal shape on the relative viscosity of magmas." *Journal of Volcanology and Geothermal Research*, 357, 128–133.

Kling, G.W., Clark, M.A., Wagner, G.N. et al. (1987). "The 1986 Lake Nyos gas disaster in Cameroon, West Africa." *Science*, 236, 169–185.

Kling, G.W., Evans, W.C., Tanyileke, G. et al. (2005). "Degassing Lakes Nyos and Monoun: Defusing certain disaster." *US National Academy of Sciences, Proceedings*, 102 (40), 14185–14190.

Klingaman, W.K. (2013). *The Year Without Summer: 1816 and the Volcano That Darkened the World and Changed History.* New York City: St. Martin's Press, 388 p.

Klug, C. and Cashman, K.V. (1994). "Vesiculation of May 18, 1980 Mount St. Helens magma." *Geology*, 22, 468–472.

Klug, C. and Cashman, K.V. (1996). "Permeability development in vesiculating magmas: Implications for fragmentation." *Bulletin Volcanologique*, 58, 87–100.

Kokelaar, B.P. (1983). "The mechanism of Surtseyan volcanism." *Journal of the Geological Society of London*, 140, 939–944.

Koppers, A.A.P., Staudigel, H., Pringle, M.S. et al. (2003). "Short-lived and discontinuous volcanism in the South Pacific: Hot spots or extensional volcanism?" *Geochemistry, Geophysics, Geosystems: The Electronic Journal of the Earth*, 4, 1089.

Koyaguchi, T. and Kaneko, K. (1999). "A two-stage thermal evolution model of magmas in continental crust." *Journal of Petrology*, 40, 241–254.

Koyaguchi, T. and Woods, A.W. (1996). "On the formation of eruption columns following the explosive mixing of magma and surface water." *Journal of Geophysical Research*, 101, 5561–5574.

Krafft, M. (1991). *Les Feux de la Terre: Histoires de Volcans.* Paris: Decouvertes Gallimard Sciences, 298 pp.

Kudo, A.M. and Weill, D. (1970). "An igneous Plagioclase geothermometer." *Contributions to Mineralogy and Petrology*, 25, 52–65.

Kunii, D. and Levenspiel, O. (1991). *Fluidization Engineering*, 2nd edition. Butterworth-Heinemann Publishing, Oxford, U.K., 520 p.

Kusakabe, M., Ohba, T., Yoshida, Y. et al. (2008). "Evolution of CO2 in lakes Monoun and Nyos, Cameroon, before and during controlled degassing." *Geochemical Journal*, 42 (1), 93–118.

Lacroix, A. (1908). *La œmontagne Pelée après ses éruptions: avec observations sur les éruptions du Vésuve en 79 et en 1906.* Paris: Masson et Cie., 136 pp.

Lamb, H.H. (1970). "Volcanic dust in the atmosphere: With a chronology and assessment of its meteorological significance." *Philosophical Transactions of the Royal Society of London*, 266 (1179), 425–533.

Lamur, A., Lavellée, Y., Iddon, F.E. et al. (2018). Disclosing the temperature of columnar jointing in lavas, *Nature Communications*, doi.org/10.1038/s41467-018-03842-4

Lassiter, J.C. (2006). "Constraints on the coupled thermal evolution of the Earth's core and mantle, the age of the inner core, and the origin of the 1860s/1880s 'core signal' in plume-derived lavas." *Earth and Planetary Science Letters*, 250, 306–317.

Latter, J.H. (1981). "Tsunamis of volcanic origin: Summary of causes, with particular reference to Krakatoa, 1883." *Bulletin Volcanologique*, 44, 467–490.

Latter, J.H. (ed.) (1989). *Volcanic Hazards: Assessment and Monitoring.* Berlin, Springer-Verlag, pp. 151–612.

Le Bas, M.J., Le Maitre, R.W., Streckheisen, A. et al. (1986). "A chemical classification of volcanic rocks based upon the total alkali-silica diagram." *Journal of Petrology*, 27, 745–750.

Le Quéré, C., Moriarty, R., Andrew, R. et al. (2014). "Global carbon budget, 2013." *Earth System Science Data*, 6, 235–263.

Lee, J.M. (2007). "Seismic tomography of magmatic systems." *Journal of Volcanology and Geothermal Research*, 167, 37–56.

Lee, W.-J. and Wyllie, P.J. (1997). "Liquid immiscibility in the join NaAlSi3O8-CaCO3 at 1Gpa: Implications for crustal carbonates." *Journal of Petrology*, 38, 1113–1135.

Legrand, D., Espíndola, J.M., Jiménez, Z. et al. (2015). "Comparison of the seismicity before and after the 1982 El Chichon Eruption." In: *Active Volcanoes of Chiapas (Mexico): El Chichón and Tacaná, Active Volcanoes of the World* (eds. T. Scolamacchia and J.L. Macías). doi 10.1007/978-3-642-25890-9_5, Springer-Verlag Berlin. Heidelberg.

Legros, F. and Kelfoun, K. (2000). "On the ability of pyroclastic flows to scale topographic barriers." *Journal of Volcanology and Geothermal Research*, 98, 235–241

LeGuern, F., Tazieff, H., and Faivre, P.R. (1982). "An example of health hazards: People killed by gas during a phreatic eruption: Dieng Plateau, Java, Indonesia." *Bulletin of Volcanology*, 45, 153–156.

LeMasurier, W. (2013). "Shield volcanoes of Marie Byrd Land, West Antractic rift: oceanic island similarities, continental signature, and tectonic controls." *Bulletin of Volcanology*, 75. doi 10.1007/s00445-013-0726.

Lénat, J.-F., Bachèlery, P., and Merle, O. (2013). "Anatomy of Piton de la Fournaise volcano (La Réunion, Indian Ocean)." *Bulletin of Volcanology*, 74, 1945–1961.

Lengliné, O., Zacharie, D., and Okubo, P.G. (2021). "Tracking dike propagation leading to the 2018 Kīlauea eruption." *Earth and Planetary Science Letters*, 553, 1 January 2021, 116653.

Lesher, C.E. and Spera, F.A. (2015). "Thermodynamic and transport properties of silicate melts and magma." Chapter 5 in *The Encyclopedia of Volcanoes* (eds. H. Sigurdsson et al.), 2nd edition, pp. 113–114, Amsterdam: Elsevier, 1421 pp.

Lewicki, J.L. and Hilley, G.E. (2014). "Multi-scale observations of the variability of magmatic CO2 emissions, Mammoth Mountain, CA, USA." *Journal of Volcanology and Geothermal Research*, 284, 1–15.

Lewis, G.B. and de Lajartre, P.E.B. (2007). *Red Volcanoes: Face to Face with Mountains of Fire.* New York: Thames and Hudson, 144 p.

Lewis, J.D. (1968). Form and structure of the Loch Ba ring-dyke, Isle of Mull. *Proceedings of Geological Society of London*, 1649, 110–111.

Li, X., Baker, D.N., and Temerin, D. et al. (1997). "Are energetic electrons in the solar wind the source of the outer radiation belt?" *Geophysical Research Letters*, 24, 923–926.

Lindsay, J.M., Schmitt, A.K., Trumbull, R.B. et al. (2001). "Magmatic evolution of the La Pacana caldera system, Central Andes, Chile: Compositional variation of two cogenetic, large-volume feslic ignimbrites." *Journal of Petrology*, 42, 459–486.

Lipman, P., Normark, W.R., Moore, J.G. et al. (1988). "The giant submarine Alika Debris Slide, Mauna Loa, Hawaii." *Journal of Geophysical Research Solid Earth*, 3, 4279–4299. https://doi.org/10.1029/JB093iB05p04279.

Lipman, P.W. (1976a). "Caldera collapse breccias in the western San Juan Mountains, Colorado." *Geological Society of America Bulletin*, 87, 1397–1410.

Lipman, P.W. (1976b). Geologic map of the Lake City Caldera area, western San Juan Mountains, southwestern Colorado: USGS Miscellaneous Investigations Series Map I-962, 1: 48,000.

Lipman, P.W. (1980). "Cenozoic volcanism in the western United States: Implications for continental tectonics." *Studies in Geophysics: Continental Tectonics*, 161–174.

Lipman, P.W. (1984). "The roots of ash flow calderas in western North America: Windows into the tops of granitic batholiths." *Journal of Geophysical Research*, 89 (B10), 8801–8841.

Lipman, P.W. (1988). "Evolution of silicic magma in the upper crust, the mid-Tertiary Latir Volcanic Field and its cogenetic granitic batholith, northern New Mexico, USA." *Transactions of the Royal Society of Edinburgh*, 79, 265–288.

Lipman, P.W. (1997). "Subsidence of ash-flow calderas: Relation to caldera size and magma-chamber geometry." *Bulletin of Volcanology*, 59, 198–218.

Lipman, P.W. (2006). Geologic Map of the Central San Juan Caldera Cluster. Southwestern Colorado, US Geological Survey Geologic Investigations Series I-2799, Scale 1:50,000.

Lipman, P.W. (2020). Geologic Map of the Bonanza Caldera Area, Northeastern San Juan Mountains, Colorado, USGS Scientific Investigations Map 3394, Scale 1:50,000.

Lipman, P.W. and Bachman, O. (2015). "Ignimbrites to batholiths: Integrating perspectives from geological, geophysical, and geochronological data." *Geosphere*, 11, 705–743. https://doi.org/10.1130/GES01091.1.

Lipman, P.W. and Banks, N.G. (1987). "'A'a flow dynamics, Mauna Loa, 1984." In: *Volcanism in Hawai'I* (eds. R.W. Decker, T.L. Wright, and P.H. Stauffer), pp. 1527–1567. US Geological Survey Professional Paper 1350.

Lipman, P.W. and McIntosh, W.C. (2008). "Eruptive and non-eruptive calderas, northeastern San Juan Mountains, Colorado: Where did the ignimbrites come from?" *Geological Society of America Bulletin*, 120 (7–8), 771–795.

Lipman, P.W. and Mullineaux, D.R. (1981). *The 1980 Eruptions of Mount St. Helens.* Washington, DC: US Geological Survey Professional Paper 1250.

Lipman, P.W. and Zimmerer, M.J. (2019). "Magmato-tectonic links: Ignimbrite calderas, regional dike swarms, and the transition from arc to rift in the Southern Rocky Mountains." *Geosphere*, 15 (6), 1893–1926. https://doi.org/10.1130/GES02068.1.

Lipman, P.W., Moore, J.G., and Swanson, D.A. (1981). "Bulging of the north flank before the 18 May eruption–geodetic data." *The 1980 Eruptions of Mount St. Helens*. Washington, DC: US Geological Survey Professional Paper 1250, pp. 143–155.

Lipman, P.W., Zimmerer, M.J., and McIntosh, W.C. (2015). "An ignimbrite caldera from the bottom up – exhumed floor and fill of the resurgent Bonanza Caldera, Southern Rocky Mountain Volcanic Field, Colorado." *Geosphere*, 11 (6), 1902–1947. https://doi.org/10.1130/GES01184.1

Lirer, L. and Vinci, A. (1991). "Grain-size distributions of pyroclastic deposits." *Sedimentology*, 38, 1075–1083.

Llewellin, E.W. and Manga, M. (2005). "Bubble suspension rheology and implications for conduit flow." *Journal of Volcanology and Geothermal Research*, 143, 205–217.

Lockwood, J., Costa, J.E., Tuttle, M.L. et al. (1988). "The potential for catastrophic dam failure at Lake Nyos Maar, Cameroon." *Bulletin of Volcanology*, 50, 340–349.

Lockwood, J.P. (1980). "Gordon Macdonald, 1911–1978." *Journal of Volcanology and Geothermal Research*, 7 (2), 177–188.

Lockwood, J.P. and Lipman, P.W. (1980). "Recovery of datable charcoal beneath young lavas: Lessons from Hawaii." *Bulletin Volcanologique*, 43 (3), 609–615.

Lockwood, J.P. and Lipman, P.W. (1987). "Holocene eruptive history of Mauna Loa." In: *Volcanism in Hawaii* (eds. R.W. Decker, T.L. Wright, and P.H. Stauffer), pp. 509–536. US Geological Survey Professional Paper 1350.

Lockwood, J.P. and Romano, R. (1985). "Diversion of lava during the 1983 eruption of Mount Etna." *Earthquake Information Bulletin*, 17 (4), 124–133.

Lockwood, J.P. and Torgerson, F.A. (1980). "Diversion of lava flows by aerial bombing: Lessons from Mauna Loa volcano, Hawaii." *Bulletin of Volcanology*, 43 (4), 727–741.

Lockwood, J.P. and Williams, I. (1978). "Lava trees and tree molds as indicators of lava flow direction." *Geological Magazine*, 115 (1), 69–74.

Lockwood, J.P. (1995). "Mauna Loa eruptive history – the preliminary radiocarbon record." In *Mauna Loa Revealed – Structure, Composition, History, and Hazards* (eds. J.M. Rhodes and J.P. Lockwood), pp. 81–94. Amer. Geophys. Union Geophysical Monograph 92, 348 pp.

Lockwood, J.P., Tilling, R.I., Holcomb, R.T. et al. (1999). "Magma Migration and resupply during the 1974 summit eruptions of Kilauea Volcano, Hawaii." US Geological Survey, United States Government Printing Office, 37p.

Lofgren, G. (1971). "Spherulitic textures in glassy and crystalline rocks." *Journal of Geophysical Research*, 76, 5635–5648.

Lopes, R. (2005). *The Volcano Adventure Guide*. Cambridge Press, 352 pp.

Lopes, M.R., Guest, J.E., and Wilson, C.J. (1980). "Origin of the Olympus Mons aureole and perimeter scarp." *The Moon and the Planets*, 22, 221–232.

Lopes, R.M.C. and Carroll, M.W. (2008). *Alien Volcanoes*. Baltimore: The Johns Hopkins University Press.

Lopez, T., Tassi, F., Aiuppa, A. et al. (2017). "Geochemical constraints on volatile sources and subsurface conditions at Mount Martin, Mount Mageik, and Trident Volcanoes, Katmai Volcanic Cluster, Alaska." *Journal of Volcanology and Geothermal Research*, 347, 64–81.

Lowe, D.J. and de Lange, W.P. (2000). "Volcano-meteorological tsunamis, the 200 CE Taupo eruption (New Zealand) and the possibility of a global tsunami." *The Holocene*, 10, 401–407.

Lowe, D., Byerly, G.R., and Kyle, F.T. (2014). Recently discovered 3.42-3.23 Ga impact layers, Barberton Belt, South Africa: 3.8 Ga detrital zircons, Archean impact history, and tectonic implications. *Geology*, 42, 747–750.

Lowe, D.R. (1982). "Sediment gravity flows II; Depositional models with special reference to the deposits of high-density turbidity currents." *Journal of Sedimentary Research*, 52, 279–297.

Lowenstern, J.B., Smith, R.B., and Hill, P.D. (2006). "Monitoring supervolcanoes: Geophysical and geochemical signals at Yellowstone and other large caldera systems." *Philosophical Transactions of the Royal Society, A*, 364, 2055–2072.

Lu, Z. and Dzurisin, D. (2011). *INSAR Imaging of Aleutian Volcanoes*. Berlin: Springer-Praxis, 300p.

Lubrich, O. and Nehrlich, T. (eds.) (2021). *Die Vulkane des William Hamilton, Naturreportagen von den Feuerbergen Ätna und Vesuv*. Darmstadt: Wissenschaftliche Buchgesellschaft, 272p. (Containing reproductions of the figures and text, translated into German by J. v. Koppenfels,of "Campi Phlegraei" by William Hamilton, 1776 and 1779).

Luedke, R.G. and Smith, R.L. (1981). "Map showing distribution, composition and age of late Cenozoic volcanic centers in California and Nevada." US Geological Survey.

Luhr, J.F. and Simkin, T. (1993). *Paricutín: The Volcano Born in a Mexican Cornfield*. Phoenix, Geoscience Press, 427p.

Lund, J.W. (2000). "World status of geothermal energy use: Overview 1995–1999." *Geothermal Resources Council Transactions*, 24, 383–388.

Luongo, G.L., Perrotta, A., and Scarpati, C. (2003). "Impact of the ad 79 explosive eruption on Pompeii, I: Relations amongst the depositional mechanisms of the pyroclastic products, the framework of the buildings and the associated destructive events." *Journal of Volcanology and Geothermal Research*, 126, 201–233.

Lutz, H. and Lorenz, V. (2013). "Early volcanological research in the Vulkaneifel, Germany, the classic region of maar-diatreme volcanoes: The years 1774–1865." *Bulletin of Volcanology*, 75, article 743. doi 10.1007/s00445-013-0743-0.

Lyell, C. (1855). *A Manual of Elementary Geology*, 5th edition. New York, D. Appleton and Company.

Macdonald, G.A. (1962). "The 1959 and 1960 eruptions of Kilauea volcano, Hawaii, and the construction of walls to restrict the spread of the lava flows." *Bulletin of Volcanology*, 24, 249–294.

Macdonald, G.A. (1972). *Volcanoes*. Englewood Cliffs: Prentice-Hall.

Macdonald, G.A. (1978). "Geologic map of the crater section of Haleakala National Park, Maui, Hawaii." I Map 1088, US Geological Survey.

Macdonald, K.C. (1982). "Mid-ocean ridges: Fine-scale tectonics, volcanic and hydrothermal processes within the plate boundary zone." *Annual Reviews of Earth and Planetary Sciences*, 10, 155–190.

Macías, J.L., Capra, L., Scott, K.M. et al. (2004). "The 26 May 1982 breakout flows derived from failure of a volcanic dam at El Chichón, Chiapas, Mexico." *GSA Bulletin*, 116 (1–2), 233–246. doi: https://doi.org/10.1130/B25318.1.

Macías, J.L., Sheridan, M.F., Espíndola, J.M. (1997). "Reappraisal of the 1982 eruptions of El Chichón Volcano, Chiapas, Mexico: New data from proximal deposits." *Bulletin of Volcanology*, 58, 459–471.

Mackaman-Lofland, C., Brand, B.D., Taddeucci, J., and Wohletz, K. (2014). "Sequential fragmentation/transport theory, pyroclast-density relationships, and the emplacement dynamics of pyroclast density currents: A case study on the Mt. St. Helens (USA) 1980 eruption." *Journal of Volcanology and Geothermal Research*, 275, 1–13.

MacLeod, N.S. and Sherrod, D.R. (1988). "Geologic evidence for a magma chamber beneath Newberry Volcano, Oregon." *Journal of Geophysical Research*, 93, B9, 10,067–10,079.

Mader, H.M., Coles, S.G., Connor, C.B. et al. (2006). *Statistics in Volcanology*. London: Geological Society of London.

Mahoney, J.J. and Coffin, M.F. (1997). *Large Igneous Provinces: Continental, Oceanic, and Planetary Flood Volcanism*. Washington, DC: American Geophysical Union.

Major, J.J. and Newhall, C.G. (1989). "Snow and ice perturbation during historical volcanic eruptions and the formation of lahars and floods." *Bulletin of Volcanology*, 52, 1–27.

Malahoff, A. (1987). "Geology of the summit of Loihi submarine volcano." Volcanism in Hawaii, The 1980 Eruptions of Mount St. Helens. US Geological Survey Professional Paper 1350, pp. 145–155.

Manga, M. and Ventura, G. (2005). Kinematics and dynamics of lava flows. *Geological Society of America Special Paper* 396, 218 p.

Mangan, M.T. and Cashman, K.V. (1993). "Vesiculation of basaltic magma during eruption." *Geology*, 21, 157–160.

Mangan, M.T. and Cashman, K.V. (1996). "The structure of basaltic scoria and reticulite and inferences for vesiculation, foam formation, and fragmentation in lava fountains." *Journal of Volcanology and Geothermal Research*, 73, 1–18.

Markhinin, Ye.K. (1980). *Вуканы ц Жцзнь [Volcanoes and Life]*. Moscow: Mysl' Publishers, 196p.

Markhinin, Ye.K. (1985). *Вуканцзм [Vulcanism]*. Moscow: Nedra Publishing.

Marland, G., Andres, R.J., Boden, T.A. et al. (1998). "Global, regional, and national CO2 emission estimates from fossil fuel burning, cement production, and gas flaring, 1751–1996, DATASET NDPO30." Carbon Dioxide

Information and Analysis Center (CDIAC), Oak Ridge National Laboratory, Oak Ridge, Tennessee.

Marsh, B.D. (1984). "On the mechanics of caldera resurgence." *Journal of Geophysical Research*, 89 (B10), 8245–8251.

Marshall, P. (1935). "Acid rocks of the Taupo-Rotorua volcanic district." *Transactions of the Royal Society of New Zealand*, 64, 323–366.

Martí, J., Aspinall, W.P., Sobradeloet, R. et al. (2008). "A long-term volcanic hazard event tree for Teide-Pico Viejo stratovolcanoes (Tenerife, Canary Islands)." *Journal of Volcanology and Geothermal Research*, 178, 543–552.

Martí, J., Geyer, A., and Folch, A. (2009). "A genetic classification of collapse calderas based on field studies, and analogue and theoretical modelling." In: *Studies in Volcanology: The Legacy of George Walker* (eds. T. Thordarson, S. Self, and G. Larsen, et al.), pp. 249–266. London: Geological Society of London.

Marti, J., Geyer, A., and Aguirre-Diaz, G. (2013). "Origin and evolution of the Deception Island caldera (South Shetland Islands, Antarctica)." *Bulletin of Volcanology*, 732. doi 10.1007/s00445-013-0732-3.

Martin, D.P. and Rose, W.I. (1981). "Behavioral patterns of Fuego volcano, Guatemala." *Journal of Volcanology and Geothermal Research*, 10, 67–81.

Martin, U., Németh, K., Lorenz, V. et al. (2007). "Introduction: Maar-diatreme volcanism." *Journal of Volcanology and Geothermal Research*, 159, 1–3.

Marty, B., Sano, Y. and France-Lanord, C. (2001). "Water-saturated oceanic lavas from the Manus Basin: Volatile behaviour during assimilation-fractional crystallisation-degassing (AFCD)." *Journal of Volcanology and Geothermal Research*, 108, 1–10.

Marzocchi, W., Newhall, C., and Woo, G. (2012). "The scientific management of volcanic crises." *Journal of Volcanology and Geothermal Research*, 247–248, 181–189.

Mason, B.G., Pyle, D.M., and Oppenheimer, C. (2004). "The size and frequency of the largest explosive eruptions on Earth." *Bulletin of Volcanology*, 66, 735–748.

Mason, T.P. and Smith, R.L. (1977). "Spectacular mobility of ash flows around Aniakchak and Fisher calderas, Alaska." *Geology*, 5, 173–176.

Masotta, M., Beier, C., and Mollo, S. (2021). "Crustal Magmatic System Evolution: Anatomy, Architecture, and Physico-Chemical Processes." American Geophysical Union Monograph 264, 239pp.

Massonet, D., Rossi, M., Carmona, C. et al. (1993). "The displacement field of the Landers earthquake mapped by radar interferometry." *Nature*, 364, 138–142.

Mastin, L.G. (1994). "Explosive tephra emissions at Mount St. Helens, 1989–1991 – the violent escape of magmatic gas following storms?" *Geological Society of America Bulletin*, 106, 175–185.

Mastin, L.G. and Pollard, D.D. (1988). "Surface deformation and shallow dike intrusion processes at Inyo Craters, Long Valley, California." *Journal of Geophysical Research*, 93, 13,221–13,235.

Mastin, L.G. and Waitt, R.B. (2000). "Glacier Peak: History and hazards of a Cascade volcano." US Geological Survey Fact Sheet 058 -00.

Mastin, L.G. and Witter, J.B. (2000). "The hazards of eruptions through lakes and seawater." *Journal of Volcanology and Geothermal Research*, 97, 195–214.

Masurenkov, Y.P. (1991). "Hypsometric and lateral patterns of active volcanoes." In: *Active Volcanoes of Kamchatka* (eds. S.A. Fedotov and Y.P. Masurenkov), 54–66. Moscow: Nauka Publishers.

Matek, B. (2015). U.S. and Global Geothermal Power Report, www.geo-energy.org/reports.aspx.

Mather, T.A., Pyle, D.M., and Allen, A.G. (2004). "Volcanic origin for fixed nitrogen in the early Earth's atmosphere." *Geology*, 3, 905–908.

Matson, M. (1984). "The 1982 El Chichon Volcano eruptions –A satellite perspective." *Journal of Volcanology and Geothermal Research*, 23, 1–10.

Matthews, A.J., Barclay, J., and Johnstone, J.E. (2009). "The fast response of volcano-seismic activity to intense precipitation: Triggering of primary volcanic activity by rainfall at Soufrière Hills volcano, Montserrat." *Journal of Volcanology and Geothermal Research*, 184, 405–415.

Mauk, F.J. and Johnston, M.J.S. (1973). "On the triggering of volcanic eruptions by Earth tides." *Journal of Geophysical Research*, 78, 3356–3362.

McBirney, A.R. (1965). "Volcanic history of Nicaragua." *University of California, Berkeley Publications in the Geological Sciences*, 55, 1–65.

McBirney, A.R. and Williams, H. (1969). *Geology and Petrology of the Galapagos Islands*. Geological Society of America, 197p.

McBirney, A.R. (2004). *Faulty Geology: Frauds, Hoaxes, and Delusions*. Eugene, Oregon: Bostok Press, 216 pp.

McCausland, W.A., Pallister, J.S., Andreastutti, S. et al. (eds.) (2019). "Lessons learned from the recent eruptions of Sinabung and Kelud volcanoes, Indonesia." *Journal of Volcanology and Geothermal Research*, 382, 310 pp.

McClelland, L., Simkin, T., Summers, M. et al. (eds.) (1989). *Global Volcanism 1975–1985*. Englewood Cliffs: Prentice-Hall and American Geophysical Union.

McCoy, F. and Heiken, G. (2000). *Volcanic Hazards and Disasters in Human Antiquity*. Boulder: Geological Society of America.

McCulloh, T.H., Fleck, R.J., Denison, R.E. et al. (2002). "Age and tectonic significance of volcanic rocks in the Northern Los Angeles Basin, California." US Geological Survey Professional Paper 1669, 24p.

McEwen, A., Keszthelyi, L., Geissler, P. et al. (1998). "Active volcanism on Io as seen by Galileo SSI 219." *Icarus*, 135–181.

McGee, K.A. and Gerlach, T.M. (1998). "Annual cycle of magmatic CO2 in a tree kill soil at Mammoth Mountain, California: Implications for soil acidification." *Geology*, 26, 463–466.

McGovern, P.J., Grosfils, E.B., Galgana, G.A. et al. (2014). Lithospheric flexure and volcano basal boundary conditions: keys to the structural evolution of large volcanic edifices on the terrestrial planets. In: *Volcanism and Tectonism Across the Inner Solar System* (eds. T. Platz, M. Massironi, P.K. Byrne, and H. Hiesingerm). Geological Society, London, Special Publications no. 401, http://dx.doi.org/10.1144/SP401.7.

McGuire, W.J. (1995). "Monitoring active volcanoes: An introduction." In: *Monitoring Active Volcanoes: Strategies, Procedures and Techniques* (eds. W. Mcguire, C. Kilburn, and J. Murray). London: UCL Press, 421p.

McGuire, W.J., Jones, A.P., and Neuberg, J. (1996). "Volcano instability on the Earth and other planets." Geological Society of London Special Paper 110, 388p.

McIntrye, D.B. and McKirdy, A. (2012). *Hutton: The Founder of Modern Geology*, 2nd edn. National Museum of Scotland Publishing, Edinburgh,80 p.

McNutt, S.R. (1994). "Volcanic tremor amplitude correlated with the volcano explosivity index and its potential use in determining ash hazards to aviation." *Acta Vulcanologia*, 5, 193–196.

McNutt, S.R. (1996). "Seismic monitoring and eruption forecasting of volcanoes: A review of the state of the art and case histories." In: *Monitoring and Mitigation of Volcano Hazards* (eds. R. Scarpa and R. Tilling), pp. 99–146. Berlin: Springer-Verlag.

McNutt, S.R. and Davis, C.M. (2000). "Lightning associated with the 1992 eruptions of Crater Peak, Mount Spurr Volcano, Alaska." *Journal of Volcanology and Geothermal Research*, 102, 45–65.

McPhie, J. and Cas, R. (2015). "Volcanic successions associated with ore deposits – facies characteristics and ore-host relationships." Ch. 49 in The Encyclopedia of Volcanoes, (eds. H. Sigurdsson et al.), 2nd edition, pp. 865–879. Amsterdam: Elsevier Press, 1421 pp.

McPhie, J., Doyle, M., and Allen, R. (1993). *Volcanic Textures: A Guide to the Interpretation of Textures in Volcanic Rocks*. Hobart, Centre for Ore Deposits and Exploration Studies, University of Tasmania, 198p.

Medina, F., González, L., Gutierrez, C. et al. (1992). Analysis of the seismic activity related to the 1982 eruptions of El Chichón Volcano, Mexico, In: *Volcanic seismology: IAVCEI Proceedings in Volcanology 3* (eds. P. Gaspariniand and K. Aki), pp. 97–108.

Medina, F., González-Morán, T., and González, L. (1990). "Gravity and seismicity analyses of the El Chichón Volcano, Chiapas, Mexico." *Pure and Applied Geophysics (PAGEOPH)*, 133, 149–165.

Medina-Martínez, F. (1982). "El Volcaán Chichón." *GEOS*, 2 (4), 19.

Medina-Martínez, F. (1983). Analysis of the eruptive history of the Volcán de Colima, Mexico (1560–1980). *Geofísica Internacional*, 22, 157–178.

Mee, K., Tuffon, H., and Gilbert, J.S. (1994). "Snow-contact volcanic facies and their use in determining past eruptive environments at Nevados del Chillán volcano, Chile." *Bulletin of Volcanology*, 68, 363–376.

Menand, T. (2007). "The mechanics of sills in layered elastic rocks and their implications for the growth of laccoliths and other igneous complexes." *Earth and Planetary Sciences Letters*, 267, 93–99.

Mencke, W., West, M., and Tolstoy, M. (2002). "Shallow-crustal magma chamber beneath the axial high of the coaxial segment of the Juan de Fuca Ridge at the source site of the 1993 eruption." *Geology*, 30, 359–362.

Mendoza-Rosas A.T. and De la Cruz-Reyna S. (2010). Hazard estimates for El Chichón volcano, Chiapas, México: a statistical approach for complex eruptive histories. *Natural Hazards and Earth System Sciences*, 10, 1159–1170. www.nat-hazards-earth-syst-sci.net/10/1159/2010/

Mercalli, G. (1907). *I Vulcani Attivi della Terra*. Milano: Ulrico Hoepli Publishing, 421 p.

Mercier-Langenin, P., Gibson, H., Hannington, M.D. et al. (2014). "A special issue on Archaean magmatism, volcanism, and ore deposits: Part 2 – Volcanogenic massive sulfide deposits, preface." *Economic Geology*, 109, 1–9.

Michaud-Dubuy, A., Carazzo, G., Kaminski, E., and Girault, F. (2018). "A revisit of the role of gas entrapment on the stability conditions of explosive volcanic columns." *Journal of Volcanology and Geothermal Research*, 357, 349–361.

Middlemost, E.A.K. (1985). *Miocene Shield Volcanoes of New South Wales*. Geological Society of Australia, New South Wales Div., Publication 1, 49–58.

Mileti, D.S., Bolton, P.A., Fernandez, G. et al. (1991). *The Eruption of Nevado del Ruiz Volcano, Colombia, South America–November 13, 1985*. Washington, DC: National Academy Press.

Miller, C.F. and Miller, J.S. (2002). "Contrasting stratified plutons exposed in tilt blocks, Eldorado Mountains, Colorado River Rift, NV, USA." *Lithos*, 61, 209–224.

Miller, C.F. and Wark, D.A. (2008). "Supervolcanoes and their explosive super-eruptions." *Elements*, 4 (1), 11–15.

Miller, C.F., Walker, B.A., Lowery, L. et al. (2005). "Construction of Plutons by Horizontal Depositional and Intrusive Sheets." *Geological Society of America Abstracts with Programs*, 37, 7, 130.

Miller, D., and Groves, D.I. (2019). *Potassic Igneous Rocks and Associated Gold-Copper Mineralization*. Mineral Resources Reviews, Springer-Verlag, 398 p.

Miller, S. and Orgel, L.E. (1974). *The Origins of Life on Earth*. Englewood Cliffs: Prentice-Hall Publishing, 229p.

Miller, S.L. (1953). "A production of amino acids under possible primitive Earth conditions." *Science*, 117, 528–529.

Miller, T.P. and Casadevall, T.J. (2000). "Volcanic ash hazards to aviation." In: *Encyclopedia of Volcanoes* (ed. H. Sigurdsson), pp. 915–930. New York: Academic Press.

Miller, T.P. and Smith, R.L. (1977). "Spectacular mobility of ash flows around Aniakchak and Fisher calderas, Alaska." *Geology*, 173–176.

Miller, W.F., Geller, R.J., and Stein, S. (1978). "Use of a bubble tiltmeter as a horizontal seismometer." *Geophysical Journal of the Royal Astronomical Society*, 54, 661–668.

Mimura, K. (1984). "Imbrication, flow direction, and possible source areas of the pumice-flow tuffs near Bend, Oregon, U.S.A." *Journal of Volcanology and Geothermal Research*, 21 (1–2), 45–60.

Minakami, T. (1956). "Report on volcanic activities and volcanological studies in Japan for the period from 1951 to 1954." *Bulletin Volcanologique*, 18, 39–76.

Minakami, T. (1960). "Fundamental research for predicting volcanic eruptions, Part I." *Bulletin of the Earthquake Research Institute*, Tokyo University, 38, 4497–4544.

Minakami, T., Ishikawa, T., and Yagi, K. (1951). "The 1944 eruption of Volcano Usu in Hokkaido, Japan." *Bulletin of Volcanology*, 11, 45–160.

Mogi, K. (1958). "Relations between eruptions of various volcanoes and the deformations of the ground surfaces around them." *Bulletin of the Earthquake Research Institute*, 36, 99–134.

Mohr, P.A. and Zanettin, B. (1988). *The Ethiopian Flood Basalt Province*. Dordrecht: Kluwer.

Montelli, R., Nolet, G., Dahlen, F.A. et al. (2003). "Finite frequency tomography reveals a variety of plumes in the mantle." *Science*, 303 (5656), 338–343.

Moore, H.J. (1987). "Preliminary estimates of the rheological properties of 1984 Mauna Loa lava." In: *Volcanism in Hawaii*, US Geological Survey Professional Paper 1350, pp. 1569–1588.

Moore, J., White, W.M., Paul, D. et al. (2011) "Evolution of shield-building and rejuvenescent volcanism of Mauritius." *Journal of Volcanology and Geothermal Research*, 207, 47–66.

Moore, J.G. (1967). "Base surge in recent volcanic eruptions." *Bulletin. Volcanologique*, 30, 337–363.

Moore, J.G. (1970). "Pillow lava in a historic lava flow from Hualalai Volcano, Hawaii." *Journal of Geology*, 78, 239–243.

Moore, J.G. (In Press). *Geosaga* Tales from the Hills and the Sea.

Moore, J.G. and Sisson, T.W. (1981). "Deposits and effects of the May 18 pyroclastic surge." In: The 1980 Eruptions of Mount St. Helens. US Geological Survey Professional Paper 1250, pp. 421–438.

Moore, J.G. and Thomas, D.M. (1988). "Subsidence of Puna, Hawaii inferred from sulfur content of drilled lava flows." *Journal of Volcanology and Geothermal Research*, 35, 165–171.

Moore, P. (2002). *Venus*. London: Cassell.

Moore, R.B. and Clague, D.A. (1991). "Geologic Map of Hualalai Volcano, Hawaii." US Geological Survey, I Map 2213.

Mooser, F., Meyer-Abich, H., and McBirney, A.R. (1958). *Central America. Catalog of Active Volcanoes of the World and Solfatara Fields*. Rome: IAVCEI, 6, 1–146.

Mora, D. and Tassara, A. (2018). "Upper crustal decompression due to deglaciation induced flexural unbending and its role in post-glacial volcanism at the southern Andes." *Geophysical Journal International*, 216 (3), 1549–1559.

Morgan, J. (1971). "Convection plumes in the lower mantle." *Nature*, 230, 42–43.

Morgan, J., Reston, T.J., and Ranero, C.R. (2004). "Contemporaneous mass extinctions, continental flood basalts, and 'impact signals': Are mantle plume-induced lithospheric gas explosions the causal link?" *Earth and Planetary Science Letters*, 217, 263–284.

Morgan, L.A, Shanks, P., and Pierce, K.L. (2009). "Hydrothermal processes above the Yellowstone magma chamber: Large hydrothermal systems and large hydrothermal explosions." Geological Society of America Sp. Paper 459, 95pp., https://doi.org/10.1130/2009.2459(01).

Mossoux, S., Saex, M., Bartolini, S. et al. (2016). Q_LASHVA: A flexible GIS plugin to simulate lava flows." *Journal of Computers and Geoscience*, 97, 98–109.

Mota, R., De la Cruz, S., and Mena, M. (1984). "Enjambres sísmicos en Chiapas: un fenómeno frecuente." *GEOS Boletin de la Unión Geofísica Mexicana*, 2, B-17.

Mouginis-Mark, J. and Robinson, M.J. (1992). "Evolution of Olympus Mons caldera, Mars." *Bulletin of Volcanology*, 54, 347–360.

Mouginis-Mark, P.J., Crisp, J.A., and Fink, J.H. (eds.) (2000). *Remote Sensing of Active Volcanism*. Washington, DC: American Geophysical Union.

Mueller, S., Scheu, B., Spieler, O. et al. (2008). "Permeability control on magma fragmentation." *Geology*, 36, 399–402.

Muffler, L.J.P., White, R.D.E., and Truesdell, A.H. (1971). "Hydrothermal explosion craters in Yellowstone National Park." *Geological Society of America Bulletin*, 82, 723–740.

Müllerried F. (1932a). "El Chichón, volcán en actividad descubierto en el Estado de Chiapas." *Mem. y Rev. Ac. Nac. Cienc. Alzate*, 53 (11 y 12), 411-416.

Müllerried F. (1932b). "Der Chichón, ein bisher undekannter tätiger Vulkan im nördlichen Chiapas. Mexiko." Mit. petr.-min. Beiträgen von O.H. Erdmannsdörffer, Heidelberg. *Z. f. Vulkan.*, XIV, 191–209.

Müllerried F. (1933). El Chichón, único volcán en actividad en el sureste de México, Revista Instituto de Geología, *UNAM México*, 33, 156–170.

Murai, I. (1961). "A study of the textural characteristics of pyroclastic flow deposits in Japan." *Bulletin of the Earthquake Research Institute of Tokyo University*, 39, 133–254.

Murphy, B., Gaines, R., and Lackey, J.S. (2016). "Co-evolution of volcanic and lacustrine systems in Pleistocene Long Valley caldera, California, U.S.A." *Journal of Sedimentary Research*, 86, 1129–1146.

Murray, J.B., Pullen, A.D., and Saunders, S.J. (1995). "Ground deformation surveying of active volcanoes." In: *Monitoring Active Volcanoes: Strategies, Procedures and Techniques* (eds. W.J. McGuire, C. Kilburn and J. Murray), pp. 113–150. London: UCL Press.

Myers, A. (2007). "Investigating volcano's plumbing system." *Bolletino del la Comunitá Scientifica in Australasia*, April, 36.

Mysen, B.O. (ed.) (1987). *Magmatic Processes: Physicochemical Principles*. The Geochemical Society, Special Publication 1.

Nairn, A.E.M., Stehli F.G., and Uyeda, S. (eds.) (1985). *The Ocean Basins and Margins: The Pacific Ocean*, vol. 7b. New York: Plenum Press, pp. 89–121.

Nairn, I.A. and Cole, J.W. (1981). "Basaltic dikes in the 1886 Tarawera Rift, New Zealand." *Journal of Geology and Geophysics*, 24, 585–592.

Nakada, S., Nagai, M., Kaneko, T. et al. (2005). "Chronology and products of the 2000 eruption of Miyakejima volcano, Japan." *Bulletin of Volcanology*, 67 (3), 205–218.

Nakada, S., Zaennudin, A., Yoshimoto, M. et al. (2019). "Growth process of the lava dome/flow complex at Sinabung Volcano during 2013-2016." *Journal of Volcanology and Geothermal Research*, 382, 120–136.

Neal, C.P., Brantley, S.R., Montgomery-Brown, E. et al. (2019). "The 2018 flank eruption of Kīlauea Volcano." *Science*, 363 (6425), 367–374.

Neall, V.E. (2001). "Volcanic landforms." In: *The Physical Environment: A New Zealand Perspective* (eds. A. Sturman and R. Spronken-Smith), pp. 39–60. South Melbourne, Australia: Oxford University Press.

Neri, A. and Dobran, F. (1994). Influence of eruption parameters on the thermofluid dynamics of collapsing volcanic columns. *Journal of Geophysical Research*, 99, 11833–11857.

Neuville, D.R., Courtail, P., Dingwell, D.B. et al. (1993). "Thermodynamic and rheological properties of rhyolite and andesite melts." *Contributions to Mineralogy and Petrology*, 113, 572–581.

Newhall, C.G. and Dzurisin, D. (1988). *Historical Unrest at Large Calderas of the World*. New York: US Geological Survey.

Newhall, C.G. and Endo, E.T. (1987). "Sudden seismic calm before eruptions: Illusory or real?" *Hawaii Symposium on How Volcanoes Work [abstracts volume]*. US Geological Survey, Hilo, Hawaii, 190.

Newhall, C.G. and Hoblitt, R.P. (2002). "Constructing event trees for volcanic crises." *Bulletin of Volcanology*, 64 (1), 3–20.

Newhall, C.G. and Palister, J.S. (2015). "Using multiple data sets to populate probabilistic volcanic event trees." Chapter 8 in *Volcanic Hazards, Risks, and Disasters* (eds. .F. Shroder and P. Papale), pp. 203–232. Amsterdam: Elsevier. https://doi.org/10.1016/B978-0-12-396453-3.00008-3.

Newhall, C.G. and Punongbayan, R.S. (eds.) (1996). *Fire and Mud: Eruptions and Lahars of Mount Pinatubo*. Philippines: Philippine Institute of Volcanology and Seismology (Quezon City) and the University of Washington Press (Seattle), 1126 pp.

Newhall, C.G. and Self, S. (1982). "The volcanic explosivity index (VEI): An estimate of explosive magnitude for historical volcanism." *Journal of Geophysical Research*, 87 (C2), 1231–1238.

Newhall, C.G., Bronto, S., Allowya, B. et al. (2000). "10,000 years of explosive eruptions of Merapi Volcano, Central Java: Archaeological and modern implications." *Journal of Volcanology and Geothermal Research*, 100, 9–50.

NIOSH (1994). *Documentation for Immediately Dangerous to Life or Health Concentrations (IDLH)*. Center for Disease Control, Atlanta, Georgia. NTIS Publication No. PB-94-195047.

Nixon, I.G. (1985). "The volcanic eruption of Thera and its effect on the Mycenean and Minoan civilizations." *Journal of Archaeological Research*, 12, 9–24.

Oddsson, B., Gudmundsson, M., Edwards, B. et al. (2016). "Subglacial lava propagation, ice melting and heat transfer during emplacement of an intermediate lava flow in the 2010 Eyjafjallajökull eruption." *Bulletin of Volcanology*, 78. doi 10.1007/s00445-016-1041-4.

Officer, C.B. and Drake, C.L. (1983). "The cretaceous-tertiary transition." *Science*, 219, 1383.

Ogburn, S.E., Loughlin, S.C., and Calder, E.S. (2015) "The association of lava dome growth with major explosive activity (VEI ≥ 4): DomeHaz, a global dataset." *Bulletin of Volcanology*, 77 (5), 40.

Ogden, D.E., Glatzmaier, G.A., and Wohletz, K.H. (2008). "Effects of vent overpressure on buoyant eruption columns: Implications for plume stability." *Earth and Planetary Science Letters*, 268, 283–292.

Ohmoto, H. and Skinner, B.J. (eds.) (1983). "The Kuroko and related volcanogenic massive sulphide deposits." *Economic Geology, Monograph*, 5, 604p.

Ohmoto, H., Graham, U., Liu, Z.-K. et al. (2021). "Discovery of a 3.46 billion-year-old impact crater in Western Australia." Earth and Space Science Open Archive, doi.org/10.1002/essoar.10505838.1.

Oikawa, T., Yoshimoto, M., Nakada, S. et al. (2016). "Reconstruction of the 2014 eruption sequence of Ontake Volcano from recorded images and interviews." *Earth Planets and Space*, 68, 79. doi 10.1186/s40623-016-0458-5.

Okubo, C.H. and Martel, S.J. (1998). "Pit crater formation on Kilauea volcano, Hawaii." *Journal of Volcanology and Geothermal Research*, 86, 1–18.

Okubo, P.G., Benz, H.M., and Chouet, B.A. (1997). "Imaging the crustal magma sources beneath Mauna Loa and Kilauea volcanoes, Hawaii." *Geology*, 25 (10), 867–870.

Okuno, M., Nakamura, T., and Kobayahi, T. (1998). "AMS 14C dating of historic eruptions of the Kirishima, Sakura-jima, and Kaimon-dake volcanoes, southern Kyushu, Japan." In: *Proceedings of the 16th International 14C Conference, Radiocarbon* (eds. W.G. Mook and J. van der Plicht), 40, pp. 825, 832.

Olafsen, E. (1774). *Des Vice-lavmands Eggert Olasens und des Landphysici Bianre Povelsens Reise durch Island, veranstaltet von der Koniglichen Societar der Wissenschaften in Kopenhagen und Leipzig*, 328 pp.

Ollier, C.D. (1969). *Volcanoes*. Boston: MIT Press.

Olson, S. (2016). *The Untold Story of Mount St. Helens*. New York: W.W. Norton, 336 pp.

Omori, F. (1914). "The Sakura-jima eruption and earthquakes." *Bulletin of the Imperial Earthquake Investigation Committee (Japan)*, 8 (1–6), 1–34.

Oppenheimer, C.M.M. (1991). "Lava flow cooling estimated from Landsat thematic mapper infrared data: The Loquimay, Chile eruption 1989." *Journal of Geophysical Research*, 96, 21856–21878.

Oreskes, N. (2001). *Plate Tectonics: An Insider's History of the Modern Theory of the Earth: Seventeen Original Essays by the Scientists Who Made Earth History*. Boulder: Westview.

Oswalt, J.S., Nichols, W., and O'Hara, J.F.O. (1996). "Meteorological observations of the 1991 Mount Pinatubo eruption." In: *Fire and Mud: Eruptions and Lahars of Mount Pinatubo, Philippines* (eds. C.G. Newhall, and R.S. Punongbayan), pp. 625–636. Quezon City, Philippine Institute of Volcanology and Seismology and University of Washington Press.

Oze, C., Cole, J., Scott, A. et al. (2013). "Corrosion of metal roof materials related to volcanic ash interactions." *Journal of Natural Hazards*, 71, 785–802. doi 10.1007/s11069-013-0943-0.

Paguican, E.M.R., van Wyk de Vries, B., and Lagmay, A.M.F. (2012). Volcano-tectonic controls and emplacement kinematics of the Iriga debris avalanches (Philippines). *Bulletin of Volcanology*, 74, 2067–2081. https://doi.org/10.1007/s00445-012-0652-7

Palgan, D., Devey, C.W., and Yeo, I.A. (2017). "Volcanism and hydrothermalism on a hotspot-influenced ridge: Comparing Reykjanes Peninsula and Reykjanes Ridge, Iceland." *Journal of Volcanology and Geothermal Research*, 348, 62–81.

Palladino, D.M. (2017). "Simply pyroclastic currents." *Bulletin of Volcanology*, 79, 53. doi 10.1007/s00445-017-1139-3.

Pallister, J.S., Hoblitt, R.P., Meeker, G.P. et al. (1996). "Magma mixing at Mount Pinatubo: Petrographic and chemical evidence from the 1991 deposits." In: *Fire and Mud: Eruptions and Lahars of Mount Pinatubo, Philippines* (eds. C.G. Newhall and R.S. Punongbayan), pp. 687–731. Quezon City, Philippine Institute of Volcanology and Seismology and University of Washington Press.

Pallister, J.S., Scneider, D.J., Griswold, J.P. et al. (2012). Merapi 2010 eruption – Chronology and extrusion rates monitored with satellite radar and used in eruption forecasting: *Journal of Volcanology and Geothermal Research*, 261, 144–152.

Papale, P. (ed.) (2015). *Volcanic Hazards, Risks, and Disasters*. Amsterdam: Elsevier Publications. 505 pp.

Papoulis, A., and Pillai, S.U. (2002). *Probability, Random Variables, and Stochastic Processes*. Boston: McGraw-Hill.

Parfitt, E.A. (2004). "A discussion of the mechanics of explosive volcanic eruptions." *Journal of Volcanology and Geothermal Research*, 134, 77–107.

Parfitt, E.A. and Wilson, L. (2008). *Fundamentals of Physical Volcanology*. Oxford: Blackwell.

Park, R.A. and MacDiarmid, C.F. (1964). *Ore Deposits*. San Francisco: W.H. Freeman, 475 p.

Passega, R. (1964). "Grain size representation by CM plots as a geological tool." *Journal of Sedimentary Petrology*, 34, 830–847.

Patrick, M., Orr, T., Sutton, A. et al. (2016). "Shallowly driven fluctuations in lava lake outgassing (gas pistoning), Kīlauea Volcano." *Earth and Planetary Science Letters*, 4533, 326–338.

Paulick, H. and Franz, G. (1997). "The color of pumice: A case study on a trachytic fall deposit, Meidob volcanic field, Sudan." *Bulletin of Volcanology*, 59, 171–185.

Peale, S.J., Cassen, P., and Reynolds, R.T. (1979). "Melting of Io by Tidal Dissipation." *Science*, 203, 892–894.

Pease, R. (2019). "Ship spies largest submarine eruption ever." *Science Journal*, doi:10.1126/science.aay1175.

Pedoja, V., Arthemayou, C., Pinegina, T. et al. (2013). "'Arc-continent collision' of the Aleutian-Komandorsky arc into Kamchatka: Insight from Quaternary tectonic segmentation through Pleistocene marine terraces and morphometric analysis of fluvial drainage." *Tectonics*, 32, 827–842.

Perez, W., Kutterolf, S., Schmincke, H.-U. and Freunt, A. (2009). The masaya triple layer; A 2100 year old basaltic multi-episodic Plinian eruption from the Masaya caldera complex (Nicaragua), *Journal of Volcanology and Geothermal Research*, 179, 171–205.

Perfit, M.R., Cann, J.R., Fornari, J. et al. (2003). "Interaction of sea water and lava during submarine eruptions at mid-ocean ridges." *Nature*, 426, 62–65.

Pering, T., McGonigle, A., James, M. et al. (2017). "The dynamics of slug trains in volcanic conduits: Evidence for expansion driven slug coalescence." *Journal of Volcanology and Geothermal Research*, 348, 26–35.

Perret, F.A. (1950). *Volcanologic Observations*. Washington, DC: Carnegie Institution. Publication 549, 162 pp.

Perret, F.A. (1924). *The Vesuvius Eruption of 1906*. Washington, DC: Carnegie Institution.

Perret, F.A. (1935). *The Eruption of Mt. Pelée, 1929–1932*. Washington, DC: Carnegie Institution.

Peterson D.W. (1988). "Volcanic hazards and public response." *Journal of Geophysical Research*, 93B5, 4161–4170.

Peterson, D.W. (1996). "Mitigation measures and preparedness plans for volcanic emergencies." In: *Monitoring and Mitigation of Volcanic Hazards* (eds. R. Scarpa and R. I. Tilling), pp. 701–718. Berlin: Springer.

Peterson, D.W. and Tilling, R.I. (1980). "Transition of basaltic lava from pahoehoe to aa, Kilauea Volcano, Hawaii: Field observations and key factors." *Journal of Volcanology and Geothermal Research*, 7, 271–293.

Pfanz, H. (1999). *Mofetten kalter atem schlafender vulkane (Mofettes–cold breath of sleeping volcanoes)*. Deutsche Vulkanologissche Gesellschaft, Mendig, 85 pp.

Pfanz, H., Vodnik, D, Wittmann, C. et al. (2004). "Plants and geothermal CO2 exhalations: Survival in and adaptation to a high CO_2 environment." In: *Progress in Botany. Progress in Botany* (eds. K. Esser, U. Lüttge, W. Beyschlag, and J. Murata), Vol. 65. Berlin, Heidelberg: Springer.

Phillips, J. (1869). *Vesuvius*. Oxford: Clarendon Press, p. 355.

Phillips, J., Humphreys, M., Daniels, K. et al. (2013). "The formation of columnar joints produced by cooling in basalt at Staffa, Scotland." *Bulletin of Volcanology*, 75, article 715. doi: 10.1007/s00445-013-0715-4.

Pierson, T.C. (1995). "Flow characteristics of large, eruption-triggered debris flows at snow-clad volcanoes: Constraints for debris-flow models." *Journal of Volcanology and Geothermal Research*, 66, 283–294.

Pierson, T.C. (2005). "Hyperconcentrated flow: Transitional process between water flow and debris flow." In: *Debris-flow Hazards and Related Phenomena* (eds. J. Matthias and H. Oldrich), pp. 159–202. Berlin: Springer-Praxis.

Pierson, T.C. and Major, J.J. (2014). "Hydrogeomorphic effects of explosive volcanic eruptions on drainage basins." *Annual Review of Earth and Planetary Sciences*, 42, 469–507. doi 10.1146/annurev-earth-060313-054913.

Pierson, T.C., Daag, A.S., Delos, P.J. et al. (1996). "Flow and deposition of posteruption hot lahars on the east side of Mount Pinatubo, July-October 1991." In: *Fire and Mud: Eruptions and Lahars of Mount Pinatubo, Philippines* (eds. C.G. Newhall and R.S. Punongbayan), pp. 921–950. Quezon City, Philippine Institute of Volcanology and Seismology and University of Washington Press.

Pierson, T.C., Janda, R.J., Thouret, J.-C. et al. (1990). "Perturbation and melting of snow and ice by the 13 November, 1985 eruption of Nevado del Ruiz, Colombia and consequent mobilization, flow, and deposition of lahars." *Journal of Volcanology and Geothermal Research*, 41, 17–66.

Pierson, T.C., Janda, R.J., Umbal, J.V. et al. (1992). "Immediate and long-term hazards from lahars and excess sedimentation in rivers draining Mt. Pinatubo, Philippines." US Geological Survey Water Resource Investigations Report 92-4039, pp. 1–35.

Pinkerton, H., James, M., and Jones, A. (2002). "Surface temperature measurements of active lava flows on Kilauea volcano, Hawaii." *Journal of Volcanology and Geothermal Research*, 112, 159–176.

Pioli, L. and Rosi, M. (2005). "Rheomorphic structures in high-grade ignimbrite: The Nuraxi Tuff, Sulcic volcanic district (SW Sardinia, Italy)." *Journal of Volcanology and Geothermal Research*, 142, 11–28.

Pistolesi, M., Cioni, R., Rosi, M. et al. (2013). "Evidence for lahar-triggering mechanisms in complex stratigraphic sequences: The post-twelfth century eruptive activity of Cotopaxi Volcano, Ecuador." *Bulletin of Volcanology*, 75. doi 10.1007/s00445-013-0698-1.

Platt, U. and Stutz, J. (2008). *Differential Optical Absorption Spectroscopy: Principles and Applications*. New York: Springer.

Poland, M.P. and Anderson, K.R. (2019). "Partly cloudy with a chance of lava flows: Forecasting volcanic eruptions in the twenty-first century." *Journal of Geophysical Research*, 125 (1), e2018JB016974, https://doi.org/10.1029/2018JB016974.

Poland, M.P. and Anderson, K.R. (2019). "Partly cloudy with a chance of lava flows: Forecasting volcanic eruptions in the twenty-first century." *Journal of Geophysical Research*, 125 (1), e2018JB016974, https://doi.org/10.1029/2018JB016974.

Pollard, D.D. and Aydin, A. (1988). "Progress in understanding jointing over the past century." *Geological Society of America Bulletin*, 100, 1184–1204.

Ponomareva, V.V., Pezner, M.M., and Melekestsev, I.V. (1998). "Large debris avalanches and associated eruptions in the Holocene eruptive history of Shiveluch Volcano, Kamchatka, Russia." *Bulletin of Volcanology*, 59, 490–505.

Porco, C.C., Helfenstein, P., Thomas, P.C. et al. (2006). "Cassini observes the active south pole of Enceladus." *Science*, 311, 1393–1401.

Porter, S.C. (1987). "Pleistocene subglacial eruptions on Mauna Kea." Volcanism in Hawaii, The 1980 Eruptions of Mount St. Helens. US Geological Survey Professional Paper 1350, pp. 587–598.

Powell, J.W. (1896). *The Physiography of the United States*. New York: National Geographic Society/American Book Company.

Powers, H.A. and Wilcox, R.E. (1964). "Volcanic ash from Mt. Mazama (Crater Lake) and from Glacier Peak." *Science*, 144 (3624), 1334–1336.

Prueher, L.M. and Rea, D.K. (2001). "Tephrochronology of the Kamchatka-Kurile and Aleutian arcs: Evidence for volcanic episodicity." *Journal of Volcanology and Geothermal Research*, 106, 67–84.

Punongbayan, R.S. (eds.) (1997). *Fire and Mud: Eruptions and Lahars of Mount Pinatubo, Philippines*. Quezon City, Philippine Institute of Volcanology and Seismology and University of Washington Press, pp. 3–20.

Pyle, D.M. (1989). "The thickness, volume, and grainsize of tephra fall deposits." *Bulletin Volcanologique*, 51, 1–15.

Pyle, D.M. (1995). "Mass and energy budgets of explosive volcanic eruptions." *Geophysical Research Letters*, 22 (5), 563–566.

Pyle, D.M. and Elliott, J.R. (2006). "Quantitative morphology, recent evolution, and future activity of the Kameni Islands volcano, Santorini, Greece." *Geosphere*, 2 (5), 253–268.

Quick, J.E., Sinigoi, S., and Denlinger, R. (1992). "Large scale evolution of the underplated igneous complex of the Ivrea-Verbano Zone." US Geological Survey Circular 1089, 14–15.

Quick, L.C., Glaaze, L.S., Baloga, S.M., and Stofan, E.R. (2016). "New approaches to inferences for steep-sided domes on Venus." *Journal of Volcanology and Geothermal; Research*, 319, 93–105.

Rader, E. and Geist, D. (2015). "Eruption conditions of spatter deposits." *Journal of Volcanology and Geothermal Research*, 304, 287–293.

Rampino, M.R. and Ambrose, S.H. (2000). "Volcanic winter in the Garden of Eden: The Toba super-eruption and the late Pleistocene human population crash." In: *Volcanic Hazards and Disasters in Human Antiquity* (eds. F.W. McCoy and G. Heiken), pp. 71–82. Geological Society of America Special Paper, 345.

Rampino, M.R. and Self, S. (1984). "The atmospheric effects of El Chichón." *Scientific American*, 250 (1), 48–57.

Rampino, M.R. and Self, S. (1993). "Climate-volcanism feedback and the Toba eruption of $\sim$ 74,000 years ago." *Quaternary Research*, 269–280.

Rampino, M.R. and Strother, R.B. (1988). "Flood basalt volcanism during the past 250 million years." *Science*, 241, 663–668.

Rapp, G.R. and Hill, C.L. (2006). *Geoarchaeology: The Earth Science Approach to Archaeological Investigations.* New Haven: Yale University Press.

Ratte, J.C. and Steven, T.A. (1967). "Ash flows and related volcanic rocks associated with the Creede caldera, San Juan Mountains, Colorado." US Geological Survey Professional Paper 524-H, 58p.

Rea, D.K. and Vallier, T.L. (1983). "Two Cretaceous volcanic episodes in the western Pacific Ocean." *Geological Society of America Bulletin*, 94, 1430–1437.

Reiche, P. (1937). "The Toreva-Block: A distinctive landslide type." *Journal of Geology*, 45, 538–548.

Reid, J.B., Jr., Murray, D.P., Hermes, O.D. et al. (1993). "Fractional crystallization in granites in the Sierra Nevada: How important is it?" *Geology*, 21, 587–590.

Reidel, S.P., Camp, V.E., Tolan, T.L., and Martin, B.S. (2013). "The Columbia River flood basalt province: Stratigraphy, areal extent, volume, and physical volcanology." In: *The Columbia River Flood Basalt Province* (eds. S.P. Reidel, V.E. Camp, M.E. Ross et al.), pp. 1–44. The Geological Society of America Special Paper 497.

Reihle, J.R., Miller, T.F., and Bailey, R.A. (1995). "Cooling, degassing, and compaction of rhyolitic lava flows: A computational model." *Bulletin of Volcanology*, 57, 319–336.

Reynolds, H.L., Gudmundsson, M.T., Högnadóttir, T., and Pálsson, F. (2018). "Thermal power of Grímsvötn, Iceland, from 1998 to 2016: Quantifying the effects of volcanic activity and geothermal anomalies." *Journal of Volcanology and Geothermal Research*, 358, 184–193.

Ricci, T., Barberi, F., Davis, M.S. et al. (2013). "Volcanic risk perception in the Campi Flegrei area." *Journal of Volcanology and Geothermal Research*, 254, 118–130.

Richards, A.F. (1959). "Geology of Islas Revillagigedo, Mexico, 1: Birth and development of Volcan Barcena, Isla San Benedicto (1)." *Bulletin Volcanologique*, 73–123.

Riker, J., Cashman, K.V., Kauahikaua, J., and Montieth, C. (2009). The length of channelized lava flows: Insight from the 1959 eruption of Mauna Loa, *U.S. Geological Survey*, doi:org/10:1016/volgeores.2009.03.002

Rittmann, A. (1936). *Vulkane und ihre T tgkeit.* Stuttgart: Ferdinand Enke Verlag, 399p.

Rittmann, A. (1958). "Cenni sulle collate di ignimbrati." *Atti della Accademia Gioenia di Scienze Naturali in Catania*, 4, 524–533.

Rittmann, A. (1962). *Volcanoes and Their Activity.* New York: John Wiley & Sons.

River, E.P. and Harris, D.V. (1999). *Geology of U.S. Parklands.* New York: Wiley.

Robert, B., Harris, A., Gurioli, L. et al. (2014). "Textural and rheological evolution of basalt flowing down a lava channel." *Bulletin of Volcanology*, 76, article 824. doi: 10.1007/s00445-014-0824-8.

Robin, C., Eissen, J.P., and Monzier, M. (1993). "Giant tuff cone and 12-mile-wide associated caldera at Ambrym Volcano (Vanuatu, New Hebrides)." *Journal of Volcanology and Geothermal Research*, 225–238.

Robinson, J.E. and Eakins, B.W. (2006). "Calculated volumes of individual shield volcanoes at the young end of the Hawaiian Ridge." *Journal of Volcanology and Geothermal Research*, 151, 309–317.

Robock, A. and Oppenheimer, C. (2003). *Volcanism and the Earth's Atmosphere.* Washington, DC: American Geophysical Union.

Roche, S.L., Davis T.L., Benson, R.D. et al. (2001). "Dynamic reservoir characterization: Application of time-lapse (4-D), multicomponent seismic to a CO2 EOR project, Vacuum Field, New Mexico." Tulsa, American Association of Petroleum Geologists.

Rodolfo, K.S. (1995). *Pinatubo and the Politics of Lahar: Eruption and Aftermath, 1991.* University of Philippines Press, Quezon City, 370 pp.

Rogers, N. and Hawkesworth, C. (2000). "Composition of magmas." In: *Encyclopedia of Volcanoes* (ed. H. Sigurdsson), 115–131. New York: Academic Press.

Roggensack, K., Hervig, R.L., McKnight, S.B. et al. (1997). "Explosive basaltic volcanism from Cerro Negro Volcano: Influence of volatiles in eruptive style." *Science*, 277, 1639–1642.

Rohrer, R. (1965). *Base Surges and Cloud Formation: Project Pre-Schooner.* Berkeley, Lawrence Livermore Laboratory, University of California, p. 10.

Romagnoli, C., Kokelaar, P., Rossi, P.L. et al. (1993). "The submarine extension of Sciara del Fuoco feature (Stromboli Is.): Morphologic characterization." *Acta Volcanologica*, 3, 91–98.

Rose, W.I., Newhall, C.G., Bornhorst, T.J. et al. (1987). "Quaternary silicic pyroclastic deposits of Atitlan Caldera, Guatemala." *Journal of Volcanology and Geothermal Research*, 33, 57–80.

Rosenbaum, J.G. and Waitt, R.B., Jr. (1981). "Summary of eyewitness accounts of the May 18 eruption." Volcanism in Hawaii. US Geological Survey Professional Paper 1350, pp. 53–68.

Rosenmüller, J.C. and Tillesius, W.G. (1799). *Beschreibung merkwurdiger Holen, ein Beitrag zur physikalischen Beschreibung der Erde. Leipzig.* Breitkopf und Hartel, 294p.

Ross, C.S. and Smith, R.L. (1961). *Ash-flow Tuffs: Their Origin, Geologic Relations, and Identification.* Washington, DC: US Geological Survey.

Rothery, D.A. (2017). "Volcanism on Mercury." In: *Oxford Research Encyclopedia of Planetary Geology* (eds. Read, P.A. et al.). Oxford: Oxford University Press. doi. 10/1093/acrefore/9780190647926.

Rouwet, D., Constantinescu, R., and Sandri, L. (2017). "Deterministic versus probabilistic volcano monitoring: not 'or' but 'and'." In: *Volcano Unrest: From Science to Society* (eds. J. Gottsmann, J. Neuberg, and B. Scheu), 35–46. Berlin: Springer Publications.

Rowe, C., Aster, R., Kyle, P.R. et al. (2000). "Seismic and acoustic observations at Mount Erebus Volcano, Ross Island, Antarctica, 1994–1998." *Journal of Volcanology and Geothermal Research*, 20, 105–128.

Rowland, I.D. (2014). *From Pompeii: The Afterlife of a Roman Town.* Cambridge, MA; London: Belknap Press, 340 p.

Rowland, S. and Walker, G. (1990). "Pahoehoe and aa in Hawaii: volumetric flow rate controls the lava structure." *Bulletin of Volcanology*, 52, 615–628.

Rowland, S.K. and Munro, D.C. (1992). "The caldera of Volcan Fernandina: A remote sensing study of its structure and recent activity." *Bulletin Volcanologique*, 55, 97–109.

Rowland, S.K., Smith, G., and Mouginis-Mark, P.J. (1993). "Preliminary ERS-1 observations of Alaskan and Aleutian volcanoes." *Remote Sensing of Environment*, 48, 358–369.

Rowley, P., Kuntz, M., and Macleod, N. (1981). "Pyroclastic flow deposits." The 1980 Eruptions of Mount St. Helens, US Geological Survey Professional Paper 1250, pp. 489–512.

Rubin, K.H., Soule, S.A., Chadwick Jr., W.W. et al. (2012). "Volcanic eruptions in the deep sea." *Oceanography* 25 (1), 142–157. http://dx.doi.org/10.5670/oceanog.2012.12.

Russell, I.C. (1902). *Geology and Water Resources of the Snake River Plains of Idaho.* Washington, DC: US Geological Survey.

Russell, J.M., III, Luo, M., Cicerone, R.J. et al. (1996). "Satellite confirmation of the dominance of chlororfluorocarbons on the global stratospheric budget." *Nature*, 379, 526.

Rutherford, M.J. (2008). "Magma ascent rates." *Reviews in Mineralogy and Geochemistry*, 69, 241–271.

Ryan, M.P. (1987a). "Neutral buoyancy and the mechanical evolution of magmatic systems." In: *Magmatic Processes – Physicochemical Principles* (ed. B.O. Mysen). The Geochemical Society, Special Publication 1.

Ryan, M.P. (1987b). "Elasticity and contractancy of Hawaiian olivine tholeiite and its role in the stability and structural evolution of subcaldera magma reservoirs and rift systems." Volcanism in Hawaii. US Geological Survey Professional Paper 1350, pp. 1395–1447.

Ryan, M.P. (1988). "The mechanics and three-dimensional internal structure of active magmatic systems: Kilauea volcano, Hawaii." *Journal of Geophysical Research*, 93 (B5), 4213–4248.

Ryan, M.P. and Sammis, C.G. (1978). "Cyclic fracture mechanisms in cooling basalt." *Geological Society of America Bulletin*, 89, 1295–1308.

Ryan, M.P., Koyanagi, R.Y., and Fiske, R.S. (1981). "Modeling the three-dimensional structure of magma transport systems: Application to Kilauea Volcano, Hawaii." *Journal of Geophysical Research*, 86, 7111–7129.

Ryder, G. (2000). Heavy bombardment on the Earth ~3.85 Ga: The search for petrographic and geochemical evidence. In: *Origin of the Earth and the Moon* (eds. R.M. Canup and K. Righter), 475–492. Tucson, Arizona: University of Arizona Press.

Rymer, H. and Williams-Jones, G. (2000). "Volcanic eruption prediction: Magma chamber physics from gravity and deformation measurements." *Geophysical Research Letters*, 27 (16), 2389–2392.

Sable, J.E., Houghton, B.F., Wilson, C.J.N. et al. (2006). "Complex proximal sedimentation from Plinian plumes: The example of Tarawera 1886." *Bulletin of Volcanology*, 9, 89–103.

Sager, W., Zhang, J., Korenaga, J. et al. (2013). "An immense shield volcano within the Shatsky Rise oceanic plateau, northwest Pacific Ocean." *Nature Geoscience*, 6, 976–981. https://doi.org/10.1038/ngeo1934.

Sager, W.W., Huang, Y., Tominaga, M. et al. (2019). "Oceanic plateau formation by seafloor spreading implied by Tamu Massif magnetic anomalies." *Nature Geoscience*, 12, 661–666.

Sahagian, D. (1985). "Bubble migration and coalescence during the solidification of basaltic lava flows." *Journal of Geology*, 93, 205–211.

Sánchez, M.C., Sarrionandia, F., and Ibarguchi, J.I.G. (2014). "Post-depositional intrusion and extrusion through a scoria and spatter cone of fountain-fed nephelinite lavas (Las Herrerías volcano, Calatrava, Spain)." *Bulletin of Volcanology*, 76. doi 10.1007/s00445-014-0860-4.

Sandri, L., Jolly, G., Lindsay, J. et al. (2012). "Combining long- and short-term probabilistic volcanic hazard assessment with cost-benefit analysis to support decision making in a volcanic crisis from the Auckland Volcanic Field, New Zealand." *Bulletin of Volcanology*, 74, 705–723.

Sanford, A.R., Olsen, K.H., and Jashka, L.H. (1979). *Seismicity of the Rio Grande Rift: Tectonics and Magmatism.* Washington, DC: American Geophysical Union, pp. 145–168.

Sarna-Wojcicki, C., Champion, D.E., and Davis, J.O. (1983). "Holocene volcanism in the conterminous United States and the role of silicic volcanic ash layers in correlation of latest-Pleistocene and Holocene deposits." In: *Late-Quaternary Environments of the United States* (ed. H.E. Wright), pp. 52–77. Minneapolis: University of Minnesota Press.

Saunders, S.J. (2001). "The shallow plumbing system of Rabaul caldera: A partially intruded ring fault?" *Bulletin of Volcanology*, 63 (6), 406–420.

Scaillet, B., Pichavant, M., and Cioni, R. (2008). "Upward migration of Vesuvius Magma Chamber over the past 20,000 Years." *Nature*, 455, 216–220.

Scandone, R. (1990). "Chaotic collapse of calderas." *Journal of Volcanology and Geothermal Research*, 42, 282–302.

Scandone, R., Arganese, G., and Galdi, F. (1993). "The evaluation of volcanic risk in the Vesuvius area." *Journal of Volcanology and Geothermal Research*, 58, 5–25.

Scarpa, R. and Tilling, R.I. (eds.) (1996). *Monitoring and Mitigation of Volcano Hazards. Bulletin of Volcanology*, v. 69. Heidelberg: Springer-Verlag, pp. 701–718, 841.

Scarth, A. (2002). *La Catastrophe: The Eruption of Mount Pelee, the Worst Volcanic Eruption of the Twentieth Century.* New York: Oxford University Press, 221 pp.

Scarth, A. (2009). *Vesuvius: A Biography.* Princeton: Princeton University Press, 352p.

Scharrer, K., Spieler, O., Meyer, Ch. et al. (2008). "Imprints of sub-glacial volcanic activity on a glacier surface: SAR study of Katla volcano, Iceland." *Bulletin of Volcanology*, 70 (4), 495–506.

Schenk, D.M. and Williams, D.A. (2004). "Potential thermal erosion channel on Io." *Geophysical Research Letters*, 31. doi./10.1027/2004GL02137.

Schenk, P., Hargitai, H., Wilson, R. et al. (2001). "The mountains of Io: Global and geological perspectives from Voyager and Galileo." *Journal of Geophysical Research*, 106, 201–233.

Schirnicke, C., Van den Bogaard, P., and Schmincke, H.U. (1999). "Cone sheet formation and intrusive growth of an oceanic island: The Miocene Tejeda complex on Gran Canaria (Canary Islands)." *Geology*, 27, 207–210. doi: 10-1130/0091-7613(1999)027<0207.CSFAIG72-2CO:2.

Schmidt, B., Dueker, K., Humphreys, E., and Hansen, S. (2012). "Hot mantle upwelling across the 660 beneath Yellowstone." *Earth and Planetary Sciences Letters*, 331–332, 224–236.

Schmidt, P., Lund, B., Hieronymus, C. et al. (2013). Effects of pre-sent day deglaciation in Iceland on mantle melt pro-duction rates. *Journal of Geophysical Research*, 118, 3366–3379.

Schmidt, R. (1981). "Descriptive nomenclature and classification of pyroclastic deposits and fragments: Recommendations of the IUGS Subcommission on the Systematics of Igneous Rocks." *Geology*, 9 (1), 41–43.

Schmincke, H.-U. (1967). "Fused tuff and peperites in south-central Washington." *Geological Society of America Bulletin*, 78, 319–330.

Schmincke, H.-U. and Swanson, D.A. (1967). Laminar viscous flowage structures in ash-flow tuffs from Gran Canaria. *Journal of Geology*, 85, 641–644.

Schmincke, H.-U., Fisher, R.V., and Waters, A.C. (1973). "Antidune and chute and pool structures in the base surge deposits of the Laacher See area, Germany." *Sedimentology*, 20 (4), 553–574.

Scholl, D.W., Kirby, S.H., and Platt, J.P. (1996). *Subduction: Top to Bottom.* American Geophysical Union Monograph Series, 96, 384 p.

Schumacher, R. and Schmincke, H.-U. (1995). "Models for the origin of accretionary lapilli." *Bulletin of Volcanology*, 56, 626–639.

Scolamacchia, T. and Dingwell, D.B. (2014). "Sulfur as a binding agent of aggregates in explosive eruptions." *Bulletin of Volcanology*, 76, 871. doi 10.1007/s00445-0871-1.

Scolamacchia, T. and Dingwell, D.B. (2014). "Sulfur as a binding agent of aggregates in explosive eruptions." *Bulletin of Volcanology*, 76, 871. doi 10.1007/s00445-0871-1.

Scolamacchia, T. and Macías, J.L. (eds.) (2015). *Active Volcanoes of Chiapas (Mexico).* El Chichón and Tacaná: Springer Publishing, New York – Active Volcanoes of the World Series, 180 pp.

Scolamacchia, T. and Schouwenaars, R. (2009). "High-speed impacts by ash particles in the 1982 eruption of El Chichón, Mexico." *Journal of Geophysical Research*, 114, B12206. doi:10.1029/2008JB005848.

Scoppola, B., Boccaletti, D., Bevis, M. et al. (2006). "The westward drift of the lithosphere: A rotational drag?" *Bulletin of the Geological Society of America*, 118, 199–209.

Scott, W.E., Hoblitt, R.P., Torres, R.C. et al. (1996). "Pyroclastic flows of the June 15, 1991, climactic eruption of Mount Pinatubo." In: *Fire and Mud: Eruptions and Lahars of Mount Pinatubo, Philippines* (eds. C.G. Newhall and R.S. Punongbayan), pp. 545–570. Quezon City: Philippine Institute of Volcanology and Seismology and University of Washington Press.

Scrope, G.P. (1862). *Volcanoes: The Character of their Phenomena, their Share in the Structure and Composition of the Surface of the Globe, and their Relation to its Internal Forces with a Descriptive Catalogue of All Known Volcanoes and Volcanic Formations.* London: Longman, Green, Longmans, and Roberts.

Scrope, G.P. (1872). *Volcanoes: The Character of their Phenomena, their Share in the Structure and Composition of the Surface of the Globe, and their Relation to its Internal Forces with a Descriptive Catalogue of All Known Volcanoes and Volcanic Formations.* 2nd edition. London: Longmans, Green, Reader and Dyer.

Segerstrom, K. (1950). *Erosion Studies at Paricutín, State of Michoacan, Mexico.* Washington DC: US Geological Survey.

Self, S. (1992). "Krakatau revisited: The course of events and interpretation of the 1883 eruption." *Geojournal*, 28, 109–121.

Self, S. (2006). "The effects and consequences of very large explosive volcanic eruptions." *Philosophical Transactions of the Royal Society*, 364, 2073–2097.

Self, S. and Sparks, R.S.J. (1978). "Characteristics of widespread pyroclastic deposits formed by the interaction of silicic magma and water." *Bulletin Volcanologique*, 41, 196–212.

Shackley, M.S. (2005). *Obsidian: Geology and Archaeology in the North American Southwest.* Tucson: University of Arizona Press, 264 p.

Shalygin, E.V., Markiewicz, W.J., Basilevsky, A.T. et al. (2015). "Active volcanism on Venus in the Ganiki Chasma rift zone." *Geophysical News Letters*, 42 (12), 4762-4769. https://doi.org/10.1002/2015GL064088.

Shane, P. (2015). "Contrasting plagioclase textures and geochemistry in response to magma dynamics in an intra-caldera rhyolite system, Okataina volcano." *Journal of Volcanology and Geothermal Research*, 297, 1–10.

Shaw, H.P. (1972). "Viscosities of magmatic silicate liquids: An empirical method of prediction." *American Journal of Science*, 272 (9), 870–893.

Shea, T., Gurioli, L., and Houghton, B.F. (2012). "Transitions between fall phases and pyroclastic density currents during the AD 79 eruption at Vesuvius: building a transient conduit model from the textural and volatile record." *Bulletin of Volcanology*, 74, 2363–2381.

Sheets, P. (2015). "Volcanoes, ancient people, and their societies." Chapter 76 in *The Encyclopedia of Volcanoes* (eds. H. Sigurdsson et al.), 2nd edn, pp. 1313–1319. Amsterdam: Academic Press, 1421 pp. https://doi.org/10.1016/B978-0-12-385938-9.00076-6.

Sheets, P.D. (1979). "Environmental and cultural effects of the Illopango eruption in Central America." In: *Volcanic Activity and Human Ecology* (eds. P.D. Sheets and D.K. Grayson), pp. 525–564. New York: Academic Press.

Sheets, P.D. (2002). *Before the Volcano Erupted: The Ancient Cerén Village in Central America.* Austin: University of Texas Press.

Sheets, P.D. and Grayson, D.K. (1979). *Volcanic Activity and Human Ecology.* New York: Academic Press.

Sheridan, M.F. and Wang, Y. (2005). "Cooling and welding history of the Bishop Tuff in Adobe Canyon and Chidago Canyon, California." *Journal of Volcanology and Geothermal Research*, 142, 119–144.

Sheridan, M.F. (1971). Particle-size characteristics of pyroclastic tuffs. *Journal of Geophysical Research*, 76, 5627–5634.

Sherrod, D.R., Scott, W.E., and Stauffer, P.H. (eds.) (2008). *A volcano rekindled; The renewed eruption of Mount St. Helens, 2004–2006.* US Geological Survey Professional Paper 1750, 856p.

Sheth, H., Ray, J., Kumar, A. et al. (2011). "Toothpaste lava from the Barren Island volcano (Andaman Sea)." *Journal of Volcanology and Geothermal Research*, 202, 73–82.

Sheth, H.C. (2007). "Large Igneous Provinces (LIPs): Definition, recommended terminology, and a hierarchical classification." *Earth-Science Reviews*, 85 (3–4), 117–124.

Shimozuru, D. (1968). "Discussion on the energy partition of volcanic eruption." *Bulletin of Volcanology*, 30, 383–394.

Siebe, C. and Macías, J.L. (2006). "Volcanic hazards in the Mexico City metropolitan area from eruptions at Popocatépetl, Nevado de Toluca, and Jocotitlán stratovolcanoes and monogenetic scoria cones in the Sierra de Chichinautzin Volcanic Field." *Special Papers-Geological Society of America*, 402, 253–329.

Siebe, C., Salinas, S., Arana-Salinas, L. et al. (2017). "The ~23,500y 14C BP White Pumice Plinian eruption and associated debris avalanche and Tochimilco lava flow of Popocatépetl volcano, México." *Journal of Volcanology and Geothermal Research*, 333, 66–95.

Sigmundsson, F., Pinel, V., Lund, B. et al. (2010) "Climate effects on volcanism: influence on magmatic systems of loading and unloading from ice mass variations, with examples from Iceland." *Philosophical Transactions of the Royal Society* A.v. 368, 2519–2534. https://doi.org/10.1098/rsta.2010.0042

Sigurdsson, H. (1999). *Melting the Earth: The History of Ideas on Volcanic Eruptions.* Oxford: Oxford University Press, 260 pp.

Sigurdsson, H. (2000). *Encyclopedia of Volcanoes.* San Diego: Academic Press.

Sigurdsson, H., Carey, S.N., Espíndola, J.M. (1984). "The 1982 eruptions of El Chichón Volcano, Mexico: Stratigraphy of pyroclastic deposits." *Journal of Volcanology and Geothermal Research*, 23 (1–2), 11–37.

Sigurdsson, H., Carey, S.N., Fisher, R.V. (1987). "The 1982 eruptions of El Chichón volcano, Mexico (3): Physical properties of pyroclastic surges." *Bulletin of Volcanology*, 49 (2), 467–488.

Sigurdsson, H., Houghton, B., McNutt, S. et al. (2015). *The Encyclopedia of Volcanoes*, 2nd edition. Amsterdam: Elsevier Press.

Silva-Mora L. (1983). "La erupcion del Volcan Chichonal, Chiapas: Una particularidad del Vulcanismo en Mexico." In: *El Volcan Chichonal. Instituto de Geologia*, Universidad Nacional Autonoma de Mexico. Mexico D.F., pp. 23–35.

Simkin, T. (1972). "Origin of some flat-topped volcanoes and guyots." *Geological Society of America Memoir*, 132, 183–193.

Simkin, T. (1984). "Geology of Galapagos Islands." In: *Galapagos* (ed. R. Perry), pp. 14–41. Oxford: Pergamon Press.

Simkin, T. and Fiske, R.S. (1983). *Krakatau, 1883: The Volcanic Eruption and its Effects.* Washington: Smithsonian Inst. Press.

Simkin, T. and Howard, K.A. (1970). "Caldera collapse in the Galápagos Islands, 1968." *Science*, 169, 429–437.

Simkin, T. and Siebert, L. (1984). "Explosive eruptions in space and time: Durations, intervals, and a comparison of the world's active volcanic belts." In: *Explosive Volcanism: Inception, Evolution, and Hazards*, N. R. C. Geophysics Study Committee. Washington, DC: National Academy Press, pp. 110–121.

Simkin, T. and Siebert, L. (1994). *Volcanoes of the World*, 2nd edition. Tucson: Geoscience Press in association with the Smithsonian Institution Global Volcanism Program.

Simkin, T., Siebert, L., and Blong, R. (2001). "Volcano fatalities: Lessons from the historical record." *Science*, 291, 5502, 255.

Simkin, T., Siebert, L., McClelland, L. et al. (1981). *Volcanoes of the World: A Regional Directory, Gazetteer, and Chronology of Volcanism During the Last 10,000 Years.* Stroudsburg: Hutchinson Ross Publishing, 240p.

Simkin, T., Tilling, R.I., Under, J.D. et al. (2006). "This dynamic planet; World map of volcanoes, earthquakes, impact craters, and plate tectonics." I Map 2800, US Geological Survey Geologic Investigations Series.

Singh, S.C., Crawford, W.C., Carton, C. et al. (2006). "Discovery of a magma chamber and faults beneath a Mid-Atlantic Ridge hydrothermal field." *Nature*, 442, 1029–1032.

Skidmore and Toya (2002). "Do natural disasters promote long-run growth." *Economic Inquiry*, 40 (4), 664–687.

Smellie, J.L. (2002). "The 1969 subglacial eruption on Deception Island (Antarctica): Events and processes during an eruption beneath a thin glacier and implications for volcanic hazards." In: *Volcano–Ice Interaction on Earth and Mars* (eds. J.L. Smellie and M.G. Chapman), p. 431. London: Geological Society of London. Special Publications.

Smith, A.L. and Roobal, M.J. (1982). "Andesitic pyroclastic flows." In: *Orogenic Andesites* (ed. R.S. Thorpe), pp. 415–433. New York: Wiley.

Smith, C.M., McNutt, S.R., and Thompson, G. (2016). "Ground-coupled airwaves at Pavlov Volcano, Alaska, and their potential for eruption monitoring." *Bulletin of Volcanology*, 78. doi 10.1007/s00445-016-1045-0.

Smith, D.R. and Leeman, W.P. (1987). "Petrogenesis of St. Helen's dacitic magmas." *Journal of Geophysical Research*, 92, 10,313–10,334.

Smith, G.A. and Fritz, W.J. (1989). "Volcanic influences on terrestrial sedimentation." *Geology*, 17, 375–376.

Smith, L.M., Barth, J.A., Kelley, D.S. et al. (2018). "The Ocean Observatories Initiative." *Oceanography*, 31 (1), 16–35. https://doi.org/10.5670/oceanog.2018.105.

Smith, R.L. (1960). "Ash flows." *GSA Bulletin*, 71 (6), 795–841.

Smith, R.L. (1979). "Ash-flow magmatism." In: *Ash-flow Tuffs* (eds. C.E. Chapin and W.E. Elston), pp. 5–27. Geological Society of America Special Paper, 180.

Smith, R.L., Friedman, I.I., and Long, W.D. (1958). "Welded tuffs, Expt. 1." *American Geophysical Union Transactions*, 39, 352–353.

Sobradelo, R. and Martí, J. (2015). "Short-term volcanic hazard assessment through Bayesian inference: Retrospective application to the Pinatubo 1991 volcanic crisis." *Journal of Volcanology and Geothermal Research*, 290, 1–11.

Soloman, S.C., Smreckar, S.E., Duane, L. et al. (1992). "Venus tectonics: An overview of Magellan observations." *Journal of Geophysical Research*, 97 (13), 199–256.

Sorey, M.L., Evans, W.C., Kennedy, B.M. et al. (1998). "Carbon dioxide and helium emissions from a reservoir of magmatic gas beneath Mammoth Mountain, California." *Journal of Geophysical Research*, 103, 15,303–15,323.

Soriano, C., Zafrilla, S., Marti, J. et al. (2002). "Welding and rheomorphism of phonolitic fallout deposits from Las Canadas caldera, Tenerife, Canary Islands." *Geological Society of America Bulletin*, 114, 883–895.

Sosa-Ceballos, G., Macías, J.L., García-Tenorio, F. et al. (2015). El Ventorrillo, a paleostructure of Popocatépetl volcano: Insights from geochronology and geochemistry. *Bulletin of Volcanology*, 77, 91.

Sottili, G., Palladino, D.M., and Zanon, V. (2004). "Plinian activity during the early eruptive history of the Sabatini Volcanic District, Central Italy." *Journal of Volcanology and Geothermal Research*, 135, 361–379.

Soule, S.A., Fornari D.J., Perfit, M.R. et al. (2007). "New insights into mid-ocean ridge volcanic processes from the 2005–2006 eruption of the East Pacific Rise, 9°46'N–9°56'N." *Geology*, 35 (12), 1079–1082.

Souther, J.G. (1990). "Volcano tectonics of Canada." In: *Volcanoes of North America: United States and Canada* (eds. C.A. Wood and J. Kienle), pp. 111–145. Cambridge: Cambridge University Press.

Sparks, R.S.J., Annen, C., Blundy, J.D. et al. (2019). "Formation and dynamics of magma reservoirs." *Philosophical Transactions of the Royal Society A: Mathematical, Physical and Engineering Sciences.* 377, 1–30, 20180019. https://doi.org/10.1098/rsta.2018.0019

Sparks, R.S.J. (1978). "The dynamics of bubble formation and growth in magmas: A review and analysis." *Journal of Volcanology and Geothermal Research*, 3, 1–37.

Sparks, R.S.J. (1986). "The dimensions and dynamics of volcanic eruption columns." *Bulletin Volcanologique*, 48, 3–15.

Sparks, R.S.J. (1988). "Petrology and geochemistry of the Loch Ba ring-dyke, Mull (N.W. Scotland): An example of the extreme differentiation of tholeiitic melts." *Contributions to Petrology and Mineralogy*, 100, 446–461.

Sparks, R.S.J. and Walker, G.P.L. (1973). "The ground surge deposit: A third type of pyroclastic rock." *Nature*, 241, 62–64.

Sparks, R.S.J. and Walker, G.P.L. (1977). The significance of vitric-enriched air fall ashes associated with crystal-enriched ignimbrites. *Journal of Volcanology and Geothermal Research*, 2, 329–341.

Sparks, R.S.J. and Wilson, L. (1976). "A model for the formation of ignimbrite by gravitational collapse." *Journal of the Geological Society of London*, 132, 441–451.

Sparks, R.S.J. and Wright, J.V. (1979). "Welded air-fall tuffs." In: *Ash-flow Tuffs* (eds. C.E. Chapin and W.E. Elston), pp. 155–166. Geological Society of America Special Paper 180.

Sparks, R.S.J. and Wilson, L. (1982). Explosive volcanic eruptions – V. Observations of plume dynamics during the 1979 Soufriére eruption, St. Vincent. *Geophysical Journal International*, 69, 551–570.

Sparks, R.S.J., Bursik, M.I., Ablay, G. et al. (1992). Sedimentation of tephra by volcanic plumes: Part 2 – Controls on thickness and grain-size variations of tephra fall deposits, *Bulletin of Volcanology*, 54, 685–695.

Sparks, R.S.J., Bursik, M.I., Carey, S.N. et al. (1997). *Volcanic Plumes*. New York, John Wiley & Sons.

Sparks, R.S.J., Moore, J., and Rice, C.J. (1986). "The initial giant umbrella cloud of the May 18th, 1980 explosive eruption of Mount St. Helens." *Journal of Volcanology and Geothermal Research*, 28, 257–274.

Sparks, R.S.J., Self, S., Grattan, J.P. et al. (2005). "Super-eruptions: Global effects and future threats." *Geological Society of London*, 28.

Sparks, R.S.J., Wilson, L., and Hulme, G. (1978). "Theoretical modeling of the generation, movement, and emplacement of pyroclastic flows by column collapse." *Journal of Geophysical Research Solid Earth*, 83, 1727–1739.

Spence, R.J.S., Pomonis, A., Naxter, P.J. et al. (1997). "Building damage caused by the Mount Pinatubo eruption of June 15." In: *Fire and Mud: Eruptions and Lahars of Mount Pinatubo* (eds. R.S. Philippines, C.G. Newhall, and R.S. Punongbayan), pp. 1055–1061. Quezon City, Philippine Institute of Volcanology and Seismology and University of Washington Press.

Spencer, J.R., Pearl, J.C., Segura, M. et al. (2006). "Cassini encounters Enceladus: Background and discovery of a south polar hot spot." *Science*, 311, 1401–1405.

Spudis, P.D. (2000). "Volcanism on the Moon." In: *Encyclopedia of Volcanoes* (ed. H. Sigurdsson), pp. 697–708. San Diego: Academic Press.

Spudis, P.D., McGovern, P.J., and Kiefer, W.S. (2013). "Large shield volcanoes on the Moon." *Journal of Geophysical Research: Planets*, 118, 1063–1081.

Stearns, H.T. (1925). "The explosive phase of Kilauea volcano, Hawaii, in 1924." *Bulletin Volcanologique*, 12 (5), 193–208.

Stearns, H.T. (1983). *Memoirs of a Geologist: From Poverty Peak to Piggery Gulch*. Honolulu: Hawaii Institute of Geophysics, 242 pp.

Stelling, P., Gardner, J.E. and Beget, J. (2005). "Eruptive history of Fisher Caldera, Alaska, USA." *Journal of Volcanology and Geothermal Research*, 139 (3–4), 163–183.

Stelten, M.E., Champion, D.E., and Kuntz, M.A. (2018). "The timing and origin of pre- and post-caldera volcanism associated with the Mesa Falls Tuff, Yellowstone Plateau volcanic field." *Journal of Volcanology and Geothermal Research*, 350, 47–60.

Stephens, C., Chouet, B.A., Page, R.A. et al. (1994). "Seismological aspects of the 1989–1990 eruptions at Redoubt Volcano, Alaska: The SSAM perspective." *Journal of Volcanology and Geothermal Research*, 62, 153–182.

Steven, T.A. and Lipman, P.W. (1976). "Calderas of the San Juan volcanic field, southwestern Colorado." US Geological Survey Professional Paper, 958, 35.

Stevenson, D.S. and Blake, S. (1998). "Modelling the dynamics and thermodynamics of volcanic degassing." *Bulletin Volcanologique*, 60, 307–317.

Stofan, E.R., Sharpton, V.L., Schubert, G. et al. (1992). "Global distribution and characteristics of coronae and related features on Venus: Implications for origin and relation to mantle processes." *Journal of Geophysical Research*, 97, 13,347–13,378.

Stoiber, R.E. and Rose, W.I., Jr. (1970). "Geochemistry of Central American volcanic gas condensates." *Geological Society of America Bulletin*, 81, 2891–2912.

Stormer, J.C.J. and Carmichael, I.S.E. (1970). "The Kudo-Weill plagioclase geothermometer and porphyritic acid glasses." *Contributions to Mineralogy and Petrology*, 28, 306–309.

Strom, R.G., Schaber, G.G., and Dawson, D.D. (1994). "The global resurfacing of Venus." *Journal of Geophysical Research*, 99, 10,899–10,926.

Strothers, R.B. (1984). "The Great Tambora eruption of 1815 and its aftermath." *Science*, 224, 1191–1198.

Sumner, J.M. (1998). "Formation of clastogenic lava flows during fissure eruption and scoria cone collapse: The 1986 eruption of Izu-Oshima Volcano, eastern Japan." *Bulletin Volcanologique*, 60, 195–212.

Surono, Jousset, P., Pallister, J. et al. (2012). "The 2010 explosive eruption of Java's Merapi volcano: A '100-year' event." *Journal of Volcanology and Geothermal Research*, 241, 121–135. doi:10.1016/j.jvolgeores.2012.06.018.

Sutton, A.J., Elias, T., Gerlach, T.M. et al. (2001). "Implications for eruptive processes as indicated by sulfur dioxide emissions from Kilauea Volcano, Hawaii, 1979–1997." *Journal of Volcanology and Geothermal Research*, 108 (1–4), 283–302.

Suzuki, T. (1977). "Volcano types and their global population percentages (in Japanese with English abstract)." *Bulletin of the Volcanological Society of Japan*, 2nd series, 27–40.

Suzuki, Y.J., and Koyaguchi, T. (2012). "3-D numerical simulations of eruption column collapse: Effects of bent size on pressure-balanced jet/plumes." *Journal of Volcanology and Geothermal Research*, 221–222, 1–13.

Swanson, D.A. (1973). "Pahoehoe flows from the 1969–1971 Mauna Ulu eruption of Kilauea volcano, Hawaii." *Geology Society America Bulletin*, 84, 615–626.

Swanson, D.A. and Holland, R.T. (1990). "Regularities in growth of the Mount St. Helens dacite dome 1980–1986." In: *Lava Flows and Domes* (ed. J.H. Fink), pp. 3–24. New York: Springer Publishing.

Swanson, D.A. and Houghton, B.F. (2018). Products, processes, and implications of Keanakāko'i volcanism, Kīlauea Volcano, Hawai'i. In: *Field Volcanology: A Tribute to the Distinguished Career of Don Swanson* (eds. Poland, M.P., Garcia, M.O., Camp, V.E., and Grunder, A.). Geological Society of America, Vol. 538.

Swanson, D.A., Casadevall, T.J., Dzurisin, D. et al. (1982). "Forecasts and predictions of eruptive activity at Mt. St. Helens, USA." *Journal of Geodynamics*, 3, 397–423.

Swanson, D.A., Rose, T.R., Mucek, A.E. et al. (2014). "Cycles of explosive and effusive eruptions at Kīlauea Volcano, Hawai'i." *Geology*, 42, 631–634.

Swanson, D.A., Weaver, S.J., and Houghton, B.F. (2015). "Reconstructing the deadly eruptive events of 1790 CE at Kīlauea Volcano, Hawai'I." *Geological Society of America Bulletin*, 127, 503–515. https://doi.org/10.1130/B31116.1

Swanson, D.A., Wright, T.L., and Helz, R.T. (1975). "Linear vent systems and estimated rates of magma production and eruption for the Yakima Basalt on the Columbia Plateau." *American Journal of Science*, 275, 877–905.

Swanson, D.A., Wright, T.L., and Hooper, P.R. (1979). "Revisions in stratigraphic nomenclature of the Columbia River Basalt Group." Washington, DC. *US Geological Survey Bulletin*, 1457-G, 59.

Symonds, G.J. (1888). *The Eruption of Krakatoa and Subsequent Phenomena*. London: Royal Society.

Symonds, R.B., Rose, W., Bluth, G.J.S. et al. (1994). "Volcanic gas studies; methods, results and applications." *Reviews in Mineralogy and Geochemistry*, 30, 1–66.

Takada, A. (1988). "Subvolcanic Structure of the Central Dike Swarm Associated with the Ring Complexes in the Shitara district, Central Japan." *Bulletin of Volcanology*, 50, 106–118.

Takahashi, J. (2003). *The Pu'u O'o-Kupaianaha Eruption of Kilauea Volcano*. Hawaii: The First 20 Years. US Geological Survey, Professional Paper, 1676, pp. 63–88.

Taran Y., Fischer T.P., Pokrovsky B. et al. (1998). "Geochemistry of the volcano-hydrothermal system of El Chichón Volcano, Chiapas, Mexico." *Bulletin of Volcanology*. 59, 436–449.

Tarduno, J.A., Duncan, R.A., Scholl, D.W. et al. (2003). "The Emperor Seamounts: southward motion of the Hawaiian hotspot plume in Earth's mantle." *Science*, 301, 1064–1069.

Taylor, B. (2006). "The single largest oceanic plateau: Ontong Java-Manahiki-Hikurangi." *Earth and Planetary Science Letters*, 241, 372–380.

Taylor, G.A. (1958). "The 1951 eruption of Mt. Lamington, Papua." *Australia, Bureau of Mineral Resources Geology and Geophysics. Bulletin*, 38, 1–117.

Tazieff, H. (1952). *Craters of Fire*. New York: Harper and Brothers, 239 pp.

Tazieff, H. (1970). "The Afar Triangle." *Scientific American*, 222 (2), 32–40.

Tazieff, H. (1977). "An exceptional eruption; Mt. Niragongo, Jan. 10th, 1977." *Bulletin of Volcanology*, 40 (4), 189–200.

Teague, A.J., Seward, T.M., and Harrison, D. (2008). "Mantle sources for Oldoinyo Lengai carbonatites: Evidence from helium isotopes in fumarole gases." *Journal of Volcanology and Geothermal Research*, 175, 366–390.

Telling, J. and Dufek, J. (2012). "An experimental evaluation of ash aggregation in explosive volcanic eruptions." *Journal of Volcanology and Geothermal Research*, 209–210, 1–8.

Tepley, F.J., III, Davidson, J.R., Tilling, R.I. et al. (2000). "Magma mixing, recharge, and eruption histories recorded in plagioclase phenocrysts from El Chichón volcano, Mexico." *Journal of Petrology*, 41, 1397–1411.

Textor, C., Graf, H.-F., Timmreck, C. et al. (2004). "Emissions from volcanoes." In: C. Granier, P. Artaxo, and C.E. Reeves (eds.), "Emissions of Atmospheric Trace Compounds," *Advances in Global Change Research*, 18, 269–303.

Thomas, R.J., and Rothery, D.A. (2019). "Volcanism on Mercury." *Elements Journal*, 15, 27–32.

Thomas, R.M.E. and Sparks, R.S.J. (1992). "Cooling of tephra during fallout from eruption columns." *Bulletin Volcanologique*, 54, 542–553.

Thorarinsson, S. (1954). "The eruption of Hekla, 1947–1948, II.3, The tephra-fall from Hekla on March 29th 1947." *Societas Scientiarum Islandica*, Reykjavik, 68p.

Thorarinsson, S. (1967). "The Surtsey eruption and related scientific work." *Polar Record*, 13 86, 571–578.

Thorarinsson, S. and Sigvaldason, G.E. (1962). "The eruption in Askja, 1961: A preliminary report." *American Journal of Science*, 260, 641–651.

Thordarson, T. (2000). "Physical volcanology of Surtsey Island: A preliminary report." *Surtsey Research*, 11, 109–126.

Thordarson, T. and Self, S. (1993). "The Laki (Skaftár Fires) and Grímsvötn eruptions in 1783–1785." *Bulletin of Volcanology*, 55, 233–263.

Thordarson, T. and Self, S. (1998). "The Roza Member, Columbia River Basalt Group: A gigantic pahoehoe lava flow field formed by endogenous processes?" *Journal of Geophysical Research*, 103, 27,411–27,445.

Thordarson, T., Miller, D.J., Larsen, G. et al. (2001). "New estimates of sulfur degassing and atmospheric mass-loading by the 934 AD Eldgja eruption, Iceland." *Journal of Volcanology and Geothermal Research*, 108, 33–54.

Thordarson, T., Self, S., Larsen, G. et al. (eds.) (2009). *Studies in Volcanology: The Legacy of George Walker*. London: Geological Society of London, pp. 249–266.

Thornber, C.R. (2003). Magma reservoir processes revealed by the geochemistry of th on-going Pu'u O'o-Kupaianaha eruption In *The Pu'u O'o-Kupaianaha eruption of Kilauea Volcano, Hawaai'i: The First 20 Years* (eds. C. Heliker, D.A. Swansonand, and T.J. Takahashi), pp. 121–136. USGS Professional Paper 1676. Washington, DC.

Thouret, J-C. and Chester, D.K. (2005). Volcanic Landforms, Processes and Hazards. International Association of Geomorphology (IAG) Working Group on Volcanic Geomorphology. *Zeitschrift für Geomorphologie, Supplementbände*, 140, 231p.

Tibaldi, A. (2001). "Multiple sector collapses of Stromboli Volcano, Italy: How they work." *Bulletin Volcanologique*, 63, 112–125.

Tilling, R.I. (1989). "Volcanic hazards and their mitigation." *Progress and problems: Reviews of Geophysics*, 27 (2), 237–269.

Tilling, R.I. (2005). "Volcano hazards." In: *Volcanoes and the Environment* (eds. J. Martí and G. Ernst), pp. 56–89. Cambridge: Cambridge University Press.

Tilling R.I. (2009). "El Chichón "surprise" eruption in 1982: Lessons for reducing volcano risk." *Geofísica Internacional*, 40 (1), 3–19.

Tilling, R.I., Christiansen, R.L., Duffield, W.A. et al. (1987). "The 1972–1974 Mauna Ulu eruption, Kilauea Volcano: An example of quasi-steady-state magma transfer." In: *Volcanism in Hawaii* (eds. R.W. Decker, T.L. Wright and P.H. Stauffer), Vol. 1, pp. 405–469. US Geological Survey Professional Paper 1350.

Tilling, R.I., Topinka, L., and Swanson, D. (1990). *Eruptions of Mount St. Helens; Past, Present, and Future*. US Geological Survey General Interest Publication, 56p.

Timms, N.E., Erickson, T.M., Zanetti, M.R. et al. (2017). "Cubic zirconia >2730°C impact melt records Earth's hottest crust." *Earth and Planetary Science Newsletter*, 477, 52–58.

Todde, A., Cioni, R., Pistolesi, M. et al. (2017). "The 1914 Taisho eruption of Sakurajima volcano: stratigraphy and dynamics of the largest explosive event in Japan in the twentieth century." *Bulletin of Volcanology*, 79, 72. doi 10.1007/s00445-017-1154-4.

Tolan, T.L., Reidel, S.P., Beeson, M.H. et al. (1989). "Revisions to the estimates of the areal extent and volume of the Columbia River Basalt Group." In: *Volcanism and Tectonism in the Columbia River Flood-Basalt Province* (eds. S.P. Reidel and P.R. Hooper), pp. 1–20. Geological Society of America Special Paper, 239.

Tomkeieff, S. (1940). "The basalt flows of the Giant's Causeway district of Northern Ireland." *Bulletin Volcanologique*, 6, 89–143.

Tootell, B. (1985). *All Four Engines Have Failed: The True and Triumphant Story of Flight BA009 and the Jakarta Incident*. London: Andre Deutsch Press.

Torres, R.C., Self, S. and Martinez, M.L. (1996). "Secondary pyroclastic flows from the June 15, 1991 ignimbrite of Mount Pinatubo." In *Fire and Mud: Eruptions and Lahars of Mount Pinatubo* (eds. C.G. Newhall and R.S. Punongbayan), pp. 665–680. Philippines: Philippine Institute of Volcanology and Seismology (Quezon City) and the University of Washington Press (Seattle), 1126 pp.

Torsvik, T.H., Doubrovine, P.V., Steinberger, B. et al. (2017). "Pacific plate motion change caused the Hawaiian-Emperor Bend." *Nature Communications*, 8, 15660. doi: 10.1038/ncomms15660.

Townsend, M., Pollard, D., Johnson, K., and Culha, C. (2015). "Jointing around magmatic dikes as a precursor to the development of volcanic plugs." *Bulletin of Volcanology*, 77, article 92. doi: 10.1007/s00445-015-0978-z.

Trial, A.F. and Spera, F.J. (1990). "Mechanisms for the generation of compositional heterogeneities in magma chambers." *Geological Society of America Bulletin*, 353–367.

Troil, U.V. (1779). *Briefe welcheeine von Herrn Dr. Uno von Troil im jahr 1772 nack Island angestellte Reise betreffen*. Magnus Swederus, Upsala nd Leipzig, 342 pp.

Trusdell, F.A. and Lockwood, J.P. (2009). "Geologic Map of the Northeast Flank of Mauna Loa Volcano, Island of Hawaii, Hawaii." *Scientific Investigations Map 2932-A*, Washington, DC, US Geological Survey.

Trusdell, F.A., Hungerford, J.D.G., Stone, J.O. et al. (2018). "Explosive eruptions at the summit of Mauna Loa: Lithology, modeling, and dating." In: *Field Volcanology: A Tribute to the Distinguished Career of Don Swanson* (eds. Poland, M.P., Garcia, M.O., Camp, V.E., and Grunder, A.), 325–349. Geological Society of America: Special Paper 538. https://doi.org/10.1130/2018.2538(15).

Tsekhmistrenko, M., Sigloch, K., Hosseini, K. et al. (2021). A tree of Indo-African mantle plumes imaged by seismic tomography. *Nature Geoscience*, 14, 612–619 https://doi.org/10.1038/s41561-021-00762-9

Tsuya, H. (1955). "Geological and petrological studies of volcano Fuji. 5. On the 1707 eruption of volcano Fuji." *Bulletin of the Earthquake Research Institute of Japan*, 33, 341–383.

Tsuya, H. and Morimoto, R. (1963). "Types of volcanic eruptions in Japan." *Bulletin Volcanologique*, 26, 209–222.

Turner, J.S. (1986). "Turbulent entrainment: The development of the entrainment assumption and its application to geophysical flows." *Journal of Fluid Mechanics*, 173, 431–471.

Turner, M.B., Cronin, S.J., Bebbington, M.S. et al. (2008). "Developing probabilistic eruption forecasts for dormant volcanoes: A case study from Mt Taranaki, New Zealand." *Bulletin of Volcanology*, 70 (4), 507–515.

Turner, M.B., Reagan, M.K., Turner, S.P. et al. (2013a). "Timescales of magma degassing: Insights from U-series disequilibria, Mount Cameroon, West Africa." *Journal of Volcanology and Geothermal Research*, 262, 38–46.

Turner, S.J., Izbekov, P., and Langmuir, C. (2013b). "The magma plumbing system of Bezymianny Volcano: Insights from a 54 year time series of trace element whole-rock geochemistry and amphibole compositions." *Journal of Volcanology and Geothermal Research*, 263, 108–121.

Uesawa, S. (2014). "A study of the Taisho lahar generated by the 1926 eruption of Tokachidake Volcano, central Hokkaido, Japan, and implications for the generation of cohesive lahars." *Journal of Volcanology and Geothermal Research*, 270, 23–34.

Ugolini, F.C. and Zasoski, R.J. (1979). "Soils derived from tephra." In: *Volcanism and Human Habitation* (eds. D. Sheets, and D.K. Grayson), pp. 83–124. New York: Academic Press.

Ui, T. (1983). "Volcanic dry avalanche deposits: Identification and comparison with nonvolcanic debris stream deposits." *Journal of Volcanology and Geothermal Research*, 18 (1), 135–50.

Ulrich, G.E. (1987). "SP Mountain cinder cone and lava flow, northern Arizona." In: *Centennial Field Guide* (ed. S.S. Beus), Vol. 2. Rocky Mountain Section of Geological Society of America Annual Meeting, Flagstaff, Arizona.

Ulrich, K. and Sindern, S. (1998). "Nd and Sr isotope signatures of fenites from Oldoinyo Lengai, Tanzania, and the genetic relationships between nephelinites, phonolites, and carbonatites." *Journal of Petrology*, 39, 1997–2004.

Urbanski, N.-A., Hort, M., and Schmincke, H.-U. (2003). "Changing eruption dynamics during the Plinian phase of the Minoan eruption, Santorini, Greece (3500 yr BP): Insight from stratigraphic variation of fall deposit and pumice texture." *Geophysical Research Abstracts*, Nice, France.

Usamah, M., and Haynes, K. (2012). "An examination of the resettlement program at Mayon Volcano: What can we learn for sustainable volcanic risk reduction?" *Bulletin of Volcanology*, 74, 839–859.

Valderrama, P., Roche, O., Samaniego, P. et al. (2016). Dynamic implications of ridges on a debris avalanche deposit at Tutupaca volcano, southern Peru." *Bulletin of Volcanology*, 78. doi 10.1007/s00445-016-1011-x.

Valentine, G.A. (1987). "Stratified flow in pyroclastic surges." *Bulletin of Volcanology*, 49, 616–630.

Valentine, G.A. (1998). "Eruption column physics." In: *From Tephra to Magma: Modelling Physical Processes of Explosive Volcanic Eruptions* (eds. A. Freundt and M. Rosi), pp. 91–138. North Holland: Elsevier.

Valentine, G.A. and Fisher, R.V. (1986). "Origin of layer 1 deposits in ignimbrites." *Geology*, 14, 146–148.

Valentine, G.A., Wohletz, K.H., and Kieffer, S.W. (1992). "Effects of topography on facies and compositional zonation in caldera-related ignimbrites." *Geological Society of America Bulletin*, 104, 154–165.

Vallance, J.W., Siebert, L., Rose, W.I., Jr. et al. (1995). "Edifice collapse and related hazards in Guatemala." *Journal of Volcanology and Geothermal Research*, 66 (1–4), 337–355.

Van der Meer, D.G., Van Hinserberg, D.J.J., and Spakman, W.Q. (2018). "Atlas of the Underworld: Slab remnants in the mantle, their sinking, history, and a new outlook on lower mantle viscosity." *Tectonophysics*, 723, 309–448.

Van der Pluijm, B. and Marshak, S. (2003). *Earth Structure: An Introduction to Structural Geology and Tectonics.* New York: W. W. Norton, 672 p.

van Otterloo, J., Cas, R.A.F., and Sheard, M.J. (2013). "Eruption processes and deposit characteristics at the monogenetic Mt. Gambier Volcanic Complex, SE Australia: implications for alternating magmatic and phreatomagmatic activity." *Bulletin of Volcanology*, 75. doi 10.1007/s00445-013-0737-y.

van Wyk de Vries, B., Self, S., Francis, P.W. et al. (2001). "A gravitational spreading origin for the Socompa debris avalanche." *Journal of Volcanology and Geothermal Research*, 105, 225–247.

Varley, N., Connor, C.B. and Komorowski, J.C. (eds.) (2019). *Volcán de Colima: Portrait of a Persistently Hazardous Volcano.* Springer, Berlin. ISBN 978-3-642-25911-1 (eBook). https://doi.org/10.1007/978-3-642-25911-1

Vásquez, R., Capra, L., Caballero, L. et al. (2014). "The anatomy of a lahar: Deciphering the 15th September 2012 lahar at Volcán de Colima, Mexico." *Journal of Volcanology and Geothermal Research*, 272, 126–136.

Veeder, G.J., Matson, D.L., Johnston, T.V. et al. (1994). "Io's heat flow from infrared radiometry: 1983–1993." *Journal of Geophysical Research, Planets*, 99, 17,095–17,162.

Verbeek, R.D.M. (1885). *Krakatau. Batavia, Dutch East Indies, Landsdrukkerij.* [Republished in 1886 as Krakatau. Batavia, Dutch East Indies, Imprimerie de l'Etat.]

Verdel, C.V., Wernicke, B.P., Hassanzadeh, J., and Guest, B. (2011). "A Paleogene arc flare-up in Iran." *Tectonics*, 30 (3). doi:10.1029/2D10TC002809.

Verhoogen, J. (1980). *Energetics of the Earth.* Washington, DC: National Academy of Sciences.

Verplank, E.P. and Duncan, R.A. (1987). "Temporal variations in plate convergence and eruption rates in the western cascades." *Tectonics*, 6, 197–209.

Vicenzi, E.P., McBirney, A.R., White, W.M. et al. (1990). "The geology and geochemistry of Isla Marchena, Galapagos Archipelago: An ocean island adjacent to a mid-ocean ridge." *Journal of Volcanology and Geothermal Research*, 40, 291–315.

Vidal, C.M. Métrich, N., Komorowsji, J.-C. et al. (2014). "The 1257 Samalas eruption (Lombok, Indonesia): the single greatest stratospheric gas release of the Common Era." *Scientific Reports*, 6, 34868; doi: 10.1038/srep34868 (2016).

Viljoen, M.J. (1999). "The nature and origin of the Merensky Reef of the western Bushveld Complex based on geological facies and geological data." *South African Journal of Geology*, 102 (3), 221–239.

Violette, S., de Marsily, G., Carbonnel, J.P. et al. (2001). "Can rainfall trigger volcanic eruptions? A mechanical stress model of an active volcano: 'Piton de la Fournaise', Reunion Island." *Terra Nova*, 13 (1), 18–24.

Vitaliano, D.B. (1973). *Legends of the Earth.* Bloomington: Indiana University Press.

Vitousek, P.M., Aplet, G.H., Raich, J.W., and Lockwood, J.P. (1995). "Biological perspectives on Mauna Loa volcano: A model system for ecological research." *American Geophysical Union Geophysical Monograph*, 92, 117–126.

Vogfjörd, K.S., Jakobsdóttir, S.S., Gudmundsson, G.B. et al. (2005). "Forecasting and monitoring a subglacial eruption in Iceland." *American Geophysical Union–EOS*, 86, 245–248.

Voight, B. (1990). "The 1985 Nevado del Ruiz volcanic catastrophe: anatomy and retrospection." *Journal of Volcanology and Geothermal Research*, 44, 349–386.

Voight, B. (2000). Structural stability of andesite volcanoes and lava domes." *Philosophical Transactions of the Royal Society of London*, 358, 1663–1703.

Vye-Brown, C., Self, S., and Barry, T. (2013). "Architecture and emplacement of flood basalt flow fields: Case studies from the Columbia River Basalt Group, NW USA." *Bulletin of Volcanology*, 75, article 697. doi: 10.1007/s00445-013-0697-2.

Wada, K. and Aomine, S. (1973). "Soil development on volcanic material during the Quaternary." *Soil Science*, 116, 170–177.

Wadge, G. (1982). "Steady-state volcanism: Evidence from eruption histories of polygenetic volcanoes." *Journal of Volcanology and Geothermal Research*, 87, 4035–4049.

Wadge, G.B., Voight, B., Sparks, R.S.J. et al. (2014). An overview of the eruption of Soufriere Hills Volcano, Montserrat from 2000 to 2010. *Geological Society London Memoirs*, 39, 1–40.

Walker, G.P.L. (1967). "Thickness and viscosity of Etnean lavas." *Nature*, 213, 484–485.

Walker, G.P.L. (1971). Grain-size characteristics of pyroclastic deposits. *Journal of Geology*, 79, 696–714.

Walker, G.P.L. (1973). "Explosive volcanic eruptions: A new classification scheme." *Geologische Rundschau*, 62, 431–446.

Walker, G.P.L. (1980). "The Taupo Pumice: Product of the most powerful known (Ultraplinian) eruption?" *Journal of Volcanology and Geothermal Research*, 8 (1), 69–94.

Walker, G.P.L. (1981). Plinian eruptions and their products. *Bulletin of Volcanology*, 44, 223–240.

Walker, G.P.L. (1983). "Ignimbrite types and ignimbrite problems." *Journal of Volcanology and Geothermal Research*, 17, 65–68.

Walker, G.P.L. (1989). "Gravitational (density) controls on volcanism, magma chambers, and intrusions." *Australian Journal of Earth Sciences*, 36, 149–165.

Walker, G.P.L. (2009). "The endogenous growth of pahoehoe lava lobes and morphology of lava-rise edges." In *Studies in Volcanology: The Legacy of George Walker* (eds. T. Thordarson, S. Self, G. Larsen et al.), pp. 17–32. London: Geological Society of London.

Walker, G.P.L. and Croasdale, R. (1972). "Characteristics of some basaltic pyroclastics." *Bulletin of Volcanology*, 35, 303–317.

Walker, G.P.L., Wilson, C.J.N., and Froggart, P.C. (1980). "Fines-depleted ignimbrite in New Zealand: The product of a turbulent pyroclastic flow." *Geology*, 8, 245–249.

Wallace, P.J., Plank, T., Edmonds, M., and Hauri, E.H. (2015). "Volatiles in magmas." Chapter 7. In: *The Encyclopedia of Volcanoes* (eds. H. Sigurdsson et al.), 2nd edition, pp. 163–183. Amsterdam: Elsevier, 1421 pp.

Wallenstein, N., Duncan, A., Coutinho, R., and Chester, D. (2018). "Origin of the term nueés ardentes and the 1580 and 1808 eruptions on São Jorge Island, Azores." *Journal of Volcanology and Geothermal Research*, 358, 165–170.

Walter, T.R. and Troll, V.R. (2001). "Formation of caldera periphery faults: An experimental study." *Bulletin of Volcanology*, 63 (2–3), 191–203.

Ward, S.N. and Day, S. (2001). "Cumbre Vieja Volcano – Potential collapse and tsunami at La Palma, Canary Islands." *Geophysical Research Letter*, 78, 3397–3400. https://doi.org/10.1029/2001GL013110

Wardman, J.B., Wilson, T.M., Bodger, P.S. et al. (2012). "Potential impacts from tephra fall to electric power systems: a review and mitigation strategy." *Bulletin of Volcanology*, 74, 2221–2241.

Wark, D.A. and Watson, E.B. (2006). "TitaniQ: A titanium-in-quartz geothermometer." *Contributions to Mineralogy and Petrology*, 152, 743–754.

Wark, D.A., Hildreth, W., Spear, F.S. et al. (2007). "Pre-eruption recharge of the Bishop magma system." *Geology*, 35, 235–238.

Watanabe, K. (1986). "Size composition of the Tosu orange pumice flow deposit from the Aso caldera." *Bulletin of the Volcanological Society of Japan*, 31, 299.

Watson, E.B., Wark, D.A., and Thomas, J.B. (2006). "Crystallization thermometers for zircon and rutile." *Contributions to Mineralogy and Petrology*, 151, 413–433.

Watt, S.F.L., Gilbert, J.S., Folch, A. et al. (2015). "An example of enhanced tephra deposition driven by topographically induced atmospheric turbulence." *Bulletin of Volcanology*, 77, 35. doi 10.1007/s00445-015-0927-x.

Webb, P.K. and Weaver, S.D. (1975). "Trachyte shield volcanoes: A new volcanic form from south Turkana, Kenya." *Bulletin Volcanologique*, 39, 294–312.

Wellman, P. (1989). "Upper mantle, crust, and geophysical volcanology of Eastern Australia." In: *Intraplate Volcanism in Eastern Australia and New Zealand* (ed. R.W. Johnson), pp. 29–37. Cambridge: Cambridge University Press.

Westervelt, W.D. (1963). *Hawaiian Legends of Volcanoes*. Rutland, Vermont: Charles E. Tuttle.

Westgate, J.A. and Gorton, M.P. (1981). "Correlation techniques in tephra studies." In: *Tephra Studies* (eds. S. Self and R.S.J. Sparks), pp. 73–93. Dordrecht: Reidel.

White, J. and Ross, P.-S. (2011). "Maar-diatreme volcanoes: A review." *Journal of Volcanology and Geothermal Research*, 201, 1–29.

White, J.C. and Urbanczyk, K.M. (2001). "Origin of a silica-oversaturated quartz trachyte-rhyolite suite through combined crystal melting and fractional crystallization: The Leyva Canyon volcano, Trans-Pecos Magmatic Province, Texas." *Journal of Volcanology and Geothermal Research*, 111, 155–182.

White, J.D.L. (1996). "Impure coolants and interaction dynamics of phreatomagmatic eruptions." *Journal of Volcanology and Geothermal Research*, 71, 155–170.

White, J.D.L., Smellie, J.L., and Clague, D.A. (eds.) (2003). *Explosive Subaqueous Volcanism*. Washington, DC: American Geophysical Union.

White, R. and McCausland, W. (2016). "Volcano-tectonic earthquakes: A new tool for estimating intrusive volumes and forecasting eruptions." *Journal of Volcanology and Geothermal Research*, 309, 139–155. https://doi.org/10.1016/j.jvolgeores.2015.10.020

Wicks, C.W., Dzurisin, D., and Lowenstern, J.B. (2020). "Magma intrusion and volatile ascent beneath Norris Geyser Basin, Yellowstone National Park." *Journal of Geophysical Research Solid Earth*, https://doi.org/10.1029/2019JB018208.

Wicks, C.W., Thatcher, W., Dzurisin, D. et al. (2006). "Uplift, thermal unrest and magma intrusion at Yellowstone caldera." *Nature*, 440, 72–75.

Wiebe, R.A., Frey, H., and Hawkins, D.P. (2001). "Pillow mounds in the Vinalhaven intrusion, Maine." *Journal of Volcanology and Geothermal Research*, 107, 171–184.

Wiesnet, D.R. and D'Aguanno, J. (1982). "Thermal imagery of Mount Erebus from the NOAA-6 satellite." *Antarctic Journal of the United States*, 17, 32–34.

Wignall, P. (2005). "The link between large igneous province eruptions and mass extinctions." *Elements*, 1 (5), 293–7.

Wilcox, R.E. (1965). *Volcanic-ash Chronology, in Quaternary of the United States*. Princeton, Princeton University Press, pp. 807–816.

Wilhelms, D.E. (1993). *To a Rocky Moon: A Geologist's View of Lunar Exploration*. Tucson: University of Arizona Press.

Wilhelms, D.E., McCauley, J.F., and Trask, N.J. (1987). *The Geologic History of the Moon*. US Geological Survey Professional Paper 1348, Washington DC, USGS.

Williams, D.A., Kerr, R.C., and Lesher, C.M. (1999). "Thermal and fluid dynamics of komatiitic lavas associated with magmatic Ni-Cu (PEG) sulphide deposits." In: *Dynamic Processes in Magmatic Ore Deposits and Their Application to Mineral Exploration* (eds. R.R. Keays et al.), pp. 367–412. Geol. Soc. Canada Short Course, 13.

Williams, H. (1941). *Calderas and their Origin*. Berkeley: University of California Press.

Williams, H. (1942). *The Geology of Crater Lake National Park, Oregon, With a Reconnaissance of The Cascade Range South to Mount Shasta*. Carnegie Institute of Washington Publication, 540, 162p.

Williams, H. and McBirney, A.R. (1979). *Volcanology*. San Francisco, Freeman Cooper.

Williams, R.S. and Moore, J.G. (1983). "Man against volcano: The eruption on Heimaey, Vestmannaeyjar, Iceland." Available at: http://pubs.usgs.gov/gip/heimaey/

Williams, R.S., Jr. and Friedman J.D. (1970). "Satellite observation of effusive volcanism." *Journal of the British Interplanetary Society*, 23, 441–450.

Williams-Jones, G. and Rymer, H. (2002). "Detecting volcanic eruption precursors: A new method using gravity and deformation measurements." *Journal of Volcanology and Geothermal Research*, 113, 379–389.

Willsey, S. (2017). *Geology Underfoot in Southern Idaho*. Missoula, Montana: Mountain Press, 304 pp, ISBN 978-0-87842-678-2.

Wilshire, H. and Kirby, S.H. (1989). "Dikes, joints, and faults in the upper mantle." *Tectonophysics*, 23–31.

Wilshire, H.G., Nielson, J.E., and Hazlett, R.W. (2008). *The American West at Risk: Science, Myths, and Politics of Land Abuse and Recovery*. Oxford: Oxford University Press.

Wilson, C.J.N. (1985). "The Taupo eruption, New Zealand, 2; The Taupo ignimbrite." *Philosophical Transactions of the Royal Society of London*, A314, 229–310.

Wilson, C.J.N. (1993). "Eruption columns." In: *Explosive Volcanism: Processes and Products* (ed. J. McPhie), pp. 4.1–4.10. Commission on Explosive Volcanism Short Course.

Wilson, C.J.N. (2001). "The 26.5 ka Oruanui eruption, New Zealand: An introduction and overview." *Journal of Volcanology and Geothermal Research*, 112, 133–174.

Wilson, C.J.N. and Hildreth, W. (1997). "Hybrid fall deposit in the Bishop Tuff, California: A novel pyroclastic depositional mechanism." *Geology*, 26, 7–10.

Wilson, C.J.N. and Hildreth, W. (2003). "Assembling an ignimbrite: Mechanical and thermal building blocks in the Bishop Tuff, California." *Journal of Geology*, 111, 635–670.

Wilson, C.J.N., Blake, S., Charlier, B.L.A. et al. (2006). "The 26.5 ka Oruanui Eruption, Taupo Volcano, New Zealand: Development, characteristics, and evacuation of a large rhyolitic magma body." *Journal of Petrology*, 47, 35–69.

Wilson, C.J.N., Houghton, B.F., McWilliams, M.O. et al. (1995). "Volcanic and structural evolution of Taupo Volcanic Zone, New Zealand: A review." *Journal of Volcanology and Geothermal Research*, 68, 1–28.

Wilson, J.T. (1963). "A possible origin of the Hawaiian Islands." *Canadian Journal of Physics*, 41, 863–870.

Wilson, L. and Head, J.W. (2017). "Lunar floor-fractured craters: Modes of dike and sill emplacement and implications of gas production and intrusion cooling on surface morphology and structure." *Icarus*, 385, 105–122.

Wilson, L. and Walker, G.P.L. (1987). Explosive volcanic eruptions – VI. Ejecta dispersal in plinian eruptions: The control of eruption conditions and atmospheric properties. *Geophysical Journal, Royal Astronomical Society*, 89, 657–679.

Wilson, L., Sparks, R.S.J., and Walker, G.P.L. (1980). Explosive volcanic eruptions – IV. The control of magma properties and conduit geometry on eruption column behavior, *Geophysical Journal International*, 63, 117–148.

Witter, J.B. and Harris, A.L. (2007). "Field measurements of heat loss from skylights and lava tube systems." *Journal of Geophysical Research, Solid Earth*, 112 (B1). doi:/10.1029/2005JB003800.

Witze, A. (2014). *Island on Fire: The Extraordinary Story of a Forgotten Volcano [Laki, Iceland] that Changed the World.* Berkeley, California: Pegasus Books, 224 p.

Witze, A. and Kanipe, J. (2014). *Island on Fire: The Extraordinary Story of a Forgotten Volcano [Laki, Iceland] that Changed the World.* Pegasus Books, 224 p.

Wohletz, K. (1985). *Volcanic Ash.* Berkeley: University of California Press, 246p .

Wohletz, K. (1986). "Explosive magma-water interactions: Thermodynamics, explosive mechanisms and field studies." *Bulletin of Volcanology*, 48, 245–264.

Wohletz, K.H. (1998). "Pyroclastic surges and compressible two-phase flow." In: *From Magma to Tephra – Modelling Volcanic Processes of Explosive Volcanic Eruptions* (eds. Freundt, A. and Rossi, M). Amsterdam: Elsevier Publishing.

Wohletz, K.H. (2003). "Water/magma interaction: Physical considerations for the deep submarine environment." In: *Geophysical Monograph 140* (eds. J.D.L. White, J.L. Smellie, and D.A. Clague), pp. 25–49. Washington, DC, American Geophysical Union.

Wohletz, K.H. and Heiken, G. (1992). *Volcanology and Geothermal Energy.* Berkeley: University of California Press.

Wohletz, K.H. and McQueen, R.G. (1984). "Experimental studies of hydromagmatic volcanism in explosive volcanism: Inception, evolution, and hazards." *Studies in Geophysics*, 158–169.

Wohletz, K.H. and Sheridan, M.F. (1979). "A model of pyroclastic surge." *Geological Society America, Special Paper*, 180, 177–193.

Wohletz, K.H., McGetchin, T.R., Sandford, M.T., II et al. (1984). "Hydrodynamic aspects of caldera-forming eruptions: Numerical models." *Journal of Geophysical Research*, 89, 8269–8285.

WoldeGabriel, G., Heiken, G., White, T.D. et al. (2000). "Volcanism, tectonism, sedimentation, and the paleo anthropological record in the Ethiopian Rift System." In: *Volcanic Hazards and Disasters in Human History* (eds. W. Floyd and G.H. McCoy), pp. 82–99. Washington, DC: Geological Society of America.

Wolf, T. (1878). "Geognostische Mittheilungen aus Ecuador Part 5, Der Cotopaxi und seine letzte Eruption am 26. Juni 1877." In: *Neues Jahrbuch für Mineralogie, Geologie und Palaeontologie* (eds. G. Leonhard and H.B. Geinitz), pp. 113–167. Stuttgart, E. Schweizerbart'sche Verlagsbuchhandlung.

Wolfe, E.W. and Hoblitt, R.P. (1996). "Overview of the eruptions." In: *Fire and Mud: Eruptions and Lahars of Mount Pinatubo* (eds. Newhall, C.G. and Punongbayan, R.S.). Philippines: Philippine Institute of Volcanology and Seismology (Quezon City) and the University of Washington Press (Seattle), 1126 pp.

Woo, G. (2008). "Probabilistic criteria for volcano evacuation decisions." *Natural Hazards*, 45, 87–97.

Woo, G. (2015). "Cost-benefit analysis in volcanic risk." Chapter 11 in *Volcanic Hazards, Risks, and Disasters* (ed. J.F. Shroder and P. Papale), pp. 289–300. Amsterdam: Elsevier. https://doi.org/10.1016/B978-0-12-396453-3.00011-3.

Wood, C. (2009). "World Heritage Volcanoes: Global Review of Volcanic World Heritage Prospects: Present Situation, Future Prospects, and Management Requirements." *International Union for the Conservation of Nature World Heritage Studies*, 8, 62 p.

Wood, C.A. (1980). "Morphometric evolution of cinder cones." *Journal of Volcanology and Geothermal Research*, 7, 387–413.

Wood, G.D. (2014). *Tambora: The Eruption that Changed the World.* Princeton: Princeton University Press, 293 p.

Woods, A.W. (1988). "A fluid dynamics and thermodynamics of eruption columns." *Bulletin of Volcanology*, 50, 169–193.

Woods, A.W. (1995). "The dynamics of explosive volcanic eruptions." *Reviews of Geophysics*, 33, 495–530.

Wooley, A.R. and Church, A.A. (2005). "Known occurrences of carbonatites." *Lithos*, 85, 1–14.

Wooster, R.J. and Rothery, D.A. (1997). "Thermal monitoring of Lascar volcano, northern Chile, using infrared data at high temporal resolution: A 1992 to 1995 time-series using the along-track scanning radiometer." *Bulletin Volcanologique*, 58, 566–579.

Workman, W.B. (1979). "The significance of volcanism in the prehistory of subarctic northwest North America." In: *Volcanic Activity and Human Ecology* (eds. P.D. Sheets and D.K. Grayson), pp. 339–371. New York: Academic Press.

Wright, J.V. (1981). "The Rio Caliente ignimbrite: Analysis of a compound intraplinian ignimbrite from a major late Quaternary Mexican eruption." *Bulletin Volcanologique*, 44, 189–212.

Wright, J.V. and Walker, G.P.L. (1977). "The ignimbrite source problem: Significance of a co-ignimbrite lag-fall deposit." *Geology*, 5, 729–732.

Wright, T.L., Mangan, M., and Swanson, D.A. (1989). "Chemical data for flows and feeder dikes of the Yakima Basalt Subgroup, Columbia River Basalt Group, Washington, Oregon, and Idaho, and their bearing on a petrogenic model." US Geological Survey Bulletin 1821, 71p.

Wu, S.M., Lin, F.C., Farrell, J. et al. (2021). "Imaging the subsurface plumbing complex of Steamboat Geyser and Cistern Spring with hydrothermal tremor migration using seismic interferometry." *Journal of Geophysical Research: Solid Earth*, doi: 10.1029/2020JB021128.

Wyllie, P.J. and Huang, W.L. (1976). "High CO2 solubilities in mantle magmas." *Geology*, 4, 21–44.

Yamada, H., Tateyama, K., Sasaki, H. et al. (2018). "Impact resistance to ballistic ejecta of wooden buildings and a simple reinforcement method using aramid fabric." *Journal of Volcanology and Geothermal Research*, 359, 37–46.

Yamagishi, H. (1985). "Growth of pillow lobes: Evidence from pillow lavas of Hokkaido, Japan and North Island, New Zealand." *Geology*, 13, 499–502.

Yamamoto, T. and Nakada, S. (2015). "Extreme volcanic risks 2: Mount Fuji." Chapter 14 in *Volcanic Hazards, Risks, and Disasters* (ed. J.F. Shroder and P. Papale), pp. 355–376. Amsterdam: Elsevier. https://doi.org/10.1016/B978-0-12-396453-3.00014-9.

Yamamoto, T., Takarada, S., and Suto, S. (1993). "Pyroclastic flows from the 1991 eruption of Unzen volcano, Japan." *Bulletin of Volcanology*, 55, 166–175. doi.org/10.1007/BF00301514.

Yamashita, S. (1999). "Experimental study of the effect of temperature on water solubility on natural rhyolite melt to 100 MPa." *Journal of Petrology*, 40, 1497–1507.

Ye, L., Kanamori, H., Lay, T. et al. (2020). The 22 December 2018 tsunami from flank collapse of Anak Krakatau volcano during eruption. *Science Advances*, 6. doi: 10.1126/sciadv.aaz1377.

Yokoyama I., De la Cruz-Reyna S., and Espíndola J.M. (1992). "Energy partition in the 1982 eruption of El Chichón volcano, Chiapas, Mexico." *Journal of Volcanology and Geothermal Research*, 51, 1–21.

Yokoyama, I. (1981). "A geophysical interpretation of the Krakatau eruption." *Journal of Volcanology and Geothermal Research*, 9, 359–378.

Young, D. (2003). *Mind over Magma: The Story of Igneous Petrology.* Princeton: Princeton University Press, 704p.

Yürür, M.T. and. Chorowicz, J. (1998). "Recent volcanism, tectonics, and plate tectonics near the junction of the African, Arabian, and Anatolian plates in the eastern Mediterranean." *Journal of Volcanology and Geothermal Research*, 85, 1–15.

Žák, J., Verner, K., Johnson, K., and Schwartz, J. (2012). "Magma emplacement process zone preserved in the roof of a large Cordillern batholith, Wallowa Mountains, northeastern Oregon." *Journal of Volcanology and Geothermal Research*, 227–228, 61–75.

Zarazúa-Carbajal, M.C. and de la Cruz-Reyna (2020). "Morpho-chronology of monogenetic scoria cones from their level contour curves. Applications to the Chichinautzin monogenetic field, Central Mexico." *Journal of Volcanology and Geothermal Research*, 407, 107093. doi: 10.1016/j.jvolgeores.2020.107093.

Zheng, L., Gordon, R.G., and Woodworth, D. (2018). "Pacific Plate apparent polar wander, hot spot frailty, and true polar wander during formation of the Hawaiian island and seamount chain from an analysis of magnetic anomaly 20r (44 M)." *Tectonics*, 37, 2094–2105.

Zimanowski, B., Wohletz, K., Dellino, P. et al. (2003). "The volcanic ash problem." *Journal of Volcanology and Geothermal Research*, 12 (1–2), 1–5.

Zimbelman, J.R. (2000). "Volcanism on Mars." In: *Encyclopedia of Volcanoes* (ed. H. Sigurdsson), pp. 771–783. San Diego: Academic Press.

Zimbelman, J.R. and Gregg, T.K.P. (eds.) (2000). *Environmental Effects on Volcanic Eruptions: From Deep Oceans to Deep Space*. Dordrecht: Kluwer Academic/Plenum.

Zlotnicki, J., Ruegg, J.C., Bacheley, P. et al. (1990). "Eruptive mechanism on Piton de la Fournaise volcano associated with the December 4, 1983, and January 18, 1984 eruptions from ground deformation monitoring and photogrammetric surveys." *Journal of Volcanology and Geothermal Research*, 40, 197–217.

APPENDIX 1 List of Prominent World Volcanoes

Volcano Global Perspectives

Map Number	Volcano Name	Smithsonian Number	Region	Primary Volcano Type	Last Known Eruption
1	Hunga Tonga–Hunga Ha'apai	243040	New Zealand to Fiji	Submarine	2022 CE
2	Home Reef	243080	New Zealand to Fiji	Submarine	2006 CE
3	West Mata	243130	New Zealand to Fiji	Submarine	2009 CE
4	Okmok	311290	Alaska	Shield	2008 CE
5	Makushin	311310	Alaska	Composite	1995 CE
6	Westdahl	311340	Alaska	Composite	1992 CE
7	Fisher	311350	Alaska	Composite	1830 CE
8	Shishaldin	311360	Alaska	Composite	2020 CE
9	Veniaminof	312070	Alaska	Composite	2021 CE
10	Aniakchak	312090	Alaska	Caldera	1931 CE
11	Koolau	332807	Hawaii and Pacific Ocean	Shield	10000 BCE
12	Ukinrek Maars	312131	Alaska	Maar(s)	1977 CE
13	Haleakala	332060	Hawaii and Pacific Ocean	Shield	1750 CE
14	Hualalai	332040	Hawaii and Pacific Ocean	Shield	1801 CE
15	Mauna Loa	332020	Hawaii and Pacific Ocean	Shield	1984 CE
16	Mauna Kea	332030	Hawaii and Pacific Ocean	Shield	2460 BCE
17	Kilauea	332010	Hawaii and Pacific Ocean	Shield	2021 CE
18	Loihi	332000	Hawaii and Pacific Ocean	Seamount	1996 CE
19	Novarupta	312180	Alaska	Caldera	1912 CE
20	Katmai	312170	Alaska	Composite	1912 CE
21	Augustine	313010	Alaska	Lava dome(s)	2006 CE
22	Iliamna	313020	Alaska	Composite	1876 CE
23	Redoubt	313030	Alaska	Composite	2009 CE

Volcanoes: Global Perspectives, Second Edition. John P. Lockwood, Richard W. Hazlett, and Servando De La Cruz-Reyna.

Companion website: www.wiley.com/go/lockwood/volcanoes2

Map Number	Volcano Name	Smithsonian Number	Region	Primary Volcano Type	Last Known Eruption
24	Mt. Spurr (Spurr)	313040	Alaska	Composite	1992 CE
25	Axial Seamount	331021	Hawaii and Pacific Ocean	Submarine	2015 CE
26	Tseax (Tseax River Cone)	320100	Canada and Western USA	Pyroclastic cone	1690 CE
27	Garibaldi	320200	Canada and Western USA	Composite	8060 BCE
28	Mt. McLoughlin (McLoughlin)	322834	Canada and Western USA	Composite	20000 BCE
29	Brown Mountain	322835	Canada and Western USA	Shield	12000 BCE
30	Shasta	323010	Canada and Western USA	Composite	1250 CE
31	St. Helens	321050	Canada and Western USA	Composite	2008 CE
32	Crater Lake (Mt. Mazama)	322160	Canada and Western USA	Caldera	2850 BCE
33	Mt. Thielsen (Thielsen)	322827	Canada and Western USA	Shield	0.29M BCE
34	Mt. Baker (Baker)	321010	Canada and Western USA	Composite	1880 CE
35	North Sister (Three Sisters)	322070	Canada and Western USA	Composite	440 CE
36	Mt. Rainier (Rainier)	321030	Canada and Western USA	Composite	1450 CE
37	Mt. Hood (Hood)	322010	Canada and Western USA	Composite	1866 CE
38	Bachelor	322090	Canada and Western USA	Composite	5800 BCE
39	Medicine Lake	323020	Canada and Western USA	Shield	1060 CE
40	Lassen Volcanic Center	323080	Canada and Western USA	Composite	1917 CE
41	Newberry	322110	Canada and Western USA	Shield	690 CE
42	Glacier Peak	321020	Canada and Western USA	Composite	1700 CE
43	Mammoth Mountain	323150	Canada and Western USA	Composite	1260 CE
44	Long Valley	323822	Canada and Western USA	Caldera	760K BCE
45	Ubehebe Craters	323160	Canada and Western USA	Maar(s)	150 BCE
46	Barcena	341020	México and Central America	Monogenic Cinder Cones	1953 CE
47	Yellowstone	325010	Canada and Western USA	Caldera	1350 BCE
48	La Garita Caledera	*	Colorado	Caldera	28M BCE
49	Jemez Caldera (Valles)	**	New Mexico	Caldera	1.25M BCE

(Continued)

Map Number	Volcano Name	Smithsonian Number	Region	Primary Volcano Type	Last Known Eruption
50	Tequila	341818	México and Central America	Composite	0.21M BCE
51	Colima	341040	México and Central America	Composite	2019 CE
52	Jorullo (Micho-acán–Guanajuato)	341060	México and Central America	Monogenic Cinder Cones	1952 CE
53	Paricutín (Micho-acan–Guanajuato)	341060	México and Central America	Monogenic Cinder Cones	1952 CE
54	Nevado de Toluca	341070	México and Central America	Composite	1350 BCE
55	Iztaccihuatl	341082	México and Central America	Composite	1868 CE
56	Popocatépetl	341090	México and Central America	Composite	2021 CE
57	El Chichón	341120	México and Central America	Composite	1982 CE
58	Santa Maria	342030	México and Central America	Composite	2021 CE
59	Volcán La Cumbre (Fernandina)	353010	South America	Shield	2020 CE
60	Los Chocoyos (Atitlan)	342060	México and Central America	Composite	1853 CE
61	Fuego	342090	México and Central America	Composite	2021 CE
62	Floreana (Charles Island)	353805	South America	Shield	1813 CE
63	Ilopango	343060	México and Central America	Caldera	1880 CE
64	Casita (San Cristobal)	344020	México and Central America	Composite	2021 CE
65	Momotombo	344090	México and Central America	Composite	2016 CE
66	Masaya	344100	México and Central America	Caldera	2021 CE
67	Tungurahua	352080	South America	Composite	2016 CE
68	Cotopaxi	352050	South America	Composite	2016 CE
69	Galeras	351080	South America	Composite	2014 CE
70	Nevado del Huila	351050	South America	Composite	2012 CE
71	Nevado del Ruiz (Ruiz)	351020	South America	Composite	2021 CE
72	Nevado Coropuna	354003	South America	Composite	Unknown
73	Chaitén	358041	South America	Caldera	2011 CE
74	Calbuco	358020	South America	Composite	2015 CE
75	Osorno	358010	South America	Composite	1869 CE
76	Villarrica	357120	South America	Composite	2021 CE
77	Llaima	357110	South America	Composite	2009 CE
78	El Misti	354010	South America	Composite	1985 CE
79	Loma caldera (Lomas Blancas)	357064	South America	Composite	Unknown
80	Huaynaputina	354030	South America	Composite	1600 CE

Map Number	Volcano Name	Smithsonian Number	Region	Primary Volcano Type	Last Known Eruption
81	Socompa	355109	South America	Composite	5250 BCE
82	Láscar	355100	South America	Composite	2017 CE
83	La Pacana Caldera	***	South America	Caldera	2 Ma
84	Soufriere Hills Montserrat	360050	West Indies	Composite	2013 CE
85	Soufriere Guadeloupe	360060	West Indies	Composite	1977 CE
86	Soufriere St. Vincent	360150	West Indies	Composite	2021 CE
87	Mt. Pelée (Pelée)	360120	West Indies	Composite	1932 CE
88	Deception Island	390030	Antarctica	Caldera	1970 CE
89	Capelinhos (Fayal)	382010	Atlantic Ocean	Composite	1958 CE
90	Surtsey (Vestmannaeyjar)	372010	Iceland and Arctic Ocean	Submarine	1973 CE
91	Skjaldbreiður (Oddnyjarhnjukur-Langjokull)	371080	Iceland and Arctic Ocean	Subglacial	950 CE
92	Hekla	372070	Iceland and Arctic Ocean	Composite	2000 CE
93	Eyjafjallajökull	372020	Iceland and Arctic Ocean	Composite	2010 CE
94	Katla	372030	Iceland and Arctic Ocean	Subglacial	1918 CE
95	Eldgjá	372030	Iceland and Arctic Ocean	Subglacial	939 CE
96	Laki	373010	Iceland and Arctic Ocean	Caldera	1784 CE
97	Cumbre Vieja (La Palma)	383010	Atlantic Ocean	Composite	2021 CE
98	Bárðarbunga	373030	Iceland and Arctic Ocean	Composite	2015 CE
99	Grímsvötn	373010	Iceland and Arctic Ocean	Caldera	2011 CE
100	Trolladyngja	373060	Iceland and Arctic Ocean	Composite	1961 CE
101	Lake Myvatn	373080	Iceland and Arctic Ocean	Caldera	1719 CE
102	Askja	373060	Iceland and Arctic Ocean	Composite	1961 CE
103	Krafla	373080	Iceland and Arctic Ocean	Caldera	1984 CE
104	Las Canadas/Teide (Tenerife)	383030	Atlantic Ocean	Composite	1909 CE
105	Laacher See	****	Germany	Caldera Lake	13,000 BCE
106	Mt. Cameroon (Cameroon)	224010	Africa and Red Sea	Composite	2000 CE
107	Lake Nyos (Oku Volcanic Field)	224030	Africa and Red Sea	Composite	Unknown
108	Campi Flegrei	211010	Mediterranean and Western Asia	Caldera	1538 CE

(Continued)

Map Number	Volcano Name	Smithsonian Number	Region	Primary Volcano Type	Last Known Eruption
109	Vesuvius	211020	Mediterranean and Western Asia	Composite	1944 CE
110	Mt. Guardia (Lipari)	211042	Mediterranean and Western Asia	Composite	1230 CE
111	Vulcano	211050	Mediterranean and Western Asia	Composite	1890 CE
112	Etna	211060	Mediterranean and Western Asia	Composite	2021 CE
113	Stromboli	211040	Mediterranean and Western Asia	Composite	2021 CE
114	Santorini	212040	Mediterranean and Western Asia	Shield(s)	1950 CE
115	Meidob Volcanic Field	225050	Africa and Red Sea	Pyro-clastic cone(s)	2950 BCE
116	Kos	212805	Mediterranean and Western Asia	Caldera	160000 BCE
117	Nyamulagira	223020	Africa and Red Sea	Shield	2021 CE
118	Nyiragongo	223030	Africa and Red Sea	Composite	2021 CE
119	Hasan Dagi	213002	Mediterranean and Western Asia	Composite	Unknown
120	Oldoinyo (Lengai, Ol Doinyo)	222120	Africa and Red Sea	Composite	2021 CE
121	Erta Ale	221080	Africa and Red Sea	Shield	2021 CE
122	Elbrus	214010	Mediterranean and Western Asia	Composite	50 CE
123	Mt. Ararat (Ararat)	213040	Mediterranean and Western Asia	Composite	1840 CE
124	Piton des Nieges (Snow Peak)	233802	Middle East and Indian Ocean	Shield	2.1M BCE
125	Piton de la Fournaise	233020	Middle East and Indian Ocean	Shield	2021 CE
126	Sinabung	261080	Indonesia	Composite	2021 CE
127	Toba	261090	Indonesia	Caldera	Unknown
128	Krakatau (Rakata)	262000	Indonesia	Caldera	2020 CE
129	Galunggung	263140	Indonesia	Composite	1984 CE
130	Dieng Volcanic Composite	263200	Indonesia	Composite	2021 CE
131	Merapi	263250	Indonesia	Composite	2021 CE
132	Kelud (Kelut)	263280	Indonesia	Composite	2014 CE
133	Samalas (Lombok Island)	*****	Indonesia	Composite	1257 CE
134	Tambora	264040	Indonesia	Composite	1967 CE
135	Pinatubo	273083	Philippines and SE Asia	Composite	1993 CE
136	Taal	273070	Philippines and SE Asia	Caldera	2021 CE
137	Mayon	273030	Philippines and SE Asia	Composite	2019 CE
138	Hibok-Hibok (Camiguin)	271080	Philippines and SE Asia	Composite	1953 CE

Map Number	Volcano Name	Smithsonian Number	Region	Primary Volcano Type	Last Known Eruption
139	Parker	271011	Philippines and SE Asia	Composite	1641 CE
140	Gamalama	268060	Indonesia	Composite	2018 CE
141	Unzen (Unzendake)	282100	Japan, Taiwan, Marianas	Composite	1996 CE
142	Sakurjima (Aira)	282080	Japan, Taiwan, Marianas	Caldera	2021 CE
143	Aso (Asosan)	282110	Japan, Taiwan, Marianas	Caldera	2020 CE
144	Ontake (Ontakesan)	283040	Japan, Taiwan, Marianas	Composite	2014 CE
145	Asama (Asamayama)	283110	Japan, Taiwan, Marianas	Composite	2019 CE
146	Mt. Fuji (Fujisan)	283030	Japan, Taiwan, Marianas	Composite	1708 CE
147	Hakone (Hakoneyama)	283020	Japan, Taiwan, Marianas	Composite	2015 CE
148	Miyake-jima	284040	Japan, Taiwan, Marianas	Composite	2010 CE
149	Myōjin-Shō	284070	Japan, Taiwan, Marianas	Submarine	1970 CE
150	Bandai (Bandaisan)	283160	Japan, Taiwan, Marianas	Composite	1888 CE
151	Nigorikawa	285805	Japan, Taiwan, Marianas	Caldera	12000 BCE
152	Usu (Toya)	285030	Japan, Taiwan, Marianas	Composite	2001 CE
153	Showa Shinzan	285030	Japan, Taiwan, Marianas	Composite	1945 CE
154	Kuttara	285034	Japan, Taiwan, Marianas	Composite	1820 CE
155	Mt. Gambier (Newer Volcanics Province)	259010	Melanesia and Australia	Maar	2900 BCE
156	Undara (McBride Volcanic Province)	259804	Melanesia and Australia	Shield	60000 BCE
157	NW Rota-1	284211	Japan, Taiwan, Marianas	Submarine	2010 CE
158	Long Island	251050	Melanesia and Australia	Composite	1993 CE
159	Mt. Lamington (Lamington)	253010	Melanesia and Australia	Composite	1956 CE
160	Rabaul (Tavurvur)	252140	Melanesia and Australia	Pyroclastic shield	2014 CE
161	Raikoke	290250	Kuril Islands	Composite	2019 CE
162	Billy Mitchell	255011	Melanesia and Australia	Pyroclastic shield	1580 CE
163	Bolshoye Tol'batchik (Kambalny)	300010	Kamchatka and Mainland Asia	Composite	2017 CE
164	Kavachi	255060	Melanesia and Australia	Submarine	2021 CE

(Continued)

Map Number	Volcano Name	Smithsonian Number	Region	Primary Volcano Type	Last Known Eruption
165	Gorelli (Gorely)	300070	Kamchatka and Mainland Asia	Composite	2010 CE
166	Koryaksky	300090	Kamchatka and Mainland Asia	Composite	2009 CE
167	Uzon	300170	Kamchatka and Mainland Asia	Caldera	200 CE
168	Ploskii Tolbatchik (Tolbachik)	300240	Kamchatka and Mainland Asia	Shield	2013 CE
169	Kronotsky	300200	Kamchatka and Mainland Asia	Composite	1923 CE
170	Bezymianny	300250	Kamchatka and Mainland Asia	Composite	2021 CE
171	Kliuchevskoi (Klyuchevskoy)	300260	Kamchatka and Mainland Asia	Composite	2021 CE
172	Shiveluch (Sheveluch)	300270	Kamchatka and Mainland Asia	Composite	2021 CE
173	Mt. Erebus (Erebus)	390020	Antarctica	Composite	2021 CE
174	Ambrym	257040	Melanesia and Australia	Pyro-clastic shield	2018 CE
175	Kuwae	257070	Melanesia and Australia	Caldera	1974 CE
176	Mt. Egmont (Taranaki)	241030	New Zealand to Fiji	Composite	1800 CE
177	Rangitoto (Auckland Volcanic Field)	241020	New Zealand to Fiji	Volcanic field	1446 CE
178	Ruapehu	241100	New Zealand to Fiji	Composite	2007 CE
179	Tongariro	241080	New Zealand to Fiji	Composite	2012 CE
180	Taupo	241070	New Zealand to Fiji	Caldera	260 CE
181	Oruanui (Maroa)	241061	New Zealand to Fiji	Caldera	180 CE
182	Mayor Island	241021	New Zealand to Fiji	Shield	5060 BCE
183	Rotorua	241816	New Zealand to Fiji	Caldera	25000 BCE
184	Tarawera (Okataina)	241050	New Zealand to Fiji	Fissure	1981 CE
185	White Island	241040	New Zealand to Fiji	Composite	2019 CE
186	Brothers	241150	New Zealand to Fiji	Submarine	Unknown

* https://en.wikipedia.org/wiki/La_Garita_Caldera

** https://en.wikipedia.org/wiki/Valles_Caldera

*** https://volcano.oregonstate.edu/sites/volcano.oregonstate.edu/files/oldroot/CVZ/lapacana/index.html

**** https://en.wikipedia.org/wiki/Laacher_See

***** https://en.wikipedia.org/wiki/1257_Samalas_eruptionlapacana/index.html

APPENDIX 2 Fun Reading

Volcanologists mostly write scientific articles and reports all their lives, although a few, like Rick, Servando, and I (JPL), also write textbooks. Some also write stories for broader audiences – either as personal memoirs about their adventurous lives, or as books describing the sociological or historical significance of volcanoes they have studied. Volcanologists have also written fictional "volcano adventure" novels, imaginative stories about future eruptions at volcanoes they know well. All of these books can be fun to read! Here are a few of them – books that offer insights into how volcanologists really think about their careers, about the social impact of their studies, and about how their fanciful imaginations can run wild to describe eruptions that will doubtless happen at their favorite volcanoes one day!

Memoirs by Volcanologists

Fisher, Richard. (2000) *Out of the Crater – Chronicles of a Volcanologist.* Princeton University Press, 180 pp. *Fisher witnessed the explosive power of atomic bombs at Bikini Atoll in 1946, and decided to spend the rest of his distinguished, adventurous career studying the much larger explosive activity and pyroclastic deposits associated with gray volcanoes in the Caribbean, Central America, China, and elsewhere.*

Hamilton, Sir William, 1768. An account of the eruption of Mount Vesuvius, in 1767. *Philosophical Transactions of the Royal Society of London* 58: 1–14. *Although his formal memoir was written by another (Veitch, 1869 – reprinted 2016), this Royal Society account best gives the flavor of his relationship with Vesuvius. Hamilton was a perceptive observer – a volcanologist before that term was ever coined!*

Jaggar, Thomas (1945) *Volcanoes Declare War. Logistics and Strategy of Pacific Volcano Science.* Honolulu: Paradise of the Pacific Press, 166 pp. *Jaggar, founder of the Hawaiian Volcano Observatory was deeply impacted by his visit to St Pierre after the 1906 Mount Pelée eruption – a disaster that caused him to devote the rest of his life to volcano studies.*

McBirney, Alexander (2004) *Faulty Geology: Frauds, Hoaxes, and Delusions.* Eugene, Oregon: Bostok Press, 216 pp. *McBirney was a first-rate volcanologist, who wrote several books about volcanology and igneous petrology, but in this "for fun," self-published book, he writes stories about little-known frauds perpetuated by geologists – including the quest for gold in the crater of a Nicaraguan volcano he knew well – Masaya.*

Krafft, Maurice. (1991) *Les Feux de la Terre: Histoires de Volcans.* Paris: Decouvertes Gallimard Sciences, 298 pp. *Maurice and his wife Katia were adventurous French volcanologists who witnessed eruptions all over the world, and whose books, films, and public lectures gave them deserved fame – especially in the French-speaking world. This fact-filled little book (in French) is richly illustrated with original illustrations from the Kraffft Collection, Unfortunately Maurice and Katia were killed at Unzen before this book was published.*

Moore, James. (In press) *Geosaga – Tales from the Hills and the Sea* [accepted for publication]. *Moore has led an adventurous life as a volcanologist – with major findings that have led to revolutionary discoveries around Hawaii and globally.*

Stearns, Harold. (1983) *Memoirs of a Geologist: From Poverty Peak to Piggery Gulch.* Honolulu: Hawaii Institute of Geophysics, 242 pp. *Stearns was an old-time volcanologist who lived an amazing life and conducted pioneering volcano mapping in Hawaii and in Idaho.*

Tazieff, Haroun. (1952) *Craters of Fire.* New York: Harper and Brothers, 239 pp. *Tazieff was a larger-than-life volcanologist, whose daredevil volcano exploits in Africa, the Caribbean and elsewhere make for exciting reading.*

Books About the Relationships Between Volcanoes and Society

Andrew, Robin George (2022) *SUPER VOLCANOES – What they reveal about Earth and the worlds beyond.* W.W. Norton Press, 372 pp. Andrews is a volcanologist who received his Ph.D from Otago University, with work on the 1886 Tarawera eruption. Rather than focus his career on academic research, Robin has chosen to devote his efforts to writing popular science articles focused on volcanology and other earth-altering phenomena. This entertaining, general interest book is an irreverent look at how volcanologists work on Earth and explore other worlds beneath the sea and on extraterrestrial planets.

Heiken, Grant. (2013) *Dangerous Neighbors – Volcanoes and Cities.* Cambridge University Press, 185 pp. *Heiken is known for classic studies at Yellowstone and elsewhere, but his work to help forge the critical links between volcanology*

Volcanoes: Global Perspectives, Second Edition. John P. Lockwood, Richard W. Hazlett, and Servando De La Cruz-Reyna.

Companion website: www.wiley.com/go/lockwood/volcanoes2

and threatened populations around volcanoes will have the greatest impact in mitigating volcanic risk.

Lopes-Gautier, Rosely. (2005) *The Volcano Adventure Guide*. Cambridge Press, 352 pp. *Lopes is mostly known for her extraterrestrial volcanology contributions, but here provides a user-friendly guide for amateur volcanologists to safely plan and execute visits to more than 40 volcanoes around the world.*

Oppenheimer, C. (2011) *Eruptions that Shook the Word*. Cambridge University Press, 392 pp. *Oppenheimer is a most talented volcanologist who has combined his perspectives on scientific, historical, and political realities to describe the greatest volcanic disasters in human history, along with sobering perspectives on the potential for future eruptions that could alter climate to the detriment of humankind.*

Volcano Fiction Written by Volcanologists

Cullen-Tanaka, Janet. (1980) *Fire Mountain*. New York: Kensington Publishing, 352 pp. *Cullen imagines the chaos and social disruption that will accompany the awakening of Mount Rainier one of these days, based on her long experience in Washington State risk mitigation work.*

Duffield, Wendell. (2003) *Pele Stirs – A Volcanic Tale of Hawaii, Hemp, and High-Jinks*. New York: Universe Press, 279 pp. *Duffield worked at the Hawaiian Volcano Observatory for several years, and his fanciful, irreverent account of future Kīlauea eruptive activity and the impact on volcanologists and various other social misfits that live on the volcano makes for a most entertaining read!*

Duffield, Wendell. (2007) *Yucca Mountain Dirty Bomb*. New York: Universe Press, 184 pp. *The "Yucca Mountain Nuclear Waste Repositor" was an incredibly complicated scheme to store nuclear waste in an area where there was low potential for disastrous volcanic activity. Duffield explores the politics involved in the promotion of this facility, and has written an exciting tale about the debate, and about what might have happened had the YMNWR been impacted by an eruption.*

Dwiggins, Toni. (2020) *Volcano Watch*. Cupertino, California: Milbank Press, 386 pp. *Dwiggins knows the Mammoth Mountain area of California well, and uses her geologic understanding of the real volcanic hazards there as a basis for an entertaining, action-packed murder mystery involving volcanologists, a "forensic geologist," the precursors to a major eruption and harrowing survival tales of* <u>*some*</u> *of the book's ash-covered characters.*

Index

Note: Page numbers followed by f indicated figure, t indicates table and n indicates notes.
As per author preferences of the key term we have retained the term with square bracket and updated with the respective page number.

Volcanoes: Global Perspectives, Second Edition. John P. Lockwood, Richard W. Hazlett, and Servando De La Cruz-Reyna.

Companion website: www.wiley.com/go/lockwood/volcanoes2

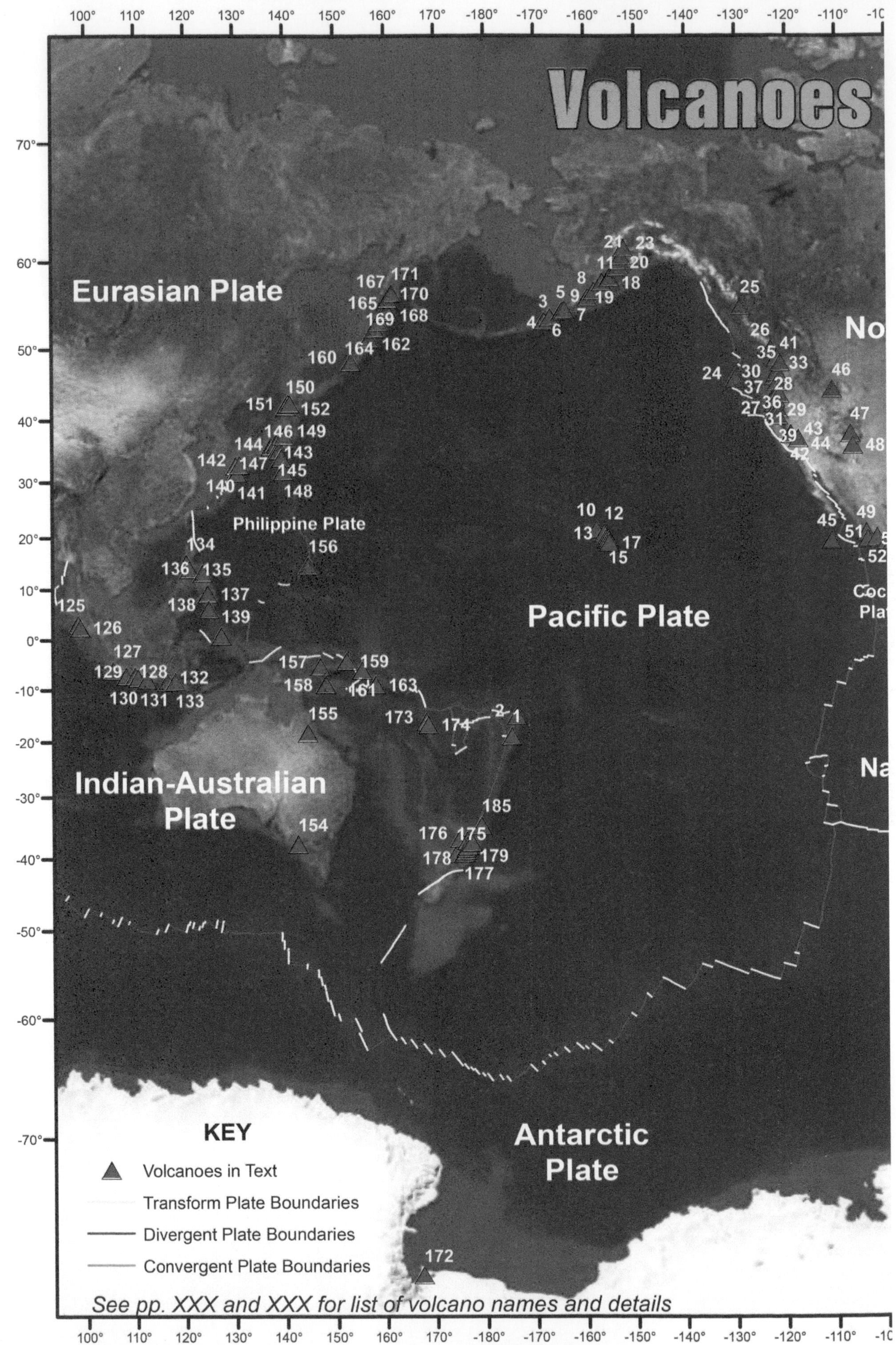

Volcanoes
Eurasian Plate
Philippine Plate
Pacific Plate
Indian-Australian Plate
Antarctic Plate
KEY
Volcanoes in Text
Transform Plate Boundaries
Divergent Plate Boundaries
Convergent Plate Boundaries
See pp. XXX and XXX for list of volcano names and details

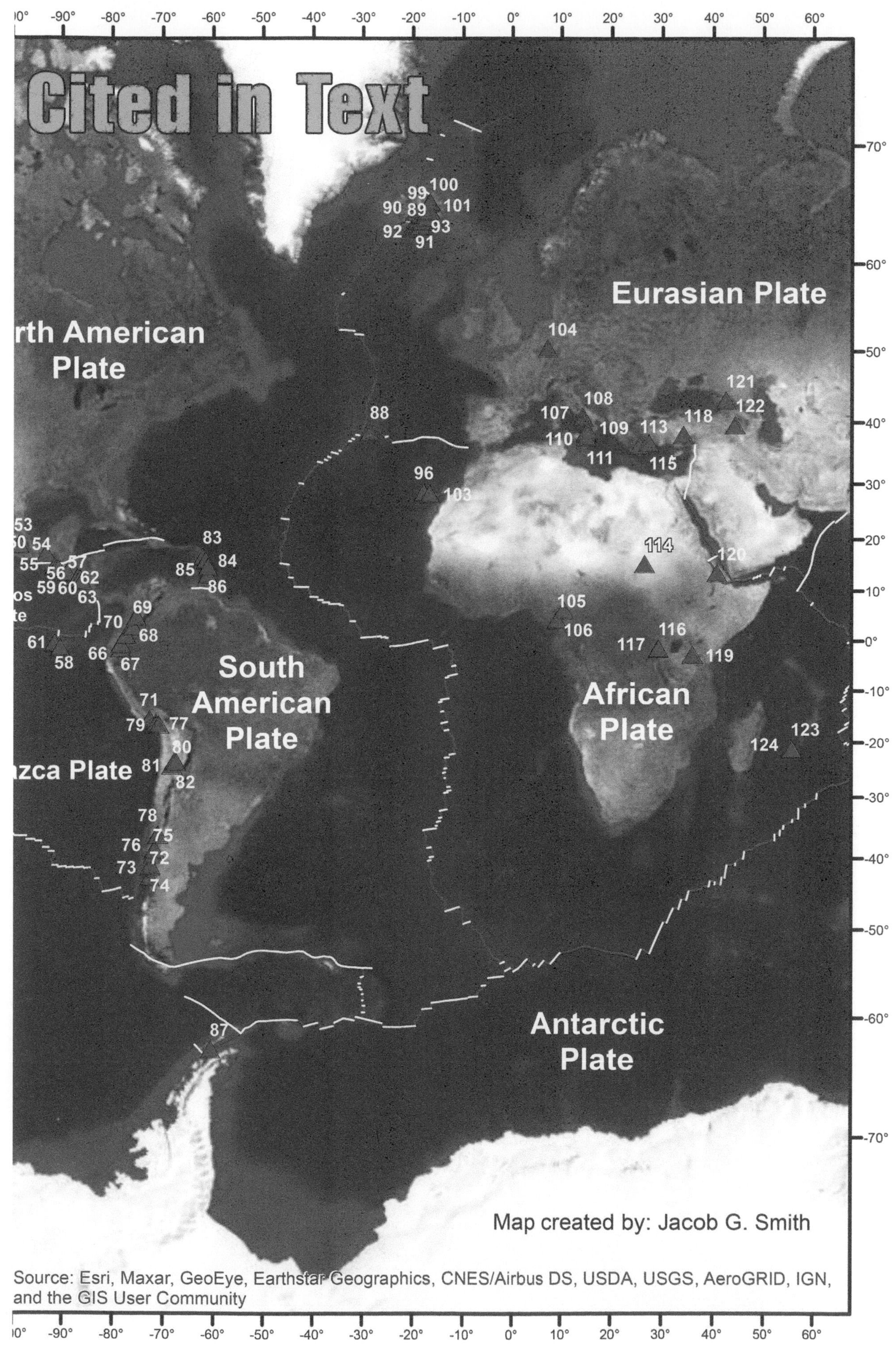
Cited in Text
rth American Plate
Eurasian Plate
South American Plate
African Plate
zca Plate
Antarctic Plate
Map created by: Jacob G. Smith
Source: Esri, Maxar, GeoEye, Earthstar Geographics, CNES/Airbus DS, USDA, USGS, AeroGRID, IGN, and the GIS User Community